The Role of PPARs in Disease II

The Role of PPARs in Disease II

Editors

Kay-Dietrich Wagner
Nicole Wagner

MDPI • Basel • Beijing • Wuhan • Barcelona • Belgrade • Manchester • Tokyo • Cluj • Tianjin

Editors
Kay-Dietrich Wagner
University Cote d'Azur
Nice, France

Nicole Wagner
University Cote d'Azur
Nice, France

Editorial Office
MDPI
St. Alban-Anlage 66
4052 Basel, Switzerland

This is a reprint of articles from the Special Issue published online in the open access journal *Cells* (ISSN 2073-4409) (available at: https://www.mdpi.com/journal/cells/special_issues/PPARs_disease_II).

For citation purposes, cite each article independently as indicated on the article page online and as indicated below:

LastName, A.A.; LastName, B.B.; LastName, C.C. Article Title. *Journal Name* **Year**, *Volume Number*, Page Range.

ISBN 978-3-0365-8110-1 (Hbk)
ISBN 978-3-0365-8111-8 (PDF)

Contents

Preface to "The Role of PPARs in Disease II"

This reprint, "The Role of PPARs in Disease II", contains all of the original and review articles published in the Special Issue of the same name in Cells. These articles contain the most recent significant contributions in the field of research on peroxisome proliferator activated receptors (PPARs) in different diseases and their pathophysiological processes. We have to thank Christina Zhang for the excellent editorial assistance. We also thank all of the contributing authors for sharing their research and thus making this Special Issue possible.

Kay-Dietrich Wagner and Nicole Wagner
Editors

cells

Editorial

Recent Insights into the Role of PPARs in Disease

Nicole Wagner * and Kay-Dietrich Wagner *

CNRS, INSERM, iBV, Université Côte d'Azur, 06107 Nice, France
* Correspondence: nwagner@unice.fr (N.W.); kwagner@unice.fr (K.-D.W.)

Peroxisome proliferator-activated receptors (PPARs) are nuclear receptors that play important roles in cell proliferation, differentiation, metabolism, and cancer [1–5]. They were originally identified more than 30 years ago [6,7] in a search for the receptors for a group of rodent hepatocarcinogens that cause the proliferation of peroxisomes. To look on the bright side of life, one of these rodent hepatocarcinogens (clofibrate) was also known to lower triglycerides and cholesterol concentrations in the plasma of patients and to be beneficial in the prevention of ischemic heart disease in a population with increased plasma cholesterol levels [6]. These effects were well known long before the cloning of the corresponding PPARs as receptors [8]. It was also known that these drugs "coincidently" induce an increased transcription of genes required for the peroxisomal β-oxidation of long-chain fatty acids and genes of the cytochrome P450 family [6,9–11]. Shortly after, it was realized that these receptors not only somehow induce genes of fatty acid metabolisms but are also activated by fatty acids [12]. After these first timid steps and with the increasingly powerful tools of modern mouse genetics and molecular and cellular biology methods, our knowledge about PPARs is increasing exponentially. Today, basic knowledge about the three different isoforms, PPARα, PPARβ/δ, and PPARγ, is well-established, and PPARα and PPARγ agonists have been in clinical use for a long time for the treatment of hyperlipidemia and type 2 diabetes, respectively. Nevertheless, the topic of PPARs attracts a lot of attention, and after a first successful Special Issue titled "The Role of PPARs in Disease" in *Cells* in 2020 [1], we decided to collect novel, exiting data and highly interesting points of view in the form of original articles and reviews for the current Special Issue, titled "The Role of PPARs in Disease II". Here, we will briefly outline and highlight the most recent insights into the roles of PPARs in disease collected in this Special Issue.

Steinke et al. describe a novel PPARβ/δ and PPARγ dual agonist, which demonstrates striking beneficial effects in a mouse model (3xTgAD) of Alzheimer's disease [13]. PPARγ agonists had been tested for this indication already before in several studies, but the effects were limited due to the poor penetration of the blood–brain barrier requiring high doses and observed severe side effects in clinical trials [14–16]. As PPARβ/δ is highly expressed in the brain compared to other isoforms and PPARβ/δ activation might counteract weight gain, the authors reasoned that a dual agonist might have additional beneficial effects compared to the PPARγ agonists reported before. They showed first that their compound AU9 activates PPARγ and PPARβ/δ. Most importantly, AU9 improves memory deficits in 3xTgAD mice, improves neurotrophin expression and spine density, reduces amyloid beta levels in the brain, and diminishes neuroinflammation. In contrast to the PPARγ agonist pioglitazone, the novel dual agonist caused less weight gain and heart hypertrophy but was still able to reduce blood glucose levels in 3xTgAD transgenic mice. Given this exiting profile of action, this novel dual agonist might represent great promises for people suffering from Alzheimer's disease. Future experiments will show whether PPARβ/δ activation by the compound is also angiogenic, as reported for other models of PPARβ/δ stimulation [17–21] and if this novel therapeutic approach for Alzheimer's disease is safe in the settings of cancer and ocular disease.

Citation: Wagner, N.; Wagner, K.-D. Recent Insights into the Role of PPARs in Disease. *Cells* **2023**, *12*, 1572. https://doi.org/10.3390/cells12121572

Received: 29 May 2023
Accepted: 4 June 2023
Published: 7 June 2023

Melissa Rayner and colleagues explored in their paper the utility of several compounds with PPARγ agonist activity in a different setting of neurological disease, i.e., peripheral nerve injury (PNI) [22]. The regeneration and remyelination of damaged nerves are essential for functional recovery, but only little progress had been made in clinical settings to stimulate repair. PPARγ acts as an inhibitor of the Rho/ROCK pathway. This inhibition enhances axon regeneration after PNI [23], making PPARγ a good candidate to stimulate peripheral nerve regeneration. Non-steroidal anti-inflammatory drugs (NSAIDs) and thiazolidinediones have shown beneficial effects on nerve regeneration, which seems to be mediated via their PPARγ agonist activity [24]. The authors combined in vitro co-culture systems of nerve and Schwann cells and in vivo models of axon regeneration in rats to explore whether the potency of different molecules to induce nerve regeneration in vitro and in vivo corresponds to their PPARγ agonist activity. All tested molecules with PPARγ agonist activity promoted to some extend neuronal outgrowth in vitro, but functional tests in the animals treated in vivo showed no significant differences. Thus, it is possible that additional mechanisms, e.g., PPARγ effects on immune system function, might modify the outcomes in the in vivo versus in vitro situation. It might be an interesting future approach to compare existing clinical data from patients with PNI treated or not treated with different NSAIDs or PPARγ agonists to answer the question of whether they might experience a functional benefit from different PPARγ agonist treatment. Of note, these analyses should be conducted by taking into account possible gender differences as it has emerged, for instance, that pioglitazone produces a female-predominant inhibition of hyperalgesia associated with peripheral nerve injury in rodents [25].

Tomczyk and colleagues report that the PPARγ agonist rosiglitazone improves cardiac and muscle function in a mouse model of Huntington's disease [26]. Huntington's disease is a rare genetic disease affecting the central nervous system, but it also negatively impacts the heart and muscle strength. In a non-diabetic Huntington's disease mouse model, the authors found that treatment with rosiglitazone improves grip strength and cardiac contractile function. These functional improvements were accompanied by an enhancement of the total adenine nucleotides pool, increased glucose oxidation, alterations in mitochondria number and function, and increased total antioxidant status. As heart failure is a frequent cause of death in Huntington's disease patients, it would be highly exciting if similar results could be observed in clinical studies. Additionally, already available data from patients with Huntington's disease might be re-analyzed for a potential medication with PPARγ agonists, which were in use for a long time for the treatment of diabetes.

Papaccio et al. explored another potential use of the PPARγ agonist pioglitazone. They treated vitiligo melanocytes and fibroblasts with pioglitazone and reported increased mRNA and protein levels of anaerobic glycolytic enzymes, restored mitochondrial membrane potential and mitochondrial DNA (mtDNA) copy number, an increase in ATP content and a decrease in reactive oxygen species (ROS) production, and a reversal of a premature senescence phenotype in vitiligo melanocytes [27]. As the current treatment options for vitiligo, an acquired pigmentation disorder of the skin, are limited, the potential use of PPARγ agonists might represent a novel therapeutic opportunity. Future clinical studies will clarify whether vitiligo patients could benefit from this alternative treatment strategy.

Grimaldi et al. explored potential beneficial effects of the PPARγ agonist rosiglitazone on the angiogenic profile of preeclampsia (PE) placentas [28]. PE is one of the most common causes of maternal-fetal morbidity and mortality. Placentas in PE are characterized by reduced PPARγ expression, disturbed trophoblast differentiation, and the abnormal secretion of angiogenic factors, which causes systemic endothelial damage and organ dysfunction. Thus, the idea of activating PPARγ to induce the normalization of these alterations seems to be straightforward. The authors cultured normal and PE placenta tissue in the presence or absence of rosiglitazone and used cell culture supernatants to characterize angiogenic properties in human umbilical vein endothelial cell (HUVEC) tube formation assays. They showed beneficial effects of rosiglitazone treatment on the angiogenic profile in the human preeclamptic placenta through a reduction in anti-angiogenic angiopoietin-

2 and soluble endoglin and the upregulation of pro-angiogenic placental growth factor, fibroblast growth factor-2, heparin-binding epidermal growth factor, and follistatin. The treatment of PE placental tissue with the PPARγ agonist enhanced the angiogenic profile of HUVECs exposed to the cell culture supernatant. Thus, it will be highly interesting to see in future studies whether rosiglitazone represents a therapeutic opportunity for PE. Besides this original investigation, the role of PPARs in PE has been reviewed recently [29].

Li and colleagues also investigated the role and potential therapeutic opportunities of PPARγ activation in the placenta in a different context [30]. It is known that exposure to the antibacterial agent triclosan (TCS), which acts also as endocrine disruptor, results in placental abnormalities, increased abortion rates, and the reduced size of fetuses and newborns. The authors show that TCS downregulates the expression of PPARγ and its downstream genes HMOX1, ANGPTL4, VEGFA, MMP-2, and MMP-9 and upregulates inflammatory genes p65, IL-6, IL-1β, and TNF-α in vitro and in vivo. The overexpression of PPARγ or activation of the receptor by rosiglitazone improved cell viability, migration, and angiogenesis and reduced the inflammatory response caused by TCS. The knockdown or inhibition of PPARγ had the opposite effects. Finally, TCS caused placenta dysfunction characterized by a significant decrease in the weight and size of the placenta and fetus, while the PPARγ agonist rosiglitazone reduced this damage in mice. Hopefully, in the future, we will be able to reduce industrial pollution with endocrine disruptors instead of treating the damage with PPARγ agonists.

From in vitro studies and pre-clinical animal models in vivo, it has been known that PPARγ activation protects kidney podocytes from injury and reduces proteinuria and glomerular diseases (reviewed in [31]). However, PPARγ signaling in podocytes seems to be different from its well-understood role in driving insulin sensitivity and adipogenesis. Bryant et al. showed in this Special Issue of *Cells* that the expression of PPARγ splice variants differ between podocytes and adipocytes and liver [32]. Podocytes express the PPARγ Var 1 (encoding γ1) but not γ2, which is expressed in adipocytes. Low levels of PPARγ Var4, Var3, Var11, VartORF4, and Var9 were also detected in podocytes. Interestingly, a distinct podocyte vs. adipocyte PPAR-promoter response element was also identified in podocytes, which puts our concept of common PPAR-response element sequences in question. This study represents a rationale for the search of novel PPARγ splice-specific agonists, which could be highly specific for targeted therapies.

Besides the multiple roles of PPARγ activation, PPARα agonists also exert several functions in addition to lipid lowering [33]. Qiu and colleagues show that the activation of PPARα ameliorates cardiac fibrosis in Dsg2-deficient arrhythmogenic cardiomyopathy [34]. Arrhythmogenic cardiomyopathy (ACM) represents a genetic disease characterized by the progressive fibro-fatty replacement of cardiac myocytes. Mutations in desmoglein-2 (Dsg2) are one of the reasons for the development of ACM. The authors showed that cardiac-specific Dsg2 knockout mice develop fibrosis, have reduced PPARα levels, and increased STAT3 and SMAD3 activity. Fenofibrate treatment as well as viral PPARα overexpression improved cardiac fibrosis and decreased the phosphorylation of STAT3, SMAD3, and AKT in cardiac-specific Dsg2 knockout mice, suggesting a novel indication for the use of PPARα agonists in ACM patients.

Adamowicz et al. postulate in this Special Issue that hepatic PPARα is suppressed in primary biliary cholangitis, which might be modulated by miR-155 [35]. The authors show that PPARα expression is reduced in human biliary cholangitis samples compared to controls. Additionally, miR-155 and miR-21 were increased in the samples from patients with primary biliary cholangitis. In human hepatocarcinoma (HepG2) and normal human cholangiocyte (NHC) cells transfected with miR-155 or miR-21 mimics, the effect on PPARα was variable. Whether these microRNAs have a direct effect on PPARα expression and whether the reduction in PPARα in primary biliary cholangitis is causative for disease progression remains to be determined.

Non-alcoholic steatohepatitis (NASH) is a stage of non-alcoholic fatty liver disease (NAFLD) which might lead to fibrosis, liver cirrhosis, and carcinomas. The only established clinical treatment is bariatric surgery, but trials with PPAR agonists, i.e., the dual PPARα/β/δ agonist elafibranor for the treatment of NASH, were conducted [36]. In the current Special Issue, Boeckmans and colleagues compared the transcriptome profiles of in vitro NASH human cell culture models with cells treated with elafibranor. Additionally, they compared the elafibranor-induced gene expression modulation to the transcriptome data of patients with improved/resolved NAFLD/NASH upon bariatric surgery [37]. The authors found a 35% overlap of transcriptome data from NASH patients with cell culture models exposed to NASH-inducing triggers. Elafibranor partially reversed the transcriptional modulations. Peroxisome Proliferator Activated Receptor Alpha, PPARG Coactivator 1 Alpha, and Sirtuin 1 were the major common upstream regulators upon exposure to elafibranor. Angiopoietin-Like 4, pyruvate dehydrogenase kinase 4, and perilipin 2 were commonly upregulated by elafibranor in the in vitro NASH models but not in patient samples after bariatric surgery. These generated large datasets are very informative. They provide evidence for the differences in the in vitro models with the ex vivo patient data and also, not unexpectedly, a different response to the dual PPAR agonist and bariatric surgery. Still a major challenge for the use of the in vitro NASH models for large-scale drug screening remains the identification of a common robust marker set, which ideally would be easy to measure and analyze.

In a highly exciting study, Murakami et al. used a combination of the selective PPARα modulator pemafibrate and sodium-glucose cotransporter 2 inhibitor tofogliflozin to treat NASH in a mouse model [38]. They carefully investigated histopathological changes in NASH animals, mice with single compound treatment, and the combination of pemafibrate and tofogliflozin compared to control animals. The authors provide evidence that the combination effectively reduces hepatocyte degeneration and improves hypertriglyceridemia, hyperglycemia, and macro vesicular steatosis. Most importantly, the combination significantly reduced the number of tumors and improved survival in the mouse NASH model. Hopefully, clinical studies in the future will provide comparable results in human patients, which would represent a major breakthrough in the field.

In the reviews as part of the current Special Issue, Kim and colleagues summarized current knowledge about the potential involvement of the different PPARs in infectious diseases [39]. Although the involvement of PPARs in metabolism and inflammatory responses is well characterized, relatively little is known about the modulatory roles of PPARs in viral, bacterial, and parasitic infections. The authors carefully summarize the current knowledge in the field and introduce future perspectives as, currently, no PPAR therapeutics are in use to treat infectious diseases.

Zhao et al. reviewed the involvement of PPARs in breast cancer. They introduce the structure of the different PPARs and the mechanisms of PPAR-mediated gene regulation and provide examples for the structure of PPAR agonists and antagonists. Afterwards, they describe in detail the knowledge of each PPAR isotype in breast cancer, including multiple observations reported from different cell lines and a potential modulation of PPAR effects by estrogen receptors, which are of special importance in breast cancer and are an established therapeutic target.

Basilotta et al. explore the potential therapeutic effects of PPAR ligands in glioblastoma [40]. Glioblastoma is the most aggressive brain tumor, with very limited therapeutic options; thus, additional therapeutic strategies are urgently needed. PPARα and PPARγ activation are thought to inhibit tumor growth, while PPARβ/δ seems to be mostly protumorigenic, although it might reduce the cardiotoxicity of doxorubicin used for glioblastoma treatment. The PPARα agonist fenofibrate has received the most attention due to its capacity to reduce the proliferation of glioblastoma cells through both PPAR-dependent and PPAR-independent mechanisms. PPAR-γ ligands have been reported to induce cell death in glioblastoma cells. Further studies are required to define the potential clinical use of PPAR modulators for glioblastoma.

Ballav and colleagues, in their review, focus on the utility of PPARγ activators for the treatment of cancer [41]. They introduce structure and functional diversity of PPARγ forms, describe various ligands and their use in different diseases, and finally focus on the use of PPARγ partial agonists for cancer treatment and provide a good overview of ongoing clinical trials.

After a successful review of PPARβ/δ in the hallmarks of cancer [42], we decided for the current Special Issue to provide a comprehensive analysis of the literature for all PPAR isoforms in relation to the hallmarks of cancer. We describe the known roles of PPARs in cancer cell proliferation, cell death, angiogenesis, invasion and metastasis, replicative immortality, tumor metabolism, and cancer immunity and graphically illustrate the signaling pathways involved therein [43]. Of note, the hallmarks of cancer are a didactic concept, and from a single positive effect on one of the hallmarks, a potential therapeutic effect of PPAR modulation cannot be predicted. Clinical studies and profound retrospective analyses of the available data are required to answer the therapeutic potential of PPAR modulation for cancer.

Mukherjee et al. summarize the role of PPARs and non-coding RNAs in non-alcoholic fatty liver disease (NAFLD) [44]. They introduce the causes and pathology of NAFLD and describe the roles of the different PPAR isoforms in this disease. Afterwards, they describe in detail knowledge about microRNAs, long non-coding RNAs, and circular RNAs in NAFLD and the potential regulation of PPARs by non-coding RNAs. As RNA-based therapies are becoming increasingly focused on, they provide an outlook of how these potential therapies might be used in the future to modify the progression of non-alcoholic fatty liver disease.

Siblini and colleagues explore the influence of methionine needs and the SIRT1/PGC-1α/PPAR-α axis on normal and cancer stem cells [45]. Normal and cancer stem cells share some common features of self-renewal and differentiation capacity. The one-carbon metabolism (OCM) plays an important role in self-renewal and differentiation through its role in the endogenous synthesis of methionine and S-adenosylmethionine (SAM), the universal donor of methyl groups in eukaryotic cells. Stem cells' reliance on methionine is linked to several mechanisms, including a high methionine flux or low endogenous methionine biosynthesis. The authors highlight the influence of SIRT1 on SAM synthesis and suggest the role of PGC-1α/PPAR-α in impaired stemness produced by methionine deprivation. Of high interest is the potential of methionine restriction in regenerative medicine and cancer treatment.

Guo et al. review the potential roles of PPARs in the fetal origin of adult diseases [46]. The fetal origin of adult disease (FOAD) hypothesis postulates that early events might predispose the development of certain diseases later in life. More than 30 years ago, it was already noted that an increased risk of death from stroke and coronary heart disease in adults was related to a low birth weight [47]. Later, the concept was expanded to different diseases, and it was even reported that transgenerational effects have their origin in early embryos [48]. The roles of PPARs in FOAD have been increasingly appreciated due to their wide variety of biological actions. Exposure to different events in early life has a significant influence on the methylation pattern of PPARs in several organs, which can affect development and health throughout the course of life. In this excellent review, the authors have compiled recent data on the role of PPARs in the fetal origin of different adult diseases and provide potential ways to prevent such diseases in the future.

In summary, this Special Issue, "The Role of PPARs in Disease II", represents an excellent collection of original articles which might open up new perspectives for therapeutic interventions and comprehensive up-to-date reviews on several different topics of PPAR signaling in different disease processes. In contrast to earlier descriptions, the roles of PPARs are not limited to metabolic alterations as many different opportunities in neurological, cardiovascular, hepatic diseases, regenerative medicine, and cancer emerge.

Author Contributions: Conceptualization, K.-D.W. and N.W.; formal analysis, K.-D.W. and N.W.; investigation, K.-D.W. and N.W.; writing—original draft preparation, K.-D.W. and N.W.; writing—review and editing, K.-D.W. and N.W.; project administration, K.-D.W. and N.W.; funding acquisition, K.-D.W. and N.W. All authors have read and agreed to the published version of the manuscript.

Funding: This research was funded by Fondation pour la Recherche Medicale, grant number FRM DPC20170139474 (K.-D.W.), Fondation ARC pour la recherche sur le cancer, grant number n°PJA 20161204650 (N.W.), Gemluc (N.W.), Plan Cancer. INSERM (K.-D.W.), Agence Nationale de la Recherche, grant R19125AA "Senage" (K.-D.W.), and Fondation ARC pour la recherche sur le cancer, grant number n°PJA 20161204650 (K.-D.W.).

Conflicts of Interest: The authors declare no conflict of interest.

References

1. Wagner, N.; Wagner, K.D. The Role of PPARs in Disease. *Cells* **2020**, *9*, 2367. [CrossRef] [PubMed]
2. Wagner, K.D.; Wagner, N. Peroxisome proliferator-activated receptor beta/delta (PPARbeta/delta) acts as regulator of metabolism linked to multiple cellular functions. *Pharmacol. Ther.* **2010**, *125*, 423–435. [CrossRef] [PubMed]
3. Fougerat, A.; Montagner, A.; Loiseau, N.; Guillou, H.; Wahli, W. Peroxisome Proliferator-Activated Receptors and Their Novel Ligands as Candidates for the Treatment of Non-Alcoholic Fatty Liver Disease. *Cells* **2020**, *9*, 1638. [CrossRef]
4. Montagner, A.; Wahli, W.; Tan, N.S. Nuclear receptor peroxisome proliferator activated receptor (PPAR) β/δ in skin wound healing and cancer. *Eur. J. Dermatol.* **2015**, *25* (Suppl. S1), 4–11. [CrossRef]
5. Michalik, L.; Wahli, W. PPARs Mediate Lipid Signaling in Inflammation and Cancer. *PPAR Res.* **2008**, *2008*, 134059. [CrossRef] [PubMed]
6. Issemann, I.; Green, S. Activation of a member of the steroid hormone receptor superfamily by peroxisome proliferators. *Nature* **1990**, *347*, 645–650. [CrossRef]
7. Dreyer, C.; Krey, G.; Keller, H.; Givel, F.; Helftenbein, G.; Wahli, W. Control of the peroxisomal beta-oxidation pathway by a novel family of nuclear hormone receptors. *Cell* **1992**, *68*, 879–887. [CrossRef]
8. Havel, R.J.; Kane, J.P. Drugs and lipid metabolism. *Annu. Rev. Pharmacol.* **1973**, *13*, 287–308. [CrossRef]
9. Reddy, J.K.; Goel, S.K.; Nemali, M.R.; Carrino, J.J.; Laffler, T.G.; Reddy, M.K.; Sperbeck, S.J.; Osumi, T.; Hashimoto, T.; Lalwani, N.D. Transcription regulation of peroxisomal fatty acyl-CoA oxidase and enoyl-CoA hydratase/3-hydroxyacyl-CoA dehydrogenase in rat liver by peroxisome proliferators. *Proc. Natl. Acad. Sci. USA* **1986**, *83*, 1747–1751. [CrossRef]
10. Hijikata, M.; Ishii, N.; Kagamiyama, H.; Osumi, T.; Hashimoto, T. Structural analysis of cDNA for rat peroxisomal 3-ketoacyl-CoA thiolase. *J. Biol. Chem.* **1987**, *262*, 8151–8158. [CrossRef]
11. Hardwick, J.P.; Song, B.J.; Huberman, E.; Gonzalez, F.J. Isolation, complementary DNA sequence, and regulation of rat hepatic lauric acid omega-hydroxylase (cytochrome P-450LA omega). Identification of a new cytochrome P-450 gene family. *J. Biol. Chem.* **1987**, *262*, 801–810. [CrossRef] [PubMed]
12. Göttlicher, M.; Widmark, E.; Li, Q.; Gustafsson, J.A. Fatty acids activate a chimera of the clofibric acid-activated receptor and the glucocorticoid receptor. *Proc. Natl. Acad. Sci. USA* **1992**, *89*, 4653–4657. [CrossRef] [PubMed]
13. Steinke, I.; Govindarajulu, M.; Pinky, P.D.; Bloemer, J.; Yoo, S.; Ward, T.; Schaedig, T.; Young, T.; Wibowo, F.S.; Suppiramaniam, V.; et al. Selective PPAR-Delta/PPAR-Gamma Activation Improves Cognition in a Model of Alzheimer's Disease. *Cells* **2023**, *12*, 1116. [CrossRef] [PubMed]
14. Pedersen, W.A.; McMillan, P.J.; Kulstad, J.J.; Leverenz, J.B.; Craft, S.; Haynatzki, G.R. Rosiglitazone attenuates learning and memory deficits in Tg2576 Alzheimer mice. *Exp. Neurol.* **2006**, *199*, 265–273. [CrossRef]
15. Escribano, L.; Simón, A.M.; Gimeno, E.; Cuadrado-Tejedor, M.; López de Maturana, R.; García-Osta, A.; Ricobaraza, A.; Pérez-Mediavilla, A.; Del Río, J.; Frechilla, D. Rosiglitazone rescues memory impairment in Alzheimer's transgenic mice: Mechanisms involving a reduced amyloid and tau pathology. *Neuropsychopharmacology* **2010**, *35*, 1593–1604. [CrossRef]
16. Geldmacher, D.S.; Fritsch, T.; McClendon, M.J.; Landreth, G. A randomized pilot clinical trial of the safety of pioglitazone in treatment of patients with Alzheimer disease. *Arch. Neurol.* **2011**, *68*, 45–50. [CrossRef]
17. Piqueras, L.; Reynolds, A.R.; Hodivala-Dilke, K.M.; Alfranca, A.; Redondo, J.M.; Hatae, T.; Tanabe, T.; Warner, T.D.; Bishop-Bailey, D. Activation of PPARbeta/delta induces endothelial cell proliferation and angiogenesis. *Arterioscler. Thromb. Vasc. Biol.* **2007**, *27*, 63–69. [CrossRef]
18. Wagner, K.-D.; Du, S.; Martin, L.; Leccia, N.; Michiels, J.-F.; Wagner, N. Vascular PPARβ/δ Promotes Tumor Angiogenesis and Progression. *Cells* **2019**, *8*, 1623. [CrossRef]
19. Wagner, K.D.; Vukolic, A.; Baudouy, D.; Michiels, J.F.; Wagner, N. Inducible Conditional Vascular-Specific Overexpression of Peroxisome Proliferator-Activated Receptor Beta/Delta Leads to Rapid Cardiac Hypertrophy. *PPAR Res.* **2016**, *2016*, 7631085. [CrossRef]
20. Wagner, N.; Jehl-Piétri, C.; Lopez, P.; Murdaca, J.; Giordano, C.; Schwartz, C.; Gounon, P.; Hatem, S.N.; Grimaldi, P.; Wagner, K.D. Peroxisome proliferator-activated receptor beta stimulation induces rapid cardiac growth and angiogenesis via direct activation of calcineurin. *Cardiovasc. Res.* **2009**, *83*, 61–71. [CrossRef]

21. Bishop-Bailey, D. A Role for PPARbeta/delta in Ocular Angiogenesis. *PPAR Res.* **2008**, *2008*, 825970. [CrossRef] [PubMed]
22. Rayner, M.L.D.; Kellaway, S.C.; Kingston, I.; Guillemot-Legris, O.; Gregory, H.; Healy, J.; Phillips, J.B. Exploring the Nerve Regenerative Capacity of Compounds with Differing Affinity for PPARγ In Vitro and In Vivo. *Cells* **2022**, *12*, 42. [CrossRef] [PubMed]
23. Hiraga, A.; Kuwabara, S.; Doya, H.; Kanai, K.; Fujitani, M.; Taniguchi, J.; Arai, K.; Mori, M.; Hattori, T.; Yamashita, T. Rho-kinase inhibition enhances axonal regeneration after peripheral nerve injury. *J. Peripher. Nerv. Syst.* **2006**, *11*, 217–224. [CrossRef] [PubMed]
24. Rayner, M.L.D.; Healy, J.; Phillips, J.B. Repurposing Small Molecules to Target PPAR-γ as New Therapies for Peripheral Nerve Injuries. *Biomolecules* **2021**, *11*, 1301. [CrossRef] [PubMed]
25. Santos, D.F.S.; Donahue, R.R.; Laird, D.E.; Oliveira, M.C.G.; Taylor, B.K. The PPARγ agonist pioglitazone produces a female-predominant inhibition of hyperalgesia associated with surgical incision, peripheral nerve injury, and painful diabetic neuropathy. *Neuropharmacology* **2022**, *205*, 108907. [CrossRef]
26. Tomczyk, M.; Braczko, A.; Mierzejewska, P.; Podlacha, M.; Krol, O.; Jablonska, P.; Jedrzejewska, A.; Pierzynowska, K.; Wegrzyn, G.; Slominska, E.M.; et al. Rosiglitazone Ameliorates Cardiac and Skeletal Muscle Dysfunction by Correction of Energetics in Huntington's Disease. *Cells* **2022**, *11*, 2662. [CrossRef]
27. Papaccio, F.; Bellei, B.; Ottaviani, M.; D'Arino, A.; Truglio, M.; Caputo, S.; Cigliana, G.; Sciuto, L.; Migliano, E.; Pacifico, A.; et al. A Possible Modulator of Vitiligo Metabolic Impairment: Rethinking a PPARγ Agonist. *Cells* **2022**, *11*, 3583. [CrossRef]
28. Grimaldi, B.; Kohan-Ghadr, H.R.; Drewlo, S. The Potential for Placental Activation of PPARγ to Improve the Angiogenic Profile in Preeclampsia. *Cells* **2022**, *11*, 3514. [CrossRef]
29. Psilopatis, I.; Vrettou, K.; Fleckenstein, F.N.; Theocharis, S. The Role of Peroxisome Proliferator-Activated Receptors in Preeclampsia. *Cells* **2023**, *12*, 647. [CrossRef]
30. Li, J.; Quan, X.; Zhang, Y.; Yu, T.; Lei, S.; Huang, Z.; Wang, Q.; Song, W.; Yang, X.; Xu, P. PPARγ Regulates Triclosan Induced Placental Dysfunction. *Cells* **2021**, *11*, 86. [CrossRef]
31. Agrawal, S.; He, J.C.; Tharaux, P.L. Nuclear receptors in podocyte biology and glomerular disease. *Nat. Rev. Nephrol.* **2021**, *17*, 185–204. [CrossRef] [PubMed]
32. Bryant, C.; Webb, A.; Banks, A.S.; Chandler, D.; Govindarajan, R.; Agrawal, S. Alternatively Spliced Landscape of PPARγ mRNA in Podocytes Is Distinct from Adipose Tissue. *Cells* **2022**, *11*, 3455. [CrossRef] [PubMed]
33. Balakumar, P.; Sambathkumar, R.; Mahadevan, N.; Muhsinah, A.B.; Alsayari, A.; Venkateswaramurthy, N.; Dhanaraj, S.A. Molecular targets of fenofibrate in the cardiovascular-renal axis: A unifying perspective of its pleiotropic benefits. *Pharmacol. Res.* **2019**, *144*, 132–141. [CrossRef]
34. Qiu, Z.; Zhao, Y.; Tao, T.; Guo, W.; Liu, R.; Huang, J.; Xu, G. Activation of PPARα Ameliorates Cardiac Fibrosis in Dsg2-Deficient Arrhythmogenic Cardiomyopathy. *Cells* **2022**, *11*, 3184. [CrossRef]
35. Adamowicz, M.; Kempinska-Podhorodecka, A.; Abramczyk, J.; Banales, J.M.; Milkiewicz, P.; Milkiewicz, M. Suppression of Hepatic PPARα in Primary Biliary Cholangitis Is Modulated by miR-155. *Cells* **2022**, *11*, 2880. [CrossRef]
36. Boeckmans, J.; Natale, A.; Rombaut, M.; Buyl, K.; Rogiers, V.; De Kock, J.; Vanhaecke, T.; Rodrigues, R.M. Anti-NASH Drug Development Hitches a Lift on PPAR Agonism. *Cells* **2019**, *9*, 37. [CrossRef] [PubMed]
37. Boeckmans, J.; Gatzios, A.; Heymans, A.; Rombaut, M.; Rogiers, V.; De Kock, J.; Vanhaecke, T.; Rodrigues, R.M. Transcriptomics Reveals Discordant Lipid Metabolism Effects between In Vitro Models Exposed to Elafibranor and Liver Samples of NAFLD Patients after Bariatric Surgery. *Cells* **2022**, *11*, 893. [CrossRef]
38. Murakami, K.; Sasaki, Y.; Asahiyama, M.; Yano, W.; Takizawa, T.; Kamiya, W.; Matsumura, Y.; Anai, M.; Osawa, T.; Fruchart, J.C.; et al. Selective PPARα Modulator Pemafibrate and Sodium-Glucose Cotransporter 2 Inhibitor Tofogliflozin Combination Treatment Improved Histopathology in Experimental Mice Model of Non-Alcoholic Steatohepatitis. *Cells* **2022**, *11*, 720. [CrossRef]
39. Kim, I.S.; Silwal, P.; Jo, E.K. Peroxisome Proliferator-Activated Receptor-Targeted Therapies: Challenges upon Infectious Diseases. *Cells* **2023**, *12*, 650. [CrossRef]
40. Basilotta, R.; Lanza, M.; Casili, G.; Chisari, G.; Munao, S.; Colarossi, L.; Cucinotta, L.; Campolo, M.; Esposito, E.; Paterniti, I. Potential Therapeutic Effects of PPAR Ligands in Glioblastoma. *Cells* **2022**, *11*, 621. [CrossRef]
41. Ballav, S.; Biswas, B.; Sahu, V.K.; Ranjan, A.; Basu, S. PPAR-γ Partial Agonists in Disease-Fate Decision with Special Reference to Cancer. *Cells* **2022**, *11*, 3215. [CrossRef] [PubMed]
42. Wagner, N.; Wagner, K.D. PPAR Beta/Delta and the Hallmarks of Cancer. *Cells* **2020**, *9*, 1133. [CrossRef] [PubMed]
43. Wagner, N.; Wagner, K.D. Peroxisome Proliferator-Activated Receptors and the Hallmarks of Cancer. *Cells* **2022**, *11*, 2432. [CrossRef] [PubMed]
44. Mukherjee, A.G.; Wanjari, U.R.; Gopalakrishnan, A.V.; Katturajan, R.; Kannampuzha, S.; Murali, R.; Namachivayam, A.; Ganesan, R.; Renu, K.; Dey, A.; et al. Exploring the Regulatory Role of ncRNA in NAFLD: A Particular Focus on PPARs. *Cells* **2022**, *11*, 3959. [CrossRef]
45. Siblini, Y.; Namour, F.; Oussalah, A.; Guéant, J.L.; Chéry, C. Stemness of Normal and Cancer Cells: The Influence of Methionine Needs and SIRT1/PGC-1α/PPAR-α Players. *Cells* **2022**, *11*, 3607. [CrossRef]
46. Guo, J.; Wu, J.; He, Q.; Zhang, M.; Li, H.; Liu, Y. The Potential Role of PPARs in the Fetal Origins of Adult Disease. *Cells* **2022**, *11*, 3474. [CrossRef]

47. Barker, D.J.; Osmond, C. Infant mortality, childhood nutrition, and ischaemic heart disease in England and Wales. *Lancet* **1986**, *1*, 1077–1081. [CrossRef]
48. Wagner, K.D.; Wagner, N.; Ghanbarian, H.; Grandjean, V.; Gounon, P.; Cuzin, F.; Rassoulzadegan, M. RNA induction and inheritance of epigenetic cardiac hypertrophy in the mouse. *Dev. Cell* **2008**, *14*, 962–969. [CrossRef]

cells

MDPI

Article

Selective PPAR-Delta/PPAR-Gamma Activation Improves Cognition in a Model of Alzheimer's Disease

Ian Steinke [1], Manoj Govindarajulu [1], Priyanka Das Pinky [1], Jenna Bloemer [2], Sieun Yoo [1], Tracey Ward [3], Taylor Schaedig [1], Taylor Young [1], Fajar Setyo Wibowo [1], Vishnu Suppiramaniam [1,4] and Rajesh H. Amin [1,*]

1 Department of Drug Discovery and Development, Auburn University, Auburn, AL 36879, USA
2 Department of Pharmaceutical and Biomedical Sciences, Touro College of Pharmacy, New York, NY 10027, USA
3 Department of Pharmaceutical Sciences, Ferris State University, Big Rapids, MI 49307, USA
4 College of Science and Mathematics, Kennesaw State University, Kennesaw, GA 31044, USA
* Correspondence: rajamin@auburn.edu

Abstract: *Background:* The continuously increasing association of Alzheimer's disease (AD) with increased mortality rates indicates an unmet medical need and the critical need for establishing novel molecular targets for therapeutic potential. Agonists for peroxisomal proliferator activating receptors (PPAR) are known to regulate energy in the body and have shown positive effects against Alzheimer's disease. There are three members of this class (delta, gamma, and alpha), with PPAR-gamma being the most studied, as these pharmaceutical agonists offer promise for AD because they reduce amyloid beta and tau pathologies, display anti-inflammatory properties, and improve cognition. However, they display poor brain bioavailability and are associated with several adverse side effects on human health, thus limiting their clinical application. *Methods*: We have developed a novel series of PPAR-delta and PPAR-gamma agonists in silico with AU9 as our lead compound that displays selective amino acid interactions focused upon avoiding the Tyr-473 epitope in the PPAR-gamma AF2 ligand binding domain. *Results*: This design helps to avoid the unwanted side effects of current PPAR-gamma agonists and improve behavioral deficits and synaptic plasticity while reducing amyloid-beta levels and inflammation in 3xTgAD animals. *Conclusions*: Our innovative in silico design of PPAR-delta/gamma agonists may offer new perspectives for this class of agonists for AD.

Keywords: peroxisomal proliferator activating receptor; in silico drug design; neurodegeneration; Alzheimer's disease; synaptic plasticity; behavioral deficits; dendritic spines

Citation: Steinke, I.; Govindarajulu, M.; Pinky, P.D.; Bloemer, J.; Yoo, S.; Ward, T.; Schaedig, T.; Young, T.; Wibowo, F.S.; Suppiramaniam, V.; et al. Selective PPAR-Delta/PPAR-Gamma Activation Improves Cognition in a Model of Alzheimer's Disease. *Cells* **2023**, *12*, 1116. https://doi.org/10.3390/cells12081116

Academic Editors: Kay-Dietrich Wagner, Nicole Wagner, Hermona Soreq and Lars Ove Brandenburg

Received: 31 December 2022
Revised: 23 March 2023
Accepted: 4 April 2023
Published: 8 April 2023

1. Introduction

Peroxisome proliferator-activated receptors (PPARs) are members of the nuclear hormone receptor superfamily that are ligand-activated transcription factors [1]. These receptors are associated with many systemic and cellular functions, including insulin sensitivity and whole-body energy regulation [2,3]. PPARα is abundantly expressed in tissues that utilize fatty acid catabolism, such as the heart, liver, brown adipose tissue, and the kidney. PPARγ, which exists in two isoforms, $\gamma 1$ and $\gamma 2$, is most abundant in expression in adipose tissue and regulates adipocyte differentiation and lipid storage [1]. PPARδ/β has a wide range of expressions and is associated with activity in the skeletal muscle, the gut, and the brain. Interestingly, all three forms are expressed in the brain; however, PPARδ/β is the most abundant in the brain, specifically in neurons and microglial cells [4]. PPARγ is less observed when compared to PPARδ/β in these cells, and PPARα is observed mainly in astrocytes. However, PPARγ agonists are the most extensively investigated form of PPAR agonist for AD therapy. They may serve as potential therapeutic targets for AD because of their positive effects against pathologies and their learning and memory-enhancing effects in transgenic AD animal models [5,6]. Orally administered rosiglitazone (Rosi) at high dosing concentrations (9–18 mg/kg) and extended times (three months) was observed to improve spatial memory and long-term potentiation (LTP) in diabetic and AD rodent models [6–8]. To date, most studies involving pharmacologically activated PPARγ have focused on anti-inflammatory mechanisms and thus altered amyloid-beta (Aβ) and Tau pathology levels. For example, pioglitazone (Pio) treatment (18 mg/kg daily) for 5 weeks improved memory in STZ-diabetic mice on high-fat diets by reducing Aβ40/Aβ42 via inhibition of NF-kB, BACE1, and RAGE in the brain, as well as attenuating hyperglycemia [9]. However, mixed results for Pio in mouse AD models make the application of PPARγ agonists for AD questionable [10]. To further dampen the enthusiasm of PPARγ agonists or thiazolidinediones (TZDs) for AD, these classes of drugs display poor blood–brain barrier (BBB) permeability and deleterious effects on human health [5,11]. Many studies utilizing Pio and Rosi require high concentrations of the drug over an extended period of time to obtain therapeutic effects for improving pathologies associated with AD. Unfortunately, higher concentrations of Rosi or Pio treatment lead to unwanted off-target effects that are life-threatening in humans. Further clinical trials investigating the efficacy of Pio with long-term treatment failed due to complaints from patients with increased weight gain and significant edema [12].

These failures to safely ameliorate or mitigate AD development in animal models and at the clinical level have quenched the clinical applicability of these agonists. Further, they have negated volumes of positive findings verifying these therapeutics' ability to reduce pathology and neurodegeneration associated with AD. Therefore, it is critical to develop novel PPAR-targeted agents that display improved bioavailability and tolerability. Second, because PPARδ is the most abundant PPAR nuclear receptor in the brain, this may offer a new therapeutic target for AD therapy. Recent work in the field of PPAR biology has focused on dual PPAR agonists for AD therapy [13]. The present manuscript discusses the design and development of a potential next-generation PPAR agonist for improving behavioral deficits, synaptic plasticity, and pathologies in a 3xTgAD mouse. We have rationally developed a novel dual PPARδ/γ agonist in silico. Twenty-three nontraditional lead compounds were designed, synthesized, and tested [14]. The design of the PPARγ compounds was based on avoiding the tyrosine-473 site in the AF2 ligand binding domain. Therefore, compounds that displayed robust PPARδ and partial PPARγ activity were then further evaluated for biological significance. The evaluation of these compounds allowed us to advance our lead compound, AU9, for further investigation for improving behavioral deficits, synaptic plasticity and reducing amyloid beta in 3xTgAD mice.

2. Materials and Methods

Animals: 3xTgAD mice (B6;129-Tg(APPSwe, tauP301L) and control (C57BL/6) female mice (wild-type) were obtained from The Jackson Laboratory (stock #008195 and #101045, respectively). According to the Jackson Laboratory, the 3xTgAD male mice exhibit fewer phenotypic traits when compared to females, and hence only female mice were utilized in the current study. All mice were group-housed (4 per cage) with free access to food and water in a temperature- and humidity-controlled colony room with a reverse 12:12 light/dark cycle, thus allowing proper timing of animal studies to occur at normal working hours. All experiments and procedures were conducted in accordance with National Institute of Health (NIH) guidelines and approved by Auburn University Institutional Animal Care and Use Committee (IACUC) and following the ARRIVE guidelines.

Drug treatment: AU9 was reconstituted in saline and administered orally (5 mg/kg) daily, starting at 9 months of age and continuing until 12 months of age. AU9 was designed to have higher water solubility and lipophilicity than traditional TZDs. The calculated partition coefficient (oil/water) of AU9 is on a scale from −2 to +2 of 1, where 2 is highest.

Cell lines: BV2: The murine microglial cell line BV-2 was purchased from ACCEGEN (Cat#ABC-TC212S) and cultured according to the manufacturer's recommendation. Briefly, cultured cells were grown in Dulbecco's Modified Eagle Medium (DMEM) supplemented with 10% fetal bovine serum and 1% penicillin-streptomycin in a humidified CO_2 incubator.

Chinese Hamster Ovary cells–CHO expressing Swedish mutant APP (APPswe) and wild type human PSEN1, were a gift from Dr. Sasha Waggen [15]. Cells were grown in Dulbecco's Modified Eagle Medium (DMEM) supplemented with 10% fetal bovine serum and 1% penicillin-streptomycin and grown in a humidified CO_2 in humidified atmosphere of 5% CO_2/95% air at 37 °C. The cells were cultured in the presence of G418 (200 µg/mL, Invitrogen) and puromycin (7.5 µg/mL, ThermoFischer Scientific, Waltham, MA, USA) to maintain selection for the expression plasmid. The cells were plated at an appropriate density according to each experimental scale.

Chemicals: AU9 was synthesized by Dr. Tracey Ward at Ferris State University. The drug synthesis scheme and validation of purity have been described previously [14].

Modeling: The Schrodinger software suite was used to perform computational analysis of the interactions of AU9, GW0742 and Pio with their respective PPAR ligand binding domains. PDB crystal structures were used to confirm ligand receptor interactions. Ligand docking studies were performed to determine the most stable docking poses determined by the ligand docking scores, which represent the free energy upon binding of the ligand to the proteins active site. Using the lowest energy conformation, a model system was built to explore the molecular dynamics of this interaction using a simulated annealing technique. For comparison, a full PPARδ agonist, GW0742, was used to illustrate key differences in our compound's ability to achieve similar transcriptional activity in vitro.

Protein Preparation: Molecular models for PPARβ/δ and γ were built using the ligand conformation obtained from X-ray crystallographic structures of GW0742 bound to PPARδ (PDB: 3TKM) and Rosiglitazone bound to PPARγ (PDB:5Y2O). Protein crystal structures were imported and prepared using the Maestro modeling software protein preparation workflow. In preprocessing of the protein structures, termini were capped and any missing chains were filled in using Prime. H-bond optimization was performed using PROPKA at a pH of 7.4. Lastly, restrained minimization was performed with convergence of heavy atoms to RMSD of 0.30 Å and deletion of all water molecules within 5 Å of the ligand utilizing the force field OPLS4.

Induced Fit Docking: All ligands were prepared using LigPrep with the OPLS4 force field. Ligands were ionized at a pH of 7.4 ± 0.2 using Epik. Prepared ligands were then subjected to induced fit docking by selecting the centroid of the workspace ligand in each protein complex. Residues were refined within 5.0 Å of ligand poses. Glide re-docking was performed using standard precision. The lowest energy docking score for each ligand was evaluated and used for further molecular modeling experiments.

Model System Generation for Molecular Dynamics: Model systems were built from the best induced fit docking poses using a predefined simple point-charge (SPC) water solvent model. An orthorhombic box shape was chosen with a salt concentration of 0.15 M. The model system was built with the force field OPLS4. The method of simulated annealing was used to evaluate molecular dynamics. Each previously built model system was loaded into the simulation from the workspace. Simulation parameters were set to have a schedule of seventeen temperature changes over the course of 1.2 ns using an NVT ensemble class at 1.01325 bar. Model systems were relaxed prior to simulation.

Behavioral Studies: Novel Object Recognition: Object recognition testing was performed as previously described [16–18]. Briefly, two days before the training, each mouse was handled gently for 5 min and then allowed to familiarize with the apparatus (a plexiglass box 40 cm × 40 cm and 15 cm high) for 10 min per day. The object recognition test consisted of two 10 min trials, one per day. This extended exposure allowed the animals to learn the task. In the first trial (T1), two identical objects were placed in the central part of the box, equally distant from the perimeter. Each mouse was placed in the apparatus and allowed to explore it. Exploration was defined as the mouse pointing its nose toward the object from a distance of no more than 2 cm (as marked by a reference circle). The mouse was then returned to its cage. The second trial (T2) was performed 24 h later to test memory retention. Mice were presented with two objects, a "familiar" (i.e., the one used for T1) and a "novel" object. The last object was placed on the left or the right side of the box in a randomly but balanced manner to minimize potential biases due to a preference for particular locations or objects. To avoid olfactory cues, the objects and the apparatus were cleaned with 70% ethanol after each trial. Recordings were measured by blind reviewers. Exploration of the objects was defined as time spent with the snout orientated toward the object at a distance of <2 cm of the object. Results were expressed as a discrimination index (DI) (T novel − T familiar)/(T familiar + T novel). The following parameters were evaluated: exploratory object preference and time of exploration of the two objects expressed as % exploration of the familiar and % exploration of the novel object; and discrimination.

Y-maze Test: Spatial recognition memory utilizing a two-trial Y-maze task was performed as previously described [16,19–22]. Briefly, the plastic Y-maze apparatus consisted of three arms, with each arm separated by 120 degrees, and visual cues were placed around the Y-maze. The two trials were separated by a 3 h inter-trial interval to evaluate spatial recognition memory. During the first trial (acquisition), mice were allowed to freely explore the two arms of the maze for 10 min. For identification, one arm was the starting arm, where the mice were initially placed, and a second arm was identified as the familiar arm; while a third (novel) arm was closed. During the second trial (retention test), mice were placed back in the starting arm and allowed to explore for 6 min with free access to all three arms (the novel arm was opened). To eliminate odors between animals, the entire Y-Maze, including the arena, was cleaned with 70% ethanol. Blinded reviewers scored recordings, and the total number of entries and time spent in each arm were measured. All data are expressed as means ± SEM. Statistical analyses were performed for all behavioral studies using Student *t*-test with Tukey's post hoc analysis for comparing specific groups. ($p < 0.05$ was considered to indicate statistical significance).

Electrophysiology studies, Hippocampal slice preparation: Animals were euthanized with carbon dioxide, and 350 μm-thick transverse slices were prepared using a Leica VT1200S Vibratome (Leica Microsystems, Wetzlar, Germany). Slices were incubated at room temperature in artificial cerebrospinal fluid (ACSF; 124 mM NaCl, 2.5 mM KCl, 1.5 mM $MgCl_2$, 2 mM $CaCl_2$, 1.25 mM NaH_2PO_4, 25 mM $NaHCO_3$, 25 mM dextrose, pH 7.4) saturated with 95% O_2/5% CO_2 until transfer to the recording chamber.

Extracellular field potential recording: Brain slices were incubated for at least two hours in ACSF and then transferred into a recording chamber for electrophysiological measurements as previously described with continuous ACSF perfusion at 34 °C [19,23,24]. A bipolar stimulating electrode (MicroProbes, Gaithersburg, MD, USA) was placed in the Schaffer collateral pathway. An extracellular recording pipette drawn with the PC-10 Dual-Stage Glass Micropipette Puller (Narishige, Amityville, NY, USA) and filled with ACSF (2–6 MΩ) was placed in the stratum radiatum of CA1 to record field excitatory postsynaptic potentials (fEPSPs). For LTP experiments, stimulus intensity was set at 50% of the amplitude, at which the preliminary population spike appeared. LTP was then induced after 10 min of stable baseline recording using a Theta Burst Stimulation (TBS) protocol (10 bursts of stimuli, each of four pulses at 100 Hz, interburst interval of 200 ms, and 20 s intervals between individual sweeps), and recording was continued for 60 min post-TBS [19,23,24]. LTP was measured as an average of fEPSP slopes from 50–60 min after the end of induction. The data were recorded online using the WinLTP software (University of Bristol, UK). Standard offline analyses of the data were conducted using Prism software (GraphPad Prism version 8, San Diego, CA, USA).

Western blot analysis: Hippocampi tissue from 3xTgAD and wild type (6 mice per group ± AU9) vehicle-control and drug-treated mice were homogenized in a neuronal lysis buffer (N-PER; Neuronal Protein Extraction Reagent, ThermoFisher Scientific) containing a protease cocktail inhibitor (Halt Protease cocktail inhibitor). Lysate was cleared by centrifugation at 4 °C at 12,000× *g* for 20 min. Cleared lysate was collected and the total protein was estimated by Nanodrop (280/260 wavelength) and stored at −80 °C until use. Lysate was mixed with 4X Laemelli buffer containing DTT and heated at 85 °C for 5 min. Protein homogenate was resolved via a 4–16% SurePAGE precast gel (GenScript Biotech), and transferred to nitrocellulose (BioRad, Hercules, CA, USA) by semi-dry techniques

(BioRad). The immuno-blots were blocked with 5% bovine serum albumin in Tris-buffered saline containing 0.1% Tween 20 (TBST) for 1 h, followed by 3x washes with TBST and incubated with primary antibodies overnight and 4 °C. The following day, blots were washed, and probed with secondary anti-rabbit or anti-mouse antibodies (Cell Signaling Technology, Danvers, MA, USA, 1:2000 in TBS-T+BSA solution) for 2 h. Immunoblots were then exposed to ECL reagent (Millipore) and imaged using a LICOR imager. The analyses of bands were based upon densities that were standardized to alpha-tubulin. All data are expressed as means ± SEM. Statistical analyses were performed using Student *t*-test with Tukey's post hoc analysis for comparing specific groups. ($p < 0.05$ was considered to indicate statistical significance).

Reporter assays: To validate the specificity of the compounds towards the activation of PPARδ and PPARγ, we utilized a PPARδ or PPARγ ligand binding domain driven GAL4 reporter HEK293 stable cell line system, purchased from Signosis (SanDiego, CA, USA). Briefly cells were plated into 24-well plates in triplicate, with 6 independent plates, followed the next day with increasing concentrations of AU9, GW074 or Pio (1 nM–20 μM). Luciferase activity was accomplished using Bright-Glo, assay system (Promega, Madison, WI, USA) and standardized to total protein concentration per well. Alternatively, compounds were tested for the capacity to bind to select PPARδ and PPARγ DNA recognition elements, Peroxisome Proliferator Response Elements (PPRE). The PPRE are unique sites located in the promoter region where PPARs bind and transcriptionally activate the target genes. AP2-PPRE is the PPARγ target involved in adipocyte growth and differentiation and a kind gift from Bruce S. Spiegelman (Addgene) [25]. To test PPARδ activity, we utilized a p4xDRE-Luc plasmid, a kind gift from Bert Vogelsein (Addgene) [26]. These vectors were co-transfected with a Renilla vector (promega) into HEK-293 cells using Jet Prime (PolyPlus, France). Relative light units (RLU) were measured using a Glomax Luminometer (Promega, Madison, WI, USA). Data were standardized to Renilla activity using Dual glo assay (Promega). VP16 vector was used for constitutively active PPARγ and was a kind gift from Mitch Lazar at the University of Pennsylvania. Mutated Human PPAR-gamma Tyr-473 to Phenylalanine plasmid was purchased from Sinobiological. Statistical analyses were performed using Student *t*-test with Tukey's post hoc analysis for comparing specific groups. ($p < 0.05$ was considered to indicate statistical significance).

Gene expression: RNA was extracted from the hippocampi regions (6 mice per group, with 2 hippocampi per mouse combined together) from mice brains using Trizol (Invitrogen). Approximately 200 ng of RNA was converted to cDNA using OneScript Plus cDNA synthesis Kit (Applied Biological Material) followed by qPCR analysis using BlasTaq 2c qPCR MasterMix. Primers used for qRT-PCR were purchased from IDT. Please see the table of primers in the Supplementary Materials. Data were represented by ΔΔCT based upon gene of interest cycle numbers standardized to beta actin ct values.

Rapid Golgi Staining Procedure: Golgi Cox staining procedure followed a previously published protocol [27]. Briefly, whole brains were harvested from mice and stained using the FD Rapid GolgiStain kit (FD NeuroTechnologies). Brains were immersed in a 1:1 mixture of FD Solution A:B for 2 weeks at room temperature in the dark and then transferred to FD Solution C and kept in the dark for an additional 48 h. Solution C was replaced after the first 24 h. Brains (6 Wt ± AU9 and 3xTgAD ± AU9) from 12-month female mice treated with AU9 for three months daily (5 mg/Kg) were cut into approximately 200 μM sections using a Leica vibratome, with no less than 10 slices from regions containing hippocampi, were transferred to gelatin coated slides onto small drops of FD Solution C and sealed using Permount mounting media (. Ten neurons per slice were imaged using a Nikon Ti inverted microscope from the CA1 and CA3 regions of the hippocampus using a z-stacking procedure with 20 slices per neuron with 0.1 μM per optical slice.

Spine length refers to the sum of the lengths of all spine branches on neurons (μm). The spatial density/volume of a spine was the smallest cubic volume that could image the entire spine (μm^3). Student *t*-test was used to analyze the differences among groups. Confidence level was set to 0.05 (p-value) and all the results are presented as the mean ± SEM.

Neurotrophin measurement: The levels of mouse neurotrophins (NGF, BDNF, NT3, and NT4/5) were measured from mice hippocampi (6 mice per group (wt and 3xTgAD) ± AU9) using a commercial ELISA kit (Biosensis) (Cat# BEK-2231). To measure neurotrophin levels, soluble proteins were extracted using a protocol based upon Kolbeck et al. [28]. Briefly, hippocampi were suspended in 20 volume/weight extraction buffer (0.05 M sodium acetate, 1 M sodium chloride, 1% Triton-X100, Roche complete inhibitor cocktail tablet) and homogenized. Protein concentrations were standardized using Nanodrop, followed by an ELISA for neurotrophins, according to the manufacturer's instructions (Biosensis). The resulting measurements (pg) were normalized per mg of total soluble protein. Hippocampal homogenate neurotrophin concentrations were based upon a standard curve from the known concentration and measured by a plated reader at 450 nm. Student t-test was used to analyze the differences among groups. Confidence level was set to 0.05 (p-value) and all the results are presented as the mean ± SEM.

Immunostain for Aβ: Cryosections (3–5 sections per sample, 10–15 μm each, six brains per Wt and 3xTgAD ± AU9) were taken from the CA3–CA1 regions of the hippocampus using a Leica cryostat and fixed with 4% formaldehyde for 10 min followed by permeabilization with 0.1% triton-100 in PBS. Sections were washed three times in PBS, 5 min each and blocked for 2 h in 5% goat serum/PNS solution. Sections were washed three times again in PBS and exposed to primary antibody overnight at 4 °C (1:500 dilution overnight with 5% goat serum overnight). Anti-6E10 antibody (Biolegend) Alexa Fluor 488, which is reactive to aa 1–16 Aβ and to APP, reacts to the abnormally processed isoforms, as well as precursor forms. The following day sections were washed three times in PBS and counterstained with Dapi (Sigma chemical) and mounted with a coverslip using an antifade solution (Molecular probes). Sections were imaged using an inverted fluorescence NikonT1 microscope. The immunofluorescence stained area was determined by the density of immunostain standardized to the total area using ImageJ software. Goat serum was used in place of the primary antibody and was used as a negative control. Images of 3xTgAD with the primary antibody were used as the baseline for time and exposure levels and were then used for all images obtained with these values to nullify background levels.

Aβ ELISA assay in mice: Mouse hippocampi from treated and untreated mice as discussed above (6 per group) Wt and 3xTgAD mice were collected to detect the secreted Aβ1–42 based on the manufacturer's protocol (R&D Systems). The Aβ1–42 concentrations were quantified using values from a standard curve associated with the ELISA kits following the manufacturer's protocol. The optical densities of each well were measured at 450 nm using a microplate reader (Agilent, Santa Clara, CA, USA) and the sample Aβ1–42 concentrations were determined by comparison with the Aβ1–42 standard curves. All readings were in the linear range of the assay. Values were standardized to total protein concentrations. Student t-test was used to analyze the differences among groups. Confidence level was set to 0.05 (p-value) and all the results are presented as the mean ± SEM.

Aβ ELISA assay in APP-Cho cells: AppCho cells were plated in triplicate with six independent plates with increasing concentrations of AU9 (1–20 μM) (24 h). Media was collected the following day to detect the secreted Aβ1–42. The Aβ1–42 concentrations were quantified using values from a standard curve associated with the ELISA kits following the manufacturer's protocol. The optical densities of each well were measured at 450 nm using a microplate reader (Agilent, Santa Clara, CA, USA) and the sample Aβ1–42 concentrations were determined by comparison with the Aβ1–42 standard curves. All readings were in the linear range of the assay. Values were based upon a known standard curve and standardized to total protein concentration. Student t-test was used to analyze the differences among groups. The confidence level was set to 0.05 (p-value) and all the results are presented as the mean ± SEM.

β-secretase activity assay: β-site-APP cleaving enzyme (BACE) or β-Secretase activity was determined fluorimetrically using an β-Secretase activity kit (BioVision, Waltham, MA, USA). APP-Cho cells were pleated in triplicate in 24-well plates with 6 independent plates for each experiment and treated with AU9, GW0742 and Pio for 24 h (10 μM). Values were determined based upon manufacturer's instructions and a standard curve. Beta-secretase activity was represented as relative fluorescence units per mg of total protein. Values were based upon known standard curve and standardized to the total protein concentration. Student *t*-test was used to analyze the differences among groups. Confidence level was set to 0.05 (*p*-value) and all the results are presented as the mean ± SEM.

Nanostring Gene Expression analysis: Supplementary Materials. Hippocampal RNA from 3xTgAD and control mice treated with either saline or AU9 were extracted and purified using an RNA Plus Universal Mini Kit (Cat. #73404, QIAGEN, Germantown, MD, USA). For nCounter analysis, total RNA was diluted to 20 ng/μL and probed using a mouse nCounter Neuropathology Panel (Nanostring Technologies, Seattle, WA, USA). Counts for target genes were normalized to the best-fitting housekeeping genes as determined by nSolver software. The tables observed in the Supplementary Materia are based on units associated with the neuropathology panel. Included are results for Nuerotransmission Supplementary Figure S1, Cellular Stress (Supplementary Figure S2), Cytokine and the associated signaling markers (Supplementary Figure S3), and markers associated with DNA damage (Supplementary Figure S4). Table for QPCR primers (Supplementary Figures S5–S7). Western Whole blots for PSD95, GluA1, GluA2, and the associated standardizing Alpha-Actin are shown in Supplemental Figure S8. Western whole blots for inflammatory markers, including IBA1 and TSPO, are shown in Supplemental Figure S9 and their associated standardizing marker alpha-actin.

Nitrite content: BV2 microglia (2×10^5 cells/mL) were seeded in the 96-well plates as triplicate in each plate and each plate was repeated six times for 12 h, followed by AU9 treatment for 12 h (5–100 μM). Media were changed and LPS (Sigma L2654,100 ng/mL) was added to the media for 24 h. Cultured supernatant was then collected, centrifuged (2500 r.p.m for 20 min) and 100 μL was added to 100 μL of Griess reagent (1% sulfanilamide and 0.1% naphthylethylenediamine dihydrochloride in 2.5% phosphoric acid; Promega G2930, Madison, WI, USA) for 10 min in the dark at room temperature. An ELISA microplate reader was used for the measurement of absorbances at 540 nm. A standard curve was generated in the same manner using $NaNO_2$ for quantitation. Student *t*-test was used to analyze the differences among groups. Confidence level was set to 0.05 (*p*-value) and all the results are presented as the mean ± SEM.

Statistical analysis: All data are expressed as means ± SEM. Statistical analyses were performed using a Student *t*-test or a two-tailed, unpaired *t*-test. Additionally, Tukey post- hoc comparisons were used to compare groups when analysis of variance indicated significant effects, except where expected effects were assessed with planned comparisons. In all cases, $p < 0.05$ was considered to indicate statistical significance. All statistical analyses were performed using the GraphPad Prism version 9 software (La Jolla, CA, USA).

3. Results

In silico design of AU9: PPARδ Site Map Description. The PPARδ ligand binding domain (LBD) consists of a Y-shaped hydrophobic cavity with three functionally different arms, identified by the computationally derived surface site map seen in Figure 1 (PDB: 3TKM). Arm 1 contains the highly conserved helix 12 (H12) C-terminus, referred to as the activation function 2 (AF2) domain [29,30]. Full agonists have been shown to form strong hydrogen bond interactions in the AF2 LBD contained in arm 1. Further stabilization of the ligand–protein complex is achieved by hydrophobic interactions in arm 2. While the AF2 domain is highly conserved across all PPAR isoforms, functional differences within the ligand binding pocket modulate substrate selectivity [31]. Observable differences in the PPARδ's site map descriptors can be seen in Figure 1, highlighting a narrow hydrophobic entrance to the AF2 domain with minimal polar contacts colored yellow and red, respectively. The strength of ligand induced activation of the AF2 domain is strictly controlled by access to the tyrosine 437 residue, colored purple in Figure 1B. The supporting polar contacts in arm 1, coming from the histidine 413 residue on H11 and histidine 287 residue on H5, restrict access of sterically hindered hydrophilic groups to tyrosine 437 (Figure 1C,D). Figure 1C demonstrates the interactions of full PPARδ agonist GW0742 bound within the active site. GW0742 has been shown to have a 300- to 1000-fold preference for PPARδ over the other PPAR isoforms [32]. The GW0742 specificity and strength of activation for PPARδ can be observed in the polar phenoxy acetic acid functional group that can extend deep into arm 1 to form a bifurcated hydrogen bond with tyrosine 437 and histidine 413 (Figure 1C). Additional hydrogen bonding to histidine 287 provides further stability and coordination to the AF2 domain for co-activator recruitment. Additionally, the hydrophobic tail of GW0742 takes advantage of arm 2 hydrophobic contacts to valine 305, tryptophan 228, and valine 312, providing stability for the heterodimerization to RXR.

3.1. PPAR Delta Induced Fit Docking and Molecular Dynamics

To further probe the interactions that AU9 has with the PPARδ LBD, the PDB crystal structure 3TKM was used, which has the PPARδ LBD with full agonist GW0742 bound. Induced fit ligand docking studies were performed to determine the lowest energy conformation, evaluated by the ligand docking scores, which represent the free energy upon binding of the ligand to the protein active site. While computationally more intensive, induced fit docking accurately accounts for both ligand and receptor flexibility [33]. Using the lowest obtained energy conformation, a model system was built to explore the molecular dynamics of this interaction using a simulated annealing technique.

For comparison, the full PPARδ agonist GW0742 was used to illustrate key differences in our compounds' ability to achieve a similar transcriptional activity in vitro. The protein–ligand contacts plot for GW0742 shows interactions occurring with twenty-three amino acid residues in the active site. The largest fraction of interactions occurring >50% throughout the simulation are hydrogen bond contacts to tyrosine 437 (H12), histidine 413 (H11) and histidine 287 (H5). This hydrogen bond network to the AF2 is characteristic of full PPAR agonists across all three isoforms, as ligand interactions with the AF2 lock the protein complex into an active conformation towards the recruitment of co-activators for gene transcription. The remaining contacts are primarily hydrophobic and occur outside of the AF2 domain, as listed in Figure 1G. The protein RMSF for GW0742 shows increased protein stability with all ligand contacts compared to the B-factor, except ligand contacts in H7. The protein–ligand contacts plot for compound AU9 shows interactions occurring with twenty-six amino acids in the active site. The largest fraction of interactions occurring >50% throughout the simulation is a mixture of hydrogen bonds, water bridges, ionic bonds, and hydrophobic bonds to lysine 331, tryptophan 228, glutamate 412, and histidine 413. Protein RMSF for AU9 shows increased stability with all ligand contacts compared to the B-factor, except contact between H1 and H2 (Figure 2B).

These results highlight the importance of PPARδ's requirement for hydrophobic contacts from the ligand to provide stabilization of the protein complex. Compared to GW0742, AU9 forms fewer hydrogen bond interactions in the PPARδ LBD AF2. However, hydrogen bonding interactions to histidine 413 and a water bridge to glutamine 412 demonstrate a comparable stabilization of the AF2 domain. Additionally, the AU9 ligand interactions extend into arm three and provide stability in other regions of the protein that differ from GW0742.

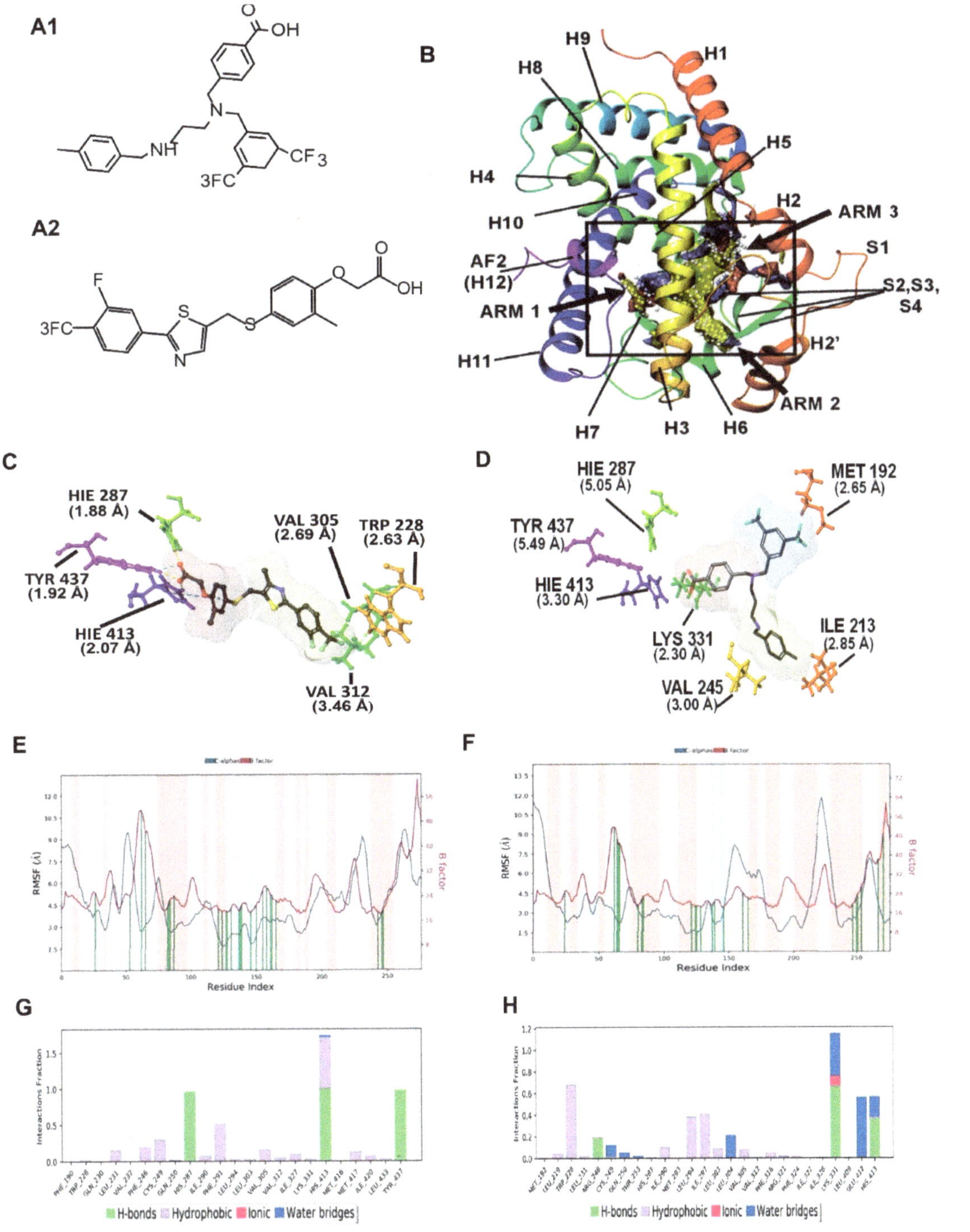

Figure 1. Molecular modeling for PPARδ showing comparison between GW0742 (PPARδ agonist) and AU9. (**A**) Chemical structure of AU9 (**A1**) and GW0742 (**A2**). (**B**) Site map analysis of PPARδ

ligand binding domain (LBD) PDB: 3TKM. Hydrophobic surface map (yellow), hydrogen-bond donor surface map (blue), and hydrogen-bond acceptor surface map (red). α-helices and β-sheets labeled H1–H12 and S1–S4, respectively, from N-terminus to C-terminus. Y-shaped ligand binding pocket where arm 1 contains the AF2 domain, arm 2 is the entrance site, and arm 3 is a secondary ligand binding pocket: (**C**) GW0742 lowest energy conformation and amino acid binding interactions with distances (angstroms). GW0742 forms a hydrogen bond network to the AF2 domain, indicating a full agonist. (**D**) AU9 lowest energy conformation and amino acid binding interactions with distances. AU9 avoids a key interaction at TYR 437 (H12/AF2) yet maintains a critical contact at HIS 413. (**E**,**F**) Molecular dynamics root-mean-square fluctuation (RMSF) plots. The blue graph represents the ligand-induced α-carbon fluctuation overlaid with the red graph experimental b-factor. α-helices and β-sheets are shaded red and blue, respectively, from N-terminus to C-terminus. Vertical green bars indicate ligand-residue contacts. (**E**) GW0742 displays ligand-induced stabilization of the AF2 domain (residue index > 250) represented by a decrease in the RMSF as compared to the b-factor plot, approximately 5-fold. (**F**) AU9 displays ligand-induced stabilization of the AF2 domain while avoiding contact to TYR 437, approximately 1.5-fold. (**G**,**H**) Protein interaction diagram categorized by the fraction and type of interactions maintained throughout the simulation. (**G**) GW0742 maintains hydrogen bonds to the AF2 residue TYR 437 with supporting hydrogen bonds to HIS 413 and HIS 287 as the major contribution of ligand–protein contacts. (**H**) AU9 maintains a mixture of hydrogen bonds/water bridges/ionic interactions at LYS 331, GLU 412, and HIS 413, which predominate through the course of simulation. Analysis of the molecular dynamic simulation are reported in a protein–ligand contacts plot (**E**–**H**), which calculates the nature and fraction of bonds formed with protein residues throughout the simulation. Stacked bar charts are normalized over the course of trajectory, and values greater than 1 indicate that the protein residue is making multiple ligand contacts of different subtypes. The contributions a ligand may have on protein stability were calculated using the protein root mean square fluctuation (RMSF) plot. The protein RMSF plot ((**E**,**F**), Figure 2E,F) characterizes the local changes along the protein chain, relative to the ligand, throughout the course of simulation and is listed on the left-hand y-axis. The y-axis is the experimental x-ray B-factor. B-factors are experimentally determined from data submitted with the PDB x-ray crystal structure and indicate the relative vibrational motion with different atoms located in the structure [34]. Ligand induced changes in the protein RMSF should approximate the experimental B-factor, otherwise significant structural changes are occurring. Green vertical lines represent ligand–protein contacts. Shaded areas represent protein secondary structures, where red are alpha helices and blue are beta-strands ((**G**,**H**) and Figure 2G,H).

PPARγ Induced Fit Docking and Molecular dynamics: The PPARγ LBD active site is similar to PPAR δ, as it also consists of a Y-shaped binding pocket. However, PPARγ has a decreased hydrophobic surface area and an increased polar surface area compared to PPARδ (Figure 3, PDB: 5Y2O). Entrance to the AF2 LBD of PPARγ can accommodate bulkier polar functional groups commonly seen in the thiazolidinedione (TZD) class of PPARγ selective agonists, thus providing substrate specificity [35]. Although full PPARγ agonists profoundly improve blood glucose levels, TZDs are associated with increased edema and heart failure [36]. Thus, AU9 was designed to avoid specific interactions in the PPARγ AF2 domain with the intention of improving clinical efficacy. Pio was used as the cognate ligand in this study to evaluate AU9′s partial agonist profile to that of a full agonist. Pio forms a strong hydrogen bond network to the AF2 domain via contacts at tyrosine 473 (H12), histidine 323 (H4), and histidine 449 (H3) (Figure 3A). Supporting hydrogen bonds from phenylalanine 282 and tyrosine 327 help to provide further stability for the AF2 domain. Pio's lipophilic tail extends into arm 2 to make hydrophobic contacts at valine 339 and isoleucine 341. In comparison, AU9 avoids contact with tyrosine 473 as its branched structure prevents extension into arm 1. AU9 forms a hydrogen bond to tyrosine 327 and lysine 367, providing partial stabilization towards the AF2 domain. Further hydrophobic contacts to glutamate 343, isoleucine 341, and isoleucine 281 provide comparable stabilization of the protein complex to that of Pio (Figure 3B).

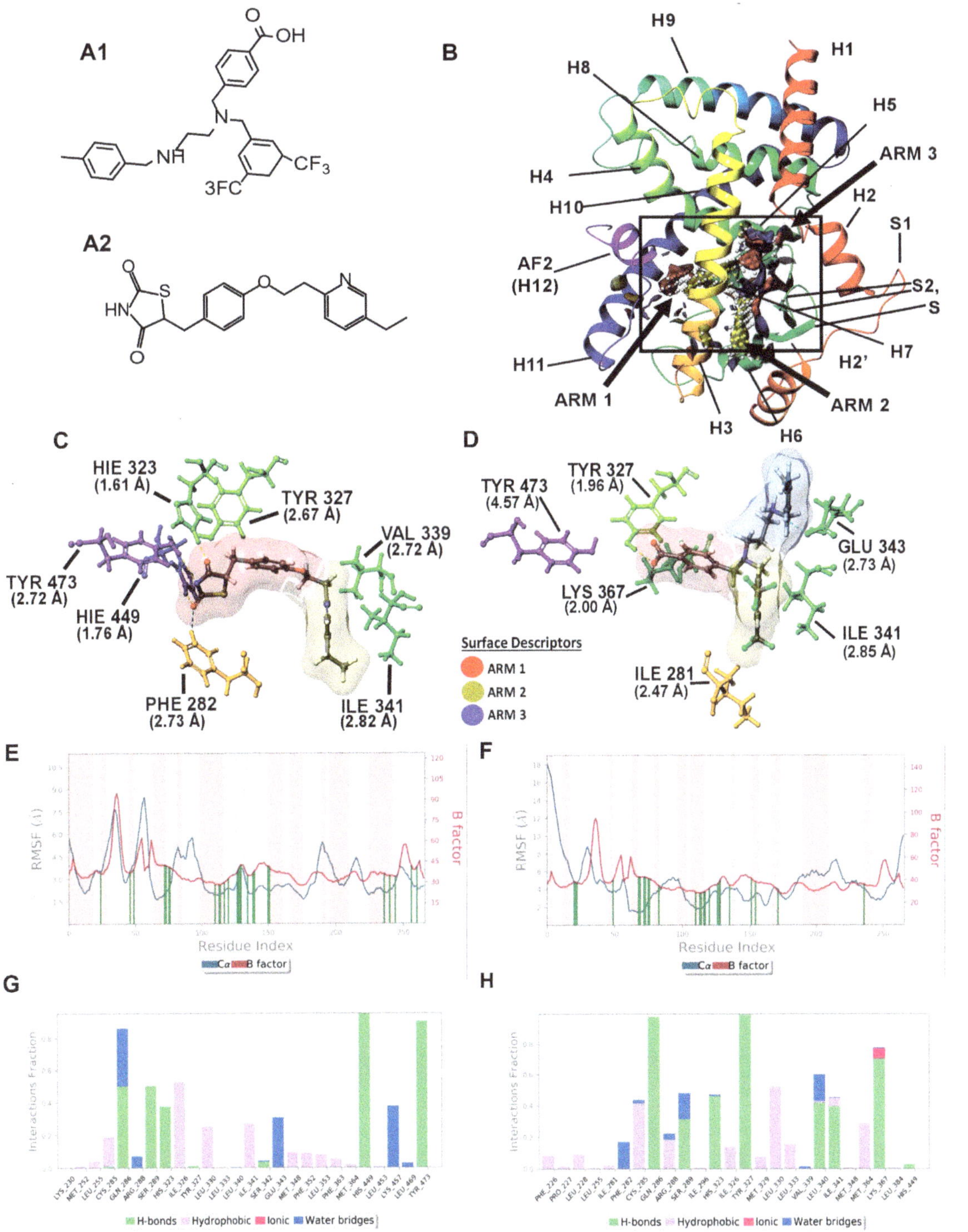

Figure 2. Molecular modeling: In silico modeling of PPARγ interactions between Pio (PPARγ agonist) and AU9. (**A**) Chemical structure of AU9 (**A1**) and Pio (**A2**). (**B**) Site Map analysis of PPARγ LBD PDB:

5Y2O. Hydrophobic surface map (yellow), hydrogen-bond donor surface map (blue), and hydrogen-bond acceptor surface map (red). α-helices and β-sheets labeled H1–H12 and S1–S4, respectively, from N-terminus to C-terminus. Y-shaped ligand binding pocket where arm 1 contains the AF2 domain, arm 2 is the entrance site, and arm 3 is a secondary ligand binding pocket. (**C**) Lowest energy conformation for Pio and amino acid binding interactions with distances (angstroms). Pio forms a hydrogen bond network to the AF2 domain indicative of a full agonist. (**D**) AU9 lowest energy conformation and amino acid binding interactions with distances. AU9 avoids a key interaction at TYR473 as its branched molecular structure inhibits extension further into the AF2 as compared to Pio. (**E**,**F**) Molecular dynamics root-mean-square fluctuation (RMSF) plots. Where the blue graph represents the ligand induced α-carbon fluctuation overlayed with red graph experimental b-factor, α-helices and β-sheets are shaded red and blue, respectively, from N-terminus to C-terminus. Vertical green bars indicate ligand–residue contacts. (**E**) Pio displays ligand-induced stabilization of the AF2 domain (residue index > 250) represented by a decrease in the RMSF as compared to the b-factor plot. (**F**) AU9 displays ligand induced stabilization in the global protein structure characterized by a reduction in RMSF as compared to the b-factor; however, avoidance of AF2 interactions cause a greater fluctuation in both the N-terminus and C-terminus indicating partial activity relative to a full agonist. (**G**,**H**) Protein interaction diagram categorized by the fraction and type of interactions maintained through the course of simulation. (**G**) Pio maintains hydrogen bonds to the AF2 residue TYR 473 with supporting interactions at residues HIS 449, HIS 323, SER 289, and GLN 286 as the major contribution of ligand-protein contacts. (**H**) AU9 maintains only minimal contact to the AF2 supporting residue HIS 449, thus contributing to the increased RMSF values observed in this region of the protein.

The protein–ligand contacts plot for Pio shows interactions occurring with twenty-six amino acid residues in the active site. The largest fraction of interactions occurring >50% throughout the simulation are hydrogen bonds to tyrosine 473 (H12/AF2), histidine 449 (H11), serine 289, glutamine 289, and a hydrophobic contact at isoleucine 326 seen in Figure 2G.

The protein–ligand contacts plot for AU9 shows interactions with twenty-five amino acids within the PPARγ LBD. The interactions occurring >50% of the simulation are hydrogen bonds to glutamine 286, tyrosine 327, lysine 367, and leucine 340 (Figure 2H).

All ligand contacts observed in the protein RMSF plots for both Pio and AU9 (Figure 2E,F) show increased stability to that of the B-factor plot, except the AF2 domain. Pio ligand contacts to the AF2 were shown to decrease the protein RMSF relative to the B-Factor. Conversely, AU9 avoids contact with the AF2 LBD tyrosine 473 residue, thereby allowing for greater AF2 flexibility and dynamic motion. As a consequence, it would be expected that AU9 would display a partial agonistic profile to that of Pio in vitro.

PPARδ and γ reporter assay: The ability of AU9 to induce PPARδ and PPARγ activity was determined by zLBD-Driven GAL4 Reporter assay (Figure 3A,D). Values obtained from the dose–response curve (Figure 3A) suggest that AU9 has PPARδ activity when compared to GW0742 (EC_{50} of AU9 is 41 nM and EC50 for GW0742 is 20 nM). Further evaluation of AU9 activating PPARδ is observed by the promoter activity (Reporter) assays involving AU9 inducing interaction of PPARδ with the PPARδ response element (DRE) similar to GW0742 (10 fold from control), a full PPARδ agonist (Figure 3B) ($p < 0.0001$). Additionally, our animal studies confirmed that AU9 induced an increase in PPARδ gene expression targets, as demonstrated by the PPARδ downstream gene expression profile observed by a bubble diagram (Figure 3C). Conversely, AU9 did not have a significant effect on PPARγ activation (Figure 3D) or the constitutively active (VP16) PPAR gamma construct (Figure 3E,F). EC50 is respectfully measured at 400 nM (AU9) and 50 nM (pioglitazone) ($p < 0.05$ and $p < 0.0001$). Further evaluation of AU9 effects on the AP2 promoter further confirmed that AU9 activates PPARγ (Figure 3E). To help explain this, our in silico design predicted that AU9 avoids Tyrosine-473 of the AF2 ligand binding domain of PPARγ. Therefore, we tested and observed that the substitution of Tyrosine-473 with phenylalanine resulted in a significant reduction of Pio-mediated activation of the PPARγ interaction with PPRE (1.9 fold) (Figure 3F,G) ($p < 0.05$ and $p < 0.0001$).

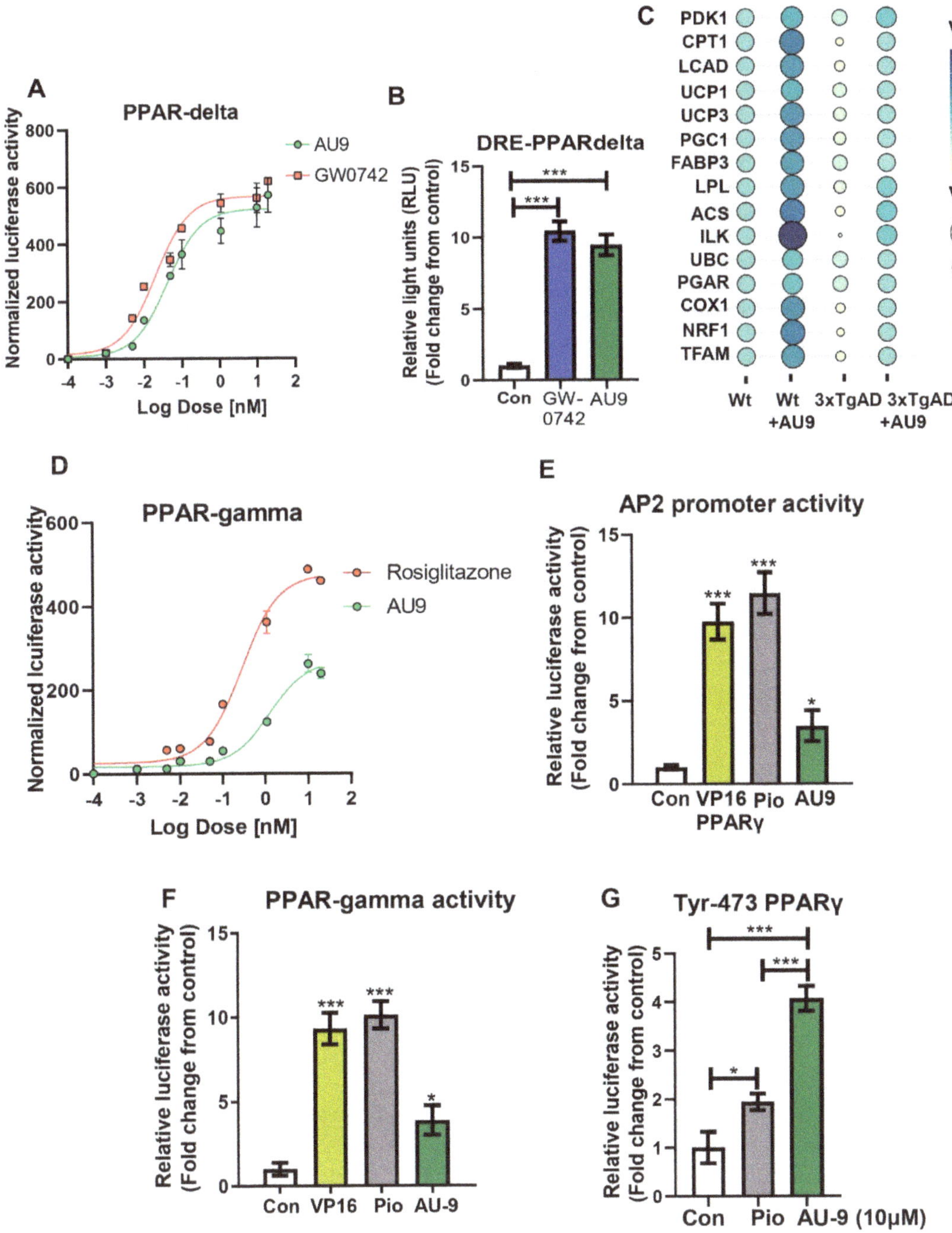

Figure 3. Reporter assays for AU9 PPARδ and PPARγ activity. Stable HEK293T cell lines expressing (**A**) PPAR-δ ligand binding domain driven GAL4 reporter assays determined AU9 activity when compared to increasing concentrations of full PPARδ agonist GW0742. (**B**) AU9 induces partial human PPAR-δ activity when compared to full PPARδ agonist GW074 by activating the PPARδ

Response Element (DRE) via transient co-transfection into HEK293 cells with human PPARδ expression vectors along with the reporter plasmid (PPRE-pk-Luc) or control reporter plasmid (pk-Luc) with *Renilla* vector for 24 h. Cells were treated with AU9 (10 μM) and Pio (10 μM) for 24 h. Luciferase activity was normalized to *Renilla* luciferase activity as described in the Methods section. Values were based upon normalized luciferase activity and ΔΔct values shown as a fold change from control. Statistical values were obtained using two-tailed, unpaired *t*-test analysis ± S.E.M. Where $n = 6$ independent experiments with three replicates per experiment, *** $p < 0.0001$. (**C**) Bubble plot of qPCR analysis (fold change from control) of wild type and 3xTgAD mice treated with and without AU9 for three months daily (5 mg/Kg). (**D**) Stable HEK293T cell lines expressing PPARγ ligand binding domain driven GAL4 reporter assays determined AU9 activity compared to increasing concentrations of full PPARγ agonist Rosi. (**E**) AU9 induces human PPARγ activity by activating the AP2 response element via co-transfection into HEK293 cells transiently with human PPARγ vector with the reporter plasmid (AP2-Luc) or control reporter plasmid (pk-Luc) with *Renilla* vector for 24 h. (**F**) AU9 induces human PPARγ activity by activating the 3XPPRE-pk-Luc response element via co-transfection into HEK293 cells transiently control reporter plasmid (pk-Luc) with *Renilla* vector for 24 h. (**G**) Human PPARγ with tyrosine-473 substituted with phenylalanine demonstrates activity using the 3XPPRE-pk-Luc response element via co-transfection into HEK293 cells transiently control reporter plasmid (pk-Luc) with *Renilla* vector for 24 h. Cells were treated with AU9 (10 μM), and Pio (10 μM) for 24 h. Luciferase activity was normalized to *Renilla* luciferase activity as described in the Methods section. Values for G and H were based upon normalized luciferase activity and fold change from control. Statistical values were obtained using two-tailed, unpaired *t*-test analysis ± S.E.M. Where $n = 6$ independent experiments with three replicates per experiment. *: $p < 0.05$ and ***: $p < 0.0001$.

AU9 improves deficits in Y-maze and NOR tests in 3xTg-AD mice: The activation of the PPARδ and PPARγ axes improves cognitive deficits in mouse models for AD [37,38]. We hypothesized that AU9 may play a role in synaptic processes and ultimately cognition and that selective activation of PPAR by AU9 may improve learning and memory deficits. To determine whether AU9 improves cognitive deficits in 3xTgAD mice, we performed novel object recognition (NOR) and Y-maze based on previously published protocols by our group and others, as discussed in the methods section [16]. In the NOR test, there was no biased exploratory preference to either object among the four groups of mice in the training session, suggesting that there was no difference in motivation and curiosity about novel objects among the groups (data not shown). In the retention session performed 24 h after training, we observed a marked decrease in the exploratory preference (Figure 4A) for novel objects, as observed in saline-treated 3xTgAD mice compared with that in control mice (reduced by ~20%, $p < 0.05$ and $p < 0.001$), indicating impaired discrimination of a novel object from a familiar one (Figure 4B). Furthermore, treatment with AU9 significantly improved the discrimination index in comparison to the vehicle-treated 3xTgAD group ($p < 0.001$). These results suggest that AU9 improves recognition memory impairment in 3xTgAD mice by two folds ($p < 0.001$).

To determine the effect of AU9 on short-term spatial recognition memory, we utilized a two-trial Y-maze task with an inter-trial interval of 3 h. The number of arm entries and the time spent in the novel arm was significantly less by the 3xTgAD mice when compared with the control mice (Figure 4D–F) ($p < 0.001$) and that AU9-treated 3xTgAD mice demonstrated an improvement in the number of entries in the novel arm (Figure 4F, $p < 0.05$). These results suggest that AU9 improves short-term memory impairment in 3xTgAD mice.

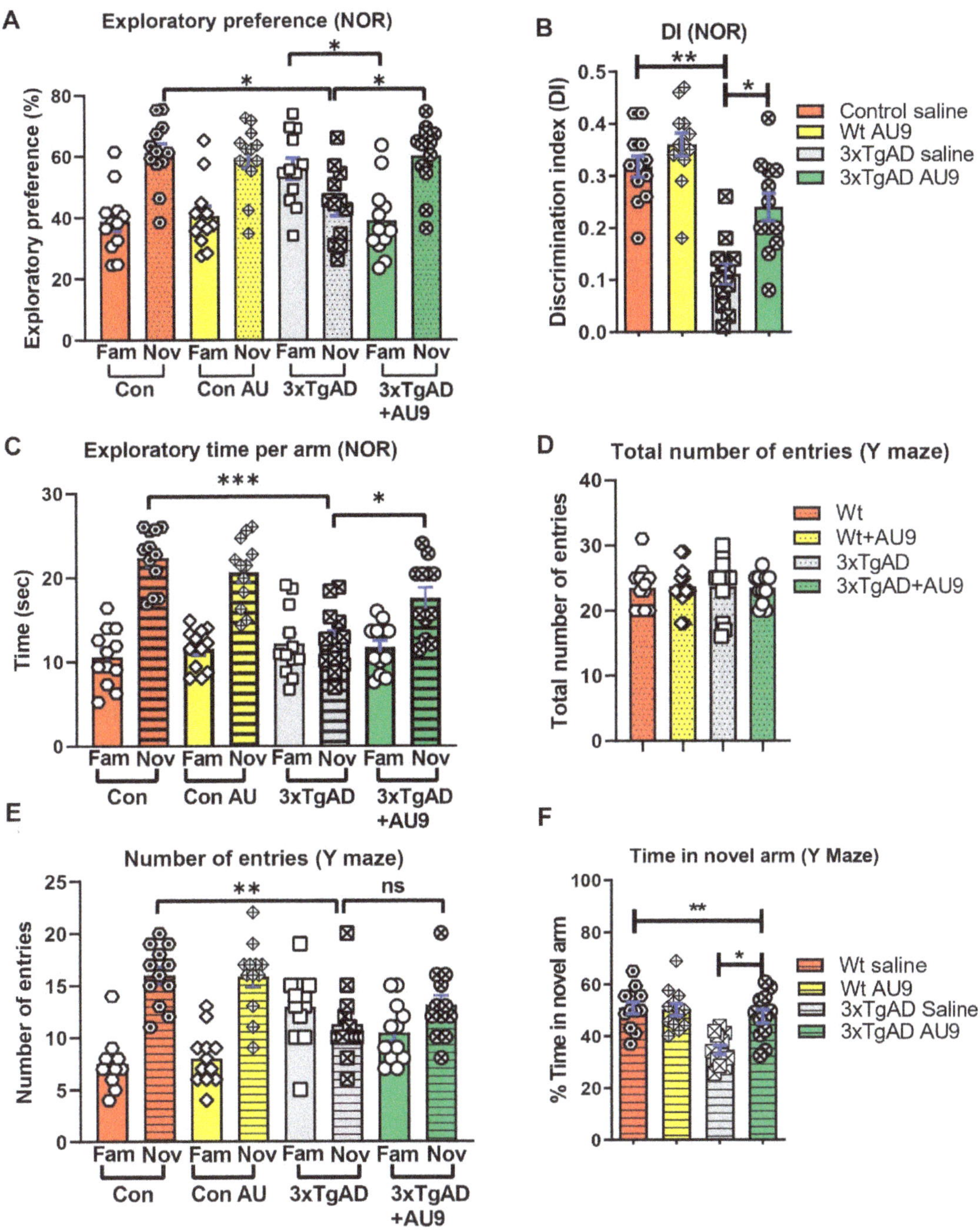

Figure 4. AU9 improves memory deficits in 3xTgAD mice. (**A**) Results for novel object recognition (NOR) tests reflect an exploratory preference for the novel object vs. the familiar object by the animals that were analyzed by a naive subject. (**B**) Discrimination index (DI), represents the recognition of memory sensitivity. Where discrimination index was calculated as (DI) (T novel − T familiar)/(T familiar + T novel). (**C**) The

exploratory time of a novel object when compared to the familiar object. (**D**) Results from Y-maze tests for the total number of entries into the novel and familiar arms. (**E**) Results from Y-maze tests for the number of times in the novel or familiar (other) arm. (**F**) The Y-maze tests for the percent time the animal spends in the novel arm. Statistical values were obtained by student t test analysis ± S.E.M. Where n = 12 mice per group and ns represents no significance, where * $p < 0.05$, ** $p < 0.001$ and *** $p < 0.0001$.

Field recordings in 3XTg-AD mice: To determine whether cognitive impairment in the 3xTgAD mice was linked to alterations in neural field transmission, hippocampal slices were used to measure the fEPSP responses at increasing stimulus intensities based on previous protocols [19,39]. We observed an alteration in fEPSPs over a range of stimuli intensities between groups. The fEPSP slope and amplitude were reduced in 3xTgAD mice compared to the controls (Figure 5A,B) ($p < 0.001$). Further, there was an improvement in transmission following AU9 treatment (Figure 5A) ($p < 0.05$). To further investigate whether the deficits in transmission in 3xTgAD mice were potentially due to alterations in presynaptic axon recruitment, we measured the fiber volley (FV) amplitude across a range of increasing stimulus intensities (amplitude) (Figure 5B). We observed that 3xTgAD mice showed a reduction in the FV amplitude compared to the wild type mice, suggesting reduced presynaptic axonal activation/recruitment. However, no improvement in FV amplitude deficit was observed in AU9 treated 3xTgAD mice (Figure 5B).

Field recordings in 3xTgAD mice: We next examined whether cognitive impairments in 3xTgAD were associated with alterations in neuronal activity by measuring LTP, based on previous methodologies [16,19,40]. Using hippocampal slices, we determined that 3xTgAD mice displayed deficits in LTP in the Schaeffer collateral pathway when compared to the control mice. The 3xTgAD mice treated with AU9 showed an improvement in the fEPSP and LTP (Figure 5C,D) ($p < 0.0048$). One possibility for the reduction LTP can be attributed to weakened signaling strength during LTP [41]. To evaluate for alterations during LTP induction (Figure 5E) ($p < 0.001$), we assessed fEPSP amplitude during theta burst stimulation and observed a significant difference between the control and 3xTgAD. Further, there was no significant difference between the control and 3xTgAD mice treated with AU9. These results suggest that AU9 improves the deficits in LTP in the hippocampus of 3xTgAD mice.

AU9 improves neurotrophin levels and spine density: Hippocampal function in the form of neuronal survival and differentiation is primarily dependent on neurotrophins, including brain-derived neurotrophic factor (BDNF) [42,43]. Neurotrophins are required for supporting the synapse-specific protein synthesis that mediates the stability of various forms of synaptic plasticity [43,44]. Several studies have indicated reduced BDNF and neurotrophin levels in the brains of patients diagnosed with AD and mild cognitive impairment (MCI) [45–47]. Similarly, reduced BDNF levels are also observed in animal models of AD [48]. Previous findings from our lab have established that the PPARγ agonist rosiglitazone promotes BDNF gene expression [40]. Hence, we sought to investigate whether AU9 could improve neurotropin levels, including BDNF expression, in 3xTgAD mice. To validate our theory., we measured neutropin levels via an ELISA and observed a significant increase in neurotropin levels following AU9 treatment in 3xTgAD mice in comparison to saline treated 3xTgAD mice (Figure 6A–D) ($p < 0.05$). Interestingly, we noted a statistically significant increase in neurotrophins in 3xTgAD treated with AU9 for 3 months including BDNF (increase by ~10 pg/mg of protein) ($p < 0.05$), Glial-derived Neurotrophic Factor (NGF) (increase by ~10 pg/mg of protein) ($p < 0.005$), and NT3 (increase by ~10 pg/mg of protein) ($p = 0.966$, and nt 4/5 ($p > 0.05$) levels (Figure 6A–D). Taken together, our data suggest that AU9 treatment improves neurotrophin levels in 3xTgAD mice. Neurotrophins can promote dendritic spine morphogenesis, including improved spine density, area, and length. We observed statistically non-significant improvement in spine density as determined from our Golgi-Cox staining results (Figure 6E–H).

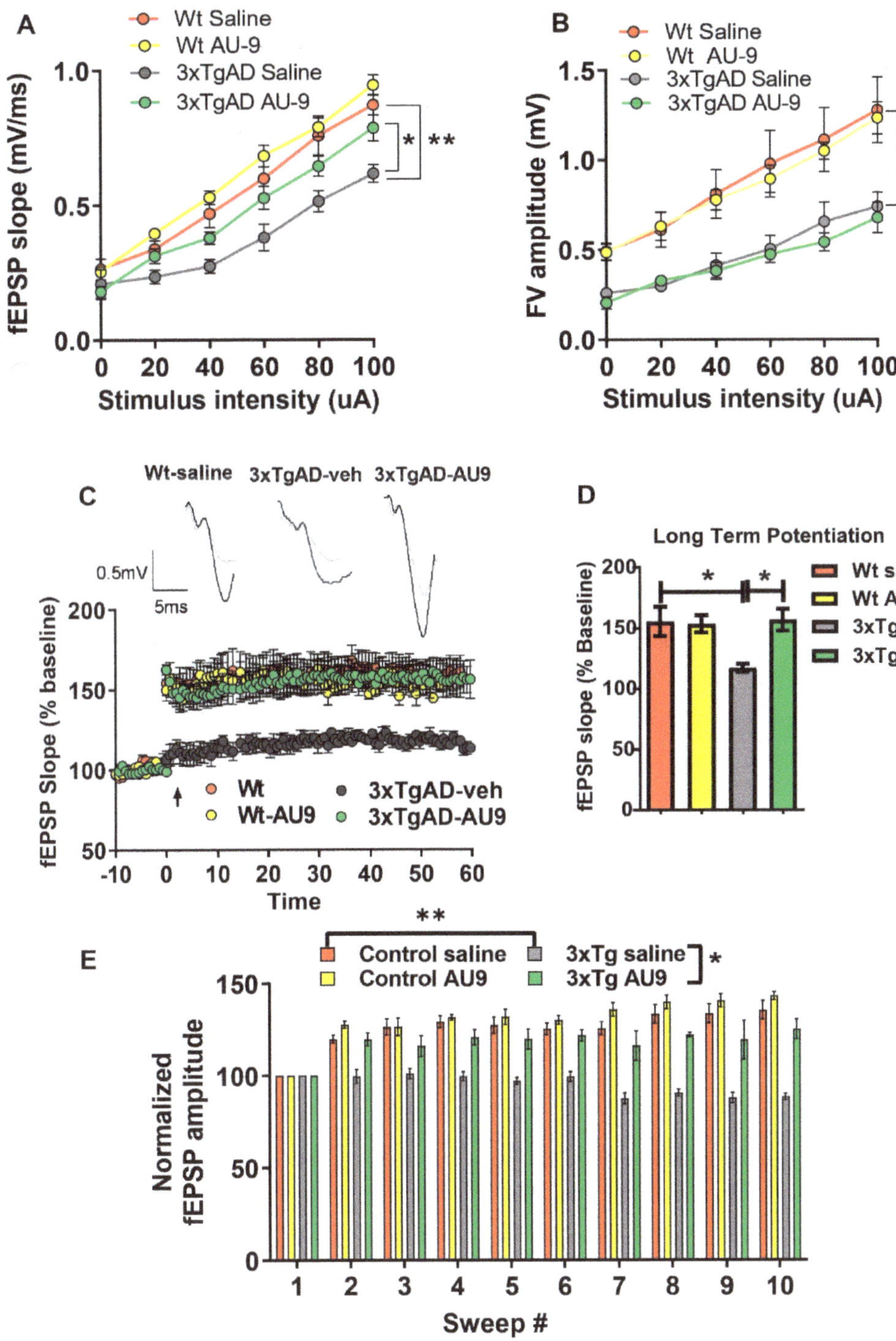

Figure 5. AU9 improves field recordings in 3xTgAD mice. 3xTgAD mice display alterations in extracellular field recordings of excitatory postsynaptic potential (EPSP). Twelve-month-aged female

wild type and 3xTgAD mice treated with AU9 or saline orally for 3 months daily (5 mg/kg). (**A**) Input–output curve of fEPSP slope recorded at increasing stimulus intensities. (**B**) Input–output curve of FV amplitude recorded at increasing stimulus intensities. (**C**) Deficits in 3xTgAD mice LTP was improved in AU9-treated 3xTgAD mice as measured by a high-frequency stimulation (3 × 100 Hz trains with a 20 s intertrain interval). LTP graphs represent fEPSP slope before and after induction by TBS. (**D**) LTP bar graphs show fEPSPs recorded during the time period 50–60 min following TBS induction normalized to baseline levels and traces before and after LTP induction. (**E**) Sweep analysis was calculated by normalizing the amplitude of the first fEPSP of sweeps 2–5 with the amplitude of the first fEPSP of sweep 1 during LTP induction. Statistical values were obtained using student *t*-test analysis ± S.E.M. Where $n = 8$ mice per group and *: $p < 0.05$, **: $p < 0.001$.

3.2. *AU9 Reduces Aβ Levels in 3xTgAD Mice*

The 3xTg-AD mice develop amyloid plaques by six months of age. The pathologies appear in a distinct pattern, with Aβ deposition starting in the neocortex and appearing later in the hippocampus [49]. Immunostaining for Aβ with 6E10 antibody revealed significantly increased Aβ deposits in the hippocampi of vehicle-treated 3xTgAD mice compared to AU9-treated mice (Figure 7A,B). Specifically, the detectable Aβ levels were markedly reduced in the hippocampi of 3xTgAD mice treated with AU9 compared with vehicle-treated mice (Figure 7A) (0.85 fold) ($p < 0.004$). Further analysis of the soluble form was measured by an Aβ1–42 enzyme-linked immunosorbent assay (ELISA). We observed a statistically significant reduction in Aβ1–42 levels in the 3xTgAD mice treated with AU9 when compared to saline-treated mice (decrease by eight pg/mg total protein). We confirmed our findings in APP-Cho cells and observed that AU9 has to reduce Aβ levels by approximately 50% (10 μM); (Figure 7C) ($p < 0.001$). Several studies have indicated that PPAR agonists reduce BACE1 expression and thereby reduce Aβ levels [24]. We, therefore, investigated the effect of AU9 on BACE1 activity and found that AU9 reduces Aβ in our APP-Cho cell line and reduces β-secretase activity with increasing concentrations of AU9 (Figure 7D) ($p < 0.001$ and $p < 0.0001$).

3.3. *AU9 Reduces Neuroinflammation*

Previous reports verify an increase in neuroinflammation associated with an increase in marker cells (microglia) as well as infiltrating macrophages. We investigated changes in gene expression patterns using gene analysis (qPCR) and Nanostring data analysis (Figure S3 in Supplementary material). Changes in gene expression patterns associated with neuroinflammation and cytokine expression verify that AU9 attenuated several markers associated with neuroinflammation in 12-month-old 3xTgAD mice (Figure 8A). Further, the markers IBA 1 and TSPO were observed to increase in 3xTgAD mice brains, approximately 0.7- and 1.75-fold increases from wild type mice ($p < 0.05$) and ($p < 0.05$), respectively. Further, AU9 (5 mg/Kg for 3 months daily) significantly reduced IBA expression (0.5-fold and 1.70-fold, respectively, $p < 0.05$) in 12-month aged 3xTgAD mice. Further nano-string analysis allowed us to determine that AU9 treatment in the same mice resulted in reduced cytokine expression cellular stress and DNA damage (Figures S4–S6 in Supplementary Material). Lastly, we measured in BV2 cells that AU9 reduced lipopolysaccharide mediated nitrite levels in a dose response manner (0–100 μM) where a 50% reduction was observed at a dose of 10 μM of AU9 ($p < 0.0001$).

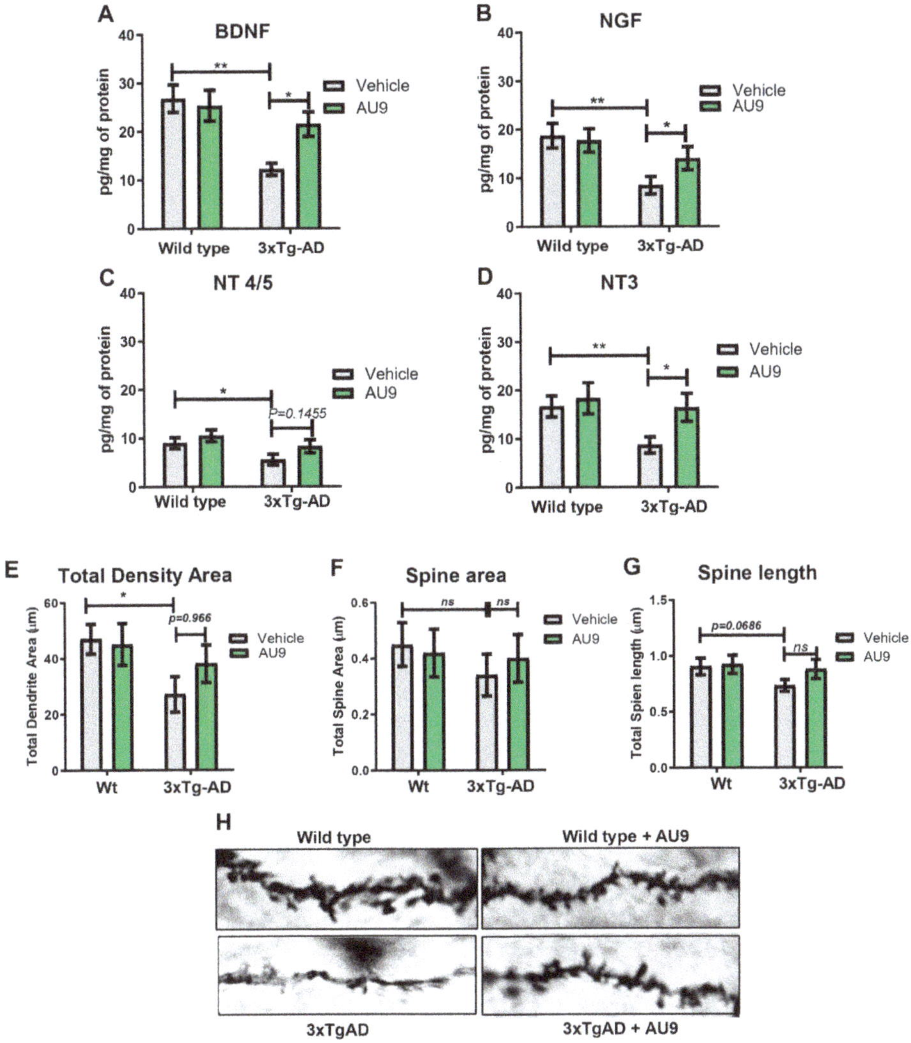

Figure 6. AU9 improves neurotrophin expression and spine density. AU9 improves neurotrophin protein expression in 3xTgAD mice treated with AU9 (3 months daily, 5 mg/kg) as determined by ELISA, (**A**) Brain-Derived Neurotrophic Factor (BDNF) (**B**) Nerve Growth Factor (NGF) expression, (**C**) Neurotrophin Factor 4/5 expression and (**D**) Neurotrophin −3 expression in 3xTgAD hippocampi. Where $n = 6$ mice per group of treatment and * $p < 0.01$, and ** $p < 0.005$. Values were based upon a normalized protein concentration, a standard curve of growth factor protein supplied in the kit. Statistical values were obtained using Student t-test analysis ± S.E.M. Where $n = 6$ mice per group and *: $p < 0.05$ and **: $p < 0.001$. AU9 improves spine density area, spine area and spine length (**E**–**H**) in 3xTgAD mice. Rapid Golgi-Cox staining was utilized to measure changes in total spine density, spine area and spine length. Overall, 200 µm sections were stained and imaged using a Z-stack procedure on a Nikon TSi microscope from a minimum of 10 slices with 10 neurons per slice from 6 mice per group using ImageJ software for measurements. Statistical values were obtained using student t-test analysis ± S.E.M. Where n = 6 mice per group and *: $p < 0.05$.

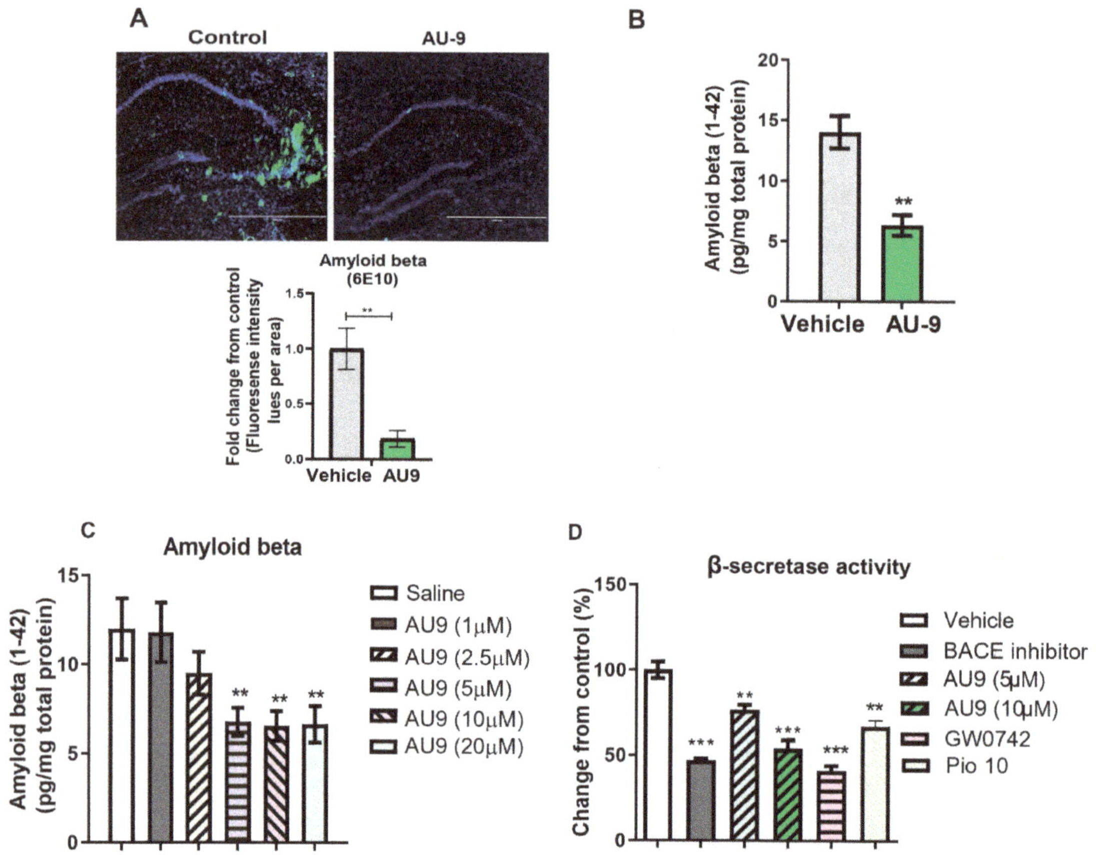

Figure 7. AU9 reduces amyloid beta (1-42) levels in 3xTgAD mice. (**A**) Immunofluorescence imaging (anti-6E10 antibody) shows a reduction in levels of all forms of Amyloid beta levels in 12-month-old 3xTgAD mice treated with AU9 orally for three months (5 mg/Kg daily). Densitometric measurements of Amyloid beta in hippocampi from six mice and four slices per mouse. Values were standardized to total area. (**B**) Elisa measurement of soluble form of Amyloid beta (1–42) from mice treated in the same manner as mice as in panel A. Amyloid beta was measured from hippocampi from six mice per group. Values were based on a standard curve of Amyloid beta 1–42 and standardized to total protein concentrations. Statistical values were obtained using Student t-test analysis ± S.E.M. Where $n = 6$ mice per group and ** $p < 0.001$. (**C**) Reduction of Amyloid-beta being secreted in media from APP-Cho cells following increasing concentrations of AU9 treatment (1, 2.5, 5, 10 and 20 μM). Values were based upon standardized curve from ELISA (R&D Systems). Statistical values were obtained using student t-test analysis ± S.E.M. Where $n = 6$ independent experiments were repeated in triplicate per group and **: $p < 0.001$. (**D**) Effects of AU9 (5 μM and 10 μM) on Beta secretase activity in APP-Cho cells. β-Secretase activity was determined fluorometrically using an β-Secretase activity kit (Biovision, Waltham, MA) and standardized to total protein concentration from APP-Cho cells. Beta-secretase activity was represented as relative fluorescence unit per mg of total protein. Values were based upon means from 6 independent repetitions with 3 replicates in each group and represented as a percent change from control. Statistical values were obtained using student t-test analysis ± S.E.M. **: $p<0.001$, ***: $p < 0.0001$.

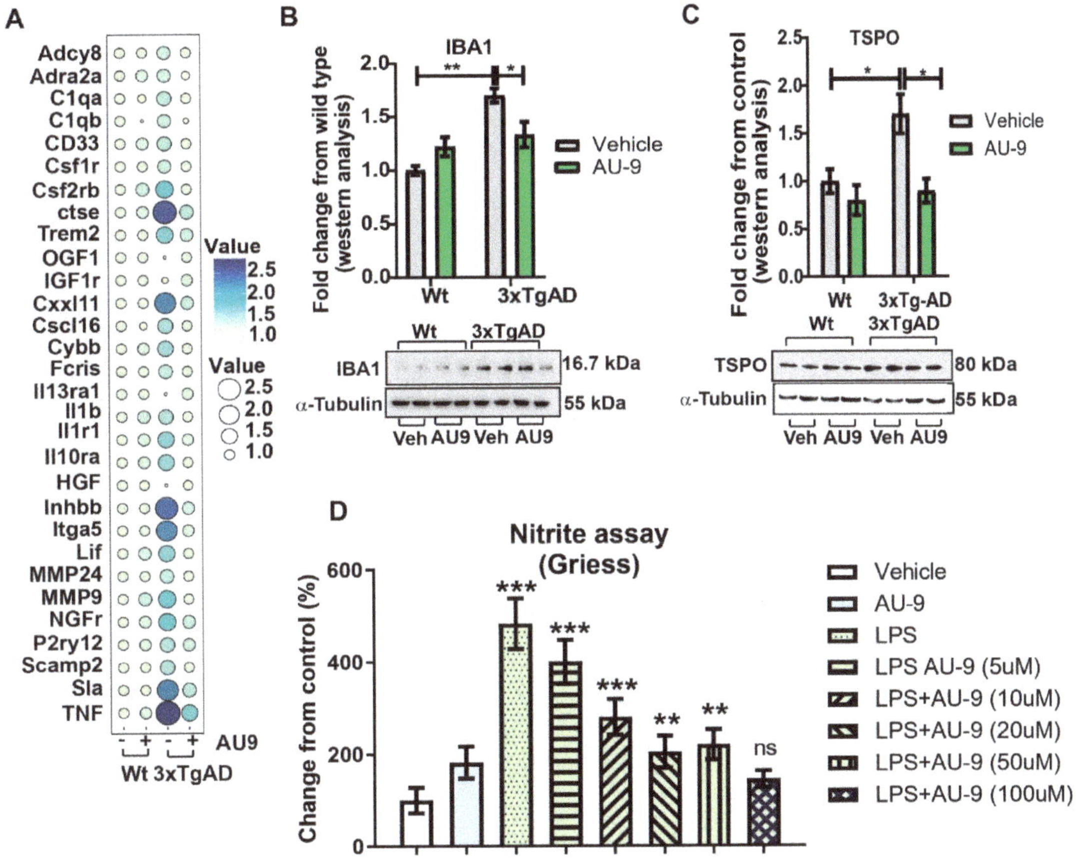

Figure 8. AU9 reduces inflammation. (**A**) Two-step qRTPCR analysis demonstrates by bubble plot that AU9 reduces inflammatory gene markers in 3xTgAD administered AU9 (5 mg/Kg daily for three months by oral gavage). (**B**,**C**) Western analysis demonstrates reduced protein expression of IBA1 and TSPO in similarly treated 3xTgAD mice as in A. Values were based on normalized protein concentrations, standardized to β-tubulin, and displayed as fold changes from control. Statistical values were obtained using Student *t*-test analysis ± S.E.M. Where $n = 6$ mice per group and *: $p < 0.05$ and **: $p < 0.001$. (**D**) Nitrite levels were measured by Griess reagent assay, where increasing concentrations of AU9 reduced LPS-mediated nitrite formation. Values were based upon triplicate readings from 6 independent assays and standardized to protein concentration. Statistical values were obtained using student *t*-test analysis ± S.E.M. *ns* = Not significant, **: $p < 0.001$, and ***: $p < 0.0001$.

3.4. Peripheral Effects of AU9

Traditional full PPARγ agonists are known to induce an increasing body weight. However, PPARδ agonists are known to improve oxidative phosphorylation and catabolic activity. After 3 months of treatment in 9- to 12-month-aged mice, we observed no significant increase in body weight in both wild type and 3xTgAD mice (Figure 9A,B). Further, it has been reported that 3xTgAD mice display elevated blood glucose levels when compared to age-matched wild type mice [50]. We observed a significant improvement in our glucose tolerance test in 12-month-old 3xTgAD mice (Figure 9C) ($p < 0.001$ and 0.05). Lastly, AU9 (10 mg/kg) did not induce a significant increase in heart weight to body weight ratio (0.25 fold increase in wild type and 0.27 in 3xTgAD mice). However, we observed a significant increase with Pio (10 mg/Kg) in wild type and 3xTgAD mice after 3 months of treatment (Figure 9D) ($p < 0.05$ and $p < 0.001$).

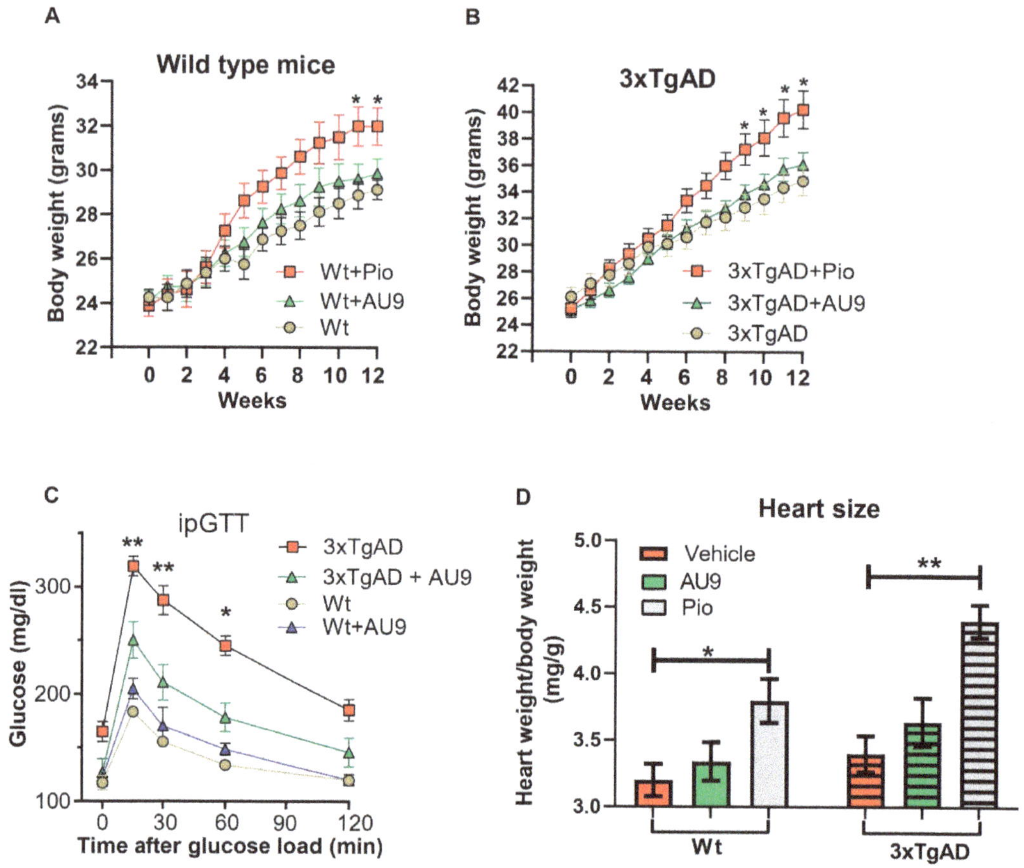

Figure 9. Physiological effects of AU9. (**A**,**B**) No significant weight change in 12-month-old wild type or 3xTgAD mice given AU9 (10 mg/Kg/day for 3 months, orally) compared to mice treated with Pio. (**C**) Intraperitoneal glucose tolerance test demonstrated that AU9 reduces circulating blood glucose in 3xTgAD mice. (**D**) Heart weight to body weight studies shows that AU9, when compared to Pio, does not induce an increase in size in age-matched wild-type and 3xTgAD mice. Statistical values were obtained using student t-test analysis ± S.E.M. Where $n = 6$ mice per group and *: $p < 0.05$ and **: $p < 0.001$.

4. Discussion

PPARγ agonists have previously been investigated as potential treatments for Alzheimer's disease; however, there have been conflicting data from preclinical studies. For example, full PPARγ agonists Pio and rosiglitazone (rosi) improved cognition in the PS1-KI (human presenilin-1 M146V knock-in mouse) mouse model of AD following a 9-month treatment (20 mg/kg) [10]. However, in 3xTgAD mice, similar effects were not observed; thus bringing into question the ability of Pio to improve memory deficits in AD [10,51]. Additionally, object recognition studies revealed a trend towards the worsening of memory in wild type male mice after Pio treatment, thus making the overall effect of Pio on cognition difficult to interpret [10]. However, evidence suggests that targeting PPARγ and/or PPARδ can improve memory deficits and/or the pathology associated with Alzheimer's disease in rodent models. Indeed, work by Searcy et al. demonstrated that pioglitazone in 10-month-old 3xTgAD mice improved learning on the active avoidance task, decreased hippocampal amyloid-β and tau deposits, and enhanced short- and long-term plasticity [52]. Interestingly, in human diabetic patients with mild cognitive deficits, Pio treatment improved peripheral insulin sensitivity, as well as plasma levels of Aβ and the insulin degrading enzyme [53]. However, findings from the TOMMORROW clinical trial utilizing pioglitazone in patients with MCI failed to improve cognitive deficits [54]. These confounding results suggested that an alternative form of PPAR agonism would have significant potential for AD. However, work by Joel Berger's group at Merck Pharmaceutical identified the significance of the physical interactions of Tyrosine-473 in the PPARγ ligand binding domain for adiposity and other biological properties [55]. Secondly, our previous work identified PPARδ as a potential therapeutic target for improving synaptic plasticity in rodent models of diabetes/AD [38]. We therefore developed a dual PPARδ-PPARγ agonist. Our in silico observations of full PPAR agonists have been shown to form strong hydrogen bond interactions with the AF2 ligand binding domain contained in arm 1. Further stabilization of the ligand–protein complex is achieved by hydrophobic interactions in arm 2. The ligand's ability to form a stable hydrogen bond network to the AF2 is representative of a strong transcriptional activation, as this leads to the displacement of co-repressor proteins and recruitment of co-activator proteins. As the AF2 adopts this active conformation, changes in the quaternary protein structure allow for new sites to become available for co-activator binding. Ligand-induced conformational changes can affect the size and residue charge distribution to accept a variety of co-activators. However, due to the functional difference in the two arms seen in PPARs, different binding modes can be adopted with some ambiguity to the specific role the arms play during transcriptional activation. Interestingly, this presents the potential for novel mechanisms of activation in PPARδ ligand binding. While AU9 does not display the typical binding profile of a full agonist, it does have the ability to provide significant stabilization of AF2 domain at histidine 413 on H11 (Figure 1B). Coordination of H11 alone can position the H12 AF2 domain for self-assembly of tyrosine 437 to histidine 287 in the absence of ligand stabilization, providing some explanation for the observed activity of AU9. Furthermore, AU9's branched structure extends deep into arms 2 and 3, providing several stabilizing hydrophobic contacts at valine 245, isoleucine 213, and methionine 192 (Figure 1B). The combined space filling of arm 2 and 3 by AU9 provides additional stabilization of the protein complex outside of the AF2, which is not observed in the PPARδ agonist GW0742.

It is interesting to note that the trifluoro side group interaction holds the ring moiety in position, making AU9 avoid contact with Tyr473, which is approximately 5Å away. This residue is crucial to the stabilization of the AF2 helix H12, which allows the binding of co-activators that lead to the activation of the genes responsible for adipogenesis [55]. Our in silico data were confirmed by transcriptional assays, which demonstrate that AU9 minimally activates the PPARγ AP2-PPRE (Figure 4A). The results appear consistent with our previously published lipid accumulation assays, where AU9 has negligible effects upon lipid accumulation in adipocytes [14].

Ligand binding affects the conformation of the AF-2 ligand binding surface, resulting in modifying the binding affinity for chromatin remodeling transcriptional co-regulator proteins and resulting in the activation or repression of selective gene transcription [56,57]. Further evidence for conformational changes associated with ligand–receptor interactions has been identified by crystal structures that define the inactive/repressive and active conformations that enable the binding of transcriptional coactivator and corepressor proteins, respectively, by stabilizing specific conformations of the AF-2 region [58]. Recent work has illuminated how ligands engage the ligand-binding domain and enter the orthosteric ligand-binding pocket, and whether ligand binding occurs through an induced fit or conformational selection mechanisms [59]. In the induced fit scenario, ligand binding selectively binds to and selects a particular conformation that is occupied within the ligand binding conformational group. In the conformational selection mechanism, ligand binding occurs through an encounter complex and promotes the ligand binding conformational group into the final ligand–bound complex.

Previously, we have observed that PPARδ agonist (GW0742) and rosiglitazone improve synaptic plasticity in db/db leptin receptor knockout mice [38,40]. The key findings of the current study are that 3xTgAD mice display cognitive deficits and impaired synaptic plasticity that can be rescued by AU9 through activation of PPARδ and partially PPARγ. This conclusion is deduced mainly from the following observations. First, AU9 treatment improved cognitive deficits, specifically impairments in working memory (assessed by NOR and Y-maze) in 3xTgAD mice. Second, hippocampal LTP was impaired in 3xTgAD mice, and AU9 improved the deficits. Third, AU9 modulated the postsynaptic receptor expression in the hippocampus of 3xTgAD mice. Importantly, AU9 improved neurotrophin levels including BDNF levels in the hippocampus. Fourth, AU9 modulated several hippocampal genes involved in synaptic plasticity and neurotransmission.

After 12 weeks of treatment, novel object recognition performance and Y-maze-dependent memory were improved in AU9-treated 3xTgAD mice. This is in agreement with prior studies, which reported that the PPARγ agonist rosiglitazone attenuated the learning and memory deficits of APP transgenic mice in the radial maze and object recognition tests after chronic administration [5,6]. However, APP mice treated with Pio for two months did not have improved spatial memory [53]. The difference in the latter study may be due to the different dosing of the agonists used and the duration of the treatment.

Several animal models of AD exhibit deficits in basal synaptic transmission, which correlate with the progression of the disease [60]. Some of these deficits include, the release of glutamate, glutamate uptake, and the expression or functionality of glutamatergic receptors [61]. Furthermore, neurotrophins, including BDNF have been shown to regulate the expression and synaptic delivery of AMPA receptor subunits in the hippocampus, which implies that neurotrophin signaling alters AMPAR trafficking in addition to influencing AMPAR activity [62]. Therefore, the improved neurotrophin expression may be responsible for the improved basal synaptic transmission in part via influencing AMPAR trafficking and function [39]. Further evidence for improved markers associated with neurotransmission can be seen in our gene analysis profile (nanostring data) in supplementary data Figure S1.

In the present study, we observed deficits in LTP, an integral component of the signaling strength and synaptic dysfunction observed in AD pathology [63]. We also found that AU9 enhances LTP in the hippocampal Schaffer collateral pathway in the 3xTg-AD mice without affecting the control LTP. Other studies have also reported an influence of PPARδ and PPARγ receptor signaling on LTP [7,38]. Previously, we observed an improvement in LTP deficits in mouse model of diabetes (leptin receptor deficient db/db) following administration of rosiglitazone via ICV and not by oral delivery [40]. PPAR agonists increase the expression of neurotrophins and transcription factors, including neurotrophic factor 1α or CREB [64], which are centrally involved in this process [40,65,66]. Neurotrophins are required for supporting synapse-specific protein synthesis that mediates the stability of long-term forms of synaptic plasticity [67]. Likewise, we observed an increase in four neurotrophins, including BDNF.

In the current study, the treatment of 3xTgAD mice with AU9 reduced hippocampal Aβ deposition compared to the control mice. We noted a significant decrease in the total Aβ plaque area and the respective staining intensity. Furthermore, soluble levels of Aβ were significantly reduced following AU9 treatment in 3xTgAD mice. Thus, AU9 may be involved in preventing the formation of Aβ deposits or augmenting the clearance of Aβ. Several lines of evidence indicate that PPARγ transcriptionally regulates the activity of beta secretase enzyme (BACE1), a key enzyme responsible for generation of Aβ peptides [68,69]. Hence, activating PPARγ with natural or synthetic ligands inhibits BACE1. We observed that AU9 inhibited BACE1 activity, which may explain the reduced Aβ levels in culture and our animal models.

Oxidative stress and damage are implicated in Alzheimer's disease and are linked to Aβ plaque formation, Alzheimer's disease pathophysiological events, and synaptic dysfunction [70]. Increased ROS occurs due to an imbalance between pro-oxidants (ROS, RNS, superoxide anion, hydroxyl radicals, and hydrogen peroxide) and antioxidants (GSH, GPX, CAT, GRx, and SOD). Down-regulation of antioxidant defense mechanisms and elevated ROS generation lead to oxidative stress-mediated neurodegeneration [71]. We observed a statistically significant increase in ROS levels in 3xTgAD mice, which were attenuated by the AU9 treatment. AU9 treatment in 3xTgAD reduced markers associated with neuroinflammation, markers of stress, and DNA damage, as seen in our supplemental data Figures S2–S5.

Our study had several shortcomings, including that our model was a moderately aged (9–12 months treatment age) 3xTgAD model. Further work on understanding long-term treatment into the late stage (16–18 months) will help understand whether AU9 can reduce or prevent the progression of the disease into late stages. Another limitation is the use of only female mice due to previous studies determining that the gene expression in males is unstable. Alternative Aβ models, such as the 5xFAD and a Tau model (P301S), may provide more insight into AU9′s ability to alter more aggressive forms of pathology. As postmortem brains from late stage AD patients indicate hyperinsulinemia and altered insulin signaling, determination of the impact of AU9 on brain glucose uptake and insulin signaling will help us better understand the impact of PPAR signaling on brain energy regulation. Our neurotrophin results offered a possible explanation for how improvement in our LTP—field recordings. Although we observed no significant influence of AU9 on presynaptic activity, further information determining the influence of AU9 on postsynaptic receptor involvement is needed. Potential studies would include patch clamp analysis and synaptosome fractional analysis for the influence of AU9 on glutamatergic and or NMDAR receptor levels in the postsynaptic region.

5. Conclusions

We have developed a novel PPARδ–partial PPARγ agonist that improves behavioral deficits and synaptic plasticity (LTP) in an AD mouse model. The anti-inflammatory effects and enhanced neurotrophin expression levels may help explain these findings. Further analysis to clarify how AU9 ameliorates amyloid beta levels needs to be further explored. Future pharmacodynamic and pharmacokinetic analyses will be beneficial to advance the clinical application of AU9. The current study's findings support the potential use of PPAR agonists in the treatment of AD.

6. Patents

Patents: 10844003 Dual PPAR-delta and PPAR-gamma agonists to Drs. Amin and Ward and are licensed to Oleolive llc.

Supplementary Materials: The following supporting information can be downloaded at: https://www.mdpi.com/article/10.3390/cells12081116/s1. Figure S1: Markers for Neurotransmission. Figure S2: Markers for cellular stress. Figure S3: Cytokines. Figure S4: Markers for DNA Damage. Figure S5: QPCR primers for Figure 3C. Figure S6: QPCR primers for Figure 8A (1). Figure S7: QPCR primers for Figure 8A (2). Figure S8: Western Whole blots for PSD95, GluA1, GluA2, and the associated standardizing Al-pha-Actin. Figure S9: Western whole blots for inflammatory markers.

Author Contributions: Conceptualization, I.S., M.G., R.H.A., P.D.P., S.Y., T.S., T.Y., F.S.W. and V.S.; methodology, I.S., M.G., R.H.A., P.D.P., S.Y., T.W., T.S., T.Y., F.S.W. and V.S.; software, I.S., M.G., P.D.P., S.Y., F.S.W., V.S. and R.H.A.; validation, I.S., M.G., R.H.A., P.D.P. and S.Y.; formal analysis, V.S., I.S., M.G., R.H.A., P.D.P., S.Y. and V.S.; investigation, I.S., M.G., R.H.A., P.D.P., S.Y., T.W., T.S., T.Y., F.S.W. and V.S.; resources, R.H.A. and V.S.; data curation, I.S., M.G., R.H.A., J.B., P.D.P., S.Y., T.W., T.S., T.Y., F.S.W. and V.S.; writing—R.H.A., I.S., M.G., P.D.P., J.B. writing—review and editing, I.S., M.G., P.D.P., F.S.W., J.B. and R.H.A.; visualization, I.S., M.G. and P.D.P.; supervision, R.H.A. and V.S.; project administration, R.H.A.; funding acquisition, R.H.A. and V.S. All authors have read and agreed to the published version of the manuscript.

Funding: This research was funded by NIH, grant numbers NIH1R15AG048643 to Amin and Suppiramaniam, NIH-1R43AG065069-01 Amin with Oleolive).

Institutional Review Board Statement: The animal study protocol was approved by the Institutional Review Board (or Ethics Committee) of Auburn University (protocol numbers 2018-3259, 2019-3612 and 2019-3566 date of approval).

Informed Consent Statement: Not applicable for studies did not involve humans in our studies.

Data Availability Statement: Not applicable.

Conflicts of Interest: The authors declare no conflict of interest. Oleolive had no role in the design of the study; in the collection, analyses, or interpretation of data; in the writing of the manuscript; or in the decision to publish the results.

References

1. Zieleniak, A.; Wojcik, M.; Wozniak, L.A. Structure and physiological functions of the human peroxisome proliferator-activated receptor gamma. *Arch. Immunol. Ther. Exp.* **2008**, *56*, 331–345. [CrossRef] [PubMed]
2. Harmon, G.S.; Lam, M.T.; Glass, C.K. PPARs and lipid ligands in inflammation and metabolism. *Chem. Rev.* **2011**, *111*, 6321–6340. [CrossRef] [PubMed]
3. Moller, D.E.; Berger, J.P. Role of PPARs in the regulation of obesity-related insulin sensitivity and inflammation. *Int. J. Obes. Relat. Metab. Disord.* **2003**, *27* (Suppl. S3), S17–S21. [CrossRef]
4. Warden, A.; Truitt, J.; Merriman, M.; Ponomareva, O.; Jameson, K.; Ferguson, L.B.; Mayfield, R.D.; Harris, R.A. Localization of PPAR isotypes in the adult mouse and human brain. *Sci. Rep.* **2016**, *6*, 27618. [CrossRef]
5. Pedersen, W.A.; McMillan, P.J.; Kulstad, J.J.; Leverenz, J.B.; Craft, S.; Haynatzki, G.R. Rosiglitazone attenuates learning and memory deficits in Tg2576 Alzheimer mice. *Exp. Neurol.* **2006**, *199*, 265–273. [CrossRef]
6. Escribano, L.; Simon, A.M.; Gimeno, E.; Cuadrado-Tejedor, M.; Lopez de Maturana, R.; Garcia-Osta, A.; Ricobaraza, A.; Perez-Mediavilla, A.; Del Rio, J.; Frechilla, D. Rosiglitazone rescues memory impairment in Alzheimer's transgenic mice: Mechanisms involving a reduced amyloid and tau pathology. *Neuropsychopharmacology* **2010**, *35*, 1593–1604. [CrossRef]
7. Costello, D.A.; O'Leary, D.M.; Herron, C.E. Agonists of peroxisome proliferator-activated receptor-gamma attenuate the Abeta-mediated impairment of LTP in the hippocampus in vitro. *Neuropharmacology* **2005**, *49*, 359–366. [CrossRef] [PubMed]
8. Pathan, A.R.; Gaikwad, A.B.; Viswanad, B.; Ramarao, P. Rosiglitazone attenuates the cognitive deficits induced by high fat diet feeding in rats. *Eur. J. Pharm. Pharmacol.* **2008**, *589*, 176–179. [CrossRef]
9. Jiang, L.Y.; Tang, S.S.; Wang, X.Y.; Liu, L.P.; Long, Y.; Hu, M.; Liao, M.X.; Ding, Q.L.; Hu, W.; Li, J.C.; et al. PPARgamma agonist pioglitazone reverses memory impairment and biochemical changes in a mouse model of type 2 diabetes mellitus. *CNS Neurosci. Ther.* **2012**, *18*, 659–666. [CrossRef]
10. Masciopinto, F.; Di Pietro, N.; Corona, C.; Bomba, M.; Pipino, C.; Curcio, M.; Di Castelnuovo, A.; Ciavardelli, D.; Silvestri, E.; Canzoniero, L.M.; et al. Effects of long-term treatment with pioglitazone on cognition and glucose metabolism of PS1-KI, 3xTg-AD, and wild-type mice. *Cell Death Dis.* **2012**, *3*, e448. [CrossRef]
11. Landreth, G.; Jiang, Q.; Mandrekar, S.; Heneka, M. PPARgamma agonists as therapeutics for the treatment of Alzheimer's disease. *Neurotherapeutics* **2008**, *5*, 481–489. [CrossRef]
12. Geldmacher, D.S.; Fritsch, T.; McClendon, M.J.; Landreth, G. A randomized pilot clinical trial of the safety of pioglitazone in treatment of patients with Alzheimer disease. *Arch. Neurol.* **2011**, *68*, 45–50. [CrossRef]
13. Chamberlain, S.; Gabriel, H.; Strittmatter, W.; Didsbury, J. An Exploratory Phase IIa Study of the PPAR delta/gamma Agonist T3D-959 Assessing Metabolic and Cognitive Function in Subjects with Mild to Moderate Alzheimer's Disease. *J. Alzheimer's Dis.* **2020**, *73*, 1085–1103. [CrossRef]
14. Gathiaka, S.; Nanayakkara, G.; Boncher, T.; Acevedo, O.; Wyble, J.; Patel, S.; Patel, A.; Shane, M.E.; Bonkowski, B.; Wieczorek, J.; et al. Design, development and evaluation of novel dual PPARdelta/PPARgamma agonists. *Bioorg. Med. Chem. Lett.* **2013**, *23*, 873–879. [CrossRef]
15. Hieke, M.; Ness, J.; Steri, R.; Dittrich, M.; Greiner, C.; Werz, O.; Baumann, K.; Schubert-Zsilavecz, M.; Weggen, S.; Zettl, H. Design, synthesis, and biological evaluation of a novel class of gamma-secretase modulators with PPARgamma activity. *J. Med. Chem.* **2010**, *53*, 4691–4700. [CrossRef] [PubMed]
16. Bloemer, J.; Pinky, P.D.; Smith, W.D.; Bhattacharya, D.; Chauhan, A.; Govindarajulu, M.; Hong, H.; Dhanasekaran, M.; Judd, R.; Amin, R.H.; et al. Adiponectin Knockout Mice Display Cognitive and Synaptic Deficits. *Front. Endocrinol.* **2019**, *10*, 819. [CrossRef] [PubMed]
17. Leger, M.; Quiedeville, A.; Bouet, V.; Haelewyn, B.; Boulouard, M.; Schumann-Bard, P.; Freret, T. Object recognition test in mice. *Nat. Protoc.* **2013**, *8*, 2531–2537. [CrossRef] [PubMed]
18. Stover, K.R.; Campbell, M.A.; Van Winssen, C.M.; Brown, R.E. Early detection of cognitive deficits in the 3xTg-AD mouse model of Alzheimer's disease. *Behav. Brain Res.* **2015**, *289*, 29–38. [CrossRef]

19. Parameshwaran, K.; Buabeid, M.A.; Karuppagounder, S.S.; Uthayathas, S.; Thiruchelvam, K.; Shonesy, B.; Dityatev, A.; Escobar, M.C.; Dhanasekaran, M.; Suppiramaniam, V. Developmental nicotine exposure induced alterations in behavior and glutamate receptor function in hippocampus. *Cell. Mol. Life Sci.* **2012**, *69*, 829–841. [CrossRef]
20. Kraeuter, A.K.; Guest, P.C.; Sarnyai, Z. The Y-Maze for Assessment of Spatial Working and Reference Memory in Mice. In *Pre-Clinical Models. Methods in Molecular Biology*; Guest, P., Ed.; Humana Press: New York, NY, USA, 2019; Volume 1916, pp. 105–111.
21. Sarnyai, Z.; Sibille, E.L.; Pavlides, C.; Fenster, R.J.; McEwen, B.S.; Toth, M. Impaired hippocampal-dependent learning and functional abnormalities in the hippocampus in mice lacking serotonin(1A) receptors. *Proc. Natl. Acad. Sci. USA* **2000**, *97*, 14731–14736. [CrossRef]
22. Conrad, C.D.; Galea, L.A.; Kuroda, Y.; McEwen, B.S. Chronic stress impairs rat spatial memory on the Y maze, and this effect is blocked by tianeptine pretreatment. *Behav. Neurosci.* **1996**, *110*, 1321–1334. [CrossRef] [PubMed]
23. Kerr, D.S.; Abraham, W.C. Cooperative interactions among afferents govern the induction of homosynaptic long-term depression in the hippocampus. *Proc. Natl. Acad. Sci. USA* **1995**, *92*, 11637–11641. [CrossRef] [PubMed]
24. Eckhardt, M.; Bukalo, O.; Chazal, G.; Wang, L.; Goridis, C.; Schachner, M.; Gerardy-Schahn, R.; Cremer, H.; Dityatev, A. Mice deficient in the polysialyltransferase ST8SiaIV/PST-1 allow discrimination of the roles of neural cell adhesion molecule protein and polysialic acid in neural development and synaptic plasticity. *J. Neurosci.* **2000**, *20*, 5234–5244. [CrossRef]
25. Graves, R.A.; Tontonoz, P.; Spiegelman, B.M. Analysis of a tissue-specific enhancer: ARF6 regulates adipogenic gene expression. *Mol. Cell. Biol.* **1992**, *12*, 1202–1208. [CrossRef] [PubMed]
26. He, T.C.; Chan, T.A.; Vogelstein, B.; Kinzler, K.W. PPARdelta is an APC-regulated target of nonsteroidal anti-inflammatory drugs. *Cell* **1999**, *99*, 335–345. [CrossRef]
27. Risher, W.C.; Ustunkaya, T.; Singh Alvarado, J.; Eroglu, C. Rapid Golgi analysis method for efficient and unbiased classification of dendritic spines. *PLoS ONE* **2014**, *9*, e107591. [CrossRef]
28. Kolbeck, R.; Bartke, I.; Eberle, W.; Barde, Y.A. Brain-derived neurotrophic factor levels in the nervous system of wild-type and neurotrophin gene mutant mice. *J. Neurochem.* **1999**, *72*, 1930–1938. [CrossRef] [PubMed]
29. Kilroy, G.E.; Zhang, X.; Floyd, Z.E. PPAR-gamma AF-2 domain functions as a component of a ubiquitin-dependent degradation signal. *Obesity* **2009**, *17*, 665–673. [CrossRef]
30. Shang, J.; Kojetin, D.J. Structural mechanism underlying ligand binding and activation of PPARgamma. *Structure* **2021**, *29*, 940–950.e4. [CrossRef]
31. Markt, P.; Schuster, D.; Kirchmair, J.; Laggner, C.; Langer, T. Pharmacophore modeling and parallel screening for PPAR ligands. *J. Comput. Aided Mol. Des.* **2007**, *21*, 575–590. [CrossRef]
32. Sznaidman, M.L.; Haffner, C.D.; Maloney, P.R.; Fivush, A.; Chao, E.; Goreham, D.; Sierra, M.L.; LeGrumelec, C.; Xu, H.E.; Montana, V.G.; et al. Novel selective small molecule agonists for peroxisome proliferator-activated receptor delta (PPARdelta)–synthesis and biological activity. *Bioorg. Med. Chem. Lett.* **2003**, *13*, 1517–1521. [CrossRef]
33. Sherman, W.; Day, T.; Jacobson, M.P.; Friesner, R.A.; Farid, R. Novel Procedure for Modeling Ligand/Receptor Induced Fit Effects. *J. Med. Chem.* **2006**, *49*, 534–553. [CrossRef]
34. Sakamoto, J.; Kimura, H.; Moriyama, S.; Odaka, H.; Momose, Y.; Sugiyama, Y.; Sawada, H. Activation of Human Peroxisome Proliferator-Activated Receptor (PPAR) Subtypes by Pioglitazone. *Biochem. Biophys. Res. Commun.* **2000**, *278*, 704–711. [CrossRef]
35. Nesto, R.W.; Bell, D.; Bonow, R.O.; Fonseca, V.; Grundy, S.M.; Horton, E.S.; Le Winter, M.; Porte, D.; Semenkovich, C.F.; Smith, S.; et al. Thiazolidinedione use, fluid retention, and congestive heart failure: A consensus statement from the American Heart Association and American Diabetes Association. *Diabetes Care* **2004**, *27*, 256–263. [CrossRef]
36. Boudina, S.; Sena, S.; Theobald, H.; Sheng, X.; Wright, J.J.; Hu, X.X.; Aziz, S.; Johnson, J.I.; Bugger, H.; Zaha, V.G.; et al. Mitochondrial energetics in the heart in obesity-related diabetes: Direct evidence for increased uncoupled respiration and activation of uncoupling proteins. *Diabetes* **2007**, *56*, 2457–2466. [CrossRef]
37. Denner, L.A.; Rodriguez-Rivera, J.; Haidacher, S.J.; Jahrling, J.B.; Carmical, J.R.; Hernandez, C.M.; Zhao, Y.; Sadygov, R.G.; Starkey, J.M.; Spratt, H.; et al. Cognitive enhancement with rosiglitazone links the hippocampal PPARgamma and ERK MAPK signaling pathways. *J. Neurosci.* **2012**, *32*, 16725–16735. [CrossRef] [PubMed]
38. Abdel-Rahman, E.A.; Bhattacharya, S.; Buabeid, M.; Majrashi, M.; Bloemer, J.; Tao, Y.X.; Dhanasekaran, M.; Escobar, M.; Amin, R.; Suppiramaniam, V. PPAR-delta Activation Ameliorates Diabetes-Induced Cognitive Dysfunction by Modulating Integrin-linked Kinase and AMPA Receptor Function. *J. Am. Coll. Nutr.* **2019**, *38*, 693–702. [CrossRef] [PubMed]
39. Watson, J.F.; Ho, H.; Greger, I.H. Synaptic transmission and plasticity require AMPA receptor anchoring via its N-terminal domain. *eLife* **2017**, *6*, e23024. [CrossRef] [PubMed]
40. Kariharan, T.; Nanayakkara, G.; Parameshwaran, K.; Bagasrawala, I.; Ahuja, M.; Abdel-Rahman, E.; Amin, A.T.; Dhanasekaran, M.; Suppiramaniam, V.; Amin, R.H. Central activation of PPAR-gamma ameliorates diabetes induced cognitive dysfunction and improves BDNF expression. *Neurobiol. Aging* **2015**, *36*, 1451–1461. [CrossRef] [PubMed]
41. Luscher, C.; Malenka, R.C. NMDA receptor-dependent long-term potentiation and long-term depression (LTP/LTD). *Cold Spring Harb. Perspect. Biol.* **2012**, *4*, a005710. [CrossRef] [PubMed]
42. Miranda, M.; Morici, J.F.; Zanoni, M.B.; Bekinschtein, P. Brain-Derived Neurotrophic Factor: A Key Molecule for Memory in the Healthy and the Pathological Brain. *Front. Cell. Neurosci.* **2019**, *13*, 363. [CrossRef]

43. Gomez-Palacio-Schjetnan, A.; Escobar, M.L. Neurotrophins and synaptic plasticity. *Curr. Top. Behav. Neurosci.* **2013**, *15*, 117–136. [CrossRef] [PubMed]
44. McAllister, A.K.; Katz, L.C.; Lo, D.C. Neurotrophins and synaptic plasticity. *Annu. Rev. Neurosci.* **1999**, *22*, 295–318. [CrossRef] [PubMed]
45. Wang, C.; Cui, Y.; Yang, J.; Zhang, J.; Yuan, D.; Wei, Y.; Li, Y.; Duo, Y.; Li, S.; Zhu, W.; et al. Combining serum and urine biomarkers in the early diagnosis of mild cognitive impairment that evolves into Alzheimer's disease in patients with the apolipoprotein E ϵ4 genotype. *Biomarkers* **2015**, *20*, 84–88. [CrossRef]
46. Siuda, J.; Patalong-Ogiewa, M.; Zmuda, W.; Targosz-Gajniak, M.; Niewiadomska, E.; Matuszek, I.; Jedrzejowska-Szypulka, H.; Lewin-Kowalik, J.; Rudzinska-Bar, M. Cognitive impairment and BDNF serum levels. *Neurol. Neurochir. Polska* **2017**, *51*, 24–32. [CrossRef] [PubMed]
47. Gezen-Ak, D.; Dursun, E.; Hanagasi, H.; Bilgic, B.; Lohman, E.; Araz, O.S.; Atasoy, I.L.; Alaylioglu, M.; Onal, B.; Gurvit, H.; et al. BDNF, TNFalpha, HSP90, CFH, and IL-10 serum levels in patients with early or late onset Alzheimer's disease or mild cognitive impairment. *J. Alzheimer's Dis.* **2013**, *37*, 185–195. [CrossRef]
48. Peng, S.; Garzon, D.J.; Marchese, M.; Klein, W.; Ginsberg, S.D.; Francis, B.M.; Mount, H.T.; Mufson, E.J.; Salehi, A.; Fahnestock, M. Decreased brain-derived neurotrophic factor depends on amyloid aggregation state in transgenic mouse models of Alzheimer's disease. *J. Neurosci.* **2009**, *29*, 9321–9329. [CrossRef]
49. Javonillo, D.I.; Tran, K.M.; Phan, J.; Hingco, E.; Kramar, E.A.; da Cunha, C.; Forner, S.; Kawauchi, S.; Milinkeviciute, G.; Gomez-Arboledas, A.; et al. Systematic Phenotyping and Characterization of the 3xTg-AD Mouse Model of Alzheimer's Disease. *Front. Neurosci.* **2021**, *15*, 1829. [CrossRef] [PubMed]
50. Vandal, M.; White, P.J.; Chevrier, G.; Tremblay, C.; St-Amour, I.; Planel, E.; Marette, A.; Calon, F. Age-dependent impairment of glucose tolerance in the 3xTg-AD mouse model of Alzheimer's disease. *FASEB J.* **2015**, *29*, 4273–4284. [CrossRef]
51. Saunders, A.M.; Burns, D.K.; Gottschalk, W.K. Reassessment of Pioglitazone for Alzheimer's Disease. *Front. Neurosci.* **2021**, *15*, 666958. [CrossRef]
52. Searcy, J.L.; Phelps, J.T.; Pancani, T.; Kadish, I.; Popovic, J.; Anderson, K.L.; Beckett, T.L.; Murphy, M.P.; Chen, K.C.; Blalock, E.M.; et al. Long-term pioglitazone treatment improves learning and attenuates pathological markers in a mouse model of Alzheimer's disease. *J. Alzheimer's Dis.* **2012**, *30*, 943–961. [CrossRef]
53. Sato, T.; Hanyu, H.; Hirao, K.; Kanetaka, H.; Sakurai, H.; Iwamoto, T. Efficacy of PPAR-gamma agonist pioglitazone in mild Alzheimer disease. *Neurobiol. Aging* **2011**, *32*, 1626–1633. [CrossRef]
54. Burns, D.K.; Alexander, R.C.; Welsh-Bohmer, K.A.; Culp, M.; Chiang, C.; O'Neil, J.; Evans, R.M.; Harrigan, P.; Plassman, B.L.; Burke, J.R.; et al. Safety and efficacy of pioglitazone for the delay of cognitive impairment in people at risk of Alzheimer's disease (TOMMORROW): A prognostic biomarker study and a phase 3, randomised, double-blind, placebo-controlled trial. *Lancet Neurol.* **2021**, *20*, 537–547. [CrossRef] [PubMed]
55. Einstein, M.; Akiyama, T.E.; Castriota, G.A.; Wang, C.F.; McKeever, B.; Mosley, R.T.; Becker, J.W.; Moller, D.E.; Meinke, P.T.; Wood, H.B.; et al. The differential interactions of peroxisome proliferator-activated receptor gamma ligands with Tyr473 is a physical basis for their unique biological activities. *Mol. Pharmacol.* **2008**, *73*, 62–74. [CrossRef] [PubMed]
56. Kojetin, D.J.; Burris, T.P. Small molecule modulation of nuclear receptor conformational dynamics: Implications for function and drug discovery. *Mol. Pharmacol.* **2013**, *83*, 1–8. [CrossRef] [PubMed]
57. Santos, G.M.; Fairall, L.; Schwabe, J.W. Negative regulation by nuclear receptors: A plethora of mechanisms. *Trends Endocrinol. Metab.* **2011**, *22*, 87–93. [CrossRef]
58. Weikum, E.R.; Liu, X.; Ortlund, E.A. The nuclear receptor superfamily: A structural perspective. *Protein Sci.* **2018**, *27*, 1876–1892. [CrossRef]
59. Zhou, H.X. From induced fit to conformational selection: A continuum of binding mechanism controlled by the timescale of conformational transitions. *Biophys. J.* **2010**, *98*, L15–L17. [CrossRef]
60. Jankowsky, J.L.; Zheng, H. Practical considerations for choosing a mouse model of Alzheimer's disease. *Mol. Neurodegener.* **2017**, *12*, 89. [CrossRef]
61. Reiner, A.; Levitz, J. Glutamatergic Signaling in the Central Nervous System: Ionotropic and Metabotropic Receptors in Concert. *Neuron* **2018**, *98*, 1080–1098. [CrossRef]
62. Reimers, J.M.; Loweth, J.A.; Wolf, M.E. BDNF contributes to both rapid and homeostatic alterations in AMPA receptor surface expression in nucleus accumbens medium spiny neurons. *Eur. J. Neurosci.* **2014**, *39*, 1159–1169. [CrossRef] [PubMed]
63. Sri, S.; Pegasiou, C.M.; Cave, C.A.; Hough, K.; Wood, N.; Gomez-Nicola, D.; Deinhardt, K.; Bannerman, D.; Perry, V.H.; Vargas-Caballero, M. Emergence of synaptic and cognitive impairment in a mature-onset APP mouse model of Alzheimer's disease. *Acta Neuropathol. Commun.* **2019**, *7*, 25. [CrossRef] [PubMed]
64. Makela, J.; Tselykh, T.V.; Kukkonen, J.P.; Eriksson, O.; Korhonen, L.T.; Lindholm, D. Peroxisome proliferator-activated receptor-gamma (PPARgamma) agonist is neuroprotective and stimulates PGC-1alpha expression and CREB phosphorylation in human dopaminergic neurons. *Neuropharmacology* **2016**, *102*, 266–275. [CrossRef] [PubMed]
65. Thouennon, E.; Cheng, Y.; Falahatian, V.; Cawley, N.X.; Loh, Y.P. Rosiglitazone-activated PPARgamma induces neurotrophic factor-alpha1 transcription contributing to neuroprotection. *J. Neurochem.* **2015**, *134*, 463–470. [CrossRef] [PubMed]

66. Benedetti, E.; D'Angelo, B.; Cristiano, L.; Di Giacomo, E.; Fanelli, F.; Moreno, S.; Cecconi, F.; Fidoamore, A.; Antonosante, A.; Falcone, R.; et al. Involvement of peroxisome proliferator-activated receptor beta/delta (PPAR beta/delta) in BDNF signaling during aging and in Alzheimer disease: Possible role of 4-hydroxynonenal (4-HNE). *Cell. Cycle* **2014**, *13*, 1335–1344. [CrossRef]
67. Sherwood, N.T.; Lo, D.C. Long-term enhancement of central synaptic transmission by chronic brain-derived neurotrophic factor treatment. *J. Neurosci.* **1999**, *19*, 7025–7036. [CrossRef] [PubMed]
68. Sastre, M.; Dewachter, I.; Rossner, S.; Bogdanovic, N.; Rosen, E.; Borghgraef, P.; Evert, B.O.; Dumitrescu-Ozimek, L.; Thal, D.R.; Landreth, G.; et al. Nonsteroidal anti-inflammatory drugs repress beta-secretase gene promoter activity by the activation of PPARgamma. *Proc. Natl. Acad. Sci. USA* **2006**, *103*, 443–448. [CrossRef]
69. Assaf, N.; El-Shamarka, M.E.; Salem, N.A.; Khadrawy, Y.A.; El Sayed, N.S. Neuroprotective effect of PPAR alpha and gamma agonists in a mouse model of amyloidogenesis through modulation of the Wnt/beta catenin pathway via targeting alpha- and beta-secretases. *Prog. Neuropsychopharmacol. Biol. Psychiatry* **2020**, *97*, 109793. [CrossRef]
70. Tonnies, E.; Trushina, E. Oxidative Stress, Synaptic Dysfunction, and Alzheimer's Disease. *J. Alzheimer's Dis.* **2017**, *57*, 1105–1121. [CrossRef]
71. Singh, A.; Kukreti, R.; Saso, L.; Kukreti, S. Oxidative Stress: A Key Modulator in Neurodegenerative Diseases. *Molecules* **2019**, *24*, 1583. [CrossRef]

cells

MDPI

Review

Peroxisome Proliferator-Activated Receptor-Targeted Therapies: Challenges upon Infectious Diseases

In Soo Kim [1,2,3], Prashanta Silwal [4] and Eun-Kyeong Jo [1,2,3,*]

[1] Department of Microbiology, College of Medicine, Chungnam National University, Daejeon 35015, Republic of Korea
[2] Department of Medical Science, College of Medicine, Chungnam National University, Daejeon 35015, Republic of Korea
[3] Infection Control Convergence Research Center, College of Medicine, Chungnam National University, Daejeon 35015, Republic of Korea
[4] Department of Orthopaedic Surgery, University of Pittsburgh School of Medicine, Pittsburgh, PA 15261, USA
* Correspondence: hayoungj@cnu.ac.kr; Tel.: +82-42-580-8243

Abstract: Peroxisome proliferator-activated receptors (PPARs) α, β, and γ are nuclear receptors that orchestrate the transcriptional regulation of genes involved in a variety of biological responses, such as energy metabolism and homeostasis, regulation of inflammation, cellular development, and differentiation. The many roles played by the PPAR signaling pathways indicate that PPARs may be useful targets for various human diseases, including metabolic and inflammatory conditions and tumors. Accumulating evidence suggests that each PPAR plays prominent but different roles in viral, bacterial, and parasitic infectious disease development. In this review, we discuss recent PPAR research works that are focused on how PPARs control various infections and immune responses. In addition, we describe the current and potential therapeutic uses of PPAR agonists/antagonists in the context of infectious diseases. A more comprehensive understanding of the roles played by PPARs in terms of host-pathogen interactions will yield potential adjunctive personalized therapies employing PPAR-modulating agents.

Keywords: peroxisome proliferator-activated receptor; infection; bacteria; virus; parasite

Citation: Kim, I.S.; Silwal, P.; Jo, E.-K. Peroxisome Proliferator-Activated Receptor-Targeted Therapies: Challenges upon Infectious Diseases. *Cells* **2023**, *12*, 650. https://doi.org/10.3390/cells12040650

Academic Editors: Nicole Wagner and Kay-Dietrich Wagner

Received: 9 December 2022
Revised: 13 February 2023
Accepted: 14 February 2023
Published: 17 February 2023

1. Introduction

Peroxisome proliferator-activated receptors (PPARs) are adopted orphan family members of the nuclear receptor group that regulates various biological functions, including glucose and lipid homeostasis, inflammation, and adipose cell differentiation [1,2]. PPARs are ligand-activated transcription factors that are subdivided into three isoforms, termed PPARα (NR1C1), PPARβ/δ (also termed PPARβ or PPARδ, or NR1C2), and PPARγ (NR1C3) [3]. The endogenous PPAR ligands include long-chain polyunsaturated fatty acids and eicosanoids, although the functions of the ligands remain largely unknown [2,4]. Each PPAR isoform evidences a distinct cellular and tissue distribution and biological functions with a focus on energy balance and inflammation [2].

PPARs feature N-terminal DNA-binding and C-terminal ligand-binding domains and form heterodimers with nuclear retinoid X receptor (RXR)-α [5,6]. After interacting with the ligands, PPAR-RXR heterodimers undergo conformational changes that allow them to regulate the transcription of many genes with peroxisome proliferator response elements (PPREs) in their promoter regions [7]. The many PPAR-mediated functions are orchestrated via recruitment of different transcriptional co-activators, including PPAR co-activator-1α, co-activator-associated proteins, and co-repressors [2,5]. Moreover, each PPAR isoform transcriptionally regulates the expression of the other PPAR isoforms via feedback control [2].

PPARα is found principally in the liver and transcriptionally regulates fatty acid oxidation, cholesterol and glycogen metabolism, gluconeogenesis, ketogenesis, and inflammation [8,9]. PPARγ is found in both hematopoietic and non-hematopoietic cells and tissues (adipose tissue and the large intestine) [10]. PPARγ modulates many biological functions, including fatty acid and glucose metabolism and anti-inflammatory signaling via nuclear factor kappa B (NF-κB); it also suppresses oxidative stress and prevents platelet-leukocyte interactions [10,11]. Recent insights into the roles played by PPAR ligands have enabled development of PPAR agonists/antagonists, which serve as candidate drugs for inflammatory, metabolic, and autoimmune diseases, as well as cancers [12]. Several PPARα ligands, including fibrates, helpfully treat dyslipidemia, while the PPARγ ligands pioglitazone and rosiglitazone are well-known anti-diabetic drugs [13]. The three PPARs play critical but distinct roles in regulating the inflammation and metabolic pathways closely associated with immune cell functions [14–16]. It is thus essential to understand how PPARs affect antimicrobial actions against diverse infections. Here, we highlight recent insights into how the PPAR isoforms and their agonists regulate antimicrobial host defenses against viral, bacterial, and parasitic diseases.

2. Overview of PPARs

2.1. Molecular Characteristics of PPARs

Peroxisomes, 0.5 μm diameter single-membrane cytoplasmic organelles, play essential roles in the oxidation of various biomolecules [17,18]. Peroxisome proliferators are multiple chemicals that increase the abundance of peroxisomes in cells [19,20]. These molecules also increase gene expression for β-oxidation of long-chain fatty acids and cytochrome P450 (CYP450) [21,22]. Given the gene transcriptional modulation of peroxisome proliferators, PPARs have been identified as nuclear receptors [23–29]. The PPAR subfamily consists of three isoforms, PPARα, PPARβ/δ, and PPARγ [30]. The three PPARs differ in tissue-specific expression patterns and ligand-biding domains, each performing distinct functions. *PPARA*, encoding PPARα, is located in chromosome 22q13.31 in humans and is mainly expressed in the liver, intestine, kidney, heart, and muscle [31,32]. PPARγ has four alternative splicing forms from *PPARG* located in chromosome 3p25.2 and is highly expressed in adipose tissue, the spleen, and intestine [33,34]. PPARδ, encoded by *PPARD*, is located in chromosome 6q21.31 and presents ubiquitously [29,35]. Thus, it is essential in the study of PPARs to consider their tissue distribution and functions.

PPAR is a nuclear receptor superfamily class II member that heterodimerizes with RXR [36,37]. The PPAR structure includes the A/B, C, D, and E domains from N-terminus to C-terminus [38]. The N-terminal A/B domain (NTD) is a ligand-independent transactivation domain containing the activator function (AF)-1 region. The NTD is targeted for variable post-translational modifications, including SUMOylation, phosphorylation, acetylation, O-GlcNAcylation, and ubiquitination, resulting in transcriptional regulatory activities [39]. DNA-binding C domain (DBD) has two DNA-binding zinc finger motifs containing cysteines, which dock to PPREs. PPARs reside upstream of RXR upon the direct repeat (DR)-1 motifs, which are composed of two hexanucleotide consensus sequences with one spacing nucleotide (AGGTCA N AGGTCA) [40]. The hinge D region is a linker between the C and E domains, which contains a nuclear localization signal, and is the site for post-translational modifications such as phosphorylation, acetylation, and SUMOylation [39]. The ligand-binding E domain (LBD) carries the hydrophobic ligand-binding pocket and the AF-2 region. The absence of agonists enables LBD to recruit co-repressors containing the CoRNR motifs [41]. Engaging agonists to LBD elicits conformational changes of AF-2 to facilitate interaction with LXXLL motifs of many co-activators [42]. Like other nuclear receptor superfamily class II members, such as thyroid hormone receptor (TR), retinoic acid receptor (RAR), and vitamin D receptor (VDR), PPARs function as heterodimers with RXR through LBD [6,43]. LBD is also targeted for SUMOylation and ubiquitination [39]. Advancement of research on the PPAR structure helps thoroughly dissect the roles of PPARs. We will discuss the roles of specific PPAR subtypes in the following subsections.

2.1.1. Roles of PPARα

PPARα is predominantly expressed in the liver but is also found in other tissues, including the heart, muscle, and kidney [4,32]. PPARα regulates the expression of genes involved in metabolism and inflammation. Activation of PPARα leads to the upregulation of genes involved in fatty acid oxidation and the downregulation of genes involved in fatty acid synthesis [8]. PPARα also modulates other genes, including genes involved in the transport and uptake of fatty acids and the synthesis and secretion of lipoproteins [4,8]. In addition, activation of PPARα has been shown to improve insulin sensitivity, reduce oxidative stress, and reduce inflammation in preclinical studies [7,8,44]. PPARα activation has been shown to modify the expression of immune response genes, including those encoding cytokines and chemokines, which are signaling molecules that regulate the immune response [44,45]. PPARα activation has also been demonstrated to reduce the production of pro-inflammatory cytokines, such as tumor necrosis factor-alpha (TNF-α) and interleukin (IL)-6 [46,47]. PPARα has been shown to interfere with the DNA binding of both AP-1 and NF-κB [45,46,48]. Thus, the roles of PPARα in infectious diseases should be studied in wide ranging aspects, including metabolism and inflammation.

In the context of infection, PPARα has been shown to play an essential role in the hepatic metabolic response to infection. During an infectious challenge, the liver coordinates several metabolic changes to support the host defense response, including the mobilization of energy stores, production of acute-phase proteins, and synthesis of new metabolites. Activation of PPARα in the liver leads to the upregulation of genes involved in fatty acid oxidation and ketogenesis with fibroblast growth factor 21 (FGF21) production [49]. FGF21 is a hormone produced by the liver that has been shown to promote ketogenesis and reduce glucose utilization [50,51]. The ketogenesis regulation of PPARα with FGF21 is essential for reacting to microbial or viral sepsis [52–54]. In conclusion, the hepatic PPARα metabolic response to infection is crucial to the host defense response.

2.1.2. Roles of PPARβ/δ

PPARβ is expressed in diverse tissues, including adipose tissue, muscle, and the liver [29,55], and is activated by multiple ligands, such as fatty acids and their derivatives [7]. PPARβ is involved in regulating lipid metabolism and energy homeostasis, as well as controlling inflammation and immune function [56]. PPARβ activation has been demonstrated to have pro- and anti-inflammatory effects based on the situation [56]. The role of PPARβ in tumorigenesis is debatable. PPARβ activation has been found in some cases to have anti-tumorigenic effects, such as causing apoptosis and inhibiting cell proliferation [57,58]. In other cases, however, activation of PPARβ has been shown to promote tumorigenesis by enhancing cell survival, promoting angiogenesis, and reducing cellular differentiation [59–62]. Overall, the role of PPARβ activation in cancer is not entirely known and is complex. Similarly, the function of PPARβ in infection is not well understood. Additional research is required to comprehend the function of PPARβ in the context of immunology against cancers and infectious diseases.

2.1.3. Roles of PPARγ

PPARγ is expressed in a variety of tissues, including adipose tissue, muscle, and the liver [33,34,55], and is activated by diverse ligands, including fatty acids and their derivatives, as well as synthetic chemicals known as thiazolidinediones [4,7]. PPARγ is responsible for regulating lipid metabolism, glucose homeostasis, and inflammation [63,64]. Numerous inflammatory mediators and cytokines are inhibited by PPARγ ligands in various cell types, including monocytes/macrophages, epithelial cells, smooth muscle cells, endothelial cells, dendritic cells, and lymphocytes. In addition, PPARγ diminishes the activities of transcription factors AP-1, STAT, NF-κB, and NFAT to adversely regulate inflammatory gene expression [65–67]. As a result, PPARγ has been demonstrated to have a protective function against infections by modulating the immune response and lowering inflammation. However, other researchers have hypothesized that PPARγ activation may impair the function of immune cells, such as macrophages, and contribute to the development of infections. Therefore, the role of PPARγ in disease is complex and context-dependent, and more research is needed to fully understand the molecular mechanisms by which PPARγ regulates the host response to infection.

2.2. Regulatory Mechanisms of PPARs

The PPAR ligand-binding pocket is large and capable of engaging diverse ligands [68,69]. Endogenous ligands vary depending on the PPAR isoform, including n-3 polyunsaturated fatty acids such as docosahexaenoic acid and eicosapentaenoic acid for all PPARs, leukotriene B4 for PPARα, carbaprostacyclin for PPARδ, and prostaglandin J2 for PPARγ [70]. Representative synthetic agonists include fibrates (PPARα agonists) and thiazolidinediones (PPARγ agonists) [7]. Fibrates, such as fenofibrate, clofibrate, and gemfibrozil, are widely used for treating dyslipidemia. Thiazolidinediones, such as rosiglitazone, pioglitazone, and lobeglitazone, improve insulin resistance [7]. Most clinical studies on PPAR actions in infectious diseases have been conducted retrospectively, and no clinical studies currently in progress are listed in ClinicalTrials.gov (https://clinicaltrials.gov/ (accessed on 13 February 2023)). Since widely used PPAR agonists exist, clinical research can be conducted through a deeper understanding of PPAR roles in infectious diseases.

PPAR-RXR heterodimerization occurs ligand-independently [6]. The heterodimer appears to exert transcriptional regulation both ligand-dependently and -independently [7]. Although LBD may interact with either co-repressor or co-activator in the state of not binding with an agonist, binding to a ligand elicits stabilized co-activator-LBD interaction, thus increasing transactivation [7,71]. Further, recent studies have shown that PPARs inhibit other transcription factors, such as NF-κB, activator protein-1 (AP-1), signal transducer and activator of transcription (STAT), and nuclear factor of activated T cells (NFAT) [44,65–67]. Recent studies revealed the possibility of forming a protein chaperone complex with PPAR-associated proteins, such as heat shock proteins (HSPs). Similar to interactions between other type I intracellular receptors and heat shock proteins, HSP90 repressed PPARα and PPARβ activities but not that of PPARγ [72]. Instead, HSP90 was required for PPARγ signaling in the nonalcoholic fatty liver disease mouse model [73]. Thus, it is necessary to study the various modes of PPAR actions. The intracellular regulatory mechanisms of PPARs are shown in Figure 1.

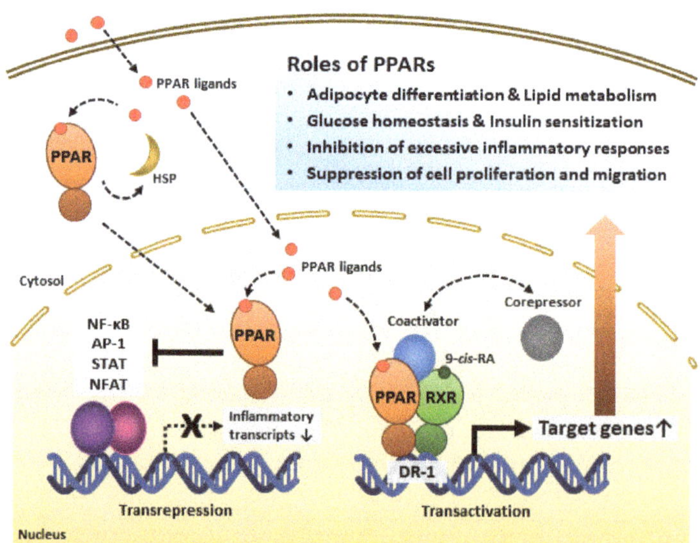

Figure 1. Roles of peroxisome proliferator-activated receptors (PPARs) and their regulatory mechanisms.

PPAR ligands bind to the PPAR ligand-binding domain and activate receptors. PPARs interact with heat shock protein (HSP) in the cytosol. PPARs inhibit inflammation-related gene transcription by interfering with transcription factors such as NF-κB, AP-1, STAT, and NFAT. PPARs form heterodimers with Retinoid X receptor (RXR), a receptor of 9-*cis*-retinoic acid (9-*cis*-RA), and bind to direct repeat 1 (DR-1), a peroxisome-proliferator-responsive element. The PPAR-RXR heterodimer complex and co-repressors represses target gene transcription. However, the complex with co-activators promotes target gene transcription. Through these mechanisms, PPARs play significant roles in energy metabolism, inflammatory modulation, and the cell cycle. AP-1, Activator protein 1; NF-κB, Nuclear factor kappa-light-chain-enhancer of activated B cells; NFAT, Nuclear factor of activated T cells; STAT, Signal transducer and activator of transcription.

3. PPARs and Viral Infections

3.1. PPARs and Respiratory Viral Infections

Many studies have shown that PPARγ controls viral replication and virus-associated inflammation by antagonizing inflammatory signaling pathways such as the NF-κB and STAT pathways [74,75]. In particular, PPARγ of alveolar macrophages critically modulates acute inflammation to promote recovery from respiratory viral infections, most of which are caused by influenza A virus (IAV) and respiratory syncytial virus (RSV) [76]. Several PPAR agonists have shown promise in terms of ameliorating virus-related cytokine storms and the damage caused by severe IAV infection [77]. Macrophage PPARγ is essential for resolving the chronic pulmonary collagen deposition and fibrotic changes that follow influenza infection [78]. Several researchers have sought new therapeutic candidates for IAV disease. A recent screening of traditional Chinese medicines showed that emodin and analogs thereof evidenced excellent anti-IAV activities mediated by activation of the PPARα/γ and adenosine monophosphate (AMP)-activated protein kinase (AMPK) pathways [79]. High-throughput screening of natural compounds and/or synthetic drugs/agents will yield new therapeutics against respiratory viral infections based on drug interactions with PPAR pathways.

A link has been suggested between severe acute respiratory syndrome coronavirus 2 (SARS-CoV-2) virus infection and PPARα activity in the context of lipid uptake, lipotoxicity, and vascular inflammation [80–82]. The PPARα agonist fenofibrate is a potential adjunctive coronavirus disease (COVID-19) therapy; the material exhibits anti-inflammatory and anti-thrombotic activities [80,82]. A study employing a public database on subjects with type 2 diabetes and COVID-19, along with animal studies, revealed that the PPARγ agonist pioglitazone may ameliorate acute lung injury and SARS-CoV-2-mediated hyperinflammation [83]. Cannabidiol working via PPARγ is proposed as a therapeutic approach for the severe form of COVID-19 [84]. A recent study demonstrated that cannabidiol attenuated inflammation and epithelial damage in colonic epithelial cells exposed to the SARS-CoV-2 spike protein through a PPARγ-dependent mechanism [85]. The natural compound γ-oryzanol may also serve as an adjunctive therapy to reduce the cytokine storm associated with COVID-19; the material stimulated PPARγ to modulate oxidative stress and the inflammatory response in adipose tissues [86]. The Middle East respiratory syndrome coronavirus (MERS-CoV)-derived S glycoprotein activates PPARγ to suppress the pathologic inflammatory responses of macrophages [87]. Further research on the modulatory roles played by PPAR agonists/antagonists in terms of virus-associated inflammation will yield novel adjunctive therapeutics to counter emerging and re-emerging viral infections. Table 1 summarizes studies on PPARs and their ligands in relation to viral infections.

Table 1. Studies on PPARs and their ligands during viral infections.

Pathogen	Study Model	Intervention	PPAR Status	Mechanism	Ref.
IAV, RSV	AMs, mice	*Pparg*$^{\Delta Lyz2}$ mice	↓	Regulation of PPARγ through STAT1 activation following IFN signaling	[76]
IAV	AMs, human lung macrophages, mice	*Pparg*$^{\Delta Lyz2}$ mice, Bleomycin	↓	Increased influenza-induced pulmonary collagen deposition in PPARγ-deficient mice	[78]
IAV	A549 cells, mice	Emodin and its analogs	↑	Activation of PPARα/γ and AMPK, decreased fatty acid biosynthesis and increased ATP level	[79]
MERS-CoV	THP-1 cells, primary human monocytes	siRNAs	↑	MERS-CoV S glycoprotein interaction with DPP4 leading to IRAK-M and PPARγ expression	[87]
CHIKV	Vero cells, RAW264.7 cells	Telmisatran, PPAR-γ antagonist GW9662	↓	Activation of PPAR-γ and inhibition of AT1 by telmisartan	[88]
HIV	Primary rat astrocytes, microglia, rats	$gp120_{ADA}$, Rosiglitazone, Pioglitazone	↓	Induction of inflammatory response and decrease in GLT-1 expression in the brain by gp120	[89]
HBV	HepG2.2.15, Huh7, HepG2-NTCP ells	OS_128167, overexpression and downregulation studies, HBV transgenic mice	-	Activation of HBV core promoter by SIRT6 through upregulation of PPARα	[90]
HCV	Huh7.5 cells	Calciterol, Linoleic acid, Ly171883, Wy14643	-	Activation of VDR but inhibition of PPARα/β/γ by calcitriol	[91]

Abbreviations: AMPK, AMP-activated protein kinase; AMs, Alveolar macrophages; AT1, Angiotensin 1; CHIKV, Chikungunya virus; DPP4, Dipeptidyl-peptidase 4; GLT-1, Glutamate transporter 1; HBV, Hepatitis B virus; HCV, Hepatitis C virus; HIV, human immunodeficiency virus; IAV, Influenza A virus; IFN, interferon; IRAK-M, Interleukin-1 receptor-associated kinase 3; MERS-CoV, Middle east respiratory syndrome corona virus; RSV, Respiratory syncytial virus; SIRT6, Sirtuin 6; STAT1, Signal transducer and activator of transcription 1; VDR; Vitamin D receptor; ↑, increase/activation; ↓, decrease/inhibition; -, not reported.

3.2. PPARs and Virus-Related Inflammation

A recent study showed that the inflammatory responses during infection with Chikungunya virus (CHIKV) involved the renin-angiotensin system (RAS) and PPARγ pathways [88]. The telmisartan-mediated suppression of CHIKV infection is at least partly mediated via activation of PPARγ; a PPARγ antagonist increased the CHIKV viral load [88]. Omeragic et al. showed that PPARγ played a critical role in terms of human immunodeficiency virus (HIV-1) ADA glycoprotein 120 (gp120)-related inflammatory marker generation was observed in primary astrocytes and microglia and also in vivo [89]. The anti-inflammatory activities induced by the PPARγ agonists rosiglitazone and pioglitazone reflected suppression of the NF-κB signaling pathway [89]. These relationships between PPARγ and viral infections are included in Table 1. Δ-9-tetrahydrocannabinol improved epithelial barrier function and thus protected colonic tissues of rhesus macaques chronically infected with simian immunodeficiency virus (SIV). This was at least partly attributable to the upregulation of PPARγ [92]. PPARα signaling is required for restoration of the intestinal barrier by the probiotic *Lactobacillus plantarum* and amelioration of gut inflammation during SIV infection [93]. Such findings strongly suggest that targeting PPARγ would both prevent and treat virus-associated inflammation of the brain, endothelial system, and intestinal tissues. The PPARγ antagonist GW9662 protected against dengue virus infection and di(2-ethylhexyl) phthalate (DEHP)-induced interleukin (IL)-23 expression, thus suppressing the viral load [94]. Therefore, future clinical trials should explore the protective effects of several possible PPAR agonists/antagonists and combinations thereof with current antivirals in patients with various viral infections.

Zika virus (ZIKV) is a serious arthropod-borne (arbovirus) pathogen that causes congenital defects and neurological diseases in both infants and adults [95]. A recent study showed that ZIKV-induced cellular responses of induced pluripotent stem cell (iPSC)-derived neural progenitor cells involved the PPAR signaling pathways, which may contribute to neurogenesis and viral replication [96]. However, further research is required.

3.3. PPARs and Hepatitis Virus Infection

The roles played by PPAR pathways in terms of hepatitis B virus (HBV) infection elimination are complex. IL-1β production induced by HBV infection of M1-like inflammatory macrophages triggered anti-HBV responses via downregulation of PPARα and forkhead box O3 (FOXO3) expression in hepatocytes [97]. OSS_128167, a sirtuin 6 inhibitor, inhibited HBV transcription and replication in hepatic cells and in vivo by targeting PPARα expression [90]. In the HBV replicative mouse model, PPAR agonists, including bezafibrate, fenofibrate, and rosiglitazone, significantly increased the serum levels of HBV antigens HBsAg, HBeAg, and HBcAg and that of HBV DNA, as well as the viral load in mouse liver [98]. Thus, patients with metabolic diseases taking PPAR-based therapeutics should take care to avoid HBV infection. However, in a retrospective study of HBV-infected patients treated with entecavir and tenofovir-disoproxil-fumarate, the drugs exerted profound extrahepatic effects on lipid metabolism, reducing serum cholesterol levels by inducing the expression of PPARα target genes such as CD36 in liver tissue and cells [99]. Thus, the PPARα-activating nucleoside analogs tenofovir-disoproxil-fumarate may usefully treat atherosclerosis and hepatocarcinogenesis, both of which are associated with dyslipidemia. This would be a new role for an anti-HBV therapeutic. However, the precise functions of PPARs during HBV infection remain unclear. The antiviral, antitumor, and extrahepatic actions of PPAR agonists vary with the clinical condition.

During hepatitis C virus (HCV) infection, PPAR-α/β/γ stimulators/agonists reduce calcitriol-mediated anti-HCV responses, presumably by counteracting the calcitriol-mediated activation of vitamin D receptor signaling and inhibiting nitrative stress [91]. Naringenin, a grapefruit flavonoid, suppressed HCV production by inhibiting viral particle assembly via PPARα activation, suggesting potential roles for PPARα agonists in the resolution of infection [100]. It is essential to perform an in-depth exploration of how the three PPARs and their signaling pathways affect the outcomes of HBV and HCV infections. Studies on PPARs and hepatitis virus infections are summarized in Table 1.

4. PPARs and Bacterial Infections

4.1. PPARs and Post-Influenza Bacterial Infections

PPARs exacerbate the severity of post-influenza bacterial infections. During *Staphylococcus aureus* superinfection following IAV infection, the levels of CYP450 metabolites, which are PPARα ligands, increase significantly and trigger receptor-interacting serine/threonine-protein kinase 3 (RIPK3)-induced necroptosis, thus exacerbating the lung pathology and increasing mortality from secondary bacterial infection [101]. The PPARγ agonist rosiglitazone reduces bacterial clearance during secondary bacterial pneumonia, which is a frequent complication of primary IAV infection [102]. Diabetic patients treated with rosiglitazone exhibited increased mortality from IAV-associated pneumonia compared to those not treated with rosiglitazone, as revealed by data from the National Health and Nutrition Examination Survey (NHANES) [102]. CYP450 metabolites reduced the protective inflammatory responses via PPARα activation, thereby increasing the susceptibility to secondary bacterial infection following IAV infection [103]. Thus, PPARα or PPARγ drives host protection but reduces bacterial clearance at different stages of IAV infection. The molecular mechanisms by which PPARα/γ mediates immune modulation during a bacterial infection following IAV infection require urgent attention. Better medicines are needed to treat the different stages of IAV-associated disease, which is often fatal in susceptible patients.

4.2. PPARs in Bacterial Infections

PPARs and agonists/antagonists thereof may modulate disease severity and outcomes in patients with bacterial infections and associated inflammation. In a model of intestinal colitis, 5-aminosalicylic acid, a PPARγ agonist, exerted therapeutic anti-inflammatory effects by activating the epithelial PPARγ signaling pathway [104]. After infection with *Klebsiella pneumoniae*, which is the respiratory Gram-negative bacterium that usually causes pneumonia, PPARγ agonists such as pioglitazone reduced proinflammatory cytokine and myeloperoxidase levels, bacterial growth in lung tissues, and bacterial dissemination to distant organs [105]. The taste receptor type-2 member 138 (TAS2R138) plays a role in neutrophil-associated host innate immune defense after *Pseudomonas aeruginosa* infection [106]. TAS2R138 mediated the degradation of lipid bodies via competitive binding to the PPARγ antagonist N-(3-oxododecanoyl)-L-homoserine lactone (AHL-12), a mediator of virulence produced by *P. aeruginosa* [106]. Although the exact roles of PPARγ in antimicrobial responses remain unclear, a study employing a model of *P. aeruginosa* infection found that the PPARγ agonist pioglitazone increased the levels of certain chemokines (*Cxcl1*, *Cxcl2*, and *Ccl20*) and cytokines (*Tnfa*, *Il6*, and *Cfs3*) in bronchial epithelial cells and suppressed inflammatory responses in bronchoalveolar lavage fluid [107]. Future studies must explore the utility of PPAR agonists/antagonists as adjuvant therapies and determine whether systemic or local treatments improve disease outcomes.

During *Chlamydia pneumoniae* infection, both PPARα and PPARγ are required to upregulate foam macrophage formation via induction of the scavenger receptor A1 (SR-A1) and the acyl-coenzyme A cholesterol acyltransferase 1 (ACAT1) involved in cholesterol esterification [108]. PPARα and PPARγ agonists, including fenofibrate and rosiglitazone, may suppress atherosclerotic plaque formation in patients with coronary heart disease infected with *C. pneumoniae* [108]. Activation of both PPARα and PPARγ by PAR5359 protected against *Citrobacter rodentium*-induced colitis. The dual agonism promoted antibacterial immunity and ameliorated the inflammatory response [109].

In contrast to studies with Gram-negative bacteria, few reports have explored the roles played by PPARs during Gram-positive infections. In a *Caenorhabditis elegans* model, induction of the gene encoding flavin-containing monooxygenase (FMO) *fmo-2/FMO5* by NHR-49/PPAR-α was critical in terms of the establishment of an effective innate host defense against *S. aureus* infection [110]. Erythropoietin limits infections caused by Gram-negative *Escherichia coli* and Gram-positive *S. aureus*; macrophage-mediated clearance of these bacteria is at least partly mediated by a PPARγ-dependent pathway [111]. Inhibition of PPARγ signaling reduced the survival of *Rickettsia conorii*, an intracellular Gram-positive bacterium, probably by reducing lipid droplet production [112]. Although PPAR-based therapeutics may counter bacterial infections, more preclinical and clinical studies are required. Table 2 summarizes the roles of PPAR ligands in bacterial infections.

4.3. PPARs and Mycobacterial Infections

Many scholars have sought to clarify the effects of PPARs in those infected with *Mycobacterium tuberculosis* (Mtb) and nontuberculous mycobacteria (NTM), which cause tuberculosis and NTM disease, respectively [113]. Although the relevant bacterial components have not been fully characterized, *M. leprae* and Mtb lead to activation of PPARs [113–115]. PPARα and PPARγ appear to play opposite roles. The virulent Mtb strain H37Rv and cell wall component lipoarabinomannan induced PPARγ expression, in turn activating IL-8 and cyclooxygenase (COX) 2 expression, but the attenuated *M. bovis* strain, termed Bacillus Calmette-Guérin (BCG), induced less PPARγ expression [115]. PPARγ activation during Mtb or BCG infection upregulates lipid body formation and increases bacterial survival in macrophages [116,117]. Either PPARγ knockdown or PPARγ antagonist GW9662 increased macrophage-mediated Mtb killing [115,117]. PPARγ activation was associated with enhanced cholesterol and triacylglycerol uptake; these materials are required for macrophage lipid body formation during mycobacterial infection [113]. Antagonists of PPARδ or PPARγ significantly inhibited lipid accumulation by cells infected with *M. leprae*, thus reducing parasitization [114,118]. Together, the data suggest that PPARγ is required for intracellular bacterial survival; PPARγ enhances lipid body formation and foam macrophage development during mycobacterial infection.

Table 2. Roles of PPAR agonists/antagonists in bacterial infections.

Pathogen	Drug/Reagent	Function	Study Model	Mechanism of Action	Ref.
Escherichia coli	5-aminosalicylic acid	PPARγ agonist	DSS-induced murine colitis model, *Pparg*-deficient mice, CaCo-2 cells	Amelioration of a respiration-dependent luminal expansion of *E. coli*	[104]
Klebsiella pneumoniae	Pioglitazone	PPARγ agonist	In vivo mouse model	Reduction of cytokines and myeloperoxidase levels in the lungs	[105]
Pseudomonas aeruginosa	Pioglitazone	PPARγ agonist	In vivo mouse model	Increased pro-inflammatory cytokines with enhanced expression of genes involved in glycolysis	[107]
Chlamydia pneumoniae	Rosiglitazone Fenofibrate GW9662 MK886	PPARγ agonist PPARα agonist PPARγ antagonist PPARα antagonist	THP-1 macrophages, HEp-2 cells	Regulation of Cpn induced macrophage-derived foam cell formation by upregulating SR-A1 an ACAT1, and downregulating ABCA1/G1 expression via PPARα/γ signaling	[108]
Citrobacter rodentium	PAR5359	PPARα/γ-dual-agonist	*Citrobacter rodentium*- and DSS-induced murine colitis model, IBD patient-derived PBMCs	Enhanced bacterial clearance, controlled production of ROS and cytokines, anti-inflammatory/healing	[109]
Rickettsia conorii	GW9662	PPARγ antagonist	THP-1 macrophages	Increased intracellular survival of bacteria	[112]

Abbreviations: ABCA1/G1, ATP binding cassette transporters A1/G1; ACAT1, acyl-coenzyme A: cholesterol acyltransferase 1; Cpn, *Chlamydia pneumonia*; DSS, Dextran sulfate sodium; IBD, Inflammatory bowel disease; PBMCs, Peripheral blood mononuclear cells; ROS, Reactive oxygen species; SR-A1, scavenger receptor A1.

In contrast, PPARα appears to enhance defenses against macrophage and lung Mtb or BCG infection in mice. PPARα-mediated antimicrobial responses are at least partly mediated via promotion of lipid catabolism and activation of the transcription factor EB (TFEB), a transcriptional factor required for lysosomal biogenesis [119]. Notably, PPARα agonists GW7647 and Wy14643 protected macrophages against Mtb or BCG infection [119]. Macrophage PPARα expression reduces inflammatory cytokine synthesis during Mtb or BCG infection [119], suggesting that PPARα ameliorates inflammation. PPARα deficiency reduced the antimicrobial response and increased lung tissue damage during pulmonary *Mycobacteroides abscessus* (Mabc) infection [120]. Gemfibrozil, a PPARα activator, reduced the in vivo Mabc load and lung inflammation during infection [120]. It is important to clarify whether PPARα modulates lipid body formation during infections with Mabc and other NTMs.

5. PPARs and Parasitic Infections

The anti-inflammatory responses of M2 macrophages and Th2 immunity protect against parasitic infections [121]. In allergic patients and those infected with the nematode *Heligmosomoides polygyrus*, PPARγ is highly expressed in Th2 cells. PPARγ affects the development of Th2-associated pathological immune responses and increases IL-33 receptor levels in Th2 cells [122]. *Neospora caninum* infection triggers maturation of M2 macrophage development via upregulation of PPARγ activity and downregulation of NF-κB signaling [123]. In a model of eosinophilic meningoencephalitis caused by the rat lungworm *Angiostrongylus cantonensis*, PPARγ played anti-inflammatory and protective roles by inhibiting NF-κB-mediated pathological inflammatory responses; the PPARγ antagonist GW9662 increased susceptibility to angiostrongyliasis [124]. In a model of cerebral malaria using clinical isolates of *Plasmodium falciparum*, dimethyl fumarate increased the expression of nuclear factor E2-related factor 2 (NRF2), in turn enhancing PPAR signaling and thus ameliorating the neuroinflammatory responses of primary human brain microvascular endothelial cells [125]. Cerebral malaria susceptibility was associated with a lack of PPARγ nuclear translocation and increased COX-2 levels in brain tissues, which was associated with higher-level parasitemia and poorer survival [126]. PPAR signaling may exert useful antiparasitic functions by attenuating inflammation.

Toxoplasma gondii, one of the most common zoonotic pathogens, infects both immunocompromised patients and healthy individuals and most commonly targets the central nervous system [127]. In *T. gondii*-infected astroglia, the PPARγ agonist rosiglitazone reduced neuroinflammation, whereas the PPARγ antagonist GW9662 increased levels of matrix metalloprotease (MMP)-2, MMP-9, and inflammatory mediators. These findings suggested that PPARγ signaling protects against *T. gondii* infection [128]. Proteomic analysis showed that the hepatic protein responses to *T. gondii* infection modulated the PPAR signaling pathways to dysregulate further liver lipid metabolism [129]. However, it remains unclear how *T. gondii*-mediated modulation of PPARγ signaling affects such metabolism and the consequence thereof.

Sometimes, PPAR signaling negatively affects host defenses against parasitic infections, particularly when M2 macrophage responses are associated with disease progression. During infection of Balb/c mice and hamsters with *Leishmania donovani*, a causative agent of visceral leishmaniasis, the mRNA expression levels of IL-4- and IL-10-driven markers increased significantly [130]. Although any IL-4-related PPARγ function remains unclear, the parasitic load correlated with the effects of IL-10 on the hamster spleen [130]. Schistosomiasis (bilharzia), caused by parasitic flatworms of the genus *Schistosoma*, is associated with inflammatory responses of the intestinal, hepato-splenic, and urogenital systems [131,132]. The Sm16/SPO-1/SmSLP protein from *S. mansoni* may allow the parasite to escape the actions of the innate immune pathway and cellular metabolism, at least partly via a PPAR-dependent pathway [133]. The *Trypanosoma cruzi* protozoan causes Chagas disease, a neglected but chronic tropical infection of great concern in Latin America [134]. During *T. cruzi* infection, both PPARα and PPARγ agonists appear to be involved in macrophage polarization from M1 to M2 types, thereby suppressing inflammation but increasing phagocytosis and macrophage parasitic loads [135]. Thus, PPAR functions may vary by parasite and experimental model. Future studies must explore whether PPARs trigger host defenses or immune evasion during parasitic infections.

Several PPAR ligands may serve as useful adjunct therapies for Chagas disease, although more preclinical and clinical data are required. The new PPARγ ligand HP24, a pyridinecarboxylic acid derivative, evidenced anti-inflammatory and pro-angiogenic activities and might serve as an adjunct therapy for Chagas disease [136]. 15-deoxy-$\Delta^{12,14}$ prostaglandin J_2 (15dPGJ2), a natural PPARγ agonist, reduced liver inflammation and fibrosis during *T. cruzi* infection [137]. However, the use of PPAR agonists/antagonists should be considered in the context of in vivo PPAR expression levels during certain parasitic infections. For example, acute *T. cruzi* mouse infections trigger significant adipose tissue loss and dysregulation of lipolytic and lipogenic enzymes, which are associated with decreased adipocyte PPAR-γ levels in vivo [138]. Given the robust PPARγ inhibition in this mouse model, PPARγ agonists were minimally effective. However, certain parasites do not specifically affect the host responses depending on PPAR down- or upregulation in target tissues or cells. After infection with the intestinal parasite *Giardia muris*, rapid PPARα induction did not affect the protective or pathological immune responses; PPARα-deficient mice cleared the parasite as did wild-type controls [139]. It is important to explore whether aberrant PPAR expression induced by different parasites improves disease status or, rather, enhances dysfunctional inflammation and infection progression. Table 3 summarizes the roles of PPAR agonists/antagonists in parasitic infections.

Table 3. Roles of PPAR agonists/antagonists in parasitic infections.

Pathogen	Drug/Reagent	Function	Study Model	Mechanism of Action	Ref.
Angiostrongylus cantonensis	GW9662	PPARγ antagonist	Mouse model of angiostrongyliasis	NF-κB activation and increase in inflammation and BBB permeability	[124]
Plasmodium falciparum	Dimethyl fumarate	-	Cerebral cortex derived HBMVECs	Upregulation of PPAR pathway, NRF2-mediated oxidative stress responses, ErbB4 signaling to downregulate the neuroinflammation	[125]
Toxoplasma gondii	Rosiglitazone	PPARγ agonist	SVG p12 cells, Hs68 cells	Decreased expression of MMP-2, MMP-9, COX-2, PGE2, iNOS and NO	[128]
	GW9662	PPARγ antagonist		Increased expression of MMP-2, MMP-9, COX-2, PGE2, iNOS and NO	
Trypanosoma cruzi	HP24	pyridinecarboxylic acid derivative	In vivo mice infection, mouse peritoneal macrophages	Induction of PI3K/Akt/mTOR signaling (pro-angiogenic), inhibition of NF-κB signaling (anti-inflammatory)	[136]
	15-deoxy-D12,14 prostaglandin J2	PPARγ agonist	In vivo mice infection	Reduction of liver inflammatory infiltrates, pro-inflammatory enzymes and cytokine expression through inhibition of NF-kB signaling, No change in parasitic load	[137]

Abbreviations: BBB, Blood-brain barrier; COX-2, Cyclooxygenase-2; ErbB4, Erb-b2 receptor tyrosine kinase 4; HP24, 1-methyl-3-hydroxy-4-pyridinecarboxylic acid derivative 24; iNOS, Inducible nitric oxide synthase; MMP, Matrix metalloproteinase; mTOR, Mammalian target of rapamycin; NF-kB, Nuclear factor-κB; NO, Nitric oxide; NRF2, Nuclear factor E2-related factor 2; PGE2, Prostaglandin E2; PI3K, Phosphoinositide 3-kinase.

6. Future Perspectives

PPARs play a wide range of roles across host metabolism, inflammation, and immune responses. Recent studies indicate that PPARs modulate the host responses to infections, such as infectious agent clearance and inflammation. Several PPAR ligands have been utilized in infection models and their functions have been investigated. However, there are no clinical trials of well-known, licensed metabolic medicines utilizing PPAR pathways for infectious diseases. PPAR-based future drugs may serve as adjuvants or components of combination therapies against infections. Understanding the fundamental processes of PPAR-mediated host immune regulation is necessary to develop the most effective treatment approaches for infectious diseases. Future research may also benefit from developing synthetic ligands that preferentially target the specific PPAR isoform implicated in immune response modification.

7. Conclusions

Accumulating evidence suggests that PPARs are involved in the host responses to infections caused by bacteria, viruses, and parasites. However, the molecular mechanisms by which PPARs modulate disease progression or protective responses remain unknown. It is essential to further explore the PPAR functions and mechanisms involved in pathogen survival, the pathological responses during different stages of infection, and the associated modulation of the distinct types of infection-associated acute and chronic inflammation. Apart from shaping the inflammatory and metabolic responses during infections, PPARs may impact disease outcomes. The PPAR signaling pathways exert potent immunomodulatory effects; pathway activation or suppression may usefully treat infectious diseases. Infectious pathogens modulate the individual and collaborative activities of PPAR(s) during infection. We speculate that aberrant PPAR expression by various parasites may contribute to inflammation-related dysfunction. It is essential to better understand the possible clinical effects of PPAR-based therapeutics in patients with various infectious diseases.

Author Contributions: Conceptualization, writing—original draft preparation, review and editing, visualization, E.-K.J., I.S.K. and P.S.; supervision, project administration, funding acquisition, E.-K.J. All authors have read and agreed to the published version of the manuscript.

Funding: This work was supported by the National Research Foundation of Korea (NRF) grant funded by the Korea government (MSIT) (grant number: 2017R1A5A2015385). Additionally, this research was supported by the Korea Health Technology R&D Project through the Korea Health Industry Development Institute (KHIDI) funded by the Ministry of Health & Welfare, Republic of Korea (grant number: HI22C1361).

Institutional Review Board Statement: Not applicable.

Informed Consent Statement: Not applicable.

Data Availability Statement: Not applicable.

Conflicts of Interest: The authors declare no conflict of interest.

References

1. Peters, J.M.; Shah, Y.M.; Gonzalez, F.J. The role of peroxisome proliferator-activated receptors in carcinogenesis and chemoprevention. *Nat. Rev. Cancer* **2012**, *12*, 181–195. [CrossRef]
2. Toobian, D.; Ghosh, P.; Katkar, G.D. Parsing the Role of PPARs in Macrophage Processes. *Front. Immunol.* **2021**, *12*, 783780. [CrossRef]
3. Schupp, M.; Lazar, M.A. Endogenous ligands for nuclear receptors: Digging deeper. *J. Biol. Chem.* **2010**, *285*, 40409–40415. [CrossRef]
4. Lamas Bervejillo, M.; Ferreira, A.M. Understanding Peroxisome Proliferator-Activated Receptors: From the Structure to the Regulatory Actions on Metabolism. *Adv. Exp. Med. Biol.* **2019**, *1127*, 39–57. [CrossRef]
5. Viswakarma, N.; Jia, Y.; Bai, L.; Vluggens, A.; Borensztajn, J.; Xu, J.; Reddy, J.K. Coactivators in PPAR-Regulated Gene Expression. *PPAR Res.* **2010**, *2010*, 250126. [CrossRef]
6. Chandra, V.; Huang, P.; Hamuro, Y.; Raghuram, S.; Wang, Y.; Burris, T.P.; Rastinejad, F. Structure of the intact PPAR-gamma-RXR-nuclear receptor complex on DNA. *Nature* **2008**, *456*, 350–356. [CrossRef]
7. Grygiel-Górniak, B. Peroxisome proliferator-activated receptors and their ligands: Nutritional and clinical implications–a review. *Nutr. J.* **2014**, *13*, 17. [CrossRef]
8. Ye, X.; Zhang, T.; Han, H. PPARα: A potential therapeutic target of cholestasis. *Front. Pharmacol.* **2022**, *13*, 916866. [CrossRef]
9. Bordet, R.; Ouk, T.; Petrault, O.; Gelé, P.; Gautier, S.; Laprais, M.; Deplanque, D.; Duriez, P.; Staels, B.; Fruchart, J.C.; et al. PPAR: A new pharmacological target for neuroprotection in stroke and neurodegenerative diseases. *Biochem. Soc. Trans.* **2006**, *34*, 1341–1346. [CrossRef]
10. Marion-Letellier, R.; Savoye, G.; Ghosh, S. Fatty acids, eicosanoids and PPAR gamma. *Eur. J. Pharmacol.* **2016**, *785*, 44–49. [CrossRef]
11. Croasdell, A.; Duffney, P.F.; Kim, N.; Lacy, S.H.; Sime, P.J.; Phipps, R.P. PPARγ and the Innate Immune System Mediate the Resolution of Inflammation. *PPAR Res.* **2015**, *2015*, 549691. [CrossRef]
12. Mirza, A.Z.; Althagafi, I.I.; Shamshad, H. Role of PPAR receptor in different diseases and their ligands: Physiological importance and clinical implications. *Eur. J. Med. Chem.* **2019**, *166*, 502–513. [CrossRef]
13. Derosa, G.; Sahebkar, A.; Maffioli, P. The role of various peroxisome proliferator-activated receptors and their ligands in clinical practice. *J. Cell. Physiol.* **2018**, *233*, 153–161. [CrossRef]
14. Diskin, C.; Ryan, T.A.J.; O'Neill, L.A.J. Modification of Proteins by Metabolites in Immunity. *Immunity* **2021**, *54*, 19–31. [CrossRef]
15. Haas, R.; Cucchi, D.; Smith, J.; Pucino, V.; Macdougall, C.E.; Mauro, C. Intermediates of Metabolism: From Bystanders to Signalling Molecules. *Trends Biochem. Sci.* **2016**, *41*, 460–471. [CrossRef]
16. Soto-Heredero, G.; Gómez de Las Heras, M.M.; Gabandé-Rodríguez, E.; Oller, J.; Mittelbrunn, M. Glycolysis—A key player in the inflammatory response. *FEBS J.* **2020**, *287*, 3350–3369. [CrossRef]
17. De Duve, C.; Baudhuin, P. Peroxisomes (microbodies and related particles). *Physiol. Rev.* **1966**, *46*, 323–357. [CrossRef]
18. Okumoto, K.; Tamura, S.; Honsho, M.; Fujiki, Y. Peroxisome: Metabolic Functions and Biogenesis. *Adv. Exp. Med. Biol.* **2020**, *1299*, 3–17. [CrossRef]
19. Reddy, J.K.; Krishnakantha, T.P.; Azarnoff, D.L.; Moody, D.E. 1-methyl-4piperidyl-bis (P-chlorophenoxy) acetate: A new hypolipidemic peroxisome proliferator. *Res. Commun. Chem. Pathol. Pharmacol.* **1975**, *10*, 589–592.
20. Reddy, J.K.; Lalwai, N.D. Carcinogenesis by hepatic peroxisome proliferators: Evaluation of the risk of hypolipidemic drugs and industrial plasticizers to humans. *Crit. Rev. Toxicol.* **1983**, *12*, 1–58. [CrossRef]

21. Reddy, J.K.; Goel, S.K.; Nemali, M.R.; Carrino, J.J.; Laffler, T.G.; Reddy, M.K.; Sperbeck, S.J.; Osumi, T.; Hashimoto, T.; Lalwani, N.D.; et al. Transcription regulation of peroxisomal fatty acyl-CoA oxidase and enoyl-CoA hydratase/3-hydroxyacyl-CoA dehydrogenase in rat liver by peroxisome proliferators. *Proc. Natl. Acad. Sci. USA* **1986**, *83*, 1747–1751. [CrossRef]
22. Hardwick, J.P.; Song, B.J.; Huberman, E.; Gonzalez, F.J. Isolation, complementary DNA sequence, and regulation of rat hepatic lauric acid omega-hydroxylase (cytochrome P-450LA omega). Identification of a new cytochrome P-450 gene family. *J. Biol. Chem.* **1987**, *262*, 801–810. [CrossRef]
23. Issemann, I.; Green, S. Activation of a member of the steroid hormone receptor superfamily by peroxisome proliferators. *Nature* **1990**, *347*, 645–650. [CrossRef]
24. Zhu, Y.; Alvares, K.; Huang, Q.; Rao, M.S.; Reddy, J.K. Cloning of a new member of the peroxisome proliferator-activated receptor gene family from mouse liver. *J. Biol. Chem.* **1993**, *268*, 26817–26820. [CrossRef]
25. Göttlicher, M.; Widmark, E.; Li, Q.; Gustafsson, J.A. Fatty acids activate a chimera of the clofibric acid-activated receptor and the glucocorticoid receptor. *Proc. Natl. Acad. Sci. USA* **1992**, *89*, 4653–4657. [CrossRef]
26. Rolls, B.J.; Hammer, V.A. Fat, carbohydrate, and the regulation of energy intake. *Am. J. Clin. Nutr.* **1995**, *62*, 1086s–1095s. [CrossRef]
27. Dreyer, C.; Krey, G.; Keller, H.; Givel, F.; Helftenbein, G.; Wahli, W. Control of the peroxisomal beta-oxidation pathway by a novel family of nuclear hormone receptors. *Cell* **1992**, *68*, 879–887. [CrossRef]
28. Greene, M.E.; Blumberg, B.; McBride, O.W.; Yi, H.F.; Kronquist, K.; Kwan, K.; Hsieh, L.; Greene, G.; Nimer, S.D. Isolation of the human peroxisome proliferator activated receptor gamma cDNA: Expression in hematopoietic cells and chromosomal mapping. *Gene Expr.* **1995**, *4*, 281–299.
29. Schmidt, A.; Endo, N.; Rutledge, S.J.; Vogel, R.; Shinar, D.; Rodan, G.A. Identification of a new member of the steroid hormone receptor superfamily that is activated by a peroxisome proliferator and fatty acids. *Mol. Endocrinol.* **1992**, *6*, 1634–1641. [CrossRef]
30. Christofides, A.; Konstantinidou, E.; Jani, C.; Boussiotis, V.A. The role of peroxisome proliferator-activated receptors (PPAR) in immune responses. *Metabolism* **2021**, *114*, 154338. [CrossRef]
31. Sher, T.; Yi, H.F.; McBride, O.W.; Gonzalez, F.J. cDNA cloning, chromosomal mapping, and functional characterization of the human peroxisome proliferator activated receptor. *Biochemistry* **1993**, *32*, 5598–5604. [CrossRef]
32. Tyagi, S.; Gupta, P.; Saini, A.S.; Kaushal, C.; Sharma, S. The peroxisome proliferator-activated receptor: A family of nuclear receptors role in various diseases. *J. Adv. Pharm. Technol. Res.* **2011**, *2*, 236–240. [CrossRef]
33. Qi, J.S.; Desai-Yajnik, V.; Greene, M.E.; Raaka, B.M.; Samuels, H.H. The ligand-binding domains of the thyroid hormone/retinoid receptor gene subfamily function in vivo to mediate heterodimerization, gene silencing, and transactivation. *Mol. Cell. Biol.* **1995**, *15*, 1817–1825. [CrossRef]
34. Evans, R.M.; Barish, G.D.; Wang, Y.X. PPARs and the complex journey to obesity. *Nat. Med.* **2004**, *10*, 355–361. [CrossRef]
35. Wang, S.; Dougherty, E.J.; Danner, R.L. PPARγ signaling and emerging opportunities for improved therapeutics. *Pharmacol. Res.* **2016**, *111*, 76–85. [CrossRef]
36. Bain, D.L.; Heneghan, A.F.; Connaghan-Jones, K.D.; Miura, M.T. Nuclear receptor structure: Implications for function. *Annu. Rev. Physiol.* **2007**, *69*, 201–220. [CrossRef]
37. Weikum, E.R.; Liu, X.; Ortlund, E.A. The nuclear receptor superfamily: A structural perspective. *Protein Sci.* **2018**, *27*, 1876–1892. [CrossRef]
38. Desvergne, B.; Wahli, W. Peroxisome proliferator-activated receptors: Nuclear control of metabolism. *Endocr. Rev.* **1999**, *20*, 649–688. [CrossRef]
39. Brunmeir, R.; Xu, F. Functional Regulation of PPARs through Post-Translational Modifications. *Int. J. Mol. Sci.* **2018**, *19*, 1738. [CrossRef]
40. Palmer, C.N.; Hsu, M.H.; Griffin, H.J.; Johnson, E.F. Novel sequence determinants in peroxisome proliferator signaling. *J. Biol. Chem.* **1995**, *270*, 16114–16121. [CrossRef]
41. Hu, X.; Lazar, M.A. The CoRNR motif controls the recruitment of corepressors by nuclear hormone receptors. *Nature* **1999**, *402*, 93–96. [CrossRef]
42. Poulsen, L.; Siersbæk, M.; Mandrup, S. PPARs: Fatty acid sensors controlling metabolism. *Semin. Cell Dev. Biol.* **2012**, *23*, 631–639. [CrossRef]
43. Chan, L.S.; Wells, R.A. Cross-Talk between PPARs and the Partners of RXR: A Molecular Perspective. *PPAR Res.* **2009**, *2009*, 925309. [CrossRef]
44. Ricote, M.; Glass, C.K. PPARs and molecular mechanisms of transrepression. *Biochim. Biophys. Acta* **2007**, *1771*, 926–935. [CrossRef]
45. Delerive, P.; Gervois, P.; Fruchart, J.C.; Staels, B. Induction of IkappaBalpha expression as a mechanism contributing to the anti-inflammatory activities of peroxisome proliferator-activated receptor-alpha activators. *J. Biol. Chem.* **2000**, *275*, 36703–36707. [CrossRef]
46. Staels, B.; Koenig, W.; Habib, A.; Merval, R.; Lebret, M.; Torra, I.P.; Delerive, P.; Fadel, A.; Chinetti, G.; Fruchart, J.C.; et al. Activation of human aortic smooth-muscle cells is inhibited by PPARalpha but not by PPARgamma activators. *Nature* **1998**, *393*, 790–793. [CrossRef]

47. Madej, A.; Okopien, B.; Kowalski, J.; Zielinski, M.; Wysocki, J.; Szygula, B.; Kalina, Z.; Herman, Z.S. Effects of fenofibrate on plasma cytokine concentrations in patients with atherosclerosis and hyperlipoproteinemia IIb. *Int. J. Clin. Pharmacol. Ther.* **1998**, *36*, 345–349.
48. Delerive, P.; Martin-Nizard, F.; Chinetti, G.; Trottein, F.; Fruchart, J.C.; Najib, J.; Duriez, P.; Staels, B. Peroxisome proliferator-activated receptor activators inhibit thrombin-induced endothelin-1 production in human vascular endothelial cells by inhibiting the activator protein-1 signaling pathway. *Circ. Res.* **1999**, *85*, 394–402. [CrossRef]
49. Kersten, S.; Seydoux, J.; Peters, J.M.; Gonzalez, F.J.; Desvergne, B.; Wahli, W. Peroxisome proliferator-activated receptor alpha mediates the adaptive response to fasting. *J. Clin. Invest.* **1999**, *103*, 1489–1498. [CrossRef]
50. Montagner, A.; Polizzi, A.; Fouché, E.; Ducheix, S.; Lippi, Y.; Lasserre, F.; Barquissau, V.; Régnier, M.; Lukowicz, C.; Benhamed, F.; et al. Liver PPARα is crucial for whole-body fatty acid homeostasis and is protective against NAFLD. *Gut* **2016**, *65*, 1202–1214. [CrossRef]
51. Fougerat, A.; Schoiswohl, G.; Polizzi, A.; Régnier, M.; Wagner, C.; Smati, S.; Fougeray, T.; Lippi, Y.; Lasserre, F.; Raho, I.; et al. ATGL-dependent white adipose tissue lipolysis controls hepatocyte PPARα activity. *Cell Rep.* **2022**, *39*, 110910. [CrossRef]
52. Wang, A.; Huen, S.C.; Luan, H.H.; Yu, S.; Zhang, C.; Gallezot, J.D.; Booth, C.J.; Medzhitov, R. Opposing Effects of Fasting Metabolism on Tissue Tolerance in Bacterial and Viral Inflammation. *Cell* **2016**, *166*, 1512–1525.e1512. [CrossRef]
53. Huen, S.C.; Wang, A.; Feola, K.; Desrouleaux, R.; Luan, H.H.; Hogg, R.; Zhang, C.; Zhang, Q.J.; Liu, Z.P.; Medzhitov, R. Hepatic FGF21 preserves thermoregulation and cardiovascular function during bacterial inflammation. *J. Exp. Med.* **2021**, *218*, e20202151. [CrossRef]
54. Paumelle, R.; Haas, J.T.; Hennuyer, N.; Baugé, E.; Deleye, Y.; Mesotten, D.; Langouche, L.; Vanhoutte, J.; Cudejko, C.; Wouters, K.; et al. Hepatic PPARα is critical in the metabolic adaptation to sepsis. *J. Hepatol.* **2019**, *70*, 963–973. [CrossRef]
55. Braissant, O.; Foufelle, F.; Scotto, C.; Dauça, M.; Wahli, W. Differential expression of peroxisome proliferator-activated receptors (PPARs): Tissue distribution of PPAR-alpha, -beta, and -gamma in the adult rat. *Endocrinology* **1996**, *137*, 354–366. [CrossRef]
56. Müller, R. PPARβ/δ in human cancer. *Biochimie* **2017**, *136*, 90–99. [CrossRef]
57. Yao, P.L.; Chen, L.; Dobrzański, T.P.; Zhu, B.; Kang, B.H.; Müller, R.; Gonzalez, F.J.; Peters, J.M. Peroxisome proliferator-activated receptor-β/δ inhibits human neuroblastoma cell tumorigenesis by inducing p53- and SOX2-mediated cell differentiation. *Mol. Carcinog.* **2017**, *56*, 1472–1483. [CrossRef]
58. Foreman, J.E.; Sharma, A.K.; Amin, S.; Gonzalez, F.J.; Peters, J.M. Ligand activation of peroxisome proliferator-activated receptor-beta/delta (PPARbeta/delta) inhibits cell growth in a mouse mammary gland cancer cell line. *Cancer Lett.* **2010**, *288*, 219–225. [CrossRef]
59. Genini, D.; Garcia-Escudero, R.; Carbone, G.M.; Catapano, C.V. Transcriptional and Non-Transcriptional Functions of PPARβ/δ in Non-Small Cell Lung Cancer. *PLoS ONE* **2012**, *7*, e46009. [CrossRef]
60. Pedchenko, T.V.; Gonzalez, A.L.; Wang, D.; DuBois, R.N.; Massion, P.P. Peroxisome proliferator-activated receptor beta/delta expression and activation in lung cancer. *Am. J. Respir. Cell Mol. Biol.* **2008**, *39*, 689–696. [CrossRef]
61. Wagner, K.D.; Benchetrit, M.; Bianchini, L.; Michiels, J.F.; Wagner, N. Peroxisome proliferator-activated receptor β/δ (PPARβ/δ) is highly expressed in liposarcoma and promotes migration and proliferation. *J. Pathol.* **2011**, *224*, 575–588. [CrossRef]
62. Bishop-Bailey, D. PPARs and angiogenesis. *Biochem. Soc. Trans.* **2011**, *39*, 1601–1605. [CrossRef]
63. Daynes, R.A.; Jones, D.C. Emerging roles of PPARs in inflammation and immunity. *Nat. Rev. Immunol.* **2002**, *2*, 748–759. [CrossRef]
64. Kostadinova, R.; Wahli, W.; Michalik, L. PPARs in diseases: Control mechanisms of inflammation. *Curr. Med. Chem.* **2005**, *12*, 2995–3009. [CrossRef]
65. Ricote, M.; Li, A.C.; Willson, T.M.; Kelly, C.J.; Glass, C.K. The peroxisome proliferator-activated receptor-gamma is a negative regulator of macrophage activation. *Nature* **1998**, *391*, 79–82. [CrossRef]
66. Chinetti, G.; Griglio, S.; Antonucci, M.; Torra, I.P.; Delerive, P.; Majd, Z.; Fruchart, J.C.; Chapman, J.; Najib, J.; Staels, B. Activation of proliferator-activated receptors alpha and gamma induces apoptosis of human monocyte-derived macrophages. *J. Biol. Chem.* **1998**, *273*, 25573–25580. [CrossRef]
67. Bao, Y.; Li, R.; Jiang, J.; Cai, B.; Gao, J.; Le, K.; Zhang, F.; Chen, S.; Liu, P. Activation of peroxisome proliferator-activated receptor gamma inhibits endothelin-1-induced cardiac hypertrophy via the calcineurin/NFAT signaling pathway. *Mol. Cell. Biochem.* **2008**, *317*, 189–196. [CrossRef]
68. Kamata, S.; Oyama, T.; Saito, K.; Honda, A.; Yamamoto, Y.; Suda, K.; Ishikawa, R.; Itoh, T.; Watanabe, Y.; Shibata, T.; et al. PPARα Ligand-Binding Domain Structures with Endogenous Fatty Acids and Fibrates. *iScience* **2020**, *23*, 101727. [CrossRef]
69. Nolte, R.T.; Wisely, G.B.; Westin, S.; Cobb, J.E.; Lambert, M.H.; Kurokawa, R.; Rosenfeld, M.G.; Willson, T.M.; Glass, C.K.; Milburn, M.V. Ligand binding and co-activator assembly of the peroxisome proliferator-activated receptor-gamma. *Nature* **1998**, *395*, 137–143. [CrossRef]
70. Krey, G.; Braissant, O.; L'Horset, F.; Kalkhoven, E.; Perroud, M.; Parker, M.G.; Wahli, W. Fatty acids, eicosanoids, and hypolipidemic agents identified as ligands of peroxisome proliferator-activated receptors by coactivator-dependent receptor ligand assay. *Mol. Endocrinol.* **1997**, *11*, 779–791. [CrossRef]
71. Aagaard, M.M.; Siersbæk, R.; Mandrup, S. Molecular basis for gene-specific transactivation by nuclear receptors. *Biochim. Biophys. Acta* **2011**, *1812*, 824–835. [CrossRef]

72. Sumanasekera, W.K.; Tien, E.S.; Davis, J.W., 2nd; Turpey, R.; Perdew, G.H.; Vanden Heuvel, J.P. Heat shock protein-90 (Hsp90) acts as a repressor of peroxisome proliferator-activated receptor-alpha (PPARalpha) and PPARbeta activity. *Biochemistry* **2003**, *42*, 10726–10735. [CrossRef]
73. Wheeler, M.C.; Gekakis, N. Hsp90 modulates PPARγ activity in a mouse model of nonalcoholic fatty liver disease. *J. Lipid Res.* **2014**, *55*, 1702–1710. [CrossRef]
74. Layrolle, P.; Payoux, P.; Chavanas, S. PPAR Gamma and Viral Infections of the Brain. *Int. J. Mol. Sci.* **2021**, *22*, 8876. [CrossRef]
75. Bassaganya-Riera, J.; Song, R.; Roberts, P.C.; Hontecillas, R. PPAR-gamma activation as an anti-inflammatory therapy for respiratory virus infections. *Viral Immunol.* **2010**, *23*, 343–352. [CrossRef]
76. Huang, S.; Zhu, B.; Cheon, I.S.; Goplen, N.P.; Jiang, L.; Zhang, R.; Peebles, R.S.; Mack, M.; Kaplan, M.H.; Limper, A.H.; et al. PPAR-γ in Macrophages Limits Pulmonary Inflammation and Promotes Host Recovery following Respiratory Viral Infection. *J. Virol.* **2019**, *93*. [CrossRef]
77. Liu, Q.; Zhou, Y.H.; Yang, Z.Q. The cytokine storm of severe influenza and development of immunomodulatory therapy. *Cell. Mol. Immunol.* **2016**, *13*, 3–10. [CrossRef]
78. Huang, S.; Goplen, N.P.; Zhu, B.; Cheon, I.S.; Son, Y.; Wang, Z.; Li, C.; Dai, Q.; Jiang, L.; Xiang, M.; et al. Macrophage PPAR-γ suppresses long-term lung fibrotic sequelae following acute influenza infection. *PLoS ONE* **2019**, *14*, e0223430. [CrossRef]
79. Bei, Y.; Tia, B.; Li, Y.; Guo, Y.; Deng, S.; Huang, R.; Zeng, H.; Li, R.; Wang, G.F.; Dai, J. Anti-influenza A Virus Effects and Mechanisms of Emodin and Its Analogs via Regulating PPARα/γ-AMPK-SIRT1 Pathway and Fatty Acid Metabolism. *Biomed. Res. Int.* **2021**, *2021*, 9066938. [CrossRef]
80. Heffernan, K.S.; Ranadive, S.M.; Jae, S.Y. Exercise as medicine for COVID-19: On PPAR with emerging pharmacotherapy. *Med. Hypotheses* **2020**, *143*, 110197. [CrossRef]
81. Dias, S.S.G.; Soares, V.C.; Ferreira, A.C.; Sacramento, C.Q.; Fintelman-Rodrigues, N.; Temerozo, J.R.; Teixeira, L.; Nunes da Silva, M.A.; Barreto, E.; Mattos, M.; et al. Lipid droplets fuel SARS-CoV-2 replication and production of inflammatory mediators. *PLoS Pathog.* **2020**, *16*, e1009127. [CrossRef]
82. Alkhayyat, S.S.; Al-Kuraishy, H.M.; Al-Gareeb, A.I.; El-Bouseary, M.M.; AboKamer, A.M.; Batiha, G.E.; Simal-Gandara, J. Fenofibrate for COVID-19 and related complications as an approach to improve treatment outcomes: The missed key for Holy Grail. *Inflamm. Res.* **2022**, *71*, 1159–1167. [CrossRef]
83. Jagat, J.M.; Kalyan, K.G.; Subir, R. Use of pioglitazone in people with type 2 diabetes mellitus with coronavirus disease 2019 (COVID-19): Boon or bane? *Diabetol. Metab. Syndr.* **2020**, *14*, 829–831. [CrossRef]
84. Nagarkatti, P.; Miranda, K.; Nagarkatti, M. Use of Cannabinoids to Treat Acute Respiratory Distress Syndrome and Cytokine Storm Associated with Coronavirus Disease-2019. *Front. Pharmacol.* **2020**, *11*, 589438. [CrossRef]
85. Corpetti, C.; Del Re, A.; Seguella, L.; Palenca, I.; Rurgo, S.; De Conno, B.; Pesce, M.; Sarnelli, G.; Esposito, G. Cannabidiol inhibits SARS-Cov-2 spike (S) protein-induced cytotoxicity and inflammation through a PPARγ-dependent TLR4/NLRP3/Caspase-1 signaling suppression in Caco-2 cell line. *Phytother. Res.* **2021**, *35*, 6893–6903. [CrossRef]
86. Francisqueti-Ferron, F.V.; Garcia, J.L.; Ferron, A.J.T.; Nakandakare-Maia, E.T.; Gregolin, C.S.; Silva, J.; Dos Santos, K.C.; Lo, Â.T.C.; Siqueira, J.S.; de Mattei, L.; et al. Gamma-oryzanol as a potential modulator of oxidative stress and inflammation via PPAR-y in adipose tissue: A hypothetical therapeutic for cytokine storm in COVID-19? *Mol. Cell. Endocrinol.* **2021**, *520*, 111095. [CrossRef]
87. Al-Qahtani, A.A.; Lyroni, K.; Aznaourova, M.; Tseliou, M.; Al-Anazi, M.R.; Al-Ahdal, M.N.; Alkahtani, S.; Sourvinos, G.; Tsatsanis, C. Middle east respiratory syndrome corona virus spike glycoprotein suppresses macrophage responses via DPP4-mediated induction of IRAK-M and PPARγ. *Oncotarget* **2017**, *8*, 9053–9066. [CrossRef]
88. De, S.; Mamidi, P.; Ghosh, S.; Keshry, S.S.; Mahish, C.; Pani, S.S.; Laha, E.; Ray, A.; Datey, A.; Chatterjee, S.; et al. Telmisartan Restricts Chikungunya Virus Infection In Vitro and In Vivo through the AT1/PPAR-γ/MAPKs Pathways. *Antimicrob. Agents Chemother.* **2022**, *66*, e0148921. [CrossRef]
89. Omeragic, A.; Hoque, M.T.; Choi, U.Y.; Bendayan, R. Peroxisome proliferator-activated receptor-gamma: Potential molecular therapeutic target for HIV-1-associated brain inflammation. *J. Neuroinflamm.* **2017**, *14*, 183. [CrossRef]
90. Jiang, H.; Cheng, S.T.; Ren, J.H.; Ren, F.; Yu, H.B.; Wang, Q.; Huang, A.L.; Chen, J. SIRT6 Inhibitor, OSS_128167 Restricts Hepatitis B Virus Transcription and Replication Through Targeting Transcription Factor Peroxisome Proliferator-Activated Receptors α. *Front. Pharmacol.* **2019**, *10*, 1270. [CrossRef]
91. Lin, Y.M.; Sun, H.Y.; Chiu, W.T.; Su, H.C.; Chien, Y.C.; Chong, L.W.; Chang, H.C.; Bai, C.H.; Young, K.C.; Tsao, C.W. Calcitriol Inhibits HCV Infection via Blockade of Activation of PPAR and Interference with Endoplasmic Reticulum-Associated Degradation. *Viruses* **2018**, *10*, 57. [CrossRef]
92. Kumar, V.; Mansfield, J.; Fan, R.; MacLean, A.; Li, J.; Mohan, M. miR-130a and miR-212 Disrupt the Intestinal Epithelial Barrier through Modulation of PPARγ and Occludin Expression in Chronic Simian Immunodeficiency Virus-Infected Rhesus Macaques. *J. Immunol.* **2018**, *200*, 2677–2689. [CrossRef]
93. Crakes, K.R.; Santos Rocha, C.; Grishina, I.; Hirao, L.A.; Napoli, E.; Gaulke, C.A.; Fenton, A.; Datta, S.; Arredondo, J.; Marco, M.L.; et al. PPARα-targeted mitochondrial bioenergetics mediate repair of intestinal barriers at the host-microbe intersection during SIV infection. *Proc. Natl. Acad. Sci. USA* **2019**, *116*, 24819–24829. [CrossRef]

94. Lin, C.Y.; Huang, C.H.; Wang, W.H.; Tenhunen, J.; Hung, L.C.; Lin, C.C.; Chen, Y.C.; Chen, Y.H.; Liao, W.T. Mono-(2-ethylhexyl) phthalate Promotes Dengue Virus Infection by Decreasing IL-23-Mediated Antiviral Responses. *Front. Immunol.* **2021**, *12*, 599345. [CrossRef]
95. Mulkey, S.B.; Arroyave-Wessel, M.; Peyton, C.; Bulas, D.I.; Fourzali, Y.; Jiang, J.; Russo, S.; McCarter, R.; Msall, M.E.; du Plessis, A.J.; et al. Neurodevelopmental Abnormalities in Children with In Utero Zika Virus Exposure Without Congenital Zika Syndrome. *JAMA Pediatr.* **2020**, *174*, 269–276. [CrossRef]
96. Thulasi Raman, S.N.; Latreille, E.; Gao, J.; Zhang, W.; Wu, J.; Russell, M.S.; Walrond, L.; Cyr, T.; Lavoie, J.R.; Safronetz, D.; et al. Dysregulation of Ephrin receptor and PPAR signaling pathways in neural progenitor cells infected by Zika virus. *Emerg. Microbes Infect.* **2020**, *9*, 2046–2060. [CrossRef]
97. Li, Y.; Zhu, Y.; Feng, S.; Ishida, Y.; Chiu, T.P.; Saito, T.; Wang, S.; Ann, D.K.; Ou, J.J. Macrophages activated by hepatitis B virus have distinct metabolic profiles and suppress the virus via IL-1β to downregulate PPARα and FOXO3. *Cell Rep.* **2022**, *38*, 110284. [CrossRef]
98. Du, L.; Ma, Y.; Liu, M.; Yan, L.; Tang, H. Peroxisome Proliferators Activated Receptor (PPAR) agonists activate hepatitis B virus replication in vivo. *Virol. J.* **2017**, *14*, 96. [CrossRef]
99. Suzuki, K.; Suda, G.; Yamamoto, Y.; Furuya, K.; Baba, M.; Nakamura, A.; Miyoshi, H.; Kimura, M.; Maehara, O.; Yamada, R.; et al. Tenofovir-disoproxil-fumarate modulates lipid metabolism via hepatic CD36/PPAR-alpha activation in hepatitis B virus infection. *J. Gastroenterol.* **2021**, *56*, 168–180. [CrossRef]
100. Goldwasser, J.; Cohen, P.Y.; Lin, W.; Kitsberg, D.; Balaguer, P.; Polyak, S.J.; Chung, R.T.; Yarmush, M.L.; Nahmias, Y. Naringenin inhibits the assembly and long-term production of infectious hepatitis C virus particles through a PPAR-mediated mechanism. *J. Hepatol.* **2011**, *55*, 963–971. [CrossRef]
101. Tam, V.C.; Suen, R.; Treuting, P.M.; Armando, A.; Lucarelli, R.; Gorrochotegui-Escalante, N.; Diercks, A.H.; Quehenberger, O.; Dennis, E.A.; Aderem, A.; et al. PPARα exacerbates necroptosis, leading to increased mortality in postinfluenza bacterial superinfection. *Proc. Natl. Acad. Sci. USA* **2020**, *117*, 15789–15798. [CrossRef]
102. Gopal, R.; Mendy, A.; Marinelli, M.A.; Richwalls, L.J.; Seger, P.J.; Patel, S.; McHugh, K.J.; Rich, H.E.; Grousd, J.A.; Forno, E.; et al. Peroxisome Proliferator-Activated Receptor Gamma (PPARγ) Suppresses Inflammation and Bacterial Clearance during Influenza-Bacterial Super-Infection. *Viruses* **2019**, *11*, 505. [CrossRef]
103. Lucarelli, R.; Gorrochotegui-Escalante, N.; Taddeo, J.; Buttaro, B.; Beld, J.; Tam, V. Eicosanoid-Activated PPARα Inhibits NFκB-Dependent Bacterial Clearance During Post-Influenza Superinfection. *Front. Cell. Infect. Microbiol.* **2022**, *12*, 881462. [CrossRef]
104. Cevallos, S.A.; Lee, J.Y.; Velazquez, E.M.; Foegeding, N.J.; Shelton, C.D.; Tiffany, C.R.; Parry, B.H.; Stull-Lane, A.R.; Olsan, E.E.; Savage, H.P.; et al. 5-Aminosalicylic Acid Ameliorates Colitis and Checks Dysbiotic Escherichia coli Expansion by Activating PPAR-γ Signaling in the Intestinal Epithelium. *mBio* **2021**, *12*. [CrossRef]
105. Ramirez-Moral, I.; Ferreira, B.L.; de Vos, A.F.; van der Poll, T. Post-treatment with the PPAR-γ agonist pioglitazone inhibits inflammation and bacterial growth during Klebsiella pneumonia. *Respir. Res.* **2021**, *22*, 230. [CrossRef]
106. Pu, Q.; Guo, K.; Lin, P.; Wang, Z.; Qin, S.; Gao, P.; Combs, C.; Khan, N.; Xia, Z.; Wu, M. Bitter receptor TAS2R138 facilitates lipid droplet degradation in neutrophils during Pseudomonas aeruginosa infection. *Signal Transduct. Target. Ther.* **2021**, *6*, 210. [CrossRef]
107. Ferreira, B.L.; Ramirez-Moral, I.; Otto, N.A.; Salomão, R.; de Vos, A.F.; van der Poll, T. The PPAR-γ agonist pioglitazone exerts proinflammatory effects in bronchial epithelial cells during acute Pseudomonas aeruginosa pneumonia. *Clin. Exp. Immunol.* **2022**, *207*, 370–377. [CrossRef]
108. Wu, X.; Cheng, B.; Guo, X.; Wu, Q.; Sun, S.; He, P. PPARα/γ signaling pathways are involved in Chlamydia pneumoniae-induced foam cell formation via upregulation of SR-A1 and ACAT1 and downregulation of ABCA1/G1. *Microb. Pathog.* **2021**, *161*, 105284. [CrossRef]
109. Katkar, G.D.; Sayed, I.M.; Anandachar, M.S.; Castillo, V.; Vidales, E.; Toobian, D.; Usmani, F.; Sawires, J.R.; Leriche, G.; Yang, J.; et al. Artificial intelligence-rationalized balanced PPARα/γ dual agonism resets dysregulated macrophage processes in inflammatory bowel disease. *Commun. Biol.* **2022**, *5*, 231. [CrossRef]
110. Wani, K.A.; Goswamy, D.; Taubert, S.; Ratnappan, R.; Ghazi, A.; Irazoqui, J.E. NHR-49/PPAR-α and HLH-30/TFEB cooperate for C. elegans host defense via a flavin-containing monooxygenase. *eLife* **2021**, *10*, e62775. [CrossRef]
111. Liang, F.; Guan, H.; Li, W.; Zhang, X.; Liu, T.; Liu, Y.; Mei, J.; Jiang, C.; Zhang, F.; Luo, B.; et al. Erythropoietin Promotes Infection Resolution and Lowers Antibiotic Requirements in E. coli- and S. aureus-Initiated Infections. *Front. Immunol.* **2021**, *12*, 658715. [CrossRef]
112. Allen, P.E.; Noland, R.C.; Martinez, J.J. Rickettsia conorii survival in THP-1 macrophages involves host lipid droplet alterations and active rickettsial protein production. *Cell. Microbiol.* **2021**, *23*, e13390. [CrossRef]
113. Tanigawa, K.; Luo, Y.; Kawashima, A.; Kiriya, M.; Nakamura, Y.; Karasawa, K.; Suzuki, K. Essential Roles of PPARs in Lipid Metabolism during Mycobacterial Infection. *Int. J. Mol. Sci.* **2021**, *22*, 7597. [CrossRef]
114. Luo, Y.; Tanigawa, K.; Kawashima, A.; Ishido, Y.; Ishii, N.; Suzuki, K. The function of peroxisome proliferator-activated receptors PPAR-γ and PPAR-δ in Mycobacterium leprae-induced foam cell formation in host macrophages. *PLoS Negl. Trop. Dis.* **2020**, *14*, e0008850. [CrossRef]

115. Rajaram, M.V.; Brooks, M.N.; Morris, J.D.; Torrelles, J.B.; Azad, A.K.; Schlesinger, L.S. Mycobacterium tuberculosis activates human macrophage peroxisome proliferator-activated receptor gamma linking mannose receptor recognition to regulation of immune responses. *J. Immunol.* **2010**, *185*, 929–942. [CrossRef]
116. Mahajan, S.; Dkhar, H.K.; Chandra, V.; Dave, S.; Nanduri, R.; Janmeja, A.K.; Agrewala, J.N.; Gupta, P. Mycobacterium tuberculosis modulates macrophage lipid-sensing nuclear receptors PPARγ and TR4 for survival. *J. Immunol.* **2012**, *188*, 5593–5603. [CrossRef]
117. Almeida, P.E.; Silva, A.R.; Maya-Monteiro, C.M.; Töröcsik, D.; D'Avila, H.; Dezsö, B.; Magalhães, K.G.; Castro-Faria-Neto, H.C.; Nagy, L.; Bozza, P.T. Mycobacterium bovis bacillus Calmette-Guérin infection induces TLR2-dependent peroxisome proliferator-activated receptor gamma expression and activation: Functions in inflammation, lipid metabolism, and pathogenesis. *J. Immunol.* **2009**, *183*, 1337–1345. [CrossRef]
118. D'Avila, H.; Melo, R.C.; Parreira, G.G.; Werneck-Barroso, E.; Castro-Faria-Neto, H.C.; Bozza, P.T. Mycobacterium bovis bacillus Calmette-Guérin induces TLR2-mediated formation of lipid bodies: Intracellular domains for eicosanoid synthesis in vivo. *J. Immunol.* **2006**, *176*, 3087–3097. [CrossRef]
119. Kim, Y.S.; Lee, H.M.; Kim, J.K.; Yang, C.S.; Kim, T.S.; Jung, M.; Jin, H.S.; Kim, S.; Jang, J.; Oh, G.T.; et al. PPAR-α Activation Mediates Innate Host Defense through Induction of TFEB and Lipid Catabolism. *J. Immunol.* **2017**, *198*, 3283–3295. [CrossRef]
120. Kim, Y.S.; Kim, J.K.; Hanh, B.T.B.; Kim, S.Y.; Kim, H.J.; Kim, Y.J.; Jeon, S.M.; Park, C.R.; Oh, G.T.; Park, J.W.; et al. The Peroxisome Proliferator-Activated Receptor α-Agonist Gemfibrozil Promotes Defense Against Mycobacterium abscessus Infections. *Cells* **2020**, *9*, 648. [CrossRef]
121. Yamanishi, Y.; Miyake, K.; Iki, M.; Tsutsui, H.; Karasuyama, H. Recent advances in understanding basophil-mediated Th2 immune responses. *Immunol. Rev.* **2017**, *278*, 237–245. [CrossRef]
122. Chen, T.; Tibbitt, C.A.; Feng, X.; Stark, J.M.; Rohrbeck, L.; Rausch, L.; Sedimbi, S.K.; Karlsson, M.C.I.; Lambrecht, B.N.; Karlsson Hedestam, G.B.; et al. PPAR-γ promotes type 2 immune responses in allergy and nematode infection. *Sci. Immunol.* **2017**, *2*. [CrossRef]
123. He, X.; Gong, P.; Wei, Z.; Liu, W.; Wang, W.; Li, J.; Yang, Z.; Zhang, X. Peroxisome proliferator-activated receptor-γ-mediated polarization of macrophages in Neospora caninum infection. *Exp. Parasitol.* **2017**, *178*, 37–44. [CrossRef]
124. Chen, K.M.; Peng, C.Y.; Shyu, L.Y.; Lan, K.P.; Lai, S.C. Peroxisome-Proliferator Activator Receptor γ in Mouse Model with Meningoencephalitis Caused by Angiostrongylus cantonensis. *J. Parasitol.* **2021**, *107*, 205–213. [CrossRef]
125. Mita-Mendoza, N.K.; Magallon-Tejada, A.; Parmar, P.; Furtado, R.; Aldrich, M.; Saidi, A.; Taylor, T.; Smith, J.; Seydel, K.; Daily, J.P. Dimethyl fumarate reduces TNF and Plasmodium falciparum induced brain endothelium activation in vitro. *Malar. J.* **2020**, *19*, 376. [CrossRef]
126. Borges, T.K.; Alves, É.A.; Vasconcelos, H.A.; Carneiro, F.P.; Nicola, A.M.; Magalhães, K.G.; Muniz-Junqueira, M.I. Differences in the modulation of reactive species, lipid bodies, cyclooxygenase-2, 5-lipoxygenase and PPAR-γ in cerebral malaria-susceptible and resistant mice. *Immunobiology* **2017**, *222*, 604–619. [CrossRef]
127. Matta, S.K.; Rinkenberger, N.; Dunay, I.R.; Sibley, L.D. Toxoplasma gondii infection and its implications within the central nervous system. *Nat. Rev. Microbiol.* **2021**, *19*, 467–480. [CrossRef]
128. Shyu, L.Y.; Chen, K.M.; Lu, C.Y.; Lai, S.C. Regulation of Proinflammatory Enzymes by Peroxisome Proliferator-Activated Receptor Gamma in Astroglia Infected with Toxoplasma gondii. *J. Parasitol.* **2020**, *106*, 564–571. [CrossRef]
129. He, J.J.; Ma, J.; Elsheikha, H.M.; Song, H.Q.; Zhou, D.H.; Zhu, X.Q. Proteomic Profiling of Mouse Liver following Acute Toxoplasma gondii Infection. *PLoS ONE* **2016**, *11*, e0152022. [CrossRef]
130. Moulik, S.; Karmakar, J.; Joshi, S.; Dube, A.; Mandal, C.; Chatterjee, M. Status of IL-4 and IL-10 driven markers in experimental models of Visceral Leishmaniasis. *Parasite Immunol.* **2021**, *43*, e12783. [CrossRef]
131. McManus, D.P.; Dunne, D.W.; Sacko, M.; Utzinger, J.; Vennervald, B.J.; Zhou, X.N. Schistosomiasis. *Nat. Rev. Dis. Prim.* **2018**, *4*, 13. [CrossRef]
132. Gryseels, B.; Polman, K.; Clerinx, J.; Kestens, L. Human schistosomiasis. *Lancet* **2006**, *368*, 1106–1118. [CrossRef]
133. Shiels, J.; Cwiklinski, K.; Alvarado, R.; Thivierge, K.; Cotton, S.; Gonzales Santana, B.; To, J.; Donnelly, S.; Taggart, C.C.; Weldon, S.; et al. Schistosoma mansoni immunomodulatory molecule Sm16/SPO-1/SmSLP is a member of the trematode-specific helminth defence molecules (HDMs). *PLoS Negl. Trop. Dis.* **2020**, *14*, e0008470. [CrossRef]
134. Rassi, A., Jr.; Rassi, A.; Marin-Neto, J.A. Chagas disease. *Lancet* **2010**, *375*, 1388–1402. [CrossRef]
135. Penas, F.; Mirkin, G.A.; Vera, M.; Cevey, Á.; González, C.D.; Gómez, M.I.; Sales, M.E.; Goren, N.B. Treatment in vitro with PPARα and PPARγ ligands drives M1-to-M2 polarization of macrophages from T. cruzi-infected mice. *Biochim. Biophys. Acta* **2015**, *1852*, 893–904. [CrossRef]
136. Penas, F.N.; Carta, D.; Cevey, Á.C.; Rada, M.J.; Pieralisi, A.V.; Ferlin, M.G.; Sales, M.E.; Mirkin, G.A.; Goren, N.B. Pyridinecarboxylic Acid Derivative Stimulates Pro-Angiogenic Mediators by PI3K/AKT/mTOR and Inhibits Reactive Nitrogen and Oxygen Species and NF-κB Activation Through a PPARγ-Dependent Pathway in T. cruzi-Infected Macrophages. *Front. Immunol.* **2019**, *10*, 2955. [CrossRef]
137. Penas, F.N.; Cevey, Á.C.; Siffo, S.; Mirkin, G.A.; Goren, N.B. Hepatic injury associated with Trypanosoma cruzi infection is attenuated by treatment with 15-deoxy-Δ(12,14) prostaglandin J(2). *Exp. Parasitol.* **2016**, *170*, 100–108. [CrossRef]

138. González, F.B.; Villar, S.R.; Toneatto, J.; Pacini, M.F.; Márquez, J.; D'Attilio, L.; Bottasso, O.A.; Piwien-Pilipuk, G.; Pérez, A.R. Immune response triggered by Trypanosoma cruzi infection strikes adipose tissue homeostasis altering lipid storage, enzyme profile and adipokine expression. *Med. Microbiol. Immunol.* **2019**, *208*, 651–666. [CrossRef]
139. Dreesen, L.; De Bosscher, K.; Grit, G.; Staels, B.; Lubberts, E.; Bauge, E.; Geldhof, P. Giardia muris infection in mice is associated with a protective interleukin 17A response and induction of peroxisome proliferator-activated receptor alpha. *Infect. Immun.* **2014**, *82*, 3333–3340. [CrossRef]

cells MDPI

Review

The Role of PPARs in Breast Cancer

Binggong Zhao [1,†], Zhiqiang Xin [2,†], Ping Ren [2,*] and Huijian Wu [1,*]

1 Key Laboratory of Protein Modification and Disease, School of Bioengineering, Dalian University of Technology, Dalian 116024, China
2 Hospital Office, The Second Hospital of Dalian Medical University, Dalian 116023, China
* Correspondence: dlrenping@163.com (P.R.); wuhj@dlut.edu.cn (H.W.)
† These authors contributed equally to this work.

Simple Summary: Breast cancer is a highly malignant tumor that threatens the health of women worldwide, with extremely high morbidity and mortality. The study of the related genes that affect the occurrence and development of breast cancer can provide more clinical evidence for its prevention and treatment. Peroxisome proliferators-activated receptors are a class of ligand-dependent nuclear receptor transcription factors discovered in 1990 that can regulate the transcription of many genes involved in various cellular physiological processes. The dysregulation of these physiological processes is highly correlated with the occurrence of various diseases, including malignant tumors. Additionally, a large number of reports have indicated that the transcriptional regulation function of peroxisome proliferator-activated receptors and its abnormal expression are related to breast cancer. This article summarizes the role of peroxisome proliferator-activated receptors and their different ligands in the progression of breast cancer since their discovery by searching relevant literature. The purpose of this review is to regard peroxisome proliferators-activated receptors as the new targets for the prevention of breast cancer and to incorporate their ligands into the new evidence for clinical drug combination therapy, especially for high-recurrence triple-negative breast cancer.

Abstract: Breast cancer is a malignant tumor with high morbidity and lethality. Its pathogenesis is related to the abnormal expression of many genes. The peroxisome proliferator-activated receptors (PPARs) are a class of ligand-dependent transcription factors in the nuclear receptor superfamily. They can regulate the transcription of a large number of target genes, which are involved in life activities such as cell proliferation, differentiation, metabolism, and apoptosis, and regulate physiological processes such as glucose metabolism, lipid metabolism, inflammation, and wound healing. Further, the changes in its expression are associated with various diseases, including breast cancer. The experimental reports related to "PPAR" and "breast cancer" were retrieved from PubMed since the discovery of PPARs and summarized in this paper. This review (1) analyzed the roles and potential molecular mechanisms of non-coordinated and ligand-activated subtypes of PPARs in breast cancer progression; (2) discussed the correlations between PPARs and estrogen receptors (ERs) as the nuclear receptor superfamily; and (3) investigated the interaction between PPARs and key regulators in several signaling pathways. As a result, this paper identifies PPARs as targets for breast cancer prevention and treatment in order to provide more evidence for the synthesis of new drugs targeting PPARs or the search for new drug combination treatments.

Keywords: breast cancer; PPARs; ligands; ERs

Citation: Zhao, B.; Xin, Z.; Ren, P.; Wu, H. The Role of PPARs in Breast Cancer. *Cells* **2023**, *12*, 130. https://doi.org/10.3390/cells12010130

Academic Editors: Kay-Dietrich Wagner and Nicole Wagner

Received: 11 November 2022
Revised: 7 December 2022
Accepted: 26 December 2022
Published: 28 December 2022

1. Introduction

Breast cancer is a highly heterogeneous tumor transformed from mammary epithelial cells. For example, it is the most common malignant tumor among female cancer patients worldwide in 2022, with the highest morbidity rate among all cancers (accounting for 31%), second only to lung cancer (15% of all cancer deaths), and the morbidity age tends to be increasingly younger [1]. On the basis of cellular gene expression profiles, 5 subtypes of breast cancer have been defined: luminal type A (ER+/progesterone receptor (PR)+/human epidermal growth factor receptor 2(HER2)-), luminal type B (ER+/PR+/HER2+), HER2-overexpression type (ER-/PR-/HER2+), basal-like type (ER-/PR-/HER2-), and normal-like type (the gene expression profile of cells is similar to that of normal breast epithelial cells, showing features of a low treatable rate via chemotherapy, a high quality prognosis, and a lower mortality rate if detected and treated early) [2,3]. In addition, the pathogenesis and progression of breast cancer are accompanied by the differential expression of many genes. Therefore, investigating the molecular mechanism of breast cancer occurrence and development and identifying valuable clinical markers and new therapeutic targets will contribute to the clinical diagnosis and drug treatment of breast cancer. It is also crucial to reducing the lethality of malignant breast cancer.

PPARs are a class of ligand-dependent nuclear transcription factors in members of the steroid hormone receptor superfamily, discovered in 1990 [4]. It is a biosensor of lipid metabolism changes in organisms, especially intracellular fatty acid levels. In addition, such lipid sensors are also involved in the regulation of cell differentiation, growth, and apoptosis in various cells of the organism. The PPARs are expressed in many species, including all mammals [5]. Moreover, the peroxisome proliferator response element (PPRE) sequences on these gene promoters were bound by the heterodimers of PPARs and retinoid X receptors (RXRs) to regulate downstream genes. In the non-ligand-bound state, the PPAR/RXR heterodimer binds to co-repressors and inhibits target gene transcription. The conformation of PPARs changes once the specific ligands are bound, which allows multicomponent complexes to release co-repressors and recruit co-activators: peroxisome proliferator-activated receptor gamma coactivators (PGCs), steroid receptor coactivators (SRCs), CREB-binding protein/p300 (CBP/p300), etc., and regulate the transcription of genes that participate in various physiological processes [6], such as lipid and glucose metabolism, inflammation, and wound healing. Additionally, the expression changes of these genes are found in many diseases, such as dyslipidemia, obesity, type 2 diabetes, metabolic syndrome, etc. [7,8]. To date, many researchers have reported that PPARs function as key players in various malignancies, including breast cancer. In this paper, we analyzed the role of PPARs in breast cancer progression by retrieving the related experimental articles from PubMed in order to provide more evidence for the prevention and treatment of breast cancer.

2. Structure of PPARs

PPARs comprise three subtypes that have a high degree of homology: PPARα, PPARβ/δ and PPARγ. The PPARs contain a modular structure consisting of an amino-terminal ligand-independent transcriptional activation A/B domain, a 70 amino acid-long DNA-binding C domain, a hinge D domain, and a carboxyl-terminal ligand-binding E/F domain composition (Figure 1) [9,10].The sequence structure of the C and E/F domains of PPARs subtypes has high homology [5].

Furthermore, the transcriptional activation of the A/B domain has phosphorylation-binding sites [11]. The phosphorylation state of this region regulates the affinity of PPARs for receptors (PPRE), ligands, and coactivators and is also a regulatory region used by PPARs to restrict the transcription of most genes [12–14]. The A/B domain is a highly variable region containing an activation function-1 (AF-1) domain, which has not been fully characterized. Additionally, the central DNA-binding C domain has two highly conserved C_4 zinc finger motifs: distal (D-box) and proximal (P-box) boxes, which confer heterodimerization and PPARs DNA binding, respectively. The C domain recognizes and binds to the PPRE motif (AGGTCANAGGTCA) on the promoter sequences of target genes. The hinge D domain supports the conformational change of PPARs upon ligand binding. The ligand-specific E/F domain is a spherical structure composed of 13 α-helices (H1–H12, H2′) and 4 short β-strands (S1–S4) [15]. On the other hand, the anti-parallel α-helical forms a sandwich structure: H3, H7, and H10/H11 form the two outer layers of the sandwich; H4, H5, H8, and H9 form the central layer of the sandwich. The central layer is mostly located in the upper half of the sphere. The lower half of the sphere consists of H3, H5, and H10, forming a very large Y-shaped cavity (~1400° A3). The three-directional arms of the Y-shaped cavity allow PPARs to be ligand-bound with various single-chain or branched structures [16]. The RXR interacts with several α-helices, including H7–H10, to form PPAR/RXR heterodimers. Further, Sheu et al. identified 10 binding "hot spots" for RXRs in PPARγ using solvent mapping techniques. Four of these spots are located within the Y-shaped cavity: two around the entry site of the Y-shaped cavity, two in the coactivator binding region, one in the dimerization domain, and one in the secondary locus [17]. The E/F domain is also a binding site for coactivators and co-repressors. The end of the E/F domain contains a domain called AF-2, which is highly conserved in all subtypes of PPARs and is closely related to the events of ligand-induced transcription. Ligand binding to the E/F domain induces a conformational change in the AF-2 domain, resulting in a suitable binding surface to recruit coactivators and promoting target gene transcription [18]. In addition, studies on the phosphorylation of PPARs have shown that phosphorylation of AF-1 could affect the activity of AF-2, revealing that the activity of PPARs is affected by intramolecular kinase cascade signaling. All domains participate in the physiological activities of PPARs as a unified whole. For example, changes in the A/B domain could affect ligand binding in the E/F domain [19] or DNA binding in the C domain [20].

The heterodimer of PPAR and RXR is considered a permissive dimer because activation of either component can activate the entire complex. The PPAR/RXR heterodimer binds to the target gene promoter, PPRE. In the non-liganded state, PPAR/RXR interacts with co-repressors such as SMRT and NCoR to recruit repressors that contain histone deacetylase (HDAC) activity, thereby inhibiting gene transcription [21]. Upon ligand stimulation, PPAR/RXR dissociates from multicomponent co-repressors, recruits RNA polymerase II and activators with histone acetyltransferase (HAT) activity, remodels chromatin structure, and promotes target gene transcription (Figure 2) [22].

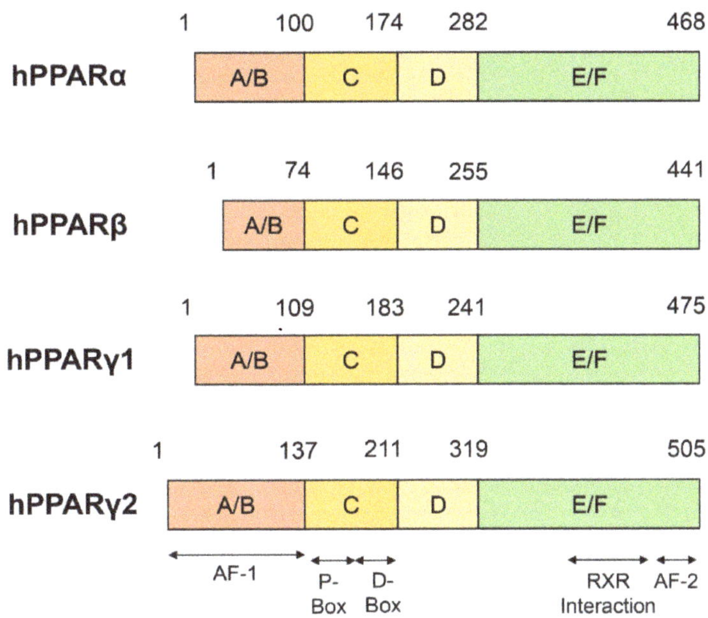

Figure 1. Schematic representation of the principal domains of PPARs. PPARα, PPARβ, and PPARγ all have a modular structure that contains four domains: A/B domain, C domain, D domain, and E/F domain. The A/B domain contains an AF-1 region involved in the regulation of PPARs phosphorylation. The C domain is the DNA binding domain. The D domain is a hinge domain. The E/F domain contains an AF-2 region and is the RXR, ligand, and cofactor binding site.

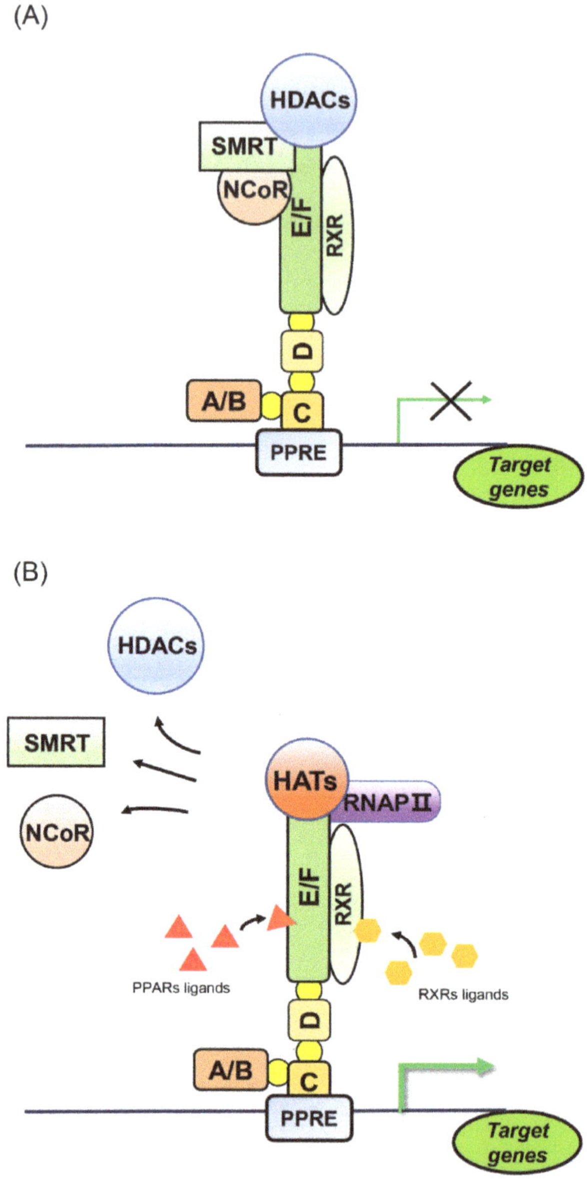

Figure 2. PPARs-mediated gene regulation. PPAR forms a heterodimer with RXR and binds to the PPRE element of the target gene promoter. In the absence of ligand binding, the heterodimer recruits transcriptional corepressors such as NCoR and SMRT, as well as HDACs, to repress target gene transcription (**A**). Upon ligand binding, PPAR changes conformation, releases transcriptional repressor complexes, and recruits transcriptional coactivators such as RNAPII and HATs to promote target gene transcription (**B**). A/B, C, D, E/F: PPAR domains; PPRE: peroxisome proliferator response element; RXR: retinoid X receptor; NCoR: nuclear receptor corepressor 1; SMRT: nuclear receptor corepressor 2; HDACs: histone deacetylases; HATs: histone acetyltransferases; RNAPII: RNA polymerase II.

3. Ligands for PPARs

The PPARs require ligand activation, such as natural and synthetic ligands, which is a characteristic of many other steroid hormone receptors [4,23,24]. The natural ligands consist of a group of endogenously secreted molecules, including various unsaturated fatty acids and their metabolic products. The specificity and activity of these molecules are not high in most circumstances. Additionally, the incubation of triglyceride-rich lipoproteins with lipoprotein lipase (LPL) produces many ligands for PPARs [25,26]. Certain prostaglandins and their metabolic derivatives are also natural ligands [27]. The structure and geometry of PPARβ/δ and PPARα are similar, while PPARγ more likely to bind long-chain polyunsaturated fatty acids [28]. At present, a variety of synthetic ligands are active on the market. These synthetic ligands often have higher PPAR subtype specificity and stronger metabolic activity than natural ligands. The synthetic ligands include agonists and antagonists (Table 1). The antagonists are also referred to as "inverse agonists" because, although they bind to the agonist binding sites of PPARs, they cause opposite pharmacological responses by stabilizing the binding state of uncoordinated PPARs and multicomponent co-repressors in order to inhibit transcriptions of downstream target genes [29,30]. Physical changes caused by ligand binding include changes in the three-dimensional structure [31,32], dissociation of heat shock proteins and chaperones [33,34], and nuclear entry [35,36] of PPARs.

The ligands of one subtype of PPARs could also act on other subtypes. For example, the natural exogenous fatty acid ombuin-3-O-β-D-glucopyranoside was shown to simultaneously activate PPARα and PPARβ/δ to reduce the expression of the lipogenic genes in hepatocytes and promote the genes' expression, which are related to reversed cholesterol transportation in macrophages so as to reduce intracellular lipid concentration [37]. This could provide dual agonists or even pan-agonists of PPARs for the clinic. The dual-agonist glitazars targeting PPARα and PPARγ, such as muraglitazar and tesaglitazar, are being tested in clinical trials and are expected to reduce cardiovascular risk. In addition, the lipid-lowering fibrate acid derivative, bezafibrate, is the first pan-agonist of PPARs that has been clinically tested with satisfactory safety levels and has become the reference for pan-agonists of PPARs [38]. Conversely, 13-HODE, an oxidized low-density lipoprotein, acts as a ligand to activate PPARγ [39]. However, it has the opposite results when it acts on PPARβ/δ. For example, when it acts on colorectal cancer cells, it is considered an antagonist that down-regulates the expression of PPARβ/δ and induces tumor cell apoptosis [40]. In preadipocytes, it is considered an agonist, activating PPARβ/δ to protect the liver from chemically induced liver injury [41]. The ligands were shown to be tissue-specific for the biological activity of PPARs, which may be due to the presence or absence of other regulatory factors in addition to known ligands. In fact, long-term bioassay studies have shown that high-affinity dual PPARα/PPARγ agonists could raise clinical safety concerns, including potential carcinogenicity, weight increase, peripheral dropsy, and a potential increased risk of heart failure in rodents [42]. Therefore, the development of dual agonists and pan-agonists of PPARs with relatively low affinity (i.e., μM or nM) is more suitable for cancer chemoprevention [43]. In addition, the use of PPARγ single agonists, thiazolidinediones (TZDs), induces bone loss in postmenopausal females and diabetic patients [44–47]. In contrast, administration of PPARα and PPARβ/δ dual agonists, linoleic acid (LA), or PPARs pan-agonist bezafibrate could upregulate bone mineral density and result in the formation of periosteal bone in male rats [48]. This suggests that dual and pan-agonists of PPARs have the potential to counteract the adverse effects elicited by the use of highly specific single agonists.

3.1. Agonists and Antagonists of PPARα

The most classic agonists of PPARα are fibrates, including bezafibrate, fenofibrate, clofibrate, gemfibrozil, and Wy-14,643 [49]. Wy-14,643, a pirinixic acid first discovered to play an effective role in anti-hypercholesterolemia [50], induces marked hepatomegaly and peroxisome proliferation in hepatocytes and reduces serum cholesterol and triglyceride levels in male mice [51]. Subsequent reports confirmed that Wy-14,643 is a specific activator of PPARα [52]. The Wy-14,643-activated PPARα regulates yes-associated protein (YAP) expression and nuclear translocation, and blockade of YAP signaling abolishes PPARα-induced hepatocyte hypertrophy and hepatocyte proliferation in mice [53]. GW9578, a urea-substituted thioisobutyric acid (TiBA), is a potent murine PPARα-selective agonist, but it has only a 20-fold selectivity for human PPARα [54]. Furthermore, GW9578 exists in the form of a viscous oil or foam, which provides a hindrance to the quantitative treatment of experiments in vitro and in vivo. Brown et al. identified GW7647 through solid-phase array synthesis to aid in identifying PPARα agonists with high selectivity and good physical properties [55]. As a thioisobutyric acid derivative, GW7647 is the first identified PPARα-specific agonist. It has a 200-fold higher specificity than PPARβ/δ and PPARγ and has lipid-lowering activity in vivo. The GW7647 is an excellent PPARα-specific agonist that could be used in experimental research since it is a powder with a melting point of 153–154 °C [28].

GW6471 is a specific antagonist of PPARα. GW6471 and PPARα could form a ternary complex with the transcriptional co-repressor SMRT, and GW6471 further strengthens the binding of the PPARα E/F domain to the SMRT co-repression motif. The co-repression motif in the ternary complex adopts a three-turn α-helix, preventing the PPARα AF-2 domain from assuming the active conformation [56]. Additionally, L-663,536 (MK-886), a leukotriene biosynthesis inhibitor, was originally identified to prevent endogenous leukotriene production during allergic reactions in guinea pigs and protect them from lethal anaphylactic shock [57]. However, it was subsequently identified as an inhibitor of the fatty acid binding protein 5-lipoxygenase-activating protein (FLAP), but the ability of L-663,536 to induce apoptosis was not mediated by FLAP [58]. The drug L-663,536 was not identified as a non-competitive antagonist of PPARα until 2001. It was then discovered to prevent the conformational change necessary for the PPARα activity formation and inhibit the PPARα target gene transcriptional activity (Figure 3) [59].

3.2. *Agonists and Antagonists of PPARβ/δ*

The first synthetic agonist was L-165,041 [60]. It is a leukotriene antagonist that can activate both the human PPARβ/δ gene and PPARγ [61]. GW501,516 is a more potent and specific PPARβ/δ agonist [62]. It has been used in a large number of experiments so far and has become the reference for PPARβ/δ agonists [63]. However, it was subsequently reported that GW501,516 had no hepatoprotective and anti-fibrotic effects in patients with chronic liver disease [64]. Further, the GW501,516 has been limited for use in clinical trials due to its potential metabolic derangement and stimulant effects and the high risk of a halt in the evolution of molecules after uncontrolled application [65]. The agonist GW0742, which was developed at the same time as GW501,516, has become a highly selective agonist of PPARβ/δ in commercial non-human experiments [66]. The most clinically used PPARβ/δ agonists are MBX-8025/RWJ80,025 and KD-3010 (Phase II trial) [67,68].

The earliest PPARβ/δ antagonist used is an irreversible PPARγ antagonist, GW9662 [69]. In 2008, GSK0660 was confirmed as the first PPARβ/δ selective antagonist [29]. However, due to its low bioavailability, the in vivo experimental effects were affected. On the other hand, SR13,904 is also a PPARβ/δ antagonist, although it also has a weak inhibiting effect on PPARγ [70]. The latest PPARβ/δ antagonist used is GSK3787 which has fair pharmacokinetics. It has been used in a large number of animal experiments due to its fine bioavailability [30,71]. The above compounds are all irreversible antagonists of PPARβ/δ, and covalently bind to the latter. DG172 and PT-S58 are currently two novel PPARβ/δ antagonists. The DG172 has high affinity and strong inhibitory ability. It recruits co-repressors, down-regulates the transcription of PPARβ/δ target genes, and still keeps mice biologically active after oral treatment [72]. In addition, PT-S58 is a cell-permeable diarylcarbonamide drug that acts directly on the PPARβ/δ ligand binding sites. It is a pure competitive specific inhibitor of PPARβ/δ (Figure 4) [73,74].

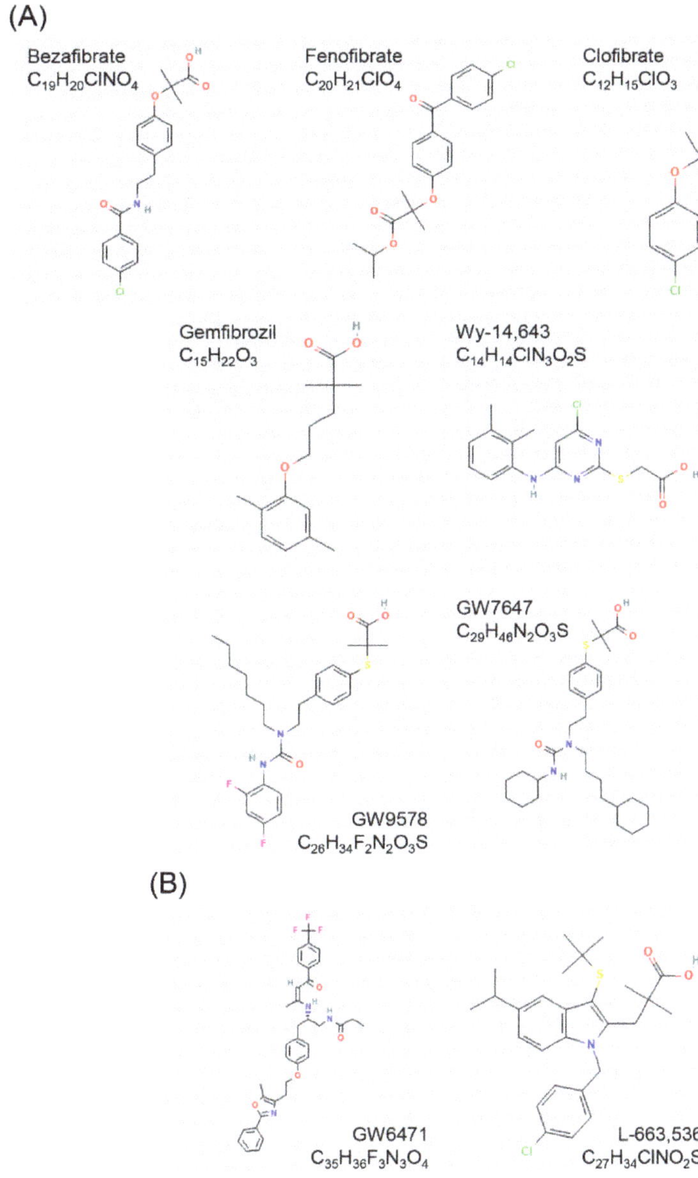

Figure 3. Agonist (**A**) and antagonist (**B**) secondary structures of PPARα.

3.3. Agonists and Antagonists of PPARγ

The most typical agonists of PPARγ are TZDs, which were the first high-affinity selective PPARγ agonists identified. The TZD family includes rosiglitazone (RGZ/BRL49,653) [75], pioglitazone (PGZ), ciglitazone (CGZ), troglitazone (TGZ), englitazone (EGZ), and balaglitazone (BGZ). They are all able to specifically activate PPARγ [76]. In addition to their ability to target PPARγ for type 2 diabetes therapy, different TZD compounds are also in clinical trials for their tumor-suppressing effects. They may become anticancer drugs in the near future. The non-TZD ligand of PPARγ, L-764,406, is the first known partial agonist of PPARγ. Covalent binding of L-764,406 to Cys313 of H3 in the E/F domain of PPARγ induces a conformational change in the receptor and specifically activates the transcriptional activity in the receptor [77]. GW0072 is the ligand of PPARγ with high affinity but is a weak partial agonist. It locates in the ligand-binding pocket, which is uncovered by X-ray crystallography, by binding to an epitope distinct from known PPARγ agonists and does not interact with AF-2 [78]. In 1999, it was first discovered that GW7845 (an L-tyrosine derivative) could be used as PPARγ activator to prevent the progression of experimental breast cancer in rats [79].

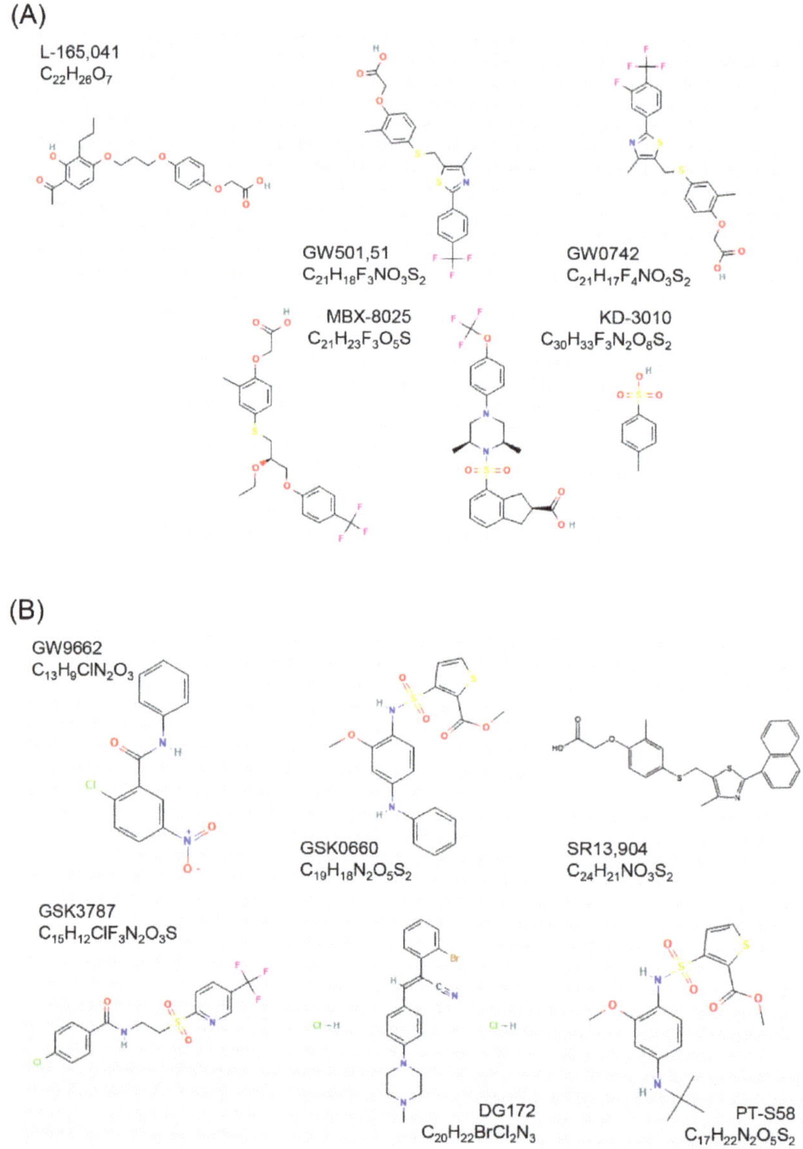

Figure 4. Agonist (**A**) and antagonist (**B**) secondary structures of PPARβ/δ.

GW9662 is an irreversible PPARγ full antagonist [80]. The GW9662 covalently binds to Cys285 of PPARγ, a residue that is highly conserved in all three PPARs. Additionally, GW9662 is 10 to 600 fold more selective for PPARγ than PPARα and PPARβ/δ in cells [81].T0,070,907, which is similar in structure to GW9662, is also a synthetic PPARγ-selective antagonist with more than 800-fold selectivity over PPARα and PPARβ/δ [82]. Bisphenol, a diglycidyl ether (BADGE), also specifically inhibits PPARγ and is a low-affinity PPARγ ligand [83]. The BADGE has been reported to antagonize PPARγ and block adipogenesis induced by BRL49,653 and insulin, under the condition that the concentration level reaches its solubility limit (100 μM) (Figure 5) [84].

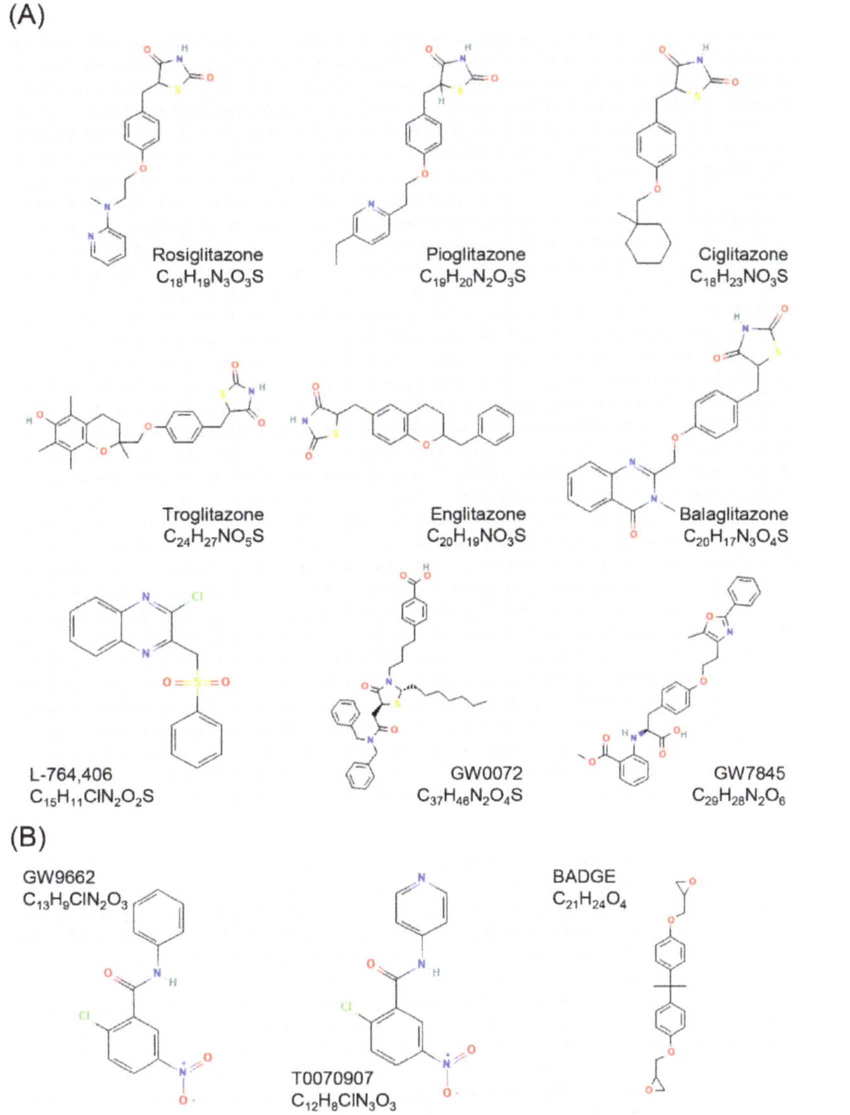

Figure 5. Agonist (**A**) and antagonist (**B**) secondary structures of PPARγ.

Table 1. Agonists and antagonists of PPARs.

PPARs		Agonists	Antagonists
PPARα	fibrates	Bezafibrate [49] Fenofibrate [49] Clofibrate [49] Gemfibrozil [49] Wy-14,643 [52]	GW6471 [56] L-663,536 [59]
		GW9578 [54] GW7647 [55]	
PPARβ/δ		L-165,041 [60] GW501,516 [62] GW0742 [66] MBX-8025/RWJ80,025 [67] KD-3010 [68]	GW9662 [69] GSK0660 [29] SR13,904 [70] GSK3787 [71] DG172 [72] PT-S58 [73,74]
PPARγ	TZDs	rosiglitazone (RGZ) [75] pioglitazone(PGZ) [76] ciglitazone(CGZ) [76] troglitazone(TGZ) [76] englitazone(EGZ) [76] balaglitazone(BGZ) [76]	GW9662 [80,81] T0,070,907 [82] BADGE [83]
		L-764,406 [77] GW0072 [78] GW7845 [79]	

3.4. Structure of PPARs Ligands

The secondary structure of PPARs ligands generally contains fluorine, chlorine, hydroxyl, aliphatic, carboxyl, and carbonyl groups. These groups can form electrophilic groups and interact with relevant sites, such as carboxyl on the E/F domain of PPARs, to form hydrogen bonds and improve the stability of the combination. For example, the carboxyl of the agonist GW409,544 forms a direct hydrogen bond with Try464 on the AF-2 domain of PPARα. GW6471, an antagonist of PPARα, replaces the carboxyl of GW409,544 with an acetamide, destroying the formation of the hydrogen bond on Try464. The GW6471 induces PPARα to recruit SMRT and enhances the binding of PPARα E/F domain to the SMRT co-repression motif, which adopts a three-turn α-helix and prevents the PPARα AF-2 domain from adopting an active conformation [56]. Several ligands contain amino, imino, or quaternary amino groups, which lead to the shift of electrons and form charge attraction with the relevant sites on the Y-shaped cavity of PPARs. The agonist bezafibrate forms a significant positive and negative charge center, which can form a strong salt bond with Lys183 on PPARα [38]. In addition to the above-mentioned intermolecular forces, some ligands can also form covalent bonds with PPARs. Covalent binding of L-764,406 to Cys313 of H3 in the PPARγ E/F domain induces a conformational change in the receptor and specifically activates its transcriptional activity [77]. GW9662, an irreversible full antagonist of PPARγ, covalently binds to Cys285 of PPARγ [81]. In addition, the molecular chains of PPARs agonists are basically long, and most of their electrophilic groups are linked to carbon atoms or small groups. On the contrary, the molecular chains of PPARs antagonists are shorter than those of agonists, and their electrophilic groups are linked to larger carbon rings, aromatic rings, or heterocyclic rings. The antagonists with relatively large molecular structures bind to the ligand-binding cavity of PPARs, resulting in steric hindrance and preventing agonists from entering, thereby inhibiting the active conformational change of PPARs [29,30]. The entrance to the Y-shaped cavity in the PPARs E/F domain includes several polar residues, and the two branches of the cavity, Arm I and Arm II, are mainly composed of hydrophobic residues, except for some moderately polar residues in Arm I. These residues play key roles in determining the interaction of agonists or antagonists with PPARs.

4. Subtypes of PPARs and Breast Cancer

The PPARα, PPARβ/δ and PPARγ express differently in different tissues, with differences in target genes, biological activities, and ligand affinities [85]. Among 225 studies of experimentally validated PPAR target genes, 83 genes were PPARα target genes, 83 were PPARβ/δ target genes, and 104 were PPARγ target genes [86]. In fact, the target genes of the three subtypes of PPARs partially overlap. For example, all three PPARs could transcriptionally activate the angiogenesis pathway-related protein Angptl4 and the lipid droplet-associated protein Plin2 after ligand activation [87]. The PPARs participate in the regulation of carbohydrate and lipid metabolism and homeostasis, as well as various physiological processes such as cell differentiation, proliferation, inflammation, and vascular biology [88]. In addition, the three subtypes of PPARs also regulate the occurrence and development of many malignant tumors via different mechanisms; breast cancer is one of them.

4.1. PPARα and Breast Cancer

PPARα, the first PPAR identified, is recognized as an orphan receptor activated by a variety of peroxisome proliferators. The PPARα was originally discovered in rodents and was named for its role in peroxisome proliferation [4]. On the other hand, PPARβ/δ and PPARγ were subsequently discovered and identified as cognate receptors that are activated by distinct peroxisome proliferators [24,52]. However, subsequent research proved that all PPARs fail to play a role in human peroxisome proliferation. PPARα is mainly expressed in metabolically vigorous cells with active fatty acid oxidation capacity, for example in skeletal muscle, brown fat, the liver, heart, and intestinal mucosal tissues [89]. PPARα is of considerable importance to glucose and lipid metabolism and the balance of transport in mammals. Its main function of maintaining lipid homeostasis is realized by increasing cell mobilization, promoting cell uptake, activation, oxidation, and decomposition of fatty acids, and generating ketone bodies for energy production [90]. The ligand-activated PPARα could also catalyze the hydroxylation of fatty acids. Hence, PPARα is the target of fibrates and hypolipidemic drugs for the treatment of abnormal lipid metabolism. The transcription of PPARα is up-regulated by fibrates, which enhance the lipolysis mediated by lipoprotein lipase, promote the oxidative decomposition of fatty acids, and achieve the curative effect of reducing total cholesterol and total triglycerides [91]. Fibrates are effective in increasing insulin sensitivity and protecting the cardiovascular system, so they are also widely used in the clinical treatment of diabetes and cardiovascular diseases [92].

In addition to regulating glucose and lipid metabolism, PPARα plays a role in various cancers. Long-term administration of PPARα agonists was reported as early as 1980 to cause liver cancer in rodents [93]. This effect of agonists was dependent on the receptor PPARα, as they (Wy-14,643 or bezafibrate) did not induce liver cancer in PPARα-null mice [94,95]. The pro-hepatocarcinogenesis effect of PPARα agonists was not evident in humans [96]. The species-specific mechanism of promoting hepatocarcinogenesis is that mouse-derived PPARα rather than human-derived PPARα down-regulated let-7C miRNA to increase the stability of its target gene MYC, an oncogenic factor. The increased expression of MYC promoted hepatocyte mitosis until carcinogenesis [97–99]. Some studies have shown increased expression of PPARα in endometrial cancer. Fenofibrate treatment significantly prevented the proliferation of endometrial cancer cells and promoted cell apoptosis [100]. However, other studies have also shown that PPARα knockdown inhibited the proliferation of endometrial cancer cells, promoted cell apoptosis, and reduced the secretion of the angiogenesis-related factor VEGF, while fenofibrate treatment also reduced the secretion of VEGF [101]. Since this contradictory phenomenon is not caused by nonspecificity to PPARα and cytotoxicity at the dose of fenofibrate [102], a possible explanation might be the biphasic response of PPARα activity, i.e., PPARα with very low activity and expression and PPARα with very high activity and expression producing the same effect, known as a U-shaped dose-response curve. PPARα was also aberrantly expressed in melanoma. Fenofibrate treatment inhibited the clone formation and migration abilities of melanoma cells and rendered them highly sensitive to staurosporine (a protein kinase C inhibitor with strong pro-apoptotic activity) [103].

Chang et al. found that, compared to adjacent normal tissues, PPARα and its natural ligand, arachidonic acid (AA), were significantly overexpressed in the tissues of breast cancer. The growth of three breast cancer cells, MDA-MB-231 (ER-), MCF7 (ER++++), and BT-474 (ER++), were stimulated by AA, with the most pronounced pro-proliferative effect on MCF7 cells, revealing a positive correlation between PPARα and the proliferation of ER+ breast cancer cells [104]. Human cytochrome P450 1B1 (CYP1B1)-mediated biotransformation of endogenous estrogens and environmental carcinogens promotes the progression of multiple hormone-dependent tumors, including breast cancer [105]. Hwang et al. found that Wy-14,643 increased CYP1B1 mRNA and protein levels in MCF7 cells and activated PPARα enhanced CYP1B1 promoter activity through directly binding to its PPRE elements [106]. In addition, Castelli et al. found that treatment of breast cancer stem cells with the specific PPARα antagonist GW6471 reduced cell proliferation, viability, and spheroid formation, resulting in metabolic dysfunction and apoptosis [107]. The above experiments in vitro all suggest that PPARα functions in promoting the development of breast cancer. However, Pighetti et al. found that treatment with Wy-14,643 inhibited the ability of DMBA to induce breast tumor formation in rats and induced tumor volume regression [108]. Chandran et al. showed that clofibrate treatment activated the PPARα transcriptional activity and exerted an anti-proliferative effect on breast cancer cells by regulating the levels of tumor suppressors, cell cycle inhibitors, and cell to cycle checkpoint kinases, causing cells to arrest in the G0/G1 phase and significantly inhibiting cell growth. In addition, activated PPARα reduced the expression of inflammatory pathway-related enzymes and their receptors, reduced the protein levels of lipogenic enzymes, regulated the fatty acid oxidation associated gene expression, and affected various lipid metabolism pathways [109]. Yin et al. found that Runt-related transcription factor 2 (RUNX2), with high expression in breast cancer, recruited metastasis-associated 1 (MTA1)/NuRD and the Cullin 4B (CUL4B)-Ring E3 ligase (CRL4B) complex to form a ternary complex. This complex catalyzed histone deacetylation and ubiquitination, inhibited the transcriptional activity of target genes, including PPARα, and promoted the proliferation and invasion of breast cancer cells in vitro. These physiological processes finally led to breast cancer occurrence, bone metastasis, and tumor stemness in vivo (Table 2) [110]. The above findings indicate that PPARα plays a role as a tumor suppressor in breast cancer.

Table 2. The effects of PPARα on breast cancer.

	The Role in Breast Cancer	Binding Ligand	The Effect on Breast Cancer
PPARα	cancer-promoting	arachidonic acid	Promoted cell growth and proliferation, especially MCF7 in cells (ER++++) [104]
		Wy-14,643	Increased target gene CYP1B1 mRNA and protein levels in MCF7 cells promoted cancer progression [106]
		GW6471	Reduced cell viability, cell proliferation, and spheroid formation lead to apoptosis and metabolic dysfunction of stem cells [107]
	cancer-suppressing	Wy-14,643	Inhibited the ability of DMBA to induce tumor formation in rats and induced tumor volume regression [108]
		clofibrate	Inhibited cell proliferation and growth, affecting various lipid metabolism pathways [109]
		–	Inhibited the proliferation and invasion of cells in vitro, inhibited cancer occurrence, bone metastasis, and tumor stemness in vivo [110]

PPARα was generally highly expressed in human primary inflammatory breast cancer cells SUM149PT (3.9-fold higher than primary human breast epithelial cells HMEC) and highly invasive breast cancer cells SUM1315MO2 (3.7-fold higher than HMEC cells) and in human breast tumor tissue (2–6-fold higher than adjacent normal tissues) [109]. The correlation between PPARα and breast cancer is worth further investigation.

4.2. PPARβ/δ and Breast Cancer

Among the three subtypes of PPARs, PPARβ/δ exhibits higher evolutionary efficiency [4]. In addition, uncoordinated PPARβ/δ also showed more potent transcriptional repression activity. Compared with uncoordinated PPARβ/δ, unligated PPARα and PPARγ do not inhibit PPRE-mediated transcription, which is possibly due to their inability to bind to the nuclear receptor corepressors such as SMRT and NCoR [111]. This relatively rapid rate of evolution and more potent transcriptional repression activity underscore the importance of investigating PPARβ/δ function. The PPARβ/δ are referred to as HUC-1 in humans [112], fatty acid-activated receptors (FAAR) in mice [113], and PPARδ in rats [114]. The PPARβ/δ are widely expressed in most tissues, and their expression level is often higher than that of PPARα and PPARγ. This widespread expression proves its importance in systemic activities and basic cell functions [52,115]. The high baseline expression of PPARγ, especially in the gastrointestinal tract and skeletal muscle, reveals the critical role of PPARβ/δ in fatty acid oxidation and obesity prevention [116]. PPARβ/δ is specific and diversified in cell fate. It can activate housekeeping genes and regulate energy metabolism. In addition, the endogenous natural ligands of PPARβ/δ are very broad and non-specific. The ability of these ligands to activate PPARβ/δ is relatively weak. Therefore, the physiological function of PPARβ/δ is difficult to simplify. Without ligand binding, PPARβ/δ degrades fast, while ligand binding inhibits ubiquitin-proteasome activity, thereby extending its half-life [117,118]. This phenomenon may also be attributed to ligand-induced PPARβ/δ expression [119]. Ligand-activated PPARβ/δ could increase the levels of serum high-density lipoprotein cholesterol, decrease the levels of serum triglycerides in mice [60], non-human primates [62], and humans [120], and improve the metabolic syndrome such as obesity and insulin resistance induced by a high-fat diet or genetic predisposition [116,121]. Inhibition of insulin resistance by activated PPARβ/δ might also improve progressive neurodegeneration and its associated learning and memory deficits and prevent Alzheimer's disease [122,123]. In addition, PPARβ/δ also have considerable preventive or therapeutic capacity against genetic [124], diet [125], or chemically induced [126] liver inflammation.

The above evidence supports the development of PPARβ/δ specific agonists acting as clinical drugs for the treatment of diseases such as obesity, diabetes, metabolic syndrome, and liver inflammation. However, the synthesis of PPARβ/δ-targeted drugs has encountered significant obstacles related to clinical safety due to substantial controversy regarding the reports on the role of PPARβ/δ in cancer [127,128]. Ligand-activated PPARβ/δ could promote terminal differentiation of keratinocytes [129], enhance lipid deposition [130], inhibit cell proliferation [131], and inhibit the progression of skin cancers such as psoriasis. However, it has also been shown that transgenic mice that induced activation of PPARβ/δ in the epidermis developed an inflammatory skin disease strikingly similar to psoriasis. These mice were characterized by hyperproliferation of keratinocytes, aggregation of dendritic cells, and endothelial cell activation. The gene dysregulation and activation of key transcriptional programs and Th17 subsets of T cells in transgenic mice were also highly similar to psoriasis [132]. In addition, PPARβ/δ activated by UV stimulation directly promoted the expression of oncogene Src and upregulated its kinase activity, enhanced the EGFR/ERK1/2 signaling pathway, and promoted epithelial-mesenchymal transition (EMT), which promotes keratinocyte differentiation and proliferation [133]. This result also reveals the cancer-promoting effect of PPARβ/δ on skin cancer. A possible and one-sided explanation for this contradiction was that activation of PPARβ/δ existed both in keratinocytes and adjacent fibroblasts. The PPARβ/δ in fibroblasts inhibited IL-1 signaling by directly upregulating the expression of secreted interleukin-1 receptor antagonist (sIL-1ra), thereby regulating keratinocyte proliferation [134]. In addition to skin cancer, the PPARβ/δ also have a controversial role in colorectal cancer [40,135,136].

Human genome PPARβ/δ is located at 6p21.2, an increased site for ER- and high-risk breast cancer [137], which reveals the correlation between PPARβ/δ and breast cancer. PPARβ/δ was highly expressed in the nucleus in human normal breast epithelial cells and weakly expressed or even absent in 92% of human breast lobular and ductal cancer cells [138–140]. The expression of PPARβ/δ in mouse malignant breast cancer cells C20 was also significantly lower than that in mouse keratinocytes (nearly 4-fold) and human normal mammary epithelial cells MCF10A (more than 2-fold) [141]. The patients' survival rate with breast cancer and the expression of PPARβ/δ have a negative correlation [142]. In 2004, Stephen et al. reported for the first time that PPARβ/δ activated by specific ligand compound F or GW501,516 could promote the proliferation of ER+ breast cancer cells MCF7 and T47D. It could also promote in T47D cells vascular endothelial growth factor α (VEGFα) and its receptor FLT-1 and encourage the proliferation of human umbilical vein endothelial cells (HUVEG) in vitro. However, activated PPARβ/δ did not exert similar effects on ER- breast cancer cells MDA-MB-231 and BT-20, revealing that the pro-proliferative and pro-angiogenic effects of PPARβ/δ on breast cancer are dependent on ER [143]. Conversely, in 2008, Girroir et al. reported that PPARβ/δ was activated by specific ligands (GW0742 or GW501,516) and inhibited the growth of MCF7 cells [144]. In 2010, Foreman et al. reported that PPARβ/δ activated by the above two ligands also inhibited proliferation and clone formation and promoted apoptosis in mouse C20 cells [141]. Additionally, in 2014, Yao et al. reported that the overexpression of PPARβ/δ prevented the proliferation of breast cancer cells, MDA-MB-231 and MCF7, while the treatment of the agonist GW0742 further inhibited the proliferation of MCF7 cells without any effect on the MDA-MB-231 cells. The overexpression of PPARβ/δ inhibited the clone formation of these two cell lines, while further treatment with GW0742 inhibited the clone formation of MDA-MB-231 cells significantly more than that of MCF7 cells. However, the overexpression or ligand-activated of PPARβ/δ did not affect apoptosis in either of the two breast cancer cell lines. Further, the overexpression of PPARβ/δ could inhibit the growth of xenograft tumor in MDA-MB-231 cells better than in MCF7 cells, and treatment with GW0742 further inhibited the volume of mouse xenografts [145]. These findings, although inconsistent with Stephen's report [143], also confirm that the effects of PPARβ/δ on ER+ and ER- breast cancer cells were different. However, by real-time analysis of cell doubling time, Palkar et al. found that neither GW0742-activated nor highly specific irreversible antagonist GSK3787 inhibited PPARβ/δ had effects on the proliferation of MCF7 cells, despite the fact that both of them had the converse effect on the mRNA level of PPARβ/δ target gene Angptl4 in vitro and in vivo [30]. Additionally, although these disparate results may be attributed to the concentration of ligands used, cell treatment time, cell proliferation assessment methods, etc., the exact function of PPARβ/δ on breast cancer cell apoptosis and proliferation remains unclarified so far. Several experiments are required to reach consensus.

Ghosh et al. obtained PPARβ/δ−/−COX-2-TG transgenic mice by crossbreeding and found that PPARβ/δ silencing antagonized cyclooxygenase-2 (COX-2)-induced mammary gland hyperplasia and tumorigenesis in mice and significantly inhibited the expression of breast epithelial cell proliferation-related genes (e.g., Ki-67, Cyclin D1, etc.), revealing that PPARβ/δ plays the role of tumor suppressor in the development of breast cancer [146]. However, Glazer's team found that treatment with GW501,516 accelerated adenosquamous carcinoma and mammary squamous cell tumor formation in mice induced with medroxyprogesterone acetate (MPA) and 7,12 dimethylbenzene(a)anthracene (DMBA). The elevated levels of PPARβ/δ were accompanied by increased activation of 3-phosphoinositide-dependent protein kinase 1 (PDK1), revealing that PPARβ/δ plays a role in promoting breast cancer development through the PDK1 signaling pathway [147]. PDK1 is a vital governor of the AGC protein kinase family, including all isoforms of the AKT/PKB, S6K, and PCK families [148]. Therefore, Glazer's team constructed MMTV-PDK1 transgenic mice and found that overexpression of PDK1 in mouse mammary epithelial cells up-regulated the levels of pT308AKT and pS9GSK3β, as well as PPARβ/δ. After induction with MPA and DMBA, GW501,516 treated wild-type and transgenic mice showed an increased formation rate of mammary tumors compared with untreated normal wild-type mice. Further, between the two types of mice, the transgenic mice showed more pronounced tumors. The GW501,516 treatment did not alter PDK1 protein levels. In addition, PDK1 overexpression also enhanced PPARβ/δ-induced energy metabolism. These results reveal that PPARβ/δ promotes breast cancer by enhancing energy metabolism, which is dependent on PDK1/AKT signaling [149]. In 2013, Glazer's team directly constructed MMTV-PPARβ/δ transgenic mice by embryo prokaryotic injection and found that overexpression of PPARβ/δ induced breast tumorigenesis and activation of the AKT/mTOR signaling pathway. The total number of mice developed invasive breast cancer within 12 months, and GW501,516 treatment strongly accelerated the oncogenic process and increased breast tumor diversity. A hallmark characteristic of MMTV-PPARβ/δ mice is the development of ER+/PR+/HER2- mammary tumors, further revealing the correlation between PPARβ/δ and ER+ ductal breast cancer [150]. The above experiments in vivo also reflect the conflicting roles of PPARβ/δ in breast cancer development, which may be attributed to the singleness of the GW501,516 therapeutic dose (0.005% (*w*/*w*)). In addition, as a specific agonist of PPARβ/δ, GW501,516 preferentially activates PPARβ/δ in human PPARs with a 667–833-fold higher affinity than the other two subtypes. However, the affinity of GW501,516 in mice is only 33–62-fold higher than that of other subtypes [151]. Therefore, this increased mammary tumorigenesis in mice treated with a single dose of GW501,516 may not be simply attributable to the activation of PPARβ/δ. However, it is undeniable that the successful construction of many transgenic mouse models is of great significance in studying the correlation between PPARβ/δ and breast cancer.

Retinoic acid (RA) as a tumor suppressor exhibits potent anticancer activity mediated by the nuclear retinoic acid receptor (RAR). The intracellular lipid-binding protein cellular retinoic acid-binding protein II (CRABP-II) targets RA to the RAR, while another lipid-binding protein, fatty acid binding protein 5 (FABP5), could deliver it to the non-canonical RA receptor PPARβ /δ. The FABP5/CRABP-II ratio determines the partition of RA between the two receptors. Noy's team constructed two breast cancer MMTV-neu transgenic mouse models expressing different FABP5/CRABP-II ratios in breast tissue. It was observed that transgenic mice with a high FABP5/CRABP-II ratio produced larger breast tumors. On the contrary, the reduction of this ratio resulted in the suppression of breast tumor growth and gene expression, including PDK1 and cell proliferation-related genes, through the transfer of RA signaling from PPARβ/δ to RAR. This study proposes a new mechanism by which PPARβ/δ promote breast cancer [152]. Additionally, the epidermal growth factor receptor (EGFR) as a tumor-promoting factor can promote breast cancer cell proliferation and induce breast tumorigenesis. Noy's team also found that treatment of MCF7 cells with the EGFR ligand heregulin-β1 could directly upregulate the expression of FABP5 and PDK1. The results indicated that FABP5 and PPARβ/δ were the key mediators of EGFR's ability to enhance cell proliferation, further confirming that PPARβ/δ acted as a tumor-promoting factor playing a role in breast cancer [153]. However, studies on human keratinocyte HaCaT found that FABP5 neither delivered RA to PPARβ/δ nor promoted anti-apoptotic activity by upregulating PDK1 levels. This phenomenon was also identified in HaCaT cells that stably overexpress PPARβ/δ [154]. The above results suggest that the cancer-promoting effect of RA-mediated PPARβ/δ may be specific to breast cancer [155]. Wang et al. found that PPARβ/δ could promote the survival of MCF7 cells under rough microenvironmental conditions by reducing oxidative stress and promoting AKT-mediated survival signaling [156]. The correlation between PPARβ/δ and PDK1 is currently controversial. Although the above studies have found that the expression levels of the two are correlated, there are also studies showing that PDK1 is not a target gene of PPARβ/δ [136,155,157]. In addition to the research around the effect of PPARβ/δ on the proliferation and apoptosis of breast cancer cells, scholars have found that PPARβ/δ also has an effect on the invasion and metastasis of breast cancer cells. Adhikary found that PPARβ/δ, specifically antagonized by ST247 and DG172, inhibited serum and transforming growth factor β (TGFβ)-induced invasion of MDA-MB-231 cells [158]. However, Wang uncovered that the PPARβ/δ expression levels in more metastatic breast cancer basal cell lines were significantly higher than those in luminal cells. Additionally, after the inoculation with MCF7 cells overexpressing PPARβ/δ, the breast tumor volume and lung metastasis of mice increased significantly (Table 3) [156]. In conclusion, the exact role of PPARβ/δ on breast cancer still requires more experimental studies.

Table 3. The effects of PPARβ/δ on breast cancer.

	The Role in Breast Cancer	Binding Ligand	The Effect on Breast Cancer
PPARβ/δ	cancer-promoting	GW501,516	Promoted the proliferation of MCF7 and T47D cells (ER+) instead of MAD-MB-231 and BT-20 cells (ER-), promoted VEGFα and FLT-1 expression [143]
		GW501,516	Accelerated adenosquamous carcinoma and mammary squamous cell tumor formation in mice, increased activation of PDK1 [147]
		GW501,516	Accelerated tumor formation, did not alter PDK1 protein levels [149]
		GW501,516	Accelerated the oncogenic process and increased tumor diversity, especially ER+/PR+/HER2- tumors [150]
		–	Promoted tumor growth and the expression of genes, including PDK1 and cell proliferation-related genes [152]
		–	Promoted the expression of FABP5 and PDK1 in MCF7 cells, promoted cell proliferation, and induced tumorigenesis [153]
		–	Promoted the survival of MCF7 cells under harsh microenvironmental conditions [156]
		ST247 or DG172	Inhibited serum and TGFβ-induced invasion of MDA-MB-231 cells [158]
		–	Increased tumor volume and lung metastasis in mice [156]
	cancer-suppressing	GW0742 or GW501,516	Inhibited the growth of MCF7 cells [144]
		GW0742 or GW501,516	Inhibited the proliferation and clone formation, and promoted apoptosis in mouse C20 cells [141]
		GW0742	Inhibited the proliferation of MCF7 cells instead of MDA-MB-231 cells, inhibited the clone formation of MDA-MB-231 cells significantly more than that of MCF7 cells, and inhibited the volume of mouse xenografts [145]
		–	Inhibited hyperplasia and tumorigenesis in mice and inhibited the expression of epithelial cell proliferation-related genes (e.g., Ki-67, Cyclin D1, etc.) [146]
	no effect	GW0742 or GSK3787	Had no effect on the proliferation of MCF7 cells, despite both of them influencing the mRNA level of the target gene Angptl4 in vitro and in vivo [30]

4.3. PPARγ and Breast Cancer

PPARγ1 and PPARγ2 are two isoforms of PPARγ, that were found in mice. The PPARγ2 mRNA was the predominant PPAR isoform in mouse mammary tissues [159]. In humans and monkeys, in addition to PPARγ1 and PPARγ2, a third isoform of PPARγ4 was found. These isoforms are the transcripts of seven mRNA spliceosomes (PPARγ1, PPARγ2, PPARγ3, PPARγ4, PPARγ5, PPARγ6, and PPARγ7) from the different transcription start sites, which are transcribed through alternative splicing of exons in the 5′-terminal region (A1, A2, B, C, and D) [160]. The PPARγ1, PPARγ3, PPARγ5, and PPARγ7 mRNAs translate into the same protein, PPARγ1, while PPARγ2 mRNA translates into PPARγ2 protein, whereas PPARγ4 and PPARγ6 mRNAs translate into the same PPARγ4 protein. PPARγ1 is expressed in almost all tissues, with the highest level in white and brown adipose tissues. Under normal physiological conditions, the larger PPARγ2 isoform (with additional amino acids at the amino-terminal of PPARγ2, 30 in mice and 28 in humans) is only expressed in brown and white adipose tissue, whereas its expression in the liver and skeletal muscle is caused by excessive caloric intake or genetic obesity. PPARγ4 is under-researched and expressed in macrophages and adipose tissues [161–163]. PPARγ widely expressed in white and brown adipose tissues, the large intestine, and the spleen. However, PPARγ is also found in the liver, pancreas, and tissues of the immune system [164]. A considerable number of studies have confirmed that ligand-activated PPARγ could regulate fat distribution and glucose and lipid metabolism [165] and reduce the inflammatory response of cardiovascular cells, especially endothelial cells [166]. Its specific agonist is relatively effective in the treatment of hyperlipidemia, hyperglycemia, and cardiovascular disease [167]. The specific agonists of PPARγ, i.e., TZDs, are clinical drugs currently on the market as insulin sensitizers for the treatment of type 2 diabetes, targeting PPARγ to exert a hypoglycemic effect. The antidiabetic activity of TZDs was first discovered in the early 1980s [168–171]. PPARγ is also involved in neural differentiation during the formation of neural precursor cells [83]. Therefore, its specific agonists could also act as protective agents for neurons, inducing synaptic plasticity and neurite outgrowth, and improving the symptoms of some neurological diseases [172]. In addition to the above effects, a large number of reports also pointed out that ligand-activated PPARγ exerts anti-tumor effects by promoting cell apoptosis and preventing cell proliferation, regulating cell metastasis, and stimulating angiogenesis, thereby inhibiting the occurrence and development of tumors of the liver [173], bladder [174], lung [175,176], brain [177], thyroid [178], esophagus [179] and colorectum [180–183].

PPARγ also plays a role in breast cancer progression. In 1998, it was reported that TZD-activated PPARγ could induce terminal differentiation of malignant mammary epithelial cells in vitro [184]. However, in 1999, researchers found that ligand-activated PPARγ could prevent the development of experimental breast cancer in vivo. The report showed that GW7845 as an activator of PPARγ significantly inhibited nitrosomethylurea (NMU)-induced mammary tumor incidence, tumor number, and tumor weight in rats [79]. Subsequent reports of ligand-activated PPARγ inhibiting breast cancer development have experienced a rise. A 2001 study showed that TGZ inhibited DMBA-induced mammary tumor progression in rats, reduced malignancy incidence, and induced regression or stasis of total tumor volume [108]. A study in 2009 showed that the conjugated fatty acid α-eleostearic acid (α-ESA) could act as an agonist of PPARγ, upregulating the level of PPARγ mRNA in MCF7 cells, upregulating PPARγ's DNA binding activity and transcriptional activity, and mediating PPARγ nuclear translocation, thereby reducing MCF7 cell viability and promoting tumor cell apoptosis. At the same time, α-ESA-induced high PPARγ expression was associated with an inhibitory effect on ERK1/2 MAPK phosphorylation activation. This suggests that pERK1/2 might play a negative regulatory role on PPARγ levels [185]. Bonofiglio's team discovered an important pathway for PPARγ in human breast cancer cell growth, cycle arrest, and apoptosis. RGZ-activated PPARγ inhibits the PI3K/AKT pathway and induces the antiproliferative effect of MCF7 cells [186]. RGZ also increased the binding of PPARγ to the NF-κB sequence on the promoter sequence of p53, upregulated the expression level of p53 in MCF7, induced caspase 9 cleavage and DNA fragmentation, triggered the apoptotic pathway, stopped the growth, and promoted apoptosis of breast cancer cells [187]. Furthermore, in several breast cancer cell lines, RGZ activated the human Fas ligand (FasL) promoter in a PPARγ-dependent manner, increased the binding of PPARγ with Sp1 to the Sp1 sequence located within the FasL promoter, and positively regulated FasL expression [188]. FasL is a type II transmembrane protein expressed on the membrane surface of activated T lymphocytes and cancer cells. By binding to its receptor Fas [189,190], it activates the cascade of caspases and induces apoptosis [191]. These studies reveal a novel molecular mechanism by which PPARγ induces growth arrest and apoptosis in breast cancer cells. An in vivo study in 2011 showed that TZD-activated PPARγ inhibited MAPK/STAT3/AKT phosphorylation-mediated leptin signaling in MCF7 cells. On one hand, this effect led to the inhibition of MCF7 xenografts through the counteraction of the stimulatory effects of leptin on estrogen signaling. On the other hand, it inhibited leptin-induced cell-cell aggregation and tumor cell proliferation, exerting pro-apoptotic and anti-proliferative effects on breast cancer cell lines [192].

Almost all experimental studies on PPARγ ligands reflect the prevention effect of these ligands on the occurrence and development of breast cancer. However, a 20-week human clinical trial found that the clinical value of TGZ treatment in patients with refractory metastatic breast cancer was not significant. All 22 patients receiving treatment displayed different levels of disease progression within 6 months. Some might even have started other systemic therapies. All patients with serum tumor marker expression above baseline had increased levels of these markers again within 8 weeks [193]. The public has been warned against TGZ by the U.S. Food and Drug Administration, and it was taken off the market in 2000 because of its specific hepatotoxicity [194]. It was subsequently withdrawn in the UK as well. In 1999 and 2000, RGZ and PGZ were marketed as targeted type 2 diabetes treatments in the US and Europe [195]. BGZ completed phase III clinical trials in 2010 and has not yet been listed [196]. However, short-term treatment with RGZ (2–6 weeks, n = 38) also did not protect tumor cell proliferation significantly in patients with an early stage of breast cancer [197]. Therefore, it is necessary to either synthesize new PPARγ activators with clinical value and few toxic side effects or find other drugs that can be used in combination with existing ligands for breast cancer treatment. In fact, as early as 1998, a study found that the combination of TGZ and all-trans-retinoic acid (ATRA) had a synergistic and irreversible inhibitory effect on the growth of MCF7 cells in vitro, induced MCF7 cell apoptosis, and was accompanied by a significant reduction of bcl-2. In vivo injection of the combined drug had no obvious toxic effects in mice. A drug combination could also significantly induce apoptosis and fibrosis-related morphological changes in breast cancer cells [198]. A 2008 study found that the PPARγ ligand N-(9-fluorenyl-methyloxycarbonyl)-l-leucine (F-L-Leu) combined with the COX-2 inhibitor celecoxib significantly delayed the median age of death in breast cancer mice. Breast cancer cell growth is also synergistically inhibited in vitro [199]. Bonofiglio's team found that combining RGZ and RXR ligand 9-cis-retinoic acid (9RA) at nanomolar levels significantly inhibited the activity of breast cancer cells and promoted endogenous apoptosis. Combined treatment with RGZ and 9RA up-regulated the mRNA and protein levels of p53 and its effector gene p21 (WAF1/Cip1) in MCF7 cells, which led to a series of programmed apoptosis events such as the disruption of mitochondrial membrane potential, the release of cytochrome c, the activation of caspase 9, and DNA fragmentation [200]. The combination of CGZ and 9RA, another compound of the TZD family, could also synergistically prevent the human colon cancer cells' Caco2 growth and induce apoptosis [201]. A 2011 study showed that the combination of TZD and the demethylating drug hydralazine could upregulate PPARγ transcriptional and translational levels in triple-negative breast cancer (TNBC) cells, thereby promoting the anti-proliferative and apoptotic effects of TNBC cells and reducing the xenograft tumor growth proliferation index [202]. In conclusion, the multi-drug combination regimen using PPARγ ligands could have a key role in the treatment of many malignant tumors, including breast cancer [203], ovarian cancer [204,205], colon cancer [206,207], and lung cancer [208,209].

In addition to its ligand-activated state, PPARγ also involves itself in the development of breast cancer in a non-ligand-independent manner. The PPARs and ERα are both members of the nuclear receptor superfamily. The ERα signaling pathway has a critical role in metabolism regulation and various physiological processes in the development of breast cancer [210,211]. Bonofiglio's team found for the first time that ERα could bind to the PPRE element to inhibit its mediated transcriptional activity independently of PPARs. Interestingly, PPAR/RXR heterodimers could also bind to the ER response element (ERE) independently of ERs [212]. PPARγ physically interacted with ERα to form a ternary complex with a regulatory subunit of PI3K and p85. PPARγ and ERα played opposite roles in the regulation of PI3K/AKT signaling, which involves cell survival and proliferation [186]. The crosstalk between the PPARγ and ERα signaling pathways revealed the important role of PPARγ in the development of ER+ breast cancer. Since PPARγ-null mice are embryonic lethal, scientists have developed other ways to create transgenic animal models that silence PPARγ. Yin et al. investigated the susceptibility of PPARγ inactivation to MPA- and DMBA-induced breast cancer in mice by constructing an MMTV-Pax8PPARγ transgenic mouse model. In the absence of induction, the mammary glands of transgenic and wild-type mice did not differ in functional development or propensity for tumor formation, a finding consistent with Cui et al.'s [213]. However, after being induced by MPA and DMBA, transgenic mice developed higher tumor diversity than wild-type mice. These tumors were predominantly ER+ ductal breast cancers, further revealing the role of PPARγ in the development of ER+ breast cancer. The decrease in PTEN expression, the induction of pERK1 and pAKT levels, and decreasing pGSK3β level, Pax8PPARγ promotes Wnt signaling [214]. However, in constructing transgenic mice with constitutively active forms of MMTV-VpPPARγ, Saez et al. found that activation of PPARγ signaling did not affect mammary gland development in transgenic mice, which had no phenotypic difference with wild-type mice. On the other hand, when such transgenic mice were crossed with breast cancer-prone transgenic MMTV-PyV mice, the progeny biogenic mice developed tumors much faster and with a higher degree of malignancy and differentiation of the tumors. This molecular mechanism for promoting breast cancer development might also be attributed to the promotion of PPARγ on the Wnt signaling pathway [215]. Tian et al. conducted a parallel experiment on immunocompetent FVB mice, with one group of implanted tumor cells transduced with wild-type PPARγ, and the other with constitutively active PPARγCA. They found that the growth of mammary tumors in mice implanted with PPARγCA-transduced cells was enhanced, which was correlated with endothelial stem cells and angiogenesis increasing. PPARγCA induced ErbB2-transformed mammary epithelial cells to secrete Angptl4 protein, which enhanced angiogenesis in vivo and promoted tumor growth [216]. The above studies based on animal models reveal the contradictory roles (either inhibiting or promoting) of PPARγ in the occurrence and development of breast cancer. The potential reasons for this discrepancy remain to be investigated. The possible causes could be traced to the differences in the construction of animal models or the difference in the length of experimental periods. In addition, a 2019 study showed that PPARγ directly bound to the PPRE element of the protein tyrosine phosphatase receptor-type F (PTPRF) promoter and recruited RNA polymerase II and H3K4me3 to promote the transcription of PTPRF. These processes inhibited breast cancer cell proliferation and migration in vitro and inhibited breast tumor growth and distant metastasis in mice [217]. A 2020 experiment in vitro showed that PPARγ, which is commonly expressed in human primary and metastatic breast cancer [218], interacted with Nur77, recruited the ubiquitin E3 enzyme Trim13 to target the ubiquitin proteasomal degradation of Nur77, and promoted breast cancer progression. Nur77, a tumor suppressor, inhibits breast cancer cells from uptaking exogenous fatty acids and blocks the accumulation of fatty acids in the tumor metabolic microenvironment by inhibiting the transcription of the transmembrane protein CD36 and the cytoplasmic fatty acid-binding protein FABP4. Therefore, blocking the interaction between PPARγ and Nur77 can be used as a clinical approach for PPARγ ligand-independent treatment of breast cancer (Table 4) [219]. However, due to the relatively high concentrations of endogenous natural ligands in cells, it remains to be verified whether these conclusions are truly ligand-independent of PPARγ.

In 2005, an immunohistochemical test of 170 patients with invasive breast cancer showed that the expression of PPARγ was negatively associated with histological grade ($p = 0.019$). PPARγ had a significantly favorable effect on recurrence-free survival in breast ductal carcinoma patients ($p = 0.027$) and was an independent prognostic factor in ductal carcinoma patients ($p = 0.039$) [220]. In 2008, a study presented that the nuclear expression of PPARγ had a preventive effect on the recurrence of female breast ductal carcinoma in situ. Its expression level was negatively correlated with tumor recurrence ($p = 0.024$) [221]. These clinical research studies and the above experimental results reveal the important function of PPARγ in the occurrence and development of breast cancer. The overexpression of PPARγ in breast tumors and the physiological effects of its ligands on breast cancer cells indicate that PPARγ will be a possible target in breast cancer clinical prevention and treatment.

Table 4. The effects of PPARγ on breast cancer.

	The Role in Breast Cancer	Binding Ligand	The Effect on Breast Cancer
PPARγ	cancer-promoting	–	Promoted Wnt signaling and induced transgenic mice to develop tumors much faster with a higher degree of malignancy and differentiation of the tumors [215]
		–	Promoted the growth of tumors and angiogenesis in mice, increasing Angptl4 expression and endothelial stem cells [216]
		–	Interacted with Nur77, recruited Trim13 to target the ubiquitin proteasomal degradation of Nur77, and promoted cancer progression [219]
	cancer-suppressing	TZD	Induced terminal differentiation of malignant mammary epithelial cells [184]
		GW7845	Inhibited NMU-induced tumor incidence, tumor number, and tumor weight in rats [79].
		TGZ	Inhibited DMBA-induced tumor progression in rats, reduced malignancy incidence, and induced regression or stasis of total tumor volume [108]
		α-eleostearic acid	Reduced MCF7 cell viability and promoted cell apoptosis [185]
		RGZ	Inhibited PI3K/AKT pathway, inhibited proliferation of MCF7 cells [186]
		RGZ	Promoted the expression of p53 in MCF7, induced caspase 9 cleavage and DNA fragmentation, and promoted cell growth arrest and apoptosis [187]
		RGZ	Promoted target gene FasL expression, activated the cascade of caspases, and induced apoptosis [191]
		TZD	Inhibited MAPK/STAT3/AKT phosphorylation-mediated leptin signaling in MCF7 cells inhibited cell proliferation and promoted cell apoptosis [192]
		BRL49,653	Inhibited the PI3K/AKT pathway and promoted PTEN expression in MCF7 cells, inhibiting cell growth [186]
		–	PPARγ silence promoted Wnt signaling and induced transgenic mice to develope higher tumor diversity, especially ER+ ductal tumors [214]
		–	Inhibited cell proliferation and migration in vitro, inhibited tumor growth, and distant metastasis in mice [217]

4.4. PPARs and TNBC

TNBC, the most aggressive subtype of breast cancer, has no effect on hormone therapy or HER2-targeted therapy due to its lack of the three receptors. Surgery or chemotherapy, the only viable option, is a systemic therapy that causes not only physical distress but a poor prognosis for TNBC patients [222]. Therefore, it is very necessary to explore new treatment methods or target drugs to improve the prognosis of TNBC. Li et al. found that the PPARα-specific agonist fenofibrate had anti-proliferative effects on breast cancer cell lines, and the top 5 most sensitive cells are all TNBC cell lines [223]. Kwong found that fatty acid binding protein 7 (FABP7) failed to induce the efficient use of glucose to generate ATP in the TNBC cell line Hs578T during serum starvation, eventually leading to cell death. This metabolic effect of FABP7 on Hs578T cells was mediated by PPARα [224]. Studies by Stephen's group showed that PPARβ/δ activated by GW501,516 could promote the proliferation of MCF7 and T47D cells, but it had no similar effect on the TNBC cell lines MDA-MB-231 and BT-20 [143]. The expression level of PPARβ/δ in highly aggressive basal cells was significantly higher than that in luminal cells [156]. In addition, Adhikary's team found that ST247 and DG172 specifically antagonized PPARβ/δ strongly inhibited the invasion ability of MDA-MB-231 cells induced by serum and TGFβ [158]. Jiang's team found that the expression of PPARγ in the breast tissues of TNBC patients was significantly lower than that of other subtype patients, and its expression in MDA-MB-231 cells was also significantly lower than that of other breast cancer cell lines. Previous studies have reported that the PPARγ-specific agonist RGZ had antitumor effects in breast cancer. However, it did not exert significant anti-proliferative effects on MDA-MB-231 cells. RGZ combined with the demethylation agent hydralazine significantly inhibited the proliferation of MDA-MB-231 cells and promoted cell apoptosis [200]. Apaya et al. showed that epoxy-eicosatrienoic acid (EET) induced the nuclear translocation of FABP4 and FABP5 in MDA-MB-231 cells, thereby promoting the nuclear accumulation of PPARγ and affecting cell proliferation and migration [225]. These results reveal the important roles of all three subtypes of PPARs and their ligands in TNBC and suggest that more attention should be directed to drug combination therapies against TNBC.

5. Discussion

PPARs are key transcription factors in the process of fatty acid oxidative decomposition. They have a key role in nutrient metabolism and lipid homeostasis. The PPARs are involved in regulating several cellular physiological functions, consisting of cell differentiation, proliferation, metabolism, apoptosis, and other activities related to tumor formation. Several controversial reports on PPARs presented in this paper suggest that their function as tumor-promoting or tumor-suppressing factors in breast cancer still remains unclear. A number of classical signaling pathways in cells as a whole affect physiological function, such as cell carcinogenesis. The complexity of the pathways regulated by PPARs provides a one-sided explanation for their different functions in breast cancer (Figure 6). For example, both silence and constitutive activation of PPARγ enhanced Wnt signaling and promoted mammary tumorigenesis in transgenic mice [214,215]. GW501,516-activated PPARβ/δ promoted increased PDK1 activation in DMBA-induced mice [147]. The overexpression of PDK1 in mouse mammary epithelial cells in turn upregulated PPARβ/δ levels and enhanced PPARβ/δ-induced energy metabolism. However, GW501,516 treatment did not alter PDK1 protein levels [149]. Although the promoting effect of PPARβ/δ on breast cancer is partially dependent on the PDK1 signaling pathway, studies showed that PDK1 is not a target gene of PPARβ/δ [136,155,157], which further reveals the correlation between the two may be mediated by some factors in other signaling pathways. Many clinical drugs targeting PPARs (such as fibrate hypolipidemic drugs and TZD hypoglycemic drugs) can treat metabolic syndromes such as diabetes, obesity, hyperlipidemia, and cardiovascular disease. Moreover, epidemiological studies have shown that metabolic disorders are often associated with the occurrence of malignant tumors, such as breast cancer [226,227]. Therefore, PPARs remain a potential target for the prevention and treatment of breast cancer.

There are many predisposing factors for breast cancer, among which long-term estrogen exposure has been confirmed to be directly associated with the malignant proliferation, invasion, and metastasis of breast cancer cells [228]. ERs are the key factors in response to estrogen stimulation and mediate signal transduction and function in cells. Additionally, together with PPARs, they are members of the nuclear receptor superfamily. This review examined numerous reports on PPARs and found that regardless of the subtypes, the effects on ER+ and ER- breast cancer cells were different. Activated PPARα had the most significant pro-proliferation effect on ER+ MCF7 cells [104].Although the effect of PPARβ/δ on the proliferation of breast cancer cells is highly controversial, its effect on ER+ and ER- cells is indeed different [143,145]. A hallmark feature of MMTV-PPARβ/δ transgenic mice constructed by embryonic pronuclear injection developed ER+/PR+/HER2- mammary tumors, directly revealing the correlation between PPARβ/δ and ER+ ductal breast cancer [150]. PPARγ and ERα physically interacted to regulate the PI3K/AKT signaling pathway, which is involved in breast cancer cell survival and proliferation [186]. Further, MMTV-Pax8PPARγ transgenic mice produce mainly ER+ ductal breast cancer under the induction of MPA and DMBA [214]. This correlation between PPARs and ERs suggests that they can be used as synergistic targets for breast cancer clinical treatment. Consequently, the molecules and signals involved in regulating estrogen and its receptor pathways are very complex. They exhibit dynamic changes with differences in the intracellular environment. The function of PPARs in breast cancer is also disputable. Therefore, more experiments are needed for the development of common target drugs in the future.

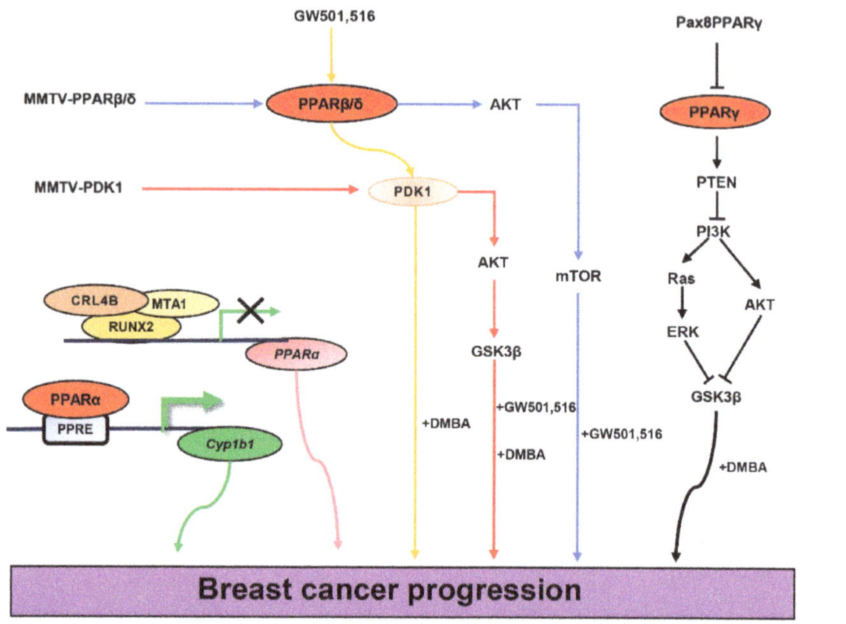

Figure 6. Schematic illustration of ligand-activated or ligand-independent PPARs affecting breast cancer progression. PPRE: peroxisome proliferator response element; *Cyp1b1*: cytochrome P450 1B1; RUNX2: Runt-related transcription factor 2; MTA1: metastasis-associated 1; CRL4B: Cullin 4B-Ring E3 ligase; PDK1: 3-phosphoinositide-dependent protein kinase 1; PTEN: phosphatase and tensin homolog; AKT: AKT serine/threonine kinase 1; GSK3β: glycogen synthase kinase 3β; mTOR: mechanistic target of rapamycin kinase; PI3K: phosphatidylinositol-4,5-bisphosphate 3-kinase; ERK: mitogen-activated protein kinase 1; DMBA: 7,12 dimethylbenzene(a)anthracene.

The selectivity and affinity of various ligands for PPARs are different between humans and other mammals. This difference might be one of the causes of the opposite results obtained from experiments in vitro and in vivo. For example, Wy-14,643, an agonist of PPARα, enhanced the transcriptional activity of the tumor-promoting factor CYP1B1 in human MCF7 cells in vitro [106]. In turn, treatment with Wy-14,643 inhibited the ability of DMBA to induce mammary tumor formation in rats [108]. The GW501,516, an agonist of PPARβ/δ, induced the proliferation of human MCF7 and T47D cells [143]. However, it inhibited the proliferation and clone formation of mouse C20 cells and promoted cell apoptosis [141]. In addition to the interspecies specificity of ligands, the presence or absence of regulatory factors such as other native natural ligands in cells or mammals may also contribute to these conflicting results [151]. In addition to acting on its specific receptors, the fact that ligands have an effect on other substances is worth investigating. In addition, the compensatory effects of living organisms and cells, ligand-related pharmacokinetic behaviors, and weak activation or antagonism of high concentrations of ligands on other subtypes are all important factors that should be considered for inclusion or exclusion in future experiments [229].

PPARα has high expression in human breast cancer cells and tissues [104,109]. The PPARβ/δ is weakly expressed or absent in human breast lobular carcinoma and ductal carcinoma [138–140], and its expression level has a negative correlation with the survival rate of breast cancer patients [142]. PPARγ is generally highly expressed in human primary and metastatic breast cancer [218]. The expression of PPARγ is inversely correlated with the histological grade of invasive breast cancer [220] and with in situ ductal breast cancer recurrence [221]. It is an independent prognostic factor in patients with ductal carcinoma.

This correlation revealed that PPARs would be potential clinical targets to prevent and treat breast cancer.

6. Conclusions

This review analyzed the roles and potential molecular mechanisms of three subtypes of PPARs in the presence or absence of ligands in breast cancer progression. In addition, the correlations between PPARs and ERs as the nuclear receptor superfamily and the investigation of the interaction between PPARs and key regulators in several signaling pathways were discussed. Furthermore, PPARs as targets for breast cancer prevention and treatment in order to provide more evidence for the synthesis of new drugs targeting PPARs or the search for new drug combination treatments. On the basis of the controversial results discovered in the review, further investigation is essential to reveal the physiological functions of PPARs.

Author Contributions: P.R. and H.W. conceived and designed the subject. B.Z. and Z.X. collected the materials. B.Z. and Z.X. wrote the paper. All authors have read and agreed to the published version of the manuscript.

Funding: This work was supported by grants (81872263) from the National Natural Science Foundation of China for Huijian Wu.

Institutional Review Board Statement: Not applicable.

Informed Consent Statement: Not applicable.

Data Availability Statement: Not applicable.

Acknowledgments: We thank Xiaoxia Shi (Dalian University of Technology, China) for revising the language of this manuscript.

Conflicts of Interest: The authors declare no conflict of interest.

References

1. Siegel, R.L.; Miller, K.D.; Fuchs, H.E.; Jemal, A. Cancer statistics, 2022. *CA Cancer J. Clin.* **2022**, *72*, 7–33. [CrossRef] [PubMed]
2. Perou, C.M.; Sørlie, T.; Eisen, M.B.; van de Rijn, M.; Jeffrey, S.S.; Rees, C.A.; Pollack, J.R.; Ross, D.T.; Johnsen, H.; Akslen, L.A.; et al. Molecular portraits of human breast tumours. *Nature* **2000**, *406*, 747–752. [CrossRef] [PubMed]
3. Sorlie, T.; Tibshirani, R.; Parker, J.; Hastie, T.; Marron, J.S.; Nobel, A.; Deng, S.; Johnsen, H.; Pesich, R.; Geisler, S.; et al. Repeated observation of breast tumor subtypes in independent gene expression data sets. *Proc. Natl. Acad. Sci. USA* **2003**, *100*, 8418–8423. [CrossRef]
4. Issemann, I.; Green, S. Activation of a member of the steroid hormone receptor superfamily by peroxisome proliferators. *Nature* **1990**, *347*, 645–650. [CrossRef] [PubMed]
5. Escher, P.; Wahli, W. Peroxisome proliferator-activated receptors: Insight into multiple cellular functions. *Mutat. Res.* **2000**, *448*, 121–138. [CrossRef] [PubMed]
6. Feige, J.N.; Gelman, L.; Michalik, L.; Desvergne, B.; Wahli, W. From molecular action to physiological outputs: Peroxisome proliferator-activated receptors are nuclear receptors at the crossroads of key cellular functions. *Prog. Lipid Res.* **2006**, *45*, 120–159. [CrossRef] [PubMed]
7. Polvani, S.; Tarocchi, M.; Tempesti, S.; Bencini, L.; Galli, A. Peroxisome proliferator activated receptors at the crossroad of obesity, diabetes, and pancreatic cancer. *World J. Gastroenterol.* **2016**, *22*, 2441–2459. [CrossRef] [PubMed]
8. Wang, Y.-X. PPARs: Diverse regulators in energy metabolism and metabolic diseases. *Cell Res.* **2010**, *20*, 124–137. [CrossRef]
9. Han, L.; Shen, W.-J.; Bittner, S.; Kraemer, F.B.; Azhar, S. PPARs: Regulators of metabolism and as therapeutic targets in cardiovascular disease. Part I: PPAR-α. *Future Cardiol.* **2017**, *13*, 259–278. [CrossRef]
10. Han, L.; Shen, W.-J.; Bittner, S.; Kraemer, F.B.; Azhar, S. PPARs: Regulators of metabolism and as therapeutic targets in cardiovascular disease. Part II: PPAR-β/δ and PPAR-γ. *Future Cardiol.* **2017**, *13*, 279–296. [CrossRef]
11. Burns, K.A.; Vanden Heuvel, J.P. Modulation of PPAR activity via phosphorylation. *Biochim. Biophys. Acta* **2007**, *1771*, 952–960. [CrossRef] [PubMed]
12. Castillo, G.; Brun, R.P.; Rosenfield, J.K.; Hauser, S.; Park, C.W.; Troy, A.E.; Wright, M.E.; Spiegelman, B.M. An adipogenic cofactor bound by the differentiation domain of PPARgamma. *EMBO J.* **1999**, *18*, 3676–3687. [CrossRef] [PubMed]
13. Hummasti, S.; Tontonoz, P. The peroxisome proliferator-activated receptor N-terminal domain controls isotype-selective gene expression and adipogenesis. *Mol. Endocrinol.* **2006**, *20*, 1261–1275. [CrossRef] [PubMed]
14. Bugge, A.; Grøntved, L.; Aagaard, M.M.; Borup, R.; Mandrup, S. The PPARgamma2 A/B-domain plays a gene-specific role in transactivation and cofactor recruitment. *Mol. Endocrinol.* **2009**, *23*, 794–808. [CrossRef] [PubMed]

15. Gampe, R.T.J.; Montana, V.G.; Lambert, M.H.; Miller, A.B.; Bledsoe, R.K.; Milburn, M.V.; Kliewer, S.A.; Willson, T.M.; Xu, H.E. Asymmetry in the PPARgamma/RXRalpha crystal structure reveals the molecular basis of heterodimerization among nuclear receptors. *Mol. Cell* **2000**, *5*, 545–555. [CrossRef] [PubMed]
16. Nolte, R.T.; Wisely, G.B.; Westin, S.; Cobb, J.E.; Lambert, M.H.; Kurokawa, R.; Rosenfeld, M.G.; Willson, T.M.; Glass, C.K.; Milburn, M.V. Ligand binding and co-activator assembly of the peroxisome proliferator-activated receptor-gamma. *Nature* **1998**, *395*, 137–143. [CrossRef] [PubMed]
17. Sheu, S.-H.; Kaya, T.; Waxman, D.J.; Vajda, S. Exploring the binding site structure of the PPAR gamma ligand-binding domain by computational solvent mapping. *Biochemistry* **2005**, *44*, 1193–1209. [CrossRef]
18. Xu, H.E.; Lambert, M.H.; Montana, V.G.; Parks, D.J.; Blanchard, S.G.; Brown, P.J.; Sternbach, D.D.; Lehmann, J.M.; Wisely, G.B.; Willson, T.M.; et al. Molecular recognition of fatty acids by peroxisome proliferator-activated receptors. *Mol. Cell* **1999**, *3*, 397–403. [CrossRef] [PubMed]
19. Shao, D.; Rangwala, S.M.; Bailey, S.T.; Krakow, S.L.; Reginato, M.J.; Lazar, M.A. Interdomain communication regulating ligand binding by PPAR-gamma. *Nature* **1998**, *396*, 377–380. [CrossRef]
20. Deeb, S.S.; Fajas, L.; Nemoto, M.; Pihlajamäki, J.; Mykkänen, L.; Kuusisto, J.; Laakso, M.; Fujimoto, W.; Auwerx, J. A Pro12Ala substitution in PPARgamma2 associated with decreased receptor activity, lower body mass index and improved insulin sensitivity. *Nat. Genet.* **1998**, *20*, 284–287. [CrossRef]
21. Lee, C.-H.; Chawla, A.; Urbiztondo, N.; Liao, D.; Boisvert, W.A.; Evans, R.M.; Curtiss, L.K. Transcriptional repression of atherogenic inflammation: Modulation by PPARdelta. *Science* **2003**, *302*, 453–457. [CrossRef] [PubMed]
22. Hall, J.M.; McDonnell, D.P. Coregulators in nuclear estrogen receptor action: From concept to therapeutic targeting. *Mol. Interv.* **2005**, *5*, 343–357. [CrossRef] [PubMed]
23. Göttlicher, M.; Widmark, E.; Li, Q.; Gustafsson, J.A. Fatty acids activate a chimera of the clofibric acid-activated receptor and the glucocorticoid receptor. *Proc. Natl. Acad. Sci. USA* **1992**, *89*, 4653–4657. [CrossRef] [PubMed]
24. Dreyer, C.; Krey, G.; Keller, H.; Givel, F.; Helftenbein, G.; Wahli, W. Control of the peroxisomal beta-oxidation pathway by a novel family of nuclear hormone receptors. *Cell* **1992**, *68*, 879–887. [CrossRef] [PubMed]
25. Ziouzenkova, O.; Perrey, S.; Asatryan, L.; Hwang, J.; MacNaul, K.L.; Moller, D.E.; Rader, D.J.; Sevanian, A.; Zechner, R.; Hoefler, G.; et al. Lipolysis of triglyceride-rich lipoproteins generates PPAR ligands: Evidence for an antiinflammatory role for lipoprotein lipase. *Proc. Natl. Acad. Sci. USA* **2003**, *100*, 2730–2735. [CrossRef] [PubMed]
26. Chawla, A.; Lee, C.-H.; Barak, Y.; He, W.; Rosenfeld, J.; Liao, D.; Han, J.; Kang, H.; Evans, R.M. PPARdelta is a very low-density lipoprotein sensor in macrophages. *Proc. Natl. Acad. Sci. USA* **2003**, *100*, 1268–1273. [CrossRef]
27. Yu, K.; Bayona, W.; Kallen, C.B.; Harding, H.P.; Ravera, C.P.; McMahon, G.; Brown, M.; Lazar, M.A. Differential activation of peroxisome proliferator-activated receptors by eicosanoids. *J. Biol. Chem.* **1995**, *270*, 23975–23983. [CrossRef]
28. Willson, T.M.; Brown, P.J.; Sternbach, D.D.; Henke, B.R. The PPARs: From orphan receptors to drug discovery. *J. Med. Chem.* **2000**, *43*, 527–550. [CrossRef]
29. Shearer, B.G.; Steger, D.J.; Way, J.M.; Stanley, T.B.; Lobe, D.C.; Grillot, D.A.; Iannone, M.A.; Lazar, M.A.; Willson, T.M.; Billin, A.N. Identification and characterization of a selective peroxisome proliferator-activated receptor beta/delta (NR1C2) antagonist. *Mol. Endocrinol.* **2008**, *22*, 523–529. [CrossRef]
30. Palkar, P.S.; Borland, M.G.; Naruhn, S.; Ferry, C.H.; Lee, C.; Sk, U.H.; Sharma, A.K.; Amin, S.; Murray, I.A.; Anderson, C.R.; et al. Cellular and pharmacological selectivity of the peroxisome proliferator-activated receptor-beta/delta antagonist GSK3787. *Mol. Pharmacol.* **2010**, *78*, 419–430. [CrossRef]
31. Dowell, P.; Peterson, V.J.; Zabriskie, T.M.; Leid, M. Ligand-induced peroxisome proliferator-activated receptor alpha conformational change. *J. Biol. Chem.* **1997**, *272*, 2013–2020. [CrossRef] [PubMed]
32. Tien, E.S.; Hannon, D.B.; Thompson, J.T.; Vanden Heuvel, J.P. Examination of Ligand-Dependent Coactivator Recruitment by Peroxisome Proliferator-Activated Receptor-alpha (PPARalpha). *PPAR Res.* **2006**, *2006*, 69612. [CrossRef] [PubMed]
33. Sumanasekera, W.K.; Tien, E.S.; Turpey, R.; Vanden Heuvel, J.P.; Perdew, G.H. Evidence that peroxisome proliferator-activated receptor alpha is complexed with the 90-kDa heat shock protein and the hepatitis virus B X-associated protein 2. *J. Biol. Chem.* **2003**, *278*, 4467–4473. [CrossRef] [PubMed]
34. Sumanasekera, W.K.; Tien, E.S.; Davis, J.W., 2nd; Turpey, R.; Perdew, G.H.; Vanden Heuvel, J.P. Heat shock protein-90 (Hsp90) acts as a repressor of peroxisome proliferator-activated receptor-alpha (PPARalpha) and PPARbeta activity. *Biochemistry* **2003**, *42*, 10726–10735. [CrossRef]
35. Akiyama, T.E.; Baumann, C.T.; Sakai, S.; Hager, G.L.; Gonzalez, F.J. Selective intranuclear redistribution of PPAR isoforms by RXR alpha. *Mol. Endocrinol.* **2002**, *16*, 707–721. [CrossRef]
36. Patel, H.; Truant, R.; Rachubinski, R.A.; Capone, J.P. Activity and subcellular compartmentalization of peroxisome proliferator-activated receptor alpha are altered by the centrosome-associated protein CAP350. *J. Cell Sci.* **2005**, *118*, 175–186. [CrossRef]
37. Malek, M.A.; Hoang, M.-H.; Jia, Y.; Lee, J.H.; Jun, H.J.; Lee, D.-H.; Lee, H.J.; Lee, C.; Lee, M.K.; Hwang, B.Y.; et al. Ombuin-3-O-β-D-glucopyranoside from Gynostemma pentaphyllum is a dual agonistic ligand of peroxisome proliferator-activated receptors α and δ/β. *Biochem. Biophys. Res. Commun.* **2013**, *430*, 1322–1328. [CrossRef]
38. Tenenbaum, A.; Motro, M.; Fisman, E.Z. Dual and pan-peroxisome proliferator-activated receptors (PPAR) co-agonism: The bezafibrate lessons. *Cardiovasc. Diabetol.* **2005**, *4*, 14. [CrossRef]

39. Nagy, L.; Tontonoz, P.; Alvarez, J.G.; Chen, H.; Evans, R.M. Oxidized LDL regulates macrophage gene expression through ligand activation of PPARgamma. *Cell* **1998**, *93*, 229–240. [CrossRef]
40. Shureiqi, I.; Jiang, W.; Zuo, X.; Wu, Y.; Stimmel, J.B.; Leesnitzer, L.M.; Morris, J.S.; Fan, H.-Z.; Fischer, S.M.; Lippman, S.M. The 15-lipoxygenase-1 product 13-S-hydroxyoctadecadienoic acid down-regulates PPAR-delta to induce apoptosis in colorectal cancer cells. *Proc. Natl. Acad. Sci. USA* **2003**, *100*, 9968–9973. [CrossRef]
41. Coleman, J.D.; Prabhu, K.S.; Thompson, J.T.; Reddy, P.S.; Peters, J.M.; Peterson, B.R.; Reddy, C.C.; Vanden Heuvel, J.P. The oxidative stress mediator 4-hydroxynonenal is an intracellular agonist of the nuclear receptor peroxisome proliferator-activated receptor-beta/delta (PPARbeta/delta). *Free Radic. Biol. Med.* **2007**, *42*, 1155–1164. [CrossRef] [PubMed]
42. Rubenstrunk, A.; Hanf, R.; Hum, D.W.; Fruchart, J.-C.; Staels, B. Safety issues and prospects for future generations of PPAR modulators. *Biochim. Biophys. Acta* **2007**, *1771*, 1065–1081. [CrossRef] [PubMed]
43. Tenenbaum, A.; Boyko, V.; Fisman, E.Z.; Goldenberg, I.; Adler, Y.; Feinberg, M.S.; Motro, M.; Tanne, D.; Shemesh, J.; Schwammenthal, E.; et al. Does the lipid-lowering peroxisome proliferator-activated receptors ligand bezafibrate prevent colon cancer in patients with coronary artery disease? *Cardiovasc. Diabetol.* **2008**, *7*, 18. [CrossRef]
44. Yaturu, S.; Bryant, B.; Jain, S.K. Thiazolidinedione treatment decreases bone mineral density in type 2 diabetic men. *Diabetes Care* **2007**, *30*, 1574–1576. [CrossRef] [PubMed]
45. Grey, A.; Bolland, M.; Gamble, G.; Wattie, D.; Horne, A.; Davidson, J.; Reid, I.R. The peroxisome proliferator-activated receptor-gamma agonist rosiglitazone decreases bone formation and bone mineral density in healthy postmenopausal women: A randomized, controlled trial. *J. Clin. Endocrinol. Metab.* **2007**, *92*, 1305–1310. [CrossRef]
46. Schwartz, A.V.; Sellmeyer, D.E. Thiazolidinedione therapy gets complicated: Is bone loss the price of improved insulin resistance? *Diabetes Care* **2007**, *30*, 1670–1671. [CrossRef]
47. Schwartz, A.V.; Sellmeyer, D.E.; Vittinghoff, E.; Palermo, L.; Lecka-Czernik, B.; Feingold, K.R.; Strotmeyer, E.S.; Resnick, H.E.; Carbone, L.; Beamer, B.A.; et al. Thiazolidinedione use and bone loss in older diabetic adults. *J. Clin. Endocrinol. Metab.* **2006**, *91*, 3349–3354. [CrossRef]
48. Still, K.; Grabowski, P.; Mackie, I.; Perry, M.; Bishop, N. The peroxisome proliferator activator receptor alpha/delta agonists linoleic acid and bezafibrate upregulate osteoblast differentiation and induce periosteal bone formation in vivo. *Calcif. Tissue Int.* **2008**, *83*, 285–292. [CrossRef]
49. Rakhshandehroo, M.; Knoch, B.; Müller, M.; Kersten, S. Peroxisome proliferator-activated receptor alpha target genes. *PPAR Res.* **2010**, *2010*, 612089. [CrossRef]
50. Santilli, A.A.; Scotese, A.C.; Tomarelli, R.M. A potent antihypercholesterolemic agent: (4-chloro-6-(2,3-xylidino)-2-pyrimidinylthio) acetic acid (Wy-14643). *Experientia* **1974**, *30*, 1110–1111. [CrossRef]
51. Reddy, J.K.; Moody, D.E.; Azarnoff, D.L.; Tomarelli, R.M. Hepatic effects of some [4-chloro-6-(2,3-xylidino)-2-pyrimidinylthio] acetic acid (WY-14,643) analogs in the mouse. *Arch. Int. Pharmacodyn. Ther.* **1977**, *225*, 51–57.
52. Kliewer, S.A.; Forman, B.M.; Blumberg, B.; Ong, E.S.; Borgmeyer, U.; Mangelsdorf, D.J.; Umesono, K.; Evans, R.M. Differential expression and activation of a family of murine peroxisome proliferator-activated receptors. *Proc. Natl. Acad. Sci. USA* **1994**, *91*, 7355–7359. [CrossRef] [PubMed]
53. Fan, S.; Gao, Y.; Qu, A.; Jiang, Y.; Li, H.; Xie, G.; Yao, X.; Yang, X.; Zhu, S.; Yagai, T.; et al. YAP-TEAD mediates PPAR α-induced hepatomegaly and liver regeneration in mice. *Hepatology* **2022**, *75*, 74–88. [CrossRef] [PubMed]
54. Brown, P.J.; Winegar, D.A.; Plunket, K.D.; Moore, L.B.; Lewis, M.C.; Wilson, J.G.; Sundseth, S.S.; Koble, C.S.; Wu, Z.; Chapman, J.M.; et al. A ureido-thioisobutyric acid (GW9578) is a subtype-selective PPARalpha agonist with potent lipid-lowering activity. *J. Med. Chem.* **1999**, *42*, 3785–3788. [CrossRef] [PubMed]
55. Brown, P.J.; Stuart, L.W.; Hurley, K.P.; Lewis, M.C.; Winegar, D.A.; Wilson, J.G.; Wilkison, W.O.; Ittoop, O.R.; Willson, T.M. Identification of a subtype selective human PPARα agonist through parallel-array synthesis. *Bioorg. Med. Chem. Lett.* **2001**, *11*, 1225–1227. [CrossRef]
56. Xu, H.E.; Stanley, T.B.; Montana, V.G.; Lambert, M.H.; Shearer, B.G.; Cobb, J.E.; McKee, D.D.; Galardi, C.M.; Plunket, K.D.; Nolte, R.T.; et al. Structural basis for antagonist-mediated recruitment of nuclear co-repressors by PPARalpha. *Nature* **2002**, *415*, 813–817. [CrossRef]
57. Guhlmann, A.; Keppler, A.; Kästner, S.; Krieter, H.; Brückner, U.B.; Messmer, K.; Keppler, D. Prevention of endogenous leukotriene production during anaphylaxis in the guinea pig by an inhibitor of leukotriene biosynthesis (MK-886) but not by dexamethasone. *J. Exp. Med.* **1989**, *170*, 1905–1918. [CrossRef]
58. Datta, K.; Biswal, S.S.; Kehrer, J.P. The 5-lipoxygenase-activating protein (FLAP) inhibitor, MK886, induces apoptosis independently of FLAP. *Biochem. J.* **1999**, *340*, 371–375. [CrossRef]
59. Kehrer, J.P.; Biswal, S.S.; La, E.; Thuillier, P.; Datta, K.; Fischer, S.M.; Vanden Heuvel, J.P. Inhibition of peroxisome-proliferator-activated receptor (PPAR)alpha by MK886. *Biochem. J.* **2001**, *356*, 899–906. [CrossRef]
60. Leibowitz, M.D.; Fiévet, C.; Hennuyer, N.; Peinado-Onsurbe, J.; Duez, H.; Bergera, J.; Cullinan, C.A.; Sparrow, C.P.; Baffic, J.; Berger, G.D.; et al. Activation of PPARdelta alters lipid metabolism in db/db mice. *FEBS Lett.* **2000**, *473*, 333–336. [CrossRef]
61. Westergaard, M.; Henningsen, J.; Svendsen, M.L.; Johansen, C.; Jensen, U.B.; Schrøder, H.D.; Kratchmarova, I.; Berge, R.K.; Iversen, L.; Bolund, L.; et al. Modulation of keratinocyte gene expression and differentiation by PPAR-selective ligands and tetradecylthioacetic acid. *J. Investig. Dermatol.* **2001**, *116*, 702–712. [CrossRef]

62. Oliver, W.R.J.; Shenk, J.L.; Snaith, M.R.; Russell, C.S.; Plunket, K.D.; Bodkin, N.L.; Lewis, M.C.; Winegar, D.A.; Sznaidman, M.L.; Lambert, M.H.; et al. A selective peroxisome proliferator-activated receptor delta agonist promotes reverse cholesterol transport. *Proc. Natl. Acad. Sci. USA* **2001**, *98*, 5306–5311. [CrossRef] [PubMed]
63. Pelton, P. GW-501516 GlaxoSmithKline/Ligand. *Curr. Opin. Investig. Drugs* **2006**, *7*, 360–370. [PubMed]
64. Iwaisako, K.; Haimerl, M.; Paik, Y.-H.; Taura, K.; Kodama, Y.; Sirlin, C.; Yu, E.; Yu, R.T.; Downes, M.; Evans, R.M.; et al. Protection from liver fibrosis by a peroxisome proliferator-activated receptor δ agonist. *Proc. Natl. Acad. Sci. USA* **2012**, *109*, E1369–E1376. [CrossRef] [PubMed]
65. Lamers, C.; Schubert-Zsilavecz, M.; Merk, D. Therapeutic modulators of peroxisome proliferator-activated receptors (PPAR): A patent review (2008-present). *Expert Opin. Ther. Pat.* **2012**, *22*, 803–841. [CrossRef] [PubMed]
66. Lin, Y.; Zhu, X.; McLntee, F.L.; Xiao, H.; Zhang, J.; Fu, M.; Chen, Y.E. Interferon regulatory factor-1 mediates PPARgamma-induced apoptosis in vascular smooth muscle cells. *Arterioscler. Thromb. Vasc. Biol.* **2004**, *24*, 257–263. [CrossRef] [PubMed]
67. Choi, Y.-J.; Roberts, B.K.; Wang, X.; Geaney, J.C.; Naim, S.; Wojnoonski, K.; Karpf, D.B.; Krauss, R.M. Effects of the PPAR-δ agonist MBX-8025 on atherogenic dyslipidemia. *Atherosclerosis* **2012**, *220*, 470–476. [CrossRef]
68. Billin, A.N. PPAR-beta/delta agonists for Type 2 diabetes and dyslipidemia: An adopted orphan still looking for a home. *Expert Opin. Investig. Drugs* **2008**, *17*, 1465–1471. [CrossRef] [PubMed]
69. Seimandi, M.; Lemaire, G.; Pillon, A.; Perrin, A.; Carlavan, I.; Voegel, J.J.; Vignon, F.; Nicolas, J.-C.; Balaguer, P. Differential responses of PPARalpha, PPARdelta, and PPARgamma reporter cell lines to selective PPAR synthetic ligands. *Anal. Biochem.* **2005**, *344*, 8–15. [CrossRef]
70. Zaveri, N.T.; Sato, B.G.; Jiang, F.; Calaoagan, J.; Laderoute, K.R.; Murphy, B.J. A novel peroxisome proliferator-activated receptor delta antagonist, SR13904, has anti-proliferative activity in human cancer cells. *Cancer Biol. Ther.* **2009**, *8*, 1252–1261. [CrossRef]
71. Shearer, B.G.; Wiethe, R.W.; Ashe, A.; Billin, A.N.; Way, J.M.; Stanley, T.B.; Wagner, C.D.; Xu, R.X.; Leesnitzer, L.M.; Merrihew, R.V.; et al. Identification and characterization of 4-chloro-N-(2-{[5-trifluoromethyl)-2-pyridyl]sulfonyl}ethyl)benzamide (GSK3787), a selective and irreversible peroxisome proliferator-activated receptor delta (PPARdelta) antagonist. *J. Med. Chem.* **2010**, *53*, 1857–1861. [CrossRef] [PubMed]
72. Lieber, S.; Scheer, F.; Meissner, W.; Naruhn, S.; Adhikary, T.; Müller-Brüsselbach, S.; Diederich, W.E.; Müller, R. (Z)-2-(2-bromophenyl)-3-{[4-(1-methyl-piperazine)amino]phenyl}acrylonitrile (DG172): An orally bioavailable PPARβ/δ-selective ligand with inverse agonistic properties. *J. Med. Chem.* **2012**, *55*, 2858–2868. [CrossRef] [PubMed]
73. Levi, L.; Lobo, G.; Doud, M.K.; von Lintig, J.; Seachrist, D.; Tochtrop, G.P.; Noy, N. Genetic ablation of the fatty acid-binding protein FABP5 suppresses HER2-induced mammary tumorigenesis. *Cancer Res.* **2013**, *73*, 4770–4780. [CrossRef] [PubMed]
74. Naruhn, S.; Toth, P.M.; Adhikary, T.; Kaddatz, K.; Pape, V.; Dörr, S.; Klebe, G.; Müller-Brüsselbach, S.; Diederich, W.E.; Müller, R. High-affinity peroxisome proliferator-activated receptor β/δ-specific ligands with pure antagonistic or inverse agonistic properties. *Mol. Pharmacol.* **2011**, *80*, 828–838. [CrossRef]
75. Lehmann, J.M.; Moore, L.B.; Smith-Oliver, T.A.; Wilkison, W.O.; Willson, T.M.; Kliewer, S.A. An antidiabetic thiazolidinedione is a high affinity ligand for peroxisome proliferator-activated receptor gamma (PPAR gamma). *J. Biol. Chem.* **1995**, *270*, 12953–12956. [CrossRef] [PubMed]
76. Mirza, A.Z.; Althagafi, I.I.; Shamshad, H. Role of PPAR receptor in different diseases and their ligands: Physiological importance and clinical implications. *Eur. J. Med. Chem.* **2019**, *166*, 502–513. [CrossRef]
77. Elbrecht, A.; Chen, Y.; Adams, A.; Berger, J.; Griffin, P.; Klatt, T.; Zhang, B.; Menke, J.; Zhou, G.; Smith, R.G.; et al. L-764406 is a partial agonist of human peroxisome proliferator-activated receptor gamma. The role of Cys313 in ligand binding. *J. Biol. Chem.* **1999**, *274*, 7913–7922. [CrossRef]
78. Oberfield, J.L.; Collins, J.L.; Holmes, C.P.; Goreham, D.M.; Cooper, J.P.; Cobb, J.E.; Lenhard, J.M.; Hull-Ryde, E.A.; Mohr, C.P.; Blanchard, S.G.; et al. A peroxisome proliferator-activated receptor gamma ligand inhibits adipocyte differentiation. *Proc. Natl. Acad. Sci. USA* **1999**, *96*, 6102–6106. [CrossRef]
79. Suh, N.; Wang, Y.; Williams, C.R.; Risingsong, R.; Gilmer, T.; Willson, T.M.; Sporn, M.B. A new ligand for the peroxisome proliferator-activated receptor-gamma (PPAR-gamma), GW7845, inhibits rat mammary carcinogenesis. *Cancer Res.* **1999**, *59*, 5671–5673.
80. Miyahara, T.; Schrum, L.; Rippe, R.; Xiong, S.; Yee, H.F.J.; Motomura, K.; Anania, F.A.; Willson, T.M.; Tsukamoto, H. Peroxisome proliferator-activated receptors and hepatic stellate cell activation. *J. Biol. Chem.* **2000**, *275*, 35715–35722. [CrossRef]
81. Leesnitzer, L.M.; Parks, D.J.; Bledsoe, R.K.; Cobb, J.E.; Collins, J.L.; Consler, T.G.; Davis, R.G.; Hull-Ryde, E.A.; Lenhard, J.M.; Patel, L.; et al. Functional consequences of cysteine modification in the ligand binding sites of peroxisome proliferator activated receptors by GW9662. *Biochemistry* **2002**, *41*, 6640–6650. [CrossRef]
82. Ji, J.; Xue, T.-F.; Guo, X.-D.; Yang, J.; Guo, R.-B.; Wang, J.; Huang, J.-Y.; Zhao, X.-J.; Sun, X.-L. Antagonizing peroxisome proliferator-activated receptor γ facilitates M1-to-M2 shift of microglia by enhancing autophagy via the LKB1-AMPK signaling pathway. *Aging Cell* **2018**, *17*, e12774. [CrossRef] [PubMed]
83. Ghoochani, A.; Shabani, K.; Peymani, M.; Ghaedi, K.; Karamali, F.; Karbalaei, K.; Tanhaie, S.; Salamian, A.; Esmaeili, A.; Valian-Borujeni, S.; et al. The influence of peroxisome proliferator-activated receptor γ(1) during differentiation of mouse embryonic stem cells to neural cells. *Differentiation* **2012**, *83*, 60–67. [CrossRef] [PubMed]

84. Wright, H.M.; Clish, C.B.; Mikami, T.; Hauser, S.; Yanagi, K.; Hiramatsu, R.; Serhan, C.N.; Spiegelman, B.M. A synthetic antagonist for the peroxisome proliferator-activated receptor gamma inhibits adipocyte differentiation. *J. Biol. Chem.* **2000**, *275*, 1873–1877. [CrossRef] [PubMed]
85. Berger, J.; Moller, D.E. The mechanisms of action of PPARs. *Annu. Rev. Med.* **2002**, *53*, 409–435. [CrossRef] [PubMed]
86. Fang, L.; Zhang, M.; Li, Y.; Liu, Y.; Cui, Q.; Wang, N. PPARgene: A Database of Experimentally Verified and Computationally Predicted PPAR Target Genes. *PPAR Res.* **2016**, *2016*, 6042162. [CrossRef] [PubMed]
87. Peters, J.M.; Foreman, J.E.; Gonzalez, F.J. Dissecting the role of peroxisome proliferator-activated receptor-β/δ (PPARβ/δ) in colon, breast, and lung carcinogenesis. *Cancer Metastasis Rev.* **2011**, *30*, 619–640. [CrossRef]
88. Tontonoz, P.; Spiegelman, B.M. Fat and beyond: The diverse biology of PPARgamma. *Annu. Rev. Biochem.* **2008**, *77*, 289–312. [CrossRef]
89. Escher, P.; Braissant, O.; Basu-Modak, S.; Michalik, L.; Wahli, W.; Desvergne, B. Rat PPARs: Quantitative analysis in adult rat tissues and regulation in fasting and refeeding. *Endocrinology* **2001**, *142*, 4195–4202. [CrossRef]
90. Pyper, S.R.; Viswakarma, N.; Yu, S.; Reddy, J.K. PPARalpha: Energy combustion, hypolipidemia, inflammation and cancer. *Nucl. Recept. Signal.* **2010**, *8*, e002. [CrossRef]
91. Staels, B.; Dallongeville, J.; Auwerx, J.; Schoonjans, K.; Leitersdorf, E.; Fruchart, J.C. Mechanism of action of fibrates on lipid and lipoprotein metabolism. *Circulation* **1998**, *98*, 2088–2093. [CrossRef] [PubMed]
92. Balfour, J.A.; McTavish, D.; Heel, R.C. Fenofibrate. A review of its pharmacodynamic and pharmacokinetic properties and therapeutic use in dyslipidaemia. *Drugs* **1990**, *40*, 260–290. [CrossRef] [PubMed]
93. Reddy, J.K.; Azarnoff, D.L.; Hignite, C.E. Hypolipidaemic hepatic peroxisome proliferators form a novel class of chemical carcinogens. *Nature* **1980**, *283*, 397–398. [CrossRef] [PubMed]
94. Peters, J.M.; Cattley, R.C.; Gonzalez, F.J. Role of PPAR alpha in the mechanism of action of the nongenotoxic carcinogen and peroxisome proliferator Wy-14,643. *Carcinogenesis* **1997**, *18*, 2029–2033. [CrossRef]
95. Hays, T.; Rusyn, I.; Burns, A.M.; Kennett, M.J.; Ward, J.M.; Gonzalez, F.J.; Peters, J.M. Role of peroxisome proliferator-activated receptor-alpha (PPARalpha) in bezafibrate-induced hepatocarcinogenesis and cholestasis. *Carcinogenesis* **2005**, *26*, 219–227. [CrossRef]
96. Peters, J.M.; Cheung, C.; Gonzalez, F.J. Peroxisome proliferator-activated receptor-alpha and liver cancer: Where do we stand? *J. Mol. Med.* **2005**, *83*, 774–785. [CrossRef]
97. Shah, Y.M.; Morimura, K.; Yang, Q.; Tanabe, T.; Takagi, M.; Gonzalez, F.J. Peroxisome proliferator-activated receptor alpha regulates a microRNA-mediated signaling cascade responsible for hepatocellular proliferation. *Mol. Cell. Biol.* **2007**, *27*, 4238–4247. [CrossRef]
98. Cheung, C.; Akiyama, T.E.; Ward, J.M.; Nicol, C.J.; Feigenbaum, L.; Vinson, C.; Gonzalez, F.J. Diminished hepatocellular proliferation in mice humanized for the nuclear receptor peroxisome proliferator-activated receptor alpha. *Cancer Res.* **2004**, *64*, 3849–3854. [CrossRef]
99. Morimura, K.; Cheung, C.; Ward, J.M.; Reddy, J.K.; Gonzalez, F.J. Differential susceptibility of mice humanized for peroxisome proliferator-activated receptor alpha to Wy-14,643-induced liver tumorigenesis. *Carcinogenesis* **2006**, *27*, 1074–1080. [CrossRef]
100. Holland, C.M.; Saidi, S.A.; Evans, A.L.; Sharkey, A.M.; Latimer, J.A.; Crawford, R.A.F.; Charnock-Jones, D.S.; Print, C.G.; Smith, S.K. Transcriptome analysis of endometrial cancer identifies peroxisome proliferator-activated receptors as potential therapeutic targets. *Mol. Cancer Ther.* **2004**, *3*, 993–1001. [CrossRef]
101. Nickkho-Amiry, M.; McVey, R.; Holland, C. Peroxisome proliferator-activated receptors modulate proliferation and angiogenesis in human endometrial carcinoma. *Mol. Cancer Res.* **2012**, *10*, 441–453. [CrossRef] [PubMed]
102. Jiao, H.; Zhao, B. Cytotoxic effect of peroxisome proliferator fenofibrate on human HepG2 hepatoma cell line and relevant mechanisms. *Toxicol. Appl. Pharmacol.* **2002**, *185*, 172–179. [CrossRef] [PubMed]
103. Grabacka, M.; Plonka, P.M.; Urbanska, K.; Reiss, K. Peroxisome proliferator-activated receptor alpha activation decreases metastatic potential of melanoma cells in vitro via down-regulation of Akt. *Clin. Cancer Res.* **2006**, *12*, 3028–3036. [CrossRef] [PubMed]
104. Chang, N.-W.; Wu, C.-T.; Chen, D.-R.; Yeh, C.-Y.; Lin, C. High levels of arachidonic acid and peroxisome proliferator-activated receptor-alpha in breast cancer tissues are associated with promoting cancer cell proliferation. *J. Nutr. Biochem.* **2013**, *24*, 274–281. [CrossRef] [PubMed]
105. Cui, J.; Meng, Q.; Zhang, X.; Cui, Q.; Zhou, W.; Li, S. Design and Synthesis of New α-Naphthoflavones as Cytochrome P450 (CYP) 1B1 Inhibitors To Overcome Docetaxel-Resistance Associated with CYP1B1 Overexpression. *J. Med. Chem.* **2015**, *58*, 3534–3547. [CrossRef]
106. Hwang, Y.P.; Won, S.S.; Jin, S.W.; Lee, G.H.; Pham, T.H.; Choi, J.H.; Kang, K.W.; Jeong, H.G. WY-14643 Regulates CYP1B1 Expression through Peroxisome Proliferator-Activated Receptor α-Mediated Signaling in Human Breast Cancer Cells. *Int. J. Mol. Sci.* **2019**, *20*, 5928. [CrossRef]
107. Castelli, V.; Catanesi, M.; Alfonsetti, M.; Laezza, C.; Lombardi, F.; Cinque, B.; Cifone, M.G.; Ippoliti, R.; Benedetti, E.; Cimini, A.; et al. PPARα-Selective Antagonist GW6471 Inhibits Cell Growth in Breast Cancer Stem Cells Inducing Energy Imbalance and Metabolic Stress. *Biomedicines* **2021**, *9*, 127. [CrossRef]

108. Pighetti, G.M.; Novosad, W.; Nicholson, C.; Hitt, D.C.; Hansens, C.; Hollingsworth, A.B.; Lerner, M.L.; Brackett, D.; Lightfoot, S.A.; Gimble, J.M. Therapeutic treatment of DMBA-induced mammary tumors with PPAR ligands. *Anticancer Res.* **2001**, *21*, 825–829.
109. Chandran, K.; Goswami, S.; Sharma-Walia, N. Implications of a peroxisome proliferator-activated receptor alpha (PPARα) ligand clofibrate in breast cancer. *Oncotarget* **2016**, *7*, 15577–15599. [CrossRef]
110. Yin, X.; Teng, X.; Ma, T.; Yang, T.; Zhang, J.; Huo, M.; Liu, W.; Yang, Y.; Yuan, B.; Yu, H.; et al. RUNX2 recruits the NuRD(MTA1)/CRL4B complex to promote breast cancer progression and bone metastasis. *Cell Death Differ.* **2022**, *29*, 2203–2217. [CrossRef]
111. Shi, Y.; Hon, M.; Evans, R.M. The peroxisome proliferator-activated receptor delta, an integrator of transcriptional repression and nuclear receptor signaling. *Proc. Natl. Acad. Sci. USA* **2002**, *99*, 2613–2618. [CrossRef] [PubMed]
112. Schmidt, A.; Endo, N.; Rutledge, S.J.; Vogel, R.; Shinar, D.; Rodan, G.A. Identification of a new member of the steroid hormone receptor superfamily that is activated by a peroxisome proliferator and fatty acids. *Mol. Endocrinol.* **1992**, *6*, 1634–1641. [CrossRef] [PubMed]
113. Amri, E.Z.; Bonino, F.; Ailhaud, G.; Abumrad, N.A.; Grimaldi, P.A. Cloning of a protein that mediates transcriptional effects of fatty acids in preadipocytes. Homology to peroxisome proliferator-activated receptors. *J. Biol. Chem.* **1995**, *270*, 2367–2371. [CrossRef] [PubMed]
114. Evans, R.M.; Barish, G.D.; Wang, Y.-X. PPARs and the complex journey to obesity. *Nat. Med.* **2004**, *10*, 355–361. [CrossRef]
115. Braissant, O.; Foufelle, F.; Scotto, C.; Dauça, M.; Wahli, W. Differential expression of peroxisome proliferator-activated receptors (PPARs): Tissue distribution of PPAR-alpha, -beta, and -gamma in the adult rat. *Endocrinology* **1996**, *137*, 354–366. [CrossRef]
116. Wang, Y.-X.; Lee, C.-H.; Tiep, S.; Yu, R.T.; Ham, J.; Kang, H.; Evans, R.M. Peroxisome-proliferator-activated receptor delta activates fat metabolism to prevent obesity. *Cell* **2003**, *113*, 159–170. [CrossRef]
117. Genini, D.; Catapano, C.V. Block of nuclear receptor ubiquitination. A mechanism of ligand-dependent control of peroxisome proliferator-activated receptor delta activity. *J. Biol. Chem.* **2007**, *282*, 11776–11785. [CrossRef]
118. Wadosky, K.M.; Willis, M.S. The story so far: Post-translational regulation of peroxisome proliferator-activated receptors by ubiquitination and SUMOylation. *Am. J. Physiol. Heart Circ. Physiol.* **2012**, *302*, H515–H526. [CrossRef]
119. Rieck, M.; Wedeken, L.; Müller-Brüsselbach, S.; Meissner, W.; Müller, R. Expression level and agonist-binding affect the turnover, ubiquitination and complex formation of peroxisome proliferator activated receptor beta. *FEBS J.* **2007**, *274*, 5068–5076. [CrossRef] [PubMed]
120. Sprecher, D.L.; Massien, C.; Pearce, G.; Billin, A.N.; Perlstein, I.; Willson, T.M.; Hassall, D.G.; Ancellin, N.; Patterson, S.D.; Lobe, D.C.; et al. Triglyceride:high-density lipoprotein cholesterol effects in healthy subjects administered a peroxisome proliferator activated receptor delta agonist. *Arterioscler. Thromb. Vasc. Biol.* **2007**, *27*, 359–365. [CrossRef]
121. Tanaka, T.; Yamamoto, J.; Iwasaki, S.; Asaba, H.; Hamura, H.; Ikeda, Y.; Watanabe, M.; Magoori, K.; Ioka, R.X.; Tachibana, K.; et al. Activation of peroxisome proliferator-activated receptor delta induces fatty acid beta-oxidation in skeletal muscle and attenuates metabolic syndrome. *Proc. Natl. Acad. Sci. USA* **2003**, *100*, 15924–15929. [CrossRef] [PubMed]
122. de la Monte, S.M.; Wands, J.R. Alzheimer's disease is type 3 diabetes-evidence reviewed. *J. Diabetes Sci. Technol.* **2008**, *2*, 1101–1113. [CrossRef]
123. de la Monte, S.M.; Tong, M.; Lester-Coll, N.; Plater, M.J.; Wands, J.R. Therapeutic rescue of neurodegeneration in experimental type 3 diabetes: Relevance to Alzheimer's disease. *J. Alzheimers Dis.* **2006**, *10*, 89–109. [CrossRef] [PubMed]
124. Gross, B.; Hennuyer, N.; Bouchaert, E.; Rommens, C.; Grillot, D.; Mezdour, H.; Staels, B. Generation and characterization of a humanized PPARδ mouse model. *Br. J. Pharmacol.* **2011**, *164*, 192–208. [CrossRef]
125. Liu, S.; Hatano, B.; Zhao, M.; Yen, C.-C.; Kang, K.; Reilly, S.M.; Gangl, M.R.; Gorgun, C.; Balschi, J.A.; Ntambi, J.M.; et al. Role of peroxisome proliferator-activated receptor {delta}/{beta} in hepatic metabolic regulation. *J. Biol. Chem.* **2011**, *286*, 1237–1247. [CrossRef] [PubMed]
126. Shan, W.; Nicol, C.J.; Ito, S.; Bility, M.T.; Kennett, M.J.; Ward, J.M.; Gonzalez, F.J.; Peters, J.M. Peroxisome proliferator-activated receptor-beta/delta protects against chemically induced liver toxicity in mice. *Hepatology* **2008**, *47*, 225–235. [CrossRef]
127. Peters, J.M.; Gonzalez, F.J. Sorting out the functional role(s) of peroxisome proliferator-activated receptor-beta/delta (PPAR-beta/delta) in cell proliferation and cancer. *Biochim. Biophys. Acta* **2009**, *1796*, 230–241. [CrossRef] [PubMed]
128. Peters, J.M.; Hollingshead, H.E.; Gonzalez, F.J. Role of peroxisome-proliferator-activated receptor beta/delta (PPARbeta/delta) in gastrointestinal tract function and disease. *Clin. Sci.* **2008**, *115*, 107–127. [CrossRef]
129. Tan, N.S.; Michalik, L.; Noy, N.; Yasmin, R.; Pacot, C.; Heim, M.; Flühmann, B.; Desvergne, B.; Wahli, W. Critical roles of PPAR beta/delta in keratinocyte response to inflammation. *Genes Dev.* **2001**, *15*, 3263–3277. [CrossRef]
130. Schmuth, M.; Haqq, C.M.; Cairns, W.J.; Holder, J.C.; Dorsam, S.; Chang, S.; Lau, P.; Fowler, A.J.; Chuang, G.; Moser, A.H.; et al. Peroxisome proliferator-activated receptor (PPAR)-beta/delta stimulates differentiation and lipid accumulation in keratinocytes. *J. Investig. Dermatol.* **2004**, *122*, 971–983. [CrossRef]
131. Kim, D.J.; Bility, M.T.; Billin, A.N.; Willson, T.M.; Gonzalez, F.J.; Peters, J.M. PPARbeta/delta selectively induces differentiation and inhibits cell proliferation. *Cell Death Differ.* **2006**, *13*, 53–60. [CrossRef] [PubMed]
132. Romanowska, M.; Reilly, L.; Palmer, C.N.A.; Gustafsson, M.C.U.; Foerster, J. Activation of PPARbeta/delta causes a psoriasis-like skin disease in vivo. *PLoS ONE* **2010**, *5*, e9701. [CrossRef] [PubMed]

133. Montagner, A.; Delgado, M.B.; Tallichet-Blanc, C.; Chan, J.S.K.; Sng, M.K.; Mottaz, H.; Degueurce, G.; Lippi, Y.; Moret, C.; Baruchet, M.; et al. Src is activated by the nuclear receptor peroxisome proliferator-activated receptor β/δ in ultraviolet radiation-induced skin cancer. *EMBO Mol. Med.* **2014**, *6*, 80–98. [CrossRef] [PubMed]
134. Chong, H.C.; Tan, M.J.; Philippe, V.; Tan, S.H.; Tan, C.K.; Ku, C.W.; Goh, Y.Y.; Wahli, W.; Michalik, L.; Tan, N.S. Regulation of epithelial-mesenchymal IL-1 signaling by PPARbeta/delta is essential for skin homeostasis and wound healing. *J. Cell Biol.* **2009**, *184*, 817–831. [CrossRef]
135. Yang, L.; Olsson, B.; Pfeifer, D.; Jönsson, J.-I.; Zhou, Z.-G.; Jiang, X.; Fredriksson, B.-A.; Zhang, H.; Sun, X.-F. Knockdown of peroxisome proliferator-activated receptor-beta induces less differentiation and enhances cell-fibronectin adhesion of colon cancer cells. *Oncogene* **2010**, *29*, 516–526. [CrossRef]
136. Marin, H.E.; Peraza, M.A.; Billin, A.N.; Willson, T.M.; Ward, J.M.; Kennett, M.J.; Gonzalez, F.J.; Peters, J.M. Ligand activation of peroxisome proliferator-activated receptor beta inhibits colon carcinogenesis. *Cancer Res.* **2006**, *66*, 4394–4401. [CrossRef]
137. Santos, G.C.; Zielenska, M.; Prasad, M.; Squire, J.A. Chromosome 6p amplification and cancer progression. *J. Clin. Pathol.* **2007**, *60*, 1–7. [CrossRef]
138. Uhlen, M.; Oksvold, P.; Fagerberg, L.; Lundberg, E.; Jonasson, K.; Forsberg, M.; Zwahlen, M.; Kampf, C.; Wester, K.; Hober, S.; et al. Towards a knowledge-based Human Protein Atlas. *Nat. Biotechnol.* **2010**, *28*, 1248–1250. [CrossRef]
139. Berglund, L.; Björling, E.; Oksvold, P.; Fagerberg, L.; Asplund, A.; Szigyarto, C.A.-K.; Persson, A.; Ottosson, J.; Wernérus, H.; Nilsson, P.; et al. A genecentric Human Protein Atlas for expression profiles based on antibodies. *Mol. Cell. Proteomics* **2008**, *7*, 2019–2027. [CrossRef]
140. Uhlén, M.; Björling, E.; Agaton, C.; Szigyarto, C.A.-K.; Amini, B.; Andersen, E.; Andersson, A.-C.; Angelidou, P.; Asplund, A.; Asplund, C.; et al. A human protein atlas for normal and cancer tissues based on antibody proteomics. *Mol. Cell. Proteomics* **2005**, *4*, 1920–1932. [CrossRef]
141. Foreman, J.E.; Sharma, A.K.; Amin, S.; Gonzalez, F.J.; Peters, J.M. Ligand activation of peroxisome proliferator-activated receptor-beta/delta (PPARbeta/delta) inhibits cell growth in a mouse mammary gland cancer cell line. *Cancer Lett.* **2010**, *288*, 219–225. [CrossRef] [PubMed]
142. Kittler, R.; Zhou, J.; Hua, S.; Ma, L.; Liu, Y.; Pendleton, E.; Cheng, C.; Gerstein, M.; White, K.P. A comprehensive nuclear receptor network for breast cancer cells. *Cell Rep.* **2013**, *3*, 538–551. [CrossRef]
143. Stephen, R.L.; Gustafsson, M.C.U.; Jarvis, M.; Tatoud, R.; Marshall, B.R.; Knight, D.; Ehrenborg, E.; Harris, A.L.; Wolf, C.R.; Palmer, C.N.A. Activation of peroxisome proliferator-activated receptor delta stimulates the proliferation of human breast and prostate cancer cell lines. *Cancer Res.* **2004**, *64*, 3162–3170. [CrossRef] [PubMed]
144. Girroir, E.E.; Hollingshead, H.E.; Billin, A.N.; Willson, T.M.; Robertson, G.P.; Sharma, A.K.; Amin, S.; Gonzalez, F.J.; Peters, J.M. Peroxisome proliferator-activated receptor-beta/delta (PPARbeta/delta) ligands inhibit growth of UACC903 and MCF7 human cancer cell lines. *Toxicology* **2008**, *243*, 236–243. [CrossRef] [PubMed]
145. Yao, P.-L.; Morales, J.L.; Zhu, B.; Kang, B.-H.; Gonzalez, F.J.; Peters, J.M. Activation of peroxisome proliferator-activated receptor-β/δ (PPAR-β/δ) inhibits human breast cancer cell line tumorigenicity. *Mol. Cancer Ther.* **2014**, *13*, 1008–1017. [CrossRef]
146. Ghosh, M.; Ai, Y.; Narko, K.; Wang, Z.; Peters, J.M.; Hla, T. PPARdelta is pro-tumorigenic in a mouse model of COX-2-induced mammary cancer. *Prostaglandins Other Lipid Mediat.* **2009**, *88*, 97–100. [CrossRef]
147. Yin, Y.; Russell, R.G.; Dettin, L.E.; Bai, R.; Wei, Z.-L.; Kozikowski, A.P.; Kopelovich, L.; Glazer, R.I. Peroxisome proliferator-activated receptor delta and gamma agonists differentially alter tumor differentiation and progression during mammary carcinogenesis. *Cancer Res.* **2005**, *65*, 3950–3957. [CrossRef] [PubMed]
148. Pearce, L.R.; Komander, D.; Alessi, D.R. The nuts and bolts of AGC protein kinases. *Nat. Rev. Mol. Cell Biol.* **2010**, *11*, 9–22. [CrossRef]
149. Pollock, C.B.; Yin, Y.; Yuan, H.; Zeng, X.; King, S.; Li, X.; Kopelovich, L.; Albanese, C.; Glazer, R.I. PPARδ activation acts cooperatively with 3-phosphoinositide-dependent protein kinase-1 to enhance mammary tumorigenesis. *PLoS ONE* **2011**, *6*, e16215. [CrossRef]
150. Yuan, H.; Lu, J.; Xiao, J.; Upadhyay, G.; Umans, R.; Kallakury, B.; Yin, Y.; Fant, M.E.; Kopelovich, L.; Glazer, R.I. PPARδ induces estrogen receptor-positive mammary neoplasia through an inflammatory and metabolic phenotype linked to mTOR activation. *Cancer Res.* **2013**, *73*, 4349–4361. [CrossRef]
151. Shearer, B.G.; Hoekstra, W.J. Recent advances in peroxisome proliferator-activated receptor science. *Curr. Med. Chem.* **2003**, *10*, 267–280. [CrossRef] [PubMed]
152. Schug, T.T.; Berry, D.C.; Toshkov, I.A.; Cheng, L.; Nikitin, A.Y.; Noy, N. Overcoming retinoic acid-resistance of mammary carcinomas by diverting retinoic acid from PPARbeta/delta to RAR. *Proc. Natl. Acad. Sci. USA* **2008**, *105*, 7546–7551. [CrossRef] [PubMed]
153. Kannan-Thulasiraman, P.; Seachrist, D.D.; Mahabeleshwar, G.H.; Jain, M.K.; Noy, N. Fatty acid-binding protein 5 and PPAR-beta/delta are critical mediators of epidermal growth factor receptor-induced carcinoma cell growth. *J. Biol. Chem.* **2010**, *285*, 19106–19115. [CrossRef]
154. Borland, M.G.; Khozoie, C.; Albrecht, P.P.; Zhu, B.; Lee, C.; Lahoti, T.S.; Gonzalez, F.J.; Peters, J.M. Stable over-expression of PPARβ/δ and PPARγ to examine receptor signaling in human HaCaT keratinocytes. *Cell Signal.* **2011**, *23*, 2039–2050. [CrossRef] [PubMed]

155. Borland, M.G.; Foreman, J.E.; Girroir, E.E.; Zolfaghari, R.; Sharma, A.K.; Amin, S.; Gonzalez, F.J.; Ross, A.C.; Peters, J.M. Ligand activation of peroxisome proliferator-activated receptor-beta/delta inhibits cell proliferation in human HaCaT keratinocytes. *Mol. Pharmacol.* **2008**, *74*, 1429–1442. [CrossRef] [PubMed]
156. Wang, X.; Wang, G.; Shi, Y.; Sun, L.; Gorczynski, R.; Li, Y.-J.; Xu, Z.; Spaner, D.E. PPAR-delta promotes survival of breast cancer cells in harsh metabolic conditions. *Oncogenesis* **2016**, *5*, e232. [CrossRef]
157. Burdick, A.D.; Bility, M.T.; Girroir, E.E.; Billin, A.N.; Willson, T.M.; Gonzalez, F.J.; Peters, J.M. Ligand activation of peroxisome proliferator-activated receptor-beta/delta(PPARbeta/delta) inhibits cell growth of human N/TERT-1 keratinocytes. *Cell Signal.* **2007**, *19*, 1163–1171. [CrossRef]
158. Adhikary, T.; Brandt, D.T.; Kaddatz, K.; Stockert, J.; Naruhn, S.; Meissner, W.; Finkernagel, F.; Obert, J.; Lieber, S.; Scharfe, M.; et al. Inverse PPARβ/δ agonists suppress oncogenic signaling to the ANGPTL4 gene and inhibit cancer cell invasion. *Oncogene* **2013**, *32*, 5241–5252. [CrossRef]
159. Gimble, J.M.; Pighetti, G.M.; Lerner, M.R.; Wu, X.; Lightfoot, S.A.; Brackett, D.J.; Darcy, K.; Hollingsworth, A.B. Expression of peroxisome proliferator activated receptor mRNA in normal and tumorigenic rodent mammary glands. *Biochem. Biophys. Res. Commun.* **1998**, *253*, 813–817. [CrossRef]
160. Fajas, L.; Auboeuf, D.; Raspé, E.; Schoonjans, K.; Lefebvre, A.M.; Saladin, R.; Najib, J.; Laville, M.; Fruchart, J.C.; Deeb, S.; et al. The organization, promoter analysis, and expression of the human PPARgamma gene. *J. Biol. Chem.* **1997**, *272*, 18779–18789. [CrossRef]
161. Fang, S.; Livergood, M.C.; Nakagawa, P.; Wu, J.; Sigmund, C.D. Role of the Peroxisome Proliferator Activated Receptors in Hypertension. *Circ. Res.* **2021**, *128*, 1021–1039. [CrossRef] [PubMed]
162. Francque, S.; Szabo, G.; Abdelmalek, M.F.; Byrne, C.D.; Cusi, K.; Dufour, J.-F.; Roden, M.; Sacks, F.; Tacke, F. Nonalcoholic steatohepatitis: The role of peroxisome proliferator-activated receptors. *Nat. Rev. Gastroenterol. Hepatol.* **2021**, *18*, 24–39. [CrossRef] [PubMed]
163. Azhar, S. Peroxisome proliferator-activated receptors, metabolic syndrome and cardiovascular disease. *Future Cardiol.* **2010**, *6*, 657–691. [CrossRef] [PubMed]
164. Takeyama, K.; Kodera, Y.; Suzawa, M.; Kato, S. [Peroxisome proliferator-activated receptor(PPAR)–structure, function, tissue distribution, gene expression]. *Nihon Rinsho.* **2000**, *58*, 357–363.
165. Janani, C.; Ranjitha Kumari, B.D. PPAR gamma gene–a review. *Diabetes Metab. Syndr.* **2015**, *9*, 46–50. [CrossRef]
166. Hamblin, M.; Chang, L.; Fan, Y.; Zhang, J.; Chen, Y.E. PPARs and the cardiovascular system. *Antioxid. Redox Signal.* **2009**, *11*, 1415–1452. [CrossRef]
167. Li, Y.; Qi, Y.; Huang, T.H.W.; Yamahara, J.; Roufogalis, B.D. Pomegranate flower: A unique traditional antidiabetic medicine with dual PPAR-alpha/-gamma activator properties. *Diabetes. Obes. Metab.* **2008**, *10*, 10–17. [CrossRef]
168. Blaschke, F.; Caglayan, E.; Hsueh, W.A. Peroxisome proliferator-activated receptor gamma agonists: Their role as vasoprotective agents in diabetes. *Endocrinol. Metab. Clin. N. Am.* **2006**, *35*, 561–574. [CrossRef]
169. Rizos, C.V.; Kei, A.; Elisaf, M.S. The current role of thiazolidinediones in diabetes management. *Arch. Toxicol.* **2016**, *90*, 1861–1881. [CrossRef]
170. Kahn, C.R.; Chen, L.; Cohen, S.E. Unraveling the mechanism of action of thiazolidinediones. *J. Clin. Investig.* **2000**, *106*, 1305–1307. [CrossRef]
171. Hauner, H. The mode of action of thiazolidinediones. *Diabetes. Metab. Res. Rev.* **2002**, *18* (Suppl. S2), S10–S15. [CrossRef] [PubMed]
172. Farshbaf, M.J.; Ghaedi, K.; Shirani, M.; Nasr-Esfahani, M.H. Peroxisome proliferator activated receptor gamma (PPARγ) as a therapeutic target for improvement of cognitive performance in Fragile-X. *Med. Hypotheses* **2014**, *82*, 291–294. [CrossRef] [PubMed]
173. Yang, Y.; Zhao, L.-H.; Huang, B.; Wang, R.-Y.; Yuan, S.-X.; Tao, Q.-F.; Xu, Y.; Sun, H.-Y.; Lin, C.; Zhou, W.-P. Pioglitazone, a PPARγ agonist, inhibits growth and invasion of human hepatocellular carcinoma via blockade of the rage signaling. *Mol. Carcinog.* **2015**, *54*, 1584–1595. [CrossRef]
174. Wang, Y.; Tan, H.; Xu, D.; Ma, A.; Zhang, L.; Sun, J.; Yang, Z.; Liu, Y.; Shi, G. The combinatory effects of PPAR-γ agonist and survivin inhibition on the cancer stem-like phenotype and cell proliferation in bladder cancer cells. *Int. J. Mol. Med.* **2014**, *34*, 262–268. [CrossRef] [PubMed]
175. Velmurugan, B.K.; Yang, H.-H.; Sung, P.-J.; Weng, C.-F. Excavatolide B inhibits nonsmall cell lung cancer proliferation by altering peroxisome proliferator activated receptor gamma expression and PTEN/AKT/NF-Kβ expression. *Environ. Toxicol.* **2017**, *32*, 290–301. [CrossRef]
176. Srivastava, N.; Kollipara, R.K.; Singh, D.K.; Sudderth, J.; Hu, Z.; Nguyen, H.; Wang, S.; Humphries, C.G.; Carstens, R.; Huffman, K.E.; et al. Inhibition of cancer cell proliferation by PPARγ is mediated by a metabolic switch that increases reactive oxygen species levels. *Cell Metab.* **2014**, *20*, 650–661. [CrossRef]
177. Pestereva, E.; Kanakasabai, S.; Bright, J.J. PPARγ agonists regulate the expression of stemness and differentiation genes in brain tumour stem cells. *Br. J. Cancer* **2012**, *106*, 1702–1712. [CrossRef]
178. Trindade-da-Silva, C.A.; Reis, C.F.; Vecchi, L.; Napimoga, M.H.; Sperandio, M.; Matias Colombo, B.F.; Alves, P.T.; Ward, L.S.; Ueira-Vieira, C.; Goulart, L.R. 15-Deoxy-Δ(12,14)-prostaglandin J2 Induces Apoptosis and Upregulates SOCS3 in Human Thyroid Cancer Cells. *PPAR Res.* **2016**, *2016*, 4106297. [CrossRef]

179. Wu, K.; Yang, Y.; Liu, D.; Qi, Y.; Zhang, C.; Zhao, J.; Zhao, S. Activation of PPARγ suppresses proliferation and induces apoptosis of esophageal cancer cells by inhibiting TLR4-dependent MAPK pathway. *Oncotarget* **2016**, *7*, 44572–44582. [CrossRef]
180. Zurlo, D.; Ziccardi, P.; Votino, C.; Colangelo, T.; Cerchia, C.; Dal Piaz, F.; Dallavalle, S.; Moricca, S.; Novellino, E.; Lavecchia, A.; et al. The antiproliferative and proapoptotic effects of cladosporols A and B are related to their different binding mode as PPARγ ligands. *Biochem. Pharmacol.* **2016**, *108*, 22–35. [CrossRef]
181. Weidner, C.; Rousseau, M.; Micikas, R.J.; Fischer, C.; Plauth, A.; Wowro, S.J.; Siems, K.; Hetterling, G.; Kliem, M.; Schroeder, F.C.; et al. Amorfrutin C Induces Apoptosis and Inhibits Proliferation in Colon Cancer Cells through Targeting Mitochondria. *J. Nat. Prod.* **2016**, *79*, 2–12. [CrossRef] [PubMed]
182. Kohno, H.; Yasui, Y.; Suzuki, R.; Hosokawa, M.; Miyashita, K.; Tanaka, T. Dietary seed oil rich in conjugated linolenic acid from bitter melon inhibits azoxymethane-induced rat colon carcinogenesis through elevation of colonic PPARgamma expression and alteration of lipid composition. *Int. J. Cancer* **2004**, *110*, 896–901. [CrossRef] [PubMed]
183. Yasui, Y.; Hosokawa, M.; Sahara, T.; Suzuki, R.; Ohgiya, S.; Kohno, H.; Tanaka, T.; Miyashita, K. Bitter gourd seed fatty acid rich in 9c,11t,13t-conjugated linolenic acid induces apoptosis and up-regulates the GADD45, p53 and PPARgamma in human colon cancer Caco-2 cells. *Prostaglandins. Leukot. Essent. Fatty Acids* **2005**, *73*, 113–119. [CrossRef] [PubMed]
184. Mueller, E.; Sarraf, P.; Tontonoz, P.; Evans, R.M.; Martin, K.J.; Zhang, M.; Fletcher, C.; Singer, S.; Spiegelman, B.M. Terminal differentiation of human breast cancer through PPAR gamma. *Mol. Cell* **1998**, *1*, 465–470. [CrossRef]
185. Moon, H.-S.; Guo, D.-D.; Lee, H.-G.; Choi, Y.-J.; Kang, J.-S.; Jo, K.; Eom, J.-M.; Yun, C.-H.; Cho, C.-S. Alpha-eleostearic acid suppresses proliferation of MCF-7 breast cancer cells via activation of PPARgamma and inhibition of ERK 1/2. *Cancer Sci.* **2010**, *101*, 396–402. [CrossRef]
186. Bonofiglio, D.; Gabriele, S.; Aquila, S.; Catalano, S.; Gentile, M.; Middea, E.; Giordano, F.; Andò, S. Estrogen receptor alpha binds to peroxisome proliferator-activated receptor response element and negatively interferes with peroxisome proliferator-activated receptor gamma signaling in breast cancer cells. *Clin. Cancer Res.* **2005**, *11*, 6139–6147. [CrossRef]
187. Bonofiglio, D.; Aquila, S.; Catalano, S.; Gabriele, S.; Belmonte, M.; Middea, E.; Qi, H.; Morelli, C.; Gentile, M.; Maggiolini, M.; et al. Peroxisome proliferator-activated receptor-gamma activates p53 gene promoter binding to the nuclear factor-kappaB sequence in human MCF7 breast cancer cells. *Mol. Endocrinol.* **2006**, *20*, 3083–3092. [CrossRef]
188. Bonofiglio, D.; Gabriele, S.; Aquila, S.; Qi, H.; Belmonte, M.; Catalano, S.; Andò, S. Peroxisome proliferator-activated receptor gamma activates fas ligand gene promoter inducing apoptosis in human breast cancer cells. *Breast Cancer Res. Treat.* **2009**, *113*, 423–434. [CrossRef]
189. Chinnaiyan, A.M.; Dixit, V.M. Portrait of an executioner: The molecular mechanism of FAS/APO-1-induced apoptosis. *Semin. Immunol.* **1997**, *9*, 69–76. [CrossRef]
190. Debatin, K.-M. Apoptosis pathways in cancer and cancer therapy. *Cancer Immunol. Immunother.* **2004**, *53*, 153–159. [CrossRef]
191. Pinkoski, M.J.; Green, D.R. Fas ligand, death gene. *Cell Death Differ.* **1999**, *6*, 1174–1181. [CrossRef] [PubMed]
192. Catalano, S.; Mauro, L.; Bonofiglio, D.; Pellegrino, M.; Qi, H.; Rizza, P.; Vizza, D.; Bossi, G.; Andò, S. In vivo and in vitro evidence that PPARγ ligands are antagonists of leptin signaling in breast cancer. *Am. J. Pathol.* **2011**, *179*, 1030–1040. [CrossRef] [PubMed]
193. Burstein, H.J.; Demetri, G.D.; Mueller, E.; Sarraf, P.; Spiegelman, B.M.; Winer, E.P. Use of the peroxisome proliferator-activated receptor (PPAR) gamma ligand troglitazone as treatment for refractory breast cancer: A phase II study. *Breast Cancer Res. Treat.* **2003**, *79*, 391–397. [CrossRef] [PubMed]
194. Gale, E.A. Lessons from the glitazones: A story of drug development. *Lancet* **2001**, *357*, 1870–1875. [CrossRef]
195. Krentz, A.J.; Bailey, C.J. Oral antidiabetic agents: Current role in type 2 diabetes mellitus. *Drugs* **2005**, *65*, 385–411. [CrossRef]
196. Garcia-Vallvé, S.; Guasch, L.; Tomas-Hernández, S.; del Bas, J.M.; Ollendorff, V.; Arola, L.; Pujadas, G.; Mulero, M. Peroxisome Proliferator-Activated Receptor γ (PPARγ) and Ligand Choreography: Newcomers Take the Stage. *J. Med. Chem.* **2015**, *58*, 5381–5394. [CrossRef]
197. Yee, L.D.; Williams, N.; Wen, P.; Young, D.C.; Lester, J.; Johnson, M.V.; Farrar, W.B.; Walker, M.J.; Povoski, S.P.; Suster, S.; et al. Pilot study of rosiglitazone therapy in women with breast cancer: Effects of short-term therapy on tumor tissue and serum markers. *Clin Cancer Res.* **2007**, *13*, 246–252. [CrossRef]
198. Elstner, E.; Müller, C.; Koshizuka, K.; Williamson, E.A.; Park, D.; Asou, H.; Shintaku, P.; Said, J.W.; Heber, D.; Koeffler, H.P. Ligands for peroxisome proliferator-activated receptorgamma and retinoic acid receptor inhibit growth and induce apoptosis of human breast cancer cells in vitro and in BNX mice. *Proc. Natl. Acad. Sci. USA* **1998**, *95*, 8806–8811. [CrossRef]
199. Mustafa, A.; Kruger, W.D. Suppression of tumor formation by a cyclooxygenase-2 inhibitor and a peroxisome proliferator-activated receptor gamma agonist in an in vivo mouse model of spontaneous breast cancer. *Clin. Cancer Res.* **2008**, *14*, 4935–4942. [CrossRef]
200. Bonofiglio, D.; Cione, E.; Qi, H.; Pingitore, A.; Perri, M.; Catalano, S.; Vizza, D.; Panno, M.L.; Genchi, G.; Fuqua, S.A.W.; et al. Combined low doses of PPARgamma and RXR ligands trigger an intrinsic apoptotic pathway in human breast cancer cells. *Am. J. Pathol.* **2009**, *175*, 1270–1280. [CrossRef]
201. Yamazaki, K.; Shimizu, M.; Okuno, M.; Matsushima-Nishiwaki, R.; Kanemura, N.; Araki, H.; Tsurumi, H.; Kojima, S.; Weinstein, I.B.; Moriwaki, H. Synergistic effects of RXR alpha and PPAR gamma ligands to inhibit growth in human colon cancer cells–phosphorylated RXR alpha is a critical target for colon cancer management. *Gut* **2007**, *56*, 1557–1563. [CrossRef]

202. Jiang, Y.; Huang, Y.; Cheng, C.; Lu, W.; Zhang, Y.; Liu, X.; Zou, L.; Ben, Q.; Shen, A. Combination of thiazolidinedione and hydralazine suppresses proliferation and induces apoptosis by PPARγ up-expression in MDA-MB-231 cells. *Exp. Mol. Pathol.* **2011**, *91*, 768–774. [CrossRef] [PubMed]
203. Girnun, G.D.; Naseri, E.; Vafai, S.B.; Qu, L.; Szwaya, J.D.; Bronson, R.; Alberta, J.A.; Spiegelman, B.M. Synergy between PPARgamma ligands and platinum-based drugs in cancer. *Cancer Cell* **2007**, *11*, 395–406. [CrossRef] [PubMed]
204. Bräutigam, K.; Biernath-Wüpping, J.; Bauerschlag, D.O.; von Kaisenberg, C.S.; Jonat, W.; Maass, N.; Arnold, N.; Meinhold-Heerlein, I. Combined treatment with TRAIL and PPARγ ligands overcomes chemoresistance of ovarian cancer cell lines. *J. Cancer Res. Clin. Oncol.* **2011**, *137*, 875–886. [CrossRef]
205. Yokoyama, Y.; Xin, B.; Shigeto, T.; Mizunuma, H. Combination of ciglitazone, a peroxisome proliferator-activated receptor gamma ligand, and cisplatin enhances the inhibition of growth of human ovarian cancers. *J. Cancer Res. Clin. Oncol.* **2011**, *137*, 1219–1228. [CrossRef] [PubMed]
206. Cesario, R.M.; Stone, J.; Yen, W.-C.; Bissonnette, R.P.; Lamph, W.W. Differentiation and growth inhibition mediated via the RXR:PPARgamma heterodimer in colon cancer. *Cancer Lett.* **2006**, *240*, 225–233. [CrossRef]
207. Desreumaux, P.; Dubuquoy, L.; Nutten, S.; Peuchmaur, M.; Englaro, W.; Schoonjans, K.; Derijard, B.; Desvergne, B.; Wahli, W.; Chambon, P.; et al. Attenuation of colon inflammation through activators of the retinoid X receptor (RXR)/peroxisome proliferator-activated receptor gamma (PPARgamma) heterodimer. A basis for new therapeutic strategies. *J. Exp. Med.* **2001**, *193*, 827–838. [CrossRef]
208. Fu, H.; Zhang, J.; Pan, J.; Zhang, Q.; Lu, Y.; Wen, W.; Lubet, R.A.; Szabo, E.; Chen, R.; Wang, Y.; et al. Chemoprevention of lung carcinogenesis by the combination of aerosolized budesonide and oral pioglitazone in A/J mice. *Mol. Carcinog.* **2011**, *50*, 913–921. [CrossRef]
209. Reddy, R.C.; Srirangam, A.; Reddy, K.; Chen, J.; Gangireddy, S.; Kalemkerian, G.P.; Standiford, T.J.; Keshamouni, V.G. Chemotherapeutic drugs induce PPAR-gamma expression and show sequence-specific synergy with PPAR-gamma ligands in inhibition of non-small cell lung cancer. *Neoplasia* **2008**, *10*, 597–603. [CrossRef]
210. Aronica, S.M.; Kraus, W.L.; Katzenellenbogen, B.S. Estrogen action via the cAMP signaling pathway: Stimulation of adenylate cyclase and cAMP-regulated gene transcription. *Proc. Natl. Acad. Sci. USA* **1994**, *91*, 8517–8521. [CrossRef]
211. Migliaccio, A.; Di Domenico, M.; Castoria, G.; de Falco, A.; Bontempo, P.; Nola, E.; Auricchio, F. Tyrosine kinase/p21ras/MAP-kinase pathway activation by estradiol-receptor complex in MCF-7 cells. *EMBO J.* **1996**, *15*, 1292–1300. [CrossRef] [PubMed]
212. Keller, H.; Givel, F.; Perroud, M.; Wahli, W. Signaling cross-talk between peroxisome proliferator-activated receptor/retinoid X receptor and estrogen receptor through estrogen response elements. *Mol. Endocrinol.* **1995**, *9*, 794–804. [CrossRef] [PubMed]
213. Cui, Y.; Miyoshi, K.; Claudio, E.; Siebenlist, U.K.; Gonzalez, F.J.; Flaws, J.; Wagner, K.-U.; Hennighausen, L. Loss of the peroxisome proliferation-activated receptor gamma (PPARgamma) does not affect mammary development and propensity for tumor formation but leads to reduced fertility. *J. Biol. Chem.* **2002**, *277*, 17830–17835. [CrossRef] [PubMed]
214. Yin, Y.; Yuan, H.; Zeng, X.; Kopelovich, L.; Glazer, R.I. Inhibition of peroxisome proliferator-activated receptor gamma increases estrogen receptor-dependent tumor specification. *Cancer Res.* **2009**, *69*, 687–694. [CrossRef] [PubMed]
215. Saez, E.; Rosenfeld, J.; Livolsi, A.; Olson, P.; Lombardo, E.; Nelson, M.; Banayo, E.; Cardiff, R.D.; Izpisua-Belmonte, J.C.; Evans, R.M. PPAR gamma signaling exacerbates mammary gland tumor development. *Genes Dev.* **2004**, *18*, 528–540. [CrossRef]
216. Tian, L.; Zhou, J.; Casimiro, M.C.; Liang, B.; Ojeifo, J.O.; Wang, M.; Hyslop, T.; Wang, C.; Pestell, R.G. Activating peroxisome proliferator-activated receptor gamma mutant promotes tumor growth in vivo by enhancing angiogenesis. *Cancer Res.* **2009**, *69*, 9236–9244. [CrossRef]
217. Xu, Y.-Y.; Liu, H.; Su, L.; Xu, N.; Xu, D.-H.; Liu, H.-Y.; Spaner, D.; Bed-David, Y.; Li, Y.-J. PPARγ inhibits breast cancer progression by upregulating PTPRF expression. *Eur. Rev. Med. Pharmacol. Sci.* **2019**, *23*, 9965–9977. [CrossRef]
218. Kotta-Loizou, I.; Giaginis, C.; Theocharis, S. The role of peroxisome proliferator-activated receptor-γ in breast cancer. *Anticancer. Agents Med. Chem.* **2012**, *12*, 1025–1044. [CrossRef]
219. Yang, P.-B.; Hou, P.-P.; Liu, F.-Y.; Hong, W.-B.; Chen, H.-Z.; Sun, X.-Y.; Li, P.; Zhang, Y.; Ju, C.-Y.; Luo, L.-J.; et al. Blocking PPARγ interaction facilitates Nur77 interdiction of fatty acid uptake and suppresses breast cancer progression. *Proc. Natl. Acad. Sci. USA* **2020**, *117*, 27412–27422. [CrossRef]
220. Papadaki, I.; Mylona, E.; Giannopoulou, I.; Markaki, S.; Keramopoulos, A.; Nakopoulou, L. PPARgamma expression in breast cancer: Clinical value and correlation with ERbeta. *Histopathology* **2005**, *46*, 37–42. [CrossRef]
221. Kulkarni, S.; Patil, D.B.; Diaz, L.K.; Wiley, E.L.; Morrow, M.; Khan, S.A. COX-2 and PPARgamma expression are potential markers of recurrence risk in mammary duct carcinoma in-situ. *BMC Cancer* **2008**, *8*, 36. [CrossRef] [PubMed]
222. Liedtke, C.; Mazouni, C.; Hess, K.R.; André, F.; Tordai, A.; Mejia, J.A.; Symmans, W.F.; Gonzalez-Angulo, A.M.; Hennessy, B.; Green, M.; et al. Response to neoadjuvant therapy and long-term survival in patients with triple-negative breast cancer. *J. Clin. Oncol.* **2008**, *26*, 1275–1281. [CrossRef] [PubMed]
223. Li, T.; Zhang, Q.; Zhang, J.; Yang, G.; Shao, Z.; Luo, J.; Fan, M.; Ni, C.; Wu, Z.; Hu, X. Fenofibrate induces apoptosis of triple-negative breast cancer cells via activation of NF-κB pathway. *BMC Cancer.* **2014**, *14*, 96. [CrossRef] [PubMed]
224. Kwong, S.C.; Jamil, A.H.A.; Rhodes, A.; Taib, N.A.; Chung, I. Metabolic role of fatty acid binding protein 7 in mediating triple-negative breast cancer cell death via PPAR-α signaling. *J. Lipid Res.* **2019**, *60*, 1807–1817. [CrossRef]
225. Apaya, M.K.; Hsiao, P.W.; Yang, Y.C.; Shyur, L.F. Deregulating the CYP2C19/Epoxy-Eicosatrienoic Acid-Associated FABP4/FABP5 Signaling Network as a Therapeutic Approach for Metastatic Triple-Negative Breast Cancer. *Cancers* **2020**, *12*, 199. [CrossRef]

226. Volkers, N. Diabetes and cancer: Scientists search for a possible link. *J. Natl. Cancer Inst.* **2000**, *92*, 192–194. [CrossRef]
227. Giovannucci, E.; Harlan, D.M.; Archer, M.C.; Bergenstal, R.M.; Gapstur, S.M.; Habel, L.A.; Pollak, M.; Regensteiner, J.G.; Yee, D. Diabetes and cancer: A consensus report. *Diabetes Care* **2010**, *33*, 1674–1685. [CrossRef]
228. Rossouw, J.E.; Anderson, G.L.; Prentice, R.L.; LaCroix, A.Z.; Kooperberg, C.; Stefanick, M.L.; Jackson, R.D.; Beresford, S.A.A.; Howard, B.V.; Johnson, K.C.; et al. Risks and benefits of estrogen plus progestin in healthy postmenopausal women: Principal results From the Women's Health Initiative randomized controlled trial. *JAMA* **2002**, *288*, 321–333. [CrossRef]
229. Youssef, J.; Badr, M. Peroxisome proliferator-activated receptors and cancer: Challenges and opportunities. *Br. J. Pharmacol.* **2011**, *164*, 68–82. [CrossRef]

cells

MDPI

Article

Exploring the Nerve Regenerative Capacity of Compounds with Differing Affinity for PPARγ In Vitro and In Vivo

Melissa L. D. Rayner [1,2,*], Simon C. Kellaway [1,2], Isabel Kingston [1,2], Owein Guillemot-Legris [1,2], Holly Gregory [1,2], Jess Healy [2] and James B. Phillips [1,2]

1 Department of Pharmacology, School of Pharmacy, University College London, London WC1N 1AX, UK
2 Centre for Nerve Engineering, University College London, London WC1N 6BT, UK
* Correspondence: m.rayner@ucl.ac.uk

Abstract: Damage to peripheral nerves can cause debilitating consequences for patients such as lifelong pain and disability. At present, no drug treatments are routinely given in the clinic following a peripheral nerve injury (PNI) to improve regeneration and remyelination of damaged nerves. Appropriately targeted therapeutic agents have the potential to be used at different stages following nerve damage, e.g., to maintain Schwann cell viability, induce and sustain a repair phenotype to support axonal growth, or promote remyelination. The development of therapies to promote nerve regeneration is currently of high interest to researchers, however, translation to the clinic of drug therapies for PNI is still lacking. Studying the effect of PPARγ agonists for treatment of peripheral nerve injures has demonstrated significant benefits. Ibuprofen, a non-steroidal anti-inflammatory drug (NSAID), has reproducibly demonstrated benefits in vitro and in vivo, suggested to be due to its agonist action on PPARγ. Other NSAIDs have demonstrated differing levels of PPARγ activation based upon their affinity. Therefore, it was of interest to determine whether affinity for PPARγ of selected drugs corresponded to an increase in regeneration. A 3D co-culture in vitro model identified some correlation between these two properties. However, when the drug treatments were screened in vivo, in a crush injury model in a rat sciatic nerve, the same correlation was not apparent. Further differences were observed between capacity to increase axon number and improvement in functional recovery. Despite there not being a clear correlation between affinity and size of effect on regeneration, all selected PPARγ agonists improved regeneration, providing a panel of compounds that could be explored for use in the treatment of PNI.

Keywords: peripheral nerve injury (PNI); peroxisome proliferator-activated receptor gamma (PPARγ); non-steroidal anti-inflammatory drugs (NSAIDs); nerve regeneration; functional recovery; drug affinity

Citation: Rayner, M.L.D.; Kellaway, S.C.; Kingston, I.; Guillemot-Legris, O.; Gregory, H.; Healy, J.; Phillips, J.B. Exploring the Nerve Regenerative Capacity of Compounds with Differing Affinity for PPARγ In Vitro and In Vivo. *Cells* **2023**, *12*, 42. https://doi.org/10.3390/cells12010042

Academic Editors: Kay-Dietrich Wagner, Nicole Wagner and Stefano Geuna

Received: 18 November 2022
Revised: 9 December 2022
Accepted: 10 December 2022
Published: 22 December 2022

1. Introduction

Even in the best-case scenario, with microsurgical therapy providing a supportive environment for regeneration, neuron growth rate is limited to ~1 mm/day [1]. Therefore, months to years can elapse before regenerating neurons reach their target organ. For motor neurons growing back to muscle, this delay leaves the muscle with no nerve stimulation over time, resulting in irreversible wasting. This means that even if the neurons eventually reach their target muscle, there is little recovery of function. The development of a drug that acts to accelerate nerve regeneration, with subsequent improvement in functional recovery, will provide a therapy for an unmet clinical need.

The role of peroxisome proliferator-activated receptor gamma (PPARγ) in nerve tissue has not been fully determined but there is evidence from rodent studies that it may be involved in neuronal development, neuronal health, and pain signalling [2]. Immunohistochemical analysis has demonstrated localisation of PPARγ in Schwann cells and endothelial cells in rat peripheral nerves [3], expression of PPARγ within Schwann cells of healthy and regenerating nerves following a crush injury [4], and an increase in expression during the inflammation process [5]. PPARγ expression has also been demonstrated in axons within a rat sciatic nerve at 2, 4, and 6 h after a nerve ligation or crush injury [6]. The presence and increased expression of PPARγ following a peripheral nerve injury (PNI) provide possible benefits for its use in developing therapies for this unmet clinical need.

PPARγ is an upstream mediator of the Rho/ROCK pathway [7] and its activation is suggested to inhibit this pathway via the upregulation of the protein tyrosine phosphatase, Src homology region 2–containing protein tyrosine phosphatase-2 (SHP-2). This cytosolic protein tyrosine phosphatase (PTP) dephosphorylates and inactivates guanine nucleotide exchange factor (GEF), Vav, which is turn suppresses the conversion of inactive guanosine diphosphate (GDP) -Rho to active guanosine triphosphate (GTP) -Rho [7]. As a consequence, ROCK is not activated and this inhibition of the Rho/ROCK pathway prevents stiffening of the actin cytoskeleton, encourages axonal elongation, and prevents growth cone collapse [8–10].

PPARγ is one of several upstream effectors of the Rho/ROCK pathway [11], making it a great target for drug therapies. Pharmacological treatment of PNI with PPARγ agonists such as non-steroidal anti-inflammatory drugs (NSAIDs) and **thiazolidinediones** has shown beneficial effects on nerve regeneration, with the opposite effect seen with the PPARγ antagonist, GW9662 [6]. Thiazolidinediones are antidiabetic agents that are potent and selective activators of PPARγ [12]. Their use in PNI has not been extensively explored but increased neurite outgrowth was seen following troglitazone-treated primary rat hippocampal cultures [13] and rosiglitazone-treated primary rat cortical neurons [6].

Ibuprofen and diclofenac are the most commonly studied NSAIDs for PNI treatment [9,14–16]. The pharmacological effects of NSAIDs are thought to be independent of their well-known anti-inflammatory activity on cyclooxygenase (COX) 1 and 2 [17,18]. The characteristic lipophilic backbone and acid moiety, most commonly a carboxylate, within the structure of NSAIDs is what may enable them to bind to PPARγ [19]. There are several classes of NSAIDs based upon their chemical structure including thiazinecarboxamides (piroxicam), derivatives of arylacetic acid (indomethacin), aminoarylcarboxylic acid (flufenamic acid), arylpropionic acid (ibuprofen and fenoprofen), and salicylic acid (aspirin) [19]. The different classes of NSAIDs have demonstrated a spectrum of agonist activity on PPARγ. Ibuprofen, fenoprofen, and flufenamic acid exhibit the greatest effect on PPARγ with indomethacin, piroxicam, and salicylic acid having little or no effect [19]. This correlated with studies identifying reduced levels of activated Rho with ibuprofen, indomethacin, and sulindac sulfide treatment, but not with piroxicam, meloxicam, and naproxen [18,20–22]. The reduced levels of activated Rho correlated to an increase in axon growth when treated with ibuprofen and indomethacin but not with naproxen [18,20,21].

Previous literature has identified and discussed the relative affinities of NSAIDs for PPARγ (sulindac sulfide > diclofenac > indomethacin > ibuprofen) [23,24]. Given the promising results already seen with ibuprofen and diclofenac, other NSAIDs could have promising effects on nerve regeneration and may even exceed those already seen. In contrast, naproxen has been shown not to activate PPARγ in neuronal cells [20], and not to have a RhoA-suppressing function [25].

Gaining an understanding of how common compounds such as **thiazolidinediones and NSAIDs** interact with PPARγ could aid the development of more potent and selective agonists, corresponding to a greater effect on nerve regeneration and functional recovery following a PNI. Therefore, the aims of this study were to investigate the regenerative capacity of a panel of PPARγ agonists and determine whether there is a correlation between a compound's affinity for PPARγ and its capacity to enhance neuronal regeneration in vitro and in vivo.

2. Materials and Methods

2.1. Cell Cultures

2.1.1. SCL4.1/F7 Schwann Cell Line

Cells from the rat Schwann cell line SCL4.1/F7 (Health Protection Agency) were grown in Dulbecco's Modified Eagle Medium (DMEM) high glucose medium supplemented with 10% *v/v* heat inactivated fetal bovine serum (FBS) and 100 U/mL of Penicillin, 100 µg/mL of Streptomycin, in standard cell culture flasks. The cultures were maintained at sub-confluency at 37 °C with 5% CO_2 and passaged when required.

2.1.2. PC12 Neuronal Cell Line

PC12 cells (Pheochromocytoma cells from rat adrenal medulla used as a neuronal cell line, 88022401; Sigma-Aldrich, St. Louis, MO, USA) were grown in suspension in RPMI 1640 medium supplemented with 100 U/mL of Penicillin, 100 µg/mL of Streptomycin, 2 mM L-glutamine, 10% *v/v* heat-inactivated horse serum and 5% *v/v* FBS in standard cell culture flasks. The cultures were maintained at sub-confluency at 37 °C with 5% CO_2 and passaged when required.

2.2. Fabrication of 3D EngNT Co-Cultures

Anisotropic cellular gels were prepared as described previously [26]. Briefly, 1 mL of solution containing 80% *v/v* type I rat tail collagen (2 mg/mL in 0.6% acetic acid), 10% *v/v* minimum essential medium, 5.8% *v/v* neutralising solution, and 4.2% Schwann cell suspension (4×10^6 SCL4.1/F7 cells per 1 mL gel) was integrated with tethering mesh at opposite ends of a rectangular mold (Dimensions 16.4 mm × 6.5 mm × 5 mm) [27]. Cellular gels were immersed in 10 mL DMEM and incubated at 37 °C with 5% CO_2 for 24 h to enable cellular self-alignment, then stabilised using plastic compression.

Plastic compression was conducted by transferring the gel to a standard blotting element, which comprised of a layer of stainless-steel support mesh covered by a layer of nylon mesh placed on top of three layers of circular Grade 1 Whatman filter paper [26]. The gel was then covered with a second layer of nylon mesh and a glass sheet. Compression was then completed by placing a 120 g stainless steel weight on top of the blotting element (stress equivalent to approximately 1.8 kN/m^2) for 1 min.

Each stabilised aligned cellular gel was cut into 4 equal segments to obtain a control and three test regions from each gel. Each gel segment was transferred to a separate well in a 24-well plate, then 100,000 PC12 cells were seeded on top of each segment for co-cultures. The gels were incubated for 1 h at 37 °C to allow attachment of neuronal cells to the collagen gel, then 1 mL of culture medium (DMEM high glucose), supplemented with 10% heat inactivated FBS and 100 U/mL of Penicillin and 100 µg/mL of Streptomycin, was added to each well. The neurons were seeded onto the top surface of the gels and neurites extended across the horizontal plane following the aligned Schwann cells.

Three-dimensional EngNT co-cultures containing SCL4.1/F7 and PC12 cells, as described previously [28], were subjected to drug treatments for 72 h before being fixed with 4% *w/v* paraformaldehyde (PFA) in PBS at 4 °C overnight for subsequent immunostaining and microscopy analysis. Drug stock solutions made up in DMSO were added directly to the media in the appropriate volume to provide the required drug concentration.

2.3. Immunocytochemistry

Following fixation with 4% (*w/v*) PFA in PBS gels were washed 3 times with PBS and permeabilised using 0.5% (*v/v*) Triton-X-100 in PBS. Washes were repeated before blocking (goat serum 1/20 in PBS). Gels were washed before the addition of anti-mouse βIII –Tubulin primary antibody (Sigma-Aldrich T8660) diluted in PBS and incubated overnight at 4 °C. Washes were repeated before adding the corresponding secondary antibody (Dylight anti-mouse IgG 488 (1:400, diluted in PBS) (Thermofisher 35502, Paisley, UK). The secondary antibody was washed, and the sample was stored at 4 °C before viewing. Omission of primary or secondary antibody was routinely used as a control. When required, Hoechst was used to stain nuclei by incubating the gel for 15 min and then washed thrice with PBS.

2.4. Surgical Nerve Injury Models In Vivo

All surgical procedures were performed in accordance with the UK Animals (Scientific Procedures) Act (1986), the European Communities Council Directives (86/609/EEC) and approved by the UCL Animal Welfare and Ethics Review Board (AWERB). Adult male Wister rats (250–300 g) (Charles River) were randomised into groups and housed in plastic cages with soft bedding and free access to food and water. The animals were deeply anaesthetised by inhalation of isoflurane, and the left sciatic nerve was exposed at mid-thigh level. This was done by making an incision (~3 cm) parallel to the femur between the knee and hip then separating muscle layers to expose the nerve. Under the microscope (Zeiss CL 1500 ECO, Carl Zeiss GmbH), the sciatic nerve was released from the surrounding tissue.

A crush injury (axonotmesis) was achieved by applying a consistent pressure with a pair of sterile TAAB type 4 tweezers closed fully on the same point of the nerve (1.5 cm distal of the femur) for 15 s. This was repeated twice more in the same location with the tweezers positioned perpendicular to the nerve and rotated through 45° between each crush application. A 10/0 epineurial suture (Ethicon) was used to mark the injury site. Following the injury an osmotic pump (Alzet, model 1004, Cupertino, CA, USA) loaded with 100 μL of drug solution in PBS was implanted locally parallel to the sciatic nerve, delivering drug at a rate of 0.11 μL/h. The overlying muscle layers were closed using two 4/0 sutures (Ethicon, W2850, London, UK) and the skin was closed using stainless steel wound clips. The animals were allowed to recover and were maintained for 28 days.

All animals received the same level of interaction throughout the study. They were handled prior to surgery and repeatedly throughout the study for training and completion of functional testing. All animals were left to settle (~5 min) before conducting any functional tests.

2.5. Nerve Tissue Harvest

Animals were culled using an overdose of anesthetic according to local regulations, and the repaired nerves (~1.5 cm) were excised under an operating microscope and cut as required for analysis. The nerve tissue was immersion-fixed in 4% (*w*/*v*) PFA in PBS at 4 °C overnight.

2.6. Nerve Tissue Analysis

2.6.1. Cryo-Sectioning

Following fixation, the nerve samples were incubated in 30% sucrose overnight and underwent subsequent snap freezing in 1:1 FSC 22 Frozen Section Media (Leica) and 30% sucrose. Transverse sections (10 μm thick) were prepared from the distal stumps, at a defined distance from the injury site, using a cryostat (Fisher scientific HM525 NX, Loughborough, UK). The sections were adhered to glass slides (Superfrost™ Plus) for histological analysis.

2.6.2. Immunohistochemistry

Nerve sections were washed in immunostaining buffer (PBS together with 0.2% Triton-X), 0.002% sodium azide, and 0.25% Bovine Serum Albumin before the addition of serum to block non-specific binding (goat serum 1/20 in immunostaining buffer) for 45 min. The blocking serum was removed, and sections were incubated with anti-mouse neurofilament-H primary antibody diluted in immunostaining buffer overnight at 4 °C. The sections were washed with immunostaining buffer before addition of the secondary antibody and incubation at room temperature for 45 min. Sections underwent a final wash with immunostaining buffer before mounting with Vectashield Hardset mounting medium with DAPI.

2.7. Image Analysis and Quantification

Fluorescence microscopy (Zeiss Axiolab A1, Axiocam Cm1, Carl Zeiss GmbH) was used to capture images of neurites from five pre-determined fields of each gel using a ×20 lens. The positions of the pre-determined fields on the gels were equally spaced (625 μm apart) in a line along the edge of the construct where alignment was greatest [27]. The length of each neurite in each field (~1–12 neurites) was measured using Fiji ImageJ [29]. Following stabilization, the gel acquired a thickness of 100 μm and the neurons extended predominantly in a single horizontal plane along the top surface, following the aligned Schwann cells [30].

Tile scans were used to capture high-magnification (×20) micrographs from the entire nerve cross-section using a Zeiss LSM 710 confocal microscope and images were analysed using Volocity™ 6.4 (PerkinElmer, Beaconsfield, UK) running automated image analysis protocols to determine the number of neurofilament-immunoreactive neurites in each transverse nerve section.

2.8. Functional Outcomes In Vivo

2.8.1. Electrophysiology

After 28 days, animals were anaesthetised using isoflurane and nerve function was assessed by electrophysiology (Synergy Ultrapro 3) by comparing the repaired nerve to the contralateral undamaged nerve in each animal. Monopolar needle electrodes were attached to the animal; a grounding electrode was placed in the tail of the animal and a reference electrode was placed above the hip bone. A stimulating electrode was placed against the proximal nerve 2 mm above the injury site and a recording needle was placed into the gastrocnemius muscle. The distance between the stimulating and recording electrodes was standardised. The nerve was stimulated using a bipolar stimulation constant voltage configuration and the muscle response recorded. The stimulation threshold was determined by increasing the stimulus amplitude in 0.1 V steps (200 μs pulse), until both a supramaximal muscle action potential was recorded, and a significant twitch of the animal's hind paw was seen. The amplitude (mV) of the compound muscle action potential (CMAP) was measured from the baseline to the peak, and the latency was measured from the time of stimulus to the first deviation from the baseline. Muscle action potential measurements were conducted in triplicate for both the injured nerve and contralateral control nerve in each animal.

2.8.2. Static Sciatic Index (SSI)

Functional recovery was analysed using the static sciatic index (SSI). The animal's hind paws were imaged from below with the animal standing on a Perspex platform and the toe spread factor (TSF), between the 1st and 5th toe, and the intermediary toe spread factor (ITSF), between the 2nd and 4th toe, were measured and Equation (1) was used to calculate SSI [31].

$$\begin{gathered} \mathrm{SSI} = (108.44 \times \mathrm{TSF}) + (31.85 \times \mathrm{ITSF}) - 5.49 \\ \mathrm{TSF} = (\mathrm{TS_{injury}} - \mathrm{TS_{control}})/\mathrm{TS_{control}} \\ \mathrm{ITSF} = (\mathrm{ITS_{injury}} - \mathrm{ITS_{control}})/\mathrm{ITS_{control}} \end{gathered} \tag{1}$$

2.8.3. Von Frey

The animals were placed on a grid and von Frey filaments made of nylon, which all have the same length but vary in diameter to provide a range of forces (0.008–300 g), were applied through the underside of the grid to stimulate the centre of the animal's hind paws. A response was measured by the retraction of the animal's paw in response to the filament stimulus. The threshold response was recorded by decreasing the stimulus until no response was detected.

2.9. Statistical Analysis

Normality tests were conducted on all data to determine appropriate statistical tests, and a one-way analysis of variance (ANOVA) was performed, as data followed a normal distribution. A one-way ANOVA was followed by a Dunnett's post hoc test. For all tests, * $p < 0.05$, ** $p < 0.01$, *** $p < 0.001$ and **** $p < 0.0001$ were considered significant.

3. Results

3.1. Regenerative Capacity of PPARγ Agonists In Vitro

PC12 neuronal cells seeded upon EngNT containing SCL4.1/F7 cells were treated with compounds sulindac sulfide, diclofenac, indomethacin, naproxen, pioglitazone, and GW1929 to modulate neurite outgrowth. Dose responses were conducted for each compound to determine their therapeutic window (Figure 1A). Doses tested in vitro were selected from previous literature to ensure non-toxic doses were used. The doses of the NSAIDs were kept the same as those previously used with ibuprofen [28] to make the results comparable across the drug class (some of these drugs had not been tested in PNI models previously and so no literature was available). Treatments with 10 µM, 100 µM, and 200 µM diclofenac and 100 µM sulindac sulfide significantly increased neurite extensions in comparison to the vehicle control (Figure 1A). Ibuprofen had previously shown to increase neurite growth and GW9662 to have no effect in the same model [28].

Table 1 shows the neurite length as a percentage increase to the experimental control, measured from co-cultures containing Schwann cells (SCL4.1/F7) and neurons (PC12s). Neurite length following the treatment with ibuprofen and GW9662 has been published previously [28]; the effect of the other compounds is presented in this study.

Table 1. The ranking order of an agonist's EC_{50} for PPARγ and neurite length increase in vitro.

Compound	PPARγ EC_{50}	Dose of Compound Tested In Vitro (µM)	Neurite Length Seen with Optimal Dose Percentage Increase to the Control (Mean ± SEM)
GW1929	13 nM [32]	10	51% ± 14.74
Pioglitazone	380 nM [33]	1	40% ± 15.58
Sulindac sulfide	1.87 µM [23]	100	106% ± 24.61
Diclofenac	1 µM [23]	100	73.93% ± 28.65
Ibuprofen	56.8 µM [23]	100	55.2% ± 17.92 [28]
Indomethacin	21 µM [23]	100	19.3% ± 17.35
Naproxen	No action [20]	100	1.8% ± 8.44
GW9662	2 nM [34] Irreversible Antagonist	100	−89% ± 10.41 [28]

Comparing a dose of 100 µM of each of the five NSAIDs and the optimal dose of pioglitazone (1 µM) and GW1929 (10 µM) indicated some correlation between the agonist's affinity for the receptor and increases in neurite outgrowth (Table 1).

Sulindac sulfide had the greatest effect on nerve regeneration, with a 106% increase in neurite length relative to the vehicle control (Figure 1B) and naproxen, which has been found not to activate PPARγ, showing minimal effect on nerve regeneration (Table 1).

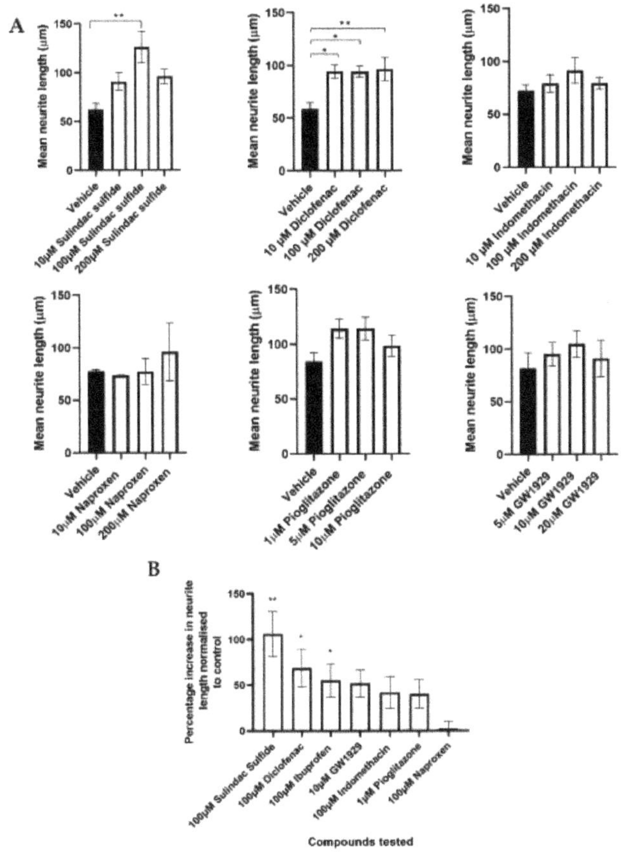

Figure 1. PC12 neuronal cells were seeded upon EngNT containing SCL4.1/F7 Schwann cells. (**A**) Significant increases in neurite length were seen in the presence of 10 μM, 100 μM, and 200 μM diclofenac and 100 μM sulindac sulfide compared to the vehicle control, after 72 h exposure. (**B**) The percentage increase in neurite length normalised to the control showed that sulindac sulfide had the greatest effect on neurite outgrowth and naproxen the least. Scale bar = 100 μm. N = 6, mean ± SEM for each condition. One-way ANOVA with Dunnett's post hoc test, * $p < 0.05$ and ** $p < 0.01$.

3.2. *Effect on Neuronal Growth of PPARγ Agonists In Vivo*

Agonists were released in a controlled manner locally at the nerve injury site using Alzet osmotic pumps. The doses delivered were based upon the dose given in vitro. Neurofilament-positive axons were quantified from 10 μm transverse rat sciatic nerve sections taken 5 mm distal to the injury site at 28 days post injury. All agonists increased the number of neurites present in comparison to the vehicle, however, statistically significant differences were only seen with 11 μg/day diclofenac, 11 μg/day sulindac sulfide, and 2 μg/day GW1929 treatment and not 12 μg/day indomethacin, 0.12 μg/day pioglitazone, and 8 μg/day naproxen (Figure 2).

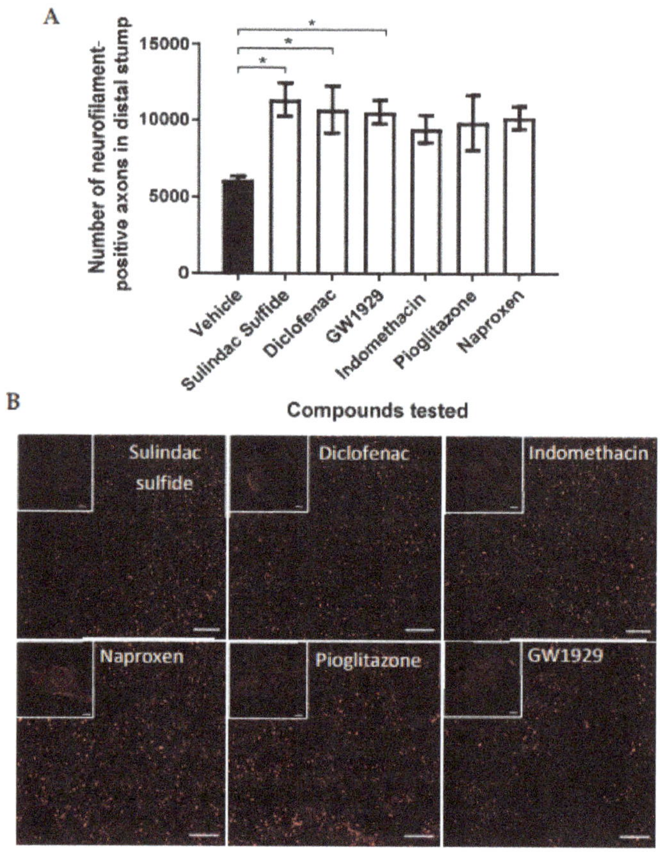

Figure 2. (**A**) The number of neurofilament-positive axons in the distal stump appeared to be elevated with all PPARγ agonists in comparison with the vehicle control, with statistical significance seen with sulindac sulfide, diclofenac, and GW1929 local treatment. Doses delivered locally to the site of injury were 11 μg/day diclofenac, 11 μg/day sulindac sulfide, 2 μg/day GW1929, 12 μg/day indomethacin, 0.12 μg/day pioglitazone, and 8 μg/day naproxen. (**B**) Micrographs are transverse sections showing neurofilament-positive neurites at 5 mm distal to the injury site. Scale bar = 100 μm. N = 6, means ± SEM. One-way ANOVA with Dunnett's post hoc test, * $p < 0.05$.

3.3. Functional Recovery Following In Vivo Treatment with PPARγ Agonists

To determine motor functional recovery, electrophysiology was used to measure the compound muscle action potential (CMAP) from the gastrocnemius muscle during electrical stimulation of the proximal nerve. CMAP amplitude and latency recordings from the contralateral nerves were consistent between animals (data not shown). Variability was seen between animals, with a trend that all PPARγ agonists increased the CMAP, except naproxen, in comparison to the vehicle group but with no statistical significance (Figure 3A). All compounds tested showed a trend towards a higher latency in comparison to no drug treatment, but no statistical significance was seen (Figure 3B).

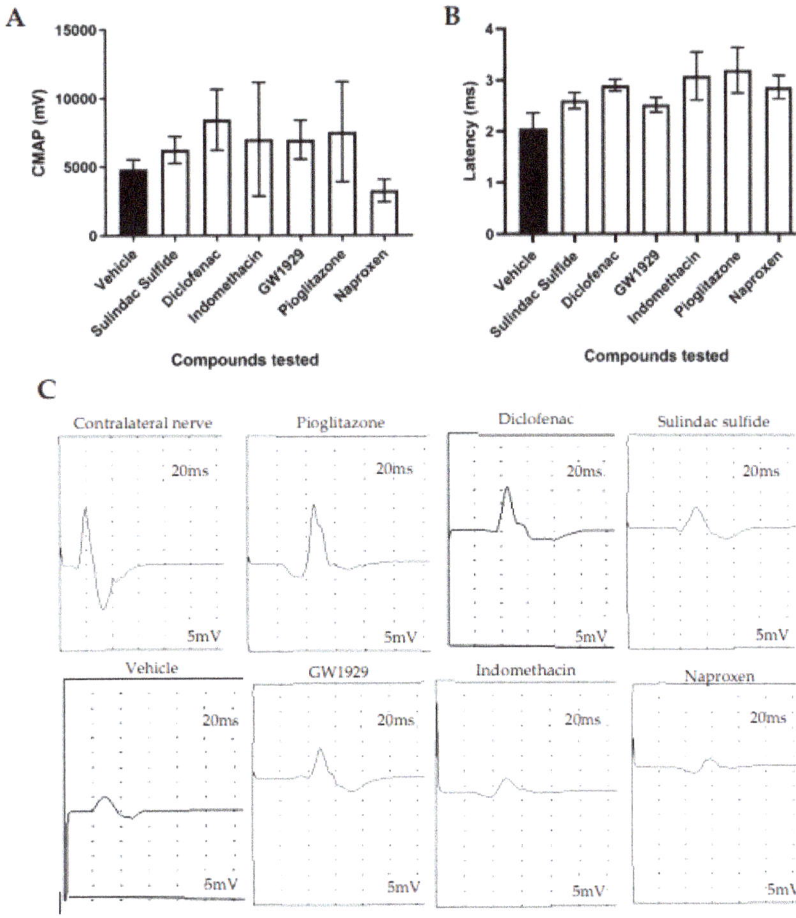

Figure 3. The sciatic nerve was stimulated proximal to the repair site and the CMAP was recorded from the gastrocnemius muscle. The CMAP amplitude reduced in the injury side in comparison with the contralateral non-injured side. (**A**) All compounds tested showed a trend towards a higher CMAP in comparison to the vehicle except naproxen, but no statistical significance was seen. (**B**) All compounds tested showed a trend towards a higher latency in comparison to the vehicle, but no statistical significance was seen. (**C**) The electrophysiological traces for the contralateral uninjured nerve and injured nerves. N = 6, means ± SEM.

To further monitor motor function, static sciatic index (SSI) was measured weekly post injury alongside von Frey to monitor sensory recovery. In both outcomes, recordings demonstrated an initial loss of function immediately after the crush injury with recovery seen in all groups returning towards the baseline by 28 days post injury. No difference was seen between the groups following SSI (Figure 4) and von Frey.

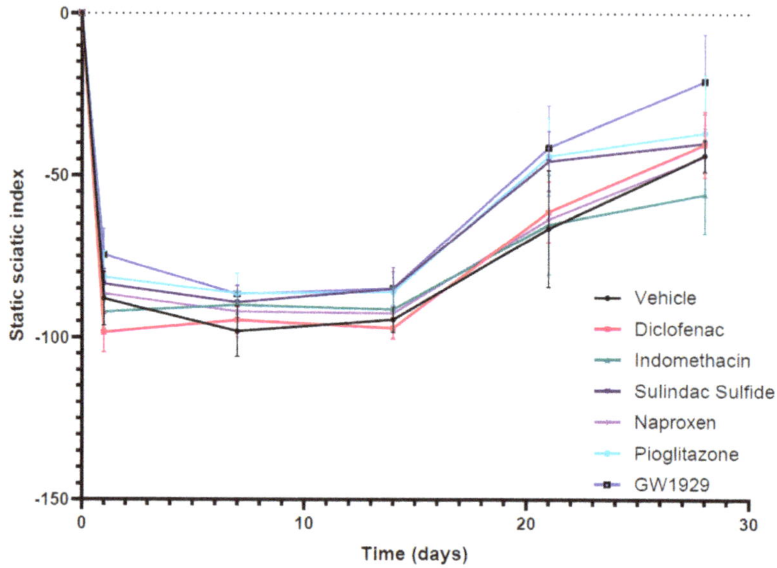

Figure 4. Hind paw images were used to conduct SSI quantitation in contralateral uninjured and injured sciatic nerves weekly until the 28-day endpoint. An initial loss of function can be seen immediately after the crush injury with recovery seen in all groups returning towards the baseline by 28 days post injury. N = 6, means ± SEM.

4. Discussion

Increased knowledge of the signaling pathways activated following a PNI, particularly the Rho/ROCK pathway, has helped the development of target-specific treatments with improved regeneration [35]. The inhibition of the Rho/ROCK pathway has been linked with the activation of PPARγ via the upregulation of the SHP-2, providing a platform for pharmacological intervention [7,20]. This study selected a panel of PPARγ agonists which were found to have a beneficial effect on nerve regeneration in vitro and in vivo. Additionally, it was found that there was some correlation between a drug's affinity for PPARγ and its capacity to promote neurite outgrowth in vitro, however, this wasn't apparent in vivo using a nerve crush model.

Using a 3D co-culture in vitro model containing neurons and Schwann cells and a sciatic nerve crush injury model in animals, we investigated the pro-regenerative effects of diclofenac, sulindac sulfide, indomethacin, naproxen, piogliazone, and GW1929. Ibuprofen has been tested in the same models in our previous studies [28]. Co-cultures comprised of both neuronal cells and Schwann cells were used to best mimic the native nerve environment following injury. There has been evidence of PPARγ activity in both neurons and Schwann cells [3–6], however, this was not explored in this study.

Ibuprofen is the most commonly studied NSAID in PNI promoting neurite growth in primary cell cultures in a dose-dependent manner [18,20,21]. In our previous study we presented a 55.2% increase in neurite growth following ibuprofen treatment compared to the vehicle in vitro. Further studies demonstrated that ibuprofen successfully increased axon number and functional recovery in animal studies [9].

Sulindac is a NSAID indicated for the relief of signs and symptoms of arthritic conditions, including osteoarthritis and rheumatoid arthritis. Sulindac is a pro-drug that undergoes reversible reduction to two metabolites: sulindac sulfide, which is of interest to us as it inhibits COX and Ras-mediated signal transduction, and sulindac sulfone [36,37]. Sulindac sulfide demonstrated the greatest effect on the neurite outgrowth in the 3D co-culture model by increasing the growth by 106% and demonstrated the highest axon number in the distal stump following the in vivo study. The drug's effect on neurite growth has not been reported before, however sulindac sulfide has previously shown to inhibit the activity of Rho in a concentration-dependent manner. The direct effect of sulindac sulfide on Rho activation was explored in SY5YAPPswe, HEK 293, and PC12 cells and the levels of active Rho-GTP were found to be reduced in all the cell lines [22]. This suggests that sulindac sulfide has a similar mechanism of action to the other NSAID by its activation of PPARγ linking to the inhibition of the Rho/ROCK pathway.

Diclofenac is another widely used NSAID as an analgesic and anti-inflammatory. In nerve regeneration, diclofenac has been found to have conflicting effects by both improving regeneration but also having harmful effects on developing nerves [38–41]. Diclofenac has been shown to be teratogenic; during the embryonic period, the number of nerve fibers and the cross-sectional area of axons in developing sciatic nerves were affected when exposed to diclofenac [39]. Prenatal administration of the drug impairs the sciatic, optic, and median nerve development in the child. Furthermore, diclofenac has been found to inhibit neuronal stem cell differentiation and reduce proliferation through the apoptotic pathway [42].

In comparison, this study demonstrated that diclofenac significantly promoted nerve regeneration in both a 3D co-culture in vitro model and crush injury animal model. This echoed another study's work that identified beneficial effects with diclofenac treatment in a transection injury on the rat sciatic nerve; by 16 weeks, increased regeneration was seen with diclofenac treatment suggested by a significant increase in gastrocnemius muscle weight, improved nerve fiber number, and axon diameter [14].

Indomethacin also increased neurite growth (19.3%) in the 3D co-culture model, however the effect was not as great as that seen with diclofenac (73%), ibuprofen (55.2%), or sulindac sulfide (106%). This mirrored the stimulation of neurite growth seen in a dorsal root ganglion (DRG) neuron culture when treated with indomethacin following exposure to inhibitory substrates [21]. Furthermore, the same study showed that indomethacin and ibuprofen blocked lysophosphatidic acid-induced Rho A activation in PC12 and DRG neuronal cells [21]. Indomethacin increased axon number in vivo but this was not significant.

The final NSAID tested was naproxen which had no effect on neurite growth in vitro. The different regenerative capacities between the five NSAIDs tested may be a result of their different chemical structures (Figure 5). This hypothesis has not been explored in this study, but it has been suggested that the PPARγ binding site cannot accommodate the CCH_3 group on the structure of naproxen [25].

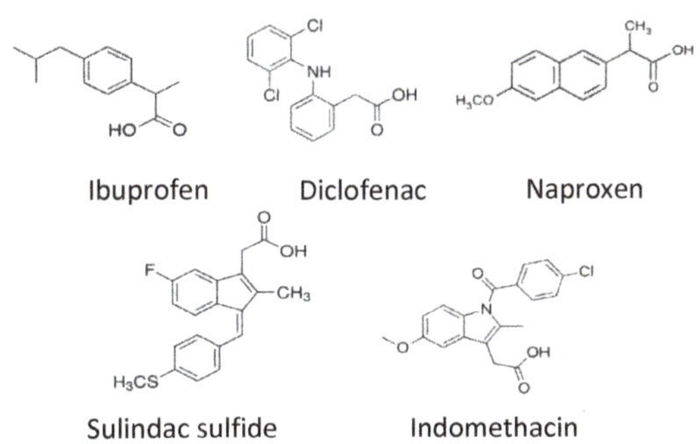

Figure 5. Chemical structures of NSAIDs.

Two selective PPARγ agonists were also tested in this study: pioglitazone and GW1929. GW1929 is a research compound and was used as a positive control. Pioglitazone is an anti-diabetic drug [43] that has previously demonstrated benefit following nervous system injury by providing neuroprotection succeeding focal cerebral ischemia in rats [44], and promoting remyelination following a sciatic nerve crush injury in mice [45]. In the 3D co-culture, pioglitazone stimulated an increase in neurite growth (40%) even at a lower dose than the other agonists.

Regeneration and functional recovery were measured 28 days post injury after the compounds were delivered locally to the injury site in a controlled manner using Alzet osmotic pumps. Histological analysis demonstrated a trend in the increase in axon number in the distal stump with indomethacin, pioglitazone, and naproxen and significant increases seen with diclofenac, sulindac sulfide, and GW1929. The extent to which the agonists improved axon number differed in vivo to what we observed in vitro. The greatest difference was seen with naproxen treatment; no regeneration was observed in vitro, but the compound increased the axon number in the distal stump in comparison to the vehicle in vivo. [21]) found that the effects of ibuprofen via the inhibition of the Rho A pathway are independent of COX action. However, for naproxen to have an effect it must be either through its inhibition of COX or another unknown mechanism. This could suggest that the NSAIDs are acting through a dual mechanism in PNI which needs to be explored further.

The electrophysiological results demonstrated a trend towards an increase in CMAP with all PPARγ agonists except naproxen, however none were significant. This suggests that there may have been an improvement in target muscle reinnervation with all treatments except naproxen. No significance was seen in latency in any of the treatment groups in comparison to the vehicle group. In all groups, the SSI and von Frey recordings returned to baseline in a similar manner over the 28-day experiment; however, no differences were seen between the groups.

The list of NSAIDs that have a demonstrated effect in PNI are not limited to the ones tested in this study. Celecoxib, a COX-2 inhibitor, was tested in a sciatic nerve crush injury in rats, which resulted in improved functional recovery. Significant improvements were observed in the sciatic functional index (SFI), which was considered to be a result of neuroprotection [46]. In the same model in mice, acetylsalicylic acid also improved functional recovery seen in the SFI, nociception, and gait [47]. In a transection injury model in a sciatic nerve being fixed to the adjacent muscle, ketoprofen improved functional recovery and enhanced regeneration of axons [48]. Exploring how other NSAIDs promote nerve regeneration could help us decipher the mechanisms through which these agonists are working and aid the development of a successful drug therapy for PNI.

5. Conclusions

In conclusion, our results indicate that the relationship between affinity of the selected agonists for PPARγ and their capacity to promote nerve regeneration is complex as they could be working through additional mechanisms. There is previous evidence of PPARγ activity in the different neural cell types—both neurons and Schwann cells [3–6]. It has also been reported that PPARγ has a role in the activation of macrophages [49], with a further study demonstrating a correlation between PPARγ activation and M2 macrophage (regenerative phenotype) marker expression [50]. Additionally, a study has explored the partial and full agonistic activity and binding properties of these compounds on PPARγ [23], which could explain the differences seen in their regenerative capacity. Further in vivo studies using various injury models over longer time periods would help us decipher these mechanisms and which cell types these compounds are working on. Additional outcome measures could be completed to broaden the understanding of how these drugs are benefiting functional recovery. Target innovation could be evaluated using, for example the foot flick test or motor unit number estimation (MUNE) and functional recovery using a rotarod test, grasping test, or algesimetry test [51].

Furthermore, additional external factors that impact nerve regeneration could be evaluated alongside drug treatments such as exercise and rehabilitation. Physical activity has demonstrated to be beneficial in many pathologies due to improvement in health and well-being and has been found to have a positive effect on axonal growth, phenotypic changes in structures, and neurotrophin levels following PNI [52].

Finally, the work presented here supports the view that PPARγ is a suitable target for drug therapies for PNI and further development in this area could prove to be promising in translating drug therapies for PNI into the clinic. Repositioning current approved drugs such as the NSAIDs could achieve this more rapidly as their safety profiles required for regulatory approval are already established.

Author Contributions: Conceptualization, M.L.D.R., J.H. and J.B.P.; Methodology, M.L.D.R. and J.B.P.; Validation, M.L.D.R., J.H. and J.B.P.; Formal Analysis, M.L.D.R., S.C.K. and I.K.; Investigation, M.L.D.R., S.C.K., I.K., O.G.-L. and H.G.; Resources, M.L.D.R. and J.B.P.; Writing—Original Draft Preparation, M.L.D.R.; Writing—Review & Editing, M.L.D.R. and J.B.P.; Visualization, M.L.D.R.; Supervision, J.H. and J.B.P.; Project Administration, M.L.D.R.; Funding Acquisition, M.L.D.R., J.H. and J.B.P. All authors have read and agreed to the published version of the manuscript.

Funding: This work was made possible through the funding from of the EPSRC (Grant EP/L01646X) and the UCL Repurposing Therapeutic Innovation Network (TIN).

Institutional Review Board Statement: All surgical procedures were performed in accordance with the UK Animals (Scientific Procedures) Act (1986), the European Communities Council Directives (86/609/EEC) and approved by the UCL Animal Welfare and Ethics Review Board (AWERB).

Informed Consent Statement: Not applicable.

Acknowledgments: Simon Kellaway and Isabel Kingston were funded by the UCL Therapeutic Acceleration Support (TAS) fund and Holly Gregory was funded by the Centre for Doctoral Advanced Therapeutics and Nanomedicine.

Conflicts of Interest: The authors declare no conflict of interest.

References

1. Hoke, A.; Brushart, T. Introduction to special issue: Challenges and opportunities for regeneration in the peripheral nervous system. *Exp. Neurol.* **2010**, *223*, 1–4. [CrossRef] [PubMed]
2. Wada, K.; Nakajima, A.; Katayama, K.; Kudo, C.; Shibuya, A.; Kubota, N.; Terauchi, Y.; Tachibana, M.; Miyoshi, H.; Kamisaki, Y.; et al. Peroxisome proliferator-activated receptor gamma-mediated regulation of neural stem cell proliferation and differentiation. *J. Biol. Chem.* **2006**, *281*, 12673–12681. [CrossRef] [PubMed]
3. Yamagishi, S.; Ogasawara, S.; Mizukami, H.; Yajima, N.; Wada, R.; Sugawara, A.; Yagihashi, S. Correction of protein kinase C activity and macrophage migration in peripheral nerve by pioglitazone, peroxisome proliferator activated-gamma-ligand, in insulin-deficient diabetic rats. *J. Neurochem.* **2008**, *104*, 491–499. [CrossRef] [PubMed]
4. Cao, Y.; Wang, Q.; Zhou, Z.; Wang, Y.; Liu, Y.; Ji, Y.; Liu, F. Changes of peroxisome proliferator-activated receptor-gamma on crushed rat sciatic nerves and differentiated primary Schwann cells. *J. Mol. Neurosci.* **2012**, *47*, 380–388. [CrossRef] [PubMed]
5. Zhang, F.; Liu, F.; Yan, M.; Ji, H.; Hu, L.; Li, X.; Qian, J.; He, X.; Zhang, L.; Shen, A.; et al. Peroxisome proliferator-activated receptor-gamma agonists suppress iNOS expression induced by LPS in rat primary Schwann cells. *J. Neuroimmunol.* **2010**, *218*, 36–47. [CrossRef]
6. Lezana, J.P.; Dagan, S.Y.; Robinson, A.; Goldstein, R.S.; Fainzilber, M.; Bronfman, F.C.; Bronfman, M. Axonal PPARgamma promotes neuronal regeneration after injury. *Dev. Neurobiol.* **2016**, *76*, 688–701. [CrossRef]
7. Wakino, S.; Hayashi, K.; Kanda, T.; Tatematsu, S.; Homma, K.; Yoshioka, K.; Takamatsu, I.; Saruta, T. Peroxisome proliferator-activated receptor gamma ligands inhibit Rho/Rho kinase pathway by inducing protein tyrosine phosphatase SHP-2. *Circ. Res.* **2004**, *95*, e45–e55. [CrossRef]
8. Chan, K.M.; Gordon, T.; Zochodne, D.W.; Power, H.A. Improving peripheral nerve regeneration: From molecular mechanisms to potential therapeutic targets. *Exp. Neurol.* **2014**, *261*, 826–835. [CrossRef]
9. Madura, T.; Tomita, K.; Terenghi, G. Ibuprofen improves functional outcome after axotomy and immediate repair in the peripheral nervous system. *J. Plast. Reconstr. Aesthet. Surg.* **2011**, *64*, 1641–1646. [CrossRef]
10. Hiraga, A.; Kuwabara, S.; Doya, H.; Kanai, K.; Fujitani, M.; Taniguchi, J.; Arai, K.; Mori, M.; Hattori, T.; Yamashita, T. Rho-kinase inhibition enhances axonal regeneration after peripheral nerve injury. *J. Peripher. Nerv. Syst.* **2006**, *11*, 217–224. [CrossRef]
11. Joshi, A.R.; Bobylev, I.; Zhang, G.; Sheikh, K.A.; Lehmann, H.C. Inhibition of Rho-kinase differentially affects axon regeneration of peripheral motor and sensory nerves. *Exp. Neurol.* **2015**, *263*, 28–38. [CrossRef] [PubMed]
12. Lehmann, J.M.; Moore, L.B.; Smith-Oliver, T.A.; Wilkison, W.O.; Willson, T.M.; Kliewer, S.A. An antidiabetic thiazolidinedione is a high affinity ligand for peroxisome proliferator-activated receptor gamma (PPAR gamma). *J. Biol. Chem.* **1995**, *270*, 12953–12956. [CrossRef] [PubMed]
13. Quintanilla, R.A.; Godoy, J.A.; Alfaro, I.; Cabezas, D.; von Bernhardi, R.; Bronfman, M.; Inestrosa, N.C. Thiazolidinediones promote axonal growth through the activation of the JNK pathway. *PLoS ONE* **2013**, *8*, e65140. [CrossRef] [PubMed]
14. Mohammadi, R.; Hirsaee, M.A.; Amini, K. Improvement of functional recovery of transected peripheral nerve by means of artery grafts filled with diclofenac. *Int. J. Surg.* **2013**, *11*, 259–264. [CrossRef]

15. Rayner, M.L.D.; Grillo, A.; Williams, G.R.; Tawfik, E.; Zhang, T.; Volitaki, C.; Craig, D.Q.M.; Healy, J.; Phillips, J.B. Controlled local release of PPARgamma agonists from biomaterials to treat peripheral nerve injury. *J. Neural Eng.* **2020**, *17*, 046030. [CrossRef] [PubMed]
16. Tamaddonfard, E.; Farshid, A.A.; Samadi, F.; Eghdami, K. Effect of vitamin B12 on functional recovery and histopathologic changes of tibial nerve-crushed rats. *Drug Res.* **2014**, *64*, 470–475. [CrossRef]
17. Kopp, M.A.; Liebscher, T.; Niedeggen, A.; Laufer, S.; Brommer, B.; Jungehulsing, G.J.; Strittmatter, S.M.; Dirnagl, U.; Schwab, J.M. Small-molecule-induced Rho-inhibition: NSAIDs after spinal cord injury. *Cell Tissue Res.* **2012**, *349*, 119–132. [CrossRef]
18. Wang, X.; Budel, S.; Baughman, K.; Gould, G.; Song, K.H.; Strittmatter, S.M. Ibuprofen enhances recovery from spinal cord injury by limiting tissue loss and stimulating axonal growth. *J. Neurotrauma* **2009**, *26*, 81–95. [CrossRef]
19. Lehmann, J.M.; Lenhard, J.M.; Oliver, B.B.; Ringold, G.M.; Kliewer, S.A. Peroxisome proliferator-activated receptors alpha and gamma are activated by indomethacin and other non-steroidal anti-inflammatory drugs. *J. Biol. Chem.* **1997**, *272*, 3406–3410. [CrossRef]
20. Dill, J.; Patel, A.R.; Yang, X.L.; Bachoo, R.; Powell, C.M.; Li, S. A molecular mechanism for ibuprofen-mediated RhoA inhibition in neurons. *J. Neurosci.* **2010**, *30*, 963–972. [CrossRef]
21. Fu, Q.; Hue, J.; Li, S. Nonsteroidal anti-inflammatory drugs promote axon regeneration via RhoA inhibition. *J. Neurosci.* **2007**, *27*, 4154–4164. [CrossRef] [PubMed]
22. Zhou, Y.; Su, Y.; Li, B.; Liu, F.; Ryder, J.W.; Wu, X.; Gonzalez-DeWhitt, P.A.; Gelfanova, V.; Hale, J.E.; May, P.C.; et al. Nonsteroidal anti-inflammatory drugs can lower amyloidogenic Abeta42 by inhibiting Rho. *Science* **2003**, *302*, 1215–1217. [CrossRef] [PubMed]
23. Puhl, A.C.; Milton, F.A.; Cvoro, A.; Sieglaff, D.H.; Campos, J.C.; Bernardes, A.; Filgueira, C.S.; Lindemann, J.L.; Deng, T.; Neves, F.A.; et al. Mechanisms of peroxisome proliferator activated receptor gamma regulation by non-steroidal anti-inflammatory drugs. *Nucl. Recept. Signal.* **2015**, *13*, e004. [CrossRef]
24. Rayner, M.L.D.; Healy, J.; Phillips, J.B. Repurposing Small Molecules to Target PPAR-gamma as New Therapies for Peripheral Nerve Injuries. *Biomolecules* **2021**, *11*, 1301. [CrossRef] [PubMed]
25. Xing, B.; Li, H.; Wang, H.; Mukhopadhyay, D.; Fisher, D.; Gilpin, C.J.; Li, S. RhoA-inhibiting NSAIDs promote axonal myelination after spinal cord injury. *Exp. Neurol.* **2011**, *231*, 247–260. [CrossRef] [PubMed]
26. Phillips, J.B.; Brown, R. Micro-structured materials and mechanical cues in 3D collagen gels. *Methods Mol. Biol.* **2011**, *695*, 183–196. [CrossRef]
27. East, E.; de Oliveira, D.B.; Golding, J.P.; Phillips, J.B. Alignment of astrocytes increases neuronal growth in three-dimensional collagen gels and is maintained following plastic compression to form a spinal cord repair conduit. *Tissue Eng. Part A* **2010**, *16*, 3173–3184. [CrossRef]
28. Rayner, M.L.D.; Laranjeira, S.; Evans, R.E.; Shipley, R.J.; Healy, J.; Phillips, J.B. Developing an In Vitro Model to Screen Drugs for Nerve Regeneration. *Anat. Rec.* **2018**, *301*, 1628–1637. [CrossRef]
29. Schindelin, J.; Arganda-Carreras, I.; Frise, E.; Kaynig, V.; Longair, M.; Pietzsch, T.; Preibisch, S.; Rueden, C.; Saalfeld, S.; Schmid, B.; et al. Fiji: An open-source platform for biological-image analysis. *Nat. Methods* **2012**, *9*, 676–682. [CrossRef]
30. Phillips, J.B. Monitoring neuron and astrocyte interactions with a 3D cell culture system. *Methods Mol. Biol.* **2014**, *1162*, 113–124. [CrossRef]
31. Bervar, M. Video analysis of standing–an alternative footprint analysis to assess functional loss following injury to the rat sciatic nerve. *J. Neurosci. Methods* **2000**, *102*, 109–116. [CrossRef] [PubMed]
32. Lamotte, Y.; Martres, P.; Faucher, N.; Laroze, A.; Grillot, D.; Ancellin, N.; Saintillan, Y.; Beneton, V.; Gampe, R.T., Jr. Synthesis and biological activities of novel indole derivatives as potent and selective PPARgamma modulators. *Bioorg. Med. Chem. Lett.* **2010**, *20*, 1399–1404. [CrossRef] [PubMed]
33. Obermoser, V.; Mauersberger, R.; Schuster, D.; Czifersky, M.; Lipova, M.; Siegl, M.; Kintscher, U.; Gust, R. Importance of 5/6-aryl substitution on the pharmacological profile of 4′-((2-propyl-1H-benzo[d]imidazol-1-yl)methyl)-[1,1′-biphenyl]-2-carboxylic acid derived PPARgamma agonists. *Eur. J. Med. Chem.* **2017**, *126*, 590–603. [CrossRef] [PubMed]
34. Quintanilla, R.A.; Jin, Y.N.; Fuenzalida, K.; Bronfman, M.; Johnson, G.V.W. Rosiglitazone treatment prevents mitochondrial dysfunction in mutant huntingtin-expressing cells: Possible role of peroxisome proliferator-activated receptor-gamma (PPARgamma) in the pathogenesis of Huntington disease. *J. Biol. Chem.* **2008**, *283*, 25628–25637. [CrossRef] [PubMed]
35. DeGeer, J.; Lamarche-Vane, N. Rho GTPases in neurodegeneration diseases. *Exp. Cell Res.* **2013**, *319*, 2384–2394. [CrossRef]
36. Lee, J.K.; Paine, M.F.; Brouwer, K.L. Sulindac and its metabolites inhibit multiple transport proteins in rat and human hepatocytes. *J. Pharmacol. Exp. Ther.* **2010**, *334*, 410–418. [CrossRef]
37. Gurpinar, E.; Grizzle, W.E.; Piazza, G.A. COX-Independent Mechanisms of Cancer Chemoprevention by Anti-Inflammatory Drugs. *Front. Oncol.* **2013**, *3*, 181. [CrossRef]
38. Ayranci, E.; Altunkaynak, B.Z.; Aktas, A.; Ragbetli, M.C.; Kaplan, S. Prenatal exposure of diclofenac sodium affects morphology but not axon number of the median nerve of rats. *Folia Neuropathol.* **2013**, *51*, 76–86. [CrossRef]
39. Canan, S.; Aktas, A.; Ulkay, M.B.; Colakoglu, S.; Ragbetli, M.C.; Ayyildiz, M.; Geuna, S.; Kaplan, S. Prenatal exposure to a non-steroidal anti-inflammatory drug or saline solution impairs sciatic nerve morphology: A stereological and histological study. *Int. J. Dev. Neurosci.* **2008**, *26*, 733–738. [CrossRef]

40. Colakoglu, S.; Aktas, A.; Raimondo, S.; Turkmen, A.P.; Altunkaynak, B.Z.; Odaci, E.; Geuna, S.; Kaplan, S. Effects of prenatal exposure to diclofenac sodium and saline on the optic nerve of 4- and 20-week-old male rats: A stereological and histological study. *Biotech. Histochem.* **2014**, *89*, 136–144. [CrossRef]
41. Keskin, I.; Kaplan, S.; Kalkan, S.; Sutcu, M.; Ulkay, M.B.; Esener, O.B. Evaluation of neuroprotection by melatonin against adverse effects of prenatal exposure to a nonsteroidal anti-inflammatory drug during peripheral nerve development. *Int. J. Dev. Neurosci.* **2015**, *41*, 1–7. [CrossRef] [PubMed]
42. Kaplan, A.A.; Yurt, K.K.; Deniz, O.G.; Altun, G. Peripheral nerve and diclofenac sodium: Molecular and clinical approaches. *J. Chem. Neuroanat.* **2018**, *87*, 2–11. [CrossRef] [PubMed]
43. Smith, U. Pioglitazone: Mechanism of action. *Int. J. Clin. Pract. Suppl.* **2001**, *121*, 13–18.
44. Zhao, Y.; Patzer, A.; Herdegen, T.; Gohlke, P.; Culman, J. Activation of cerebral peroxisome proliferator-activated receptors gamma promotes neuroprotection by attenuation of neuronal cyclooxygenase-2 overexpression after focal cerebral ischemia in rats. *FASEB J.* **2006**, *20*, 1162–1175. [CrossRef]
45. Eto, M.; Sumi, H.; Fujimura, H.; Yoshikawa, H.; Sakoda, S. Pioglitazone promotes peripheral nerve remyelination after crush injury through CD36 upregulation. *J. Peripher. Nerv. Syst.* **2008**, *13*, 242–248. [CrossRef]
46. Camara-Lemarroy, C.R.; Guzman-de la Garza, F.J.; Barrera-Oranday, E.A.; Cabello-Garcia, A.J.; Garcia-Tamez, A.; Fernandez-Garza, N.E. Celecoxib accelerates functional recovery after sciatic nerve crush in the rat. *J. Brachial Plex. Peripher. Nerve Inj.* **2008**, *3*, 25. [CrossRef]
47. Subbanna, P.K.; Prasanna, C.G.; Gunale, B.K.; Tyagi, M.G. Acetyl salicylic acid augments functional recovery following sciatic nerve crush in mice. *J. Brachial Plex. Peripher. Nerve Inj.* **2007**, *2*, 3. [CrossRef]
48. Mohammadi, R.; Mehrtash, M.; Nikonam, N.; Mehrtash, M.; Amini, K. Ketoprofen combined with artery graft entubulization improves functional recovery of transected peripheral nerves. *J. Cranio-Maxillo-Facial Surg.* **2014**, *42*, 2076–2081. [CrossRef]
49. Tontonoz, P.; Spiegelman, B.M. Fat and beyond: The diverse biology of PPARgamma. *Annu. Rev. Biochem.* **2008**, *77*, 289–312. [CrossRef]
50. Bouhlel, M.A.; Derudas, B.; Rigamonti, E.; Dievart, R.; Brozek, J.; Haulon, S.; Zawadzki, C.; Jude, B.; Torpier, G.; Marx, N.; et al. PPARgamma activation primes human monocytes into alternative M2 macrophages with anti-inflammatory properties. *Cell Metab.* **2007**, *6*, 137–143. [CrossRef]
51. Navarro, X. Functional evaluation of peripheral nerve regeneration and target reinnervation in animal models: A critical overview. *Eur. J. Neurosci.* **2016**, *43*, 271–286. [CrossRef] [PubMed]
52. Maugeri, G.; D'Agata, V.; Trovato, B.; Roggio, F.; Castorina, A.; Vecchio, M.; Di Rosa, M.; Musumeci, G. The role of exercise on peripheral nerve regeneration: From animal model to clinical application. *Heliyon* **2021**, *7*, e08281. [CrossRef] [PubMed]

cells

Review

Exploring the Regulatory Role of ncRNA in NAFLD: A Particular Focus on PPARs

Anirban Goutam Mukherjee [1], Uddesh Ramesh Wanjari [1], Abilash Valsala Gopalakrishnan [1,*], Ramkumar Katturajan [1], Sandra Kannampuzha [1], Reshma Murali [1], Arunraj Namachivayam [1], Raja Ganesan [2], Kaviyarasi Renu [3], Abhijit Dey [4], Balachandar Vellingiri [5] and Sabina Evan Prince [1]

1 Department of Biomedical Sciences, School of Biosciences and Technology, Vellore Institute of Technology (VIT), Vellore 632014, Tamil Nadu, India
2 Institute for Liver and Digestive Diseases, Hallym University, Chuncheon 24252, Republic of Korea
3 Centre of Molecular Medicine and Diagnostics (COMManD), Department of Biochemistry, Saveetha Dental College & Hospitals, Saveetha Institute of Medical and Technical Sciences, Saveetha University, Chennai 600077, Tamil Nadu, India
4 Department of Life Sciences, Presidency University, Kolkata 700073, West Bengal, India
5 Stem Cell and Regenerative Medicine/Translational Research, Department of Zoology, School of Basic Sciences, Central University of Punjab (CUPB), Bathinda 151401, Punjab, India
* Correspondence: abilash.vg@vit.ac.in

Citation: Mukherjee, A.G.; Wanjari, U.R.; Gopalakrishnan, A.V.; Katturajan, R.; Kannampuzha, S.; Murali, R.; Namachivayam, A.; Ganesan, R.; Renu, K.; Dey, A.; et al. Exploring the Regulatory Role of ncRNA in NAFLD: A Particular Focus on PPARs. *Cells* **2022**, *11*, 3959. https://doi.org/10.3390/cells11243959

Academic Editors: Kay-Dietrich Wagner and Nicole Wagner

Received: 27 September 2022
Accepted: 2 December 2022
Published: 7 December 2022

Abstract: Liver diseases are responsible for global mortality and morbidity and are a significant cause of death worldwide. Consequently, the advancement of new liver disease targets is of great interest. Non-coding RNA (ncRNA), such as microRNA (miRNA) and long ncRNA (lncRNA), has been proven to play a significant role in the pathogenesis of virtually all acute and chronic liver disorders. Recent studies demonstrated the medical applications of miRNA in various phases of hepatic pathology. PPARs play a major role in regulating many signaling pathways involved in various metabolic disorders. Non-alcoholic fatty liver disease (NAFLD) is the most prevalent form of chronic liver disease in the world, encompassing a spectrum spanning from mild steatosis to severe non-alcoholic steatohepatitis (NASH). PPARs were found to be one of the major regulators in the progression of NAFLD. There is no recognized treatment for NAFLD, even though numerous clinical trials are now underway. NAFLD is a major risk factor for developing hepatocellular carcinoma (HCC), and its frequency increases as obesity and diabetes become more prevalent. Reprogramming anti-diabetic and anti-obesity drugs is an effective therapy option for NAFLD and NASH. Several studies have also focused on the role of ncRNAs in the pathophysiology of NAFLD. The regulatory effects of these ncRNAs make them a primary target for treatments and as early biomarkers. In this study, the main focus will be to understand the regulation of PPARs through ncRNAs and their role in NAFLD.

Keywords: PPARs; NAFLD; ncRNA; liver; NASH

1. Introduction

Non-alcoholic fatty liver disease (NAFLD) already affects 25% of the worldwide adult population and is steadily increasing [1]. Hepatic steatosis without a high alcohol intake is defined as NAFLD. Although there is undoubtedly a genetic component to NAFLD susceptibility, increased insulin resistance is strongly linked to the onset and development of this disease. It includes many liver diseases, including steatosis with inflammation through NASH, fibrosis, cirrhosis, and hepatocellular cancer [2]. Insulin resistance is the pathophysiological hallmark of the metabolic syndrome, and NAFLD is regarded as its hepatic component [3]. The "two-hit" concept, where steatosis is considered the first hit and oxidative stress and injury as the second hit, can explain the process of NAFLD [4]. Based on the underlying pathogenesis, NAFLD can be classified as a primary or secondary metabolic syndrome, and insulin resistance is linked to primary NAFLD. The processes of advancement from steatosis to more severe liver inflammation and fibrosis have been explained in recent years, given that the pathophysiology of NAFLD/NASH is poorly understood [5].

The processes of advancement from steatosis to more severe liver inflammation and fibrosis have been explained in recent years, even though the pathophysiology of NAFLD/NASH is still poorly understood. Steatosis is considered the first setting for the progression of NASH, although the second hit is necessary to recruit inflammation [6]. Hepatic steatosis mainly results from the overaccumulation of triglyceride (TG) fat vacuoles in the liver due to improper lipid metabolism. Compared to steatosis patients and healthy controls, hepatic CYP2E1 expression levels were elevated in NASH patients, and expression was localized in steatosis regions [7]. Another important event in steatosis and NASH may be an increased FAO in cytochromes, the increased ROS generated by the CYP enzymes enhancing hepatic oxidative stress, and worsening liver damage [8]. Another factor that increases hepatic steatosis is de novo lipogenesis. It has been observed that plasma glucose and insulin concentrations lead to de novo lipogenesis. It ultimately dysregulated the β-oxidation process, leading to hepatic steatosis's progression [9]. During NAFLD progression, several factors, such as gut-derived microbial components, cytokines, and lipotoxicity, arise from different tissues, including adipose tissues [10]. Inflammatory responses in the adipose tissues eventually result in elevated levels of several pro-inflammatory cytokines, TNF, IL-6, IL-1, and CCL2, which are responsible for insulin resistance [10].

Additionally, endocrine conditions such as polycystic ovarian disorders, hypothyroidism, and growth hormone deficit may exist with NAFLD. The underlying endocrine pathology affects the treatment, which could change the prognosis of NAFLD [11]. According to recent studies, dehydroepiandrosterone sulfate insufficiency has been linked to the emergence of advanced NASH [12]. A bidirectional link connecting liver and endocrine processes is also shown by the growing reports of adrenal failure in individuals with end-stage liver disease and in those who have had liver transplants [13]. Endocrine dysfunction may result in metabolic liver disease in some people who may later be diagnosed with cryptogenic cirrhosis because endocrine hormones regulate cellular functions and the distribution of body fat [13]. ncRNAs were discovered in endocrine organs, and their functions in the growth and functions in endocrine tissues, as well as the possible associations of lncRNAs with particular disorders such as diabetes mellitus, were explained [14].

Contemporary living's high-power consumption and sedentary habits, which fuel several harmful lifestyle-related disorders, particularly NAFLD, are reflected in the rising incidence rate [15]. The second reason for liver transplantation in the USA is NAFLD, an epidemic liver disease that affects roughly one-fourth of the global total [16]. Additionally, although around 20–30% of patients with normal weight have NAFLD, 75–100% of people with obesity have the disease [17]. Insulin resistance, dyslipidemia, and a pro-inflammatory state are the defining features of NAFLD. The core of current NAFLD treatment is weight loss through lifestyle adjustment, which is challenging for most patients to accomplish and maintain [18]. Few pharmacological alternatives are available, and patients with progressing NASH have generally been the focus of treatment [19]. PPARs regulate dyslipidemia, inflammation, atherogenesis, and glucose homeostasis (insulin-sensitizing characteristics) [20]. As a result, these substances should have diverse effects on NAFLD etiology, making them a prime candidate for therapeutic development.

PPARs are a subclass of nuclear receptors known as peroxisome proliferator-activated receptors [21–23]. The nuclear receptor superfamily, of which PPARs are a part, has been shown to govern distinct stages of NAFLD through its many subtypes, making it an appealing target for therapeutic intervention [24]. For several metabolic illnesses, including diabetes, dyslipidemia, and cardiovascular diseases, including NAFLD, PPARs are thought to be potential therapeutic targets [25]. The activation of hepatic stellate cells (HSCs), inflammation, and glucolipid metabolism intimately associated with NAFLD are regulated by all PPARs isotypes [26]. The nuclear receptor family comprises PPARγ, PPARα, and

PPARδ (also known as PPARβ) [27]. In adipocytes, PPARγ is highly expressed in the adipocytes and liver and acts as an inducer of adipocyte development [28]. PPARβ/δ is mainly expressed in skeletal muscles, with skin and adipose tissue expressing it to a lesser extent. Adipose tissue has high levels of PPARs, which are transcriptionally upregulated in steatosis, activate lipogenic enzymes, and worsen steatosis [29].

In the prognosis and progression of liver diseases, lncRNAs act as potential markers and serve as a direct target for therapeutic purposes. Numerous lncRNAs have been demonstrated to be related to liver disorders [30,31]. miRNAs are endogenous, tiny non-coding RNAs that play a crucial role in regulating the expression of mRNA and proteins of the target genes. The 3′-untranslated regions (3′-UTRs), which typically contain specific stability elements (including miRNAs binding sites), are the targets of the miRNAs at the post-transcriptional level [32–34]. In numerous pathophysiological processes, microRNAs play a crucial role as post-transcriptional regulators. Usually, they impact mRNA stability or translation, which represses the expression of the target gene. It is believed that microRNAs regulate gene expression in metabolic-related disorders, including NAFLD [35].

ncRNAs have emerged as important gene expression regulators at transcriptional, post-transcriptional, and post-translational levels. A plethora of research over the last several decades has provided evidence that ncRNAs coordinate most activities of liver metabolism. They are implicated in various illnesses and cellular processes, and there is evidence that they cooperate to form a dynamic regulatory network [36]. lncRNAs have been linked to various physiological and pathological processes, including cell growth, death, metastasis, angiogenesis, and liver function. LncRNAs, which act as drivers of NAFLD development, are overexpressed in NAFLD [37]. In this review, we will focus on the role of microRNAs, lncRNAs, and circular RNAs as functional coordinators of NAFLD progression, as well as the current understanding of how their dysregulation contributes to aberrant metabolism and pathology in insulin resistance and other associated features related to NAFLD. We will also focus on the role of PPARs in the ncRNA pathway regulation. With an emphasis on the relationship between ncRNAs and PPARs in the formation of NAFLD, we seek to present an overview of the interplay of these events in this review.

2. Activation and Regulation PPARs

PPARs regulate the homeostasis of lipids. There are generally three PPARs subtypes, known as PPARα, PPARγ, and PPARβ/δ [22]. The liver, which significantly impacts overall body nutrient/energy balance, is where PPARα governs lipid metabolism. PPARδ is also upregulated in oxidative tissues, affecting genes associated with substrate delivery, oxidative phosphorylation (OXPHOS), and energy homeostasis [27,38]. The stimulation of PPAR β/δ, which is crucial for regulating lipid and glucose metabolism in the liver, may slow the development of NAFLD [23,38–41].

2.1. *PPARα*

Although PPARs are distributed everywhere, the liver is where it is most prominently expressed. Fatty acid absorption, β-oxidation, ketogenesis, bile acid production, and TG turnover are all crucially regulated [42,43]. Due to its distinct roles in controlling lipid metabolism, encouraging TG oxidation, and suppressing hepatic inflammation, nuclear receptor PPARα has recently attracted special attention as a promising drug target for treating type 2 diabetes and associated illnesses. It has sparked interest in PPAR agonists as therapeutic targets for NAFLD [44].

2.2. *PPARγ*

The liver is significantly protected by PPARγ from oxidation, inflammation, fibrosis, fatty liver, and tumors [45]. Adipose tissue is where PPARγ is highly expressed and plays a crucial part in controlling adipocyte differentiation, adipogenesis, and lipid metabolism [46]. Additionally, it should be mentioned that NAFLD patients and experimental animals have dramatically increased hepatic PPARγ expression [42]. In addition to being closely associated with adipogenesis in mature adipocytes, PPARγ can enhance pre-adipocyte differentiation into mature adipocytes [47]. In dormant hepatic stellate cells (HSCs), PPARγ is highly expressed, however, PPARγ is repressed during the fibrosis process. According to studies, activating PPAR prevents HSCs from becoming active and lessens the amount of collagen deposited during hepatic fibrogenesis. PPARγ is, therefore, a valuable target in the treatment of fibrosis [48,49].

2.3. PPAR β/δ

Although PPARβ/δ is present in many tissues, alcoholic liver disease is also attributed to the noticeably increased expression of PPARβ/δ in the gut, liver, and keratinocytes [29]. Additionally, the connection between PPARβ/δ and hepatic lipid buildup in alcoholic liver disease has been researched [50]. Alcohol-induced hepatic lipid accumulation is suppressed by PPARβ/δ ligand activation [51]. PPARα also controls the production of carnitine-acylcarnitine translocase (CAC T) [52], a protein that aids in transporting fatty acids across the mitochondrial membrane [53]. Additionally, it can control how fatty acids are metabolized in peripheral tissues such as the liver. In line with these results, PPARβ/δ ablation worsens the hepatic TG response to alcohol consumption [54]. Additionally, numerous models prove that PPARβ/δ ligand activation blocks several molecular targets that stop NAFLD [55]. Additionally, various models suggest that PPARs may play a similar function in hepatotoxicity as a preventative and treatment for alcoholic liver disease (Figure 1) [56].

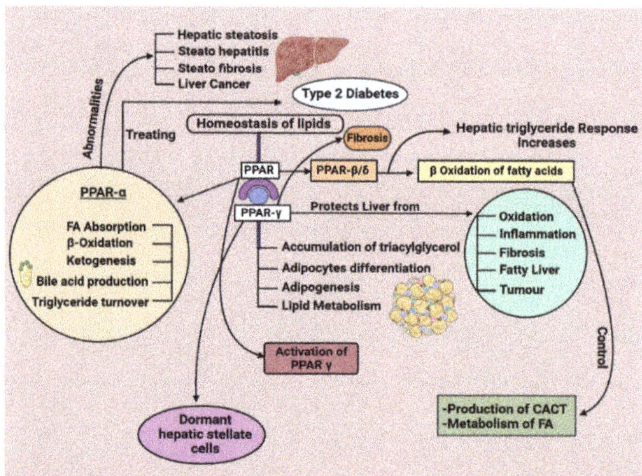

Figure 1. This illustration represents the regulatory mechanisms of PPARs involved in NAFLD. It gives a complete overview of the various PPARs (PPARα, PPARγ, and PPARβ/δ) and their critical roles and mechanisms in the body.

3. A Relationship between PPAR and NAFLD

The activation of total PPARγ can have detrimental effects by stimulating lipogenesis in the liver. However, there are advantages to partial PPARγ activation, mostly brought about by elevated adiponectin levels and lower insulin resistance. Adipocytes respond to PPARγ by secreting adiponectin, inhibiting resistin and visfatin, and differentiating into adipocytes [57]. Additionally, it blocks pro-inflammatory signaling molecules such as inducible nitric oxide synthase (iNOS), NF-κB, TNF-α, and IL-6, causing inflammatory cascades to be suppressed. The insulin-sensitizing activity of PPARγ ligands is mediated by all these effects [58].

Paradoxically, there is a strong correlation between the onset of NAFLD and hepatocyte-specific PPARγ expression. Numerous clinical research found that NAFLD patients had higher levels of hepatic PPARγ expression. Although PPARγ activation in adipose tissue offers a unique method for managing NAFLD, the side effect of activation of PPARγ will be hepatic steatosis [59]. Even though PPARγ's primary function is in adipose tissue, where the highest expression levels are seen, patients with NAFLD have much higher hepatic expression levels of PPARγ, reflecting the fact that different tissues and cell types play different roles for PPARγ in different contexts. As a result, NAFLD patients using PPARγ agonists may see a different response in their liver than in their adipose tissue. Due to the various functions of PPAR-γ, new treatment strategies include the development of drugs that harness these positive effects while limiting adverse effects such as adipogenesis that can lead to weight gain. Additionally, preclinical research on dual agonists targeting two or more PPARs has yielded promising results, and some of these agents are now moving toward clinical trials [60].

Even though excessive adiposity and obesity are frequently linked to diabetes and insulin resistance, PPARγ works as an insulin sensitizer by encouraging the development of adipose tissue [61]. This apparent disparity is due to the development of metabolically abnormal adipocytes during chronic dyslipidemia, which is frequently accompanied by obesity. Hypertrophic adipocytes that produce TNF-α and have increased rates of lipolysis due to insulin resistance are characteristics of dysfunctional adipose tissue [62]. Enhancing lipolysis leads to the release of free fatty acids (FFA), which are ectopically stored in organs such as the liver, resulting in steatosis and lipotoxicity, ultimately leading to NASH and cirrhosis [63,64].

PPARγ has distinct functions in liver cells. PPARγ plays a steatogenic role in hepatocytes by regulating the expression of adipogenic genes, including Ap2 and CD36, which increases FFA uptake. The simultaneous activation of FAS and ACC1 facilitate the buildup of TG inside cells. It lowers the release of inflammatory cytokines (such as TNF-α and MCP1) and growth factors (such as TGFβ), which reduces inflammation and activates hepatic stellate cells, which in turn lessens fibrosis. A quiescent phenotype of the liver stellate cells (HSC) is also associated with PPAR-γ, limiting their activation and fibrosis [60].

The expected result of the elevated hepatic PPARγ expression observed in NAFLD patients would be activation of DNL in the liver cells and increased expression of adipogenic genes, exacerbating steatosis [65]. However, clinical trials suggest that the PPARγ agonists rosiglitazone and pioglitazone significantly reduce hepatic steatosis in NAFLD patients [66]. This mitigation is probably brought about by changes in the adipose tissue, where activation of PPARγ promotes the growth of healthy adipose tissue, limiting the shunting of extra lipids to the liver and the development of defective adipocytes [42]. The weight gain observed in TZD-treated NAFLD patients is consistent with an enhanced formation of adipose tissue [67]. Although liver PPARγ levels are elevated in NAFLD patients, the same factors that led to this upregulation throughout the illness are yet unknown. PPARγ may be activated in reaction to lipid build-up in hepatocytes, or it may be activated in response to stimuli before the build-up of lipids [68].

4. ncRNA Regulation in NAFLD

The molecular mechanisms underlying the initiation and development of NAFLD fibrosis are increasingly attributed to lncRNAs as significant contributors. There have been reports of specific potential processes by which lncRNAs may contribute to fibrogenesis, but much more research is still required to fully comprehend the role of lncRNAs in the progression of NAFLD fibrosis. Aptr, Malat1, Neat1, and HOTAIR are some of the lncRNAs found in the CCl_4 mouse model that may also be relevant to human NAFLD fibrosis [69–71]. To understand the pathogenic mechanisms of NAFLD and establish miRNAs as diagnostic biomarkers, potential therapeutic targets at the early stage of NASH, and potential predictors of HCC, the emerging studies in NAFLD/NASH have used miRNA analysis as a starting point [72]. Several pieces of research on NASH have consistently found overexpression of miR-34a and downregulation of the liver-specific miR-122 [73,74]. CircRNAs_0046366 and 0046367 deficiency have been identified as NAFLD features, and restoration of these circRNAs reduces oxidative stress, lipotoxicity, and the severity of the illness. Circ_0071410 silencing has been demonstrated to reduce HSC activation, a crucial stage in developing liver cirrhosis. While circMTO1 adversely affects the progression of HCC, CDR1 and circ_0067934 can enhance the invasion and metastasis in HCC [75].

4.1. miRNAs

4.1.1. miR-122

By increasing Sirt1 and turning on the AMPK pathway, miR-122 inhibition protects hepatocytes from lipid metabolic diseases such as NAFLD and inhibits lipogenesis [76]. By controlling its targets, such as FASN, ACC, SCD1, and SREBPs, miR-122 is crucial for the metabolism of fatty acids, TGs, and cholesterol, as well as for the terminal differentiation of hepatocytes [76–79]. Additionally, animals with mir-122a deletions triggered steatosis that resulted in NASH, fibrosis, and HCC, indicating that this miRNA is crucial for developing NAFLD. Mice lacking miR-122 had higher liver cholesterol and TG levels but lower serum cholesterol and TG levels [79]. Changes in serum and hepatic cholesterol and TG levels are caused by changes in very-low-density lipoprotein assembly and secretion in a way that is miR-122-dependent [80,81]. This disagreement has been seen in numerous research studies and may result from applying various models and inhibitory techniques [78]. Additionally, miR-122 participates in the NAFLD fibrogenic and carcinogenic-signaling pathways. miR-122 is a vital regulator of the epithelial-to-mesenchymal transition, a crucial step in developing chronic inflammation, fibrosis, and metastasis. When miR-122 expression was downregulated, MEKK-3, vimentin, and hypoxia-inducible factor-1 were activated [77,79,82].

4.1.2. miR-21

According to Becker et al.'s analysis of the serum profiles of two cohorts consisting of 137 NAFLD/NASH patients and 61 healthy controls, the patients with NASH had higher levels of miR-21 in their blood than the NAFLD patients and healthy controls [83]. According to scientists, it may be caused by increased fibrosis in NASH patients [84,85].

4.1.3. miR-34a

miR-34a is another miRNA that appears to play a role in the emergence of NAFLD. For instance, a study that involved 34 patients found that this miRNA was exclusively found in the serum of people with NAFLD/NASH and not in healthy individuals. This finding has also been verified by Liu et al. [86,87]. This abnormal increase has a deleterious influence on the signaling of fibroblast growth factors 19 and 21 and is strongly linked with BMI in patients with obesity [88,89]. Lipids also drive the expression of miR-34a, and since sirtuin 1 has been identified as one of its targets, this miRNA appears to have a role in exacerbating the symptoms of NAFLD and NASH, mostly by raising p53 acetylation and inducing hepatocyte death [79,90,91].

4.1.4. miR-192

MiR-192 stands out among the miRNAs involved in NAFLD because it affects lipid synthesis by targeting stearoyl-CoA desaturase 1, making its upregulation a potential treatment for the condition [87,92–95].

4.1.5. miR-370

The FA oxidation enzyme CPT1A is a direct target of miR-370. Fascinatingly, miR-370 may influence miR-122 expression, contributing to TG buildup in the liver. Additionally, miR-370 overexpression in HepG2 cells activates genes involved in lipogenesis, including FAS and ACC1, by altering the expression of SREBP-1c [35,96].

4.1.6. Other miRNAs

By inhibiting cholesterol export and FA oxidation, miR-33 regulates cholesterol homeostasis [97]. Its blockage boosts reverse cholesterol transport, decreases LDL and TGs, and raises plasma HDH [98]. However, mice on a high-fat diet experience a faster progression of fatty liver and hepatic steatosis when miR-33 is knocked out. It is because miR-33 increases the number of lipids in the blood, particularly TGs [99,100]. According to prior studies, the upregulation of the miR-29 family inhibits insulin-stimulated glucose absorption through the Akt pathway by obstructing insulin signaling. In 2011, researchers found that the miR29 family is upregulated in HepG2 cells, causing insulin resistance that connects all components of the metabolic syndrome and is the most likely risk factor for developing NAFLD [101,102]. In the livers of humans and murine with NAFLD/NASH, the miR-873-5p increased, inhibiting GNMT. In in vivo NAFLD mouse models, its blockage decreases liver steatosis, inflammation, and fibrosis [103]. Several studies have demonstrated that miR-221 and its paralog miRNA-222 worsen hepatocarcinogenesis by targeting apoptosis-related proteins such as p53, p53 upregulated modulator of apoptosis (PUMA), NF-κB, and signal transducer and activator of transcription 3 (STAT3) [104,105].

Extracellular matrix (ECM) buildup in excess alters the liver's typical architecture, causing pathophysiological damage that eventually leads to liver fibrosis–cirrhosis. ECM's increased synthesis and deposition in response to a fibrotic stimulus have been linked to activated hepatic stellate cells (HSC) [106,107]. The main factor leading to liver fibrosis is the activation of HSCs. In stimulating LX-2 cells, miR-21 plays a vital role as a mediator through the PTEN/Akt pathway. miR-21's fibrogenic effects on the activation of LX-2 cells are mediated. miR-21 may be a potential new molecular target for liver fibrosis [108]. The miR-27a and miR-199a are recognized as HSC activators during fibrogenesis, whereas miRNAs, including miR-335, miR-150, and miR-194, have been identified as HSC modulatory molecules [109]. However, more research is needed to determine their significance in fibrogenesis and to evaluate their relationship with TGF-β/Smad signaling [110]. A difference in HSC activation stages may cause a discrepancy between studies. An HSC is exposed to a variety of additional stimuli from surrounding cells. It may lead to in vivo activation of HSC with a different outcome. The expression and role of miR-192 in the activation of HSC in human livers with various pathologies, mouse livers with liver fibrosis, or HSC, must thus be further investigated [111].

4.2. lncRNAs

lncRNAs may serve as prospective indicators for the prognosis and development of liver disorders and direct treatment targets. Several lncRNAs have been linked to liver disorders. In this review, we will discuss different ncRNAs and their role in the progression of NAFLD.

4.2.1. MALAT1

The significance of lncRNA metastasis-associated lung adenocarcinoma transcript 1 (MALAT1) in increased proliferation and inflammation has received the most attention in studying a wide range of illnesses [112]. It is related to lung, liver, heart, and kidney fibrosis and can control gene expression at several molecular levels [113]. According to research on NASH fibrosis, MALAT1 is essential in modulating the chemokine CXCL5 [114]. According to a study by Xiang et al., in NAFLD, MALAT1 expression increased in vitro and in vivo, and MALAT1 knockdown prevented FFA from causing hepatocytes to accumulate lipids. Additionally, ARNT's interaction with the PPARs' promoter could limit PPARs' expression. PPARs knockdown reversed these phenomena, whereas MALAT1 knockdown greatly increased PPARs' levels and decreased CD36 expression. By controlling the miR-206/ARNT axis, MALAT1 regulated PPARs/CD36-mediated hepatic lipid accumulation in NAFLD. Consequently, MALAT1/miR-206/ARNT may be a therapeutic target for NAFLD [115].

4.2.2. HOTAIR

During liver fibrosis, lncRNA and HOTAIR inhibits the expression of PTEN in hepatic stellate cells and is involved in the dysregulation of liver lipids. In the study by Li et al., they looked at whether HOTAIR could be a possible mediator of PTEN downregulation and lipid accumulation in hepatic cells at the onset of NAFLD. In HepG2 cells, exposure to FFAs increased TG accumulation by markedly boosting HOTAIR expression and suppressing PTEN expression (both at mRNA and protein levels). After stopping the FFAs therapy, the impacts on HOTAIR and PTEN expressions disappeared. In HepG2 cells treated with FFAs, PTEN downregulation and TG accumulation were inhibited by SiRNA-mediated HOTAIR knockdown, and they were also blocked by CAPE (an NF-Bp65 inhibitor). In HepG2 cells, FFAs may cause the upregulation of HOTAIR, which is likely dependent on NF-kB signaling. It would reduce PTEN expression and encourage the accumulation of TGs [116]. Interestingly, lncRNA HOTAIR suppressed another lncRNA, MEG3, through epigenetic pathways in a study on fibrotic animals. The subsequent reduction in MEG3 expression and promotion of liver fibrosis resulted from these promoter changes [117,118].

4.2.3. APTR

The lncRNA Alu-mediated p21 transcriptional regulator (APTR), which is putatively involved in liver fibrogenesis, is increased in fibrotic liver samples. By preventing TGF-β-dependent induction of α-SMA in vivo, the reduction in APTR prevents collagen formation [119,120]. In fibrosis patients and animal models, high expression of APTR was seen [119]. Extracellular matrix protein (ECM) buildup and HSC activation were prevented by silencing APTR expression [119,121]. APTR was overexpressed and activated the hepatic stellate cells in fibrotic liver tissues. The activation of HSCs in vitro was prevented, and the build-up of collagen in vivo was reduced by APTR knockdown. In primary HSCs, p21 siRNA1 reduced the effects of APTR knockdown on the cell cycle and cell proliferation [120]. In vivo COL1A accumulation was reduced, and in vitro HSC activation was inhibited by APTR knockdown. Finally, elevated levels of APTR in the serum of patients with liver cirrhosis point to APTR as a potential biomarker for the disease [119]. Additional research on APTR in sera from large cohorts will likely provide more insight into the significance and role of APTR in fibrosis linked to NAFLD [122].

4.2.4. PVT1

It was shown that Plasmacytoma Variant Translocation 1 (PVT1), whose function was more prominent in different malignancies, played a role in the development of fibrotic liver tissues by suppressing the expression of PTCH1 and stimulating the Hedgehog pathway. These pathways are essential for liver fibrosis and collagen deposition [123]. Patients with NAFLD had elevated relative expression levels of the lncRNA-PVT1. Additionally, patients with complex cirrhosis and hepatocellular carcinoma (HCC) have considerably greater levels of lncRNAPVT1 in the advanced stages of NAFLD. Given this, the lncRNA-PVT1 levels might be a helpful diagnostic biomarker for identifying patients with advanced NAFLD stages [124]. miR-152 was a driver of EMT and HSC activation through the suppression of Patchd1 (PTCH1) methylation and activation of the Hedgehog pathway during the characterization of a putative signaling network [69].

4.2.5. lncRNA COX2

lncRNAs are becoming recognized as important players in the molecular mechanisms underlying the development and progression of NAFLD fibrosis [69]. Numerous lncRNAs are expressed as a result of germline-encoded receptors such as the Toll-like receptors. One of these, lincRNA-Cox2, activates and suppresses various immunological gene subclasses [125]. These genes appear to be co-regulated because the most strongly stimulated lncRNAs tended to be found in chromosomal locations where immune gene expression was also elevated. One of the most strongly stimulated lncRNAs was lincRNA-Cox2, located close to the prostaglandin-endoperoxide synthase 2 (Ptgs2/Cox2) gene [125]. In

CCl4-induced fibrotic mice, lncRNA-Cox2, a lncRNA close to Cox2/Ptgs2, was examined since there were indications that it was involved in regulating inflammatory genes [126].

4.2.6. NEAT1

By controlling heterogeneous nuclear ribonucleoprotein A2, NEAT1 expression was elevated in HCC, whereas its knockdown decreased HCC cell proliferation, invasion, and migration [127]. When Neat1 was knocked down using an adenovirus, the CCl4-induced liver fibrosis in these animals was reduced. Neat1 expression was also higher in the entire livers and primary HSCs produced from CCl4-treated mice than in oil-fed controls [128]. By modulating the c-Jun/SREBP1c axis by sponging miR-139-5p, NEAT1 exacerbated the FFA-induced lipid accumulation in hepatocytes, suggesting its potential as a novel therapeutic target for NAFLD [114]. For the prevention and treatment of NAFLD, lncRNA NEAT1 may be a potential biological target. Further evidence that the control of lncRNA NEAT1 expression in NAFLD subjects is connected to liver function and lipid metabolism comes from the robust correlation between the lncRNA NEAT1 expression in the peripheral blood of NAFLD patients and the ALT, GGT, TC, and TG levels [129]. NEAT1 increases the development of hepatocellular carcinoma, NAFLD, and liver fibrosis while acting as a preventative in the pathogenesis of acute-on-chronic liver failure by suppressing the inflammatory response [130,131]. Adipogenesis has been reported to need miR-140, and NEAT1 contains a particular binding site for this miRNA. NEAT1's expression and stability are improved due to the interaction between miR-140 and NEAT1. Additionally, research has shown that adipogenesis needs NEAT1 to be activated in a miR-140-dependent manner [132].

4.2.7. SRA

The steroid receptor RNA activator (SRA) was initially reported as a lncRNA that increases the expression of nuclear receptors' steroid-dependent genes [131]. A study looking at the involvement of SRA in NAFLD in SRA knockout mice showed that the lack of SRA upregulates the expression of hepatic ATGL. Additionally, SRA inhibits the expression of ATGL in hepatocytes, preventing FFA from being oxidized. Although forced SRA expression suppresses ATGL expression and FFA-oxidation, loss of SRA in primary hepatocytes or a hepatocyte cell line upregulates hepatocyte function. SRA blocks the forkhead box protein O1 (FoxO1) transcription factor's ordinary inductive functions to limit ATGL promoter activity [133].

4.2.8. UC372

One of the ultra-conserved lncRNAs, the ultra-conserved element (UC372), has 100% identity in the rat, mouse, and human genomes [134]. The upregulation of UC372 in mice with type 2 diabetes (db/db mice), mice on a high-fat diet (HFD), and NAFLD patients suggest that this lncRNA has a role in liver steatosis and fatty liver. The prevention of miR-195/miR-4668-related target genes, such as fatty acid synthase (FASN), acetyl-CoA carboxylase (ACC), stearoyl-CoA desaturase 1 (SCD1), and lipid uptake-related genes such as CD36, has been proposed as a mechanism by which UC372 causes hepatic steatosis and causes the buildup of hepatic lipids [135].

4.2.9. lncARSR

A previous study suggests that lncARSR may be connected to hepatic steatosis. It is less apparent how lncARSR affects NAFLD. A prior study showed that over-expressing the lncARSR gene accelerated the formation of liver fat both in vivo and in culture, suggesting that lncARSR may play a role in NAFLD and serve as a potential new treatment target for the disease [136]. The studies showed that lncARSR silencing might treat NAFLD by shutting down the IRS2/AKT pathway through YAP1. YAP1 has already been seen to rise in cases of liver disease as the severity of the liver damage increases. Another study revealed that LATS2 controlled YAP1's phosphorylation and regulation in NAFLD. It has been demonstrated that lncARSR is linked with YAP1 and encourages YAP import into the nucleus [31].

4.2.10. APOA4-AS

A plasma lipoprotein called APOA4 is involved in the control of several metabolic processes, including lipid and glucose metabolism. In rats, the liver and small intestine manufacture APOA4 before secreting it into the blood [137]. An in vivo investigation has demonstrated that APOA4 increases hepatic TG output [137]. According to a recent study, the antisense lncRNA APOA4-AS, produced from the APOA4 gene's reverse strand and partially overlaps with the gene's 3 ends, is crucial in controlling the expression of the APOA4 gene. This correlation could be a therapeutic target for treating NAFLD-affected patients [76].

4.2.11. lncRNA H19

A nearby reciprocally imprinted gene for Igf2 and H19 is found in a highly conserved gene cluster. Steatosis and NAFLD caused by a high-fat diet (HFD) increased H19 expression. H19 silencing lowered the lipid accumulation in hepatocytes, while H19 overexpression caused lipid accumulation and upregulated several genes involved in lipid metabolism. By upregulating MLXIPL/ChREBP and silencing, Mlxipl reduced H19-induced hepatic steatosis. H19 silencing decreased the mTORC1 signaling complex, upregulated by H19 overexpression in hepatocytes. H19 increased hepatic steatosis by upregulating mTORC1 and MLXIPL in hepatocytes, according to hepatocyte implantation experiments [138,139].

4.2.12. lncRNA NONMMUT010685 and NONMMUT050689

LncRNA NONMMUT010685 was identified to control the XBP1 gene in a co-expression system. The unrelated process of typical liver fatty acid production needed XBP1, a crucial regulator of the unfolded protein response. XBP1 plays a major role in human dyslipidemias [140]. Patients with NASH who have inadequate protein degradation in response to XBP1 may be more likely to develop cirrhosis [141]. In the co-expression network of the lncRNA-mRNA study, it was discovered that the lncRNA NONMMUT050689 might control the RIPK1 gene. It is known that RIPK1′s kinase activity promotes RIPK3-mediated necroptosis and that RIPK1 is involved in the inflammatory and cell death pathways. During NASH, it was discovered that RIPK1 in hepatocytes prevented the advancement of liver fibrosis [99,142].

4.2.13. Other lncRNAs

LncRNAs can be employed as prospective targets for the diagnosis and therapy of NAFLD even though the molecular mechanism of the disease has not yet been fully understood. Upregulation of the lncRNA Gm15622 in NAFLD caused by HFD was examined in a study that discovered that Gm15622 could sponge miR-742-3p, boosting the quantity of SREBP-1 protein, a transcription factor that controls the expression of genes that govern the production of fatty acids, lipids, and cholesterol. Gm15622 overexpression can increase SREBP-1c expression, encouraging hepatic fat build-up [143]. In addition to being a possible prognostic biomarker and therapeutic target for PC, RUNX1-IT1 is an essential oncogenic lncRNA that controls and recruits RUNX1 to trigger c-FOS expression [144]. Inflammation, fibrosis, and NASH activity scores were all substantially linked with RUNX1 expression in NASH patients. Studies suggest a correlation between the lncRNA RUNX1-IT1 and its role in NAFLD [145] (Table 1).

Table 1. Table showing the research studies completed on different miRNAs and lncRNAs and the underlying mechanism.

ncRNA	Cells Involved	Animal Model Used	Genes Downregulated/Upregulated	Mechanism	Pathway Activated	Reference
miR-122	HepG2 and Huh-7 cells	16 C57BL/6 mice (males, 8 weeks old)	Downregulation of Sirt1	Binds to 3′-UTR and substantial mRNA degradation downregulates sirt1.	AMPK pathway	[146]
miR-21	CD45$^+$ cells, T cells	Eight-week-old male Ldlr$^{-/-}$ mice, nine-week-old male WT or miR-21$^{-/-}$ mice	Downregulation of ß-oxidation genes Cpt1 and Acox1	Through normalizing liver PPARα, miR-21 inhibition and suppression significantly reduced liver damage, inflammation, and fibrogenesis.	PPARα pathway	[84]
miR-34a	Kupffer cells	miR-34a$^{fl/fl}$ mice, albumin-Cre (Alb-Cre) mice, and C57BL/6J mice	Induces (Cyp7a1 and Cyp8b1)	miR-34a promotes the induction of Kupffer cell activation/inflammation, lipotoxicity, and apoptosis; boosts CYP7A1 and CYP8B1 expression, which causes a temporary drop in free cholesterol (FC) levels and the subsequent induction of SREBP2 processing and its target genes, including HMGCR. It enhances the conversion of FC to bile acids.	miR-34a-sirtuin 1 (SIRT1)-SREBP1c pathway activates hepatic TGFb signaling,	[147]
miR-192	HepG2 cells	Male C57BL/6 mice	Downregulation of SREBF1 target genes, including FASN and PPARG i	miR192 acts directly on the 3′UTR of SREBF1, which results in the dysregulation of lipid homeostasis in the hepatocytes.	Suppress SBREPF1	[148]
lncRNA MALAT1	HepG2 and AR42J	Male C57BL/6 mice (8 weeks old, $n = 6$/group)	——	PPARα expression was significantly increased, and CD36 levels were significantly decreased when MALAT1 was knocked down.	MALAT1 could regulate PPARα/CD36 through the mediation of the miR-206/ARNT axis	[115]
lncRNA HOTAIR	HepG2 cells	Adult (8-week-old) male C57BL/6J mice	HOTAIR knockout increases the expression of genes	HOTAIR enhances the accumulation of PRC2 and H3K27 trimethylation to the MEG3 promoter	Regulates the expression of the DNMT1/MEG3/p53 pathways	[117]
lncRNA APTR	Hepatic stellate cells	(CCl_4)-treated mice	Attenuates TGFB1-induced upregulation of ACTA2	The overexpression of α-SMA in HSCs brought on by TGF-β1 is prevented by suppressing APTR.	Abrogate TGF-β_1-induced upregulation of α-SMA	[69,149]
lncRNA PVT1	Primary HSCs cells	Eight-week-old male C57BL/6J mice	Regulates the expression of PTCH1, SMO, and GLI2	Chromatin modification, transcriptional regulation, and post-transcriptional regulation	PVT1 silencing inhibits the Hh pathway	[150]
lncRNA COX2	IEC cell line (IEC4.1)	C57BL/6J mice	Enhances the transcription of Il12b	By controlling Mi-2/NuRD-mediated epigenetic histone changes, LincRNA-Cox2 affects the transcription of the Il12b gene in intestinal epithelial cells stimulated by TNF.	Activates NF-kB signaling pathway	[126]
lncRNA NEAT 1	HepG2 cells and LO2 cells	——	Activates lipogenesis-related genes, such as ACC and FAS	By targeting miR-139-5p, NEAT1 knockdown reduces lipid buildup in the NAFLD cellular model.	Regulates the c-Jun/SREBP1c pathway	[151]

Table 1. *Cont.*

ncRNA	Cells Involved	Animal Model Used	Genes Downregulated/Upregulated	Mechanism	Pathway Activated	Reference
lncRNA UC372	HepG2 cells	C57BL/6J mice	Regulates the expression of genes related to lipid synthesis and uptake, including ACC, FAS, SCD1, and CD36	By inhibiting miR-195/miR-4668 maturation from reversing miR-195/miR-4668-mediated suppression of functional target gene expression, uc.372 promotes hepatic steatosis.	—	[135]
lncARSR	human HCC cells (HepG2)	C57BL/6 male mice (aged 6 weeks)	——	lncARSR inhibits YAP1, which inhibits phosphorylation nuclear translocation	Suppresses IRS2/AKT pathway	[31]
lncRNA APOA4-AS	Primary hepatocytes cells	C57BL/6 mice	Decreases APOA4 mRNA and protein levels, whereas expression of other lipid metabolism-associated genes (e.g., FASN, SCD1, APOB, MTP, CPT1α, and MCAD) were not altered	HuR, a protein that stabilizes mRNA, and APOA4-AS interact to stabilize APOA4 mRNA.	Regulates many metabolic pathways	[76]
lncRNA H19	HepG2 and Huh-7 cells	C57BL/6 mice (males, 8 weeks old)	H19 and PPARγ were upregulated	H19 exerts its biological functions in NAFLD by using a ceRNA mechanism to control the expression of miR-130a's downstream genes.	Regulates miR-130a/PPARγ pathway	[152]

4.3. Circular RNAs

Circular RNA (circRNA), a family of ncRNAs, was formerly considered a nonfunctional result of mRNA splicing [153]. Nevertheless, the properties of circRNA, such as tissue and development-specific expression, enrichment of miRNA response element (MRE), and tolerance to both RNase R and RNA exonuclease, indicate its putative function in eukaryotic gene regulation [154]. circRNA is a kind of rediscovered endogenous ncRNA that forms a covalently closed continuous loop via a specific splicing process and is regarded as the predominant subtype in gene transcription [155]. It demonstrates an extensive array of physiological and pathological activity. It has been demonstrated that circRNAs play crucial implications in cancer biology, but their implications in NASH remain unknown. circRNA, which exhibits tissue and pathology-specific expression, has been elucidated to influence circRNA–miRNA interactions [156]. Hepatic steatosis is a miRNA-related degenerative disease characterized by TG buildup and lipid peroxidation, which develops into non-alcoholic steatohepatitis, liver fibrosis/cirrhosis, and even hepatocellular cancer [72,157].

4.3.1. *circRNA_0046367 and circRNA_0046366/miR-34a/PPARα*

circRNA_0046367 normalization decreased miR-34a's inhibition activity on PPARα via preventing the miRNA/mRNA interaction with miRNA response elements during hepatocellular steatosis in vivo and in vitro (MREs). A unique epigenetic mechanism behind hepatic steatosis and accompanying oxidative stress is represented by the circRNA_0046367/miR-34a/PPARα regulatory system. In contradiction to its expression being reduced during steatogenesis, normalization of circRNA_0046367 eliminates miR-34a-induced PPARα inhibition and hepatic steatosis (Figure 2) [158].

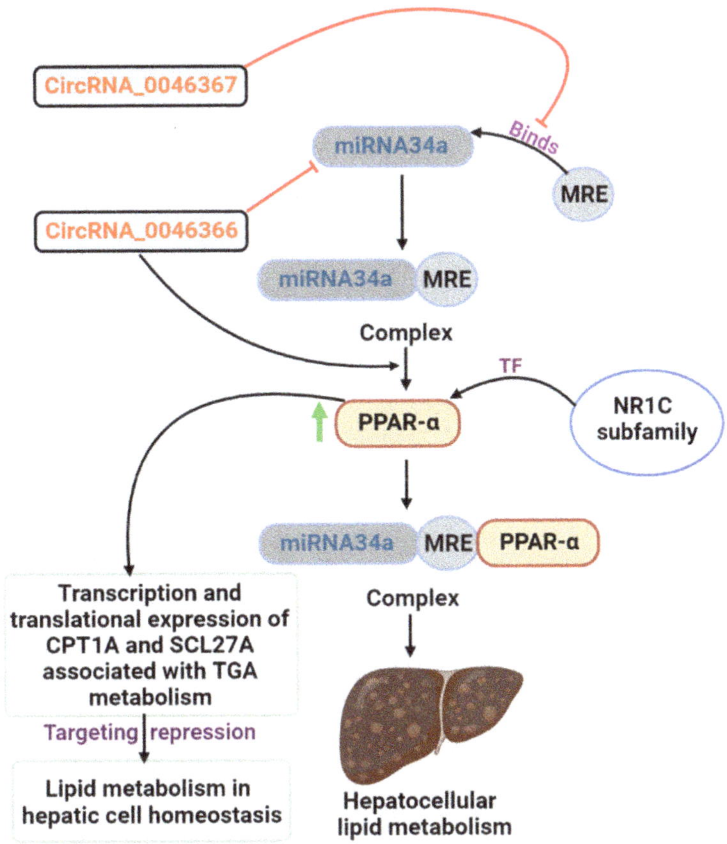

Figure 2. The figure depicts the inhibitory role of circRNA_0046367, preventing MRE from binding miRNA34a, and the role of circRNA_0046366 in inhibiting miRNA34a.

circRNA_0046366 exerts its influence mostly on metabolic processes, particularly lipid metabolism. Recent research identifies PPARα, an NR1C nuclear receptor subfamily transcription factor, as a direct target of miR-34a [120]. This type of liver-specific, ligand-activated PPARs isoform stimulates the expression of several genes with lipometabolic properties [159]. In obese patients, the absence of PPARα confers a high risk of insulin resistance, depletion of n-3 long-chain polyunsaturated fatty acids, and liver steatogenesis [160]. Accumulating data show a miR-34a/PPARα regulatory mechanism, which is the essential mediator of circRNA_0046366 effects on hepatic lipid metabolism (Figure 2). circRNA_0046366, whose expression was reduced in HepG2-based hepatocellular steatosis, inhibits the function of miR-34a. Silencing of miR-34a eliminates its inhibitory effect on PPARα and restores PPARα expression [161]. PPARα restoration enhances the expression of TG metabolism-related downstream genes [i.e., carnitine palmitoyl transferase 1A (CPT1A) and solute carrier family 27A (SLC27A)] at the transcriptional and translational levels. miR-34a is highly upregulated in multiple rodent hepatic steatosis models [162,163], and its expression persistently corresponds well with the clinical presentation of hepatic steatosis in Chinese [87], Japanese [164], and Philippine populations [165]. Targeted inhibition of PPARα underlines the steatosis-related action of miR-34a [120]. Targetome and targetome-based pathway analyses of miR-34a were merged to identify the miRNA-dependent downstream signaling and activities of circRNA_0046366 [166,167]. Using databases of ncRNA (circBase and miRBase) and aircRNA–miRNA interaction, algorithms to elucidate the circRNA–miRNA connection underpinning hepatic steatosis, expression patterns of circRNA_0046366 and miR-34a were examined in a HepG2-based steatosis model stimulated by high levels of fat [161].

4.3.2. *circRNA_021412/miR-1972/LPIN1*

Guo et al., 2017, identified Lipin 1 (LPIN1) as a transcriptional regulator of circRNAs on metabolic pathways. The circRNA-miRNA-mRNA network subsequently revealed the signaling cascade of circRNA_021412/miR-1972/LPIN1, characterized by a lower level of circRNA _021412 and miR-1972-based regulation of LPIN1. Hepatic steatosis was caused by the LPIN1-induced suppression of long-chain acyl-CoA synthetases (ACSLs) expression [168].

LPIN1 could be the critical mediator of the transcriptional regulatory action of circRNAs on metabolic pathways. In the circRNA–miRNA–mRNA network, Has_circRNA_021412 and miR-1972 with LPIN1-regulatory activity and LPIN1 were identified as essential nodes. During hepatic steatosis, the interplay of the circRNA, miRNA, and LPIN1 exposes several unexpected yet crucial pathways [168].

LPIN1 may preferentially activate fatty acid oxidation and mitochondrial oxidative phosphorylation, as demonstrated by gain-of-function and loss-of-function techniques [169,170]. Furthermore, downregulated LPIN1 inhibits fatty acid breakdown in an ACSL-dependent manner. In addition, LPIN1 is confirmed to be the crucial element of multiple signaling transductions that are deeply involved in lipid homeostasis, such as SIRT1/AMPK signaling [171], mammalian target of rapamycin complex 1 (mTORC1)/SREBP signaling [172], NF-E2-related factor 1 (Nrf1) signaling [173], and hepatocyte nuclear factor 4 (HNF4) signaling [168,174].

4.3.3. *circRNA_002581/miR-122/SLC1A5, PLP2, CPEB1*

circRNA has been shown to play a role in liver regeneration [175] and to serve as a molecular marker and therapeutic target for liver cancer [176]. miRNAs serve crucial roles in a variety of liver diseases [177]. MiR-122 levels were considerably reduced in NASH patients [178], whereas liver-specific and germline miR-122 deletion mice developed hepatic steatosis in early adulthood, followed by NASH, liver fibrosis, and eventually HCC as they aged [80].

circRNA_002581 (also known as circ_0001351 in circBase) is situated in the exonic region of the mouse chromosome 5 between positions 66,753,956 and 66,756,359, with a complete length of 2403 bp and a spliced length of 275 bp. The circRNA_002581–miR-122–CPEB1 pathway is implicated in NASH etiology [179,180]. By sponging miR-122, circRNA_002581 overexpression significantly alleviated the inhibitory effect of miR-122 on its target CPEB1. circRNA_002581 knockdown significantly attenuated lipid droplet accumulation, decreased alanine aminotransferase (ALT) levels, aspartate aminotransferase (AST), pro-inflammatory cytokines, apoptosis, and hydrogen peroxide, and raised the level of ATP in both mouse and cellular models of NASH (Figure 3) [179].

4.3.4. *circRNA_0067835/miR-155/FOXO3a*

circRNAs serve primarily as miRNA sponges, binding to functional miRNAs and then regulating gene expression. By functioning as a sponge of miR155 to boost FOXO3a expression, circRNA_0067835 modulated hepatic fibrosis development, suggesting that it may offer targeted therapy for patients with liver fibrosis [181]. The FOXO family of Forkhead transcription factors has been discovered to regulate HSC proliferation via the PI3K/Akt pathway [182]. FOXO3a is essential for the integration of various pathways. FOXO3a proteins are crucial in maintaining intracellular redox equilibrium under diverse environmental stresses [183]. In the study by Lili zhu et al., bioinformatics analysis and subsequent luciferase reporter experiments were used to identify the FOXO3a gene as a direct target of miR-155. FOXO3a and AKT expression levels were considerably reduced in miR-155-overexpressing cells and elevated in miR-155-silenced cells, indicating that miR-155 modulated AKT/FOXO3a signaling (Figure 4). There are correlations between AKT/FOXO3a signaling and liver fibrosis since AKT/FOXO3a expression was considerably enhanced in HSCs and CCl4-treated liver [181]. LX 2 cells are the predominant cell type involved in liver fibrosis [157,160,161].

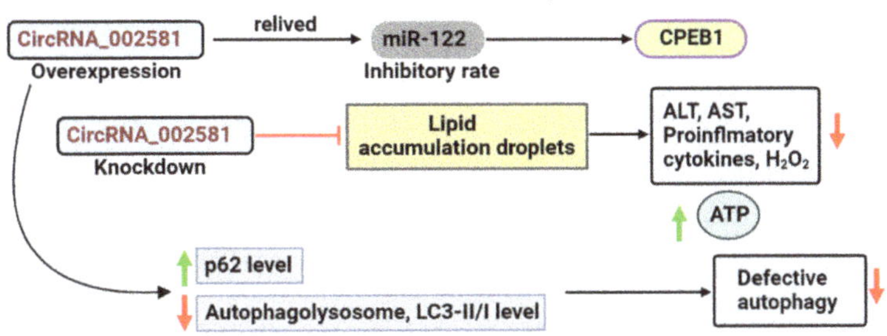

Figure 3. This figure portrays the result of overexpression and the knockdown of the circRNA_002581.

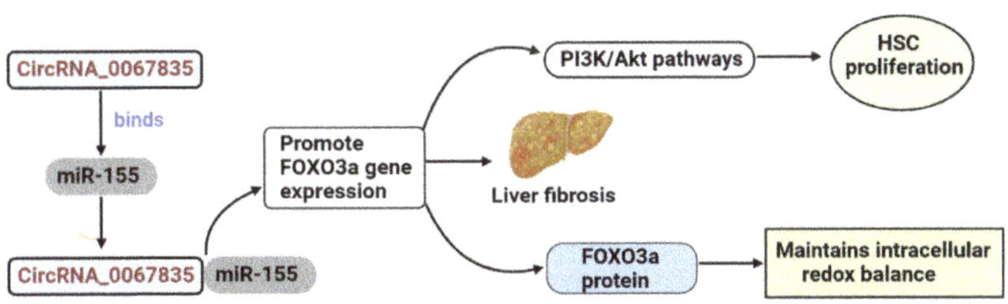

Figure 4. This figure depicts the role of circRNA_0067835 in promoting the FOXO3a gene expression.

4.3.5. *circRNA_0074410/miR-9-5p/KEGG Pathway*

The profiling of circRNAs in fibrotic HSCs indicated that 179 and 630 circRNAs were elevated. Recent research revealed that circ_0074410 inhibited the expression of miR-9-5p and increased HSC activation via the α-SMA protein. Suppression of hsa_circ_0071410 also increased the expression of miR-9-5p, resulting in a decrease in irradiation-induced HSC activity [122,184].

4.3.6. *circRNA_34116/miR-22-3p/BMP7*

In a mouse model of liver fibrosis generated by CCl4, microarray analysis found 10,389 circRNAs, of which 69 were differentially expressed in fibrotic liver tissues; 55 were negatively regulated, and 14 were activated [185]. An in silico study anticipated the existence of miR-22 MRE on circRNA_34116 and suggested that this circRNA can link to miR-22-3p competitively and consequently influence the transcription of its target gene bone morphogenetic protein 7 (BMP7) [186]. In conclusion, networks between circRNAs and miRNAs have evolved as an entirely new strategy for gene expression regulation. They may facilitate the identification of new treatment targets and enhance our knowledge of the molecular regulation of disease advancement and progression [122].

5. ncRNA Regulation of PPARs in NAFLD

ncRNA may have significant regulatory functions in the initiation and progression of NAFLD. This class of molecules, including the lncRNA, circRNA, and miRNA, does not code for proteins but still affects gene expression [187]. miRNAs have many functions, including regulating posttranscriptional gene expression, adipocyte differentiation, lipid metabolism, cholesterol metabolism, insulin resistance, and immune responses [188,189]. lncRNAs are 200 nucleotides in size and act as a transcriptional regulators of gene activation or silencing through chromatin modification [190]. There are many conserved binding sites on circRNA, which function as a "miRNA sponge" by inhibiting miRNA activity by interacting with miRNA AGO proteins [191].

5.1. miRNAs

5.1.1. miR-124-3p

According to accumulating data, the development and progression of NAFLD are attributed to miRNAs. However, little is known about the molecular model of miRNA in this NAFLD [122]. PPARα is known to be a target of many miR-124. A study has reported the role of miR-124-3p in the progression of NAFLD. The study showed that miR-124-3p targets Pref-1, the preadipocyte factor 1 [192]. During a high-fat diet, miR-124-3p activity in the liver rises. Suppressing the expression of miR-124-3p lowers hepatocyte lipid levels and inflammatory markers, and vice versa. As a result, miR-124-3p is an important regulator of the liver's lipid homeostasis. It was also noted that when Pref-1 expression is inhibited in preadipocytes, it increases PPARs' expression, and preadipocytes eventually differentiate into adipocytes [193].

Another study reported that when Pref-1 was downregulated, and PPARs were increased by blocking Notch signaling, it improved adipogenic differentiation [194,195]. It is consistent with lower PPAR activity, which has already been linked to decreased TGHs ATGL and CES1 expression [196].

5.1.2. miR-21

Liver steatosis is one of the features found in NAFLD patients. miR-21 targets PPARs during the development of NAFLD [197]. In NASH patients and diseased mice, miR-21 levels were elevated in the liver and muscle, which reduced the PPAR-α expression [85]. An investigation that was recently completed revealed this observation. This study also showed that β-oxidation in hepatocytes is known to be controlled by PPARs, which also have anti-inflammatory effects on non-parenchymal cells. Consequently, it is possible that the pro-metabolic and anti-inflammatory activities of PPARs in various liver cells contributed partially to the enhanced behavior of miR-21 KO mice. The production of C/EBP and PPARs is suppressed by β-catenin when the classical WNT/β-catenin signaling pathway is activated, which prevents pre-adipocyte differentiation [198]. It was reported that by targeting low-density lipoprotein-related receptor 6 (LRP6) to promote the WNT signaling pathway, miR-21 production could be inhibited to treat NAFLD [199]. LRP6 is the target of miR-21, which controls the classic WNT signaling pathway in NAFLD. By regulating the miR-21-5p/SFRP5 pathway, PPARs were reported in another study as being able to reduce oxidative stress and inflammation in NASH, suggesting a potential designated target for NAFLD therapy [147].

5.1.3. miR-122

Hepatic cholesterol, as well as the metabolism of lipids, are regulated by miR-122, considered the most prevalent miR in the liver [200]. Its role in NAFLD has also been reported and showed an increased expression pattern in the affected individuals [94]. It has been demonstrated that miR-122 is necessary for translating respiratory proteins, increasing the respiratory enzyme activity of mitochondria, and improving mitochondrial proteostasis in the liver [201]. Multiple signaling pathways are targeted by miR-122 to control lipid metabolism. FFAs increased the expression of miR-122, facilitated its release into the bloodstream, and inhibited TG production by targeting Agpat1 and Dgat1 and boosting β-oxidation. The scientists also discovered that blocking miR-122 caused FOXO1 to be downregulated and PPARs to be upregulated, showing that miR-122 is implicated in several pathways relevant to lipid metabolism [202,203].

5.1.4. miR-34a

Several studies have reported the role of miR-34a in NAFLD and hepatic steatosis [163]. Numerous investigations found that people with NAFLD had greater levels of miR-34a. With increasing disease severity, miR-34a, apoptosis, and acetylated p53 levels grew [120]. Hepatic PPARs and SIRT1, which are miR-34a's major targets, were significantly suppressed due to the upregulation of miR-34a. When miR-34a was silenced, the expression of PPARs, SIRT1 and PPAR's downstream genes initially rose [147]. Additionally, there was a rise in AMPK activation, the primary metabolic sensor. The miR-34a inhibitor reduced the level of steatosis and reduced lipid build-up. A recent study investigated the role of the SIRT1 signaling pathway in miR-34a by inducing resveratrol in high-fat diet-fed rats [204].

5.1.5. miR-130a and miR-130b

In a study conducted in 2017, it was reported that the hepatic expression of miR-130a-3p was significantly reduced in mouse models with fibrotic steatohepatitis. NASH fibrosis has a complicated pathophysiology. There are few effective treatments, and researchers do not fully comprehend how liver fibrosis works. miRNAs are crucially implicated in the many phases of liver fibrosis, particularly HSC activation, proliferation, and ECM synthesis [205]. HSC apoptosis is a crucial component of NASH-related liver fibrosis that could be associated with the severity of NAFLD. Increased production of the initiator caspase-9 and effector caspase-3, which miR-130a-3p triggers, causes the proteolytic cleavage of PARP, which results in cellular disintegration and apoptosis [206]. In a study conducted in rat HSC-T6 cells, miR-130a and miR-130b reduced PPAR expression by specifically binding to the 3′-UTR of PPARs mRNA. In cell culture, miR-130a and miR-130b overexpression, PPARs' downregulation, and ECM gene overexpression may be mediated by TGF-1. These results imply that PPARs are downregulated in liver fibrosis by miR-130a and miR-130b [207]. In another study completed in HepG2 cell lines and primary mouse hepatocytes, miR-130a-3p overexpression enhances insulin signaling, whereas miR-130a-3p silencing has the opposite effect. MiR-130a-5p has little impact on the control of insulin signaling [208]. It has been demonstrated that miR-130a-3p effectively inhibited the production of PPARs-regulated genes by targeting the mRNA coding and 3′UTRs. It is believed that miR-130-3p effects on (human fatty acid synthase) FAS expression may be indirectly mediated by reducing PPARs expression since PPARs might control FAS expression [208], thus, playing a role in NAFLD.

5.1.6. miR-155

The metabolism of lipids plays a major role in the pathogenesis of NAFLD. C/EBP, SIRT1, and PPARs are considered to be targets of miR-155. PPAR α promotes utilization in regulating fat metabolism, whereas PPARγ activation increases storage [209]. They are also known to act as an anti-fibrotic gene which is a direct target of miR-155 [210]. Variations in steatosis, inflammation and fibrosis are regulated by signals in normal lipid and insulin regulation and the control of inflammatory responses due to miR-155′s effect on PPARs signaling. It is possible that miR-155 functions as a negative regulator of adipogenesis because the miR-155 expression is lowered during in vitro adipogenesis and miR-155 overexpression inhibits the production of PPARs and cEBP (Figure 5) [211].

5.2. lncRNAs

5.2.1. lncRNA FLRL6/FLRL2

In NAFLD, it was noted that seven lncRNAs and five mRNAs expression patterns were associated with circadian rhythm. In the pathway analysis, the central molecule in the circadian rhythm, Per2, was found to be a target of lncRNA FLRL6, and it was also observed that the level of this lncRNA increased about 3.3-fold along with which the level of Per2 was also elevated by about 3.5-fold [212,213]. A study reported that FLR2 lncRNA, an upstream lncRNA of another circadian-rhythm target aryl hydrocarbon-receptor nuclear translocator-like (ARNTL), was downregulated in the rodent models affected with NAFLD. In the previous study, several lncRNAs, including FLRL8, FLRL3, and FLRL7, were associated with lipogenesis via proteins in the PPARs signaling pathway, suggesting their potential regulatory involvement in lipid metabolism [212].

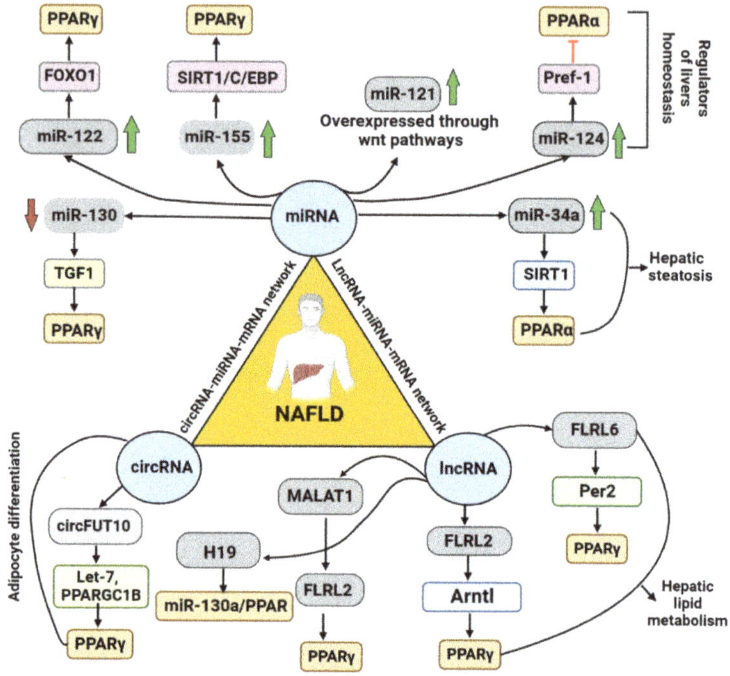

Figure 5. Role of ncRNAs and their respective molecular factors found to be involved in regulating NAFLD progression via targeting PPARs. This figure depicts the suspected pathway and molecules involved in regulating different ncRNA and the type of PPARs involved.

5.2.2. lncRNA H19

One of the first and most extensively studied lncRNAs in liver illness is H19. Hepatocellular carcinoma patients had higher levels of H19, which inhibited tumor metastasis in a miR-220-dependent manner [214]. In in vitro and in vivo models of NAFLD study, H19 has a molecular function in controlling NAFLD by directly modulating the miR-130a/PPARs axis, inhibiting hepatic lipogenesis [152]. Fatty acids elevate the expression of H19 in hepatocytes and diet-induced fatty liver, and H19 overexpression might promote steatosis and increase lipid accumulation. Obstructive cholestatic liver fibrosis quickly developed due to hepatic overexpression of H19 [215]. These suggested that H19 could play a crucial role in NAFLD and act as a lipid sensor to regulate hepatic metabolic balance.

5.2.3. MALAT 1

In vivo and in vitro studies have shown the role of lncRNA MALAT-1 in NAFLD. Patients with NAFLD were observed to have elevated MALAT1, which increased the development of liver fibrosis [216]. According to previous research and bioinformatics analysis, a correlation between the binding site of miR-206 and MALAT1 or aryl hydrocarbon-receptor nuclear translocator (ARNT) was found [115]. More studies are required to thoroughly analyze and study the exact involvement of PPARs in NAFLD by regulating lncRNA MALAT 1.

5.3. CircRNAs

A study noted that circRNA expression patterns in developing bovine adipose tissue in animals showed that circFUT10 overexpression strongly reduces PPARs and C/EBPa expression [217]. By targeting PPARs, GC1B, circFUT10 acts as a ceRNA for miRNA let-7c/let-e to control the differentiation of bovine adipocytes. Adipocyte differentiation and tissue-specific adipose deposition are also reported to be influenced by circLCLAT1, circFNDC3AL, circCLEC19A, and circARMH1 through the PPARs pathway. circ-PLXNA1 is primarily expressed in the liver and adipose tissue of ducks. The development of duck adipose cells is restricted by circ-PLXNA1 inhibition [218]. Studies have also observed and predicted that a possible target for the therapy of hepatic steatosis is circRNA 0046366 (Table 2) [161].

Table 2. Pathway and proteins involved in PPARs-related ncRNA in NAFLD and the possible treatment strategy.

ncRNA	Expression Pattern	Type of PPAR	PPARs Associated Molecular Factor	Type of Disease	Possible Treatment and Advantages	References
MiR-124	Elevated	PPARα	Pref 1, Notch signaling,	NAFLD—Hepatic lipid metabolism	Antisense miR-124 can be used. MiR-124 acts as a biomarker in NAFLD progression	[192]
MiR-21	Elevated	PPARα	LRP6, WNT/β-catenin signaling, SFRP5 pathway	Liver steatosis	Antisense MiR-21 therapy and biomarker for early diagnosis	[116]
MiR-122	Elevated	PPARγ	FoxO1	Hepatic cholesterol	Antisense MiR-122, which inhibits the production	[203]
MiR-34a	Elevated	PPARα	SIRT1, AMPK activation	Hepatic steatosis	Antisense MiR-34a	[120]
MiR-130	Reduced	PPARγ	TGF-1	Fibrotic steatohepatitis, insulin signaling	MiR-130 can be administered	[206]
MiR-155	Elevated	PPARγ	C/EBP, SIRT1	Adipogenesis	Antisense-155	[219]
circFUT10	Elevated	PPARγ	Let-7, PPARGC1B	Adipocyte Differentiation	-	[217]
circRNA_ 0046367/ circRNA_ 0046366	Reduced	PPARα	CPT2, ACBD3, SLC27A	Hepatic steatosis	circRNA_0046367/ circRNA_0046366	[158]
lncRNA FLRL6/FLRL2	Elevated	PPARγ PPARγ	circadian rhythm- Per2, Arntl	Hepatic lipid metabolism	-	[158]
LncRNA HI9	Elevated	PPARγ	miR-130a/PPAR axis	Liver fibrosis	-	[152]
MALAT 1	Elevated	PPARα	miR-206/ARNT	Liver fibrosis	-	[115]

6. Role of lincRNA Paral1 in Adipogenesis and Activation of PPARγ

Adipocyte differentiation and functioning depend on a complex network of interrelated transcription factors focused on PPARγ [220]. lincRNA promotes adipocyte development and coactivates the master adipogenic regulator PPARγ via association with the paraspeckle component and hnRNP-like RNA binding protein 14 (RBM14/NCoAA); hence, it was named PPARγ-activator RNA-Binding Motif Protein 14 (RBM14)-associated lncRNA (Paral1) [221,222]. According to a study by Firmin et al. (2017), the expression of Paral1 and PPARγ rises simultaneously during adipocyte development [222]. The presence of a PPARγ binding site upstream of the Paral1 transcription start site indicates that PPARγ may regulate the production of a coactivating RNA. It would create a favorable feedforward loop on PPARγ expression, prompting the inquiry into the effect of PPARγ agonism on the expression of Parall. The molecular mechanism for Parall′s pro-adipogenic action is its ability to engage with and enhance the coactivating capacity of RBM14, whose expression tracks Parall′s throughout adipocyte maturation. This newly found function of RBM14 is likely the result of its capacity to operate as an indirect coactivator via synergistic effects with nuclear receptor coactivators. The expression of Parall is confined to adipocytes and decreases with a rising body mass index in humans [223,224].

7. Future Perspectives

The interaction of pharmacologic drugs with PPARs' modulation resulting from the environment of the individual patients is a facet of PPARs-targeted therapy that requires additional research. Lifestyle modifications may have an immediate impact on PPARs-directed environmental parameters. Numerous metabolic processes are modulated by PPARs, which also cause pleiotropic effects in numerous tissues [225]. Specialized effects can be evoked and combined using compounds with different activity patterns on various PPARs isotypes, further adding to the complexity [226]. It offers significant prospects and problems for studying PPAR treatments for NAFLD.

8. Conclusions

Several ncRNAs have been involved in the etiology of NASH, making them attractive targets for RNA-based therapies. Antisense oligonucleotides (ASOs) are typically 14–20 single-stranded nucleotides intended to bind and suppress complementary RNA transcripts [227,228]. Most ASOs have phosphorothioate (PS) backbone linkages to improve their pharmacokinetic characteristics in vivo [227,228]. The FDA-approved ASO-based drug eteplirsen for the therapy of Duchenne muscular dystrophy [229] and nusinersen for diagnosing the neurodegenerative disease spinal muscular atrophy in both cases [230] demonstrate the high clinical efficacy of chemically altered ASOs as RNA-targeted therapeutics [231].

In order to distinguish between the various stages of NAFLD, ncRNAs are emerging as helpful biomarkers. The majority of clinical development will be concentrated on drugs that bind to miRNAs or have an impact on how splicing functions. Our knowledge of how lncRNAs and other ncRNAs function will probably improve over time. New categories of possible non-coding targets might also appear. The design of efficient development programs will gain a more robust foundation as our knowledge of ncRNAs, and their mechanisms improves, and the likelihood of clinical success will increase [232].

Today, the prevalence of NAFLD and NASH is rising, which correlates favorably with the rate of obesity and diabetes. NAFLD and NASH are the leading contributors to the advancement of HCC, the predominant form of liver cancer. Furthermore, there is no effective cure for NAFLD and NASH. For the future diagnosis of NAFLD, new noninvasive diagnostic markers such as miRNAs have been investigated. Since PPARs are one of the primary regulators and targets of ncRNA, several studies have focused on the treatment strategies for NAFLD via ncRNA-PPARs regulation. As the number of lncRNAs is greater than that of protein-coding genes, a more focused, detailed system of lncRNAs, mainly using genetic manipulation and knockdown by ASOs, is necessary to verify prior reports on lncRNAs in the liver-directed at elucidating their physiological functions and to mentor new treatment strategies to fight life-threatening liver diseases, such as NASH. The rising significance of ncRNAs as therapeutic agents constitutes a landmark in developing innovative treatment techniques. The fascinating field of lncRNA is still in its infancy, and additional experimental research is required to comprehend their potential as therapeutic targets. Understanding the cellular circuits and network-associated miRNA necessitates comprehending the various miRNA processes, including the decoy function and 5′UTR regulatory activity.

Author Contributions: Conceptualization, S.E.P., A.D., B.V., R.G. and A.V.G.; resources and data curation, A.G.M., U.R.W., R.K., S.K., R.M. and A.N.; writing— original draft preparation, A.G.M., U.R.W., R.K., S.K., R.M. and A.N.; writing—review and editing, K.R., R.K., A.G.M. and U.R.W.;

visualization, A.V.G. and R.G.; supervision, A.V.G. and R.G.; project administration, A.V.G., S.E.P. and B.V. All authors have read and agreed to the published version of the manuscript.

Funding: This research received no external funding.

Institutional Review Board Statement: Not applicable.

Informed Consent Statement: Not applicable.

Data Availability Statement: Data are available from the authors on request (AVG).

Acknowledgments: The authors thank the VIT, Vellore, Tamilnadu, India, for supporting this work.

Conflicts of Interest: The authors declared no potential conflict of interest concerning this article's research, authorship, and publication.

References

1. Mitra, S.; De, A.; Chowdhury, A. Epidemiology of non-alcoholic and alcoholic fatty liver diseases. *Transl. Gastroenterol. Hepatol.* **2020**, *5*, 16. [CrossRef] [PubMed]
2. Yatsuji, S.; Hashimoto, E.; Tobari, M.; Taniai, M.; Tokushige, K.; Shiratori, K. Clinical features and outcomes of cirrhosis due to non-alcoholic steatohepatitis compared with cirrhosis caused by chronic hepatitis C. *J. Gastroenterol. Hepatol.* **2009**, *24*, 248–254. [CrossRef] [PubMed]
3. Bugianesi, E.; Moscatiello, S.; Ciaravella, M.F.; Marchesini, G. Insulin resistance in nonalcoholic fatty liver disease. *Curr. Pharm. Des.* **2010**, *16*, 1941–1951. [CrossRef]
4. Paschos, P.; Paletas, K. Non alcoholic fatty liver disease two-hit process: Multifactorial character of the second hit. *Hippokratia* **2009**, *13*, 128.
5. Dowman, J.K.; Tomlinson, J.; Newsome, P. Pathogenesis of non-alcoholic fatty liver disease. *QJM Int. J. Med.* **2009**, *103*, 71–83. [CrossRef]
6. Farrell, G.C.; Larter, C.Z. Nonalcoholic fatty liver disease: From steatosis to cirrhosis. *Hepatology* **2006**, *43*, S99–S112. [CrossRef]
7. Orellana, M.; Rodrigo, R.; Varela, N.; Araya, J.; Poniachik, J.; Csendes, A.; Smok, G.; Videla, L.A. Relationship between in vivo chlorzoxazone hydroxylation, hepatic cytochrome P450 2E1 content and liver injury in obese non-alcoholic fatty liver disease patients. *Hepatol. Res.* **2006**, *34*, 57–63. [CrossRef]
8. Nakamuta, M.; Kohjima, M.; Higuchi, N.; Kato, M.; Kotoh, K.; Yoshimoto, T.; Yada, M.; Yada, R.; Takemoto, R.; Fukuizumi, K.; et al. The significance of differences in fatty acid metabolism between obese and non-obese patients with non-alcoholic fatty liver disease. *Int. J. Mol. Med.* **2008**, *22*, 663–667.
9. Postic, C.; Girard, J. Contribution of de novo fatty acid synthesis to hepatic steatosis and insulin resistance: Lessons from genetically engineered mice. *J. Clin. Investig.* **2008**, *118*, 829–838. [CrossRef] [PubMed]
10. Tilg, H. Editorial [Hot topic: Adipocytokines in Nonalcoholic Fatty Liver Disease: Key Players Regulating Steatosis, Inflammation and Fibrosis (Executive Editor: Herbert Tilg)]. *Curr. Pharm. Des.* **2010**, *16*, 1893–1895. [CrossRef] [PubMed]
11. Von-Hafe, M.; Borges-Canha, M.; Vale, C.; Leite, A.R.; Neves, J.S.; Carvalho, D.; Leite-Moreira, A. Nonalcoholic Fatty Liver Disease and Endocrine Axes—A Scoping Review. *Metabolites* **2022**, *12*, 298. [CrossRef] [PubMed]
12. Charlton, M.; Angulo, P.; Chalasani, N.; Merriman, R.; Viker, K.; Charatcharoenwitthaya, P.; Sanderson, S.; Gawrieh, S.; Krishnan, A.; Lindor, K. Low circulating levels of dehydroepiandrosterone in histologically advanced nonalcoholic fatty liver disease. *Hepatology* **2007**, *47*, 484–492. [CrossRef]
13. Loria, P.; Carulli, L.; Bertolotti, M.; Lonardo, A. Endocrine and liver interaction: The role of endocrine pathways in NASH. *Nat. Rev. Gastroenterol. Hepatol.* **2009**, *6*, 236–247. [CrossRef] [PubMed]
14. Knoll, M.; Lodish, H.F.; Sun, L. Long non-coding RNAs as regulators of the endocrine system. *Nat. Rev. Endocrinol.* **2015**, *11*, 151–160. [CrossRef]
15. Cotter, T.G.; Rinella, M. Nonalcoholic Fatty Liver Disease 2020: The State of the Disease. *Gastroenterology* **2020**, *158*, 1851–1864. [CrossRef]
16. Lin, H.; Yip, T.C.F.; Zhang, X.; Li, G.; Tse, Y.K.; Hui, V.W.K.; Liang, L.Y.; Lai, J.C.T.; Chan, S.L.; Chan, H.L.Y. Age and the relative importance of liver-related deaths in nonalcoholic fatty liver disease. *Hepatology* **2022**. [CrossRef]
17. Eslam, M.; El-Serag, H.B.; Francque, S.; Sarin, S.K.; Wei, L.; Bugianesi, E.; George, J. Metabolic (dysfunction)-associated fatty liver disease in individuals of normal weight. *Nat. Rev. Gastroenterol. Hepatol.* **2022**, *19*, 638–651. [CrossRef] [PubMed]
18. Kumar, D.P.; Caffrey, R.; Marioneaux, J.; Santhekadur, P.K.; Bhat, M.; Alonso, C.; Koduru, S.V.; Philip, B.; Jain, M.R.; Giri, S.R.; et al. The PPAR α/γ Agonist Saroglitazar Improves Insulin Resistance and Steatohepatitis in a Diet Induced Animal Model of Nonalcoholic Fatty Liver Disease. *Sci. Rep.* **2020**, *10*, 1–14. [CrossRef]
19. Polyzos, S.A.; Kang, E.S.; Boutari, C.; Rhee, E.-J.; Mantzoros, C.S. Current and emerging pharmacological options for the treatment of nonalcoholic steatohepatitis. *Metabolism* **2020**, *111*, 154203. [CrossRef] [PubMed]
20. Negi, C.K.; Babica, P.; Bajard, L.; Bienertova-Vasku, J.; Tarantino, G. Insights into the molecular targets and emerging pharmacotherapeutic interventions for nonalcoholic fatty liver disease. *Metabolism* **2021**, *126*, 154925. [CrossRef] [PubMed]

21. Qaoud, M.T.; Almasri, I.; Önkol, T. Peroxisome Proliferator-Activated Receptors as Superior Targets for Treating Diabetic Disease, Design Strategies—Review Article. *Turk. J. Pharm. Sci.* **2022**, *19*, 353–370. [CrossRef] [PubMed]
22. Monroy-Ramirez, H.; Galicia-Moreno, M.; Sandoval-Rodriguez, A.; Meza-Rios, A.; Santos, A.; Armendariz-Borunda, J. PPARs as Metabolic Sensors and Therapeutic Targets in Liver Diseases. *Int. J. Mol. Sci.* **2021**, *22*, 8298. [CrossRef] [PubMed]
23. Francque, S.; Szabo, G.; Abdelmalek, M.F.; Byrne, C.D.; Cusi, K.; Dufour, J.-F.; Roden, M.; Sacks, F.; Tacke, F. Nonalcoholic steatohepatitis: The role of peroxisome proliferator-activated receptors. *Nat. Rev. Gastroenterol. Hepatol.* **2020**, *18*, 24–39. [CrossRef] [PubMed]
24. Puengel, T.; Liu, H.; Guillot, A.; Heymann, F.; Tacke, F.; Peiseler, M. Nuclear Receptors Linking Metabolism, Inflammation, and Fibrosis in Nonalcoholic Fatty Liver Disease. *Int. J. Mol. Sci.* **2022**, *23*, 2668. [CrossRef]
25. Lange, N.F.; Graf, V.; Caussy, C.; Dufour, J.-F. PPAR-Targeted Therapies in the Treatment of Non-Alcoholic Fatty Liver Disease in Diabetic Patients. *Int. J. Mol. Sci.* **2022**, *23*, 4305. [CrossRef] [PubMed]
26. Fougerat, A.; Montagner, A.; Loiseau, N.; Guillou, H.; Wahli, W. Peroxisome Proliferator-Activated Receptors and Their Novel Ligands as Candidates for the Treatment of Non-Alcoholic Fatty Liver Disease. *Cells* **2020**, *9*, 1638. [CrossRef] [PubMed]
27. Christofides, A.; Konstantinidou, E.; Jani, C.; Boussiotis, V.A. The role of peroxisome proliferator-activated receptors (PPAR) in immune responses. *Metabolism* **2020**, *114*, 154338. [CrossRef]
28. Fali, T.; Aychek, T.; Ferhat, M.; Jouzeau, J.-Y.; Busslinger, M.; Moulin, D.; Eberl, G. Metabolic regulation by PPARγ is required for IL-33-mediated activation of ILC2s in lung and adipose tissue. *Mucosal Immunol.* **2020**, *14*, 585–593. [CrossRef]
29. Palomer, X.; Barroso, E.; Pizarro-Delgado, J.; Peña, L.; Botteri, G.; Zarei, M.; Aguilar, D.; Montori-Grau, M.; Vázquez-Carrera, M. PPAR β/δ: A Key Therapeutic Target in Metabolic Disorders. *Int. J. Mol. Sci.* **2018**, *19*, 913. [CrossRef]
30. Takahashi, K.; Yan, I.; Haga, H.; Patel, T. Long noncoding RNA in liver diseases. *Hepatology* **2014**, *60*, 744–753. [CrossRef]
31. Chi, Y.; Gong, Z.; Xin, H.; Wang, Z.; Liu, Z. Long noncoding RNA lncARSR promotes nonalcoholic fatty liver disease and hepatocellular carcinoma by promoting YAP1 and activating the IRS2/AKT pathway. *J. Transl. Med.* **2020**, *18*, 126. [CrossRef] [PubMed]
32. Pasquinelli, A.E. MicroRNAs and their targets: Recognition, regulation and an emerging reciprocal relationship. *Nat. Rev. Genet.* **2012**, *13*, 271–282. [CrossRef] [PubMed]
33. Lakner, A.M. microRNAs: Fad or future of liver disease. *World J. Gastroenterol.* **2011**, *17*, 2536–2542. [CrossRef] [PubMed]
34. Siomi, H.; Siomi, M.C. On the road to reading the RNA-interference code. *Nature* **2009**, *457*, 396–404. [CrossRef]
35. Ceccarelli, S.; Panera, N.; Gnani, D.; Nobili, V. Dual Role of MicroRNAs in NAFLD. *Int. J. Mol. Sci.* **2013**, *14*, 8437–8455. [CrossRef]
36. Eddy, S.R. Non–coding RNA genes and the modern RNA world. *Nat. Rev. Genet.* **2001**, *2*, 919–929. [CrossRef]
37. Cui, J.; Wang, Y.; Xue, H. Long non-coding RNA GAS5 contributes to the progression of nonalcoholic fatty liver disease by targeting the microRNA-29a-3p/NOTCH2 axis. *Bioengineered* **2022**, *13*, 8370–8381. [CrossRef]
38. Wang, Y.; Nakajima, T.; Gonzalez, F.J.; Tanaka, N. PPARs as Metabolic Regulators in the Liver: Lessons from Liver-Specific PPAR-Null Mice. *Int. J. Mol. Sci.* **2020**, *21*, 2061. [CrossRef]
39. Zarei, M.; Aguilar-Recarte, D.; Palomer, X.; Vázquez-Carrera, M. Revealing the role of peroxisome proliferator-activated receptor β/δ in nonalcoholic fatty liver disease. *Metabolism* **2020**, *114*, 154342. [CrossRef]
40. Bellanti, F.; Villani, R.; Facciorusso, A.; Vendemiale, G.; Serviddio, G. Lipid oxidation products in the pathogenesis of non-alcoholic steatohepatitis. *Free Radic. Biol. Med.* **2017**, *111*, 173–185. [CrossRef]
41. Berthier, A.; Johanns, M.; Zummo, F.P.; Lefebvre, P.; Staels, B. PPARs in liver physiology. *Biochim. et Biophys. Acta (BBA) - Mol. Basis Dis.* **2021**, *1867*, 166097. [CrossRef] [PubMed]
42. Liss, K.H.H.; Finck, B.N. PPARs and nonalcoholic fatty liver disease. *Biochimie* **2017**, *136*, 65–74. [CrossRef] [PubMed]
43. Roth, K.; Yang, Z.; Agarwal, M.; Liu, W.; Peng, Z.; Long, Z.; Birbeck, J.; Westrick, J.; Liu, W.; Petriello, M.C. Exposure to a mixture of legacy, alternative, and replacement per- and polyfluoroalkyl substances (PFAS) results in sex-dependent modulation of cholesterol metabolism and liver injury. *Environ. Int.* **2021**, *157*, 106843. [CrossRef] [PubMed]
44. Yu, D.D.; Van Citters, G.; Li, H.; Stoltz, B.M.; Forman, B.M. Discovery of novel modulators for the PPARα (peroxisome proliferator activated receptor α): Potential therapies for nonalcoholic fatty liver disease. *Bioorganic Med. Chem.* **2021**, *41*, 116193. [CrossRef]
45. Yang, X.; Gonzalez, F.J.; Huang, M.; Bi, H. Nuclear receptors and non-alcoholic fatty liver disease: An update. *Liver Res.* **2020**, *4*, 88–93. [CrossRef]
46. Sun, C.; Mao, S.; Chen, S.; Zhang, W.; Liu, C. PPARs-Orchestrated Metabolic Homeostasis in the Adipose Tissue. *Int. J. Mol. Sci.* **2021**, *22*, 8974. [CrossRef]
47. Zhao, L.; Yagiz, Y.; Xu, C.; Lu, J.; Chung, S.; Marshall, M.R. Muscadine grape seed oil as a novel source of tocotrienols to reduce adipogenesis and adipocyte inflammation. *Food Funct.* **2015**, *6*, 2293–2302. [CrossRef]
48. Wu, L.; Guo, C.; Wu, J. Therapeutic potential of PPARγ natural agonists in liver diseases. *J. Cell. Mol. Med.* **2020**, *24*, 2736–2748. [CrossRef]
49. Zhang, F.; Kong, D.; Lu, Y.; Zheng, S. Peroxisome proliferator-activated receptor-γ as a therapeutic target for hepatic fibrosis: From bench to bedside. *Cell Mol. Life Sci.* **2012**, *70*, 259–276. [CrossRef]
50. Chen, J.; Montagner, A.; Tan, N.S.; Wahli, W. Insights into the Role of PPARβ/δ in NAFLD. *Int. J. Mol. Sci.* **2018**, *19*, 1893. [CrossRef]

51. Ramirez, T.; Tong, M.; Chen, W.C.; Nguyen, Q.-G.; Wands, J.R.; De La Monte, S.M. Chronic alcohol-induced hepatic insulin resistance and endoplasmic reticulum stress ameliorated by peroxisome-proliferator activated receptor-δ agonist treatment. *J. Gastroenterol. Hepatol.* **2012**, *28*, 179–187. [CrossRef] [PubMed]
52. Gutgesell, A.; Wen, G.; König, B.; Koch, A.; Spielmann, J.; Stangl, G.I.; Eder, K.; Ringseis, R. Mouse carnitine–acylcarnitine translocase (CACT) is transcriptionally regulated by PPARα and PPARδ in liver cells. *Biochim. et Biophys. Acta (BBA) - Gen. Subj.* **2009**, *1790*, 1206–1216. [CrossRef] [PubMed]
53. Giudetti, A.M.; Stanca, E.; Siculella, L.; Gnoni, G.V.; Damiano, F. Nutritional and Hormonal Regulation of Citrate and Carnitine/Acylcarnitine Transporters: Two Mitochondrial Carriers Involved in Fatty Acid Metabolism. *Int. J. Mol. Sci.* **2016**, *17*, 817. [CrossRef] [PubMed]
54. Goudarzi, M.; Koga, T.; Khozoie, C.; Mak, T.D.; Kang, B.-H.; Fornace, A.J., Jr.; Peters, J.M. PPARβ/δ modulates ethanol-induced hepatic effects by decreasing pyridoxal kinase activity. *Toxicology* **2013**, *311*, 87–98. [CrossRef] [PubMed]
55. Pawlak, M.; Lefebvre, P.; Staels, B. Molecular mechanism of PPARα action and its impact on lipid metabolism, inflammation and fibrosis in non-alcoholic fatty liver disease. *J. Hepatol.* **2015**, *62*, 720–733. [CrossRef] [PubMed]
56. Koga, T.; Peters, J.M. Targeting Peroxisome Proliferator-Activated Receptor-β/δ (PPARβ/δ) for the Treatment or Prevention of Alcoholic Liver Disease. *Biol. Pharm. Bull.* **2021**, *44*, 1598–1606. [CrossRef] [PubMed]
57. Polyzos, S.A.; Kountouras, J.; Mantzoros, C.S. Adipokines in nonalcoholic fatty liver disease. *Metabolism* **2015**, *65*, 1062–1079. [CrossRef]
58. Ipsen, D.H.; Lykkesfeldt, J.; Tveden-Nyborg, P. Molecular mechanisms of hepatic lipid accumulation in non-alcoholic fatty liver disease. *Cell. Mol. Life Sci.* **2018**, *75*, 3313–3327. [CrossRef]
59. Ahmadian, M.; Suh, J.M.; Hah, N.; Liddle, C.; Atkins, A.R.; Downes, M.; Evans, R.M. PPARγ signaling and metabolism: The good, the bad and the future. *Nat. Med.* **2013**, *19*, 557–566. [CrossRef]
60. Skat-Rørdam, J.; Højland Ipsen, D.; Lykkesfeldt, J.; Tveden-Nyborg, P. A role of peroxisome proliferator-activated receptor γ in non-alcoholic fatty liver disease. *Basic Clin. Pharmacol. Toxicol.* **2019**, *124*, 528–537. [CrossRef]
61. Kahn, S.E.; Hull, R.L.; Utzschneider, K.M. Mechanisms linking obesity to insulin resistance and type 2 diabetes. *Nature* **2006**, *444*, 840–846. [CrossRef] [PubMed]
62. Okuno, A.; Tamemoto, H.; Tobe, K.; Ueki, K.; Mori, Y.; Iwamoto, K.; Umesono, K.; Akanuma, Y.; Fujiwara, T.; Horikoshi, H.; et al. Troglitazone increases the number of small adipocytes without the change of white adipose tissue mass in obese Zucker rats. *J. Clin. Investig.* **1998**, *101*, 1354–1361. [CrossRef] [PubMed]
63. Machado, M.V.; Diehl, A.M. Pathogenesis of Nonalcoholic Steatohepatitis. *Gastroenterology* **2016**, *150*, 1769–1777. [CrossRef] [PubMed]
64. Tontonoz, P.; Spiegelman, B.M. Fat and Beyond: The Diverse Biology of PPARγ. *Annu. Rev. Biochem.* **2008**, *77*, 289–312. [CrossRef]
65. Pan, X.; Wang, P.; Luo, J.; Wang, Z.; Song, Y.; Ye, J.; Hou, X. Adipogenic changes of hepatocytes in a high-fat diet-induced fatty liver mice model and non-alcoholic fatty liver disease patients. *Endocrine* **2014**, *48*, 834–847. [CrossRef]
66. Sanyal, A.; Abdelmalek, M.; Diehl, A.; Caldwell, S.; Shiffman, M.; Ghalib, R.; Lawitz, E.; Rockey, D.; Schall, R.; Jia, C.; et al. Efficacy and safety of simtuzumab for the treatment of nonalcoholic steatohepatitis with bridging fibrosis or cirrhosis: Results of two phase 2b, dose-ranging, randomized, placebo-controlled trials. *J. Hepatol.* **2017**, *66*, S54. [CrossRef]
67. Cusi, K.; Orsak, R.B.; Bril, F.; Lomonaco, R.; Hecht, R.J.; Ortiz-Lopez, C.; Tio, F.; Hardies, J.; Darland, R.C.; Musi, N.; et al. Long-Term Pioglitazone Treatment for Patients With Nonalcoholic Steatohepatitis and Prediabetes or Type 2 Diabetes Mellitus. *Ann. Intern. Med.* **2016**, *165*, 305–315. [CrossRef]
68. Pettinelli, P.; Videla, L.A. Up-Regulation of PPAR-γ mRNA Expression in the Liver of Obese Patients: An Additional Reinforcing Lipogenic Mechanism to SREBP-1c Induction. *J. Clin. Endocrinol. Metab.* **2011**, *96*, 1424–1430. [CrossRef]
69. Hanson, A.; Wilhelmsen, D.; DiStefano, J.K. The Role of Long Non-Coding RNAs (lncRNAs) in the Development and Progression of Fibrosis Associated with Nonalcoholic Fatty Liver Disease (NAFLD). *Non-Coding RNA* **2018**, *4*, 18. [CrossRef]
70. Delire, B.; Stärkel, P.; Leclercq, I. Animal Models for Fibrotic Liver Diseases: What We Have, What We Need, and What Is under Development. *J. Clin. Transl. Hepatol.* **2015**, *3*, 53–66. [CrossRef]
71. Dong, S.; Chen, Q.-L.; Song, Y.-N.; Sun, Y.; Wei, B.; Li, X.-Y.; Hu, Y.-Y.; Liu, P.; Su, S.-B. Mechanisms of CCl4-induced liver fibrosis with combined transcriptomic and proteomic analysis. *J. Toxicol. Sci.* **2016**, *41*, 561–572. [CrossRef] [PubMed]
72. Gerhard, G.S. Micro RNAs in the development of non-alcoholic fatty liver disease. *World J. Hepatol.* **2014**, *7*, 226–234. [CrossRef] [PubMed]
73. Su, Q.; Kumar, V.; Sud, N.; Mahato, R.I. MicroRNAs in the pathogenesis and treatment of progressive liver injury in NAFLD and liver fibrosis. *Adv. Drug Deliv. Rev.* **2018**, *129*, 54–63. [CrossRef] [PubMed]
74. Ardekani, A.M.; Naeini, M.M. The Role of MicroRNAs in Human Diseases. *Avicenna J. Med Biotechnol.* **2010**, *2*, 161–179. [PubMed]
75. Chien, Y.; Tsai, P.-H.; Lai, Y.-H.; Lu, K.-H.; Liu, C.-Y.; Lin, H.-F.; Huang, C.-S.; Wu, W.-W.; Wang, C.-Y. CircularRNA as novel biomarkers in liver diseases. *J. Chin. Med Assoc.* **2019**, *83*, 15–17. [CrossRef]
76. Qin, W.; Li, X.; Xie, L.; Li, S.; Liu, J.; Jia, L.; Dong, X.; Ren, X.; Xiao, J.; Yang, C.; et al. A long non-coding RNA, APOA4-AS, regulatesAPOA4expression depending on HuR in mice. *Nucleic Acids Res.* **2016**, *44*, 6423–6433. [CrossRef]
77. Valdmanis, P.N.; Kim, H.K.; Chu, K.; Zhang, F.; Xu, J.; Munding, E.M.; Shen, J.; Kay, M.A. miR-122 removal in the liver activates imprinted microRNAs and enables more effective microRNA-mediated gene repression. *Nat. Commun.* **2018**, *9*, 5321. [CrossRef]

78. Ye, D.; Zhang, T.; Lou, G.; Xu, W.; Dong, F.; Chen, G.; Liu, Y. Plasma miR-17, miR-20a, miR-20b and miR-122 as potential biomarkers for diagnosis of NAFLD in type 2 diabetes mellitus patients. *Life Sci.* **2018**, *208*, 201–207. [CrossRef]
79. López-Pastor, A.R.; Infante-Menéndez, J.; Gómez-Hernández, A. miRNA Dysregulation in the Development of Non-Alcoholic Fatty Liver Disease and the Related Disorders Type 2 Diabetes Mellitus and Cardiovascular Disease. *Front. Med.* **2020**, *7*, 527059. [CrossRef]
80. Tsai, W.-C.; Hsu, S.-D.; Hsu, C.-S.; Lai, T.-C.; Chen, S.-J.; Shen, R.; Huang, Y.; Chen, H.-C.; Lee, C.-H.; Tsai, T.-F.; et al. MicroRNA-122 plays a critical role in liver homeostasis and hepatocarcinogenesis. *J. Clin. Investig.* **2012**, *122*, 2884–2897. [CrossRef]
81. Esau, C.; Davis, S.; Murray, S.F.; Yu, X.X.; Pandey, S.K.; Pear, M.; Watts, L.; Booten, S.L.; Graham, M.; McKay, R.; et al. miR-122 regulation of lipid metabolism revealed by in vivo antisense targeting. *Cell Metab.* **2006**, *3*, 87–98. [CrossRef] [PubMed]
82. Csak, T.; Bala, S.; Lippai, D.; Satishchandran, A.; Catalano, D.; Kodys, K.; Szabo, G. microRNA-122 regulates hypoxia-inducible factor-1 and vimentin in hepatocytes and correlates with fibrosis in diet-induced steatohepatitis. *Liver Int.* **2014**, *35*, 532–541. [CrossRef] [PubMed]
83. Becker, P.P.; Rau, M.; Schmitt, J.; Malsch, C.; Hammer, C.; Bantel, H.; Mullhaupt, B.; Geier, A. Performance of Serum microRNAs -122, -192 and -21 as Biomarkers in Patients with Non-Alcoholic Steatohepatitis. *PLoS ONE* **2015**, *10*, e0142661. [CrossRef]
84. Loyer, X.; Paradis, V.; Hénique, C.; Vion, A.-C.; Colnot, N.; Guerin, C.L.; Devue, C.; On, S.; Scetbun, J.; Romain, M.; et al. Liver microRNA-21 is overexpressed in non-alcoholic steatohepatitis and contributes to the disease in experimental models by inhibiting PPARα expression. *Gut* **2015**, *65*, 1882–1894. [CrossRef] [PubMed]
85. Rodrigues, P.M.; Rodrigues, C.; Castro, R.E. Modulation of liver steatosis by miR-21/PPARα. *Cell Death Discov.* **2018**, *4*, 9. [CrossRef]
86. Cermelli, S.; Ruggieri, A.; Marrero, J.A.; Ioannou, G.N.; Beretta, L. Circulating MicroRNAs in Patients with Chronic Hepatitis C and Non-Alcoholic Fatty Liver Disease. *PLoS ONE* **2011**, *6*, e23937. [CrossRef] [PubMed]
87. Liu, X.-L.; Pan, Q.; Zhang, R.-N.; Shen, F.; Yan, S.-Y.; Sun, C.; Xu, Z.-J.; Chen, Y.-W.; Fan, J.-G. Disease-specific miR-34a as diagnostic marker of non-alcoholic steatohepatitis in a Chinese population. *World J. Gastroenterol.* **2016**, *22*, 9844–9852. [CrossRef] [PubMed]
88. Choi, S.-E.; Fu, T.; Seok, S.; Kim, D.-H.; Yu, E.; Lee, K.-W.; Kang, Y.; Li, X.; Kemper, B.; Kemper, J.K. Elevated microRNA-34a in obesity reduces NAD^+ levels and SIRT1 activity by directly targeting NAMPT. *Aging Cell* **2013**, *12*, 1062–1072. [CrossRef] [PubMed]
89. Fu, T.; Kemper, J. MicroRNA-34a and Impaired FGF19/21 Signaling in Obesity. In *Vitamins & Hormones*; Litwack, G., Ed.; Academic Press: Cambridge, MA, USA, 2016; Volume 101, pp. 175–196. [CrossRef]
90. Castro, R.E.; Ferreira, D.M.; Afonso, M.B.; Borralho, P.M.; Machado, M.V.; Cortez-Pinto, H.; Rodrigues, C.M. miR-34a/SIRT1/p53 is suppressed by ursodeoxycholic acid in the rat liver and activated by disease severity in human non-alcoholic fatty liver disease. *J. Hepatol.* **2013**, *58*, 119–125. [CrossRef]
91. Lee, J.; Padhye, A.; Sharma, A.; Song, G.; Miao, J.; Mo, Y.-Y.; Wang, L.; Kemper, J.K. A Pathway Involving Farnesoid X Receptor and Small Heterodimer Partner Positively Regulates Hepatic Sirtuin 1 Levels via MicroRNA-34a Inhibition. *J. Biol. Chem.* **2010**, *285*, 12604–12611. [CrossRef]
92. Liu, X.-L.; Cao, H.-X.; Wang, B.-C.; Xin, F.-Z.; Zhang, R.-N.; Zhou, D.; Yang, R.-X.; Zhao, Z.-H.; Pan, Q.; Fan, J.-G. miR-192-5p regulates lipid synthesis in non-alcoholic fatty liver disease through SCD-1. *World J. Gastroenterol.* **2017**, *23*, 8140–8151. [CrossRef]
93. Yu, Y.; Zhu, J.; Liu, J.; Huang, M.; Wan, J. Identification of 8-miRNAs as biomarkers for nonalcoholic fatty liver disease. *J. Cell. Physiol.* **2019**, *234*, 17361–17369. [CrossRef]
94. Pirola, C.J.; Fernandez Gianotti, T.; Castano, G.O.; Mallardi, P.; San Martino, J.; Mora Gonzalez Lopez Ledesma, M.; Flichman, D.; Mirshahi, F.; Sanyal, A.J.; Sookoian, S. Circulating microRNA signature in non-alcoholic fatty liver disease: From serum non-coding RNAs to liver histology and disease pathogenesis. *Gut* **2015**, *64*, 800–812. [CrossRef]
95. Braza-Boïls, A.; Marí-Alexandre, J.; Molina, P.; Arnau, M.A.; Barceló-Molina, M.; Domingo, D.; Girbes, J.; Giner, J.; Martínez-Dolz, L.; Zorio, E. Deregulated hepatic microRNAs underlie the association between non-alcoholic fatty liver disease and coronary artery disease. *Liver Int.* **2016**, *36*, 1221–1229. [CrossRef]
96. Iliopoulos, D.; Drosatos, K.; Hiyama, Y.; Goldberg, I.J.; Zannis, V.I. MicroRNA-370 controls the expression of MicroRNA-122 and Cpt1α and affects lipid metabolism. *J. Lipid Res.* **2010**, *51*, 1513–1523. [CrossRef]
97. Gerin, I.; Clerbaux, L.-A.; Haumont, O.; Lanthier, N.; Das, A.K.; Burant, C.F.; Leclercq, I.A.; MacDougald, O.A.; Bommer, G.T. Expression of miR-33 from an SREBP2 Intron Inhibits Cholesterol Export and Fatty Acid Oxidation*. *J. Biol. Chem.* **2010**, *285*, 33652–33661. [CrossRef]
98. Rayner, K.; Sheedy, F.; Esau, C.C.; Hussain, F.N.; Temel, R.E.; Parathath, S.; van Gils, J.; Rayner, A.J.; Chang, A.N.; Suarez, Y.; et al. Antagonism of miR-33 in mice promotes reverse cholesterol transport and regression of atherosclerosis. *J. Clin. Investig.* **2011**, *121*, 2921–2931. [CrossRef] [PubMed]
99. Ghareghani, P.; Shanaki, M.; Ahmadi, S.; Khoshdel, A.R.; Rezvan, N.; Meshkani, R.; Delfan, M.; Gorgani-Firuzjaee, S. Aerobic endurance training improves nonalcoholic fatty liver disease (NAFLD) features via miR-33 dependent autophagy induction in high fat diet fed mice. *Obes. Res. Clin. Pr.* **2018**, *12*, 80–89. [CrossRef] [PubMed]
100. Goedeke, L.; Salerno, A.; Ramírez, C.; Guo, L.; Allen, R.M.; Yin, X.; Langley, S.R.; Esau, C.; Wanschel, A.; A Fisher, E.; et al. Long-term therapeutic silencing of miR-33 increases circulating triglyceride levels and hepatic lipid accumulation in mice. *EMBO Mol. Med.* **2014**, *6*, 1133–1141. [CrossRef] [PubMed]

101. Jampoka, K.; Muangpaisarn, P.; Khongnomnan, K.; Treeprasertsuk, S.; Tangkijvanich, P.; Payungporn, S. Serum miR-29a and miR-122 as Potential Biomarkers for Non-Alcoholic Fatty Liver Disease (NAFLD). *MicroRNA* **2018**, *7*, 215–222. [CrossRef]
102. Pandey, A.K.; Verma, G.; Vig, S.; Srivastava, S.; Srivastava, A.K.; Datta, M. miR-29a levels are elevated in the db/db mice liver and its overexpression leads to attenuation of insulin action on PEPCK gene expression in HepG2 cells. *Mol. Cell. Endocrinol.* **2011**, *332*, 125–133. [CrossRef]
103. Fernández-Tussy, P.; Fernández-Ramos, D.; Lopitz-Otsoa, F.; Simón, J.; Barbier-Torres, L.; Gomez-Santos, B.; Nuñez-Garcia, M.; Azkargorta, M.; Juan, V.G.-D.; Serrano-Macia, M.; et al. miR-873-5p targets mitochondrial GNMT-Complex II interface contributing to non-alcoholic fatty liver disease. *Mol. Metab.* **2019**, *29*, 40–54. [CrossRef]
104. Oura, K.; Morishita, A.; Masaki, T. Molecular and Functional Roles of MicroRNAs in the Progression of Hepatocellular Carcinoma—A Review. *Int. J. Mol. Sci.* **2020**, *21*, 8362. [CrossRef] [PubMed]
105. Li, Y.; Di, C.; Li, W.; Cai, W.; Tan, X.; Xu, L.; Yang, L.; Lou, G.; Yan, Y. Oncomirs miRNA-221/222 and Tumor Suppressors miRNA-199a/195 Are Crucial miRNAs in Liver Cancer: A Systematic Analysis. *Am. J. Dig. Dis.* **2016**, *61*, 2315–2327. [CrossRef]
106. Friedman, S.L. Molecular Regulation of Hepatic Fibrosis, an Integrated Cellular Response to Tissue Injury. *J. Biol. Chem.* **2000**, *275*, 2247–2250. [CrossRef] [PubMed]
107. Maher, J.J.; McGuire, R.F. Extracellular matrix gene expression increases preferentially in rat lipocytes and sinusoidal endothelial cells during hepatic fibrosis in vivo. *J. Clin. Investig.* **1990**, *86*, 1641–1648. [CrossRef] [PubMed]
108. Wei, J.; Feng, L.; Li, Z.; Xu, G.; Fan, X. MicroRNA-21 activates hepatic stellate cells via PTEN/Akt signaling. *Biomed. Pharmacother.* **2013**, *67*, 387–392. [CrossRef]
109. Noetel, A.; Kwiecinski, M.; Elfimova, N.; Huang, J.; Odenthal, M. microRNA are Central Players in Anti- and Profibrotic Gene Regulation during Liver Fibrosis. *Front. Physiol.* **2012**, *3*, 49. [CrossRef]
110. Davoodian, P.; Ravanshad, M.; Hosseini, S.Y.; Khanizadeh, S.; Almasian, M.; Nejati Zadeh, A.; Esmaiili Lashgarian, H. Effect of TGF-β/smad signaling pathway blocking on expression profiles of miR-335, miR-150, miR-194, miR-27a, and miR-199a of hepatic stellate cells (HSCs). *Gastroenterol. Hepatol. Bed Bench* **2017**, *10*, 112–117.
111. Bae, M.; Kim, M.-B.; Lee, J.-Y. Astaxanthin Attenuates the Changes in the Expression of MicroRNAs Involved in the Activation of Hepatic Stellate Cells. *Nutrients* **2022**, *14*, 962. [CrossRef]
112. Cai, L.-J.; Tu, L.; Huang, X.-M.; Huang, J.; Qiu, N.; Xie, G.-H.; Liao, J.-X.; Du, W.; Zhang, Y.-Y.; Tian, J.-Y. LncRNA MALAT1 facilitates inflammasome activation via epigenetic suppression of Nrf2 in Parkinson's disease. *Mol. Brain* **2020**, *13*, 1–15. [CrossRef]
113. Xiong, L.; Zhou, B.; Liu, H.; Cai, L. Comprehensive Review of Cadmium Toxicity Mechanisms in Male Reproduction and Therapeutic Strategies. *Rev. Environ. Contam. Toxicol.* **2021**, *258*, 151–193. [CrossRef] [PubMed]
114. Leti, F.; Legendre, C.; Still, C.D.; Chu, X.; Petrick, A.; Gerhard, G.S.; DiStefano, J.K. Altered expression of MALAT1 lncRNA in nonalcoholic steatohepatitis fibrosis regulates CXCL5 in hepatic stellate cells. *Transl. Res.* **2017**, *190*, 25–39. [CrossRef] [PubMed]
115. Xiang, J.; Deng, Y.-Y.; Liu, H.-X.; Pu, Y. LncRNA MALAT1 Promotes PPARα/CD36-Mediated Hepatic Lipogenesis in Nonalcoholic Fatty Liver Disease by Modulating miR-206/ARNT Axis. *Front. Bioeng. Biotechnol.* **2022**. [CrossRef]
116. Wang, H.-F.; Chang, M.; Peng, T.-T.; Yang, Y.; Li, N.; Luo, T.; Cheng, Y.-M.; Zhou, M.-Z.; Zeng, X.-H.; Zheng, L.-P. Exposure to Cadmium Impairs Sperm Functions by Reducing CatSper in Mice. *Cell. Physiol. Biochem.* **2017**, *42*, 44–54. [CrossRef] [PubMed]
117. Bian, E.-B.; Wang, Y.-Y.; Yang, Y.; Wu, B.-M.; Xu, T.; Meng, X.-M.; Huang, C.; Zhang, L.; Lv, X.-W.; Xiong, Z.-G.; et al. Hotair facilitates hepatic stellate cells activation and fibrogenesis in the liver. *Biochim. et Biophys. Acta (BBA) - Mol. Basis Dis.* **2017**, *1863*, 674–686. [CrossRef] [PubMed]
118. Sulaiman, S.A.; Muhsin, N.I.A.; Jamal, R. Regulatory Non-coding RNAs Network in Non-alcoholic Fatty Liver Disease. *Front. Physiol.* **2019**, *10*, 279. [CrossRef]
119. Negishi, M.; Wongpalee, S.; Sarkar, S.; Park, J.; Lee, K.Y.; Shibata, Y.; Reon, B.J.; Abounader, R.; Suzuki, Y.; Sugano, S.; et al. A New lncRNA, APTR, Associates with and Represses the CDKN1A/p21 Promoter by Recruiting Polycomb Proteins. *PLoS ONE* **2014**, *9*, e95216. [CrossRef]
120. Ding, J.; Li, M.; Wan, X.; Jin, X.; Chen, S.; Yu, C.; Li, Y. Effect of miR-34a in regulating steatosis by targeting PPARα expression in nonalcoholic fatty liver disease. *Sci. Rep.* **2015**, *5*, 13729. [CrossRef]
121. Dorairaj, V.; Sulaiman, S.A.; Abu, N.; Murad, N.A.A. Nonalcoholic Fatty Liver Disease (NAFLD): Pathogenesis and Noninvasive Diagnosis. *Biomedicines* **2021**, *10*, 15. [CrossRef]
122. Khalifa, O.; Errafii, K.; Al-Akl, N.S.; Arredouani, A. Noncoding RNAs in Nonalcoholic Fatty Liver Disease: Potential Diagnosis and Prognosis Biomarkers. *Dis. Markers* **2020**, *2020*, 1–16. [CrossRef]
123. Di Mauro, S.; Scamporrino, A.; Petta, S.; Urbano, F.; Filippello, A.; Ragusa, M.; Di Martino, M.T.; Scionti, F.; Grimaudo, S.; Pipitone, R.M.; et al. Serum coding and non-coding RNAs as biomarkers of NAFLD and fibrosis severity. *Liver Int.* **2019**, *39*, 1742–1754. [CrossRef]
124. Rashad, N.M.; Khalil, U.A.; Ahmed, S.M.; Hussien, M.H.; Sami, M.M.; Wadea, F.M. The Expression Level of Long Non-Coding RNA PVT1 as a Diagnostic Marker for Advanced Stages in Patients with Nonalcoholic Fatty Liver Disease. *Egypt. J. Hosp. Med.* **2022**, *87*, 1825–1834. [CrossRef]
125. Carpenter, S.; Aiello, D.; Atianand, M.K.; Ricci, E.P.; Gandhi, P.; Hall, L.L.; Byron, M.; Monks, B.; Henry-Bezy, M.; Lawrence, J.B.; et al. A Long Noncoding RNA Mediates Both Activation and Repression of Immune Response Genes. *Science* **2013**, *341*, 789–792. [CrossRef]

126. Tong, Q.; Gong, A.; Zhang, X.; Lin, C.; Ma, S.; Chen, J.; Hu, G.; Chen, X. LincRNA-Cox2 modulates TNF-α-induced transcription of Il12b gene in intestinal epithelial cells through regulation of Mi-2/NuRD-mediated epigenetic histone modifications. *FASEB J.* **2015**, *30*, 1187–1197. [CrossRef] [PubMed]
127. Mang, Y.; Li, L.; Ran, J.; Zhang, S.; Liu, J.; Li, L.; Chen, Y.; Liu, J.; Gao, Y.; Ren, G. Long noncoding RNA NEAT1 promotes cell proliferation and invasion by regulating hnRNP A2 expression in hepatocellular carcinoma cells. *OncoTargets Ther.* **2017**, *10*, 1003–1016. [CrossRef]
128. Yu, F.; Jiang, Z.; Chen, B.; Dong, P.; Zheng, J. NEAT1 accelerates the progression of liver fibrosis via regulation of microRNA-122 and Kruppel-like factor 6. *Klin. Wochenschr.* **2017**, *95*, 1191–1202. [CrossRef]
129. Zhou, W.; Qiu, K. The correlation between lncRNA NEAT1 and serum hepcidin in the peripheral blood of non-alcoholic fatty liver disease patients. *Am. J. Transl. Res.* **2022**, *14*, 2593–2599.
130. Bu, F.; Wang, A.; Zhu, Y.; You, H.; Zhang, Y.; Meng, X.; Huang, C.; Li, J.; Zhu, Y. LncRNA *NEAT1*: Shedding light on mechanisms and opportunities in liver diseases. *Liver Int.* **2020**, *40*, 2612–2626. [CrossRef] [PubMed]
131. Wang, X. Down-regulation of lncRNA-NEAT1 alleviated the non-alcoholic fatty liver disease via mTOR/S6K1 signaling pathway. *J. Cell. Biochem.* **2017**, *119*, 1567–1574. [CrossRef] [PubMed]
132. Gernapudi, R.; Wolfson, B.; Zhang, Y.; Yao, Y.; Yang, P.; Asahara, H.; Zhou, Q. MicroRNA 140 Promotes Expression of Long Noncoding RNA NEAT1 in Adipogenesis. *Mol. Cell. Biol.* **2016**, *36*, 30–38. [CrossRef] [PubMed]
133. Chen, G.; Yu, D.; Nian, X.; Liu, J.; Koenig, R.J.; Xu, B.; Sheng, L. LncRNA SRA promotes hepatic steatosis through repressing the expression of adipose triglyceride lipase (ATGL). *Sci. Rep.* **2016**, *6*, 35531. [CrossRef] [PubMed]
134. Bejerano, G.; Pheasant, M.; Makunin, I.; Stephen, S.; Kent, W.J.; Mattick, J.S.; Haussler, D. Ultraconserved Elements in the Human Genome. *Science* **2004**, *304*, 1321–1325. [CrossRef] [PubMed]
135. Guo, J.; Fang, W.; Sun, L.; Lu, Y.; Dou, L.; Huang, X.; Tang, W.; Yu, L.; Li, J. Ultraconserved element uc.372 drives hepatic lipid accumulation by suppressing miR-195/miR4668 maturation. *Nat. Commun.* **2018**, *9*, 612. [CrossRef] [PubMed]
136. Jeong, S.-H.; Kim, H.-B.; Kim, M.-C.; Lee, J.-M.; Lee, J.H.; Kim, J.-H.; Kim, J.W.; Park, W.-Y.; Kim, S.-Y.; Kim, J.B.; et al. Hippo-mediated suppression of IRS2/AKT signaling prevents hepatic steatosis and liver cancer. *J. Clin. Investig.* **2018**, *128*, 1010–1025. [CrossRef] [PubMed]
137. Wang, F.; Kohan, A.B.; Lo, C.-M.; Liu, M.; Howles, P.; Tso, P. Apolipoprotein A-IV: A protein intimately involved in metabolism. *J. Lipid Res.* **2015**, *56*, 1403–1418. [CrossRef]
138. Wang, H.; Cao, Y.; Shu, L.; Zhu, Y.; Peng, Q.; Ran, L.; Wu, J.; Luo, Y.; Zuo, G.; Luo, J.; et al. Long non-coding RNA (lncRNA) H19 induces hepatic steatosis through activating MLXIPL and mTORC1 networks in hepatocytes. *J. Cell. Mol. Med.* **2019**, *24*, 1399–1412. [CrossRef]
139. Wang, Y.; Liu, W.; Liu, Y.; Cui, J.; Zhao, Z.; Cao, H.; Fu, Z.; Liu, B. Long noncoding RNA H19 mediates *LCoR* to impact the osteogenic and adipogenic differentiation of mBMSCs in mice through sponging miR-188. *J. Cell. Physiol.* **2018**, *233*, 7435–7446. [CrossRef]
140. Lee, A.-H.; Scapa, E.F.; Cohen, D.E.; Glimcher, L.H. Regulation of Hepatic Lipogenesis by the Transcription Factor XBP1. *Science* **2008**, *320*, 1492–1496. [CrossRef]
141. Fuchs, M.; Sanyal, A.J. Lipotoxicity in NASH. *J. Hepatol. 2012*(56), 291–293.
142. Farooq, M.; Piquet-Pellorce, C.; Dion, S.; Eugenio, M.; Santamaria, K.; Filliol, A.; Dimanche-Boitrel, M.-T.; Samson, M.; Le Seyec, J. RIPK1 depletion exacerbates progression of liver fibrosis in high fat diet induced non-alcoholic steatohepatitis (NASH) in mice. *J. Hepatol.* **2018**, *68*, S345. [CrossRef]
143. Alakunle, E.; Moens, U.; Nchinda, G.; Okeke, M. Monkeypox Virus in Nigeria: Infection Biology, Epidemiology, and Evolution. *Viruses* **2020**, *12*, 1257. [CrossRef]
144. Liu, S.; Zhang, J.; Yin, L.; Wang, X.; Zheng, Y.; Zhang, Y.; Gu, J.; Yang, L.; Yang, J.; Zheng, P.; et al. The lncRNA RUNX1-IT1 regulates C-FOS transcription by interacting with RUNX1 in the process of pancreatic cancer proliferation, migration and invasion. *Cell Death Dis.* **2020**, *11*, 412. [CrossRef]
145. Kaur, S.; Rawal, P.; Siddiqui, H.; Rohilla, S.; Sharma, S.; Tripathi, D.M.; Baweja, S.; Hassan, M.; Vlaic, S.; Guthke, R.; et al. Increased Expression of RUNX1 in Liver Correlates with NASH Activity Score in Patients with Non-Alcoholic Steatohepatitis (NASH). *Cells* **2019**, *8*, 1277. [CrossRef] [PubMed]
146. Long, J.-K.; Dai, W.; Zheng, Y.-W.; Zhao, S.-P. miR-122 promotes hepatic lipogenesis via inhibiting the LKB1/AMPK pathway by targeting Sirt1 in non-alcoholic fatty liver disease. *Mol. Med.* **2019**, *25*, 26. [CrossRef] [PubMed]
147. Xu, Y.; Zhu, Y.; Hu, S.; Pan, X.; Bawa, F.C.; Wang, H.H.; Wang, D.Q.-H.; Yin, L.; Zhang, Y. Hepatocyte miR-34a is a key regulator in the development and progression of non-alcoholic fatty liver disease. *Mol. Metab.* **2021**, *51*, 101244. [CrossRef]
148. Lin, Y.; Ding, D.; Huang, Q.; Liu, Q.; Lu, H.; Lu, Y.; Chi, Y.; Sun, X.; Ye, G.; Zhu, H.; et al. Downregulation of miR-192 causes hepatic steatosis and lipid accumulation by inducing SREBF1: Novel mechanism for bisphenol A-triggered non-alcoholic fatty liver disease. *Biochim. et Biophys. Acta (BBA) - Mol. Cell Biol. Lipids* **2017**, *1862*, 869–882. [CrossRef] [PubMed]
149. Yu, F.; Zheng, J.; Mao, Y.; Dong, P.; Li, G.; Lu, Z.; Guo, C.; Liu, Z.; Fan, X. Long non-coding RNA APTR promotes the activation of hepatic stellate cells and the progression of liver fibrosis. *Biochem. Biophys. Res. Commun.* **2015**, *463*, 679–685. [CrossRef] [PubMed]
150. Zheng, J.; Yu, F.; Dong, P.; Wu, L.; Zhang, Y.; Hu, Y.; Zheng, L. Long non-coding RNA PVT1 activates hepatic stellate cells through competitively binding microRNA-152. *Oncotarget* **2016**, *7*, 62886–62897. [CrossRef]

151. Jin, S.-S.; Lin, C.-J.; Lin, X.-F.; Zheng, J.-Z.; Guan, H.-Q. Silencing lncRNA NEAT1 reduces nonalcoholic fatty liver fat deposition by regulating the miR-139-5p/c-Jun/SREBP-1c pathway. *Ann. Hepatol.* **2021**, *27*, 100584. [CrossRef] [PubMed]
152. Liu, J.; Tang, T.; Wang, G.-D.; Liu, B. LncRNA-H19 promotes hepatic lipogenesis by directly regulating miR-130a/PPARγ axis in non-alcoholic fatty liver disease. *Biosci. Rep.* **2019**, 39. [CrossRef]
153. Hansen, T.B.; Jensen, T.I.; Clausen, B.H.; Bramsen, J.B.; Finsen, B.; Damgaard, C.K.; Kjems, J. Natural RNA circles function as efficient microRNA sponges. *Nature* **2013**, *495*, 384–388. [CrossRef] [PubMed]
154. Memczak, S.; Jens, M.; Elefsinioti, A.; Torti, F.; Krueger, J.; Rybak, A.; Maier, L.; Mackowiak, S.D.; Gregersen, L.H.; Munschauer, M.; et al. Circular RNAs are a large class of animal RNAs with regulatory potency. *Nature* **2013**, *495*, 333–338. [CrossRef] [PubMed]
155. Yin, Y.; Long, J.; He, Q.; Li, Y.; Liao, Y.; He, P.; Zhu, W. Emerging roles of circRNA in formation and progression of cancer. *J. Cancer* **2019**, *10*, 5015–5021. [CrossRef]
156. Jin, X.; Feng, C.-Y.; Xiang, Z.; Chen, Y.-P.; Li, Y.-M. CircRNA expression pattern and circRNA-miRNA-mRNA network in the pathogenesis of nonalcoholic steatohepatitis. *Oncotarget* **2016**, *7*, 66455–66467. [CrossRef] [PubMed]
157. Benedict, M.; Zhang, X. Non-alcoholic fatty liver disease: An expanded review. *World J. Hepatol.* **2017**, *9*, 715–732. [CrossRef] [PubMed]
158. Guo, X.-Y.; Chen, J.-N.; Sun, F.; Wang, Y.-Q.; Pan, Q.; Fan, J.-G. circRNA_0046367 Prevents Hepatoxicity of Lipid Peroxidation: An Inhibitory Role against Hepatic Steatosis. *Oxidative Med. Cell. Longev.* **2017**, *2017*, 1–16. [CrossRef] [PubMed]
159. Kanehisa, M.; Goto, S.; Sato, Y.; Furumichi, M.; Tanabe, M. KEGG for integration and interpretation of large-scale molecular data sets. *Nucleic Acids Res.* **2011**, *40*, D109–D114. [CrossRef]
160. Pettinelli, P.; del Pozo, T.; Araya, J.; Rodrigo, R.; Araya, A.V.; Smok, G.; Csendes, A.; Gutierrez, L.; Rojas, J.; Korn, O.; et al. Enhancement in liver SREBP-1c/PPAR-α ratio and steatosis in obese patients: Correlations with insulin resistance and n-3 long-chain polyunsaturated fatty acid depletion. *Biochim. et Biophys. Acta (BBA) - Mol. Basis Dis.* **2009**, *1792*, 1080–1086. [CrossRef]
161. Guo, X.-Y.; Sun, F.; Chen, J.-N.; Wang, Y.-Q.; Pan, Q.; Fan, J.-G. circRNA_0046366 inhibits hepatocellular steatosis by normalization of PPAR signaling. *World J. Gastroenterol.* **2018**, *24*, 323–337. [CrossRef]
162. Feng, Y.Y.; Xu, X.Q.; Ji, C.B.; Shi, C.M.; Guo, X.R.; Fu, J.F. Aberrant Hepatic MicroRNA Expression in Nonalcoholic Fatty Liver Disease. *Cell. Physiol. Biochem.* **2014**, *34*, 1983–1997. [CrossRef] [PubMed]
163. Pogribny, I.P.; Starlard-Davenport, A.; Tryndyak, V.; Han, T.; A Ross, S.; Rusyn, I.; Beland, F. Difference in expression of hepatic microRNAs miR-29c, miR-34a, miR-155, and miR-200b is associated with strain-specific susceptibility to dietary nonalcoholic steatohepatitis in mice. *Lab. Investig.* **2010**, *90*, 1437–1446. [CrossRef] [PubMed]
164. Yamada, H.; Suzuki, K.; Ichino, N.; Ando, Y.; Sawada, A.; Osakabe, K.; Sugimoto, K.; Ohashi, K.; Teradaira, R.; Inoue, T.; et al. Associations between circulating microRNAs (miR-21, miR-34a, miR-122 and miR-451) and non-alcoholic fatty liver. *Clin. Chim. Acta* **2013**, *424*, 99–103. [CrossRef]
165. Salvoza, N.C.; Klinzing, D.C.; Gopez-Cervantes, J.; Baclig, M.O. Association of Circulating Serum miR-34a and miR-122 with Dyslipidemia among Patients with Non-Alcoholic Fatty Liver Disease. *PLoS ONE* **2016**, *11*, e0153497. [CrossRef] [PubMed]
166. Kozomara, A.; Griffiths-Jones, S. miRBase: Annotating high confidence microRNAs using deep sequencing data. *Nucleic Acids Res.* **2013**, *42*, D68–D73. [CrossRef] [PubMed]
167. Huang, D.W.; Sherman, B.T.; Tan, Q.; Kir, J.; Liu, D.; Bryant, D.; Guo, Y.; Stephens, R.; Baseler, M.W.; Lane, H.C.; et al. DAVID Bioinformatics Resources: Expanded annotation database and novel algorithms to better extract biology from large gene lists. *Nucleic Acids Res.* **2007**, *35*, W169–W175. [CrossRef]
168. Guo, X.-Y.; He, C.-X.; Wang, Y.-Q.; Sun, C.; Li, G.-M.; Su, Q.; Pan, Q.; Fan, J.-G. Circular RNA Profiling and Bioinformatic Modeling Identify Its Regulatory Role in Hepatic Steatosis. *BioMed Res. Int.* **2017**, *2017*, 1–13. [CrossRef]
169. Lehman, J.J.; Barger, P.M.; Kovacs, A.; Saffitz, J.E.; Medeiros, D.M.; Kelly, D.P. Peroxisome proliferator–activated receptor γ coactivator-1 promotes cardiac mitochondrial biogenesis. *J. Clin. Investig.* **2000**, *106*, 847–856. [CrossRef]
170. Vega, R.B.; Huss, J.M.; Kelly, D.P. The Coactivator PGC-1 Cooperates with Peroxisome Proliferator-Activated Receptor α in Transcriptional Control of Nuclear Genes Encoding Mitochondrial Fatty Acid Oxidation Enzymes. *Mol. Cell. Biol.* **2000**, *20*, 1868–1876. [CrossRef] [PubMed]
171. Everitt, H.; Hu, M.; Ajmo, J.M.; Rogers, C.Q.; Liang, X.; Zhang, R.; Yin, H.; Choi, A.; Bennett, E.S.; You, M. Ethanol administration exacerbates the abnormalities in hepatic lipid oxidation in genetically obese mice. *Am. J. Physiol. Liver Physiol.* **2013**, *304*, G38–G47. [CrossRef]
172. Peterson, T.R.; Sengupta, S.S.; Harris, T.E.; Carmack, A.E.; Kang, S.A.; Balderas, E.; Guertin, D.A.; Madden, K.L.; Carpenter, A.E.; Finck, B.N.; et al. mTOR Complex 1 Regulates Lipin 1 Localization to Control the SREBP Pathway. *Cell* **2011**, *146*, 408–420. [CrossRef]
173. Hirotsu, Y.; Hataya, N.; Katsuoka, F.; Yamamoto, M. NF-E2-Related Factor 1 (Nrf1) Serves as a Novel Regulator of Hepatic Lipid Metabolism through Regulation of the *Lipin1* and *PGC-1* β Genes. *Mol. Cell. Biol.* **2012**, *32*, 2760–2770. [CrossRef]
174. Chen, Z.; Gropler, M.C.; Mitra, M.S.; Finck, B.N. Complex Interplay between the Lipin 1 and the Hepatocyte Nuclear Factor 4 α (HNF4α) Pathways to Regulate Liver Lipid Metabolism. *PLoS ONE* **2012**, *7*, e51320. [CrossRef] [PubMed]
175. Li, L.; Guo, J.; Chen, Y.; Chang, C.; Xu, C. Comprehensive CircRNA expression profile and selection of key CircRNAs during priming phase of rat liver regeneration. *BMC Genom.* **2017**, *18*, 80. [CrossRef]
176. Lin, X.; Chen, Y. Identification of potentially functional CircRNA-miRNA-mRNA regulatory network in hepatocellular carcinoma by integrated microarray analysis. *Med. Sci. Monit. Basic Res.* **2018**, *24*, 70–78. [CrossRef]

177. Szabo, G.; Sarnow, P.; Bala, S. MicroRNA silencing and the development of novel therapies for liver disease. *J. Hepatol.* **2012**, *57*, 462–466. [CrossRef] [PubMed]
178. Cheung, O.; Puri, P.; Eicken, C.; Contos, M.J.; Mirshahi, F.; Maher, J.W.; Kellum, J.M.; Min, H.; Luketic, V.A.; Sanyal, A.J. Nonalcoholic steatohepatitis is associated with altered hepatic MicroRNA expression. *Hepatology* **2008**, *48*, 1810–1820. [CrossRef]
179. Jin, X.; Gao, J.; Zheng, R.; Yu, M.; Ren, Y.; Yan, T.; Huang, Y.; Li, Y. Antagonizing circRNA_002581-miR-122-CPEB1 axis alleviates NASH through restoring PTEN-AMPK-mTOR pathway regulated autophagy. *Cell Death Dis.* **2020**, *11*, 123. [CrossRef]
180. Musso, G.; Cassader, M.; Gambino, R. Non-alcoholic steatohepatitis: Emerging molecular targets and therapeutic strategies. *Nat. Rev. Drug Discov.* **2016**, *15*, 249–274. [CrossRef] [PubMed]
181. Zhu, L.; Ren, T.; Zhu, Z.; Cheng, M.; Mou, Q.; Mu, M.; Liu, Y.; Yao, Y.; Cheng, Y.; Zhang, B.; et al. Thymosin-β4 Mediates Hepatic Stellate Cell Activation by Interfering with CircRNA-0067835/miR-155/FoxO3 Signaling Pathway. *Cell. Physiol. Biochem.* **2018**, *51*, 1389–1398. [CrossRef]
182. Kao, Y.-H.; Chen, P.-H.; Wu, T.-Y.; Lin, Y.-C.; Tsai, M.-S.; Lee, P.-H.; Tai, T.-S.; Chang, H.-R.; Sun, C.-K. Lipopolysaccharides induce Smad2 phosphorylation through PI3K/Akt and MAPK cascades in HSC-T6 hepatic stellate cells. *Life Sci.* **2017**, *184*, 37–46. [CrossRef]
183. Tu, W.; Ye, J.; Wang, Z.-J. Embryonic liver fordin is involved in glucose glycolysis of hepatic stellate cell by regulating PI3K/Akt signaling. *World J. Gastroenterol.* **2016**, *22*, 8519–8527. [CrossRef]
184. Chen, Y.; Yuan, B.; Wu, Z.; Dong, Y.; Zhang, L.; Zeng, Z. Microarray profiling of circular RNAs and the potential regulatory role of hsa_circ_0071410 in the activated human hepatic stellate cell induced by irradiation. *Gene* **2017**, *629*, 35–42. [CrossRef]
185. Zhou, Y.; Lv, X.; Qu, H.; Zhao, K.; Fu, L.; Zhu, L.; Ye, G.; Guo, J. Preliminary screening and functional analysis of circular RNAs associated with hepatic stellate cell activation. *Gene* **2018**, *677*, 317–323. [CrossRef] [PubMed]
186. Long, J.; Badal, S.S.; Wang, Y.; Chang, B.H.J.; Rodriguez, A.; Danesh, F.R. MicroRNA-22 Is a Master Regulator of Bone Morphogenetic Protein-7/6 Homeostasis in the Kidney. *J. Biol. Chem.* **2013**, *288*, 36202–36214. [CrossRef] [PubMed]
187. Esteller, M. Non-coding RNAs in human disease. *Nat. Rev. Genet.* **2011**, *12*, 861–874. [CrossRef]
188. Fernández-Hernando, C.; Ramírez, C.M.; Goedeke, L.; Suárez, Y. MicroRNAs in Metabolic Disease. *Arter. Thromb. Vasc. Biol.* **2013**, *33*, 178–185. [CrossRef] [PubMed]
189. Rottiers, V.; Näär, A.M. MicroRNAs in metabolism and metabolic disorders. *Nat. Rev. Mol. Cell Biol.* **2012**, *13*, 239–250. [CrossRef] [PubMed]
190. Ponting, C.P.; Oliver, P.L.; Reik, W. Evolution and Functions of Long Noncoding RNAs. *Cell* **2009**, *136*, 629–641. [CrossRef] [PubMed]
191. Huang, R.; Duan, X.; Fan, J.; Li, G.; Wang, B. Role of Noncoding RNA in Development of Nonalcoholic Fatty Liver Disease. *BioMed Res. Int.* **2019**, *2019*, 1–9. [CrossRef] [PubMed]
192. Wang, G.; Zou, H.; Lai, C.; Huang, X.; Yao, Y.; Xiang, G. Repression of MicroRNA-124-3p Alleviates High-Fat Diet–Induced Hepatosteatosis by Targeting Pref-1. *Front. Endocrinol.* **2020**, *11*, 589994. [CrossRef]
193. Hudak, C.S.; Sul, H.S. Pref-1, a Gatekeeper of Adipogenesis. *Front. Endocrinol.* **2013**, *4*, 79. [CrossRef]
194. Huang, Y.; Yang, X.; Wu, Y.; Jing, W.; Cai, X.; Tang, W.; Liu, L.; Liu, Y.; Grottkau, B.E.; Lin, Y. γ-secretase inhibitor induces adipogenesis of adipose-derived stem cells by regulation of Notch and PPAR-γ. *Cell Prolif.* **2010**, *43*, 147–156. [CrossRef] [PubMed]
195. Shaw, T.A.; Singaravelu, R.; Powdrill, M.H.; Nhan, J.; Ahmed, N.; Özcelik, D.; Pezacki, J.P. MicroRNA-124 Regulates Fatty Acid and Triglyceride Homeostasis. *iScience* **2018**, *10*, 149–157. [CrossRef] [PubMed]
196. Xu, J.; Li, Y.; Chen, W.-D.; Xu, Y.; Yin, L.; Ge, X.; Jadhav, K.; Adorini, L.; Zhang, Y. Hepatic carboxylesterase 1 is essential for both normal and farnesoid X receptor-controlled lipid homeostasis. *Hepatology* **2014**, *59*, 1761–1771. [CrossRef] [PubMed]
197. Kida, K.; Nakajima, M.; Mohri, T.; Oda, Y.; Takagi, S.; Fukami, T.; Yokoi, T. PPARα Is Regulated by miR-21 and miR-27b in Human Liver. *Pharm. Res.* **2011**, *28*, 2467–2476. [CrossRef]
198. Ackers, I.; Malgor, R. Interrelationship of canonical and non-canonical Wnt signalling pathways in chronic metabolic diseases. *Diabetes Vasc. Dis. Res.* **2017**, *15*, 3–13. [CrossRef]
199. Wang, X.; Wang, X.; Huang, Y.; Chen, X.; Shi, L.; Li, C. Role and mechanisms of action of microRNA-21 as regards the regulation of the WNT/β-catenin signaling pathway in the pathogenesis of non-alcoholic fatty liver disease. *Int. J. Mol. Med.* **2019**, *44*, 2201–2212. [CrossRef]
200. Krützfeldt, J.; Rajewsky, N.; Braich, R.; Rajeev, K.G.; Tuschl, T.; Manoharan, M.; Stoffel, M. Silencing of microRNAs in vivo with 'antagomirs'. *Nature* **2005**, *438*, 685–689. [CrossRef]
201. Zhang, R.; Wang, X.; Qu, J.-H.; Liu, B.; Zhang, P.; Zhang, T.; Fan, P.-C.; Wang, X.-M.; Xiao, G.-Y.; Su, Y.; et al. Caloric Restriction Induces MicroRNAs to Improve Mitochondrial Proteostasis. *iScience* **2019**, *17*, 155–166. [CrossRef]
202. Chai, C.; Rivkin, M.; Berkovits, L.; Simerzin, A.; Zorde-Khvalevsky, E.; Rosenberg, N.; Klein, S.; Yaish, D.; Durst, R.; Shpitzen, S.; et al. Metabolic Circuit Involving Free Fatty Acids, microRNA 122, and Triglyceride Synthesis in Liver and Muscle Tissues. *Gastroenterology* **2017**, *153*, 1404–1415. [CrossRef] [PubMed]
203. Fang, Z.; Dou, G.; Wang, L. MicroRNAs in the Pathogenesis of Nonalcoholic Fatty Liver Disease. *Int. J. Biol. Sci.* **2021**, *17*, 1851–1863. [CrossRef] [PubMed]
204. BinMowyna, M.N.; AlFaris, N.A.; Al-Sanea, E.A.; AlTamimi, J.Z.; Aldayel, T.S. Resveratrol attenuates against high-fat-diet-promoted non-alcoholic fatty liver disease in rats mainly by targeting the miR-34a/SIRT1 axis. *Arch. Physiol. Biochem.* **2022**, 1–16. [CrossRef] [PubMed]

205. Kumar, V.; Mondal, G.; Dutta, R.; Mahato, R.I. Co-delivery of small molecule hedgehog inhibitor and miRNA for treating liver fibrosis. *Biomaterials* **2016**, *76*, 144–156. [CrossRef] [PubMed]
206. Wang, Y.; Du, J.; Niu, X.; Fu, N.; Wang, R.; Zhang, Y.; Zhao, S.; Sun, D.; Nan, Y. MiR-130a-3p attenuates activation and induces apoptosis of hepatic stellate cells in nonalcoholic fibrosing steatohepatitis by directly targeting TGFBR1 and TGFBR2. *Cell Death Dis.* **2017**, *8*, e2792. [CrossRef]
207. Lu, L.; Wang, J.; Lu, H.; Zhang, G.; Liu, Y.; Wang, J.; Zhang, Y.; Shang, H.; Ji, H.; Chen, X.; et al. MicroRNA-130a and -130b enhance activation of hepatic stellate cells by suppressing PPARγ expression: A rat fibrosis model study. *Biochem. Biophys. Res. Commun.* **2015**, *465*, 387–393. [CrossRef]
208. Xiao, F.; Yu, J.; Liu, B.; Guo, Y.; Li, K.; Deng, J.; Zhang, J.; Wang, C.; Chen, S.; Du, Y.; et al. A Novel Function of MicroRNA 130a-3p in Hepatic Insulin Sensitivity and Liver Steatosis. *Diabetes* **2014**, *63*, 2631–2642. [CrossRef]
209. Auguet, T.; Berlanga, A.; Guiu-Jurado, E.; Porras, J.A. Molecular pathways in non-alcoholic fatty liver disease. *Clin. Exp. Gastroenterol.* **2014**, *7*, 221–239. [CrossRef]
210. Bala, S.; Csak, T.; Saha, B.; Zatsiorsky, J.; Kodys, K.; Catalano, D.; Satishchandran, A.; Szabo, G. The pro-inflammatory effects of miR-155 promote liver fibrosis and alcohol-induced steatohepatitis. *J. Hepatol.* **2016**, *64*, 1378–1387. [CrossRef]
211. Skårn, M.; Namløs, H.M.; Noordhuis, P.; Wang, M.-Y.; Meza-Zepeda, L.A.; Myklebost, O. Adipocyte Differentiation of Human Bone Marrow-Derived Stromal Cells Is Modulated by MicroRNA-155, MicroRNA-221, and MicroRNA-222. *Stem Cells Dev.* **2012**, *21*, 873–883. [CrossRef]
212. Chen, Y.; Huang, H.; Xu, C.; Yu, C.; Li, Y. Long Non-Coding RNA Profiling in a Non-Alcoholic Fatty Liver Disease Rodent Model: New Insight into Pathogenesis. *Int. J. Mol. Sci.* **2017**, *18*, 21. [CrossRef] [PubMed]
213. Grimaldi, B.; Bellet, M.M.; Katada, S.; Astarita, G.; Hirayama, J.; Amin, R.H.; Granneman, J.G.; Piomelli, D.; Leff, T.; Sassone-Corsi, P. PER2 Controls Lipid Metabolism by Direct Regulation of PPARγ. *Cell Metab.* **2010**, *12*, 509–520. [CrossRef] [PubMed]
214. Zhang, L.; Yang, F.; Yuan, J.-H.; Yuan, S.-X.; Zhou, W.-P.; Huo, X.-S.; Xu, D.; Bi, H.-S.; Wang, F.; Sun, S.-H. Epigenetic activation of the MiR-200 family contributes to H19-mediated metastasis suppression in hepatocellular carcinoma. *Carcinogenesis* **2013**, *34*, 577–586. [CrossRef] [PubMed]
215. Li, J.-Z.; Ye, L.-H.; Wang, D.-H.; Zhang, H.-C.; Li, T.-Y.; Liu, Z.-Q.; Dai, E.-H.; Li, M.-R. The identify role and molecular mechanism of the MALAT1/hsa-mir-20b-5p/TXNIP axis in liver inflammation caused by CHB in patients with chronic HBV infection complicated with NAFLD. *Virus Res.* **2021**, *298*, 198405. [CrossRef] [PubMed]
216. Sookoian, S.; Flichman, D.; Garaycoechea, M.E.; Martino, J.S.; Castaño, G.O.; Pirola, C.J. Metastasis-associated lung adenocarcinoma transcript 1 as a common molecular driver in the pathogenesis of nonalcoholic steatohepatitis and chronic immune-mediated liver damage. *Hepatol. Commun.* **2018**, *2*, 654–665. [CrossRef] [PubMed]
217. Jiang, R.; Li, H.; Yang, J.; Shen, X.; Song, C.; Yang, Z.; Wang, X.; Huang, Y.; Lan, X.; Lei, C.; et al. circRNA Profiling Reveals an Abundant circFUT10 that Promotes Adipocyte Proliferation and Inhibits Adipocyte Differentiation via Sponging let-7. *Mol. Ther. - Nucleic Acids* **2020**, *20*, 491–501. [CrossRef]
218. Zhang, M.; Han, Y.; Zhai, Y.; Ma, X.; An, X.; Zhang, S.; Li, Z. Integrative analysis of circRNAs, miRNAs, and mRNAs profiles to reveal ceRNAs networks in chicken intramuscular and abdominal adipogenesis. *BMC Genom.* **2020**, *21*, 594. [CrossRef]
219. Karkeni, E.; Astier, J.; Tourniaire, F.; El Abed, M.; Romier, B.; Gouranton, E.; Wan, L.; Borel, P.; Salles, J.; Walrand, S.; et al. Obesity-associated Inflammation Induces microRNA-155 Expression in Adipocytes and Adipose Tissue: Outcome on Adipocyte Function. *J. Clin. Endocrinol. Metab.* **2016**, *101*, 1615–1626. [CrossRef]
220. Eeckhoute, J.; Oger, F.; Staels, B.; Lefebvre, P. Coordinated Regulation of PPAR Expression and Activity through Control of Chromatin Structure in Adipogenesis and Obesity. *PPAR Res.* **2012**, *2012*, 1–9. [CrossRef]
221. Perron, U.; Provero, P.; Molineris, I. In silico prediction of lncRNA function using tissue specific and evolutionary conserved expression. *BMC Bioinform.* **2017**, *18*, 144. [CrossRef]
222. Firmin, F.F.; Oger, F.; Gheeraert, C.; Dubois-Chevalier, J.; Vercoutter-Edouart, A.-S.; Alzaid, F.; Mazuy, C.; Dehondt, H.; Alexandre, J.; Derudas, B.; et al. The RBM14/CoAA-interacting, long intergenic non-coding RNA Paral1 regulates adipogenesis and coactivates the nuclear receptor PPARγ. *Sci. Rep.* **2017**, *7*, 1–16. [CrossRef]
223. Iwasaki, T.; Chin, W.W.; Ko, L. Identification and Characterization of RRM-containing Coactivator Activator (CoAA) as TRBP-interacting Protein, and Its Splice Variant as a Coactivator Modulator (CoAM). *J. Biol. Chem.* **2001**, *276*, 33375–33383. [CrossRef] [PubMed]
224. Auboeuf, D.; Hönig, A.; Berget, S.M.; O'Malley, B.W. Coordinate Regulation of Transcription and Splicing by Steroid Receptor Coregulators. *Science* **2002**, *298*, 416–419. [CrossRef]
225. Corrales, P.; Vidal-Puig, A.; Medina-Gómez, G. PPARs and Metabolic Disorders Associated with Challenged Adipose Tissue Plasticity. *Int. J. Mol. Sci.* **2018**, *19*, 2124. [CrossRef]
226. Tyagi, S.; Gupta, P.; Saini, A.S.; Kaushal, C.; Sharma, S. The peroxisome proliferator-activated receptor: A family of nuclear receptors role in various diseases. *J. Adv. Pharm. Technol. Res.* **2011**, *2*, 236–240. [CrossRef] [PubMed]
227. Crooke, S.T.; Wang, S.; Vickers, T.A.; Shen, W.; Liang, X.-H. Cellular uptake and trafficking of antisense oligonucleotides. *Nat. Biotechnol.* **2017**, *35*, 230–237. [CrossRef] [PubMed]
228. Khvorova, A.; Watts, A.K.J.K. The chemical evolution of oligonucleotide therapies of clinical utility. *Nat. Biotechnol.* **2017**, *35*, 238–248. [CrossRef]

229. Lim, K.R.Q.; Maruyama, R.; Yokota, T. Eteplirsen in the treatment of Duchenne muscular dystrophy. *Drug Des. Dev. Ther.* **2017**, *ume11*, 533–545. [CrossRef]
230. Corey, D.R. Nusinersen, an antisense oligonucleotide drug for spinal muscular atrophy. *Nat. Neurosci.* **2017**, *20*, 497–499. [CrossRef]
231. Uchida, S.; Kauppinen, S. Long Non-Coding RNAs in Liver Cancer and Nonalcoholic Steatohepatitis. *Non-Coding RNA* **2020**, *6*, 34. [CrossRef]
232. Matsui, M.; Corey, D.R. Non-coding RNAs as drug targets. *Nat. Rev. Drug Discov.* **2016**, *16*, 167–179. [CrossRef] [PubMed]

cells

MDPI

Review

Stemness of Normal and Cancer Cells: The Influence of Methionine Needs and SIRT1/PGC-1α/PPAR-α Players

Youssef Siblini [1], Farès Namour [1,2], Abderrahim Oussalah [1,2], Jean-Louis Guéant [1,2,*,†] and Céline Chéry [1,2,†]

[1] INSERM, UMR_S1256, NGERE—Nutrition, Genetics, Environmental Risk Exposure, University of Lorraine, 54500 Vandoeuvre-lès-Nancy, France

[2] Reference Centre of Inborn Metabolism Diseases and Department of Molecular Medicine, University Hospital Center, 54500 Vandoeuvre-lès-Nancy, France

* Correspondence: jean-louis.gueant@univ-lorraine.fr

† These authors contributed equally to this work.

Abstract: Stem cells are a population of undifferentiated cells with self-renewal and differentiation capacities. Normal and cancer stem cells share similar characteristics in relation to their stemness properties. One-carbon metabolism (OCM), a network of interconnected reactions, plays an important role in this dependence through its role in the endogenous synthesis of methionine and S-adenosylmethionine (SAM), the universal donor of methyl groups in eukaryotic cells. OCM genes are differentially expressed in stem cells, compared to their differentiated counterparts. Furthermore, cultivating stem cells in methionine-restricted conditions hinders their stemness capacities through decreased SAM levels with a subsequent decrease in histone methylation, notably H3K4me3, with a decrease in stem cell markers. Stem cells' reliance on methionine is linked to several mechanisms, including high methionine flux or low endogenous methionine biosynthesis. In this review, we provide an overview of the recent discoveries concerning this metabolic dependence and we discuss the mechanisms behind them. We highlight the influence of SIRT1 on SAM synthesis and suggest a role of PGC-1α/PPAR-α in impaired stemness produced by methionine deprivation. In addition, we discuss the potential interest of methionine restriction in regenerative medicine and cancer treatment.

Keywords: normal stem cells; cancer stem cells; methionine; methionine dependence; stemness; sirtuin 1; peroxisome proliferator-activated receptor gamma coactivator 1-alpha; peroxisome proliferator-activated receptor alpha; S-adenosylmethionine

Citation: Siblini, Y.; Namour, F.; Oussalah, A.; Guéant, J.-L.; Chéry, C. Stemness of Normal and Cancer Cells: The Influence of Methionine Needs and SIRT1/PGC-1α/PPAR-α Players. *Cells* **2022**, *11*, 3607. https://doi.org/10.3390/cells11223607

Academic Editors: Kay-Dietrich Wagner and Nicole Wagner

Received: 8 October 2022
Accepted: 9 November 2022
Published: 15 November 2022

1. Introduction

Stem cells are an unspecialized population of cells that exhibit self-renewal capacity and can generate other cell types [1]. The stemness level defines the capacity of the cells to produce other differentiated cells. For instance, the stemness level of the cell varies from a totipotent stem cell that can give rise to a whole organism to a pluripotent stem cell, such as embryonic stem cells (ESCs) or their induced counterpart, induced pluripotent stem cells (iPSCs), that can give rise to all cell types of the three germ layers (ectoderm, endoderm, and mesoderm), down to a unipotent cell that only gives rise to one type of cells [1]. On the other hand, there is a cancer subpopulation that shares similar characteristics with normal stem cells, frequently called cancer stem cells (CSCs) or tumor-initiating cells (TICs) [2]. This subpopulation demonstrates upregulated pluripotency gene expression, can give rise to other cancerous cell types, and has been blamed for cancer relapse and treatment resistance [3–6]. In addition, recent discoveries shed light on a new similarity in the stemness capacity between cancer stem cells and normal stem cells. This similarity lies in a metabolic dependence on methionine (Met) [3–5,7–12]. Met is an essential amino acid that is fundamental for protein synthesis. The Met cycle is linked to one-carbon metabolism, which is essential for many biological processes, including methylation and nucleic/amino acid synthesis [13]. In this review, we described recent studies concerning the stemness capacity of cells and Met, highlighting the influence of sirtuin 1 (SIRT1) on SAM synthesis and suggesting the role of peroxisome proliferator-activated receptor gamma coactivator 1-alpha (PGC-1α) and peroxisome proliferator-activated receptor alpha (PPAR-α) in impaired stemness produced by methionine deprivation. We also discussed potential strategies that have taken advantage of this dependence on stem cells for the progress of regenerative medicine and to help treat recurrent cancer.

2. Methionine Cycle and One-Carbon Metabolism

One-carbon metabolism is central to many biological processes that are crucial for cell development and survival. This metabolic network allows the synthesis of nucleic acid (purines and thymidine), maintains the homeostasis of the amino acid (glycine, serine, and Met), and permits epigenetic regulations [13]. The Met cycle plays a crucial role in epigenomic mechanisms. Met is converted into S-adenosyl methionine (SAM) through methionine adenosyltransferase (MAT). SAM is the universal methyl donor for methylation reactions, including histone and DNA methylation [14]. SAM is then converted into S-adenosylhomocysteine (SAH), after giving away its methyl group through histone methyltransferase (HMT) or DNA methyltransferase (DNMT) [14]. SAH is, in turn, converted to homocysteine (Hcy) that can either pass through the transsulfuration pathway or be converted back to Met by methionine synthase (MS) [14]. The Met cycle is tightly linked to the folate cycle. Met, folate, and cobalamin are required in the interconnected reactions [13]. For this reason, the one-carbon metabolism is understood as a potential link between environmental factors, such as nutrition and epigenomic mechanisms [15,16]. The Met salvage pathway illustrates the metabolic importance of Met, also called the 5′-methylthioadenosine (MTA) cycle. In this pathway, SAM is decarboxylated by adenosylmethionine decarboxylase 1 (AMD1), which is then used to donate the propylamine group to polyamines (putrescine and spermidine) by spermidine synthase (SRM) and spermine synthase (SMS). This results in the formation of spermidine and spermine, respectively, with methylthioadenosine (MTA) as a by-product. MTA is further processed by methylthioadenosine phosphorylase (MTAP) to regenerate Met [17] (Figure 1).

The one-carbon unit production is compartmentalized between the mitochondria, cytosol, and during the S- and G2/M-phases of the cell cycle in the nucleus [13,18]. The mitochondrial one-carbon units are produced by the glycine decarboxylase (GLDC) reaction [19]. GLDC is a rate-limiting enzyme in the serine–glycine pathway that catalyzes glycine degradation into ammonium, CO_2, and 5,10-methylenetetrahydrofolate (methylene-THF) [19]. On the same note, ALDH1L2, the mitochondrial form of 10-formyltetrahydrofolate dehydrogenase (FDH), plays an essential role in the distribution of one-carbon units between the mitochondria and cytosol via an NADP (+)-dependent reaction [20]. On the other hand, the cytosolic source of one-carbon units is obtained through the serine hydroxymethyltransferase 1 (SHMT1) rate-limiting reaction [13,19]. SHMT1 catalyzes the conversion of serine and tetrahydrofolate to glycine and 5,10-methylene tetrahydrofolate [13,19].

In the Met cycle, the concentration of SAM in the cells is highly buffered by glycine N-methyltransferase (GNMT), an enzyme that uses the methyl group from SAM to convert glycine into sarcosine (monomethylglycine), thus regulating the SAM/SAH ratio [21–23]. Excess Met intake is counteracted by increased MAT and GNMT reactions [24]. Although GNMT is mainly expressed in the liver, there is some evidence of its expression in stem cells [25].

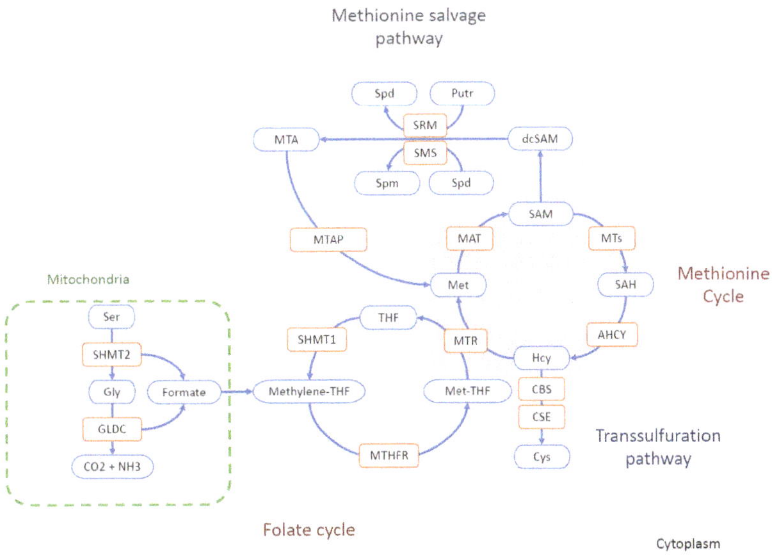

Figure 1. Simplified chart of the one-carbon metabolism. In the methionine cycle, methionine is converted into SAM via MAT. SAM is a universal methyl donor. SAM is converted to SAH in transmethylation reactions by methyltransferases. The resulting SAH is converted to homocysteine, which can either pass through the transsulfuration pathway, forming cystein in 2 steps catalyzed by CBS and CSE, or remethylated to methionine by MS, using the methyl group provided by Met-THF. The remethylation regenerates THF, which is then used to produce methylene-THF and Met-THF via SHMT1 and MTHFR, respectively. Methylene-THF can also be generated in the mitochondria by pathways that involve SHMT2 and GLDC. In the methionine salvage pathway, SAM decarboxylation produces an aminopropyl group donor for Putr and Spd synthesis. The donation of the aminopropyl group is catalyzed by aminopropyl transferases and produces MTA. MTA is converted back to SAM via MTAP. Abbreviations: AHCY, adenosylhomocysteinase; CBS, cystathionine beta synthase; CO2, carbon dioxide; CSE, cystathionine γ-lyase; Cys, cysteine; dcSAM, decarboxylated S-adenosylmethionine; GLDC, glycine decarboxylase; Gly, glycine; Hcy, homocysteine; MAT, methionine adenosyltransferase; methylene-THF, methylenetetrahydrofolate; Met-THF, methyltetrahydrofolate; MTA, methylthioadenosine; MTAP, methylthioadenosine phosphorylase; MTHFR, methylenetetrahydrofolate reductase; MTR, methionine synthase; MTs, methyltransferases; NH3, ammonia; Putr, putrescine; SAH, S-adenosylhomocysteine; SAM, S-adenosylmethionine; Ser, serine; SHMT1, serine hydroxymethyltransferase 1; SHMT2, serine hydroxymethyltransferase 2; SMS, spermine synthase; Spd, spermidine; Spm, spermine; SRM, spermidine synthase;; THF, tetrahydrofolate.

3. Methionine and Stemness

Several one-carbon metabolism genes, such as MAT2A, GLDC, SHMT1, and ALDH1L2, have been shown to demonstrate differential expression in stem cells compared to their differentiated counterparts [3,4,12,26]. In addition, Met restriction or inhibition of the upregulated genes showed a close link with the stemness capacities of cells, paving the way to novel strategies by exploiting this metabolic dependence in stem cells [4,5,8–10,12].

Several studies have shown the importance of Met in the maintenance of the self-renewal capacities and stemness of normal and cancerous stem cells. [3–5,8–10,12]. We review the recent studies that show the link between Met and the stemness of normal and cancer stem cells.

3.1. Methionine in Normal Stem Cells

The work of Shiraki et al. on human embryonic stem cells (hESCs) and human induced pluripotent stem cells (hiPSCs) highlighted the high-Met metabolic state of these cells compared to differentiated cells and their dependence on Met in a concentration-dependent manner. Their work was inspired by the specialized metabolic state of stem cells and the dependence of mouse ESCs on threonine catabolism [27]. Mouse embryonic stem cells demonstrate a high-flux metabolic state with a high expression of threonine dehydrogenase (TDH). The TDH enzyme catalyses a rate-limiting step in the mitochondrial conversion of threonine into glycine and acetyl-coenzyme A (CoA), which are essential for the folate cycle and tricarboxylic acid (TCA) cycle, respectively. This makes mouse ESCs critically dependent on threonine for their survival, pluripotency maintenance, and differentiation [27]. Given that threonine dehydrogenase is a non-functional pseudogene in humans, Shiraki et al. sought an equivalent approach to detect whether a similar effect could be observed with other amino acids in human cells by depriving them of single amino acids and measuring the total number of cells 48 h after deprivation. The most striking results were observed in the Met-deprived condition [12]. Furthermore, they showed that short Met deprivation led to a fast decrease in intracellular SAM with a consequent decrease in H3K4me3, and a decrease in NANOG expression, along with an increase in the differentiation potency in the three germ layers [12]. It is also interesting to note that the supplementation of Met increased NANOG expression and decreased the proportion of p53+ cells in a concentration-dependent manner [12]. In contrast, the prolonged Met deprivation effect on the cells was irreversible, with G0/G1 phase arrest triggering apoptosis. This leads to the conclusion that hESCs/hiPSCs do not rely on threonine-derived one-carbon metabolism for self-renewal and pluripotency, and alternatively use Met metabolism to achieve the same result [11,12]. Recently, the same team showed the involvement of zinc signaling in the regulation of stem cell pluripotency and differentiation. Culturing PSCs in Zn-deprived medium partially mimicked methionine deprivation (e.g., potentiated differentiation), showing altered methionine metabolism-related metabolite profiles. Likewise, the depletion of methionine reduced protein-bound Zn, which includes MS, through displacement with homocysteine [28].

The increase in Met requirements in the stemness and self-renewal capacity of ESCs/iPSCs is related to SAM levels and the synthesis of SAM by MAT2A and MAT2B. Knocking down MAT2A or MAT2B hampers the cell's ability to transform Met into SAM and leads to a decreased self-renewal capacity that can be rescued with SAM addition. The Met salvage pathway seems to play a limited role in the adaption of ESCs/iPSCs cells to their increased Met needs. The knocking down of SMS, the enzyme synthesizing spermine, impairs the Met salvage pathway, but does not phenocopy the Met depletion condition [12]. In addition, cycloleucine, an analog of Met that specifically inhibits MAT, decreases SAM levels and cell growth without affecting Met or SAH levels. Met restriction induces the upregulation of MAT2A expression as a way for the cells to possibly increase SAM levels and cope with the restrictive conditions [12]. These results indicate that SAM, rather than Met itself, is essential for the self-renewal and survival of stem cells. However, in cells with knocked down SIRT1, MAT2A is downregulated, and the upregulation induced through Met restriction leads to a limited increase in MAT2A level, compared to the wild-type (WT) condition. This downregulation is triggered by c-Myc, a proliferation proto-oncogene transregulator under the control of SIRT1 through a deacetylation mechanism that increases its stability and activity [10].

Adenosyl homocysteinase (AHCY) is another target enzyme of the Met cycle, which may influence the relation between SAM and cell stemness. AHCY catalyzes the hydrolysis of S-adenosylhomocysteine (SAH) to generate adenine and homocysteine as a part of the Met cycle [8]. Inducing the differentiation of mESCs decreases the mRNA levels of AHCY, MAT2A, MAT2B SAM and SAH, and homocysteine metabolites of the Met cycle. Depleting AHCY using shRNA leads to a decrease in the SAM/SAH ratio and pluripotency markers, notably Oct4 and Nanog, and an increase in the differentiation markers. Furthermore, the depletion results in an increase in cell number in the G1 phase, accompanied by a reduction in cells in the S phase and activation of the p53-dependent signaling pathway, leading to increased apoptosis [8]. The decrease in the SAM/SAH ratio leads to a decrease in H3K4me3 levels, notably at the Pou5f1 and Nanog loci and the O-GlcNAc post-translational modifications on serine or threonine residues of nucleocytoplasmic proteins [8]. O-GlcNAcylation is a nutrient-responsive modification with a pivotal role in stem cell biology [29–31]. This post-translational modification modulates enzyme activities. AHCY undergoes T136 O-GlcNAcylation, which promotes its activity by increasing its tetrameric assembly and its affinity with Hcy [8]. Inducing mESC differentiation leads to a gradual decrease in AHCY O-GlcNAcylation with reduced enzyme activity, thus regulating mESC pluripotency and self-renewal capacity [8] (Figure 2).

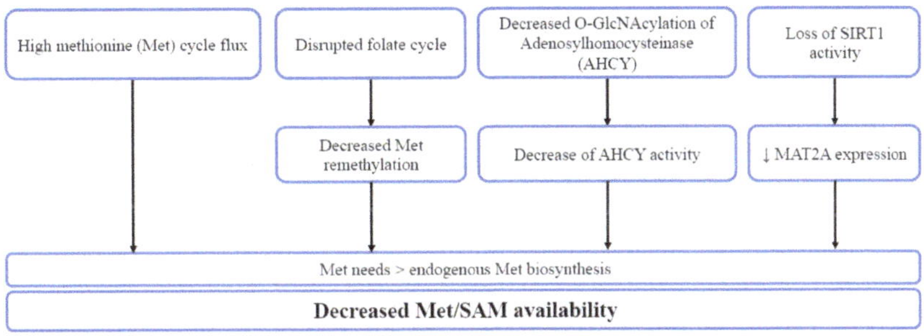

Figure 2. The diverse mechanisms leading to decreased Met/SAM availability. High methionine cycle flux, disrupted folate cycle, decreased O-GlcNAcylation of adenosylhomocysteinase (AHCY) or loss of SIRT1 activity lead to higher methionine needs in comparison to endogenous synthesis, thus decreasing methionine/SAM availability.

3.2. *Methionine in Cancer Stem Cells*

The growth of many cancer cells depends on Met cellular availability [32–34]. The Met dependency of cancer cells is defined by the inability of cells to proliferate in a medium deprived of Met, even when the metabolic precursor of Met, Hcy, is present [32–34]. Recent studies have shown that cancer stem cells are Met-dependent. The disruption of one-carbon metabolism enzymes or the reduction in Met hamper the self-renewal capacity and pluripotency of cancer stem cells [3–5,7,9].

Zgheib et al. found evidence of Met dependency in glioblastoma cancer stem cells, but not in the related adherent differentiated cells. The cells' tumorsphere formation capacities are recovered upon adding folate or MeTHF major molecules in the folate cycle. Glioblastoma stem cells demonstrated a disrupted folate cycle and could not furnish the necessary one-carbon unit to convert it into Met and SAM [3]. In contrast, the Met dependency stems from the high need for SAM in lung TICs [4]. Lung TICs have a high Met cycle flux, leading to their dependency on exogenous Met, as observed for normal stem cells. Compared to their differentiated counterpart, these cells demonstrate high GLDC expression and activity [4]. The high activity of GLDC in TICs redirects the flux of one-carbon units towards the increased demand of the Met cycle [4,35]. Knocking down this gene decreases Met and SAH levels to a level equivalent to differentiated cells, decreases histone methylation, and hampers the tumor-initiating capacities of these cells, further demonstrating the importance of Met [4]. Interestingly, recent studies have shown the importance of GLDC in maintaining and inducing pluripotency in ESCs and iPSCs through fueling H3K4me3 modification, while knocking down GLDC suppressed the pluripotency state [26,36,37]. Transient (48 h) Met starvation of lung TICs led to more striking results, with a decrease in SAM, SAH, and histone methylation levels and again hampered tumor-initiating capacities [4]. Supplementing cells with Hcy does not allow the cells to recover from the effect of Met starvation, in agreement with the Met dependency definition described above. Adding SAM or replating cells in Met-supplemented media allowed the cells to recover and regain their tumor-initiating ability [4]. Moreover, three other enzymes, MAT2A, MTHFR, and SHMT2, are also elevated in TICs compared to differentiated cells, further confirming the impact of GLDC upregulation of TICs on Met cycle flux. Furthermore, in cases where any of these genes are knocked down, the cell's ability to form tumors is hampered, similarly knocking down GLDC [4]. Both MAT2A and MTHFR were found to be upregulated in lung tumors, but only MAT2A was found to be upregulated in high-grade primary tumors or metastasis, highlighting the possibility of using MAT2A inhibitors, such as FIDAS-5, to treat patients with aggressive lung tumors. Wang et al. found that it hampers the tumor-initiating ability of TICs [4]. The mechanisms of Met dependency observed in TICs from glioblastoma and lung cancer cells is, thus, due to the disruption of the balance between folate and methionine cycles, leading to excess or impaired production of methyl-THF and methionine, which limits their growth [34]. Overall, these results suggest a prominent role of one-carbon metabolism in conferring stemness in cancer cells [3,34,38].

The dependence on Met in glioblastoma TICs seems to have a bell shape depending on the concentration used, with an optimal concentration. Therefore, lower or higher concentrations lead to lower tumorsphere formation capacity [3]. Such a correlation was not investigated in normal stem cells, as the limit of the tested concentration of Met was approximately 120 μM in comparison to an upper concentration of 500 μM in the study on glioblastoma TICs [3,7,12]. In normal stem cells, the concentration of Met was positively correlated with the increase in pluripotency gene expression and lower mortality rates were observed with a higher concentration of Met [12]. Cultivating breast cancer stem cells in Met-deprived conditions mirrors the results obtained in normal stem cells cultured in similar conditions [5,9]. The mammosphere formation assay shows the significantly lower capacity of cancer stem cells to form in those conditions, possibly through the same mechanism observed in ESCs/iPSCs with a recovered capacity upon adding SAM [5,9]. In addition, the Met-deprived condition leads to the increase in MAT2A levels as a homeostatic response and a decrease in H3K4me3 levels with lower levels of SOX9 markers and CD44+/CD24−. Intriguingly, these conditions potentiate the knocking down effect of MAT2A or its inhibition by MAT2A inhibitors. Similarly, mice fed a Met-restricted diet and treated with MAT2A inhibitor show a potentiated result, with increased apoptosis, lower tumor volume, and metastatic capacity compared to the control conditions [5]. Breast cancer stem cells were also found to highly express the O-GlcNAc transferase enzyme (OGT) and O-GlcNAcylation [9]. In line with what was found in normal stem cells, altering OGT and O-GlcNAc levels hampers stem cells' ability to form mammospheres in vitro, tumor formation in vivo, and lowered the levels of CD44+CD24− [9]. In addition, OGT seems to be a regulator of epithelial–mesenchymal transition (EMT) and cancer stem-like cell markers, including CD44, NANOG, and c-Myc [9]. Likewise, O-GlcNAcylation is involved in regulating stem cell marker expression in colon cancer cells [39]. Despite the resemblance in the importance of the O-GlcNAc effect on the stemness between normal and cancer stem cells, they are not linked to AHCY. Thus, the influence of O-GlcNAc on the Met cycle in cancer stem cells needs to be assessed in future studies [9].

3.3. *The Influence of SIRT1 and PPAR-Alpha/PGC1-Alpha Pathway*

SIRT1 and PPAR-α are key players in the links between Met metabolism and cell stemness. SIRT1 is one of the seven mammalian proteins that belong to the sirtuin family [14,40,41]. It catalyzes histone and non-histone lysine deacetylation in a NAD+-dependent manner [14,40,41]. Previous work in our lab showed the role of SIRT1 in regulating energy metabolism through PPAR-α and PGC-1α in a methyl-deficient diet [14,40–43]. PGC-1α is a master regulator of lipid metabolism and fatty acid oxidation. It is regulated by methylation and acetylation. The deacetylation of the PGC-1α protein leads to its activation and is known to coactivate PPAR-α to enhance the expression of fatty acid oxidation genes, antioxidant enzymes, and mitochondrial biogenesis [40]. Methyl-deficient diets decrease the expression of SIRT1 and subsequent activation of PGC-1α through imbalanced acetylation and methylation of the latter dysregulating energy metabolism [14,40,41]. The impaired expression and/or activity of methionine synthase in fibroblasts from patients with mutations in MTR and/or other inherited disorders of vitamin B12 metabolism also result in decreased protein expression of SIRT1, which plays a key role in the underlying pathological mechanisms of these disorders [40].

Sirtuin 1 (SIRT1) is involved in iPSC formation [44]. SIRT1 knockdown decreased, while resveratrol (RSV) increased the efficiency of iPSCs. SIRT1 enhances iPSC generation through deacetylation of p53, inhibition of p21 and enhancement of Nanog expression [45]. SIRT1 has been demonstrated to interact and acetylate Oct4 to maintain the stemness of naive pluripotent stem cells [46]. Sirt1 also deacetylates Sox2 through direct interaction with Oct4 [47]. The B12 and folate deficiency decreases the endogenous synthesis of methionine and decreases the brain expression of miR-34a in pups from deficient mother rats [48]. Of note, miR-34a reduces the reprogramming efficiency through inhibition of SIRT1 expression [45]. Decreased SIRT1 leads to the stabilization and increased activity of the P53 protein through its increased acetylation at the K120 and K164 sites [49]. This SIRT1-dependent upregulation of P53 activity is effective in undifferentiated hESCs, but not in other cell types. SIRT1 plays a role in DNA damage repair that is crucial for hESCs' fast mitotic division, which is prone to replication-related DNA errors [50]. It leads to programmed cell death through decreased expression of DNA repair enzymes, such as MSH2, MSH6 and APEX1, in hESCs [50]. Furthermore, SIRT1 is essential for telomere elongation during the iPSC generation process [51]. The level of expression of SIRT1 has been demonstrated to be elevated in normal and cancer stem cells (CSC), compared to their differentiated counterparts [45,46,50–54]. However, there is a debate on the role of SIRT1 in CSCs with its double functionality as a tumor suppressor and promoter [55,56]. SIRT 1 expression is increased, and its activity is critical for stemness and cell survival in cancer stem cells from glioma, colon and liver cancer, and leukemia [57]. SIRT1 inhibits DNMT3A and promotes the expression of SOX2 through promoter-reduced methylation [54]. It also increases the expression of other stemness-associated genes, including Oct4, Nanog, Cripto, Tert and Lin28, in colon cancer stem cells [58].

Several studies have investigated the influence of Met restriction on the effects of SIRT1 on pluripotency. However, whether SIRT1 influences stemness through its decreased expression produced by the impaired remethylation pathway of Met metabolism is unknown. Kilberg et al. showed that SIRT1 knock-out produces effects similar to those caused by Met restriction observed by Shiraki et al., in mouse embryonic stem cells (mESCs). The SIRT1 knock-out (KO) mESCs specifically impaired Met metabolism [10]. The metabolomic analysis of the SIRT1 KO cells cultured in complete media showed an elevated Met level, along with a decrease in SAM levels. The restriction of Met led to elevated differentiation markers and induced apoptosis with high sensitivity to Met deprivation, compared to other amino acids [10]. Interestingly, culturing SIRT1 KO mESCs in a complete medium with normal levels of Met demonstrated SAM levels similar to those of WT mESCs after Met restriction, suggesting that knocking out SIRT1 is somewhat equivalent to Met restriction, even with normal Met levels in culture medium [10]. Furthermore, SIRT1 KO reduced NANOG and OCT4 expression, with a marked decrease in NANOG expression when cultured in Met-restricted medium. In line with these results, the relative levels of H3K4me3 at the transcription starting site of the Nanog gene were reduced by both SIRT1 deletion and Met restriction, possibly through the same mechanism observed in hESCs/hiPSCs and steered by SAM levels [10].

PPAR-α is a key player in stemness. It triggers the expression of key genes of pluripotency reprogramming and enables pluripotent cells to adapt to their metabolic needs [59–62]. A Food and Drug Administration (FDA)-approved PPAR-α agonist was found to facilitate iPSC generation and enhance their programming efficiency by increasing the expression of pluripotency genes, including Nanog, Nr5A2, Oct4, and Rex1 [60]. PPAR-α knockdown of human glioma stem cells by shRNA reduces in vitro proliferation and inhibits orthotopic xenograft tumor growth [59]. Furthermore, PPAR-alpha was shown to play an important role in promoting mammosphere formation by modulating the expression of stem cell genes, including Jagged1, via the NF-κB/IL6 axis [62]. The PPARα-specific agonist treatment increases the number of mammospheres [62]. On the other hand, PPARα siRNA conditions decrease the number of mammospheres [62]. Whether the decreased activity of PPAR-α is related to the inactivation of PGC-1α by SIRT1 in stem cells has not been considered in experimental studies. However, this hypothesis is consistent with the role of SIRT1 activity in stem cells and its relationship with the cellular synthesis of SAM.

Given the link between SIRT1, PGC-1α and PPAR-α and their relation to stemness, we speculate that depriving cells of methionine would decrease SIRT1 levels and lead to the subsequent dysregulation of the energy metabolism through PGC-1α/PPAR-α [42], thus, participating in the hampered stemness induced by the decreased SAM levels. The data found in the literature point out the interplay between methionine restriction, SIRT1 and the PGC-1α/PPAR-α axis in stemness reprogramming and the need for further studies to produce a more integrated and mechanistic view of this interplay (Figure 3).

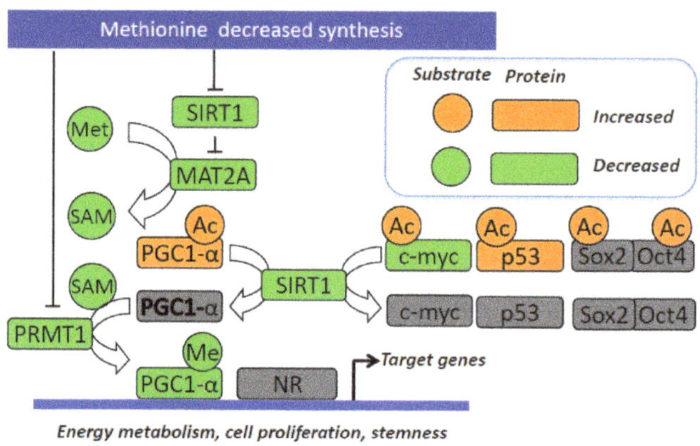

Figure 3. Summary of the links between decreased methionine endogenous synthesis and stemness. The decrease in methionine synthesis reduces SAM availability and SIRT1 and PRMT1 expression, which lead to decreased methylation and increased acetylation of PGC-1α and the subsequent impaired coactivation of nuclear receptors and other target genes of energy metabolism and pluripotency. The decreased SIRT1 reduces MAT2A expression, thus contributing to the decrease in the SAM levels. The decreased SIRT1 activity also leads to increased acetylation and reduced stability of p53 and increased acetylation of c-Myc and interacting Sox2 and Oct4, which are the three key genes of pluripotency reprograming. These mechanisms suggest that methionine endogenous synthesis/availability influence stemness capacity through its effects on the expression on key actors of pluripotency reprogramming and energy metabolic needs of stem cells. Abbreviations: Ac, acetyl group; c-Myc, MYC proto-oncogene; MAT2A, methionine adenosyltransferase 2A; Me, methyl group; Met, methionine; NR, nuclear receptor; PGC-1α, peroxisome proliferator-activated receptor gamma coactivator 1 alpha; PRMT1, protein arginine methyltransferase 1; SAM, S-adenosylmethionine; SIRT1, sirtuin 1.

4. Targeting Methionine Diet and/or Metabolism in Therapeutic Strategies

The metabolic dependence of stem cells on Met paved the way for new strategies that can be used in regenerative medicine and novel therapeutic approaches for cancer treatment. For instance, targeting Met dependency in iPSCs can eliminate leftover iPSCs after cardiac differentiation in transplanted cells to prevent tumor formation upon engraftment [63,64]. Cultivating engineered cardiac tissue in Met-free culture conditions at 42 °C led to the decreased expression of Lin28, OCT3/4, and NANOG without negatively impacting the tissue. Cardiac tissue showed spontaneous and synchronous beating, while maintaining or upregulating the expression of various cardiac and extracellular matrix genes [63,64].

Furthermore, a recently established protocol used the deprivation of both methionine and zinc to generate functional endocrine β cells [28].

A Met-restricted diet or treatment with an MAT2A inhibitor is a potential way to treat cancer by the targeted elimination of the cancer stem cell population [4–6,34]. Several of the previously mentioned studies have tested the impact of using Met restriction/deprivation or MAT2A inhibitors on the cancer stem cell population. Met restriction could be a metabolic primer for cancer cell death in combination with other strategies [65]. The targetable vulnerabilities have been studied thoroughly in preclinical studies with an expanding list of targets [66]. For example, Met restriction enhances the chemotherapeutic sensitivity of colorectal cancer stem cells by the miR-320d/c-Myc axis [67]. In addition, a recent clinical study showed that dietary Met restriction for patients with adjuvant transarterial chemoembolization (TACE) might be beneficial, as the pro-stemness capacities may be attributed to the activation of the Met cycle [68]. In addition, L-methioninase, an enzyme that catalyzes the degradation of l-Met to methanethiol, α-ketobutyrate, and ammonia, could be an additional therapeutic approach [69]. Recombinant methioninase inhibits the self-renewal and proliferation of gastric cancer stem cells. In vivo experiments demonstrated that HA-coated nanoparticles that co-encapsulated plasmid methioninase and 5-Fu enhance the targeting ability and promote the inhibition effects on tumor growth in gastric cancer [70]. These therapeutic approaches raise the question of their influence on normal stem cell pools in the human body, although normal stem cells have been shown to partially utilize Hcy to recover a part of the Met pool [12]. Future studies are needed to better understand Met dependency as a common hallmark of normal and cancer stem cells and help to build personalized approaches for cancer treatment [4,5,71].

5. Conclusions

Met dependence is a common feature of normal and cancer stem cells produced by diverse mechanisms specific to certain cell lines, including the influence of decreased synthesis of endogenous Met, and increased flux of the Met cycle. Convergent evidence shows that SIRT1 and PPAR-α/PGC-1α could be involved in the cause and/or consequences of Met dependency and stemness capacities (Figures 2 and 3). Cultivating stem cells in Met-restricted/deprived conditions alters their stemness capacity by halting SAM generation, which decreases histone methylation levels, notably H3K4me3, and alters stemness gene expression (Tables 1 and 2). This suggests that targeting the Met dependence mechanism could be useful in regenerative medicine and recurrent cancer treatment.

Table 1. Highlights of recent publications in relation to methionine restriction outcome and SIRT1/PGC-1α/PPAR-α players in the stemness of normal stem cells/embryos.

Normal Stem Cells				
Cell Type	Experimental Condition	Observed Changes: Increased	Observed Changes: Decreased	Reference
hESCs/iPSCs	Short Met deprivation	Differentiation potency Salvage pathway MAT2A expression p53-p38 signaling	SAM H3K4me3 mark NANOG Homocysteine	[12]
	Prolonged Met deprivation	Apoptosis (if not exposed to differentiation signals) G1-G0 arrest		

Table 1. *Cont.*

<table>
<tr><th colspan="5">Normal Stem Cells</th></tr>
<tr><th rowspan="2">Cell Type</th><th rowspan="2">Experimental Condition</th><th colspan="2">Observed Changes</th><th rowspan="2">Reference</th></tr>
<tr><th>Increased</th><th>Decreased</th></tr>
<tr><td rowspan="2">mESCs</td><td>SIRT1 KO or KD</td><td>Differentiation
Sensitivity to methionine restriction (mimics methionine restriction)</td><td>MAT2A
SAM
H3K4me3 + other histone marks
NANOG and OCT4</td><td rowspan="4">[10]</td></tr>
<tr><td>SIRT1 KO or KD + Met restriction</td><td>Differentiation
Apoptosis</td><td>MAT2A expression
SAM levels
H3K4me3 + other histone marks
NANOG expression</td></tr>
<tr><td rowspan="2">Mice embryos</td><td>SIRT1 KO embryos</td><td>Sensitivity to maternal methionine restriction</td><td>Mat2a expression
SAM levels
H3K4me3 mark</td></tr>
<tr><td>SIRT1 KO embryos + Met restriction</td><td>Developmental defects</td><td>Growth (retardation)
Survival rate</td></tr>
<tr><td rowspan="2">mESCs</td><td>Depletion of AHCY</td><td>Differentiation
p53-dependent signaling pathway
Apoptosis</td><td>SAM
H3K4me3
NANOG and OCT4</td><td rowspan="2">[8]</td></tr>
<tr><td>Blocking O-GlcNAcylation of AHCY</td><td>Teratomas formation in vivo
Differentiation</td><td>AHCY activity
SAM levels
H3K4me3 mark
NANOG and OCT4 expression</td></tr>
<tr><td rowspan="2">iPSCs</td><td>SIRT1 knockdown</td><td></td><td>iPSC formation (during the initiation phase of reprogramming)</td><td rowspan="2">[45]</td></tr>
<tr><td>Resveratrol (SIRT1 activation)</td><td>iPSC formation (acts on the initiation phase of reprogramming)</td><td></td></tr>
<tr><td>iPSCs</td><td>Sirt1 KO</td><td>Chromosome/chromatid breaks</td><td>Telomeres length after several cycles</td><td>[51]</td></tr>
<tr><td>mESCs</td><td>Sirt1 KO</td><td>Acetylation of Oct4
Fgf5 and Otx2 expression
Maintenance of Oct4 expression
Primed pluripotency network</td><td>Nanog and Klf2</td><td>[46]</td></tr>
<tr><td>hESCs</td><td>SIRT1 inhibition</td><td>DNA damage
p53 activation
Cell death</td><td>DNA repair enzyme levels (such as MSH2, MSH6, and APEX1)</td><td>[50]</td></tr>
<tr><td>iPSC</td><td>PPARα agonist</td><td>Nanog expression (reprogramming-promoting effect)</td><td></td><td>[60]</td></tr>
</table>

Table 2. Highlights of recent publications in relation to methionine restriction outcome and SIRT1/PGC-1α/PPAR-α players in the stemness of cancer stem cells.

Cancer Stem Cells				
Cell Type	**Experimental Condition**	**Observed Changes**		**Reference**
		Increased	**Decreased**	
Triple-negative breast CSCs	Met restriction	MAT2A Sensitivity to MAT2A inhibition	Mammosphere formation CD44(hi)/C24(low) CSC population Sox9 expression H3K4me3 mark	[5]
Lung CSCs	Met restriction		SAM levels H3K4me3 and other histone marks Colony-forming abilities in vitro Tumorigenic potential in vivo Cell-surface expression of CD166	[4]
Glioblastoma CSCs	Standard limiting dilution of Met	Mitochondrial SHMT2 and ALDH1L2 SOX2, OCT4, NANOG	Cytoplasmic SHMT1, MTHFD1 and DHFR	[3]
Breast CSCs	Inhibition of OGT (potential relation to methionine cycle)		Mammosphere formation CD44(hi)/C24(low) CSC population NANOG+ population ALDH+ population c-Myc+ population	[9]
Colorectal carcinoma CSCs	SIRT1 knockdown/inhibition		Stemness-associated genes (including Oct4, Nanog, Cripto, Tert and Lin28) Abilities of colony and sphere formation Percentage of CD133+ cells Tumorigenicity in vivo	[58]
Liver CSCs	SIRT1 knockdown/inhibition		Cell growth of liver CSCs Sphere and clone formation efficiencies in vitro Tumorigenic potential in vivo SOX2, Nanog and Oct4 expression levels	[54]
Liver CSCs	Overexpression of exogenous SIRT1	Self-renewal of liver non-CSCs Clone and sphere formation efficiencies Tumorigenic potential in vivo		[54]
Glioma CSCs	PPARα KD	Astrocytic differentiation	Tumorigenicity of in vivo Proliferative capacity and clonogenic potential in vitro Tumorigenicity of orthotopic xenografts Stem cell markers (SOX2, c-Myc and nestin)	[59]
Liver CSCs	SIRT1 inhibition	Susceptibility to chemotherapeutic drugs Senescence via activation of p53-p21 and p16 pathway	Stemness-associated genes (including NANOG, SOX2, OCT4, CD13, CD44 and EpCAM) Spheroid formation Tumorigenicity in vivo	[52]
Breast CSCs	PPARα agonist	Mammosphere formation NF-κB/IL6 axis Mammosphere regulatory genes		[62]

Author Contributions: Main drafting of the manuscript: Y.S., C.C., J.-L.G.; Partial drafting of specific parts: F.N.; Critical revision of the manuscript: All authors; Study supervision: J.-L.G. All authors have read and agreed to the published version of the manuscript.

Funding: The EpiMet project received grants from La Ligue Contre le Cancer (Grand Est/Bourgogne Franche Comté, France) and from the University of Lorraine (Pôle Biologie Médecine Santé, AAP-BMS_003_211).

Institutional Review Board Statement: Not applicable.

Informed Consent Statement: Not applicable.

Data Availability Statement: Not applicable.

Conflicts of Interest: The authors declare no conflict of interest.

Abbreviations

AHCY, adenosylhomocysteine; AMD1, adenosylmethionine decarboxylase 1; CoA, coenzyme A; CSCs, cancer stem cells; DNMT, DNA methyltransferase; EMT, epithelial-mesenchymal transition; ESCs, embryonic stem cells; FDA, Food and Drug Administration; FDH, 10-formyltetrahydrofolate dehydrogenase; GLDC, glycine decarboxylase; GNMT, glycine N-methyltransferase; Hcy, homocysteine; hESCs, human embryonic stem cells; hiPSCs, human induced pluripotent stem cells; HMT, histone methyltransferase; iPSCs, induced pluripotent stem cells; KO, knock-out; MAT, methionine adenosyltransferase; mESCs, mouse embryonic stem cells; Met, methionine; methylene-THF, 5,10-methylenetetrahydrofolate; MS, methionine synthase; MTA, 5′-methylthioadenosine; MTAP, methylthioadenosine phosphorylase; OGT, O-GlcNAc transferase; PGC-1α, peroxisome proliferator-activated receptor gamma coactivator 1 alpha; PPAR-α, peroxisome proliferator-activated receptor alpha; SAH, S-adenosylhomocysteine; SAM, S-adenosylmethionine; SHMT1, serine hydroxymethyltransferase 1; SIRT1, sirtuin 1; SMS, spermine synthase; SRM, spermidine synthase; TACE, transarterial chemoembolization; TCA, tricarboxylic acid; TDH, threonine dehydrogenase; TICs, tumor-initiating cells; WT, wild type.

References

1. Zakrzewski, W.; Dobrzynski, M.; Szymonowicz, M.; Rybak, Z. Stem cells: Past, present, and future. *Stem Cell Res. Ther.* **2019**, *10*, 68. [CrossRef] [PubMed]
2. Shackleton, M. Normal stem cells and cancer stem cells: Similar and different. *Semin. Cancer Biol.* **2010**, *20*, 85–92. [CrossRef] [PubMed]
3. Zgheib, R.; Battaglia-Hsu, S.F.; Hergalant, S.; Quere, M.; Alberto, J.M.; Chery, C.; Rouyer, P.; Gauchotte, G.; Gueant, J.L.; Namour, F. Folate can promote the methionine-dependent reprogramming of glioblastoma cells towards pluripotency. *Cell Death Dis.* **2019**, *10*, 596. [CrossRef] [PubMed]
4. Wang, Z.; Yip, L.Y.; Lee, J.H.J.; Wu, Z.; Chew, H.Y.; Chong, P.K.W.; Teo, C.C.; Ang, H.Y.; Peh, K.L.E.; Yuan, J.; et al. Methionine is a metabolic dependency of tumor-initiating cells. *Nat. Med.* **2019**, *25*, 825–837. [CrossRef] [PubMed]
5. Strekalova, E.; Malin, D.; Weisenhorn, E.M.M.; Russell, J.D.; Hoelper, D.; Jain, A.; Coon, J.J.; Lewis, P.W.; Cryns, V.L. S-adenosylmethionine biosynthesis is a targetable metabolic vulnerability of cancer stem cells. *Breast Cancer Res. Treat* **2019**, *175*, 39–50. [CrossRef]
6. Batlle, E.; Clevers, H. Cancer stem cells revisited. *Nat. Med.* **2017**, *23*, 1124–1134. [CrossRef]
7. Altundag, O.; Canpinar, H.; Celebi-Saltik, B. Methionine affects the expression of pluripotency genes and protein levels associated with methionine metabolism in adult, fetal, and cancer stem cells. *J. Cell. Biochem.* **2022**, *123*, 406–416. [CrossRef]
8. Zhu, Q.; Cheng, X.; Cheng, Y.; Chen, J.; Xu, H.; Gao, Y.; Duan, X.; Ji, J.; Li, X.; Yi, W. O-GlcNAcylation regulates the methionine cycle to promote pluripotency of stem cells. *Proc. Natl. Acad. Sci. USA* **2020**, *117*, 7755–7763. [CrossRef]
9. Akella, N.M.; Le Minh, G.; Ciraku, L.; Mukherjee, A.; Bacigalupa, Z.A.; Mukhopadhyay, D.; Sodi, V.L.; Reginato, M.J. O-GlcNAc Transferase Regulates Cancer Stem-like Potential of Breast Cancer Cells. *Mol. Cancer Res.* **2020**, *18*, 585–598. [CrossRef]
10. Tang, S.; Fang, Y.; Huang, G.; Xu, X.; Padilla-Banks, E.; Fan, W.; Xu, Q.; Sanderson, S.M.; Foley, J.F.; Dowdy, S.; et al. Methionine metabolism is essential for SIRT1-regulated mouse embryonic stem cell maintenance and embryonic development. *EMBO J.* **2017**, *36*, 3175–3193. [CrossRef]
11. Kilberg, M.S.; Terada, N.; Shan, J. Influence of Amino Acid Metabolism on Embryonic Stem Cell Function and Differentiation. *Adv. Nutr.* **2016**, *7*, 780S–789S. [CrossRef] [PubMed]
12. Shiraki, N.; Shiraki, Y.; Tsuyama, T.; Obata, F.; Miura, M.; Nagae, G.; Aburatani, H.; Kume, K.; Endo, F.; Kume, S. Methionine metabolism regulates maintenance and differentiation of human pluripotent stem cells. *Cell Metab.* **2014**, *19*, 780–794. [CrossRef] [PubMed]
13. Ducker, G.S.; Rabinowitz, J.D. One-Carbon Metabolism in Health and Disease. *Cell Metab.* **2017**, *25*, 27–42. [CrossRef]
14. Gueant, J.L.; Namour, F.; Gueant-Rodriguez, R.M.; Daval, J.L. Folate and fetal programming: A play in epigenomics? *Trends Endocrinol. Metab.* **2013**, *24*, 279–289. [CrossRef]
15. Lyon, P.; Strippoli, V.; Fang, B.; Cimmino, L. B Vitamins and One-Carbon Metabolism: Implications in Human Health and Disease. *Nutrients* **2020**, *12*, 2867. [CrossRef] [PubMed]
16. Clare, C.E.; Brassington, A.H.; Kwong, W.Y.; Sinclair, K.D. One-Carbon Metabolism: Linking Nutritional Biochemistry to Epigenetic Programming of Long-Term Development. *Annu. Rev. Anim. Biosci.* **2019**, *7*, 263–287. [CrossRef]
17. Pirkov, I.; Norbeck, J.; Gustafsson, L.; Albers, E. A complete inventory of all enzymes in the eukaryotic methionine salvage pathway. *FEBS J.* **2008**, *275*, 4111–4120. [CrossRef]

18. Field, M.S.; Kamynina, E.; Chon, J.; Stover, P.J. Nuclear Folate Metabolism. *Annu. Rev. Nutr.* **2018**, *38*, 219–243. [CrossRef]
19. Meiser, J.; Vazquez, A. Give it or take it: The flux of one-carbon in cancer cells. *FEBS J.* **2016**, *283*, 3695–3704. [CrossRef]
20. Krupenko, N.I.; Dubard, M.E.; Strickland, K.C.; Moxley, K.M.; Oleinik, N.V.; Krupenko, S.A. ALDH1L2 is the mitochondrial homolog of 10-formyltetrahydrofolate dehydrogenase. *J. Biol. Chem.* **2010**, *285*, 23056–23063. [CrossRef]
21. Obata, F.; Miura, M. Enhancing S-adenosyl-methionine catabolism extends Drosophila lifespan. *Nat. Commun.* **2015**, *6*, 8332. [CrossRef] [PubMed]
22. Frau, M.; Feo, F.; Pascale, R.M. Pleiotropic effects of methionine adenosyltransferases deregulation as determinants of liver cancer progression and prognosis. *J. Hepatol.* **2013**, *59*, 830–841. [CrossRef]
23. Takata, Y.; Huang, Y.; Komoto, J.; Yamada, T.; Konishi, K.; Ogawa, H.; Gomi, T.; Fujioka, M.; Takusagawa, F. Catalytic mechanism of glycine N-methyltransferase. *Biochemistry* **2003**, *42*, 8394–8402. [CrossRef] [PubMed]
24. Stipanuk, M.H. Metabolism of Sulfur-Containing Amino Acids: How the Body Copes with Excess Methionine, Cysteine, and Sulfide. *J. Nutr.* **2020**, *150*, 2494S–2505S. [CrossRef] [PubMed]
25. Steele, W.; Allegrucci, C.; Singh, R.; Lucas, E.; Priddle, H.; Denning, C.; Sinclair, K.; Young, L. Human embryonic stem cell methyl cycle enzyme expression: Modelling epigenetic programming in assisted reproduction? *Reprod. Biomed. Online* **2005**, *10*, 755–766. [CrossRef]
26. Tian, S.; Feng, J.; Cao, Y.; Shen, S.; Cai, Y.; Yang, D.; Yan, R.; Wang, L.; Zhang, H.; Zhong, X.; et al. Glycine cleavage system determines the fate of pluripotent stem cells via the regulation of senescence and epigenetic modifications. *Life Sci. Alliance* **2019**, *2*, e201900413. [CrossRef]
27. Wang, J.; Alexander, P.; Wu, L.; Hammer, R.; Cleaver, O.; McKnight, S.L. Dependence of mouse embryonic stem cells on threonine catabolism. *Science* **2009**, *325*, 435–439. [CrossRef]
28. Sim, E.Z.; Enomoto, T.; Shiraki, N.; Furuta, N.; Kashio, S.; Kambe, T.; Tsuyama, T.; Arakawa, A.; Ozawa, H.; Yokoyama, M.; et al. Methionine metabolism regulates pluripotent stem cell pluripotency and differentiation through zinc mobilization. *Cell Rep.* **2022**, *40*, 111120. [CrossRef]
29. Hanover, J.A.; Krause, M.W.; Love, D.C. Bittersweet memories: Linking metabolism to epigenetics through O-GlcNAcylation. *Nat. Rev. Mol. Cell Biol.* **2012**, *13*, 312–321. [CrossRef]
30. Sheikh, M.A.; Emerald, B.S.; Ansari, S.A. Stem cell fate determination through protein O-GlcNAcylation. *J. Biol. Chem.* **2021**, *296*, 100035. [CrossRef]
31. Wells, L.; Vosseller, K.; Hart, G.W. Glycosylation of nucleocytoplasmic proteins: Signal transduction and O-GlcNAc. *Science* **2001**, *291*, 2376–2378. [CrossRef] [PubMed]
32. Sorin, M.; Watkins, D.; Gilfix, B.M.; Rosenblatt, D.S. Methionine dependence in tumor cells: The potential role of cobalamin and MMACHC. *Mol. Genet. Metab.* **2021**, *132*, 155–161. [CrossRef] [PubMed]
33. Kaiser, P. Methionine Dependence of Cancer. *Biomolecules* **2020**, *10*, 568. [CrossRef] [PubMed]
34. Gueant, J.L.; Oussalah, A.; Zgheib, R.; Siblini, Y.; Hsu, S.B.; Namour, F. Genetic, epigenetic and genomic mechanisms of methionine dependency of cancer and tumor-initiating cells: What could we learn from folate and methionine cycles. *Biochimie* **2020**, *173*, 123–128. [CrossRef] [PubMed]
35. Zhang, W.C.; Shyh-Chang, N.; Yang, H.; Rai, A.; Umashankar, S.; Ma, S.; Soh, B.S.; Sun, L.L.; Tai, B.C.; Nga, M.E.; et al. Glycine decarboxylase activity drives non-small cell lung cancer tumor-initiating cells and tumorigenesis. *Cell* **2012**, *148*, 259–272. [CrossRef]
36. Tsogtbaatar, E.; Landin, C.; Minter-Dykhouse, K.; Folmes, C.D.L. Energy Metabolism Regulates Stem Cell Pluripotency. *Front. Cell Dev. Biol.* **2020**, *8*, 87. [CrossRef]
37. Kang, P.J.; Zheng, J.; Lee, G.; Son, D.; Kim, I.Y.; Song, G.; Park, G.; You, S. Glycine decarboxylase regulates the maintenance and induction of pluripotency via metabolic control. *Metab. Eng.* **2019**, *53*, 35–47. [CrossRef]
38. Chisari, A.; Golan, I.; Campisano, S.; Gelabert, C.; Moustakas, A.; Sancho, P.; Caja, L. Glucose and Amino Acid Metabolic Dependencies Linked to Stemness and Metastasis in Different Aggressive Cancer Types. *Front. Pharmacol.* **2021**, *12*, 723798. [CrossRef]
39. Fuentes-Garcia, G.; Castaneda-Patlan, M.C.; Vercoutter-Edouart, A.S.; Lefebvre, T.; Robles-Flores, M. O-GlcNAcylation Is Involved in the Regulation of Stem Cell Markers Expression in Colon Cancer Cells. *Front. Endocrinol.* **2019**, *10*, 289. [CrossRef]
40. Kosgei, V.J.; Coelho, D.; Gueant-Rodriguez, R.M.; Gueant, J.L. Sirt1-PPARS Cross-Talk in Complex Metabolic Diseases and Inherited Disorders of the One Carbon Metabolism. *Cells* **2020**, *9*, 1882. [CrossRef]
41. Gueant, J.L.; Elakoum, R.; Ziegler, O.; Coelho, D.; Feigerlova, E.; Daval, J.L.; Gueant-Rodriguez, R.M. Nutritional models of foetal programming and nutrigenomic and epigenomic dysregulations of fatty acid metabolism in the liver and heart. *Pflugers Arch.* **2014**, *466*, 833–850. [CrossRef] [PubMed]
42. Gueant, J.L.; Gueant-Rodriguez, R.M.; Kosgei, V.J.; Coelho, D. Causes and consequences of impaired methionine synthase activity in acquired and inherited disorders of vitamin B12 metabolism. *Crit. Rev. Biochem. Mol. Biol.* **2022**, *57*, 133–155. [CrossRef] [PubMed]
43. Garcia, M.M.; Gueant-Rodriguez, R.M.; Pooya, S.; Brachet, P.; Alberto, J.M.; Jeannesson, E.; Maskali, F.; Gueguen, N.; Marie, P.Y.; Lacolley, P.; et al. Methyl donor deficiency induces cardiomyopathy through altered methylation/acetylation of PGC-1alpha by PRMT1 and SIRT1. *J. Pathol.* **2011**, *225*, 324–335. [CrossRef]

44. Chen, A.C.H.; Peng, Q.; Fong, S.W.; Lee, K.C.; Yeung, W.S.B.; Lee, Y.L. DNA Damage Response and Cell Cycle Regulation in Pluripotent Stem Cells. *Genes* **2021**, *12*, 1548. [CrossRef]
45. Lee, Y.L.; Peng, Q.; Fong, S.W.; Chen, A.C.; Lee, K.F.; Ng, E.H.; Nagy, A.; Yeung, W.S. Sirtuin 1 facilitates generation of induced pluripotent stem cells from mouse embryonic fibroblasts through the miR-34a and p53 pathways. *PLoS ONE* **2012**, *7*, e45633. [CrossRef] [PubMed]
46. Williams, E.O.; Taylor, A.K.; Bell, E.L.; Lim, R.; Kim, D.M.; Guarente, L. Sirtuin 1 Promotes Deacetylation of Oct4 and Maintenance of Naive Pluripotency. *Cell Rep.* **2016**, *17*, 809–820. [CrossRef] [PubMed]
47. Mu, W.L.; Wang, Y.J.; Xu, P.; Hao, D.L.; Liu, X.Z.; Wang, T.T.; Chen, F.; Chen, H.Z.; Lv, X.; Liu, D.P. Sox2 Deacetylation by Sirt1 Is Involved in Mouse Somatic Reprogramming. *Stem Cells* **2015**, *33*, 2135–2147. [CrossRef]
48. Geoffroy, A.; Saber-Cherif, L.; Pourie, G.; Helle, D.; Umoret, R.; Gueant, J.L.; Bossenmeyer-Pourie, C.; Daval, J.L. Developmental Impairments in a Rat Model of Methyl Donor Deficiency: Effects of a Late Maternal Supplementation with Folic Acid. *Int. J. Mol. Sci.* **2019**, *20*, 973. [CrossRef]
49. Zhang, Z.N.; Chung, S.K.; Xu, Z.; Xu, Y. Oct4 maintains the pluripotency of human embryonic stem cells by inactivating p53 through Sirt1-mediated deacetylation. *Stem Cells* **2014**, *32*, 157–165. [CrossRef]
50. Jang, J.; Huh, Y.J.; Cho, H.J.; Lee, B.; Park, J.; Hwang, D.Y.; Kim, D.W. SIRT1 Enhances the Survival of Human Embryonic Stem Cells by Promoting DNA Repair. *Stem Cell Rep.* **2017**, *9*, 629–641. [CrossRef]
51. De Bonis, M.L.; Ortega, S.; Blasco, M.A. SIRT1 is necessary for proficient telomere elongation and genomic stability of induced pluripotent stem cells. *Stem Cell Rep.* **2014**, *2*, 690–706. [CrossRef] [PubMed]
52. Wang, M.J.; Chen, J.J.; Song, S.H.; Su, J.; Zhao, L.H.; Liu, Q.G.; Yang, T.; Chen, Z.; Liu, C.; Fu, Z.R.; et al. Inhibition of SIRT1 Limits Self-Renewal and Oncogenesis by Inducing Senescence of Liver Cancer Stem Cells. *J. Hepatocell Carcinoma* **2021**, *8*, 685–699. [CrossRef]
53. Shi, L.; Tang, X.; Qian, M.; Liu, Z.; Meng, F.; Fu, L.; Wang, Z.; Zhu, W.G.; Huang, J.D.; Zhou, Z.; et al. A SIRT1-centered circuitry regulates breast cancer stemness and metastasis. *Oncogene* **2018**, *37*, 6299–6315. [CrossRef] [PubMed]
54. Liu, L.; Liu, C.; Zhang, Q.; Shen, J.; Zhang, H.; Shan, J.; Duan, G.; Guo, D.; Chen, X.; Cheng, J.; et al. SIRT1-mediated transcriptional regulation of SOX2 is important for self-renewal of liver cancer stem cells. *Hepatology* **2016**, *64*, 814–827. [CrossRef]
55. Carafa, V.; Altucci, L.; Nebbioso, A. Dual Tumor Suppressor and Tumor Promoter Action of Sirtuins in Determining Malignant Phenotype. *Front. Pharmacol.* **2019**, *10*, 38. [CrossRef] [PubMed]
56. Lim, C.S. SIRT1: Tumor promoter or tumor suppressor? *Med. Hypotheses* **2006**, *67*, 341–344. [CrossRef]
57. Garcia-Peterson, L.M.; Li, X. Trending topics of SIRT1 in tumorigenicity. *Biochim. Biophys. Acta Gen. Subj.* **2021**, *1865*, 129952. [CrossRef]
58. Chen, X.; Sun, K.; Jiao, S.; Cai, N.; Zhao, X.; Zou, H.; Xie, Y.; Wang, Z.; Zhong, M.; Wei, L. High levels of SIRT1 expression enhance tumorigenesis and associate with a poor prognosis of colorectal carcinoma patients. *Sci. Rep.* **2014**, *4*, 7481. [CrossRef]
59. Haynes, H.R.; Scott, H.L.; Killick-Cole, C.L.; Shaw, G.; Brend, T.; Hares, K.M.; Redondo, J.; Kemp, K.C.; Ballesteros, L.S.; Herman, A.; et al. shRNA-mediated PPARalpha knockdown in human glioma stem cells reduces in vitro proliferation and inhibits orthotopic xenograft tumour growth. *J. Pathol.* **2019**, *247*, 422–434. [CrossRef]
60. Lee, J.; Lee, J.; Cho, Y.S. Peroxisome Proliferator-Activated Receptor alpha Agonist and Its Target Nanog Cooperate to Induce Pluripotency. *J. Clin. Med.* **2018**, *7*, 488. [CrossRef]
61. Papi, A.; Storci, G.; Guarnieri, T.; De Carolis, S.; Bertoni, S.; Avenia, N.; Sanguinetti, A.; Sidoni, A.; Santini, D.; Ceccarelli, C.; et al. Peroxisome proliferator activated receptor-alpha/hypoxia inducible factor-1alpha interplay sustains carbonic anhydrase IX and apoliprotein E expression in breast cancer stem cells. *PLoS ONE* **2013**, *8*, e54968. [CrossRef] [PubMed]
62. Papi, A.; Guarnieri, T.; Storci, G.; Santini, D.; Ceccarelli, C.; Taffurelli, M.; De Carolis, S.; Avenia, N.; Sanguinetti, A.; Sidoni, A.; et al. Nuclear receptors agonists exert opposing effects on the inflammation dependent survival of breast cancer stem cells. *Cell Death Differ.* **2012**, *19*, 1208–1219. [CrossRef]
63. Matsuura, K.; Ito, K.; Shiraki, N.; Kume, S.; Hagiwara, N.; Shimizu, T. Induced Pluripotent Stem Cell Elimination in a Cell Sheet by Methionine-Free and 42 degrees C Condition for Tumor Prevention. *Tissue Eng. Part C Methods* **2018**, *24*, 605–615. [CrossRef] [PubMed]
64. Matsuura, K.; Kodama, F.; Sugiyama, K.; Shimizu, T.; Hagiwara, N.; Okano, T. Elimination of remaining undifferentiated induced pluripotent stem cells in the process of human cardiac cell sheet fabrication using a methionine-free culture condition. *Tissue Eng. Part C Methods* **2015**, *21*, 330–338. [CrossRef] [PubMed]
65. Strekalova, E.; Malin, D.; Rajanala, H.; Cryns, V.L. Preclinical Breast Cancer Models to Investigate Metabolic Priming by Methionine Restriction. *Methods Mol. Biol.* **2019**, *1866*, 61–73. [CrossRef]
66. Yamamoto, J.; Han, Q.; Simon, M.; Thomas, D.; Hoffman, R.M. Methionine Restriction: Ready for Prime Time in the Cancer Clinic? *Anticancer Res.* **2022**, *42*, 641–644. [CrossRef]
67. Liu, C.; Wang, J.L.; Wu, D.Z.; Yuan, Y.W.; Xin, L. Methionine restriction enhances the chemotherapeutic sensitivity of colorectal cancer stem cells by miR-320d/c-Myc axis. *Mol. Cell Biochem.* **2022**, *477*, 2001–2013. [CrossRef]
68. Huang, L.; Xu, D.; Qian, Y.; Zhang, X.; Guo, H.; Sha, M.; Hu, R.; Kong, X.; Xia, Q.; Zhang, Y. A gene signature is critical for intrahepatic cholangiocarcinoma stem cell self-renewal and chemotherapeutic response. *Stem Cell Res. Ther.* **2022**, *13*, 292. [CrossRef]

69. Kreis, W.; Hession, C. Isolation and purification of L-methionine-alpha-deamino-gamma-mercaptomethane-lyase (L-methioninase) from Clostridium sporogenes. *Cancer Res.* **1973**, *33*, 1862–1865.
70. Yang, W.; Zhang, H.; Xin, L. A novel design of HA-coated nanoparticles co-encapsulating plasmid METase and 5-Fu shows enhanced application in targeting gastric cancer stem cells. *Biol. Chem.* **2018**, *399*, 293–303. [CrossRef]
71. Lamb, R.; Harrison, H.; Smith, D.L.; Townsend, P.A.; Jackson, T.; Ozsvari, B.; Martinez-Outschoorn, U.E.; Pestell, R.G.; Howell, A.; Lisanti, M.P.; et al. Targeting tumor-initiating cells: Eliminating anabolic cancer stem cells with inhibitors of protein synthesis or by mimicking caloric restriction. *Oncotarget* **2015**, *6*, 4585–4601. [CrossRef] [PubMed]

cells MDPI

Article

A Possible Modulator of Vitiligo Metabolic Impairment: Rethinking a PPARγ Agonist

Federica Papaccio [1,*], Barbara Bellei [1], Monica Ottaviani [1], Andrea D'Arino [1], Mauro Truglio [1], Silvia Caputo [1], Giovanni Cigliana [2], Lorenzo Sciuto [1], Emilia Migliano [3], Alessia Pacifico [4], Paolo Iacovelli [4] and Mauro Picardo [1,*]

1 Cutaneous Physiopathology and Integrated Center of Metabolomics Research, San Gallicano Dermatological Institute, IRCCS, 00144 Rome, Italy
2 Clinical Pathology Unit, Department of Research, Advanced Diagnostics, and Technological Innovation, Translational Research Area, IRCCS Regina Elena National Cancer Institute, 00144 Rome, Italy
3 Plastic and Regenerative Surgery, San Gallicano Dermatological Institute, IRCCS, 00144 Rome, Italy
4 Phototherapy Unit, San Gallicano Dermatological Institute, IRCCS, 00144 Rome, Italy
* Correspondence: federica.papaccio@ifo.it (F.P.); mauro.picardo@hotmail.it (M.P.)

Abstract: Vitiligo is a complex disease wherein derangements in multiple pathways determine the loss of functional melanocytes. Since its pathogenesis is not yet completely understood, vitiligo lacks a definitive safe and efficacious treatment. At present, different therapies are available; however, each modality has its baggage of disadvantages and side effects. Recently we have described several metabolic abnormalities in cells from pigmented skin of vitiligo patients, including alterations of glucose metabolism. Therefore, we conducted a study to evaluate the effect of Pioglitazone (PGZ), a Peroxisome proliferator-activated receptor-γ (PPARγ) agonist, on cells from pigmented vitiligo skin. We treated vitiligo melanocytes and fibroblasts with low doses of PGZ and evaluated the effects on mitochondrial alterations, previously reported by our and other groups. Treatment with PGZ significantly increased mRNA and protein levels of several anaerobic glycolytic enzymes, without increasing glucose consumption. The PGZ administration fully restored the metabolic network, replacing mitochondrial membrane potential and mitochondrial DNA (mtDNA) copy number. These effects, together with a significant increase in ATP content and a decrease in reactive oxygen species (ROS) production, provide strong evidence of an overall improvement of mitochondria bioenergetics in vitiligo cells. Moreover, the expression of HMGB1, Hsp70, defined as a part of DAMPs, and PD-L1 were significantly reduced. In addition, PGZ likely reverts premature senescence phenotype. In summary, the results outline a novel mode of action of Pioglitazone, which may turn out to be relevant to the development of effective new vitiligo therapeutic strategies.

Keywords: vitiligo; PPARγ; pioglitazone; melanocytes; fibroblasts; cellular metabolism; glucose metabolism; metabolic impairment

Citation: Papaccio, F.; Bellei, B.; Ottaviani, M.; D'Arino, A.; Truglio, M.; Caputo, S.; Cigliana, G.; Sciuto, L.; Migliano, E.; Pacifico, A.; et al. A Possible Modulator of Vitiligo Metabolic Impairment: Rethinking a PPARγ Agonist. *Cells* **2022**, *11*, 3583. https://doi.org/10.3390/cells11223583

Academic Editors: Nicole Wagner and Kay-Dietrich Wagner

Received: 7 October 2022
Accepted: 7 November 2022
Published: 12 November 2022

1. Introduction

Vitiligo is an acquired chronic pigmentation disorder characterized by the appearance of white spots on the skin due to melanocyte loss. Affecting 0.5–1% of the population worldwide, it is the most diffused depigmentation disorder [1,2].

The etiology of the disease is multifactorial and presents different manifestations, progression rates, and responses to treatment [3]. Multiple mechanisms are supposed to be involved in melanocyte disappearance. These include genetic predisposition, autoimmune responses, oxidative stress, environmental triggers, metabolic abnormalities, impaired renewal of melanocytes, and altered inflammatory state [4]. However, the overall contribution of each of these processes is still under debate, suggesting that multiple mechanisms may work jointly in vitiligo to contribute to the destruction of melanocytes.

We previously demonstrated that melanocytes, obtained from pigmented vitiligo skin, show cellular and molecular alterations. In particular, we described a constitutive activation of antioxidant enzyme genes and defects in mitochondrial metabolism, reflected in an altered expression and activity of complex I, increased generation of reactive oxygen species, low ATP production, and modified expression of some glycolytic enzymes [5,6]. The imbalance in the pro and anti-oxidative state is responsible for the increased susceptibility to external stressful stimuli and for the premature senescence of the skin characterized by the production of different proteins of the so-called Senescence-Associated Secretory Phenotype (SASP) [5] and increased cholesterol content in vitiligo melanocytes cultures [5].

Recently, our group has shown that some of the described alterations are not exclusive to melanocytes but can be observed in other skin cells obtained from pigmented areas, suggesting that vitiligo leads to the degeneration of the entire skin [7,8]. Fibroblasts exhibit oxidative stress, overexpression of p53, and a senescent phenotype [7]. These features are the basis for the altered secretion of soluble growth factors supporting melanocyte survival and homeostasis. The consequent alteration of the dermal-epidermal network could be the basis for melanocyte detachment [7–9]. Moreover, our group recently demonstrated abnormalities in the keratinocyte differentiation process and consequently an inappropriate assembly of the epidermal layers which, following stressful events, can induce an inflammatory reaction capable of activating an immune response targeting melanocytes [8]. These findings further strengthen the pivotal role of metabolic alterations in vitiligo onset and development.

Lastly, our vitiligo unit carried out a retrospective analysis underlining that vitiligo patients frequently display modest alterations of the spectrum of the metabolic syndrome [10]. Specifically, while in vitiligo patients, lower levels of total cholesterol were measured overall, the distribution of lipid subpopulations was less favorable with respect to the healthy controls. In particular, triglycerides and LDL cholesterol showed greater values while HDL cholesterol levels were consistently lower than healthy controls. Likewise, fasting blood glucose levels were significantly higher in vitiligo patients, even if below the diabetes threshold and within the range of impaired fasting glucose [10]. PPARγ is a ligand-activated transcription factor, belonging to the PPARs nuclear receptor superfamily, and is considered one of the master modulators of mitochondrial biogenesis and function [11]. After the heterodimerization with the retinoid X receptors (RXRs), PPARγ produces functional transcription factors contributing to the trans-activation of key genes involved in energy homeostasis and cellular differentiation [12]. A crucial role for PPARγ in fat cell differentiation, lipid storage, vascular function, and energy metabolism has been identified [13]. Thiazolidinediones (TZDs) represent a class of compounds that are high-affinity ligands for the transcription factor PPARγ [14]. Among the TZD, PGZ is commercially available and selectively stimulates PPARγ [15]. PGZ is best known for its use in improving glycemic control in adults with type 2 diabetes mellitus and reducing insulin resistance. Moreover, it can normalize hyperglycemia induced by intracellular ROS and mitochondrial ROS (mtROS) production [16]. PGZ is effective in many important biologic processes, including inflammation [17] and beneficial effects have been described in the treatment of psoriatic patients, decreasing the production of inflammatory mediators, such as Interleukin 6 (IL-6) [18]. PGZ can act as an immune-modulating agent inhibiting the proliferation and differentiation of keratinocytes in psoriasis [18]. According to the idea that vitiligo cells carry metabolic alterations, PGZ was tested on melanocytes and fibroblasts from pigmented skin of vitiligo subjects, in order to evaluate whether PPARγ activator is able to rescue metabolic impairment in vitiligo and consequently improve the biological behavior of melanocytes. Our study also provides proof-of-concept evidence that PGZ contributes to counteracting the premature aging phenotype, characteristic of vitiligo skin. This study aims to clarify that increased awareness of the metabolic aspects of vitiligo could be considered crucial to advance novel and alternative therapeutic target options in vitiligo treatment.

2. Materials and Methods

2.1. Ethic Statement

The Declaration of Helsinki Principles was followed, and patients were given written informed consent. The Institute's Research Ethics Committee (Regina Elena Institute and San Gallicano Dermatological Institute) approval was obtained to collect samples of human material for research (Prot CE/751/16 approved on 12 January 2016).

2.2. Skin Biopsies and Cell Cultures

The cell lines obtained from 10 vitiligo and 10 normal subjects, age and sex-matched, were used in the study. The control samples, normal human primary epidermal melanocytes (NHM), were obtained from subjects who underwent plastic surgery for diseases unrelated to pigmentation disorders. The primary epidermal melanocytes (VHM) and fibroblasts (VHF) from vitiligo subjects were isolated from 1 cm^2 skin biopsy in a non-lesional area. Briefly, the skin was cut into approximately 4 mm^2 sized pieces and incubated overnight at 4 °C with dispase (2.5 mg/mL) to separate the epidermis from the dermis. The dermis was digested with collagenase 0.35% for 1 h at 37 °C. Isolated NHM and VHM were cultured in 254 Medium (Cascade Biologics, Portland, OR, USA; ThermoFisher, Waltham, MA, USA) supplemented with a specific Growth Factors cocktail (Cascade Biologics) and penicillin/streptomycin (Gibco Hyclone Laboratories, South Logan, UT, USA). Isolated VHF were cultured in DMEM (EuroClone S.p.A., Milan, Italy) supplemented with 10% fetal bovine serum (FBS) (EuroClone S.p.A., Milan, Italy) and antibiotics (Hyclone Laboratories, South Logan, UT, USA). All the aforementioned analyses were performed between 2 and 7 culture passages.

2.3. Pioglitazone (PGZ) Treatment

PGZ (Merck, Sigma-Aldrich, Merck KGaA, Darmstadt, Germany) was dissolved in dimethyl sulfoxide (DMSO) (Merck, Sigma-Aldrich, Merck KGaA, Darmstadt, Germany) to a stock solution of 20 mM and added to the cell growth medium at the final concentrations of 2μM for periods from 6 h to 10 days, according to the type of parameter evaluated. In untreated control cells, an equal volume of DMSO was added.

2.4. Proliferation Assay

In this case, 10^5 of vitiligo melanocytes were seeded in 6-well plates and treated with 2 μM of PGZ or DMSO. After 48 h and 10 days, cells were washed with PBS, trypsinized, centrifuged at 800 rpm, resuspended and then counted by TC20TM Automated Cell Counter (BIO-RAD).

2.5. Semi-Quantitative Real-Time Polymerase Chain Reaction (RT-PCR)

The total RNA was extracted from each cell line using Aurum Total mini kit (Bio-Rad Laboratories, Milan, Italy). Here, cDNA was synthesized from 1 μg of total RNA using the PrimeScriptTM RT Master Mix (Takara Bio Inc., Beijing, China) according to the manufacturer's instructions. Quantitative real-time RT-PCR was performed in a reaction mixture containing SYBR qPCR Master Mix (Vazyme Biotech Co., Ltd., Nanjing, China) and 25 pmol of forward and reverse primers. The reactions were carried out using a CFX96 Real-Time System (Bio-Rad Laboratories). All samples were run in triplicate. The amplification of the β-Actin (β-act) from each sample has been used as the internal control. For each gene, the assessment of quality was performed by examining PCR melt curves after quantitative (q)RT-PCR to ensure product specificity. Supplementary Table S1 shows the oligonucleotide sequences used to detect the expression of reported target genes.

2.6. Western Blot Analysis

The cell extracts were prepared with RIPA buffer containing proteases and phosphatase inhibitors. The proteins were separated on SDS-polyacrylamide gels, transferred to nitrocellulose membranes and then treated with the primary antibodies reported in Table S2.

Horseradish peroxide-conjugated goat anti-mouse or goat anti-rabbit secondary antibody complexes were detected by chemiluminescence (Cell Signaling Technology, Danvers, MA, USA). Imaging and densitometry analyses were performed with the UVITEC Mini HD9 acquisition system (Alliance UVItec Ltd., Cambridge, UK).

2.7. ATP Determination

The intracellular level of ATP was measured using a commercial fluorimetric kit (ThermoFisher Scientific) according to the manufacturer's instructions. The results were normalized for the number of cells contained in each sample and reported as µM. The measurement was performed in duplicate for each sample and the experiments were repeated twice.

2.8. Glucose Determination

The extracellular level of glucose was measured using Roche/Hitachi cobas c 503 according to the manufacturer's instructions. The results were reported as mg/dL mean value.

2.9. mtDNA Quantification

The total DNA was prepared from melanocytes using DNeasy Blood and Tissue (Qiagen GmbH, Hiden, Germany) according to the manufacturer's recommendations and stored at −20 °C. mtDNA content was measured by real-time PCR using a CFX96 Real-Time System (Bio-Rad Laboratories). The amplification conditions were as follows: 5 min at 95 °C, then 45 cycles of 15 s at 95 °C and 1 min at 58 °C. A dissociation curve was also calculated for each sample to ensure the presence of a single PCR product. The experiment was performed in triplicate. The relative quantification of mitochondrial DNA (mtDNA) over nuclear DNA (nuDNA) levels was determined using the difference in the threshold cycle values of the nuclear TATA-box-binding protein region on chromosome 6 and the mitochondrial non-coding control region D-loop (ΔCt, namely, CtmtDNA−CtnuDNA). The relative abundance of the mitochondrial genome was reported as 2−ΔCt. The primers used were the following:

mtDNA forward, GATTTGGGTACCACCCAAGTATTG (SEQ ID NO:15);
reverse, GTACAATATTCATGGTGGCTGGCA (SEQ ID NO:16);
nuDNA forward, TTCCACCCAAGTATTG (SEQ ID NO:17);
reverse, TGTTCCATGCAGGGGAAAACAAGC (SEQ ID NO:18)

2.10. Protein Determination by Sandwich Enzyme-Linked Immunosorbent Assay (ELISA)

IL-6 determination in the supernatants of treated and untreated vitiligo melanocytes was quantified by ELISA assay (Aviva System Biology, San Diego, CA, USA) according to the manufacturer's protocol. The supernatants were collected after 48 h of treatment. The results were normalized for the number of cells contained in each sample and were expressed as picograms per milliliter (pg/mL). The measurement was performed in duplicate for each sample and the experiments were repeated twice.

2.11. Detection of Intracellular ROS Levels

The production of ROS has been assessed with the fluorescent dye 2′7′-dichlorodihydrofluorescein diacetate (H_2DCFDA; Sigma-Aldrich). Cell permeable, non-fluorescent H_2DCF is oxidized to highly fluorescent dye 2′7′-dichlorofluorescein (DCF) in the presence of intracellular ROS. The cells were incubated with 2.5 µmol L−1 H_2DCF for 30 min at 37 °C and 5% CO_2 in phenol red-free full-starved medium in the dark. After removing the probe solution, cells were washed with PBS, trypsinized, centrifuged at 800 rpm, and then resuspended in PBS. After the oxidation of H_2DCF into fluorescent DCF by ROS, signals were measured by MACSQuant Analyzer 10 Flow Cytometer (Miltenyi Biotec, Bergisch Gladbach, Germany). The data were collected from three independent experiments.

2.12. Assessment of Mitochondrial Membrane Potential (ΔΨ)

The cells were incubated with 2 μM of dye JC-1 (5′,6,6′-tetrachloro-1,1′,3,3′-tetraethylbenzimidazolylcarbocyanine iodide) (ThermoFisher Scientific) for 30 min at 37 °C and 5% CO_2 in phenol red-free full-starved medium in the dark. After removing the probe solution, the cells were washed with PBS, trypsinized, centrifuged at 800 rpm, and then resuspended in PBS. The double fluorescence staining of mitochondria by JC-1, either as green fluorescent J-monomers or as red fluorescent J-aggregates, was used for monitoring the mitochondrial membrane potential. The signals were measured by MACSQuant Analyzer 10 Flow Cytometer (Miltenyi Biotec). The data were collected from three independent experiments.

2.13. Mitochondrial Mass Measurement

The cells were incubated with 0.1 μM of MitoTracker® probe (ThermoFisher Scientific) for 30 min at 37 °C and 5% CO_2 in phenol red-free full-starved medium in the dark. After removing the probe solution, cells were washed with PBS, trypsinized, centrifuged at 800 rpm, and then resuspended in PBS. Fluorescence signals were measured by MACSQuant Analyzer 10 Flow Cytometer (Miltenyi Biotec). The data were collected from three independent experiments.

2.14. PCA Analysis

The Principal Component Analysis (PCA) was performed on the gene expression profiles, using Singular Value Decomposition (SVD) of the data to project it to a lower dimensional space. The input data were centered but not scaled for each feature before applying the SVD, using the Linear Algebra PACKage (LAPACK) implementation in the Python library scikit-learn. PCA analysis was performed on the genes reported in Supplementary Table S1.

2.15. Statistical Analysis

The results in the figures are representative of several experiments performed with at least six cell lines from independent donors. Student t-test was used to assess the statistical significance with thresholds of * $p \leq 0.05$ and ** $p \leq 0.01$.

3. Results

3.1. Effects of Pioglitazone on Melanocytes Glucose Metabolism

Since prior observations reported that glucose uptake is greatly increased in vitiligo melanocytes [6], we measured the glucose amount in the medium of cultured vitiligo melanocytes both treated with or without PGZ. As shown, the glucose medium content was increased by treatment; consequently, PGZ-treated vitiligo melanocytes presented a lower glucose consumption compared to the untreated ones after 48 h (Figure 1A).

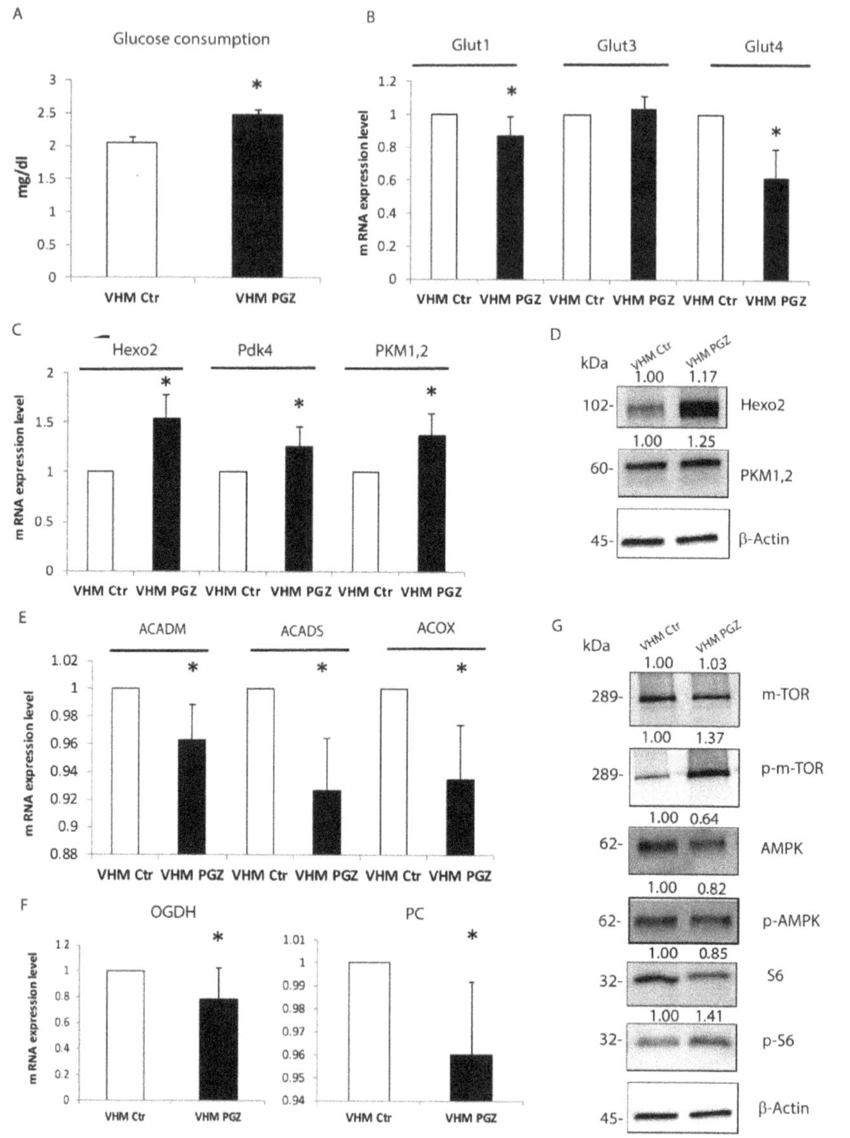

Figure 1. Pioglitazone modulates glucose metabolism in vitiligo melanocytes. (**A**) Two days of PGZ significantly increased glucose content in the vitiligo melanocytes medium. (**B**) PGZ reduced the gene expression of glucose transporters Glut1 and Glut4, but not Glut 3. (**C**,**D**) The transcriptional and protein levels of several anaerobic glycolytic enzymes were significantly induced by PGZ. mRNA levels of enzymes involved in β-oxidation (**E**) and Krebs cycle (**F**) were lowered by PGZ. (**G**) The representative Western blot of mTOR, pmTOR, AMPK, pAMPK, and S6, pS6, normalized to β-Actin, in vitiligo melanocytes following 2 days of treatment. The results are the mean ± SD of at least 3 independent experiments, * $p \leq 0.05$.

Consistent with this decrease, PGZ diminished the gene expression levels of glucose transporters Glut1 and Glut4 after 6 h, while did not seem to modify the expression of Glut3 (Figure 1B). We next investigated whether the expression of enzymes involved in glucose metabolism was adjusted by PGZ. Using the real-time evaluation, we analyzed the gene expression levels of some glycolytic enzymes involved in anaerobic respiration after 6 h of PGZ treatment. In sharp contrast, all the different enzymes evaluated were induced by treatment. Hexokinase2 (Hexo2), Pyruvate kinase isozymes M1/M2 (Pkm1,2), and subunit 4 of Pyruvate dehydrogenase complex (Pdk4) resulted significantly upregulated by PGZ (Figure 1C). Furthermore, Western blot analysis confirmed the induction of glycolytic enzymes by PGZ treatment (Figure 1D). To further corroborate our analysis, we conducted a measure of β-oxidation key enzymes, such as Acyl-CoA Dehydrogenase medium-chain (ACADM), Acyl-CoA Dehydrogenase (ACADS), Peroxisomal Acyl-CoA Oxidase (ACOX). A slight significant threshold decrease in the gene expression levels of this class of enzymes was observed (Figure 1E). Similarly, we measured a substantial reduction of enzymes involved in the Krebs cycle, as well as Pyruvate Carboxylase (PC) and Alpha-ketoglutarate (OGDH) (Figure 1F), confirming PGZ's selective activity on pathways implicated in ATP production. It is well documented that the mammalian target of rapamycin (mTOR) and AMP-activated protein kinase (AMPK) signaling pathways play a central role in response to ATP levels [19]. Recently, our group demonstrated that, in vitiligo melanocytes, the expression of AMPK phosphorylated (pAMPK) was upregulated, whereas the phosphorylation level of the downstream target of mTORC1, S6 kinase (pS6K), was reduced [20], further supporting the impaired metabolic condition. To better affirm the effect of Pioglitazone on cellular metabolism, we evaluated the effect of treatment on these pathways. Thus, 48 h of treatment with PGZ induced dephosphorylation of AMPK leading to higher phosphorylation of mTOR and S6 (Figure 1G). Taken together, these findings highlight that both acute and prolonged Pioglitazone administration improves glucose metabolism without an effectively increased consumption. Moreover, PGZ seems to be able to revert signaling pathways that regulate the intracellular metabolic state.

3.2. *Pioglitazone Attenuates Altered Melanocytes Mitochondrial Condition*

Starting from previously and currently reported mitochondrial alterations, we examined the role of PGZ in the recovery of mitochondrial structural and functional parameters. RT-PCR analysis showed a significant increase in the mtDNA copies in treated vitiligo melanocytes. The relative increment of mtDNA content was normalized on the nuDNA (Figure 2A).

Similarly, we observed that the gene expression of PGC1α, a key factor in mitochondrial biogenesis, was induced by treatment (Figure 2B). As previously reported, vitiligo melanocytes are characterized by a higher mitochondrial mass, as a compensatory mechanism to obtain a sufficient energetic level in a steady-state condition [6]. Here, we did not measure a mitochondrial mass decrease after 10 days of treatment (Figure 2C). Instead, q-RT-PCR underlined a reduction of voltage-dependent anion-selective channel 1 (VDAC1) (Figure 2D). Additionally, we did not find increased protein levels of total mitochondrial OXPHOS complexes (complex I, complex II, complex III, complex IV, and complex V) (Figure 2E). Our findings highlighted the ability of PGZ to modulate the processes that vitiligo cells act to compensate for mitochondrial defects. To further strengthen the effective ability of PGZ to reverse the altered phenotype of vitiligo melanocytes, we performed a principal component analysis (PCA). PCA was generated by comparing 6 primary treated vitiligo melanocytes with the same number of primary untreated vitiligo and primary untreated normal melanocytes. The PCA plot was based on the fold change of all the genes surveyed in the present study. As shown in Figure 2F, the melanocytes from normal individuals clustered closely, whereas the melanocytes from vitiligo patients are spread out, representing the heterogeneous signatures of these cells [21]. Moreover, a clear similarity is evident between treated vitiligo melanocytes and the normal controls. Collectively, our evidence supports the idea that aberrant mitochondrial components could be restored by Pioglitazone.

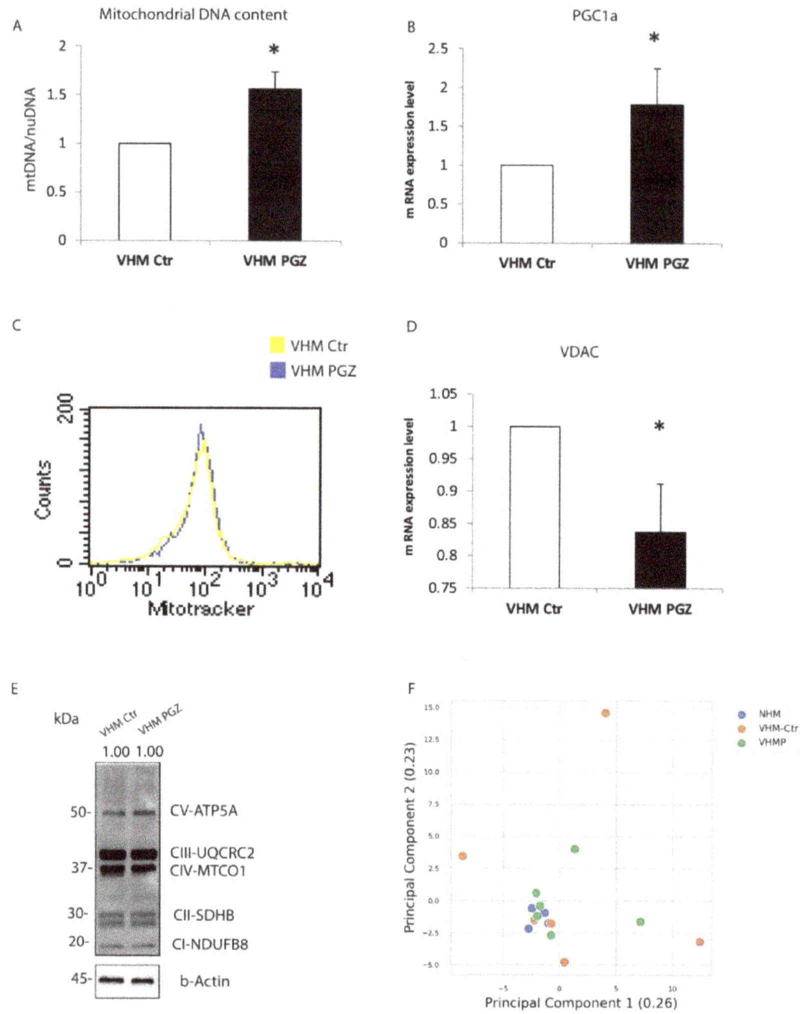

Figure 2. Pioglitazone increases the mitochondrial DNA integrity and copy number in vitiligo melanocytes. (**A**) Treatment with PGZ at 2 μM for 10 days significantly increased the mtDNA copy number. Nuclear DNA was used as an internal reference for data normalization. (**B**) RT-PCR underlined an upregulation of the PGC1α gene expression level. (**C**) The cytomic approach did not demonstrate any mitochondrial mass modification after PGZ treatment. (**D**) mRNA of VDAC1 was reduced by PGZ. (**E**) The representative immuno-blot did not show any modification of total OXPHOS complexes. (**F**) Principal component analysis (PCA) was generated by comparing 6 primary treated vitiligo melanocytes with the same number of primary untreated vitiligo and primary normal melanocytes. The data are mean ± SD of at least 3 independent experiments, * $p \leq 0.05$.

3.3. *Pioglitazone Improves the Whole Energetic Status in Vitiligo Melanocytes*

The cumulative evidence from our and other research groups demonstrated that melanocytes from pigmented skin of vitiligo patients present some structural and functional alterations and are characterized by an impaired energetic metabolism [5,6,22]. The bioenergetic deficit involves both oxidative phosphorylation and ATP synthesis due to a defect in the mitochondrial complex activity [6]. Once verified that 2 μM of PGZ was the lowest concentration able to induce a significant increase in cell proliferation rate after 48 h and 10 days of treatment (Figure 3A), this concentration was used in all experiments.

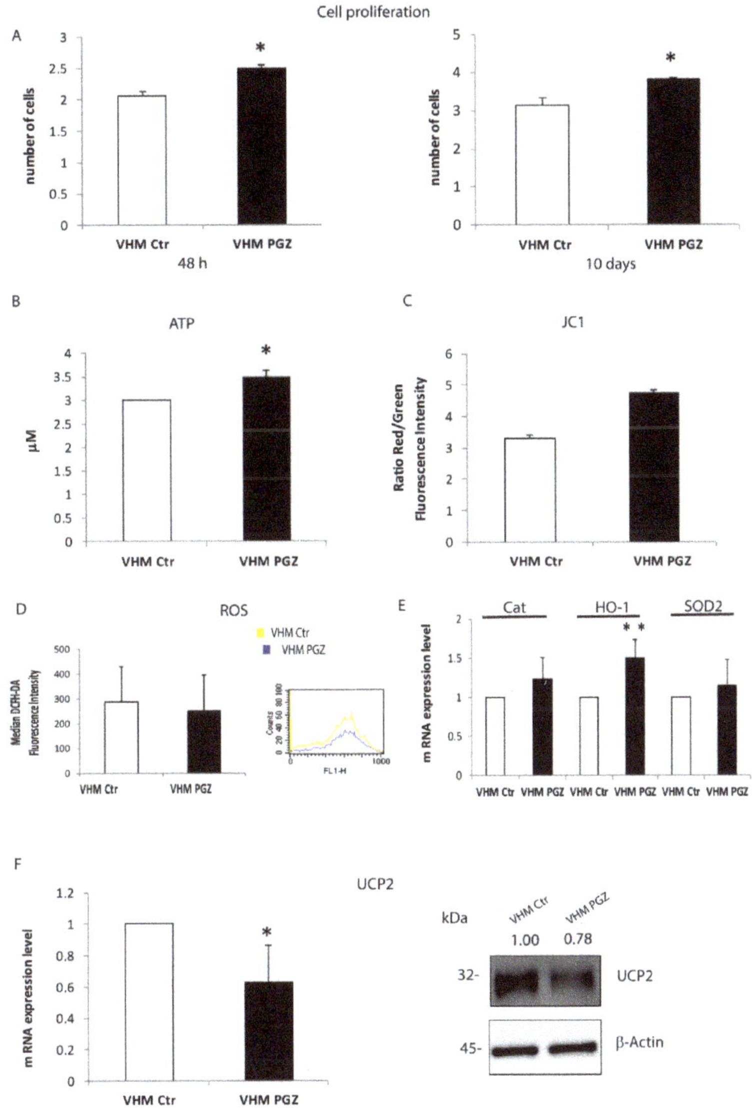

Figure 3. The impact of Pioglitazone on vitiligo melanocytes energetic status. (**A**) PGZ improved vitiligo melanocytes proliferation rate, evaluated by automatic cell counter, after 2 and 10 days of treatment. (**B**) Mitochondria-targeted luciferase assay revealed an increased ATP production in response to 10 days of Pioglitazone. (**C**) Flow cytometric analysis was used to quantify the improvement of the mitochondrial transmembrane potential of vitiligo cells. (**D**) The cytofluorimetric analysis highlighted a ROS reduction in treated vitiligo melanocytes. (**E**) Gene expression analysis of Cat, HO-1 and SOD2 was performed after 10 days of PGZ treatment. β-Actin expression was used to normalize the cDNA concentration. (**F**) RT-PCR analysis showed a significant down-modulation of UCP2, confirmed by Western blot analysis. The values are the mean ± SD of at least 3 independent experiments, * $p \leq 0.05$ and ** $p \leq 0.01$.

Firstly, we evaluated the effect of the treatment on mitochondrial bioenergetic status. We measured ATP content in treated and untreated cells. An increase in ATP levels was detected in melanocytes treated for 10 days (Figure 3B). Maintaining the mitochondrial membrane potential ($\Delta\Psi m$) is required for ATP production. To evaluate if PGZ is able to restore mitochondrial membrane potential, which is altered in vitiligo melanocytes, we compared treated vitiligo melanocytes with untreated controls (Figure 3C). Following 10 days of incubation with PGZ, flow cytometry analysis revealed an increase in mitochondrial membrane potential expressing as a ratio green/red. Since lower $\Delta\Psi$ and decreased activity of the respiratory chain is observed with a simultaneous increase in reactive oxygen species production, we investigated the effects of PGZ on intracellular ROS in vitiligo cells using the redox-sensitive probe DCFH-DA. DCFH-associated fluorescence was decreased in vitiligo melanocytes after treatment with PGZ for 10 days (Figure 3D). Parallelly, the mRNA of antioxidant enzymes such as Catalase (Cat), heme oxygenase (HO-1), and superoxide dismutase 2 (SOD-2) were increased by PGZ, even without being statically significant, possibly due to the variability of the basal level of expression (Figure 3E). The upregulation of mitochondrial uncoupling proteins is another potential explanation for the reduced mitochondrial efficiency [23]. Thus, we evaluated the effect of PGZ on Uncoupling protein 2 (UCP2), a member of the uncoupling mitochondrial transmembrane proteins family. RT-PCR analysis revealed that Pioglitazone is capable to reduce UCP2 gene expression already after 6 h (Figure 3F). This modulation was then confirmed at the protein level (Figure 3F). These findings suggest that the impaired functionality of mitochondria in vitiligo could be ameliorated by a low dose of Pioglitazone.

3.4. *Pioglitazone Reduces MITF and Senescence-Associated Markers Expression in Vitiligo Melanocytes*

An additional intrinsic abnormality in vitiligo melanocytes, compared to the normal counterpart, is the marked higher expression of the Microphthalmia-associated transcription factor (MITF), which is associated with a differentiated status (Figure 4A).

Here, we showed that PGZ significantly decreased gene expression of MITF after 6 h (Figure 4B). However, we did not detect a similar reduction at the protein level, after 48h (Figure 4C). Many reports have suggested that the ERK signaling pathway has a critical role in the regulation of melanogenesis and it has been reported that ERK activation leads to MITF phosphorylation and its subsequent degradation [24–26]. To better describe the effect of PGZ on MITF expression, we examined ERK phosphorylation levels up to 48 h of treatment. As shown in Figure 4D, PGZ induced de-phosphorylation of ERK and consequently MITF activation, contrary to the gene expression level. Kim et al. reported that inhibition of ROS and ERK abolished the degradation of MITF [27], similarly the inhibitory effect on ROS production by PGZ treatment could lead to MITF upregulation in our model, explaining its dual action. In vitiligo melanocytes, the premature senescence phenotype is characterized by the production of different proteins of the so-called senescence-associated Secretory Phenotype (SASP), including cytokines, growth factors, and molecules implicated in cell adhesion and tissue remodeling [5]. Next, we tested whether PGZ treatment was capable to alter the premature senescence phenotype in vitiligo melanocytes. Our results indicated that PGZ significantly reduced insulin growth factor binding proteins 3 (IGFBP3), both at mRNA and protein levels (Figure 4E). Additionally, insulin growth factor binding proteins 7 (IGFBP7) and p16 gene expression levels were diminished by PGZ (Figure 4E). Furthermore, the ELISA assay verified that PGZ also inhibited interleukin 6 levels (IL-6), a senescence-associated inflammatory mediator (Figure 4F). Thus, Cyclooxygenase-2 (Cox-2), an inducible enzyme, involved in stress-induced senescence and upregulated in vitiligo melanocytes [5], was feebly, but significantly, downregulated by PGZ (Figure 4G).

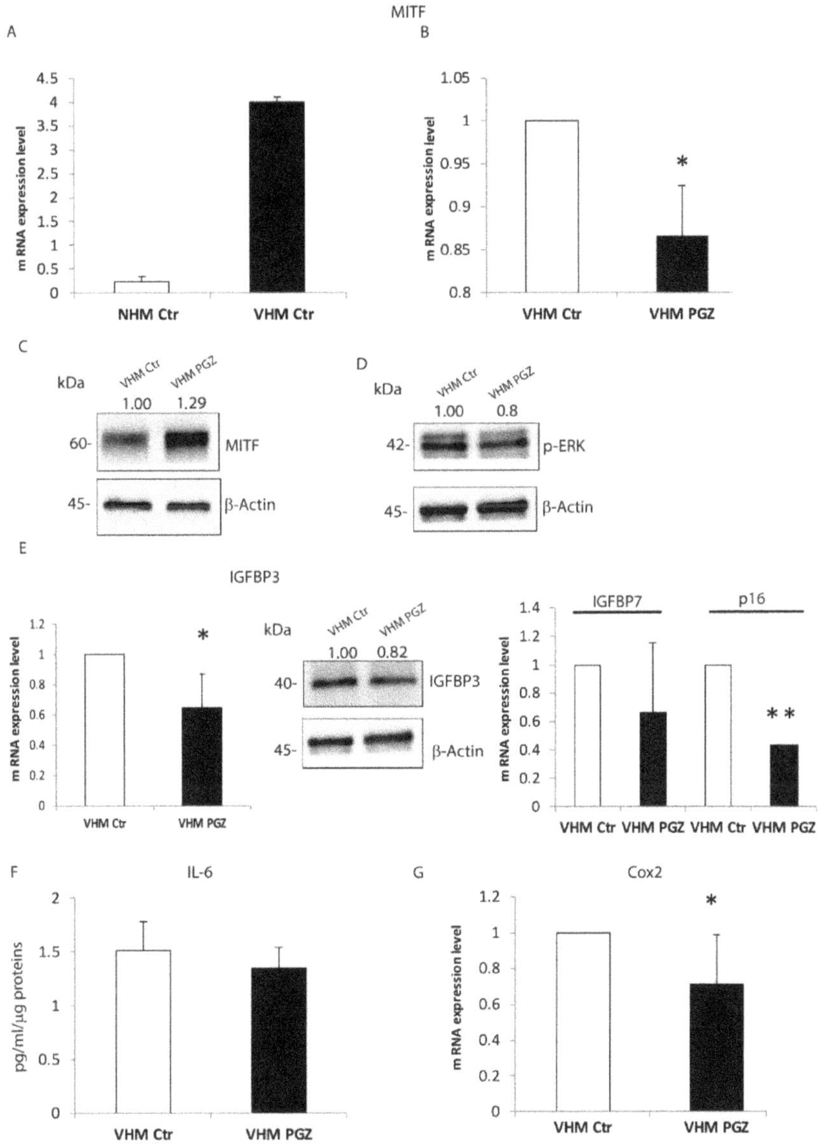

Figure 4. Pioglitazone modifies the expression of MITF and SASPs in vitiligo melanocytes. (**A**) Semi-quantitative real-time PCR was used to measure MITF mRNA expression in NHM and VHM. VHM showed a higher MITF expression level indicating a high grade of differentiation, which was reverted by PGZ treatment at mRNA (**B**), but not at protein level (**C**). (**D**) Representative Western blot of pErk. β-Actin was used as a loading control. (**E**) PGZ modulated the expression of senescence-associated markers, such as IGFBP3, both mRNA and protein levels, and IGFBP7 and p16 at mRNA level. (**F**) In vitiligo melanocytes, PGZ induced a reduction in IL-6 expression quantified by ELISA assay. (**G**) The relative Cox2 mRNA expression was evaluated after 6 h of treatment with PGZ (2 µM) and measured by qRT-PCR upon normalization to a reference gene (β-Actin), * $p \leq 0.05$ and ** $p \leq 0.01$.

3.5. Pioglitazone Modulates the Pro-Inflammatory Mediators in Vitiligo Melanocytes

Vitiligo pathogenesis involves complex combinatorial factors, including dysregulation of both innate and adaptive immune responses [2,28]. Several groups have demonstrated that inhibition of the immune response could appear as a promising strategy for vitiligo treatment [28]. Increased oxidative stress can induce the production of pro-inflammatory cytokines and activation of signals important for the induction of the immune system in vitiligo. Regarding the adaptive immune response, we have detected that PGZ reduced the gene expression of CXCR3B-isoform of the chemokine receptor (Figure 5A), whose expression is upregulated in vitiligo melanocytes [29].

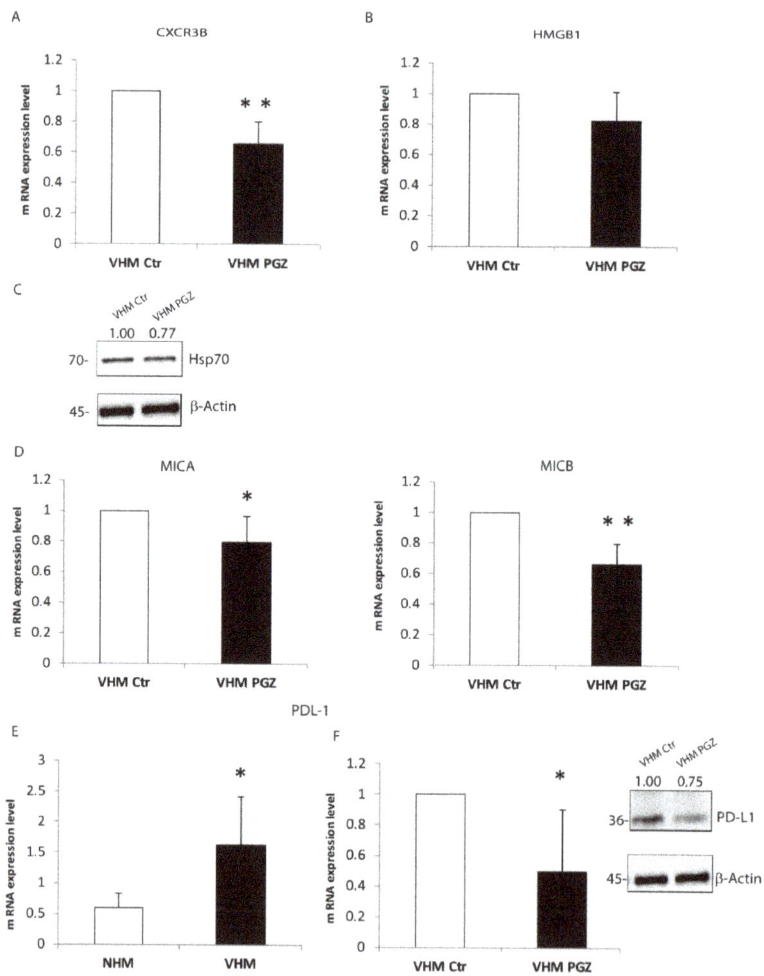

Figure 5. The mediation of pro-inflammatory agents by PGZ in vitiligo melanocytes. The mRNA level of chemokine receptor CXCR3B (**A**) and the inflammatory mediator HMGB1 (**B**) were investigated by RT-PCR after 6 h of treatment. (**C**) The representative protein appraisal of Hsp70. (**D**) mRNA evaluation of MICA and MICB, significantly downregulated by PGZ. (**E**) PDL-1 gene expression comparison between normal and vitiligo melanocytes. (**F**) mRNA downregulation of PDL-1 in vitiligo melanocytes by PGZ treatment, confirmed by immunoblotting assay. The data are mean ± SD of at least 3 independent experiments, * $p \leq 0.05$ and ** $p \leq 0.01$.

Notably, PGZ also inhibited the release of HMGB1, which is capable of activating the immune response (Figure 5B). In vitiligo, melanocytes increase the expression of Heat shock protein 70 (Hsp70) [30,31], a stress-inducible protein belonging to the heat shock protein family. Here we detected a downregulation of Hsp70 mediated by PGZ at the protein level (Figure 5C). Moreover, PGZ significantly lowered gene expression levels of stress molecule MHC class 1 chain-related protein A and B (MICA/MICB) (Figure 5D), NKG2D ligands, whose role was recently described in vitiligo pathogenesis [32]. The PD-1/PD-L1 pathway is classically associated with the modulation of the immune response of T cells, and the binding of PD-1 to a ligand can inhibit the T-cells proliferation and secretion of cytokines [33]. Interestingly, we measured, in vitiligo melanocytes compared to the normal controls, an overexpression of PD-L1 (Figure 5E), which can be expressed by melanocytes, especially in inflammatory environments [34]. As well, PGZ treatment seems to significantly reduce PD-L1 mRNA and protein levels (Figure 5F). The data collected propose a novel possible PGZ target, highlighting the synergistic interplay between metabolic alteration and inflammatory process in vitiligo pathogenesis.

3.6. Pioglitazone Ameliorates Intrinsic Alterations of Vitiligo Fibroblasts

Our earlier studies revealed functional and metabolic alteration in dermal fibroblasts, as well as myofibroblast and premature senescence phenotype [5,7]. To further characterize PGZ action in vitiligo cells, we performed a similar analysis on vitiligo fibroblasts. In treated vitiligo fibroblasts, we confirmed an increase in anaerobic glycolysis processes (Figure 6A) without an increment of glucose consumption, as proved by the glucose increased amount in fibroblasts treated medium (Figure 6B).

Furthermore, we detected a significant reduction of IGFBP3 gene expression and protein levels after PGZ treatment (Figure 6C). Similarly, mRNA levels of IGFBP5, p16 and p21 appeared downregulated by PGZ administration (Figure 6D). Most recently, it was reported that perturbation of glycolysis and fatty acid oxidation (FAO) pathway enzymes reveals their reciprocal effects in extracellular matrix components (ECM) upregulation and downregulation [35]. Instead, an enhanced release of growth factors and messengers that are part of a paracrine signaling network that controls melanocyte function is ascribed to vitiligo fibroblasts [7]. Starting from these two different pieces of evidence, to better elucidate a potential effect of PGZ in vitiligo cells, we focused our analysis on peculiar vitiligo fibroblast characteristics. We verified that PGZ treatment decreased the expression of several growth factors as well as Hepatocyte growth factor (HGF), both mRNA and protein levels (Figure 6E), and gene expression of Stem cell factor (SCF) (Figure 6E) and Vascular endothelial growth factor (VEGF) (Figure 6E). Furthermore, the Western blot analysis also revealed a decreased protein level of basic fibroblast growth factor (b- FGF) (Figure 6E). Conversely, the gene expression and protein levels of Fibronectin resulted increased by treatment (Figure 6F). Here, we proved that the modulation of metabolism could modify cellular phenotype and that PGZ could revert the altered expression of growth factors, involved in melanocyte homeostasis. The expression of transmembrane glycoprotein CD36, which imports long-chain fatty acids intracellularly for FAO, has been demonstrated to be inversely correlated with ECM abundance in normal skin and be downregulated in skin fibrosis [36]. Therefore, we evaluated whether CD36 could be modified by PGZ in vitiligo cells. In contrast with this, the CD36 gene and protein expressions were upregulated by PGZ in vitiligo fibroblasts and melanocytes (Figure 6G). These data show that PGZ action on the anaerobic glycolytic pathway could adjust aberrant vitiligo fibroblasts phenotype, also modulating the production of paracrine signals which contribute to melanocytes functionality.

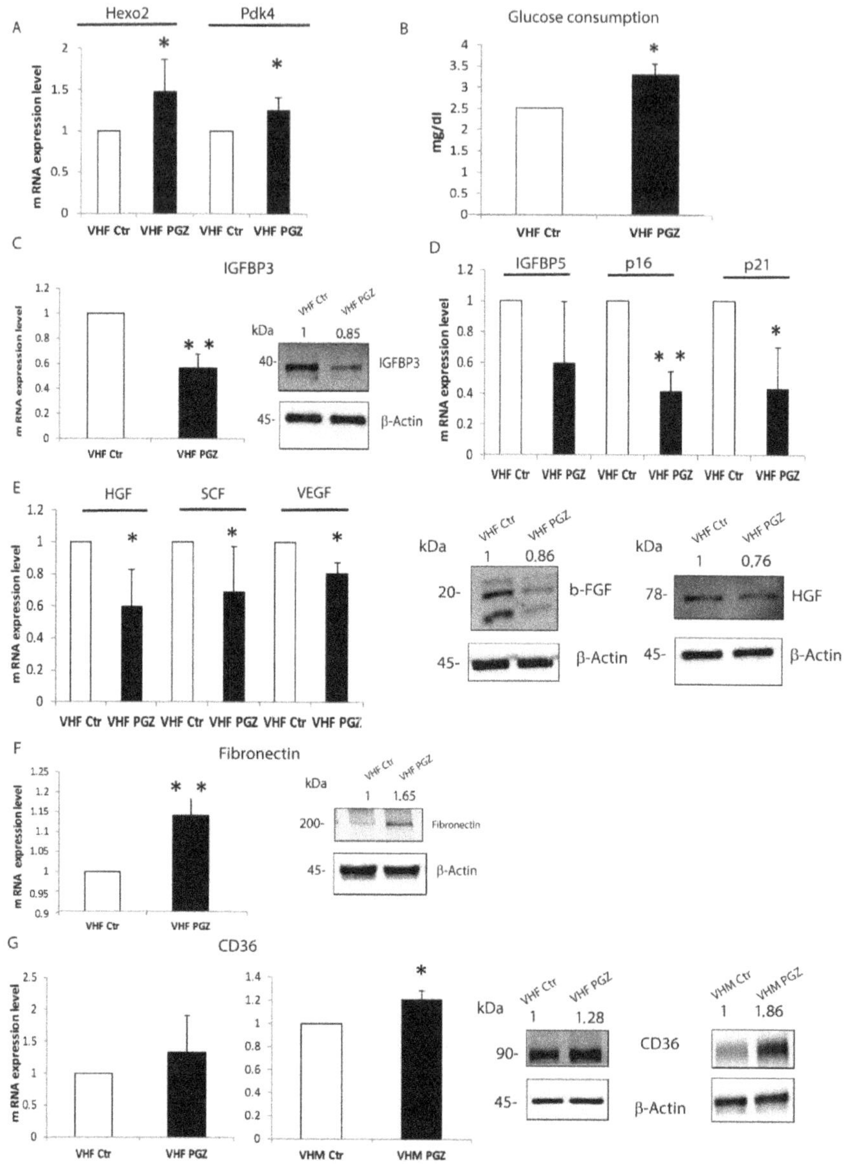

Figure 6. PGZ treatment is also effective on the dermal component. (**A**) The evaluation of gene expression of several anaerobic glycolytic enzymes induced by PGZ. (**B**) Significant glucose increase content in the medium of vitiligo fibroblasts. (**C**) IGFBP3 was downregulated in vitiligo fibroblasts by PGZ, both mRNA and protein levels. (**D**) Gene expression downregulation of senescence markers after 6 h of PGZ treatment. (**E**) The growth factors gene expression was analyzed by RT-PCR. The results were shown as a fold increase above the control. b-FGF and HGF were also assessed by Western blot. (**F**) The overexpression of Fibronectin was detected by RT-PCR and Western blot in vitiligo fibroblasts, respectively, after 6 and 48 h of PGZ treatment. (**G**) CD36 gene expression in vitiligo fibroblasts and melanocytes after PGZ treatment. The representative Western blot of CD36 total protein in vitiligo cell lines. All data are expressed as means ± SD of at least 3 independent experiments, * $p \leq 0.05$ and ** $p \leq 0.01$.

4. Discussion

Our data clearly demonstrate that, in vitro, the Peroxisome proliferator-activated receptors-γ agonist, Pioglitazone, significantly embanks the underlying cellular damage in non-depigmented skin of vitiligo patients. Compelling evidence, both at a cellular and molecular level, indicates that cumulative defects cause sickness in the skin [37]. Among the various pathogenetic hypotheses, our research has been particularly effective in the explanation of non-immunological factors, demonstrating the presence of metabolic impairment in almost all epidermal cells [5,7]. In accordance with our data, vitiligo could be defined as a mitochondrial disease leading to an aging phenomenon and to an alteration of the skin barrier, which could be an initial event in the development of the inflammatory process [5,6,8,38]. The regulation of energy metabolism is critical in maintaining intracellular redox equilibrium, strictly related to mitochondria efficiency. PPARs are nuclear receptors, which have an essential role in the mammalian physiological system [39]. PPARγ, the most widely investigated subtype, is involved in regulating glucose and lipid metabolism. In the skin, it manages the expression of a class of genes involved in cell proliferation, differentiation, and inflammatory responses [40]. Studies investigating its regulatory role on inflammatory markers such as IL-6, IL-17 and TNF-α, suggest that PGZ may be an effective candidate for autoimmune diseases and inflammatory skin disorders [41]. While the anti-diabetic effects of PPARγ agonists thiazolidinediones are well established, the full spectrum of action is not fully understood, especially when the effects on mitochondrial function are considered. To that end, we evaluated the drug's effect on the metabolic features of vitiligo cells, trying to highlight whether an improvement of the cellular metabolic status was associated with a significant modification of the biological behavior. All data presented, have precisely clarified how the known activity of PGZ is capable of shifting the metabolism of vitiligo cells towards the aerobic end of the spectrum with a consequent appropriate use of glucose, reverting the aberrant metabolic abnormalities observed. Specifically, we verified that acute PGZ administration reduced glucose consumption, meanwhile clearly increasing the expression of several anaerobic glycolytic enzymes. Likewise, the prolonged treatment decreased reactive oxygen species content and restored the physiological membrane potential. The acute metabolic effects of PGZ also reduced the phosphorylation of AMPK, a key cellular energetic sensor [42]. When cellular energy is low, as well as in vitiligo cells, AMPK is activated and targets a range of physiological processes, which finally concur in increased energy production and a coordinated decrease in ATP usage [43]. Moreover, it is known that hyperactive AMPK has been linked to age-related diseases, such as Alzheimer's disease and other cognitive dysfunctions [44]. Therefore, the inactivation of AMPK signaling, induced by PGZ treatment, confirmed the possibility of reversing the senescent changes we described in previous works. As mentioned above, Dell'Anna et al., stated a mitochondria failure associated with the peroxidation of the mitochondrial membrane due to the ROS generation and the overexpression of glycolytic enzymes to try to compensate for the decreased ATP generation via mitochondrial respiration. The disclosed increase of PGC1α and mitochondrial DNA copies suggest a mitochondria genesis induced by PGZ with the rescue of the ATP generation. In addition, our study confirmed previous reports [45,46], proving the anti-oxidative capability of PGZ. Collectively, our experimental evidence showed, that by raising metabolic spectrum, diminishing oxidative stress, and improving ATP production, PGZ seems to render vitiligo cells more similar to normal controls, as also confirmed by PCA analysis performed. MITF is a chief transcriptional regulator of melanogenesis and is crucial for developing melanocytes [47]. In vitiligo melanocytes, we observed MITF gene overexpression, indicating the activation of the differentiation processes. However, the protein level of MITF increased after 48 h of treatment. Numerous previous studies documented that several signaling pathways regulate MITF expression. MITF is phosphorylated by ERK and S6 [26,47]. Activation of ERK induces phosphorylation of MITF at Ser73, which leads to ubiquitination and degradation of the transcription factor [48]. In accordance with these previous reports, in treated vitiligo melanocytes, we found that phosphorylation of ERK induces MITF activation at the protein level. The dysfunction of intracellular metabolism in vitiligo may perturb and influence immune response. Vitiligo melanocytes release multiple cytokines and danger-associated molecular patterns, such as Hsp70 and HMGB1, in response to environmental signals relevant to the initiation of autoimmunity [5,49]. The relationships between the inflammatory milieu and metabolic impairment have always represented a key to understanding the complicated pathogenesis of vitiligo [28]. Furthermore, the stimulation of the immunological process may converge in terminal autoimmunity-mediated melanocyte loss. The amount of growing appreciation of similarities between vitiligo (autoimmunity) and melanoma (tumor immunity), has indicated the PD-1/PD-L1 axis as a promising target for future immunotherapies in human vitiligo [50]. A very novel observation of our current work is the capability of Pioglitazone to modulate several actors of the autoimmunity process. For the first time, here we reported a significant downregulation of PDL-1 by a PPARγ activator. Accordingly, PGZ reduces the expression of chemokine receptor CXCR3B, decreases the release of several DAMPs as well as, HMGB1 and Hsp70, and significantly lowers the gene expression levels of MICA and MICB. During the last years, our group reported that metabolic abnormalities are common to the entire vitiligo skin [7,8]. Starting from these considerations, we assayed that, in fibroblasts isolated from vitiligo patients, PGZ affected the release of growth factors and improved the production of extracellular components, such as Fibronectin. Furthermore, similar to precedent studies, we demonstrated that PPARγ agonist enhanced glycolysis and reverted the myofibroblast phenotype, characteristic of vitiligo dermal components. Moreover, PGZ treatment upregulated the gene and protein expression of CD36, a mediator of fatty acid uptake in both melanocytes and fibroblasts. This response can be explained by the fact that PGZ induces the accumulation of free fatty acids, as reported by Baranowski et al. [51]. The recent evaluation of the metabolic comorbidities in vitiligo patients [10], together with the known presence of pro-inflammatory cytokines linked to insulin resistance [52], defends the hypothesis of considering metabolic derangements as a new target to improve the quality of life of vitiligo patients. Unfortunately, vitiligo remains a difficult disease to treat. Taking into account all the different pathogenetic factors occurring in vitiligo onset, a therapeutical approach with a mechanism addressing the entire microenvironment seems to be needed.

5. Conclusions

The encouraging results discussed above, including the tendency to restore the deranged metabolic profile, the decrease in senescence-related markers, and the amelioration trend of the inflammatory process, lead us to think of the possible use of a PPARγ agonist in the treatment of vitiligo. With in-depth research on targeted intrinsic metabolic abnormalities, we believe that the present research could represent an alternative therapeutic option to existing vitiligo treatments, generally providing only modest efficacy. Future studies will clarify our experimental furtherance in clinical practices.

Supplementary Materials: The following supporting information can be downloaded at: https://www.mdpi.com/article/10.3390/cells11223583/s1, Table S1: Primers for Real-Time Quantitative Polymerase Chain Reaction; Table S2: Primary Antibodies.

Author Contributions: F.P. designed and performed experiments, conducted acquisition, analysis and interpretation of data, and wrote the manuscript. B.B. performed cytofluorimetric analysis and discussed the data. M.O. performed biochemical analysis and contributed to the data interpretation. M.T. performed PCA analysis. G.C. conducted serum parameters analysis. L.S. contributed to performing biochemical analysis. S.C. isolated cell lines from biopsies. E.M. collected normal skin biopsies. A.D., A.P. and P.I. collected vitiligo skin biopsies and clinical information. M.P. coordinated research, drafting, and revising the article. All authors have read and agreed to the published version of the manuscript.

Funding: The research was supported by a grant from the Italian Ministry of Health.

Institutional Review Board Statement: The study was conducted in accordance with the Declaration of Helsinki, and approved by the Ethics Committee of Regina Elena Institute and San Gallicano Dermatological Institute (Protocol code CE/751/16 approved on 12 January 2016).

Informed Consent Statement: Informed consent was obtained from all subjects involved in the study.

Data Availability Statement: Not applicable.

Acknowledgments: We want to thank Antonella Vento for her technical support.

Conflicts of Interest: The authors declare no conflict of interest.

References

1. Bergqvist, C.; Ezzedine, K. Vitiligo: A Review. *Dermatology* **2020**, *236*, 571–592. [CrossRef] [PubMed]
2. Picardo, M.; Dell'Anna, M.L.; Ezzedine, K.; Hamzavi, I.; Harris, J.E.; Parsad, D.; Taieb, A. Vitiligo. *Nat. Rev. Dis. Prim.* **2015**, *1*, 15011. [CrossRef] [PubMed]
3. Seneschal, J.; Boniface, K.; D'Arino, A.; Picardo, M. An update on Vitiligo pathogenesis. *Pigment Cell Melanoma Res.* **2021**, *34*, 236–243. [CrossRef]
4. Abdel-Malek, Z.A.; Jordan, C.; Ho, T.; Upadhyay, P.R.; Fleischer, A.; Hamzavi, I. The enigma and challenges of vitiligo pathophysiology and treatment. *Pigment Cell Melanoma Res.* **2020**, *33*, 778–787. [CrossRef]
5. Bellei, B.; Pitisci, A.; Ottaviani, M.; Ludovici, M.; Cota, C.; Luzi, F.; Dell'Anna, M.L.; Picardo, M. Vitiligo: A possible model of degenerative diseases. *PLoS ONE* **2013**, *8*, e59782. [CrossRef] [PubMed]
6. Dell'Anna, M.L.; Ottaviani, M.; Kovacs, D.; Mirabilii, S.; Brown, D.A.; Cota, C.; Migliano, E.; Bastonini, E.; Bellei, B.; Cardinali, G.; et al. Energetic mitochondrial failing in vitiligo and possible rescue by cardiolipin. *Sci. Rep.* **2017**, *7*, 13663. [CrossRef]
7. Kovacs, D.; Bastonini, E.; Ottaviani, M.; Cota, C.; Migliano, E.; Dell'Anna, M.L.; Picardo, M. Vitiligo Skin: Exploring the Dermal Compartment. *J. Investig. Dermatol.* **2018**, *138*, 394–404. [CrossRef] [PubMed]
8. Kovacs, D.; Bastonini, E.; Briganti, S.; Ottaviani, M.; D'Arino, A.; Truglio, M.; Sciuto, L.; Zaccarini, M.; Pacifico, A.; Cota, C.; et al. Altered epidermal proliferation, differentiation, and lipid composition: Novel key elements in the vitiligo puzzle. *Sci. Adv.* **2022**, *8*, eabn9299. [CrossRef]
9. Bastonini, E.; Bellei, B.; Filoni, A.; Kovacs, D.; Iacovelli, P.; Picardo, M. Involvement of non-melanocytic skin cells in vitiligo. *Exp. Dermatol.* **2019**, *28*, 667–673. [CrossRef]
10. D'Arino, A.; Picardo, M.; Truglio, M.; Pacifico, A.; Iacovelli, P. Metabolic Comorbidities in Vitiligo: A Brief Review and Report of New Data from a Single-Center Experience. *Int. J. Mol. Sci.* **2021**, *22*, 8820. [CrossRef]
11. Tyagi, S.; Gupta, P.; Saini, A.S.; Kaushal, C.; Sharma, S. The peroxisome proliferator-activated receptor: A family of nuclear receptors role in various diseases. *J. Adv. Pharm. Technol. Res.* **2011**, *2*, 236–240. [CrossRef] [PubMed]
12. Hassan, F.U.; Nadeem, A.; Li, Z.; Javed, M.; Liu, Q.; Azhar, J.; Rehman, M.S.; Cui, K.; Rehman, S.U. Role of Peroxisome Proliferator-Activated Receptors (PPARs) in Energy Homeostasis of Dairy Animals: Exploiting Their Modulation through Nutrigenomic Interventions. *Int. J. Mol. Sci.* **2021**, *22*, 12463. [CrossRef] [PubMed]
13. Ahmadian, M.; Suh, J.M.; Hah, N.; Liddle, C.; Atkins, A.R.; Downes, M.; Evans, R.M. PPARγ signaling and metabolism: The good, the bad and the future. *Nat. Med.* **2013**, *19*, 557–566. [CrossRef]
14. Villacorta, L.; Schopfer, F.J.; Zhang, J.; Freeman, B.A.; Chen, Y.E. PPARgamma and its ligands: Therapeutic implications in cardiovascular disease. *Clin. Sci.* **2009**, *116*, 205–218. [CrossRef] [PubMed]
15. Yew, T.; Toh, S.A.; Millar, J.S. Selective peroxisome proliferator-activated receptor-γ modulation to reduce cardiovascular risk in patients with insulin resistance. *Recent Pat. Cardiovasc. Drug Discov.* **2012**, *7*, 33–41. [CrossRef]
16. Fujisawa, K.; Nishikawa, T.; Kukidome, D.; Imoto, K.; Yamashiro, T.; Motoshima, H.; Matsumura, T.; Araki, E. TZDs reduce mitochondrial ROS production and enhance mitochondrial biogenesis. *Biochem. Biophys. Res. Commun.* **2009**, *379*, 43–48. [CrossRef]
17. Grygiel-Górniak, B. Peroxisome proliferator-activated receptors and their ligands: Nutritional and clinical implications–A review. *Nutr. J.* **2014**, *13*, 17. [CrossRef] [PubMed]
18. El-Gharabawy, R.M.; Ahmed, A.S.; Al-Najjar, A.H. Mechanism of action and effect of immune-modulating agents in the treatment of psoriasis. *Biomed. Pharmacother.* **2017**, *85*, 141–147. [CrossRef]
19. González, A.; Hall, M.N.; Lin, S.C.; Hardie, D.G. AMPK and TOR: The Yin and Yang of Cellular Nutrient Sensing and Growth Control. *Cell Metab.* **2020**, *31*, 472–492. [CrossRef]
20. Bastonini, E.; Kovacs, D.; Raffa, S.; Delle Macchie, M.; Pacifico, A.; Iacovelli, P.; Torrisi, M.R.; Picardo, M. A protective role for autophagy in vitiligo. *Cell Death Dis.* **2021**, *12*, 318. [CrossRef]
21. Ezzedine, K.; Vance, T.M.; Hamzavi, I.H.; Qureshi, A.A. Vitiligo: Targeted Therapies Add Color to Disease Pathophysiology. *J. Investig. Dermatol.* **2020**, *140*, 1498–1500. [CrossRef] [PubMed]
22. Qiao, Z.; Wang, X.; Xiang, L.; Zhang, C. Dysfunction of Autophagy: A Possible Mechanism Involved in the Pathogenesis of Vitiligo by Breaking the Redox Balance of Melanocytes. *Oxidative Med. Cell. Longev.* **2016**, *2016*, 3401–3570. [CrossRef] [PubMed]

23. Lee, T.W.; Bai, K.J.; Lee, T.I.; Chao, T.F.; Kao, Y.H.; Chen, Y.J. PPARs modulate cardiac metabolism and mitochondrial function in diabetes. *J. Biomed. Sci.* **2017**, *24*, 5. [CrossRef] [PubMed]
24. Hemesath, T.J.; Price, E.R.; Takemoto, C.; Badalian, T.; Fisher, D.E. MAP kinase links the transcription factor Microphthalmia to c-Kit signalling in melanocytes. *Nature* **1998**, *391*, 298–301. [CrossRef] [PubMed]
25. Wu, M.; Hemesath, T.J.; Takemoto, C.M.; Horstmann, M.A.; Wells, A.G.; Price, E.R.; Fisher, D.Z.; Fisher, D.E. c-Kit triggers dual phosphorylations, which couple activation and degradation of the essential melanocyte factor Mi. *Genes Dev.* **2000**, *14*, 301–312. [CrossRef] [PubMed]
26. Xu, W.; Gong, L.; Haddad, M.M.; Bischof, O.; Campisi, J.; Yeh, E.T.; Medrano, E.E. Regulation of microphthalmia-associated transcription factor MITF protein levels by association with the ubiquitin-conjugating enzyme hUBC9. *Exp. Cell Res.* **2000**, *255*, 135–143. [CrossRef] [PubMed]
27. Kim, E.S.; Park, S.J.; Goh, M.J.; Na, Y.J.; Jo, D.S.; Jo, Y.K.; Shin, J.H.; Choi, E.S.; Lee, H.K.; Kim, J.Y.; et al. Mitochondrial dynamics regulate melanogenesis through proteasomal degradation of MITF via ROS-ERK activation. *Pigment Cell Melanoma Res.* **2014**, *27*, 1051–1062. [CrossRef]
28. Boniface, K.; Seneschal, J.; Picardo, M.; Taïeb, A. Vitiligo: Focus on Clinical Aspects, Immunopathogenesis, and Therapy. *Clin. Rev. Allergy Immunol.* **2018**, *54*, 52–67. [CrossRef]
29. Tulic, M.K.; Cavazza, E.; Cheli, Y.; Jacquel, A.; Luci, C.; Cardot-Leccia, N.; Hadhiri-Bzioueche, H.; Abbe, P.; Gesson, M.; Sormani, L.; et al. Innate lymphocyte-induced CXCR3B-mediated melanocyte apoptosis is a potential initiator of T-cell autoreactivity in vitiligo. *Nat. Commun.* **2019**, *10*, 2178. [CrossRef]
30. Mosenson, J.A.; Flood, K.; Klarquist, J.; Eby, J.M.; Koshoffer, A.; Boissy, R.E.; Overbeck, A.; Tung, R.C.; Le Poole, I.C. Preferential secretion of inducible HSP70 by vitiligo melanocytes under stress. *Pigment Cell Melanoma Res.* **2014**, *27*, 209–220. [CrossRef]
31. Doss, R.W.; El-Rifaie, A.A.; Abdel-Wahab, A.M.; Gohary, Y.M.; Rashed, L.A. Heat Shock Protein-70 Expression in Vitiligo and its Relation to the Disease Activity. *Indian J. Dermatol.* **2016**, *61*, 408–412. [CrossRef] [PubMed]
32. Raam, L.; Kaleviste, E.; Šunina, M.; Vaher, H.; Saare, M.; Prans, E.; Pihlap, M.; Abram, K.; Karelson, M.; Peterson, P.; et al. Lymphoid Stress Surveillance Response Contributes to Vitiligo Pathogenesis. *Front. Immunol.* **2018**, *9*, 2707. [CrossRef] [PubMed]
33. Qin, W.; Hu, L.; Zhang, X.; Jiang, S.; Li, J.; Zhang, Z.; Wang, X. The Diverse Function of PD-1/PD-L Pathway Beyond Cancer. *Front. Immunol.* **2019**, *10*, 2298. [CrossRef] [PubMed]
34. Speeckaert, R.; van Geel, N. Targeting CTLA-4, PD-L1 and IDO to modulate immune responses in vitiligo. *Exp. Dermatol.* **2017**, *26*, 630–634. [CrossRef] [PubMed]
35. Zhao, X.; Psarianos, P.; Ghoraie, L.S.; Yip, K.; Goldstein, D.; Gilbert, R.; Witterick, I.; Pang, H.; Hussain, A.; Lee, J.H.; et al. Metabolic regulation of dermal fibroblasts contributes to skin extracellular matrix homeostasis and fibrosis. *Nat. Metab.* **2019**, *1*, 147–157. [CrossRef]
36. Glatz, J.F.C.; Luiken, J.J.F.P. Dynamic role of the transmembrane glycoprotein CD36 (SR-B2) in cellular fatty acid uptake and utilization. *J. Lipid Res.* **2018**, *59*, 1084–1093. [CrossRef]
37. Rashighi, M.; Harris, J.E. Vitiligo Pathogenesis and Emerging Treatments. *Dermatol. Clin.* **2017**, *35*, 257–265. [CrossRef]
38. Dell'anna, M.L.; Picardo, M. A review and a new hypothesis for non-immunological pathogenetic mechanisms in vitiligo. *Pigment Cell Res.* **2006**, *19*, 406–411. [CrossRef]
39. Karásek, D. Pioglitazone. *Vnitr. Lek.* **2020**, *66*, 121–125. [CrossRef]
40. Ramot, Y.; Mastrofrancesco, A.; Camera, E.; Desreumaux, P.; Paus, R.; Picardo, M. The role of PPARγ-mediated signalling in skin biology and pathology: New targets and opportunities for clinical dermatology. *Exp. Dermatol.* **2015**, *24*, 245–251. [CrossRef]
41. Silva-Abreu, M.; Espinoza, L.C.; Rodríguez-Lagunas, M.J.; Fábrega, M.J.; Espina, M.; García, M.L.; Calpena, A.C. Human Skin Permeation Studies with PPARγ Agonist to Improve Its Permeability and Efficacy in Inflammatory Processes. *Int. J. Mol. Sci.* **2017**, *18*, 2548. [CrossRef] [PubMed]
42. Hardie, D.G. AMP-activated protein kinase: An energy sensor that regulates all aspects of cell function. *Genes Dev.* **2011**, *25*, 1895–1908. [CrossRef] [PubMed]
43. Hardie, D.G.; Ashford, M.L. AMPK: Regulating energy balance at the cellular and whole body levels. *Physiology* **2014**, *29*, 99–107. [CrossRef] [PubMed]
44. Garcia, D.; Shaw, R.J. AMPK: Mechanisms of Cellular Energy Sensing and Restoration of Metabolic Balance. *Mol. Cell* **2017**, *66*, 789–800. [CrossRef]
45. Sun, L.; Yuan, Q.; Xu, T.; Yao, L.; Feng, J.; Ma, J.; Wang, L.; Lu, C.; Wang, D. Pioglitazone, a Peroxisome Proliferator-Activ ated Receptor x03B3; Agonist, Ameliorates Chronic Kidney Disease by Enhancing Antioxidative Capacity and Attenuating Angiogenesis in the Kidney of a 5/6 Nephrectomized Rat Model. *Cell. Physiol. Biochem.* **2016**, *38*, 1831–1840. [CrossRef]
46. Zhang, Z.; Zhang, X.; Meng, L.; Gong, M.; Li, J.; Shi, W.; Qiu, J.; Yang, Y.; Zhao, J.; Suo, Y.; et al. Pioglitazone Inhibits Diabetes-Induced Atrial Mitochondrial Oxidative Stress and Improves Mitochondrial Biogenesis, Dynamics, and Function Through the PPAR-γ/PGC-1α Signaling Pathway. *Front. Pharmacol.* **2021**, *12*, 658362. [CrossRef]
47. Levy, C.; Khaled, M.; Fisher, D.E. MITF: Master regulator of melanocyte development and melanoma oncogene. *Trends Mol. Med.* **2006**, *12*, 406–414. [CrossRef]
48. Hsiao, J.J.; Fisher, D.E. The roles of microphthalmia-associated transcription factor and pigmentation in melanoma. *Arch. Biochem. Biophys.* **2014**, *563*, 28–34. [CrossRef]

49. Richmond, J.M.; Frisoli, M.L.; Harris, J.E. Innate immune mechanisms in vitiligo: Danger from within. *Curr. Opin. Immunol.* **2013**, *25*, 676–682. [CrossRef]
50. Willemsen, M.; Melief, C.J.M.; Bekkenk, M.W.; Luiten, R.M. Targeting the PD-1/PD-L1 Axis in Human Vitiligo. *Front. Immunol.* **2020**, *11*, 579022. [CrossRef]
51. Baranowski, M.; Blachnio-Zabielska, A.; Zabielski, P.; Gorski, J. Pioglitazone induces lipid accumulation in the rat heart despite concomitant reduction in plasma free fatty acid availability. *Arch. Biochem. Biophys.* **2008**, *477*, 86–91. [CrossRef] [PubMed]
52. Bastard, J.P.; Maachi, M.; Van Nhieu, J.T.; Jardel, C.; Bruckert, E.; Grimaldi, A.; Robert, J.J.; Capeau, J.; Hainque, B. Adipose Tissue IL-6 Content Correlates with Resistance to Insulin Activation of Glucose Uptake Both in vivo and in vitro. *J. Clin. Endocrinol. Metab.* **2002**, *87*, 2084–2089. [CrossRef] [PubMed]

cells

MDPI

Article

The Potential for Placental Activation of PPARγ to Improve the Angiogenic Profile in Preeclampsia

Brooke Grimaldi [1], Hamid-Reza Kohan-Ghadr [1] and Sascha Drewlo [2,3,*]

1 Department of Obstetrics, Gynecology and Reproductive Biology, College of Human Medicine, Michigan State University, Grand Rapids, MI 49503, USA
2 Biological Sciences Platform, Sunnybrook Health Sciences Centre, Sunnybrook Research Institute, Toronto M4N 3M5, Canada
3 Department of Obstetrics and Gynecology, Temerty Faculty of Medicine, University of Toronto, Toronto M5G 1E2, Canada
* Correspondence: sascha.drewlo@sri.utoronto.ca

Abstract: Preeclampsia (PE) is one of the most common causes of maternal-fetal morbidity and mortality world-wide. While the underlying causes of PE remain elusive, aberrant trophoblast differentiation and function are thought to cause an imbalance of secreted angiogenic proteins resulting in systemic endothelial dysfunction and organ damage in the mother. The placental dysfunction is also characterized by a reduction of the transcription factor, peroxisome proliferator activated receptor γ (PPARγ) which normally promotes trophoblast differentiation and healthy placental function. This study aimed to understand how placental activation of PPARγ effects the secretion of angiogenic proteins and subsequently endothelial function. To study this, healthy and PE placental tissues were cultured with or without the PPARγ agonist, Rosiglitazone, and a Luminex assay was performed to measure secreted proteins from the placenta. To assess the angiogenic effects of placental activation of PPARγ, human umbilical vein endothelial cells (HUVECs) were cultured with the placental conditioned media and the net angiogenic potential of these cells was measured by a tube formation assay. This is the first study to show PPARγ's beneficial effect on the angiogenic profile in the human preeclamptic placenta through the reduction of anti-angiogenic angiopoietin-2 and soluble endoglin and the upregulation of pro-angiogenic placental growth factor, fibroblast growth factor-2, heparin-binding epidermal growth factor, and follistatin. The changes in the angiogenic profile were supported by the increased angiogenic potential observed in the HUVECs when cultured with conditioned media from rosiglitazone-treated preeclamptic placentas. The restoration of these disrupted pathways by activation of PPARγ in the preeclamptic placenta offers potential to improve placental and endothelial function in PE.

Keywords: placenta; preeclampsia; PPARγ; angiogenesis

Citation: Grimaldi, B.; Kohan-Ghadr, H.-R.; Drewlo, S. The Potential for Placental Activation of PPARγ to Improve the Angiogenic Profile in Preeclampsia. *Cells* **2022**, *11*, 3514. https://doi.org/10.3390/cells11213514

Academic Editors: Kay-Dietrich Wagner and Nicole Wagner

Received: 13 October 2022
Accepted: 1 November 2022
Published: 6 November 2022

1. Introduction

The placenta has a significant role in establishing and maintaining a healthy pregnancy. Abnormal placental development and function is implicated in preeclampsia (PE), which is the leading cause of maternal and fetal morbidity and mortality worldwide [1]. PE clinically manifests around 20 weeks of gestation and is diagnosed based on new onset of maternal hypertension and proteinuria [2]. There is no treatment for PE, and often requires preterm delivery (<37 weeks) as the only treatment option in severe cases. This poses a significant risk to a newborn's health and is associated with extensive neonatal intensive care costs [3,4]. If untreated, PE can impair maternal hepatic and coagulation systems causing seizures, brain damage, or maternal death.

Although the etiology of PE remains elusive, evidence suggests insufficient placental perfusion is a major cause of increased inflammation and oxidative stress in the placenta [5].

This environment promotes abnormal villous trophoblast (VT) differentiation, and trophoblast immaturity [6] that further causes abnormal secretion of placental proteins causing an anti-angiogenic state of the placenta [7–13] and contributing to villous vascular dysfunction [14]. These conditions contribute to systemic endothelial dysfunction which result in the onset of clinical symptoms in the mother. The systemic anti-angiogenic environment in PE additionally poses life-long maternal complications. Nearly half of all women with PE have high blood pressure through 12 weeks post-partum [15,16] and are at risk of developing chronic hypertension within a few years after giving birth [17]. PE poses an even greater risk for cardiovascular disease than smoking [17], and women in the United States who are diagnosed with PE have a 9.4-fold increased risk for cardiovascular-related deaths if the infant is born prior to 34 weeks of gestation [17]. Thus, there is a great need to identify mechanisms of the placental contribution to PE and to establish interventions to dampen maternal sequalae.

Peroxisome proliferator activated receptor-γ (PPARγ) is a widely studied transcription factor, prominently known for its roles in metabolism and adipocyte differentiation [18–21]. It is also well known that PPARγ acts upstream of several pathways that regulate cell metabolism, anti-inflammatory pathways, and oxidative stress response in the placenta [22]. Our group and others have been uncovering PPARγ's roles in the placenta, specifically in the regulation of VT differentiation and turnover [22–29]. PPARγ expression and activity is significantly reduced in the PE placenta, and is therefore thought to contribute to PE pathogenesis [25]. Studies have also suggested that activating PPARγ can restore placental function as a potential treatment for PE [23]. Our group has postulated that there could be a connection between the aberrant VT differentiation and the imbalance of secreted proteins from the placenta in PE which cause this anti-angiogenic state. We hypothesize that PPARγ may contribute to maintaining the angiogenic balance in the placenta, and this notion is based on our prior reports showing that activation of PPARγ and its downstream pathways in the human placenta can improve the placental angiogenic environment through down-regulating the anti-angiogenic molecule, Soluble fms-like tyrosine kinase 1 [25]. This study further investigates the potential for placental activation of PPARγ to affect the secretion of angiogenic and other growth factor proteins from the placenta that subsequently impact the surrounding endothelium.

To investigate this, we cultured healthy and PE human placentas in the presence or absence of a PPARγ agonist, Rosiglitazone, and we measured expression of several angiogenic, metabolic, and growth factor proteins (Angiopoeitin-2 (Ang-2), soluble Endoglin (sEng), and Endothelin-1 (ET-1), Placental growth factor (PlGF), Fibroblast growth factor-2 (FGF-2), Epidermal growth factor (EGF), Heparin-binding growth factor (HB-EGF), Follistatin (FST), and Leptin). To better understand if placental activation of PPARγ exerts an effect on the surrounding endothelium, we recapitulated the maternal endothelial response to the placental protein secretion by culturing human umbilical vein endothelial cells (HUVECs) with placental conditioned media. This study emphasizes the benefits of targeting of PPARγ-related pathways as a novel interventional strategy for PE. Although likely other molecules are involved in the process, this first study to report placental activation of PPARγ improves the net placental angiogenic profile and subsequently increases angiogenic potential of the endothelium. Improving maternal endothelial function by managing the negative protein secretion would be beneficial in reducing maternal symptoms in PE.

2. Materials and Methods

2.1. Tissue Collection

Placentas for this study were provided from two locations at the Research Centre for Women's and Infants' Health (RCWIH) BioBank program of Mount Sinai Hospital in Toronto, Canada and from the Women's Health Center at Spectrum Hospital in Grand Rapids, MI. All placentas collected from Mount Sinai Hospital were done so in accordance with the policies of the Mount Sinai Hospital Research Ethics Board. All placentas collected from Spectrum Health were approved by the IRB waiver of parental consent. Available patient clinical information is listed Table 1. Healthy control placentas (n = 10) were collected from un-complicated pregnancies that delivered by either Cesarean section or vaginal birth between weeks 34–39 of gestation. Placentas collected before 37 weeks were from idiopathic preterm births and the placentas did not contain histological evidence of chorioamnionitis. Pathologic placentas were collected from PE pregnancies (n = 10, Gestation age = 31–37 weeks) or severe PE pregnancies (n = 4; Gestation age = 37–39 weeks) and were delivered either by Cesarean section or vaginal birth. Placentas selected for this study must have met the inclusion criteria for PE/sPE, which is based on the current guidelines for diagnosis of PE, briefly maternal blood pressure >140/90 mm Hg on two occasions longer than 6 h apart, with or without proteinuria and fetal growth restriction [2].

Table 1. Clinical Information.

Diagnosis	Gestation Age	Mode of Delivery	Maternal Age
PE	37wks	C-section	32
PE	39wks	C-section	33
PE	37wks	Vaginal	32
PE	38wks	C-section	22
PE	37wks	Vaginal	26
PE	37wks	Vaginal	33
PE	35wks	Vaginal	29
PE	37wks	Vaginal	37
PE	37wks	C-section	29
PE	31wks	C-section	19
sePE	39wks	C-section	37
sePE	37wks	Vaginal	20
sePE	39wks	Vaginal	20
sePE	36wks	Vaginal	19
Healthy control	40wks	Vaginal	33
Healthy control	39wks	Vaginal	25
Healthy control	39wks	Vaginal	28
Healthy control	39wks	Vaginal	28
Healthy control	36wks	C-section	29
Healthy control	36wks	Vaginal	28
Healthy control	35wks	Vaginal	33
Healthy control	36wks	C-section	34
Healthy control	37wks	Vaginal	36
Healthy control	37wks	C-section	26

2.2. *Explant Culture*

Placentas were collected using our published methods [25]. In brief, within 2 h post-birth, four 1 cm^3 cuboidal sections of the placenta were collected and transported to the laboratory in ice-cold HBSS (Hank's Balanced Salt Solution). The placental sections were immediately processed by rinsing in HBSS and dissected into 20–30 mg sized pieces. The placental tissues were cultures overnight 8% O_2 with 5% CO_2 at 37 °C in 500 µL of Dulbecco's modified Eagle's medium/Ham's F-12 nutrient mixture (DMEM/F-12; 1:1; Life Technologies; Grand Island, NY, USA) containing 1% Gibco™ antibiotic-antimycotic. The following day, drug treatments were administered for 18–24 h; 10 µM Rosiglitazone (Selleckchem, Houston, TX, USA) dissolved in dimethyl sulfoxide (DMSO, Sigma Life Sciences, Burlington, MA, USA), DMSO alone was used as a vehicle control. After the culture period for each treatment, the placental conditioned media was collected, snap frozen and stored at −80 °C. Separate conditioned media controls were generated by culturing DMEM/F-12 media overnight at 8% O_2 with 5% CO_2 at 37 °C with or without 10 µM Rosiglitazone or DMSO to show the effects of placenta conditioned media. These conditioned media controls are used for controls in the tube formation assay.

2.3. Luminex Assay

The angiogenesis Luminex assay (HAGP1MAG, Millipore Sigma, Burlington, MA, USA) is a multiplex antibody-coated bead based fluorescent assay that allows for quantitative assessment of several proteins using only 25 µL of placental conditioned media. The assay was performed according to manufacturer's instruction. Briefly, the placental conditioned media was centrifuged at 4500× *g* at 4 °C to pellet any tissue/cell debris and the supernatant was used for the assay. The individual antibody-bead vials were all combined into one solution and pipetted into a 96-well plate and incubated with the conditioned media samples, internal control standard or analyte standards overnight with rocking at 4 °C. The following day, the supernatant contents were removed (analyte:antibody:bead mixtures were contained in the 96-well plate with a plate magnet during removal of supernatant/wash steps). The plate was washed 3 times with wash buffer and then the plate was incubated with the detection antibodies for 1 h at room temperature. Following this, Streptavidin-phycoerythrin was added to each well and incubated for another 30 min at room temperature. The plate was washed 3 times with wash buffer then the sheath fluid was added to each well and the plate was inserted into the Luminex 200 machine. The machine detects the median fluorescent intensity (MFI) of each analyte to measure how much material is present in each sample (pg/mL). The Luminex 200 software performs a 5-point logistic curve analysis of each sample which is compared relative the standard curve. Data from the internal controls were compared to the expected values provided by the kit to ensure the assay was performed properly.

2.4. Human Umbilical Vein Endothelial Cell Culture and Tube Formation Assay

Human umbilical vein endothelial cells (HUVECs) were purchased from American Type Culture Collection (ATCC, Manassas, VA, USA) and cultured according to manufacture instruction. Briefly, frozen cells were thawed and seeded in a T-75 flask with F-12K medium (ATCC, Manassas, VA, USA) containing 10% fetal bovine serum (FBS; Life Technologies, Carlsbad, CA, USA), 1% Gibco™ antibiotic-antimycotic, 0.1 mg/mL Heparin and 30 µg/mL Endothelial Cell Growth Supplement. Cell culture medium was changed every 2 days and cells were passaged at 60–70% confluency using 0.25% Trypsin-EDTA solution and 10,000 cells were re-seeded into a new flask. The Tube formation assay was performed based on published protocols [30–36]. Briefly, the HUVECs were serum-starved for 6 h prior to the tube formation assay and 2 × 96-well plates were coated with 50 µL of Growth-Factor Reduced Matrigel (Corning, Corning, NY, USA) added to each well and incubated at 37 °C for a minimum of 30 min until gel was solidified. HUVECs were removed from the cell culture flasks with trypsin and counted. A 7 mL solution was generated which contained 500,000 cells per mL. 200 µL of this cell suspension was added to separate 1.5 mL Eppendorf tubes, each tube corresponded to a different treatment, listed in Table 2. 800 µL of the control medium or placental conditioned medium (spun at 4500× *g* at 4 °C for 10 min) was added to the 1.5 mL Eppendorf tubes that each contained 100,000 HUVECs. 100 µL of the HUVEC-conditioned media solution was added per well in the 96-well plate. There was a minimum of 7 wells (technical replicates) for each treatment. The plates were incubated for 18 h in 20% O_2 with 5% CO_2 at 37 °C then phase contrast images were captured on an inverted microscope using the 4× objective. Images were imported into the ImageJ software [37] and the images were analyzed using the Angiogenesis Analyzer macro plugin [30]. To quantify angiogenesis, the images were segmented and skeletonized, and the trees were analyzed to provide quantitative assessment of the number of nodes, junctions, meshes and total branching length present in the HUVECs.

Table 2. Experimental Conditions and Controls for HUVEC Tube Formation Assay.

Experimental Treatment	Culture Conditions
HUVEC Only	Standard culture medium

Table 2. *Cont.*

Experimental Treatment	Culture Conditions
Vehicle Control	Placental culture medium (without human tissue) supplemented with DMSO and cultured for 24 h
Rosiglitazone Control	Placental culture medium (without human tissue) supplemented with Rosiglitazone and cultured for 24 h
Conditioned medium from non-treated preeclamptic placentas	Preeclamptic tissue cultured in placental culture medium for 24 h
Conditioned medium from preeclamptic placentas treated with vehicle	Preeclamptic tissue cultured in placental culture medium for 24 h with DMSO
Conditioned medium from preeclamptic placentas treated with Rosiglitazone	Preeclamptic tissue cultured in placental culture medium for 24 h with Rosiglitazone
Conditioned medium from non-treated healthy control placentas	Healthy placentas cultured in placental culture medium from 24 h

2.5. Statistical Analysis

All statistical analysis was performed with GraphPad Prism 7.0 software. Raw expressions were analyzed by student's *t*-test, after determination if samples are normally distributed and an F-test was applied to determine variances between groups which was then used in the parameters for the *t*-test. $p < 0.05$ is considered significant and is indicated with (*) on each graph. Data is reported as Mean ± S.E.M. All sample numbers are reported as per group, for example, n = 6 designates 6 samples per treatment/group.

3. Results

3.1. Rosiglitazone Has a Significant Impact on Angiogenic and Growth Factor Protein Secretion from the Preeclamptic Placenta

To test if placental activation of PPARγ influences angiogenic protein secretion, preeclamptic placentas treated with or without Rosiglitazone and vehicle, as well as non-treated healthy placentas were cultured for 24 h, and the conditioned medium was used for Luminex assay. Our luminex data shows no significant differences in Angiopoietin-2 (Ang-2) secretion levels between the non-treated control and non-treated PE placentas (2532 ± 517.6 pg/mL vs. 2365 ± 303.7 pg/mL, n > 10, $p > 0.05$, Figure 1A). However, we observed a significant reduction in Ang-2 secretion in Rosiglitazone-treated PE placentas when compared to the vehicle (1671 ± 223 pg/mL vs. 3086 ± 407 pg/mL, n = 14, $p = 0.04$, Figure 1B).

We observed a significant upregulation of soluble Endoglin (sEng) from the preeclamptic placentas in comparison to the healthy control placentas (2017 ± 364 pg/mL vs. 836 ± 135 pg/mL, n > 10, $p = 0.0164$, Figure 2A). We also observed a reduction of sEng in the Rosiglitazone-treated PE placentas in comparison to the vehicle control however it was not statistically significant (745 ± 161.5 pg/mL vs. 1242 ± 179 pg/mL, n = 14, $p = 0.0537$, Figure 2B).

We did not observe a significant difference in Endothelin-1 (ET-1) secretion from control compared to preeclamptic placentas (2.2 ± 0.2 pg/mL vs. 1.9 ± 0.25 pg/mL, n > 10, $p > 0.05$, Figure 3A). Further, there was no change in ET-1 secretion from Rosiglitazone-treated PE placentas compared to the vehicle control (1.83 ± 0.15 pg/mL vs. 1.74 ± 0.19 pg/mL, n = 14, $p > 0.05$, Figure 3B).

We observed a decreasing trend of placental growth factor (PlGF) secretion from PE placentas however, this was not statistically significant in comparison to the control placentas (2.913 ± 0.82 pg/mL vs. 7.12 ± 2.3 pg/mL, n > 10, $p = 0.13$, Figure 4A). We observed an increase of PlGF secretion in the Rosiglitazone-treated PE placentas however, it was not statistically significant in comparison to the vehicle (4.765 ± 1 pg/mL vs. 2.668 ± 0.6 pg/mL, n = 14, $p = 0.07$, Figure 4B).

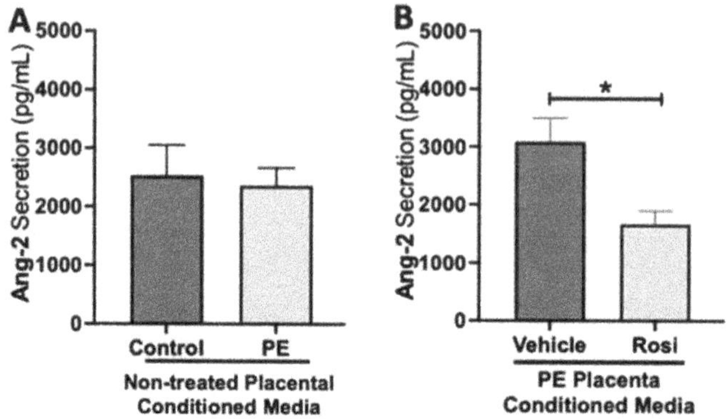

Figure 1. Angiopoeitin-2 secretion is reduced in Rosiglitazone-treated preeclamptic placentas. Angiopoeitin-2 (Ang-2) levels were measured via Luminex assay from conditioned media from non-treated control and preeclamptic (PE) placentas (**A**) and vehicle- or Rosiglitazone (Rosi)-treated PE placentas (**B**). There was no significant difference in Ang-2 secretion between PE and control placentas (**A**, n > 10). Rosi-treated PE placentas show a significant reduction of Ang-2 secretion in comparison to the vehicle control (**B**, n = 14). (Protein secretion was measured by a Luminex assay where experimental values were determined relative to a standard curve. Statistical analysis was performed by student's *t*-test to determine significant differences between groups, * $p < 0.05$, bar plots and data reported are reported as mean pg/mL values ± SEM).

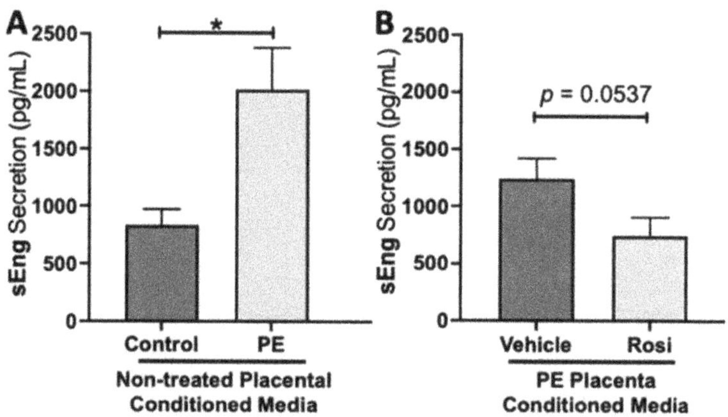

Figure 2. Soluble Endoglin secretion is increased in the preeclamptic placenta but reduced after Rosiglitazone treatment. Soluble Endoglin (sEng) levels were measured via Luminex assay from conditioned media from non-treated control and preeclamptic (PE) placentas (**A**) and vehicle- or Rosiglitazone (Rosi)-treated PE placentas (**B**). There was a significant upregulation of sEng secretion in PE compared to control placentas (**A**, n > 10). Rosi-treated PE placentas show a reduction of sEng secretion however this was not statistically significant when compared to the vehicle control (**B**, n = 14). (Protein secretion was measured by a Luminex assay where experimental values were determined relative to a standard curve. Statistical analysis was performed by student's *t*-test to determine significant differences between groups, * $p < 0.05$, bar plots and data reported are reported as mean pg/mL values ± SEM).

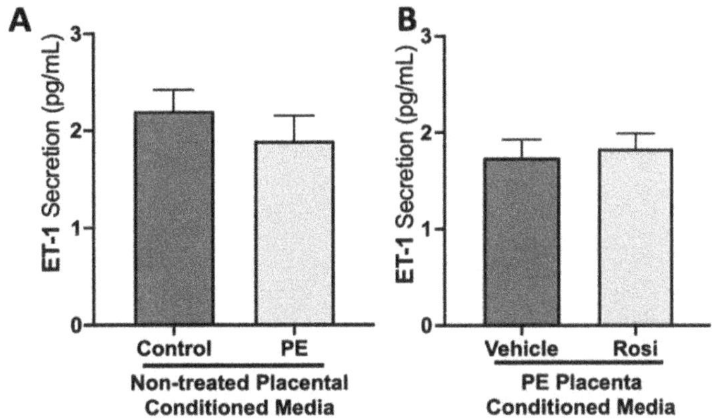

Figure 3. There are no significant differences in Endothelin-1 secretion from healthy or preeclamptic placentas with or without drug treatment. Secretion of Endothelin-1 was measured via Luminex assay from conditioned media from non-treated control and preeclamptic (PE) placentas (**A**) and vehicle- or Rosiglitazone (Rosi)-treated PE placentas (**B**). There was no significant change in ET-1 secretion between PE and control placentas (**A**, n > 10) and between vehicle and Rosi-treated PE placentas in the PE compared to control placentas however this was not statistically significant (**B**, n = 14). (Protein secretion was measured by a Luminex assay where experimental values were determined relative to a standard curve. Statistical analysis was performed by student's *t*-test to determine significant differences between groups, bar plots and data reported are reported as mean pg/mL values ± SEM).

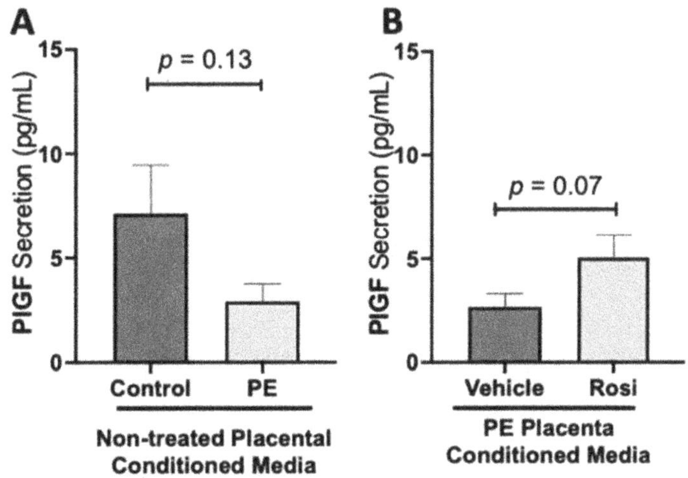

Figure 4. There is a decreasing trend in placental growth factor secretion in the preeclamptic placenta that is partially rescued by Rosiglitazone treatment. Secretion of Placental growth factor (PlGF) were measured via Luminex assay from conditioned media from non-treated control and preeclamptic (PE) placentas (**A**) and vehicle- or Rosiglitazone (Rosi)-treated PE placentas (**B**). There was a trending decrease of PlGF secretion in the PE compared to control placentas however this was not statistically significant (**A**, n > 10). Rosi-treated PE placentas show an increase of PlGF secretion however this was not statistically significant when compared to the vehicle control (**B**, n = 14). (Protein secretion was measured by a Luminex assay where experimental values were determined relative to a standard curve. Statistical analysis was performed by student's *t*-test to determine significant differences between groups, bar plots and data reported are reported as mean pg/mL values ± SEM).

While we observed a decreasing trend of fibroblast growth factor (FGF-2) secretion from preeclamptic placentas, this was not statistically significant change in FGF-2 secretion in preeclamptic compared to healthy control placentas (977 ± 266 pg/mL vs. 646 ± 96 pg/mL, n > 10, p = 0.12, Figure 5A). However, Rosiglitazone caused a significant increase of FGF-2 secretion in PE placentas compared to vehicle treatment (1041 ± 121 pg/mL vs. 649 ± 97 pg/mL, n = 14, p = 0.04, Figure 5B).

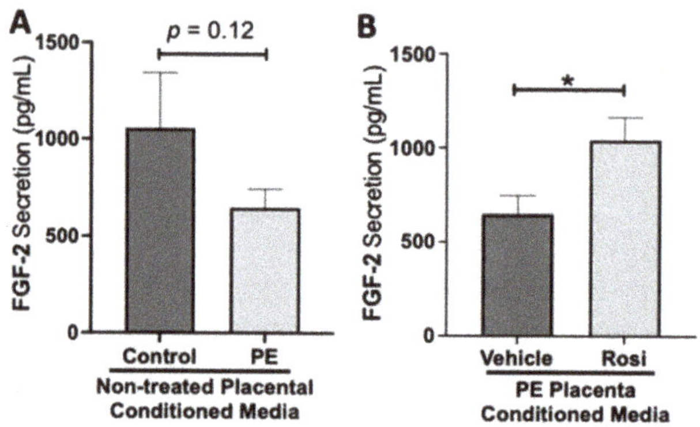

Figure 5. Fibroblast Growth Factor 2 shows reduced secretion from the preeclamptic placenta that is reversed by Rosiglitazone treatment. Secretion of Fibroblast Growth Factor 2 (FGF-2) was measured via Luminex assay from conditioned media from non-treated control and preeclamptic (PE) placentas (**A**) and vehicle- or Rosiglitazone (Rosi)-treated PE placentas (**B**). FGF-2 secretion appears to be reduced in preeclamptic placenta however this decrease is not statistically different from control placentas (**A**, n > 10). Rosi-treated PE placentas show a significant increase in FGF-2 secretion compared to the vehicle control (**B**, n = 14). (Protein secretion was measured by a Luminex assay where experimental values were determined relative to a standard curve. Statistical analysis was performed by student's t-test to determine significant differences between groups, * $p < 0.05$, bar plots and data reported are reported as mean pg/mL values ± SEM).

There was no significant change in epidermal growth factor (EGF) secretion between preeclamptic and control placentas (1.38 ± 0.15 pg/mL vs. 1.84 ± 0.27 pg/mL, n > 10, p > 0.05, Figure 6A). There were also no measurable changes of EGF secretion from Rosiglitazone-treated PE placentas in comparison to vehicle-treated PE placentas (2.1 ± 0.25 pg/mL vs. 1.8 ± 0.22 pg/mL, n = 14, p > 0.05, Figure 6B).

Our data indicates that secretion of heparin-binding epidermal growth factor (HB-EGF) is significantly reduced in the PE placenta in comparison to controls (46.5 ± 6.4 pg/mL vs. 119 ± 38 pg/mL, n > 10, p = 0.05, Figure 7A). However, HB-EGF secretion was restored when treated with Rosiglitazone in comparison to the vehicle control (71.6 ± 4.9 pg/mL vs. 44.8 ± 6 pg/mL, n = 14, p = 0.0027, Figure 7B).

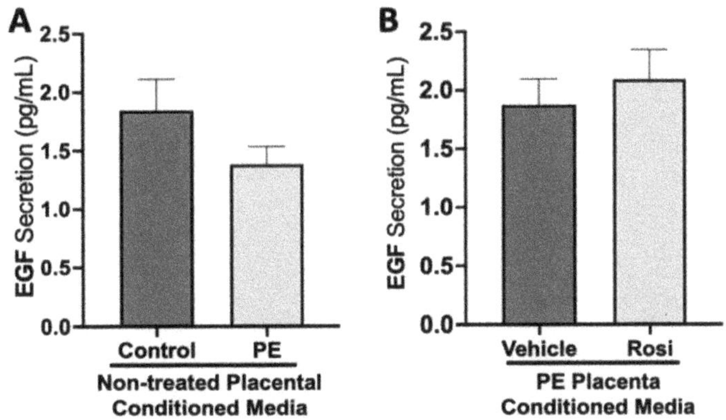

Figure 6. There are no significant changes in Epidermal Growth Factor secretion between healthy and preeclamptic placentas treated with or without Rosiglitazone. Secretion of Epidermal Growth Factor (EGF) was measured via Luminex assay from conditioned media from non-treated control and preeclamptic (PE) placentas (**A**) and vehicle- or Rosiglitazone (Rosi)-treated PE placentas (**B**). There was no significant change in EGF secretion between PE and control placentas (**A**, n > 10) and between vehicle and Rosi-treated PE placentas in the PE compared to control placentas however this was not statistically significant (**B**, n = 14). (Protein secretion was measured by a Luminex assay where experimental values were determined relative to a standard curve. Statistical analysis was performed by student's t-test to determine significant differences between groups, bar plots and data reported are reported as mean pg/mL values ± SEM).

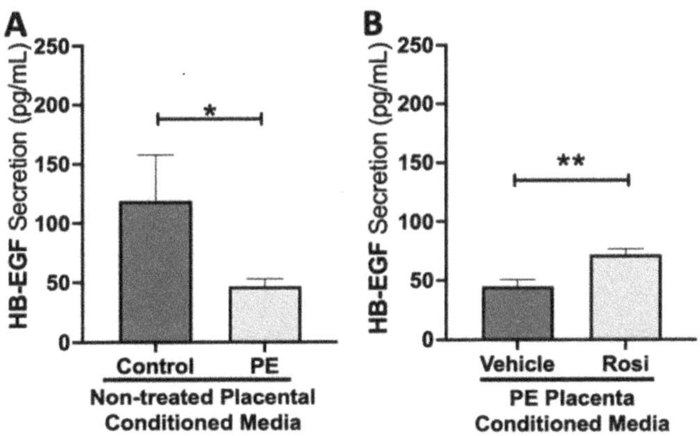

Figure 7. Heparin-Binding Epidermal Growth Factor shows reduced secretion from the preeclamptic placenta but is reversed by Rosiglitazone treatment. Secretion of Heparin-Binding Epidermal Growth Factor (HB-EGF) was measured via Luminex assay from conditioned media from non-treated control and preeclamptic (PE) placentas (**A**) and vehicle- or Rosiglitazone (Rosi)-treated PE placentas (**B**). There was a significant reduction of HB-EGF secretion from the PE placenta compared to control (**A**, n > 6). Rosi-treated PE placentas show a significant increase in HB-EGF secretion compared to the vehicle control (**B**, n = 10). (Protein secretion was measured by a Luminex assay where experimental values were determined relative to a standard curve. Statistical analysis was performed by student's *t*-test to determine significant differences between groups, * $p < 0.05$, ** $p < 0.01$, bar plots and data reported are reported as mean pg/mL values ± SEM).

Our data shows a significant reduction of Follistatin (FST) secretion from the PE placenta in comparison to control (89 ± 9 pg/mL vs. 63 ± 6 pg/mL, n > 10, $p = 0.034$,

Figure 8A). Rosiglitazone caused a significant increase in FST secretion from the PE placenta in comparison to the vehicle control (89 ± 10 pg/mL vs. 63 ± 5 pg/mL, n = 14, $p = 0.044$, Figure 8B).

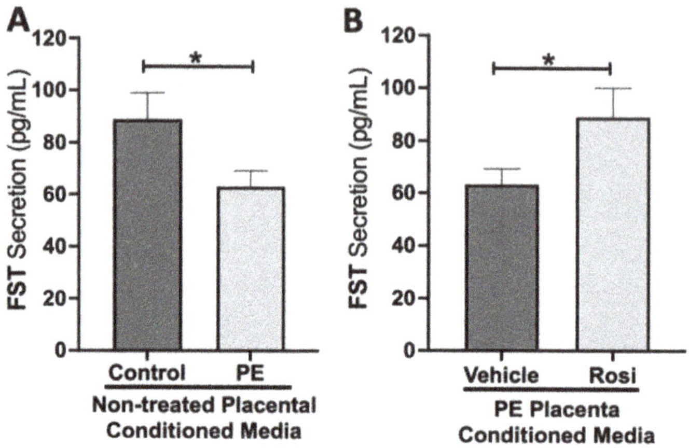

Figure 8. Follistatin shows reduced secretion from the preeclamptic placenta but is reversed by Rosiglitazone. Secretion of Follistatin (FST) was measured via Luminex assay from conditioned media from non-treated control and preeclamptic (PE) placentas (**A**) and vehicle- or Rosiglitazone (Rosi)-treated PE placentas (**B**). FST secretion was significantly reduced in the PE placenta compared to control placentas (**A**, n > 10) however, Rosi treatment led to a significantly increased secretion of FST from the PE placenta compared to vehicle-treated PE placentas (**B**, n = 14). (Protein secretion was measured by a Luminex assay where experimental values were determined relative to a standard curve. Statistical analysis was performed by student's t-test to determine significant differences between groups, * $p < 0.05$, bar plots and data reported are reported as mean pg/mL values ± SEM).

There appeared to be a greater secretion of Leptin from the PE placenta in comparison to the control placentas although this was not statistically significant (2889 ± 1047 pg/mL vs. 1517 ± 514 pg/mL, n > 10, $p > 0.05$, Figure 9A). Rosiglitazone-treated PE placentas did not show a significant change in Leptin secretion in comparison to the vehicle-treated PE placentas (4146 ± 1320 pg/mL vs. 3376 ± 1651 pg/mL, n = 14, $p > 0.05$, Figure 9B).

3.2. Tube Formation Assays Reveals a Pro-Angiogenic Effect from Rosiglitazone-Treated Preeclamptic Placentas

The endothelial tube formation is an assessment of angiogenesis through the measurement of nodes, junctions, meshes, and total branching length of the HUVECs when cultured with placental conditioned media or control media. The nodes from the HUVEC structure represent the location of two endothelial branches. Junctions are determined when a node has three or more branches that are intersecting. The total branching length is quantified as the sum of all the branch lengths per image. The 'mesh' is used to describe the general HUVEC structure (typically appearing in a spider-web like structure) and is calculated by measuring the areas enclosed by the branches. In total, these measurements can relate to the potential for the HUVECs to undergo tube formation, which we refer to as 'angiogenic potential'.

We observed a significant reduction in the number of nodes present in the HUVECs treated with PE conditioned media compared to control placenta conditioned media (165 ± 15 vs. 243 ± 14, n = 6, $p = 0.0004$, Figure 10A). However, there was a statistically significant increase in the number of nodes present in the HUVECs cultured with Rosiglitazone-treated placental conditioned media in comparison to the conditioned media from vehicle-treated placentas (271 ± 30 vs. 165 ± 17, n = 6, $p = 0.0032$, Figure 10B). We observed no significant changes in the number of nodes produced from the HUVEC positive control when compared to the HUVECs cultured with the vehicle and Rosiglitazone controls (139 ± 8 vs. 167 ± 20 vs. 16 9± 12, n = 6, $p > 0.05$, Figure 10C).

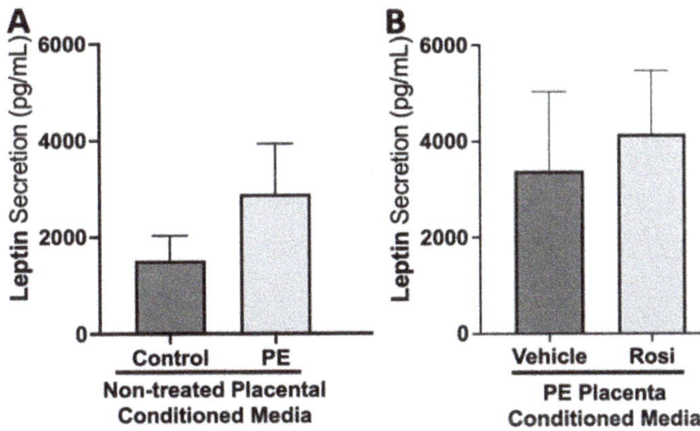

Figure 9. There are no significant changes in Leptin secretion between healthy and preeclamptic placentas with or without drug treatment. Secretion of Leptin was measured via Luminex assay from conditioned media from non-treated control and preeclamptic (PE) placentas (**A**) and vehicle- or Rosiglitazone (Rosi)-treated PE placentas (**B**). Although there was an increasing trend of Leptin secretion from PE placentas, this was not statistically different from the control placentas (**A**, n > 10). There were no significant changes in Leptin secretion between vehicle and Rosi-treated PE placentas (**B**, n = 14). (Protein secretion was measured by a Luminex assay where experimental values were determined relative to a standard curve. Statistical analysis was performed by student's *t*-test to determine significant differences between groups, bar plots and data reported are reported as mean pg/mL values ± SEM).

The number of junctions among the HUVECs was also measured and is shown to be significantly reduced when cultured with conditioned media from non-treated PE compared to control placentas (50 ± 4 vs. 64 ± 3, n = 6, $p = 0.02$, Figure 11A). There was a significant increase in the number of junctions in the HUVECs after culture with conditioned media from Rosiglitazone-treated PE placentas as compared to vehicle-treated PE placentas (87 ± 8 vs. 45 ± 4, n = 6, $p < 0.0001$, Figure 11B). There were no significant differences in the number of junctions shown in the HUVEC Only control as compared to the vehicle and Rosiglitazone conditioned media controls (37 ± 3 vs. 47 ± 6 vs. 46 ± 3, n = 6, $p > 0.05$, Figure 11C).

Our data shows that HUVECs cultured with conditioned media from PE placentas show significantly reduced total branching length in comparison to the conditioned media from the control placentas (5085 ± 414 vs. 6074 ± 257 relative pixel values, n = 6, $p = 0.03$, Figure 12A). However, the conditioned medium from the Rosiglitazone-treated PE placentas led to a significant increase in the total branching length as compared to the conditioned media from the vehicle-treated PE placentas (7994 ± 662 vs. 5538 ± 337 relative pixel values, n = 6, $p = 0.0013$, Figure 12B). There was no significant difference in total branching length between the HUVEC Only control, the vehicle conditioned media, and Rosiglitazone conditioned media controls (5238 ± 621, 6401 ± 554, 5885 ± 336 relative pixel values, n = 6, $p > 0.05$, Figure 12C).

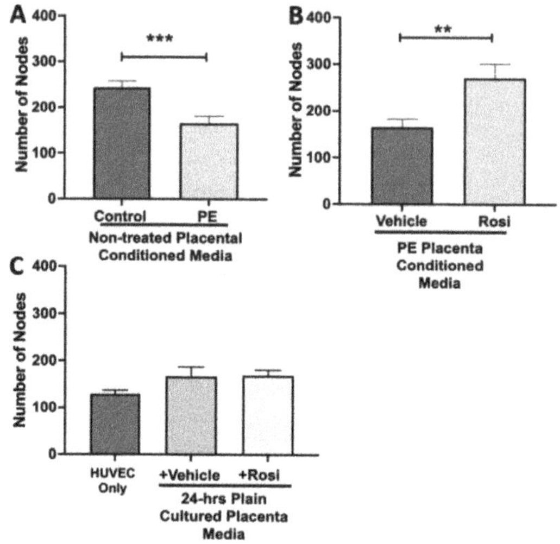

Figure 10. There is a significant reduction in the number of nodes present in the HUVECs cultured with preeclamptic conditioned media, but this is reversed in Rosiglitazone-treated placentas. HUVECs were cultured with conditioned media on matrigel, and the number of nodes present was calculated by the Image J Angiogenesis Analyzer tool (30). HUVECs cultured with conditioned media from preeclamptic (PE) placentas show significantly reduced number of nodes as compared to healthy control placentas (**A**, n =6). Conditioned media from Rosiglitazone (Rosi)-treated PE placentas led to a significant increase in the number of nodes present in the HUVECs compared to the conditioned media from the vehicle-treated PE placentas (**B**, n = 6). Rosi and vehicle were cultured in placental media without any tissues for 24 h then applied to the HUVECs. Additionally, HUVECs cultured with standard full-serum media (HUVEC Only) all served as controls for this experiment. There were no significant changes in the number of nodes present among the HUVEC Only, Rosi, and vehicle conditioned media controls (**C**, n = 6). (Statistical analysis was performed by student's *t*-test to determine significant differences between groups, ** $p < 0.01$, *** $p < 0.001$, bar plots and data reported are reported as numerical values ± SEM).

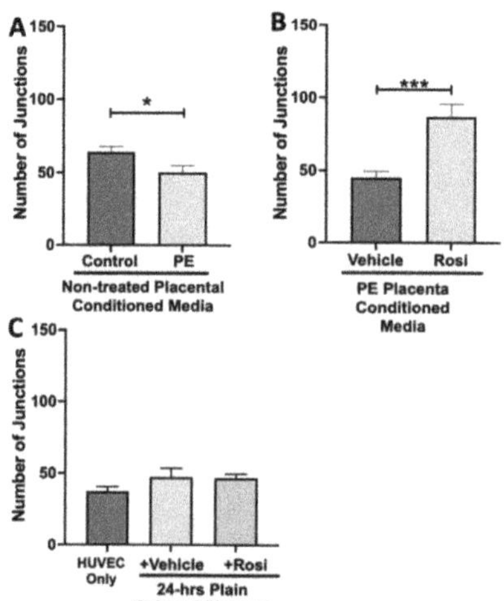

Figure 11. There is a reduction of junctions present in HUVECs cultured with preeclamptic placental

conditioned media and this was significantly increased after culture with Rosiglitazone-treated preeclamptic placentas. HUVECs were cultured with conditioned media on matrigel, and the number of junctions present was calculated by the Image J Angiogenesis Analyzer tool (30). Conditioned media from preeclamptic (PE) placentas show significantly reduced number of junctions as compared to healthy control placentas (**A**, n = 6). Conditioned media from Rosiglitazone (Rosi)-treated PE placentas led to a significant increase in the number of junctions present in the HUVECs compared to the conditioned media from the vehicle-treated PE placentas (**B**, n = 6). Rosi and vehicle were cultured in placental media without any tissues for 24 h then applied to the HUVECs. Additionally, HUVECs cultured with standard full-serum media (HUVEC Only) all as controls for this experiment. There were no significant changes in the number of junctions present among the HUVEC Only, Rosi, and vehicle conditioned media controls (**C**, n = 6). (Statistical analysis was performed by student's t-test to determine significant differences between groups, * $p < 0.05$, *** $p < 0.001$, bar plots and data reported are reported as numerical values ± SEM).

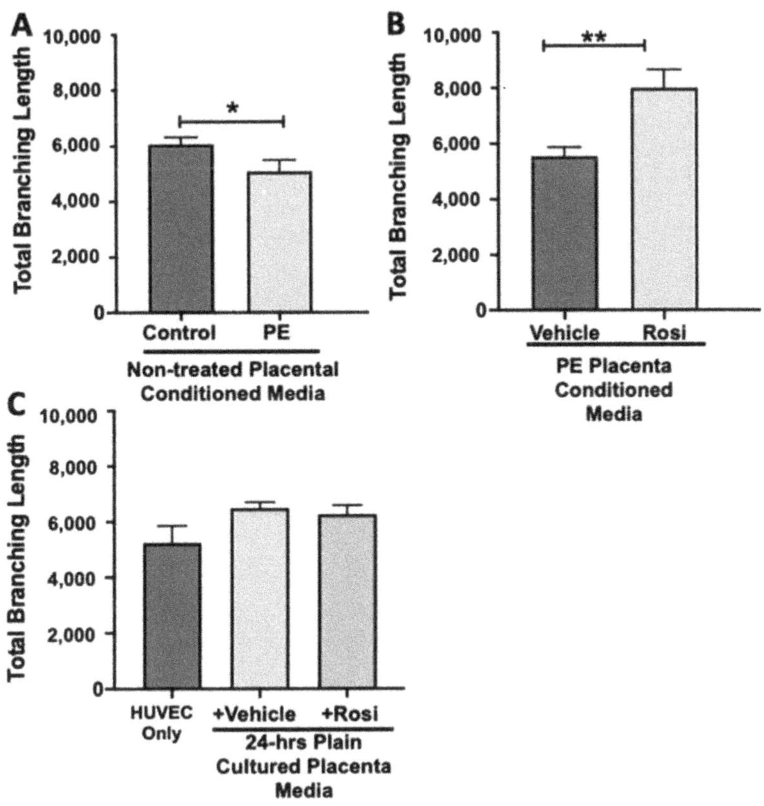

Figure 12. There is a reduction of the total branching length present in HUVECs cultured with preeclamptic placental conditioned media, but this was significantly increased after culture with Rosiglitazone-treated preeclamptic placentas. HUVECs were cultured with conditioned media on matrigel, and the total branching length was calculated by the Image J Angiogenesis Analyzer tool (30). Conditioned media from preeclamptic (PE) placentas show significantly reduced total branching length as compared to healthy control placentas (**A**, n = 6). Conditioned media from Rosiglitazone (Rosi)-treated PE placentas led to a significant increase in the total branching length present in the HUVECs compared to the conditioned media from the vehicle-treated PE placentas (**B**, n = 6). Rosiglitazone and vehicle were cultured in placental media without any tissues for 24 h along with

HUVECs cultured with standard full-serum media (HUVEC Only) all served as controls for this experiment. There were not any significant changes in the total branching length among the HUVEC Only control, Rosi, and vehicle conditioned media controls (**C**, n = 6). (Statistical analysis was performed by student's *t*-test to determine significant differences between groups, * $p < 0.05$, ** $p < 0.01$, bar plots and data reported are reported as numerical values ± SEM).

The number of meshes appear to be significantly reduced when HUVECs were cultured with conditioned media from PE placentas as compared to the conditioned media from control placentas (10 ± 1.5 vs. 15.5 ± 1.6, n = 6, $p = 0.0165$, Figure 13A). Conditioned media from Rosiglitazone-treated PE placentas led to a remarkable increase in the number of meshes present in the HUVECs, as compared to the conditioned media from the vehicle-treated PE placentas (24.5 ± 4 vs. 9.6 ± 1.7, n = 6, $p = 0.0032$, Figure 13B). There were no significant changes in the number of meshes present in the among the HUVEC Only control and the vehicle and Rosiglitazone conditioned media controls (7 ± 2.5, 7 ± 3, 9.7 ± 2.3, n = 6, $p > 0.05$, Figure 13C).

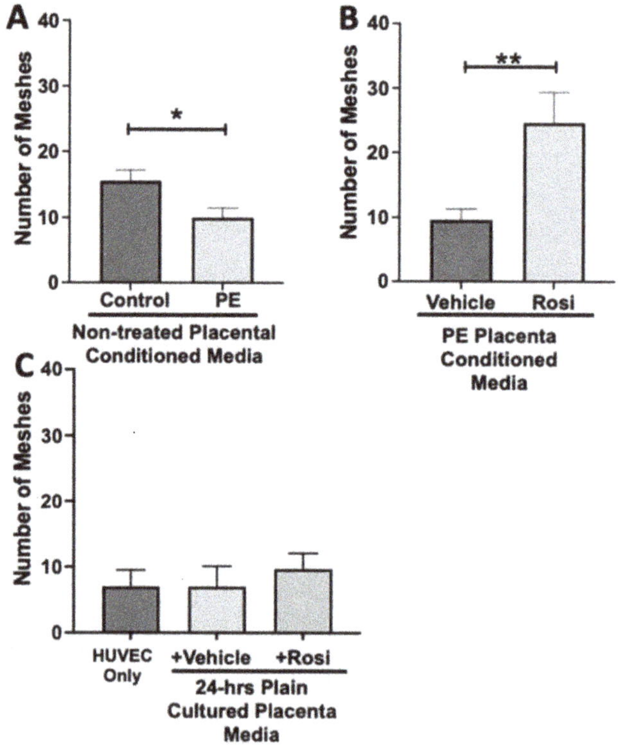

Figure 13. There is a reduction of the total number of meshes present in HUVECs from culture with conditioned media from preeclamptic placentas, but this was significantly increased after culture with Rosiglitazone-treated preeclamptic placentas. HUVECs were cultured with conditioned media on matrigel, and the number of meshes was calculated by the Image J Angiogenesis Analyzer tool (30).

Conditioned media from preeclamptic (PE) placentas show significantly reduced the number of meshes as compared to healthy control placentas (**A**, n = 6). Conditioned media from Rosiglitazone (Rosi)-treated PE placentas led to a significant increase in the number of meshes present in the HUVECs compared to the conditioned media from the vehicle-treated PE placentas (**B**, n = 6). Rosi and vehicle were cultured in placental media without any tissues for 24 h along with HUVECs cultured with standard full-serum media all served as controls for this experiment. There were not any significant changes in the number of meshes among the HUVEC Only control, Rosi, and vehicle conditioned media controls (**C**, n = 6). (Statistical analysis was performed by student's *t*-test to determine significant differences between groups, * $p < 0.05$, ** $p < 0.01$, bar plots and data reported are reported as numerical values ± SEM).

Representative images of the HUVECs with each treatment correlate with the reduction of angiogenic potential that was observed in the cells incubated with the conditioned media from PE placentas in comparison to the HUVECs cultured with conditioned media from control placentas (Figure 14A,B). There is a visible increase of tube formation observed in the HUVECs that were cultured with conditioned media from Rosiglitazone-treated PE placentas as compared to HUVEC culture with conditioned media from vehicle-treated PE placentas (Figure 14C,D). There appears to be no visible differences in tube formation in the HUVEC Only control or in the HUVECs cultured with either the vehicle-conditioned media control or Rosiglitazone-conditioned media control (Figure 14E–G).

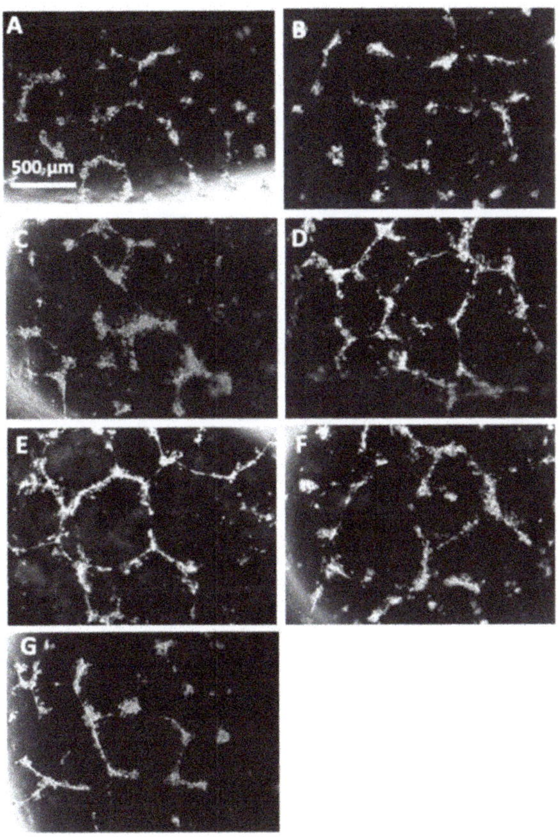

Figure 14. Representative images of HUVEC tube formation assays. Conditioned media from non-treated

control (**A**) and preeclamptic (PE) (**B**) placentas, vehicle-treated PE placentas (**C**), Rosiglitazone (Rosi)-treated PE placentas (**D**), HUVEC Only control (**E**) and Rosi (**F**) and vehicle conditioned media controls (**G**) were cultured with HUVECs on matrigel and images were captured after 18 h of culture. Images were captured using a 4× magnification and then uploaded to the ImageJ Angiogenesis Analyzer to measure various parameters of tube formation to indicate which conditions permit the greatest angiogenic potential of the HUVECs. Scale bar represents 500 µm.

4. Discussion

Endothelial dysfunction is a major hallmark of PE which can cause severe symptoms in the mother that pose long-term risk of cardiovascular disease. In early onset PE and severe PE, the hypoxic and ischemic nature of the placenta is hypothesized to be a major contribution to the aberrant secretion of angiogenic and growth factor proteins that result in endothelial dysfunction [12,38–40]. There has been considerable evidence to show that PPARγ not only improves trophoblast function in ischemic placentas [22,23,41–44], but it can also influence the secretion of proteins that are important for maintaining an angiogenic balance, such as sFLT1 [25].

We questioned whether altered PPARγ activity and expression also drives aberrant placental protein secretion that leads to an overall anti-angiogenic state in the surrounding endothelium. We hypothesized through restoring placental expression of PPARγ, this could rescue the imbalance of angiogenic/growth factor proteins secreted from the PE placenta to subsequently lead to improved angiogenesis in the endothelium. To test this, we measured multiple angiogenic/growth factor proteins from healthy control and PE placentas as well as PE placentas that were treated with Rosiglitazone or vehicle to better understand the secretory profile in control versus PE placentas and to learn how these factors are influenced from placental activation of PPARγ. We further cultured human umbilical vein endothelial cells with conditioned media from these pregnancies to understand the overall 'angiogenic potential' of the secreted factors.

Angiopoietin 2 (Ang-2) is an important angiogenic protein that belongs to the angiopoietin/Tie2 pathway and is necessary for endothelial cell (EC) survival, maturation and morphogenesis [45]. Ang-2 is most known for its anti-angiogenic roles through serving as an antagonist to Angiopoeitin-1 (Ang-1) by competing for interaction with the EC surface receptor, Tie2 [45–47]. Ang-1 is very important for the reorganization of ECs, promoting structural integrity of the blood vessels and preventing EC leakage and migration of leukocytes to surrounding tissues by inhibiting EC activation [45], whereas blocking these effects through Ang-2 can largely contribute to vascular disease. In pregnancy, Ang-2 is mainly produced by the placenta and regulates EC survival, angiogenic sprouting and vascular regression [45,46]. Our Luminex data shows there is not much change in Ang-2 secretion between control and PE placentas. These results do not clarify prior reports in the literature. Some studies report that maternal plasma Ang-2 levels are increased in healthy pregnancies, as compared to non-pregnancy and post-partum women [48]. However, there have been conflicting evidence reporting Ang-2 levels in PE. Some studies show that maternal blood plasma Ang-2 levels are decreased in PE [48] while others describe an increase of Ang-2 placental mRNA expression and higher maternal Ang-2 plasma levels in PE compared to healthy pregnancies [49]. Furthermore, additional research suggests that measuring the Ang-1/Ang-2 ratio could be a method for predicting sePE onset, as the Ang-1/Ang-2 ratio has been shown to decrease during 25–28 weeks of gestation in women who later developed sePE [50]. Prior to this study, there has been little investigation on the role for PPARγ in regulating the expression of Ang-2. Our study is the first to report that placental activation of PPARγ by Rosiglitazone leads to a significant reduction of Ang-2 protein secretion. While this may seem like an exciting result, some studies have shown that intrauterine growth restricted (IUGR) pregnancies are associated with reduced Ang-2 levels, which could potentially interfere with placental angiogenesis [46]. More research is needed to understand the effects of PPARγ on Ang-2 and to determine healthy versus pathologic expression of Ang-2 in pregnancy.

Endothelial vasodilation is a crucial aspect to maintaining a steady and low-pressure flow of maternal blood to the implantation site that, when disrupted, can contribute to placental ischemia [51]. Soluble Endoglin (sEng) significantly impairs vasodilation and our Luminex data is in accordance with the literature in showing that there is a significant upregulation of sEng from the preeclamptic placentas in comparison to the healthy control placentas [52,53]. We excitedly found that sEng secretion is dampened in the PE placenta through PPARγ activation. This could be very beneficial in PE due to the significant contributions of sEng on endothelial dysfunction in PE.

Endothelin-1 (ET-1) is another potent vasoconstrictive molecule that is known to be elevated in PE and contribute to endothelial dysfunction in multiple cardiovascular diseases [54–56]. We surprisingly did not observe a significant difference in ET-1 secretion from control placentas compared to preeclamptic placentas. One explanation for a lack of increased secretion in PE could be due to the placenta not serving as the primary source of ET-1 during pregnancy. Due to Rosiglitazone-treated PE placentas not showing any change in ET-1 levels, this result could suggest that placental PPARγ activation does not affect ET-1 secretion. There is little research that has investigated PPARγ regulation of ET-1 in the human placenta. However, there are reports in the literature that suggest in endothelial cells, PPARγ can regulate upstream pathways of ET-1 such as though increasing the expression of ET-1-inhibiting miRNAs, which result in reduced ET-1 expression [57]. It has also been shown that treatment with PPARγ agonists led to inactivation of the Activating Protein-1 (AP1) pathway which then led to transcriptional inactivation of ET-1 and subsequent reduction of ET-1 secretion [58]. It is possible that PPARγ activation could affect ET-1 at the transcript levels, but this may not be great enough to measure changes in protein secretion.

Placental growth factor (PlGF) is another major factor in the regulation of angiogenesis [59]. While our data is not statically significant, it does follow a pattern of reduced PlGF secretion from the PE placenta and is increased in PE placentas treated with Rosiglitazone. Our data supports findings in the literature that discuss PlGF being downregulated in PE [60]. There is considerable evidence to suggest PPARγ may be able to indirectly regulate PlGF, due to the role for PPARγ to regulate the expression of GCM1 which acts directly upstream of PlGF to induce transcription [61–63].

Fibroblast growth factor-2 (FGF-2) has significant roles in regulating angiogenesis both in the endothelium and in the placenta [64]. FGF-2 has a direct role in the production of NO, which is very important since NO is one of the main vasodilatory agents in the placenta that contributes to regulation of trophoblast invasion, uterine vascular remodeling, and placental perfusion [65]. Our data shows that FGF-2 is decreased in PE. Although this result was not statistically significant, it does correlate with reports in the literature stating that women with sPE are shown to have significantly reduced blood serum levels of FGF-2 compared to women of healthy pregnancies [66]. Given the many roles of FGF-2 in the placenta and endothelium, increasing its production in PE would likely have many beneficial effects. Previous studies have shown that other PPARγ agonists in the TZD drug-family are shown to increase FGF-2 secretion from osteoblasts [67], however our study still brings novel information from the upregulation of FGF-2 secretion in the placenta from Rosiglitazone treatment.

Placental function is also mediated through epidermal growth factor (EGF) and heparin-binding epidermal growth factor (HB-EGF) which both act on EGF receptors on the trophoblast and in the decidua [68–71]. There were no significant changes in EGF secretion between PE and control placentas, which is contradictory to reports in the literature that state EGF is decreased in PE [72]. It is however possible that the placenta may not be the main source of EGF production which could explain our results. We did observe significant reductions of HB-EGF from the PE placenta which is in accordance with reports in the literature [68,70]. The reduction of HB-EGF in term PE placentas may contribute to the enhanced trophoblast apoptosis. Fortunately, our data shows that Rosiglitazone treatment significantly increase HB-EGF secretion from the placenta, which could help to promote trophoblast cell survival in the PE placenta. This is an exciting finding can also confirm data reported by Kushwaha et al., who previously found that Rosiglitazone can increase HB-EGF in astrocytes [73].

Follistatin (FST) has important roles throughout the menstrual cycle such as through preventing hormone release to prevent follicular development [74–76]. Throughout pregnancy, FST levels generally increase then decrease towards normal levels within a few days following parturition [76]. We observed a significant reduction of FST from the PE placenta. While there is little data to show FST levels in during PE, it is known that FST levels are reduced during miscarriages [77,78] and implantation failure in IVF [79]. We observed a significant increase in FST levels when the PE placenta was treated with Rosiglitazone. This was a surprising finding, due to reports stating that PPARγ activation downregulates FST in intestinal epithelial cells [80]. The potential for PPARγ-upregulation of follistatin should be further investigated not only in the PE placenta, but also in the first trimester based on the known importance of FST functions in early pregnancy.

Leptin is an important metabolic molecule that is known to be increased throughout healthy pregnancy [81,82]. We did not observe significant changes in leptin expression between PE and control placentas. Reports of leptin measurements in PE pregnancies do not all follow one pattern. Some studies mention there is an increase of leptin in women with late on-set PE, while others mention that early on-set PE have greater leptin expression even compared to late on-set PE [83]. There were no significant changes for leptin secretion based on Rosiglitazone treatment, which is surprising because other studies have reported that PPARγ and leptin can both enact on each other to reduce each other's expressions in chondrocytes [84].

Collectively, the Luminex assay results supports the notion that the PE placenta exhibits greater secretions of anti-angiogenic proteins compared to controls, evidenced by the increase in sEng secretion and the decrease in PlGF, FGF-2 and HB-EGF secretion. We found that placental activation of PPARγ has an overall beneficial effect on the angiogenic profile through the reduction of Ang-2 and sEng and the upregulation of PlGF, FGF-2, HB-EGF and FST. To greater understand the impact of the angiogenic secretory profile influenced by placental activation of PPARγ, we used the conditioned medium from these placentas in culture with HUVECs.

Endothelial cells undergo angiogenesis to form new blood vessels from existing blood vessels which occur in multiple conditions, such as hypoxia and during wound healing [30]. Angiogenesis must be initiated by a stimulus, often VEGF-A, which causes endothelial activation, degradation of the basement membrane, proliferation, and migration of the cells to form into tube-like structures. Using the Angiogenesis Analyzer tool [30], the phase contrast images captured from the tube formation assay are transformed and characteristic points and elements from the images are extracted and quantified. The nodes from the HUVEC structure represent the location of two branches and junctions are determined as a node with three or more branches that are intersecting. The total branching length is quantified as the sum of all the branch lengths per image. The 'mesh' is used to describe the capillary-like HUVEC structure and is measured by the areas enclosed by the branches. These measurements are often used in the literature to describe the potential for the HUVECs to undergo tube formation [85], which we refer to as 'angiogenic potential'.

Our data shows a very clear pattern of reduced number of nodes, junctions, total branching length and meshes in the HUVECs which were cultured with conditioned media from preeclamptic placentas. These data are not a surprise and confirm the claims presented in the literature that state the preeclamptic placenta causes an overall anti-angiogenic state. Remarkably, we saw that the Rosiglitazone-treated PE placentas cause overall greater number of nodes, junctions, total branching length and meshes, in comparison to the vehicle-treated PE placentas. This finding further validated our Luminex findings which had suggested an increase towards pro-angiogenic state of the PE placenta when treated with Rosiglitazone.

We used several controls to ensure the changes in the HUVECs were due to indirect effects from the human placenta and not from residual drugs retained in the conditioned media (please see Table 2). To determine if there could be an effect on the HUVECs from any potential Rosiglitazone or vehicle retained in the placental conditioned media, we generated drug controls that consisted of human placenta culture media (without tissue) that was cultured overnight with Rosiglitazone or vehicle and then was applied to the HUVECs. We did not observe significant effects on the HUVECs from these conditioned media drug controls, thus we are confident the effects we did observe were due to proteins secreted from the human placenta. Moreover, HUVECs were serum-starved for 6 h prior to culture with the placental conditioned media, which was intended to halt HUVEC cell growth and proliferation, which further supports our findings that there was a placental-dependent effect on the HUVECs. In the 'HUVEC Only' samples, we applied the standard HUVEC culture medium that contained 10 % FBS and the required growth supplements, and we naturally observed that these cells formed capillary-tube-like structures.

To our knowledge, we report novel findings of PPARγ actions in the placenta which have dramatic indirect effects on the endothelium. Future studies should follow up on the ability for PPARγ to modulate secreted proteins from the placenta such as Ang-2, sEng, PlGF, FGF-2, HB-EGF and FST. While VEGF-A was in our panel of markers to investigate, we did not obtain data that was within the standard curve from the Luminex assay. Due to the significant impacts of VEGF-A on the endothelium and initiation of angiogenesis, it would be helpful to know if placental activation of PPARγ effects VEGF-A secretion and thus should be investigated in a future follow-up study. Moreover, more detailed studies investigating the overall impact on the endothelial cells are warranted. For example, performing RNA-sequencing of the HUVEC cells would provide significant detail on the molecular mechanisms that are altered in these cells to permit the increased angiogenic potential after exposure to the human placental conditioned media. Lastly, it is important to note that the placenta secretes many proteins and that the combination of a variety of proteins determines the net effect on the endothelial cells. Therefore, we are unable to determine at this point if one or multiple proteins are responsible for the described phenotype. Nevertheless, the overall finding that the preeclamptic placenta can be treated ex vivo to impact the surrounding angiogenic potential deserves further attention due to the clinical significance of dampening endothelial dysfunction in the mother.

Author Contributions: All authors listed contributed significantly to this manuscript. Specific roles include conceptualization, B.G.; methodology, B.G.; formal analysis, B.G.; writing—original draft preparation, B.G.; writing—review and editing, B.G., H.-R.K.-G. and S.D.; supervision, H.-R.K.-G. and S.D.; funding acquisition, S.D. All authors have read and agreed to the published version of the manuscript.

Funding: This research was supported by: National Heart Lung and Blood Institute under grant number R01-HL128628-01 awarded to S.D.; Department of Obstetrics, Gynecology and Reproductive Biology at Michigan State University College of Human Medicine; March of Dimes Foundation Award provided to S.D.; Research support for B.G.'s work in this publication was provided by the Eunice Kennedy Shriver National Institute of Child Health and Human Development Health under Award Number T32HD087166 awarded to Michigan State University AgBio Research. The content is solely the responsibility of the authors and does not necessarily represent the official views of the National Institutes of Health Support.

Institutional Review Board Statement: The Institutional Review Board of Michigan State University approved all consent forms and protocols used in this study, which abide by the National Institutes of Health research guidelines.

Informed Consent Statement: Informed consent was obtained from all subjects involved in the study.

Data Availability Statement: Not applicable.

Acknowledgments: We thank the women who have graciously donated their tissues for this study. We also thank the Spectrum Health Accelerator of Research Excellence (SHARE) biorepository (Spectrum Health, Grand Rapids, MI, USA) for their work in collecting and providing de-identified biospecimens for this study.

Conflicts of Interest: The authors declare no conflict of interest.

References

1. Jeyabalan, A. Epidemiology of preeclampsia: Impact of obesity. *Nutr. Rev.* **2013**, *71*, S18–S25. [CrossRef] [PubMed]
2. ACOG Practice Bulletin No. 202: Gestational Hypertension and Preeclampsia. *Obstet. Gynecol.* **2019**, *133*, e1–e25. [CrossRef]
3. Toal, M.; Chan, C.; Fallah, S.; Alkazaleh, F.; Chaddha, V.; Windrim, R.C.; Kingdom, J.C. Usefulness of a placental profile in high-risk pregnancies. *Am. J. Obstet. Gynecol.* **2007**, *196*, 363.e1–363.e7. [CrossRef] [PubMed]
4. Sibai, B.M. Evaluation and management of severe preeclampsia before 34 weeks' gestation. *Am. J. Obstet. Gynecol.* **2011**, *205*, 191–198. [CrossRef]
5. Staff, A.C. The two-stage placental model of preeclampsia: An update. *J. Reprod. Immunol.* **2019**, *134-135*, 1–10. [CrossRef] [PubMed]
6. Fantone, S.; Mazzucchelli, R.; Giannubilo, S.R.; Ciavattini, A.; Marzioni, D.; Tossetta, G. AT-rich interactive domain 1A protein expression in normal and pathological pregnancies complicated by preeclampsia. *Histochem. Cell Biol.* **2020**, *154*, 339–346. [CrossRef]
7. Regnault, T.; Galan, H.; Parker, T.; Anthony, R. Placental Development in Normal and Compromised Pregnancies—A Review. *Placenta* **2002**, *23*, S119–S129. [CrossRef]
8. Phipps, E.A.; Thadhani, R.; Benzing, T.; Karumanchi, S.A. Pre-eclampsia: Pathogenesis, novel diagnostics and therapies. *Nat. Rev. Nephrol.* **2019**, *15*, 275–289. [CrossRef]
9. Pereira, R.D.; De Long, N.E.; Wang, R.C.; Yazdi, F.T.; Holloway, A.C.; Raha, S. Angiogenesis in the Placenta: The Role of Reactive Oxygen Species Signaling. *BioMed Res. Int.* **2015**, *2015*, 814543. [CrossRef]
10. O'Brien, M.; Baczyk, D.; Kingdom, J.C. Endothelial Dysfunction in Severe Preeclampsia is Mediated by Soluble Factors, Rather than Extracellular Vesicles. *Sci. Rep.* **2017**, *7*, 5887. [CrossRef]
11. McCarthy, F.P.; Drewlo, S.; English, F.A.; Kingdom, J.; Johns, E.J.; Kenny, L.C.; Walsh, S.K. Evidence Implicating Peroxisome Proliferator-Activated Receptor-γ in the Pathogenesis of Preeclampsia. *Hypertension* **2011**, *58*, 882–887. [CrossRef]
12. Maynard, S.E.; Karumanchi, S.A. Angiogenic Factors and Preeclampsia. *Semin. Nephrol.* **2011**, *31*, 33–46. [CrossRef]
13. Armistead, B.; Kadam, L.; Drewlo, S.; Kohan-Ghadr, H.-R. The Role of NFκB in Healthy and Preeclamptic Placenta: Trophoblasts in the Spotlight. *Int. J. Mol. Sci.* **2020**, *21*, 1775. [CrossRef] [PubMed]
14. Opichka, M.A.; Rappelt, M.W.; Gutterman, D.D.; Grobe, J.L.; McIntosh, J.J. Vascular Dysfunction in Preeclampsia. *Cells* **2021**, *10*, 3055. [CrossRef]
15. Ditisheim, A.; Wuerzner, G.; Ponte, B.; Vial, Y.; Irion, O.; Burnier, M.; Boulvain, M.; Pechère-Bertschi, A. Prevalence of Hypertensive Phenotypes After Preeclampsia: A Prospective Cohort Study. *Hypertension* **2018**, *71*, 103–109. [CrossRef] [PubMed]
16. Goel, A.; Maski, M.R.; Bajracharya, S.; Wenger, J.B.; Zhang, D.; Salahuddin, S.; Shahul, S.S.; Thadhani, R.; Seely, E.W.; Karumanchi, S.A.; et al. Epidemiology and Mechanisms of De Novo and Persistent Hypertension in the Postpartum Period. *Circulation* **2015**, *132*, 1726–1733. [CrossRef]
17. Lindheimer, M.D. Preface. In *Chesley's Hypertensive Disorders in Pregnancy*, 4th ed.; Ix-X; Academic Press: Cambridge, MA, USA, 2015.
18. Wang, S.; Dougherty, E.J.; Danner, R.L. PPARγ signaling and emerging opportunities for improved therapeutics. *Pharmacol. Res.* **2016**, *111*, 76–85. [CrossRef] [PubMed]
19. Janani, C.; Ranjitha Kumari, B.D. PPAR gamma gene—A review. *Diabetes Metab. Syndr.* **2015**, *9*, 46–50. [CrossRef]
20. Ahmadian, M.; Suh, J.M.; Hah, N.; Liddle, C.; Atkins, A.R.; Downes, M.; Evans, R.M. PPARγ signaling and metabolism: The good, the bad and the future. *Nat. Med.* **2013**, *19*, 557–566. [CrossRef]
21. Abdelrahman, M.; Sivarajah, A.; Thiemermann, C. Beneficial effects of PPAR-gamma ligands in ischemia-reperfusion injury, inflammation and shock. *Cardiovasc. Res.* **2005**, *65*, 772–781. [CrossRef]
22. Kadam, L.; Kohan-Ghadr, H.R.; Drewlo, S. The balancing act - PPAR-gamma's roles at the maternal-fetal interface. *Syst. Biol. Reprod. Med.* **2015**, *61*, 65–71. [CrossRef] [PubMed]
23. McCarthy, F.P.; Drewlo, S.; Kingdom, J.; Johns, E.J.; Walsh, S.K.; Kenny, L.C. Peroxisome Proliferator-Activated Receptor-gamma as a Potential Therapeutic Target in the Treatment of Preeclampsia. *Hypertension* **2011**, *58*, 280–286. [CrossRef] [PubMed]
24. Levytska, K.; Drewlo, S.; Baczyk, D.; Kingdom, J. PPAR-γ Regulates Trophoblast Differentiation in the BeWo Cell Model. *PPAR Res.* **2014**, *2014*, 637251. [CrossRef] [PubMed]
25. Armistead, B.; Kadam, L.; Siegwald, E.; McCarthy, F.P.; Kingdom, J.C.; Kohan-Ghadr, H.-R.; Drewlo, S. Induction of the PPARγ (Peroxisome Proliferator-Activated Receptor γ)-GCM1 (Glial Cell Missing 1) Syncytialization Axis Reduces sFLT1 (Soluble fms-Like Tyrosine Kinase 1) in the Preeclamptic Placenta. *Hypertension* **2021**, *78*, 230–240. [CrossRef]
26. Drewlo, S.; Czikk, M.; Baczyk, D.; Lye, S.; Kingdom, J. Glial cell missing-1 mediates over-expression of tissue inhibitor of metalloproteinase-4 in severe pre-eclamptic placental villi. *Hum. Reprod.* **2011**, *26*, 1025–1034. [CrossRef]

27. Baczyk, D.; Drewlo, S.; Proctor, L.M.; Dunk, C.; Lye, S.; Kingdom, J. Glial cell missing-1 transcription factor is required for the differentiation of the human trophoblast. *Cell Death Differ.* **2009**, *16*, 719–727. [CrossRef]
28. Baczyk, D.; Dunk, C.; Huppertz, B.; Maxwell, C.; Reister, F.; Giannoulias, D.; Kingdom, J. Bi-potential Behaviour of Cytotrophoblasts in First Trimester Chorionic Villi. *Placenta* **2006**, *27*, 367–374. [CrossRef]
29. Baczyk, D.; Satkunaratnam, A.; Nait-Oumesmar, B.; Huppertz, B.; Cross, J.; Kingdom, J. Complex Patterns of GCM1 mRNA and Protein in Villous and Extravillous Trophoblast Cells of the Human Placenta. *Placenta* **2004**, *25*, 553–559. [CrossRef]
30. Carpentier, G.; Berndt, S.; Ferratge, S.; Rasband, W.; Cuendet, M.; Uzan, G.; Albanese, P. Angiogenesis Analyzer for ImageJ—A comparative morphometric analysis of "Endothelial Tube Formation Assay" and "Fibrin Bead Assay". *Sci. Rep.* **2020**, *10*, 11568. [CrossRef]
31. Wu, C.C.; Chen, Y.C.; Chang, Y.C.; Wang, L.W.; Lin, Y.C.; Chiang, Y.L.; Ho, C.J.; Huang, C.C. Human umbilical vein endothelial cells protect against hypoxic-ischemic damage in neonatal brain via stromal cell-derived factor 1/C-X-C chemokine receptor type 4. *Stroke* **2013**, *44*, 1402–1409. [CrossRef]
32. Wada, Y.; Otu, H.; Wu, S.; Abid, R.; Okada, H.; Libermann, T.; Kodama, T.; Shih, S.; Minami, T.; Aird, W.C. Preconditioning of primary human endothelial cells with inflammatory mediators alters the "set point" of the cell. *FASEB J.* **2005**, *19*, 1914–1916. [CrossRef] [PubMed]
33. Labitzke, R.; Friedl, P. A serum-free medium formulation supporting growth of human umbilical cord vein endothelial cells in long-term cultivation. *Cytotechnology* **2001**, *35*, 87–92. [CrossRef] [PubMed]
34. Potapova, I.A.; Gaudette, G.R.; Brink, P.R.; Robinson, R.B.; Rosen, M.R.; Cohen, I.S.; Doronin, S.V. Mesenchymal Stem Cells Support Migration, Extracellular Matrix Invasion, Proliferation, and Survival of Endothelial Cells In Vitro. *Stem Cells* **2007**, *25*, 1761–1768. [CrossRef] [PubMed]
35. Shi, H.; Zuo, Y.; Navaz, S.; Harbaugh, A.; Hoy, C.K.; Gandhi, A.A.; Sule, G.; Yalavarthi, S.; Gockman, K.; Madison, J.A.; et al. Endothelial cell-activating antibodies in COVID-19. *medRxiv* **2021**. [CrossRef] [PubMed]
36. Angiogenesis Protocols. Third Edition. *Anticancer Res.* **2016**, *36*, 4370.
37. Schneider, C.A.; Rasband, W.S.; Eliceiri, K.W. NIH Image to ImageJ: 25 Years of image analysis. *Nat. Methods* **2012**, *9*, 671–675. [CrossRef]
38. Soto, E.; Romero, R.; Kusanovic, J.P.; Ogge, G.; Hussein, Y.; Yeo, L.; Hassan, S.S.; Kim, C.J.; Chaiworapongsa, T. Late-onset preeclampsia is associated with an imbalance of angiogenic and anti-angiogenic factors in patients with and without placental lesions consistent with maternal underperfusion. *J. Matern. Neonatal Med.* **2011**, *25*, 498–507. [CrossRef]
39. Owaki, Y.; Watanabe, K.; Iwasaki, A.; Saitou, T.; Matsushita, H.; Wakatsuki, A. Placental hypoplasia and maternal organic vascular disorder in pregnant women with gestational hypertension and preeclampsia. *J. Matern. Neonatal Med.* **2019**, *34*, 353–359. [CrossRef]
40. Rana, S.; Lemoine, E.; Granger, J.P.; Karumanchi, S.A. Preeclampsia: Pathophysiology, Challenges, and Perspectives. *Circ. Res.* **2019**, *124*, 1094–1112. [CrossRef]
41. Kohan-Ghadr, H.-R.; A Kilburn, B.; Kadam, L.; Johnson, E.; Kolb, B.L.; Rodriguez-Kovacs, J.; Hertz, M.; Armant, D.R.; Drewlo, S. Rosiglitazone augments antioxidant response in the human trophoblast and prevents apoptosis†. *Biol. Reprod.* **2018**, *100*, 479–494. [CrossRef]
42. Kohan-Ghadr, H.-R.; Kadam, L.; Jain, C.; Armant, D.R.; Drewlo, S. Potential role of epigenetic mechanisms in regulation of trophoblast differentiation, migration, and invasion in the human placenta. *Cell Adhes. Migr.* **2016**, *10*, 126–135. [CrossRef] [PubMed]
43. Kadam, L.; Kilburn, B.; Baczyk, D.; Kohan-Ghadr, H.R.; Kingdom, J.; Drewlo, S. Rosiglitazone blocks first trimester in-vitro placental injury caused by NF-κB-mediated inflammation. *Sci. Rep.* **2019**, *9*, 2018. [CrossRef] [PubMed]
44. McCarthy, F.P.; Walsh, S.K.; Drewlo, S.; Kingdom, J.C.; Johns, E.J.; Kenny, L.C. Peroxisome Proliferator Activated Receptor Gamma Critically Regulates the Risk of Preeclampsia in Rodent Gestation. *Reprod. Sci.* **2011**, *18*, 285.
45. Kappou, D.; Sifakis, S.; Konstantinidou, A.; Papantoniou, N.; Spandidos, D. Role of the angiopoietin/Tie system in pregnancy (Review). *Exp. Ther. Med.* **2015**, *9*, 1091–1096. [CrossRef]
46. Akwii, R.G.; Sajib, M.S.; Zahra, F.T.; Mikelis, C.M. Role of Angiopoietin-2 in Vascular Physiology and Pathophysiology. *Cells* **2019**, *8*, 471. [CrossRef]
47. David, S.; Kümpers, P.; Hellpap, J.; Horn, R.; Leitolf, H.; Haller, H.; Kielstein, J.T. Angiopoietin 2 and Cardiovascular Disease in Dialysis and Kidney Transplantation. *Am. J. Kidney Dis.* **2009**, *53*, 770–778. [CrossRef]
48. Hirokoshi, K.; Maeshima, Y.; Kobayashi, K.; Matsuura, E.; Sugiyama, H.; Yamasaki, Y.; Masuyama, H.; Hiramatsu, Y.; Makino, H. Increase of Serum Angiopoietin-2 During Pregnancy Is Suppressed in Women with Preeclampsia. *Am. J. Hypertens.* **2005**, *18*, 1181–1188. [CrossRef]
49. Han, S.Y.; Jun, J.K.; Lee, C.-H.; Park, J.S.; Syn, H.C. Angiopoietin-2: A Promising Indicator for the Occurrence of Severe Preeclampsia. *Hypertens. Pregnancy* **2010**, *31*, 189–199. [CrossRef]
50. Bolin, M.; Wiberg-Itzel, E.; Wikström, A.K.; Goop, M.; Larsson, A.; Olovsson, M.; Akerud, H. Angiopoietin-1/angiopoietin-2 ratio for prediction of preeclampsia. *Am. J. Hypertens.* **2009**, *22*, 891–895. [CrossRef]
51. Myatt, L.; Webster, R.P. Vascular biology of preeclampsia. *J. Thromb. Haemost.* **2009**, *7*, 375–384. [CrossRef]
52. Ahmed, A. New insights into the etiology of preeclampsia: Identification of key elusive factors for the vascular complications. *Thromb. Res.* **2011**, *127*, S72–S75. [CrossRef]

53. Huynh, D.T.N.; Heo, K.-S. Therapeutic targets for endothelial dysfunction in vascular diseases. *Arch. Pharm. Res.* **2019**, *42*, 848–861. [CrossRef] [PubMed]
54. Saleh, L.; Verdonk, K.; Visser, W.; van den Meiracker, A.H.; Danser, A.H.J. The emerging role of endothelin-1 in the pathogenesis of pre-eclampsia. *Ther. Adv. Cardiovasc. Dis.* **2016**, *10*, 282–293. [CrossRef] [PubMed]
55. Marasciulo, M.M.A.M.A.P.F.L.; Montagnani, M.; Potenza, M.A. Endothelin-1: The Yin and Yang on Vascular Function. *Curr. Med. Chem.* **2006**, *13*, 1655–1665. [CrossRef] [PubMed]
56. Lankhorst, S.; Danser, A.H.; van den Meiracker, A.H. Endothelin-1 and antiangiogenesis. *Am. J. Physiol. Regul. Integr. Comp. Physiol.* **2016**, *310*, R230–R234. [CrossRef] [PubMed]
57. Kang, B.-Y.; Park, K.K.; Kleinhenz, J.M.; Murphy, T.C.; Green, D.E.; Bijli, K.M.; Yeligar, S.M.; Carthan, K.A.; Searles, C.D.; Sutliff, R.L.; et al. Peroxisome Proliferator–Activated Receptor γ and microRNA 98 in Hypoxia-Induced Endothelin-1 Signaling. *Am. J. Respir. Cell Mol. Biol.* **2016**, *54*, 136–146. [CrossRef] [PubMed]
58. Delerive, P.; Martin-Nizard, F.; Chinetti, G.; Trottein, F.; Fruchart, J.-C.; Najib, J.; Duriez, P.; Staels, B. Peroxisome proliferator-activated receptor activators inhibit thrombin-induced endothelin-1 production in human vascular endothelial cells by inhibiting the activator protein-1 signaling pathway. *Circ. Res.* **1999**, *85*, 394–402. [CrossRef]
59. Kliche, S.; Waltenberger, J. VEGF Receptor Signaling and Endothelial Function. *IUBMB Life* **2001**, *52*, 61–66. [CrossRef]
60. McLaughlin, K.; Snelgrove, J.W.; Audette, M.C.; Syed, A.; Hobson, S.R.; Windrim, R.C.; Melamed, N.; Carmona, S.; Kingdom, J.C. PlGF (Placental Growth Factor) Testing in Clinical Practice: Evidence from a Canadian Tertiary Maternity Referral Center. *Hypertension* **2021**, *77*, 2057–2065. [CrossRef]
61. Chang, M.; Mukherjea, D.; Gobble, R.M.; Groesch, K.A.; Torry, R.J.; Torry, D.S. Glial Cell Missing 1 Regulates Placental Growth Factor (PGF) Gene Transcription in Human Trophoblast1. *Biol. Reprod.* **2008**, *78*, 841–851. [CrossRef]
62. Li, S.; Roberson, M.S. Dlx3 and GCM-1 functionally coordinate the regulation of placental growth factor in human trophoblast-derived cells. *J. Cell Physiol.* **2017**, *232*, 2900–2914. [CrossRef]
63. Chiu, Y.-H.; Yang, M.-R.; Wang, L.-J.; Chen, M.-H.; Chang, G.-D.; Chen, H. New insights into the regulation of placental growth factor gene expression by the transcription factors GCM1 and DLX3 in human placenta. *J. Biol. Chem.* **2018**, *293*, 9801–9811. [CrossRef] [PubMed]
64. Arany, E.; Hill, D. Fibroblast growth factor-2 and fibroblast growth factor receptor-1 mRNA expression and peptide localization in placentae from normal and diabetic pregnancies. *Placenta* **1998**, *19*, 133–142. [CrossRef]
65. Özkan, S.; Vural, B.; Fi̇li̇z, S.; Coştur, P.; Dalçik, H. Placental expression of insulin-like growth factor-I, fibroblast growth factor-basic, and neural cell adhesion molecule in preeclampsia. *J. Matern. Neonatal Med.* **2008**, *21*, 831–838. [CrossRef] [PubMed]
66. Hohlagschwandtner, M.; Knöfler, M.; Ploner, M.; Zeisler, H.; Joura, E.A.; Husslein, P. Basic Fibroblast Growth Factor and Hypertensive Disorders in Pregnancy. *Hypertens. Pregnancy* **2002**, *21*, 235–241. [CrossRef]
67. Yasuda, E.; Tokuda, H.; Ishisaki, A.; Hirade, K.; Kanno, Y.; Hanai, Y.; Nakamura, N.; Noda, T.; Katagiri, Y.; Kozawa, O. PPAR-gamma ligands up-regulate basic fibroblast growth factor-induced VEGF release through amplifying SAPK/JNK activation in osteoblasts. *Biochem. Biophys. Res. Commun.* **2005**, *328*, 137–143. [CrossRef]
68. E Leach, R.; Romero, R.; Kim, Y.M.; Chaiworapongsa, T.; Kilburn, B.; Das, S.K.; Dey, S.K.; Johnson, A.; Qureshi, F.; Jacques, S.; et al. Pre-eclampsia and expression of heparin-binding EGF-like growth factor. *Lancet* **2002**, *360*, 1215–1219. [CrossRef]
69. Leach, R.E.; Khalifa, R.; Ramirez, N.D.; Das, S.K.; Wang, J.; Dey, S.K.; Romero, R.; Armant, D.R. Multiple Roles for Heparin-Binding Epidermal Growth Factor-Like Growth Factor Are Suggested by Its Cell-Specific Expression during the Human Endometrial Cycle and Early Placentation1. *J. Clin. Endocrinol. Metab.* **1999**, *84*, 3355–3363. [CrossRef]
70. Jessmon, P.; Leach, R.E.; Armant, D.R. Diverse functions of HBEGF during pregnancy. *Mol. Reprod. Dev.* **2009**, *76*, 1116–1127. [CrossRef]
71. Imudia, A.; Kilburn, B.; Petkova, A.; Edwin, S.; Romero, R.; Armant, D. Expression of Heparin-binding EGF-like Growth Factor in Term Chorionic Villous Explants and Its Role in Trophoblast Survival. *Placenta* **2008**, *29*, 784–789. [CrossRef]
72. Armant, D.; Fritz, R.; Kilburn, B.; Kim, Y.; Nien, J.; Maihle, N.; Romero, R.; Leach, R. Reduced expression of the epidermal growth factor signaling system in preeclampsia. *Placenta* **2014**, *36*, 270–278. [CrossRef] [PubMed]
73. Kushwaha, R.; Mishra, J.; Gupta, A.P.; Gupta, K.; Vishwakarma, J.; Chattopadhyay, N.; Gayen, J.R.; Kamthan, M.; Bandyopadhyay, S. Rosiglitazone up-regulates glial fibrillary acidic protein via HB-EGF secreted from astrocytes and neurons through PPARγ pathway and reduces apoptosis in high-fat diet-fed mice. *J. Neurochem.* **2019**, *149*, 679–698. [CrossRef] [PubMed]
74. Köninger, A.; Schmidt, B.; Damaske, D.; Birdir, C.; Enekwe, A.; Kimmig, R.; Strowitzki, T.; Gellhaus, A. Follistatin during pregnancy and its potential role as an ovarian suppressing agent. *Eur. J. Obstet. Gynecol. Reprod. Biol.* **2017**, *212*, 150–154. [CrossRef] [PubMed]
75. Fullerton, P.T.; Monsivais, D.; Kommagani, R.; Matzuk, M.M. Follistatin is critical for mouse uterine receptivity and decidualization. *Proc. Natl. Acad. Sci. USA* **2017**, *114*, E4772–E4781. [CrossRef] [PubMed]
76. Petraglia, F. Inhibin, activin and follistatin in the human placenta—A new family of regulatory proteins. *Placenta* **1997**, *18*, 3–8. [CrossRef]
77. Prakash, A.; Laird, S.; Tuckerman, E.; Li, T.C.; Ledger, W.L. Inhibin A and activin A may be used to predict pregnancy outcome in women with recurrent miscarriage. *Fertil. Steril.* **2005**, *83*, 1758–1763. [CrossRef]
78. Prakash, A.; Li, T.C.; Tuckerman, E.; Laird, S.; Wells, M.; Ledger, W.L. A study of luteal phase expression of inhibin, activin, and follistatin subunits in the endometrium of women with recurrent miscarriage. *Fertil. Steril.* **2006**, *86*, 1723–1730. [CrossRef]

79. Prakash, A.; Tuckerman, E.; Laird, S.; Ola, B.; Li, T.C.; Ledger, W. A preliminary study comparing the endometrial expression of inhibin, activin and follistatin in women with a history of implantation failure after IVF treatment and a control group. *BJOG: Int. J. Obstet. Gynaecol.* **2008**, *115*, 532–537. [CrossRef]
80. Necela, B.M.; Su, W.; Thompson, E.A. Peroxisome proliferator-activated receptor gamma down-regulates follistatin in intestinal epithelial cells through SP1. *J. Biol. Chem.* **2008**, *283*, 29784–29794. [CrossRef]
81. Ladyman, S.R.; Augustine, R.A.; Grattan, D.R. Hormone interactions regulating energy balance during pregnancy. *J. Neuroendocrinol.* **2010**, *22*, 805–817. [CrossRef]
82. Armistead, B.; Johnson, E.; Vanderkamp, R.; Kula-Eversole, E.; Kadam, L.; Drewlo, S.; Kohan-Ghadr, H.-R. Placental Regulation of Energy Homeostasis During Human Pregnancy. *Endocrinology* **2020**, *161*, bqaa076. [CrossRef] [PubMed]
83. Taylor, B.D.; Ness, R.B.; Olsen, J.; Hougaard, D.M.; Skogstrand, K.; Roberts, J.M.; Haggerty, C. Serum Leptin Measured in Early Pregnancy Is Higher in Women with Preeclampsia Compared with Normotensive Pregnant Women. *Hypertension* **2015**, *65*, 594–599. [CrossRef] [PubMed]
84. Wang, L.; Shao, Y.Y.; Ballock, R.T. Leptin Antagonizes Peroxisome Proliferator-Activated Receptor-γ Signaling in Growth Plate Chondrocytes. *PPAR Res.* **2012**, *2012*, 756198. [CrossRef] [PubMed]
85. Ma, H.; Jiang, S.; Du, L.; Liu, J.; Xu, X.; Lu, X.; Ma, L.; Zhu, H.; Wei, J.; Yu, Y. Conditioned medium from primary cytotrophoblasts, primary placenta-derived mesenchymal stem cells, or sub-cultured placental tissue promoted HUVEC angiogenesis in vitro. *Stem Cell Res. Ther.* **2021**, *12*, 141. [CrossRef]

cells

MDPI

Review

The Potential Role of PPARs in the Fetal Origins of Adult Disease

Jun Guo [2,†], Jue Wu [3,†], Qinyuan He [4], Mengshu Zhang [1], Hong Li [1,*] and Yanping Liu [1,*]

1 Center of Reproduction and Genetics, The Affiliated Suzhou Hospital of Nanjing Medical University, Suzhou Municipal Hospital, Gusu School, Nanjing Medical University, Suzhou 215006, China
2 Institute for Fetology, The First Affiliated Hospital of Soochow University, Suzhou 215006, China
3 Department of Obstetrics and Gynecology, Suzhou Dushu Lake Hospital, Clinical College of Soochow University, Suzhou 215006, China
4 Department of Obstetrics & Gynecology, The Second Hospital of Nanjing, Nanjing University of Chinese Medicine, Nanjing 210000, China
* Correspondence: hongli@njmu.edu.cn (H.L.); liuyanping_629@163.com(Y.L.); Tel.: +86-512-6236-2065 (H.L. & Y.L.)
† These authors contributed equally to this work.

Abstract: The fetal origins of adult disease (FOAD) hypothesis holds that events during early development have a profound impact on one's risk for the development of future adult disease. Studies from humans and animals have demonstrated that many diseases can begin in childhood and are caused by a variety of early life traumas, including maternal malnutrition, maternal disease conditions, lifestyle changes, exposure to toxins/chemicals, improper medication during pregnancy, and so on. Recently, the roles of Peroxisome proliferator-activated receptors (PPARs) in FOAD have been increasingly appreciated due to their wide variety of biological actions. PPARs are members of the nuclear hormone receptor subfamily, consisting of three distinct subtypes: PPARα, β/δ, and γ, highly expressed in the reproductive tissues. By controlling the maturation of the oocyte, ovulation, implantation of the embryo, development of the placenta, and male fertility, the PPARs play a crucial role in the transition from embryo to fetus in developing mammals. Exposure to adverse events in early life exerts a profound influence on the methylation pattern of PPARs in offspring organs, which can affect development and health throughout the life course, and even across generations. In this review, we summarize the latest research on PPARs in the area of FOAD, highlight the important role of PPARs in FOAD, and provide a potential strategy for early prevention of FOAD.

Keywords: fetal origins of adult disease; PPARs; early development; metabolic; epigenetic

Citation: Guo, J.; Wu, J.; He, Q.; Zhang, M.; Li, H.; Liu, Y. The Potential Role of PPARs in the Fetal Origins of Adult Disease. *Cells* **2022**, *11*, 3474. https://doi.org/10.3390/cells11213474

Academic Editors: Kay-Dietrich Wagner and Nicole Wagner

Received: 25 September 2022
Accepted: 27 October 2022
Published: 2 November 2022

1. Introduction

In 1989, David Barker and his colleagues performed an epidemiological survey. They found that both newborn deaths and the increased risk of death from stroke and coronary heart disease in adults were related to low birth weight [1]. Later studies have confirmed that low birth weight is linked to a variety of chronic disorders, such as hypertension [2], type 2 diabetes (T2DM) [3], autoimmune thyroid disease [4], and chronic bronchitis [5]. This led to the fetal origin of adult diseases (FOAD) hypothesis that the roots of adult metabolic and cardiovascular disorders lay in the effects of malnutrition in fetal life and early infancy [6].

The FOAD hypothesis builds on the "developmental plasticity" that the organisms exhibit plastic or sensitivity in response to environmental influence during critical developmental periods to improve the match between phenotype and environment [7]. For example, a fetus will undergo the process of remodeling and altering the structure or function of various organs, which is critical for survival as well as neurodevelopment when confronted with the adversity of malnutrition [8]. However, it is important to recognize that a person's response to environmental stimulation or pathological conditions can be limited, and such an evolutionary advantage of "plasticity" is lost over time [9]. This phenomenon called "programming" shows how early-life stimuli may lead to lifelong and irreversible changes [10]. The FOAD hypothesis attracted a lot of attention in the field of developmental plasticity.

With the expansion and deep-going of research, the recognition that "programming" occurs not only during the fetal period but also during the whole process of life development, including the early embryonic period, infancy, and early childhood [11]. FOAD hypothesis has been expanded and recognized as the Developmental Origins of Health and Diseases (DOHaD). The DOHaD theory states that the interplay between genes and environments (nutrition, stress, or environmental chemicals) from fertilization to the neonatal stage affects the disease risks related to lifestyle in later periods of life [12]. Research regarding the potential mechanisms of adverse stimuli in utero or early stage of life increases the risk of diseases later in life has been a focus of the various current animal and clinical studies [13]. One of the most exciting emerging themes in the DOHaD field is epigenetics [14]. Epigenetic mechanisms typically include DNA methylation, histone modifications, and non-coding RNAs (ncRNAs) [11]. These epigenetic modifications may have long-term consequences for gene expression and may be involved in the occurrence and/or progression of various diseases in postnatal life.

Peroxisome proliferator-activated receptors (PPARs) belong to the nuclear receptor superfamily and perform a broad range of physiological functions, including cellular development, differentiation, energy homeostasis, and metabolism [15]. Numerous studies have shed light on the involvement of the PPARs in multiple system impairments or protective effects against impairment, such as the nervous system, cardiovascular system, and metabolism system [16].

To date, three members of PPARs (PPARα, PPARβ/δ, and PPARγ) have been identified [17]. These nuclear receptors play important roles in cell differentiation, development, and reproduction [18]. All of the PPAR isoforms are identified in the rat ovary [19]. PPARα and PPARβ/δ are present primarily in the theca and stroma. PPARγ is localized mainly in the granulosa cells surrounding and supporting the oocyte meiotic maturation [20]. PPARγ expression increases from the primary/secondary follicle stage to the large follicle stage [21]. However, one study found that its expression level remains consistent during follicle development [20]. The absence of PPARα has no discernible impact on the fertility of mice, whereas the deletion of PPARγ and PPARβ/δ does [22]. Chaffin et al. held the opposite opinion; activation of PPARγ appears to play an inhibitory role in follicular growth and differentiation, according to their research [23]. The three PPAR isoforms are expressed in both somatic and germ cells in the testis [24]. Although the action of PPARs in testis development is still unclear [25], several published studies suggest that male fertility may be influenced by PPARs-regulated lipid metabolism, particularly the β-oxidation of fatty acids [25].

The PPARs isotypes are expressed in the placenta and play an important role in modulating embryo implantation and placental development [26]. Mutation of PPARβ/δ drastically influences placenta development and even embryonic death [27]. PPARγ is necessary for the formation of the labyrinthine layer of the placenta. Mice with deletion of the PPARγ gene exhibit defective placental vascular discourse and embryonic lethality [28]. During fetal development, the interaction of PPARα and its ligands in the liver may be important for the nutrient supply that the fetus may encounter after birth [29]. In addition, there is strong evidence that PPARβ/δ and PPARγ regulate the expression of genes involved in sarcoplasmic and adipose tissue production [30]. Remarkably, PPARs play a crucial role in metabolism during early life, and alterations in PPAR metabolic pathways could be one candidate mechanism contributing to the FOAD [31]. The recent work in developmental epigenetics has significantly expanded our understanding of this interaction. Exposure to adverse events in early life can affect the methylation pattern of PPARs in multiple organs, such as the brain, lungs, heart, blood vessels, liver, and skeletal muscles. Various chronic adult diseases, especially diabetes, cardiovascular disease, and chronic lung disease, show a clear association with PPARs (Figure 1). This review summarizes the contributions of PPARs to the potential mechanisms involved in the FOAD in order to provide a new theoretical direction for the early prevention and even treatment of fetal origin diseases.

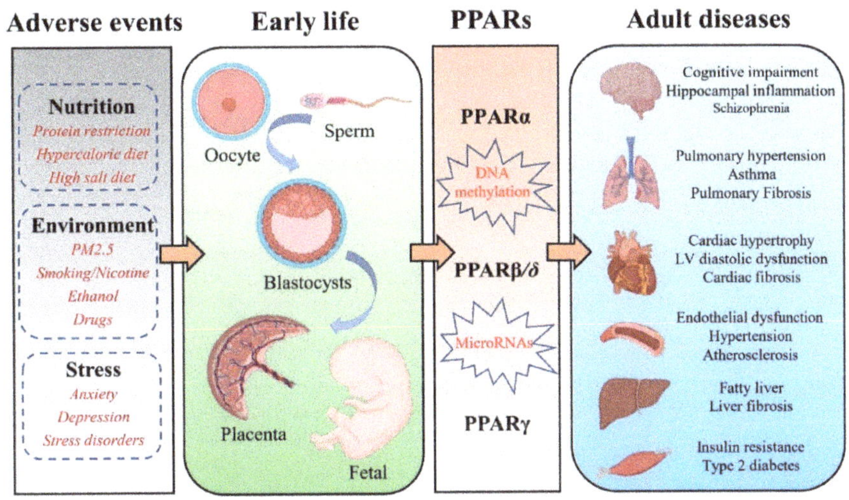

Figure 1. Schematic illustration of potential mechanisms to support the association between early-life adverse events exposure and epigenetic regulation of PPARs that promote chronic diseases in adulthood.

2. Early Life Adverse Exposure and Future Disease Risk

Epidemiological studies in humans have demonstrated an association between the quality of the early life environment and future disease risk [32]. Animal studies provided insight into the potential mechanisms for these observations by highlighting the environmentally induced changes to epigenetic marks during development [33]. The FOAD theory goes beyond nutrition assumptions and links fetal development to many other exposure factors, such as obesity [34], prenatal maternal stress [35], and environment [36]. Notably, a growing body of research indicates that the paternal environment and dietary habits influence disease onset in offspring [37].

2.1. Nutrition

The investigations have shown a strong correlation between maternal food and nutritional status and fetal development and child health [38]. Maternal diet has often been shown to affect subsequent phenotypic [39]. Many diseases such as type 2 diabetes [40], cardiovascular disease [41], and certain cancers [42] are related to low birth weight. Early epidemiological research used data from several well-documented famines and historical cohorts [43]. The most well-known is the Dutch famine cohort. Studies on the Dutch Hunger Winter have provided convincing evidence. Prenatal exposure to famine, especially in the third trimester, has been reported to be associated with decreased glucose tolerance in adults [44]. Even if the effect on fetal growth is minimal, malnutrition in utero may result in long-term alterations in insulin-glucose metabolism [44]. Additionally, individuals whose mothers were exposed to the Dutch Famine before or during gestation were almost three times more likely to develop hypertension in adulthood than unexposed adults [43]. A study of 290 men born in East Hertfordshire during 1920–1930 showed that the risk of coronary heart disease is increased in children with low birth weight [45]. Apart from the above, the studies of the 1959–1961 mass famine in China and the 1944–1945 Dutch

Hunger Winter have found a link between poor nutrition early in life and mental health and cognitive development [46].

Early research focused on evidence linking maternal protein restriction or malnutrition to the long-term health of offspring. Nowadays, a fat-rich diet is prevalent around the world, and 50% of women of childbearing age are overweight or obese in the US [47]. The impact of high-energy-dense, high-glucose, and high-salt foods during pregnancy on the phenotype of the offspring is being studied using animal models. Offspring whose mothers received these various diets showed persistent metabolic changes that were comparable to human cardio-metabolic disorders such as hypertension, insulin resistance, and obesity [48].

Obesity is also highly prevalent among adult men. The prevalence of overweight or obesity (BMI $\geq$ 25) is 72.1% in men and 61.2% in women, according to recent national statistics on the US population [49]. In 2000, Figueroa et al. published the first research revealing the parental influence on child health in humans [50]. They found that fathers' total and percentage body fat were predictors of changes in body fat of premenarcheal girls during a 2.7-y period [50]. In 2010, Ng et al. reported that a high-fat diet in male rats resulted in β-cell dysfunction in F1 female offspring [51].

2.2. Environment

In modern society, humans are exposed to a wide range of environmental chemicals, such as endocrine disruptors and other toxins from lifestyle habits [52]. Numerous epidemiological studies have shown that prenatal exposure to multiple environmental pollutants has an impact on fetal development [53]. There have been many investigations that found a connection between four major environmental pollutants (perfluorinated compounds, polyhalogenated aromatic hydrocarbons, heavy metals, and air pollutants) and impaired fetal development and lower birth weight in humans [54]. A study of 1277 children from the European HELIX (Human Early Life Exposure Group) cohort reveals that BP in children may be influenced by early exposure to some substances, as well as the built environment and climatic conditions [55]. Miguel et al. summarize the influence of the early environment on the structural and functional development of children's brains in their review [56]. For instance, most drugs of abuse (e.g., ecstasy, opiates) can readily cross the placenta and impact fetal brain development [57]. The genes associated with brain growth, myelination, and neuronal migration were down-regulated in the brain of a fetus exposed to tobacco in utero [58]. Several large population-based cohort studies have shown that prenatal exposure to maternal smoking during pregnancy or smoking cessation in early pregnancy was significantly associated with childhood $\approx$ [59]. Prenatal ethanol exposure (PE) impairs dopaminergic (DA) neuron function in the midbrain [60]. Air pollution can affect the anatomy and physiology of the umbilical cord and placenta [61]. Particles induce antiangiogenesis, resulting in the thinner and less voluminous umbilical cord in mouse models, which affects oxygen transport [62] and replicates in humans [63]. A meta-analysis of epidemiological studies suggests that exposure to air pollution increases the risk of pregnancy-induced hypertensive disorders [64].

A cross-sectional study of 67 men in North Carolina indicated that exposure to environmental chemicals/factors (organophosphates) could alter DNA methylation in human sperm cells, thereby affecting the health of offspring [65]. It has also been shown that human exposure to bisphenol A affects the global methylation of sperm DNA [66]. Environmental toxins also include lifestyle habits such as smoking and alcohol intake. Chronic consumption of smoking and alcohol was associated with epigenetic abnormalities and altered miRNA expression in spermatozoa [67].

2.3. Stress

In addition to physical status, the effects of altered maternal mental health and psychological stress during pregnancy on the offspring have been extensively documented in the literature. Maternal anxiety, depression, and stress disorders are common in pregnant women. A wide range of acute and chronic maternal stress exposures, such as daily hassles [68], life event stress [69], and unusual and extremely stressful events [70], have a negative impact on child development [71]. Low birth weight in infants is linked to chronic maternal stress, racism exposure, and depressive symptoms during pregnancy [72]. Accumulating research indicates that prenatal stress and depression during pregnancy are associated with cognitive and academic performance difficulties [73]. Maternal anxiety during pregnancy is associated with subsequent infant development, increased risk of behavioral/emotional disorders, and depression later in children [74]. According to electroencephalography and MRI research results, infants whose mothers had prenatal anxiety may have less volume and/or thickness in their frontal, temporal, and limbic regions and more frontal activity [75]. Natural changes in maternal care during the first day of life are associated with long-term changes in stress reactivity and hippocampal morphology and function in rodent studies [76]. These effects are mediated by epigenetic changes in the promoter of the progeny hippocampal glucocorticoid receptor gene [77]. In humans, childhood abuse was similarly associated with increased DNA methylation and decreased hippocampal glucocorticoid receptor expression [78]. Prenatal stress exposure has been linked to neurodevelopment and the risk of neuropsychiatric disorders in offspring [79]. Retrospective epidemiological studies have provided compelling evidence linking lifetime stress exposure in men with disease risk in offspring [80]. The rodent studies have demonstrated the susceptibility of germ cells to stressful environments throughout the paternal lifetime [81].

3. Peroxisome Proliferator-Activated Receptors

Peroxisome proliferator-activated receptors (PPARs) belong to nuclear hormone receptors (NRs) and ligand-activated transcription factors that regulate genes crucial for cell differentiation and a variety of metabolic processes such as glucose and lipid homeostasis. The PPAR family consists of three different isoforms: PPARα, PPARβ/δ, and PPARγ. These three isotypes have different tissue distribution, biological activity, and affinity for ligands [82]. The essential roles of PPARs in regulating mitochondrial function and energy metabolism have been clearly established. Notably, all three PPAR subtypes have overlapping and also distinct functions in regulating metabolic processes. Six functional domains (from A to F) make up the PPARs [83]. A C structural domain is present at the N-terminus of PPAR, also known as the DNA binding domain (DBD). The DNA sequence in the promoter region of genes, called the peroxisome proliferator response element (PPRE), is recognized by DBD. On the other hand, a ligand-binding domain (LBD) in the C-terminus is responsible for the specificity of the ligand and dimerization of the receptor with the retinoid X receptors (RXR) [84]. PPARs translocate to the nucleus after interacting with specific ligands (synthetic or non-synthetic) [85]. PPARs interact with RXR, peroxisome proliferator-activated receptor gamma-coactivators (PGC), steroid receptor coactivators, and CREB binding protein (CBP/p300) after translocating to the nucleus, then bind to the sequences of PPRE, which subsequently initiate the transcription of target genes involved in different physiological processes [86]. The target genes are involved primarily in the metabolism of fat, as well as in cellular proliferation and differentiation, protein and glucose, inflammation, and tumorigenesis [86]. Their aberrant expression is related to a variety of disorders, such as neurodegenerative disorders, cardiovascular disease, obesity, type 2 diabetes, pancreatic cancer, and so on [16].

Given their central roles in regulating metabolic flexibility, it is essential to understand the manner in which PPARs regulate gene expression. The function of PPARs is principally modulated by ligand binding, which induces structural changes, further recruiting co-activator or co-repressor complexes, which stimulate or inhibit their functions [87]. In addition to ligand binding, post-translational modifications of PPARs are emerging as one such way PPARs are regulated, including phosphorylation, ubiquitination, SUMOylation, acetylation, and O-GlcNAcylation, which contribute to fine-tuning of the transcriptional activities [88]. Recent studies have suggested that post-translational modifications are observed in all three PPAR isoforms [87,88]. A detailed view of the functional regulation of PPARs through post-translational modifications can be found in a recent review by Xu et al. [87].

3.1. PPARα

PPARα is known to be important for regulating the transcriptional expression of key enzymes that are involved in mitochondrial dynamics and metabolic functions, including glucose metabolism, fatty acids β-oxidation, and fatty acid transport [89]. Moreover, PPARα receptors are found largely in metabolically active tissues, such as brown adipose, skeletal muscle, heart, liver, and intestinal mucosa tissues. Natural ligands for the PPAR receptor include saturated, monounsaturated, and polyunsaturated fatty acids and their metabolites, such as leukotrienes B4, oxidized phospholipids, lipolytic lipoprotein products, etc. Nature ligands bind to PPARα and activate PPAR-responsive genes, increasing hepatic intracellular fatty acid absorption [90]. PPARα also plays an important role in extracellular lipid homeostasis by modulating the transcriptional regulation of major very-low-density and high-density apolipoproteins [91]. Furthermore, PPARα seems to modulate the bioactivity of leptin in the liver and adipose tissue [92].

The transcriptional activity of PPARα is enhanced by binding to the ligands, after which transcriptional coactivators contribute to the activation of target genes [93]. In addition, PPARα trans-activity is regulated by post-translational modifications such as phosphorylation, SUMOylation, and ubiquitination. As a phosphoprotein, PPARα is phosphorylated exclusively on serine residues in vivo [94]. It was reported that treatment with insulin or ciprofibrate (a PPARα agonist) increased the phosphorylation of PPARα [95,96]. SUMOylation is a reversible post-translational modification that has been established as one of the key regulatory protein modifications in eukaryotic cells. Two lysine residues of PPARα, K185, and K358, have been reported to be modified by SUMOylation [97,98]. Moreover, several studies have shown that the ubiquitin–proteasome system is involved in the regulation of PPARα activity. These studies suggest the ubiquitination of PPARα in a ligand-dependent manner, and that effect of ubiquitination on PPARα activity depends on the systems studied [87,88].

3.2. PPARβ/δ

PPARβ/δ is generally expressed in nearly all tissues, such as the brain, skin, liver, skeletal muscle, heart, and various types of cancer [99]. Polyunsaturated fatty acids (arachidonic and linoleic acids) and their metabolites, such as prostacyclin PGI2, are suggested as natural ligands. Similar to the other PPAR family members, it mainly participates in the oxidation of fatty acids and affects lipid metabolism, both reducing fat and hence preventing the development of obesity and controlling blood sugar and cholesterol levels in the heart and skeletal muscle [100]. Overall, PPARβ/δ has significant functional overlap with PPARα in most tissues. For example, PPARβ/δ also stimulates fatty acid oxidation in muscle and heart [101]. However, PPARβ/δ and PPARα carry out different roles in regulating hepatic energy metabolism. Unlike PPARα, PPARβ/δ regulates gene expression associated with lipogenesis and glucose utilization rather than inducing fatty acid oxidation [102]. Additionally, several investigations have shown a large expression of PPARβ/δ in the central nervous system (CNS) [103]. PPARβ/δ may affect the differentiation of neural and glial cells and alter cholesterol metabolism in the brain. It is well known that PPARs regulate inflammatory processes associated with lipid signaling pathways [104]. Inhibiting inflammatory processes in the CNS may reduce brain damage and enhance motor and cognitive outcomes [105]. Comparatively, PPARβ/δ is the least reported PPAR family member in terms of a post-translational modification. To date, we are aware of only one previous study showing that SUMOization of PPARβ/δ at K104 is removed by SENP2 and promotes FAO gene expression in muscle [106].

3.3. PPARγ

PPARγ is widely expressed in brown and white adipose tissue, spleen, and large intestine. PPARγ has two isoforms in mice and four different isoforms in humans [107]. Unsaturated fatty acids and their metabolites are the primary natural modulators of PPARγ. Activated PPARγ by these natural ligands regulates adipogenesis and fat distribution, the levels of adipokines such as adiponectin, which involve insulin sensitivity and lipid and glucose metabolism [108]. PPARγ are ligand-inducible transcription factors involved in regulating the expression of genes, including glucose sensitivity (IRS-1, IRS-2, GLUT-4, and PI3K), fatty acid uptake and mobilization (FAT/CD36, FABPs, and LPL) and triglyceride synthesis (ACSL, GK, and PEPCK) [102]. In addition, PPARγ also induces the expression of mitochondrial proteins, such as CPT-1 and UCPs, which play an important role in the regulation of mitochondrial metabolism. PPARγ is associated with the pathology of many diseases, such as obesity, atherosclerosis, diabetes, and cancer. PPARγ agonists, including troglitazone, rosiglitazone, ciglitazone, and pioglitazone, have been used in the treatment of hyperlipidemia and hyperglycemia [109]. The role of PPARγ in cancer initiation/progression is contradictory. Numerous studies show that PPARγ has a tumor-promoting effect. Conversely, some literature has reported that PPARγ plays a key role in tumorigenesis as a tumor suppressor. PPARγ activation by many agonists has been demonstrated to have antiproliferative and proapoptotic actions in prostate, thyroid, and lung cancers [110].

Thiazolidinediones, such as rosiglitazone, pioglitazone, and lobeglitazone, are PPARγ agonists that modulate the transcriptional activity of PPARγ. Like PPARα, PPARγ activity is also regulated by post-translational modifications. PPARγ is now known to be phosphorylated upon stimulation of the MAPK activation pathway [87]. A variety of stimuli (growth factors, platelet-derived growth factors, transforming growth factor beta, insulin, and prostaglandin F2 alpha, etc.) can activate PPARγ phosphorylation via specific activation of MAPKs [87]. Moreover, PPARγ is regulated by SUMO1 and SUMO2 sumoylation. The targeted lysine residue was identified as K107, K33, K64, K68, and K77, respectively [111,112]. In addition to this, other post-translational modifications of PPARγ, such as acetylation, ubiquitination, and O-GlcNAcylation, have also been reviewed by Xu et al. [87].

4. Effects of PPARs in the Placenta and the Fetus

During pregnancy, the placental metabolism can adapt to the environment throughout pregnancy to adapt to the maternal nutritional status and meet the demands of the fetus [31]. All three PPAR isoforms are expressed in the placenta [26,113]. The PPARs promote placental developmental plasticity by regulating lipid, hormone, and glucose metabolic pathways, including lipidogenesis, steroidogenesis, glucose transporters, and placental signaling pathways [114]. Although the role of each PPAR in placental function has not been fully determined, unique and common functions between these isoforms have been observed. Among the PPAR-isoforms, PPARγ appears to be a major regulator of the mammalian placenta [115]. PPARγ was the first to be demonstrated in the placenta [116]. In rodent placenta, PPARγ is largely expressed in the trophoblastic layer of the labyrinth zone [117,118]. In human placenta, PPARγ is present in villous trophoblast and extravillous trophoblast [119,120]. There is some evidence suggesting that PPARγ modulates villous trophoblast differentiation, oxidative pathways, inflammatory response, and barrier formation [121,122]. Furthermore, dysregulation of both PPARα and PPARγ in the placenta has been implicated in common complications of pregnancy, such as gestational diabetes mellitus, intrauterine growth restriction, and preeclampsia [123]. Their expression pattern is regulated at least partially by DNA methylation in the placenta, and the involvement of other PPAR-regulated processes, such as placenta-specific miRNAs, has just been discovered [124]. Placental epigenetic regulation appears to provide a plausible connection between environmental exposures and fetal development. Studies have shown that changes in placental DNA methylation patterns have been associated with fetal growth after exposure to maternal risk conditions such as GDM, obesity, and preeclampsia [125,126].

During the development of the human embryo and fetus, three isoforms are expressed in cells of the endoderm and mesoderm at early time points in gestation [127]. PPARβ/δ was the first allotype to be expressed during embryonic development in rodents [128]. PPARα and PPARγ are expressed first in the placenta and then in the fetus [128,129]. The role of PPAR in development has been revealed by studies in PPAR knockout mice [130]. The important role of PPARα in lipid catabolism in the fetal liver and heart is consistent with the function of PPARα in adult tissues [131–133]. Knockout of PPARα in mice causes a high miscarriage rate, hepatic lipid accumulation, obesity, and prolonged inflammation [134,135]. In the early stages of organogenesis in the rat embryo, only the PPARβ/δ isotype is expressed [128] and plays an important role in the closure of the neural tube [136]. The signaling pathway involved in PPARβ/δ activation associated with nervous system development is profoundly altered by maternal diabetes [136]. PPARγ null mutations are lethal. The developmental defects in the placenta occur in parallel to developmental defects in the embryo [137]. In fetuses from diabetic rats, the concentration of PPARγ endogenous is reduced [138]. The capacity of PPARγ endogenous to prevent the overproduction of both NO and MMPs in fetuses from diabetic rats demonstrates its anti-inflammatory effects [138]. PPARγ activation increases lipid concentrations in midgestation fetuses from diabetic rats [139]. Collectively, these data indicate that PPARs-mediated mechanisms are involved in the fetal origins of health and disease.

5. PPARs and FOAD

It is now well recognized that adverse events exposure in early life contribute to the development of the chronic diseases of adulthood, including hypertension, type 2 diabetes, stroke, cognitive impairment, and pulmonary hypertension. Additionally, the role of PPARs in numerous chronic diseases such as diabetes, cardiovascular diseases, autoimmune diseases, chronic fatigue, depression, and neurodegenerative diseases is well established. PPARs are ubiquitously expressed in almost all mammalian tissues and organs. Altering PPARs methylation patterns during early development may be maintained throughout the life course and even across generations [31]. In the following sections, we review the expression pattern of PPARs in various organs, including the brain, lung, heart, vessel, liver, and skeletal muscle, and discuss the potential roles of PPARs in FOAD (Table 1).

Table 1. Summary of studies on PPARs in the fetal origins of adult disease.

Organs	Adverse Factors	PPARs	Adverse Outcomes/Phenotype	Reference
Brain	Maternal dietary restriction	PPARα	Abnormal sleep homeostatic regulation	[140]
	Maternal immune activation		Disruption of dopamine function	[141]
	Maternal vitamin D deficiency	PPARγ	Angiogenesis impairment	[142]
	Maternal immune activation		Cognitive impairments and anxiety behaviors	[143]
	Maternal high fructose		Hippocampal neuroinflammation	[144]
	Intrauterine growth restriction		Neurodevelopment and neurocognitive impairment	[145]
Lung	Perinatal nicotine exposure	PPARγ	Asthma	[146]
	Perinatal nicotine exposure		Lung dysplasia	[147]
	Perinatal nicotine exposure		Lung mitochondrial dysfunction	[148]
	Intrauterine growth restriction		Impairment of lung development	[149]
Heart	Maternal protein restriction	PPARα	Dysregulation of lipid metabolism	[150]
	Maternal diabetes		Fetal hypertrophic cardiomyopathy	[132]
	Maternal diabetes		Cardiac oxidative stress	[151]
	Maternal obesity	PPARγ	Fetal cardiac dysfunction	[152]
	Maternal protein restriction		Cardiac fibrosis	[153]
	Maternal nutrient restriction		Myocardial lipid deposition	[154]

Table 1. *Cont.*

Organs	Adverse Factors	PPARs	Adverse Outcomes/Phenotype	Reference
Vessel	Preeclampsia	PPARβ/δ	Endothelial dysfunction	[155]
	Maternal protein restriction	PPARγ	Aortic remodeling	[156]
Liver	Maternal exposure to PFOA Maternal high-fat diet Maternal nicotine exposure Unbalanced folates/vitamin B12 diet Maternal high-fat diet	PPARα	Liver damage Non-alcoholic fatty liver disease Metabolic-associated fatty liver disease Lipid metabolism impairment Obesity	[157] [158] [159] [160] [161]
Liver	Maternal ethanol exposure Paternal hyperglycemia	PPARα	Non-alcoholic fatty liver disease Hepatic steatosis	[162] [163]
	Maternal high-fat feeding Maternal bisphenol A exposure	PPARγ	Metabolic dysfunction Non-alcoholic fatty liver disease	[164] [165]
Skeletal muscle	Maternal protein restriction	PPARα	Metabolic inflexibility	[166]
	Intrauterine growth retardation Maternal/Paternal type 2 diabetes	PPARβ/δ	Insulin resistance Insulin resistance	[167] [168]
	Intrauterine growth retardation Maternal cafeteria diet	PPARγ	Insulin resistance Skeletal muscle development and metabolic disorders	[169] [170]

PFOA: perfluorooctanoic acid.

5.1. Brain

PPARα is expressed in several regions of the central nervous system (CNS), and its specific biological function remains unclear [171]. Various inflammatory parameters were significantly enhanced in PPARα KO mice [172]. Neuroinflammation is considered a cause and/or contributing factor to neuronal degeneration [173]. It suggests that PPARα attenuates the inflammatory response after ischemia/brain injury [174]. Moreover, the activation of PPARα has anti-inflammatory properties and a beneficial impact on certain neurologic diseases, including Alzheimer's disease (AD) [175], Multiple sclerosis (MS) [176], Huntington's disease (HD) [177], and Parkinson's disease (PD) [178]. Malnutrition during pregnancy affects sleep homeostasis and increases sleep pressure in offspring, which may be related to the increased PPARα mRNA expression in the hypothalamus [140]. In a study by Felice et al., it was found that prenatal administration of fenofibrate (PPARα agonist) reduced the risk of schizophrenia-like behavior in male offspring of maternal immune activation (MIA) and emphasizes PPARα as a possible target for schizophrenia therapies [141].

Although PPARβ/δ is the most abundant PPAR subtype in the CNS, its role is rarely studied [171]. It has been suggested that the most important roles of PPARβ/δ in brain cells are antioxidant and anti-inflammatory effects [179]. There also a study identified that the differentiation of neural and glial cells might be impacted by PPARβ/δ, which may also affect the metabolism of cholesterol in the brain [103]. One study found that prenatal exposure to a high-fat diet increased the density of cells immunoreactive for PPARβ/δ in the hypothalamic paraventricular nucleus, perifornical lateral hypothalamus, and central nucleus of the amygdala [180]. However, the clinical significance of this change and the potential role of PPARβ/δ in fetal origins of CNS diseases remains unclear.

PPARγ is the most studied subgroup of the PPAR family and has an important role in the CNS, including relieving endoplasmic reticulum stress and the inflammatory response [181], the balance of cerebral metabolite [182] and the maintenance of glucose homeostasis [183]. Animal studies have demonstrated that maternal vitamin D deficiency leads to decreased PPARγ levels in the offspring's brain and affects angiogenesis in the brain [142]. Fetal hippocampal inflammation is significantly increased in immune-activated mothers, followed by cognitive deficits, which are highly correlated with hippocampal neurogenesis disorders in pre-pubertal male offspring. The PPARγ agonist pioglitazone improves neurogenesis, cognitive impairment, and anxious behavior in MIA offspring [143]. Maternal high-fructose-induced hippocampal neuroinflammation in the adult female offspring. Adult female offspring exposed to high maternal fructose have decreased levels of PPARγ and endogenous antioxidant expression in the hippocampus, which leads to hippocampal neuroinflammation. An oral dose of pioglitazone (PPARγ agonist) effectively increases the expression of antioxidants and blocks neuroinflammation [144]. Based on the findings described above, synthetic PPARγ agonists have been suggested as therapeutic medicines for the treatment of CNS diseases such as PD [184], AD [185], HD, and Autism spectrum disorder [186].

5.2. Lung

PPARα has been implicated in the control of airway inflammation, but as yet, little is known about its role in lung disease. There is a mouse model of pulmonary fibrosis suggesting that PPARα regulates fibrosis [187]. A study by Genovese et al. revealed that endogenous and exogenous PPARα ligands reduced bleomycin-induced lung injury in mice [188]. Liu et al. found that the activity of PPARα was inhibited in lipopolysaccharide (LPS) induced acute lung injury (ALI) [189]. By reducing oxidative stress and inflammation, which are both directly related to the activation of PPARα, eupatilin has a protective function in ALI [190]. Taken together, they proposed that PPARα could be a potential therapeutic target for lung injury.

PPARβ/δ agonist inhibited the proliferation of lung fibroblasts and enhanced the antifibrotic properties of PPARγ agonist [187]. The role of PPARβ/δ in pulmonary hypertension and lung cancer has received attention in recent years. According to epidemiological and experimental animal studies, prenatal hypoxia, intrauterine growth restriction (IUGR), and obesity raise the risk of pulmonary hypertension in offspring [191]. Prostacyclin and prostacyclin mimetics are the cornerstone of treatment for patients with pulmonary arterial hypertension (PAH) [192]. One study suggests that PPARβ/δ may be a potent target for prostacyclin mimics in the treatment of pulmonary hypertension. PPARβ/δ agonist (GW0742) mediates vascular relaxation and prevents the right heart from hypertrophy associated with pulmonary arterial hypertension [193]. The role of PPARβ/δ in the negative growth regulation of lung cancer cells was first reported in an in vitro study [194]. Using a variety of lung cancer models, one research group demonstrated that increased synthesis of the PPARβ/δ agonist (prostacyclin) inhibited lung tumorigenesis [195]. These findings imply that PPARβ/δ may play a protective function in PAH and lung cancer.

PPARγ is expressed in many lung cells, including bronchial epithelial cells, airway smooth muscle (HASM) cells, fibroblasts, alveolar type II pneumocytes, and mononuclear phagocytes [187,196]. The activation of PPARγ signaling is involved in the paracrine effect of interstitial fibroblasts and alveolar type II (ATII) cells, which is necessary to maintain alveolar homeostasis [197]. The PPARγ gene depends on developmentally specific transcription of mRNA variants and epigenetics for normal tissue. Therefore, it is susceptible to epigenetic changes [198]. There is evidence that perinatal damage, including exposure to nicotine or maternal tobacco smoke (MTS), IUGR, and preterm delivery, altered both epigenetic determinants and gene expression in the lung [198]. It has been demonstrated that IUGR caused epigenetic modifications to the PPARγ gene in rat lungs [199]. The levels of PPARγ mRNA variants, PPARγ protein, and downstream targets were decreased in the lung of neonatal rats [149]. Numerous studies have shown an increase in asthma in offspring whose mothers smoked during pregnancy [200]. Perinatal smoke/nicotine exposure is a recognized factor that affects lung growth and differentiation by down-regulating the expression of PPARγ [201]. Downregulation of PPARγ causes lipid-rich alveolar mesenchymal fibroblasts to transdifferentiate into myofibroblasts, which is the cellular hallmark of chronic lung diseases such as asthma [202,203]. PPARγ agonist (rosiglitazone) can effectively block asthma induced by perinatal smoke exposure [148].

5.3. Heart

PPARs are the physiological master switches of the heart, which guide the energy metabolism of cardiomyocytes, thereby influencing pathological heart failure and diabetic cardiomyopathy [204]. However, the roles of PPARs in heart function and the results of their respective agonists differ greatly in preclinical animal models and clinical studies [205]. PPARα is highly expressed in the heart and can affect the expression of numerous genes implicated in the uptake and oxidation of cellular fatty acid (FA) [206]. Therefore, it plays a major role in cardiac fatty acid homeostasis. Down-regulation of PPARα expression altered the expression of fatty acid-metabolizing proteins that lead to myocardial damage and fibrosis [207]. The expression of fetal cardiac lipid metabolism genes (PPARα, fatty acid translocase, lipoprotein lipase, etc.) was reduced in offspring from mothers with high blood glucose levels, not accompanied by cardiac triglyceride deposition or cardiac hypertrophy [132]. However, it has subsequently been suggested that the heart of adult offspring from diabetic rats showed increased lipid concentrations. The increased expression of PPARα in offspring from diabetic rats can prevent toxic lipid accumulation in the heart [208]. There is also solid evidence that PPARα exerts a protective effect on cellular oxidative damage [209]. Thus, chronic deactivation of the PPARα signaling pathway may disrupt the balance between oxidant production and antioxidant defenses and ultimately contribute to heart damage [210]. In the 2-day-old and pre-pubertal stage progeny from diabetic rats, there was an increase in the expression of prooxidative/proinflammatory markers and PPARα protein expression in the hearts. Maternal treatment with mitochondrial antioxidants led to reductions in PPARα protein expression and pro-oxidant/ro-inflammatory markers and prevented the adverse programming of heart alterations in prepubertal offspring from diabetic rats [151]. Both neonatal and adult hearts from the offspring of maternal protein restriction (PR) during pregnancy showed a reduction in the level of PPARα promoter methylation and an increase in PPARα mRNA expression [150]. The possible implication of these findings is that the enhanced capacity of fatty acid β-oxidation leads to an increased risk of oxidative damage to offspring hearts.

PPARγ is expressed at very low levels in the adult heart [211]. PPARγ activation in cardiomyocytes is associated with impaired cardiac function due to its lipogenic effect [211]. Maternal obesity leads to cardiac hypertrophy, and left ventricular diastolic dysfunction in offspring might be related to persistent upregulation of PPARγ expression [152]. In rat offspring programmed by the reduced protein diet during gestation, the PPARγ agonist (rosiglitazone) was shown to have beneficial effects by reducing cardiac fibrosis and enhancing myocardial vascularization [153]. PPARγ activator therapy has a beneficial impact on risk factors for cardiovascular disease, but it also appears to have adverse effects on the cardiovascular system. It has been reported that treatment with rosiglitazone is associated with an increase in myocardial infarction (MI) or heart failure in humans [212].

5.4. Vessel

Studies have shown that PPARs are present in all essential vascular cells, including monocyte-macrophages, endothelial cells, and vascular smooth muscle cells [213]. PPARs influence lipid metabolism and vascular diseases such as atherosclerosis and hypertension [214]. PPARα has been implicated in blood pressure regulation and vascular inflammation [215]. PPARα was expressed in both vascular endothelial cells and vascular smooth muscle cells [216]. Activation of PPARα blocks multiple pathways such as NF-κB and MAPK, which in turn inhibit the expression of many genes involved in vascular inflammation, oxidative stress, and cell growth and migration [217]. In experimental hypertension models, PPARα ligands can reportedly lower blood pressure [218]. PPARα was also associated with atherosclerotic processes [219]. The administration of the fibrate class of PPARα agonists to patients with type 2 diabetes or dyslipidemia significantly slowed the development of atherosclerosis and reduced their risk of cardiovascular events [220], but surprisingly, high-fat diet PPARα-null mice are more responsive to insulin, have lower blood pressures, and develop less atherosclerosis [219].

Activation of PPARβ/δ has a significant effect on anti-hypertension [221]. However, it is argued that PPARβ/δ agonist acts via interference with the ET-1 signaling and lower blood pressure through a PPARβ/δ-independent mechanism [222]. Moreover, the reduction of vascular oxidative stress markers and improvement of endothelial dysfunction were observed after a high dose of the PPARβ/δ antagonist GSK0660 [223]. It has been shown that the offspring of rats with maternal diabetes have abnormal fetal programming of vascular endothelial function, which is linked to increased ER stress and may be attributed to the down-regulation of the AMPK/PPAR signaling cascade [224].

Whether PPARγ is hypotensive or hypertensive is still under debate so far [225]. Genetic studies showed impaired vascular smooth muscle contraction in response to alpha-adrenergic drugs and hypotension in a generalized PPARγ knockout mouse model [226], which is very well in agreement with the findings by Tontonoz [227]. These findings suggest that PPARγ has a hypertensive function in controlling blood pressure. However, activation of PPARγ has beneficial effects on hypertension in a number of animal and human studies [228]. PPARγ activation may regulate blood pressure via modulating endothelial vasoactive factors such as prostacyclin, nitric oxide, and endothelin-1. Additionally, PPARγ may also be involved in vessel tone regulation by down-regulating ANG II receptor 1 (AT1-R) in vascular smooth muscle cells [229]. Angiotensin II-induced endothelial dysfunction in adult offspring of pregnancy complicated with hypertension is associated with impaired endothelial PPARγ [155]. Rosiglitazone (a PPARγ agonist) reduced blood pressure and attenuated vascular remodeling in perinatal low-protein offspring rats [156]. Chronic treatment with rosiglitazone has also been shown to prevent impaired nitric oxide synthase-dependent responses induced by prenatal alcohol exposure [230]. Collectively, it is widely believed that activation of PPARγ can moderately lower blood pressure and plays a protective role in endothelial dysfunction, vascular inflammation, and other pathological processes that lead to atherosclerosis [231].

5.5. Liver

The liver is a major organ that regulates whole-body nutrient and energy homeostasis. PPARs are involved in the regulation of adipogenesis, lipid metabolism, inflammation, and metabolic homeostasis [232]. PPARα is a major regulator of lipid metabolism in the liver, especially at fasting. In addition to fatty acid oxidation and ketogenesis, PPARα controls the expression of almost all genes involved in lipid metabolism in the liver [233]. Free fatty acids and other lipids are known to activate PPARα to increase lipid clearance in the liver [234]. In the liver of the PPARα-null mice (lacking the PPARα gene), constitutive mitochondrial β-oxidative activity was significantly reduced [235]. Polyunsaturated fatty Acids (PUFAs) are endogenous PPARα activators. Mice on a high-fat diet supplemented with PUFAs showed enhanced hepatic FA β-oxidation and ameliorated fatty liver [236]. Maternal exposure to perfluorooctanoic acid (PFOA) significantly decreased the expression of the PPARα gene in female offspring mice, leading to reduced fatty acid oxidation and histone acetylation and increased liver oxidative stress [157]. Other authors have found a lower expression of PPARα in the liver of rat offspring exposed to vitamin B12 deficient diets before and during pregnancy due to increased global methylation levels. [237]. The offspring born to an obese mother has a greater likelihood of progression to the fatty liver, which may be associated with PPARα dysfunction [238]. Similar works showed that a high-fat diet during pregnancy impairs the demethylation of PPARα, therefore inducing lipid metabolism disorders and obesity in offspring [161]. Maternal high-fat diet decreased the expression of PPARα and genes for fatty acid oxidation, which contributes to nonalcoholic fatty liver disease (NAFLD) in offspring [158]. Prenatal 1,2-Cyclohexane dicarboxylic acid diisononyl ester (DINCH) plasticizer exposure downregulates PPARα expression, which, in turn, affects the liver function of offspring [239]. Maternal nicotine exposure leads to lipid metabolism disorders and insulin resistance by activating PI3K/Akt signaling, inhibiting PPARα protein expression, and promoting the progression of MAFLD in adult offspring [159].

PPARγ is expressed at much lower levels in the liver and muscle than in adipose tissue and macrophages [240]. Many studies have demonstrated a link between elevated PPARγ expression and hepatic steatosis [241]. Specific disruption of liver PPARγ in mice can effectively improve fatty liver [242]. Overexpression of PPARγ in mouse liver can lead to the development of adipogenic hepatic steatosis [243]. Activation of PPARγ is steatogenic. Paradoxically, treatment of PPARγ-null mice with PPARγ ligands protects other tissues from TAG accumulation and insulin resistance [244]. In A-ZIP/F-1 mice, disrupting hepatocyte PPARγ reduced hepatic steatosis but worsened hyperlipidemia and muscle insulin resistance [244]. PPARγ has anti-inflammatory effects; PPARγ activation decreases inflammatory response by negatively interfering with NF-κB and signal transducers and transcriptional activators [245]. PPARγ agonists may have potential in the prevention of liver fibrosis/cirrhosis [246]. The NAFLD induced by gestational BPA exposure in male offspring may be related to the dysregulation of the HNF1b/PPARγ pathway [165]. Co-agonists of PPARα and PPARγ attenuated liver and white adipose tissue inflammation in male offspring of obese mothers [247]. The reduction of PPARγ level plays a crucial role in arsenic-induced hepatic autophagy in progeny [248]. Metabolic and reproductive disturbances in the female offspring of polycystic ovary syndrome may be associated with the upregulation of PPARγ in the liver [249]. Prenatal exposure to a low-protein diet exhibited a lower expression of PPARγ and hepatic steatosis [250].

5.6. Skeletal Muscle

Skeletal muscle is a metabolic organ that accounts for 40% of the total body weight in a healthy person. It produces adenosine triphosphate (ATP) through insulin-mediated glucose uptake, stores excess glucose as glycogen, and is involved in fatty acid oxidation. All three PPAR isotypes have significant effects on muscle homeostasis, either directly or indirectly. PPARα participates in glucose metabolism and fatty acid catabolism, which is crucial in regulating inflammation and energy expenditure [251]. PPARβ/δ is the major PPAR isotype in skeletal muscle. It is involved in lipid and glucose metabolism, energy expenditure, inflammation, tissue repair and regeneration, and muscle fiber type switching associated with physical exercise [252]. One of the main functions of PPARγ in skeletal muscle is fat deposition [253]. Several observations suggest that PPAR isotypes are at least partially related and overlapping in muscle.

Maternal protein restriction impaired the expression of genes that increased the ability to oxidize fat in response to fasting and exhibited an enhanced expression of PPARα in adult offspring [250]. The study by Zhou et al. showed that miR-29a was upregulated in the skeletal muscles of IUGR offspring. The direct interaction between miR-29a and PPARβ/δ inhibited the expression of PPARβ/δ, which was associated with the progression of insulin resistance (IR) [167]. The reduced mitochondrial content in the muscle of IR offspring may be in part due to decreased PPARβ/δ activation [168]. Maternal cafeteria diet during gestation and lactation Maternal cafeteria diets during pregnancy and lactation were associated with the increased PPARγ mRNA level in pups [170]. The adult mice suffered from maternal caloric restriction during late pregnancy, and a post-weaning high-fat diet, the expressions of PPARγ in their skeletal muscle tissue were significantly increased [169]. PPARγ agonist can improve skeletal muscle insulin sensitivity in the pregestational intrauterine growth-restricted rat offspring [254]. In conclusion, skeletal muscle insulin resistance and impaired fat or glucose metabolism may be closely related to PPARs changes in offspring exposed to adverse factors during pregnancy.

6. Conclusions and Outlook

The concept of FOAD (or DOHaD) has provided new insights into the origin of lifestyle diseases, and the field of FOAD has grown rapidly to high prominence in biomedical science and public health. This review is concerned with understanding how stressful environmental conditions during sensitive periods of early development influence the risk of chronic disease later in life, particularly the role of PPARs in this process. Notably, agonists of PPARs have been intensively evaluated as a potential strategy for the early prevention of FOAD. A growing body of exciting evidence demonstrates that PPAR activators reverse some of the adverse effects of adverse exposure during pregnancy on offspring. These data provide important proof that the epigenetic state of a particular gene can be modified. It provides a novel therapeutic strategy to prevent or delay the fetal origin of adult diseases through epigenetic regulation of metabolic genes. Here, we briefly summarize the relevant studies in Table 2. Nevertheless, studies on PPARs in the area of FOAD are currently in the nascent stage, especially the application of PPARs agonists in the primary prevention and treatment of FOAD remains controversial. Therefore, further research is necessary to enhance our understanding of the PPAR-mediated mechanisms involved in the fetal origins of health and disease. Connecting early-life adverse events exposures and PPARs epigenomic measures with later-life health outcomes is a proven strategy for investigating such underlying mechanisms. Recent research has begun to identify features of the PPARs-related regulation of non-coding RNAs, histone modification, and DNA methylation in FOAD. These advances drive the development of the complex transcriptional and epigenetic regulation of PPARs in FOAD. We believe that the studies of such new perspectives will open up new avenues in FOAD research, as well as potential strategies for early prevention of FOAD.

Table 2. Summary of studies on PPAR agonists in the fetal origins of adult disease.

Types	Drugs	Rescued Phenotype	Reference
PPARα agonist	Clofibrate	Fatty liver disease	[238]
	Fenofibrate	Disruption of dopamine function	[141]
	WY-14643	Obesity	[255]
PPARβ/δ agonist	GW1516	Endothelial dysfunction	[224]
PPARγ agonist	Rosiglitazone	Asthma	[256]
	Pioglitazone	Neuroinflammation and oxidative stress	[257]
	Pioglitazone	Learning and memory abilities impairment	[258]
	Rosiglitazone	Cardiac adverse remodeling	[153]
	Rosiglitazone	Skeletal muscle insulin sensitivity	[254]
	Rosiglitazone	Blood pressure and aortic structure	[156]

Author Contributions: Conceptualization: Y.L., J.G. and J.W.; Writing—original draft preparation: J.G., J.W., Q.H. and M.Z.; writing—review and editing: J.G., H.L. and Y.L.; supervision: H.L. and Y.L. All authors have read and agreed to the published version of the manuscript.

Funding: This research was funded by the Special Scientific Research Cultivation of Gusu School of Nanjing Medical University (GSKY20220523, Szslyyrc2022003), the Suzhou introduce expert team of clinical medicine (SZYJTD201708).

Institutional Review Board Statement: Not applicable.

Informed Consent Statement: Not applicable.

Data Availability Statement: Not applicable.

Conflicts of Interest: The authors declare no conflict of interest.

References

1. Barker, D.J.; Osmond, C. Infant mortality, childhood nutrition, and ischaemic heart disease in England and Wales. *Lancet* **1986**, *1*, 1077–1081. [CrossRef]
2. Barker, D.J.; Osmond, C. Low birth weight and hypertension. *BMJ* **1988**, *297*, 134–135. [CrossRef] [PubMed]
3. Hales, C.N.; Barker, D.J. Type 2 (non-insulin-dependent) diabetes mellitus: The thrifty phenotype hypothesis. *Diabetologia* **1992**, *35*, 595–601. [CrossRef] [PubMed]
4. Phillips, D.I.; Cooper, C.; Fall, C.; Prentice, L.; Osmond, C.; Barker, D.J.; Rees Smith, B. Fetal growth and autoimmune thyroid disease. *Q. J. Med.* **1993**, *86*, 247–253.

5. Barker, D.J.; Osmond, C.; Law, C.M. The intrauterine and early postnatal origins of cardiovascular disease and chronic bronchitis. *J. Epidemiol. Community Health* **1989**, *43*, 237–240. [CrossRef] [PubMed]
6. Barker, D.J. The origins of the developmental origins theory. *J. Intern. Med.* **2007**, *261*, 412–417. [CrossRef] [PubMed]
7. Bateson, P.; Barker, D.; Clutton-Brock, T.; Deb, D.; D'Udine, B.; Foley, R.A.; Gluckman, P.; Godfrey, K.; Kirkwood, T.; Lahr, M.M.; et al. Developmental plasticity and human health. *Nature* **2004**, *430*, 419–421. [CrossRef]
8. Calkins, K.; Devaskar, S.U. Fetal origins of adult disease. *Curr. Probl. Pediatric Adolesc. Health Care* **2011**, *41*, 158–176. [CrossRef]
9. Hochberg, Z.; Feil, R.; Constancia, M.; Fraga, M.; Junien, C.; Carel, J.C.; Boileau, P.; Le Bouc, Y.; Deal, C.L.; Lillycrop, K.; et al. Child health, developmental plasticity, and epigenetic programming. *Endocr. Rev.* **2011**, *32*, 159–224. [CrossRef]
10. Koletzko, B. Developmental origins of adult disease: Barker's or Dorner's hypothesis? *Am. J. Hum. Biol.* **2005**, *17*, 381–382. [CrossRef]
11. Hoffman, D.J.; Reynolds, R.M.; Hardy, D.B. Developmental origins of health and disease: Current knowledge and potential mechanisms. *Nutr. Rev.* **2017**, *75*, 951–970. [CrossRef] [PubMed]
12. Arima, Y.; Fukuoka, H. Developmental origins of health and disease theory in cardiology. *J. Cardiol.* **2020**, *76*, 14–17. [CrossRef] [PubMed]
13. Krontira, A.C.; Cruceanu, C.; Binder, E.B. Glucocorticoids as Mediators of Adverse Outcomes of Prenatal Stress. *Trends Neurosci.* **2020**, *43*, 394–405. [CrossRef]
14. Bird, A. DNA methylation patterns and epigenetic memory. *Genes Dev.* **2002**, *16*, 6–21. [CrossRef] [PubMed]
15. Cheng, H.S.; Tan, W.R.; Low, Z.S.; Marvalim, C.; Lee, J.Y.H.; Tan, N.S. Exploration and Development of PPAR Modulators in Health and Disease: An Update of Clinical Evidence. *Int. J. Mol. Sci.* **2019**, *20*, 5055. [CrossRef]
16. Wójtowicz, S.; Strosznajder, A.K.; Jeżyna, M.; Strosznajder, J.B. The Novel Role of PPARα in the Brain: Promising Target in Therapy of Alzheimer's Disease and Other Neurodegenerative Disorders. *Neurochem. Res.* **2020**, *45*, 972–988. [CrossRef]
17. Han, S.W.; Roman, J. Anticancer actions of PPARγ ligands: Current state and future perspectives in human lung cancer. *World J. Biol. Chem.* **2010**, *1*, 31–40. [CrossRef]
18. Jia, Z.; Xinhua, X.; Qian, Z.; Miao, Y.; Jianping, X.; Zhixin, W.; Yijing, L.; Mingmin, L. PPARγ links maternal malnutrition and abnormal glucose and lipid metabolism in the offspring of mice. *Yi Chuan* **2015**, *37*, 70–76. [CrossRef]
19. Braissant, O.; Foufelle, F.; Scotto, C.; Dauça, M.; Wahli, W. Differential expression of peroxisome proliferator-activated receptors (PPARs): Tissue distribution of PPAR-α, -β, and -γ in the adult rat. *Endocrinology* **1996**, *137*, 354–366. [CrossRef]
20. Komar, C.M.; Braissant, O.; Wahli, W.; Curry, T.E., Jr. Expression and localization of PPARs in the rat ovary during follicular development and the periovulatory period. *Endocrinology* **2001**, *142*, 4831–4838. [CrossRef]
21. Froment, P.; Fabre, S.; Dupont, J.; Pisselet, C.; Chesneau, D.; Staels, B.; Monget, P. Expression and functional role of peroxisome proliferator-activated receptor-gamma in ovarian folliculogenesis in the sheep. *Biol. Reprod.* **2003**, *69*, 1665–1674. [CrossRef] [PubMed]
22. Cui, Y.; Miyoshi, K.; Claudio, E.; Siebenlist, U.K.; Gonzalez, F.J.; Flaws, J.; Wagner, K.U.; Hennighausen, L. Loss of the peroxisome proliferation-activated receptor gamma (PPARγ) does not affect mammary development and propensity for tumor formation but leads to reduced fertility. *J. Biol. Chem.* **2002**, *277*, 17830–17835. [CrossRef] [PubMed]
23. Lovekamp-Swan, T.; Chaffin, C.L. The peroxisome proliferator-activated receptor gamma ligand troglitazone induces apoptosis and p53 in rat granulosa cells. *Mol. Cell. Endocrinol.* **2005**, *233*, 15–24. [CrossRef] [PubMed]
24. Bhattacharya, N.; Dufour, J.M.; Vo, M.N.; Okita, J.; Okita, R.; Kim, K.H. Differential effects of phthalates on the testis and the liver. *Biol. Reprod.* **2005**, *72*, 745–754. [CrossRef]
25. Froment, P.; Gizard, F.; Defever, D.; Staels, B.; Dupont, J.; Monget, P. Peroxisome proliferator-activated receptors in reproductive tissues: From gametogenesis to parturition. *J. Endocrinol.* **2006**, *189*, 199–209. [CrossRef]
26. Fournier, T.; Tsatsaris, V.; Handschuh, K.; Evain-Brion, D. PPARs and the placenta. *Placenta* **2007**, *28*, 65–76. [CrossRef]
27. Nadra, K.; Anghel, S.I.; Joye, E.; Tan, N.S.; Basu-Modak, S.; Trono, D.; Wahli, W.; Desvergne, B. Differentiation of trophoblast giant cells and their metabolic functions are dependent on peroxisome proliferator-activated receptor beta/delta. *Mol. Cell. Biol.* **2006**, *26*, 3266–3281. [CrossRef]
28. Asami-Miyagishi, R.; Iseki, S.; Usui, M.; Uchida, K.; Kubo, H.; Morita, I. Expression and function of PPARγ in rat placental development. *Biochem. Biophys. Res. Commun.* **2004**, *315*, 497–501. [CrossRef]
29. Rees, W.D.; McNeil, C.J.; Maloney, C.A. The Roles of PPARs in the Fetal Origins of Metabolic Health and Disease. *PPAR Res.* **2008**, *2008*, 459030. [CrossRef]
30. Rosen, E.D.; MacDougald, O.A. Adipocyte differentiation from the inside out. *Nat. Rev. Mol. Cell Biol.* **2006**, *7*, 885–896. [CrossRef]
31. Lendvai, Á.; Deutsch, M.J.; Plösch, T.; Ensenauer, R. The peroxisome proliferator-activated receptors under epigenetic control in placental metabolism and fetal development. *Am. J. Physiol. Endocrinol. Metab.* **2016**, *310*, E797–E810. [CrossRef] [PubMed]
32. Tobi, E.W.; Slieker, R.C.; Luijk, R.; Dekkers, K.F.; Stein, A.D.; Xu, K.M.; Slagboom, P.E.; van Zwet, E.W.; Lumey, L.H.; Heijmans, B.T. DNA methylation as a mediator of the association between prenatal adversity and risk factors for metabolic disease in adulthood. *Sci. Adv.* **2018**, *4*, eaao4364. [CrossRef] [PubMed]
33. Rinaudo, P.; Wang, E. Fetal programming and metabolic syndrome. *Annu. Rev. Physiol.* **2012**, *74*, 107–130. [CrossRef] [PubMed]
34. Sanchez, C.E.; Barry, C.; Sabhlok, A.; Russell, K.; Majors, A.; Kollins, S.H.; Fuemmeler, B.F. Maternal pre-pregnancy obesity and child neurodevelopmental outcomes: A meta-analysis. *Obes. Rev.* **2018**, *19*, 464–484. [CrossRef] [PubMed]

35. Lipner, E.; Murphy, S.K.; Ellman, L.M. Prenatal Maternal Stress and the Cascade of Risk to Schizophrenia Spectrum Disorders in Offspring. *Curr. Psychiatry Rep.* **2019**, *21*, 99. [CrossRef] [PubMed]
36. Araya, F.; Stingone, J.A.; Claudio, L. Inequalities in Exposure to Ambient Air Neurotoxicants and Disparities in Markers of Neurodevelopment in Children by Maternal Nativity Status. *Int. J. Environ. Res. Public Health* **2021**, *18*, 7512. [CrossRef] [PubMed]
37. Houfflyn, S.; Matthys, C.; Soubry, A. Male Obesity: Epigenetic Origin and Effects in Sperm and Offspring. *Curr. Mol. Biol. Rep.* **2017**, *3*, 288–296. [CrossRef]
38. Berti, C.; Agostoni, C.; Davanzo, R.; Hyppönen, E.; Isolauri, E.; Meltzer, H.M.; Steegers-Theunissen, R.P.; Cetin, I. Early-life nutritional exposures and lifelong health: Immediate and long-lasting impacts of probiotics, vitamin D, and breastfeeding. *Nutr. Rev.* **2017**, *75*, 83–97. [CrossRef]
39. Lillycrop, K.A.; Burdge, G.C. Maternal diet as a modifier of offspring epigenetics. *J. Dev. Orig. Health Dis.* **2015**, *6*, 88–95. [CrossRef]
40. Zhang, Z.; Kris-Etherton, P.M.; Hartman, T.J. Birth weight and risk factors for cardiovascular disease and type 2 diabetes in US children and adolescents: 10 year results from NHANES. *Matern. Child Health J.* **2014**, *18*, 1423–1432. [CrossRef]
41. Liang, J.; Xu, C.; Liu, Q.; Fan, X.; Xu, J.; Zhang, L.; Hang, D.; Shang, H.; Gu, A. Association between birth weight and risk of cardiovascular disease: Evidence from UK Biobank. *Nutr. Metab. Cardiovasc. Dis. NMCD* **2021**, *31*, 2637–2643. [CrossRef] [PubMed]
42. Yang, T.O.; Reeves, G.K.; Green, J.; Beral, V.; Cairns, B.J. Birth weight and adult cancer incidence: Large prospective study and meta-analysis. *Ann. Oncol.* **2014**, *25*, 1836–1843. [CrossRef] [PubMed]
43. Stein, A.D.; Zybert, P.A.; van der Pal-de Bruin, K.; Lumey, L.H. Exposure to famine during gestation, size at birth, and blood pressure at age 59 y: Evidence from the Dutch Famine. *Eur. J. Epidemiol.* **2006**, *21*, 759–765. [CrossRef] [PubMed]
44. Ravelli, A.C.; van der Meulen, J.H.; Michels, R.P.; Osmond, C.; Barker, D.J.; Hales, C.N.; Bleker, O.P. Glucose tolerance in adults after prenatal exposure to famine. *Lancet* **1998**, *351*, 173–177. [CrossRef]
45. Fall, C.H.; Vijayakumar, M.; Barker, D.J.; Osmond, C.; Duggleby, S. Weight in infancy and prevalence of coronary heart disease in adult life. *BMJ* **1995**, *310*, 17–19. [CrossRef]
46. St Clair, D.; Xu, M.; Wang, P.; Yu, Y.; Fang, Y.; Zhang, F.; Zheng, X.; Gu, N.; Feng, G.; Sham, P.; et al. Rates of adult schizophrenia following prenatal exposure to the Chinese famine of 1959–1961. *JAMA* **2005**, *294*, 557–562. [CrossRef]
47. Mirpuri, J. Evidence for maternal diet-mediated effects on the offspring microbiome and immunity: Implications for public health initiatives. *Pediatric Res.* **2021**, *89*, 301–306. [CrossRef]
48. Ozkan, H.; Topsakal, S.; Ozmen, O. Investigation of the diabetic effects of maternal high-glucose diet on rats. *Biomed. Pharmacother.* **2019**, *110*, 609–617. [CrossRef]
49. Soubry, A. POHaD: Why we should study future fathers. *Environ. Epigenetics* **2018**, *4*, dvy007. [CrossRef]
50. Figueroa-Colon, R.; Arani, R.B.; Goran, M.I.; Weinsier, R.L. Paternal body fat is a longitudinal predictor of changes in body fat in premenarcheal girls. *Am. J. Clin. Nutr.* **2000**, *71*, 829–834. [CrossRef]
51. Ng, S.F.; Lin, R.C.; Laybutt, D.R.; Barres, R.; Owens, J.A.; Morris, M.J. Chronic high-fat diet in fathers programs β-cell dysfunction in female rat offspring. *Nature* **2010**, *467*, 963–966. [CrossRef] [PubMed]
52. Hoffman, K.; Butt, C.M.; Webster, T.F.; Preston, E.V.; Hammel, S.C.; Makey, C.; Lorenzo, A.M.; Cooper, E.M.; Carignan, C.; Meeker, J.D.; et al. Temporal Trends in Exposure to Organophosphate Flame Retardants in the United States. *Environ. Sci. Technol. Lett.* **2017**, *4*, 112–118. [CrossRef] [PubMed]
53. Zheng, T.; Zhang, J.; Sommer, K.; Bassig, B.A.; Zhang, X.; Braun, J.; Xu, S.; Boyle, P.; Zhang, B.; Shi, K.; et al. Effects of Environmental Exposures on Fetal and Childhood Growth Trajectories. *Ann. Glob. Health* **2016**, *82*, 41–99. [CrossRef] [PubMed]
54. Song, Q.; Li, R.; Zhao, Y.; Zhu, Q.; Xia, B.; Chen, S.; Zhang, Y. Evaluating effects of prenatal exposure to phthalates on neonatal birth weight: Structural equation model approaches. *Chemosphere* **2018**, *205*, 674–681. [CrossRef] [PubMed]
55. Warembourg, C.; Maitre, L.; Tamayo-Uria, I.; Fossati, S.; Roumeliotaki, T.; Aasvang, G.M.; Andrusaityte, S.; Casas, M.; Cequier, E.; Chatzi, L.; et al. Early-Life Environmental Exposures and Blood Pressure in Children. *J. Am. Coll. Cardiol.* **2019**, *74*, 1317–1328. [CrossRef] [PubMed]
56. Miguel, P.M.; Pereira, L.O.; Silveira, P.P.; Meaney, M.J. Early environmental influences on the development of children's brain structure and function. *Dev. Med. Child Neurol.* **2019**, *61*, 1127–1133. [CrossRef]
57. Ross, E.J.; Graham, D.L.; Money, K.M.; Stanwood, G.D. Developmental consequences of fetal exposure to drugs: What we know and what we still must learn. *Neuropsychopharmacology* **2015**, *40*, 61–87. [CrossRef]
58. Salihu, H.M.; Paothong, A.; Das, R.; King, L.M.; Pradhan, A.; Riggs, B.; Naik, E.; Siegel, E.M.; Whiteman, V.E. Evidence of altered brain regulatory gene expression in tobacco-exposed fetuses. *J. Perinat. Med.* **2017**, *45*, 1045–1053. [CrossRef]
59. Dong, T.; Hu, W.; Zhou, X.; Lin, H.; Lan, L.; Hang, B.; Lv, W.; Geng, Q.; Xia, Y. Prenatal exposure to maternal smoking during pregnancy and attention-deficit/hyperactivity disorder in offspring: A meta-analysis. *Reprod. Toxicol.* **2018**, *76*, 63–70. [CrossRef]
60. Aghaie, C.I.; Hausknecht, K.A.; Wang, R.; Dezfuli, P.H.; Haj-Dahmane, S.; Kane, C.J.M.; Sigurdson, W.J.; Shen, R.Y. Prenatal Ethanol Exposure and Postnatal Environmental Intervention Alter Dopaminergic Neuron and Microglia Morphology in the Ventral Tegmental Area During Adulthood. *Alcohol. Clin. Exp. Res.* **2020**, *44*, 435–444. [CrossRef]
61. Ghosh, R.; Causey, K.; Burkart, K.; Wozniak, S.; Cohen, A.; Brauer, M. Ambient and household PM2.5 pollution and adverse perinatal outcomes: A meta-regression and analysis of attributable global burden for 204 countries and territories. *PLoS Med.* **2021**, *18*, e1003718. [CrossRef] [PubMed]

62. Veras, M.M.; Guimarães-Silva, R.M.; Caldini, E.G.; Saldiva, P.H.; Dolhnikoff, M.; Mayhew, T.M. The effects of particulate ambient air pollution on the murine umbilical cord and its vessels: A quantitative morphological and immunohistochemical study. *Reprod. Toxicol.* **2012**, *34*, 598–606. [CrossRef] [PubMed]
63. van den Hooven, E.H.; Pierik, F.H.; de Kluizenaar, Y.; Hofman, A.; van Ratingen, S.W.; Zandveld, P.Y.; Russcher, H.; Lindemans, J.; Miedema, H.M.; Steegers, E.A.; et al. Air pollution exposure and markers of placental growth and function: The generation R study. *Environ. Health Perspect.* **2012**, *120*, 1753–1759. [CrossRef] [PubMed]
64. Pedersen, M.; Stayner, L.; Slama, R.; Sørensen, M.; Figueras, F.; Nieuwenhuijsen, M.J.; Raaschou-Nielsen, O.; Dadvand, P. Ambient air pollution and pregnancy-induced hypertensive disorders: A systematic review and meta-analysis. *Hypertension* **2014**, *64*, 494–500. [CrossRef] [PubMed]
65. Soubry, A.; Hoyo, C.; Butt, C.M.; Fieuws, S.; Price, T.M.; Murphy, S.K.; Stapleton, H.M. Human exposure to flame-retardants is associated with aberrant DNA methylation at imprinted genes in sperm. *Environ. Epigenetics* **2017**, *3*, dvx003. [CrossRef]
66. Miao, M.; Zhou, X.; Li, Y.; Zhang, O.; Zhou, Z.; Li, T.; Yuan, W.; Li, R.; Li, D.K. LINE-1 hypomethylation in spermatozoa is associated with Bisphenol A exposure. *Andrology* **2014**, *2*, 138–144. [CrossRef]
67. Marczylo, E.L.; Amoako, A.A.; Konje, J.C.; Gant, T.W.; Marczylo, T.H. Smoking induces differential miRNA expression in human spermatozoa: A potential transgenerational epigenetic concern? *Epigenetics* **2012**, *7*, 432–439. [CrossRef]
68. Huizink, A.C.; Robles de Medina, P.G.; Mulder, E.J.; Visser, G.H.; Buitelaar, J.K. Stress during pregnancy is associated with developmental outcome in infancy. *J. Child Psychol. Psychiatry* **2003**, *44*, 810–818. [CrossRef]
69. Paarlberg, K.M.; Vingerhoets, A.J.; Passchier, J.; Dekker, G.A.; Heinen, A.G.; van Geijn, H.P. Psychosocial predictors of low birthweight: A prospective study. *Br. J. Obstet. Gynaecol.* **1999**, *106*, 834–841. [CrossRef]
70. King, S.; Mancini-Marïe, A.; Brunet, A.; Walker, E.; Meaney, M.J.; Laplante, D.P. Prenatal maternal stress from a natural disaster predicts dermatoglyphic asymmetry in humans. *Dev. Psychopathol.* **2009**, *21*, 343–353. [CrossRef]
71. Korja, R.; Nolvi, S.; Grant, K.A.; McMahon, C. The Relations Between Maternal Prenatal Anxiety or Stress and Child's Early Negative Reactivity or Self-Regulation: A Systematic Review. *Child Psychiatry Hum. Dev.* **2017**, *48*, 851–869. [CrossRef] [PubMed]
72. Dunkel Schetter, C.; Tanner, L. Anxiety, depression and stress in pregnancy: Implications for mothers, children, research, and practice. *Curr. Opin. Psychiatry* **2012**, *25*, 141–148. [CrossRef] [PubMed]
73. Wood, S.J.; Yung, A.R.; Pantelis, C. Cognitive precursors of severe mental disorders. *Cogn. Neuropsychiatry* **2013**, *18*, 1–8. [CrossRef] [PubMed]
74. Davis, E.P.; Sandman, C.A. Prenatal psychobiological predictors of anxiety risk in preadolescent children. *Psychoneuroendocrinology* **2012**, *37*, 1224–1233. [CrossRef] [PubMed]
75. Adamson, B.; Letourneau, N.; Lebel, C. Prenatal maternal anxiety and children's brain structure and function: A systematic review of neuroimaging studies. *J. Affect. Disord.* **2018**, *241*, 117–126. [CrossRef]
76. Meaney, M.J. Epigenetics and the biological definition of gene x environment interactions. *Child Dev.* **2010**, *81*, 41–79. [CrossRef]
77. Weaver, I.C.; Cervoni, N.; Champagne, F.A.; D'Alessio, A.C.; Sharma, S.; Seckl, J.R.; Dymov, S.; Szyf, M.; Meaney, M.J. Epigenetic programming by maternal behavior. *Nat. Neurosci.* **2004**, *7*, 847–854. [CrossRef]
78. McGowan, P.O.; Sasaki, A.; D'Alessio, A.C.; Dymov, S.; Labonté, B.; Szyf, M.; Turecki, G.; Meaney, M.J. Epigenetic regulation of the glucocorticoid receptor in human brain associates with childhood abuse. *Nat. Neurosci.* **2009**, *12*, 342–348. [CrossRef]
79. Chan, J.C.; Nugent, B.M.; Bale, T.L. Parental Advisory: Maternal and Paternal Stress Can Impact Offspring Neurodevelopment. *Biol. Psychiatry* **2018**, *83*, 886–894. [CrossRef]
80. González, C.R.; González, B. Exploring the Stress Impact in the Paternal Germ Cells Epigenome: Can Catecholamines Induce Epigenetic Reprogramming? *Front. Endocrinol.* **2020**, *11*, 630948. [CrossRef]
81. Ly, L.; Chan, D.; Trasler, J.M. Developmental windows of susceptibility for epigenetic inheritance through the male germline. *Semin. Cell Dev. Biol.* **2015**, *43*, 96–105. [CrossRef] [PubMed]
82. Grygiel-Górniak, B. Peroxisome proliferator-activated receptors and their ligands: Nutritional and clinical implications—A review. *Nutr. J.* **2014**, *13*, 17. [CrossRef] [PubMed]
83. Lazennec, G.; Canaple, L.; Saugy, D.; Wahli, W. Activation of peroxisome proliferator-activated receptors (PPARs) by their ligands and protein kinase A activators. *Mol. Endocrinol.* **2000**, *14*, 1962–1975. [CrossRef] [PubMed]
84. Basilotta, R.; Lanza, M.; Casili, G.; Chisari, G.; Munao, S.; Colarossi, L.; Cucinotta, L.; Campolo, M.; Esposito, E.; Paterniti, I. Potential Therapeutic Effects of PPAR Ligands in Glioblastoma. *Cells* **2022**, *11*, 621. [CrossRef]
85. Chan, L.S.; Wells, R.A. Cross-Talk between PPARs and the Partners of RXR: A Molecular Perspective. *PPAR Res.* **2009**, *2009*, 925309. [CrossRef]
86. Yousefnia, S.; Momenzadeh, S.; Seyed Forootan, F.; Ghaedi, K.; Nasr Esfahani, M.H. The influence of peroxisome proliferator-activated receptor γ (PPARγ) ligands on cancer cell tumorigenicity. *Gene* **2018**, *649*, 14–22. [CrossRef]
87. Brunmeir, R.; Xu, F. Functional Regulation of PPARs through Post-Translational Modifications. *Int. J. Mol. Sci.* **2018**, *19*, 1738. [CrossRef]
88. Wadosky, K.M.; Willis, M.S. The story so far: Post-translational regulation of peroxisome proliferator-activated receptors by ubiquitination and SUMOylation. *Am. J. Physiol. Heart Circ. Physiol.* **2012**, *302*, H515–H526. [CrossRef]
89. Takeyama, K.; Kodera, Y.; Suzawa, M.; Kato, S. Peroxisome proliferator-activated receptor(PPAR)—Structure, function, tissue distribution, gene expression. *Nihon Rinsho Jpn. J. Clin. Med.* **2000**, *58*, 357–363.

90. Tan, N.S.; Michalik, L.; Desvergne, B.; Wahli, W. Multiple expression control mechanisms of peroxisome proliferator-activated receptors and their target genes. *J. Steroid Biochem. Mol. Biol.* **2005**, *93*, 99–105. [CrossRef]
91. Lee, Y.; Yu, X.; Gonzales, F.; Mangelsdorf, D.J.; Wang, M.Y.; Richardson, C.; Witters, L.A.; Unger, R.H. PPAR alpha is necessary for the lipopenic action of hyperleptinemia on white adipose and liver tissue. *Proc. Natl. Acad. Sci. USA* **2002**, *99*, 11848–11853. [CrossRef] [PubMed]
92. Storlien, L.H.; Kraegen, E.W.; Chisholm, D.J.; Ford, G.L.; Bruce, D.G.; Pascoe, W.S. Fish oil prevents insulin resistance induced by high-fat feeding in rats. *Science* **1987**, *237*, 885–888. [CrossRef] [PubMed]
93. Feige, J.N.; Gelman, L.; Michalik, L.; Desvergne, B.; Wahli, W. From molecular action to physiological outputs: Peroxisome proliferator-activated receptors are nuclear receptors at the crossroads of key cellular functions. *Prog. Lipid Res.* **2006**, *45*, 120–159. [CrossRef] [PubMed]
94. Burns, K.A.; Vanden Heuvel, J.P. Modulation of PPAR activity via phosphorylation. *Biochim. Et Biophys. Acta* **2007**, *1771*, 952–960. [CrossRef] [PubMed]
95. Shalev, A.; Siegrist-Kaiser, C.A.; Yen, P.M.; Wahli, W.; Burger, A.G.; Chin, W.W.; Meier, C.A. The peroxisome proliferator-activated receptor alpha is a phosphoprotein: Regulation by insulin. *Endocrinology* **1996**, *137*, 4499–4502. [CrossRef]
96. Passilly, P.; Schohn, H.; Jannin, B.; Cherkaoui Malki, M.; Boscoboinik, D.; Dauça, M.; Latruffe, N. Phosphorylation of peroxisome proliferator-activated receptor alpha in rat Fao cells and stimulation by ciprofibrate. *Biochem. Pharmacol.* **1999**, *58*, 1001–1008. [CrossRef]
97. Pourcet, B.; Pineda-Torra, I.; Derudas, B.; Staels, B.; Glineur, C. SUMOylation of human peroxisome proliferator-activated receptor alpha inhibits its trans-activity through the recruitment of the nuclear corepressor NCoR. *J. Biol. Chem.* **2010**, *285*, 5983–5992. [CrossRef]
98. Leuenberger, N.; Pradervand, S.; Wahli, W. Sumoylated PPARalpha mediates sex-specific gene repression and protects the liver from estrogen-induced toxicity in mice. *J. Clin. Investig.* **2009**, *119*, 3138–3148. [CrossRef]
99. Hong, F.; Pan, S.; Guo, Y.; Xu, P.; Zhai, Y. PPARs as Nuclear Receptors for Nutrient and Energy Metabolism. *Molecules* **2019**, *24*, 2545. [CrossRef]
100. Rogue, A.; Lambert, C.; Jossé, R.; Antherieu, S.; Spire, C.; Claude, N.; Guillouzo, A. Comparative gene expression profiles induced by PPARγ and PPARα/γ agonists in human hepatocytes. *PLoS ONE* **2011**, *6*, e18816. [CrossRef]
101. Cheng, L.; Ding, G.; Qin, Q.; Huang, Y.; Lewis, W.; He, N.; Evans, R.M.; Schneider, M.D.; Brako, F.A.; Xiao, Y.; et al. Cardiomyocyte-restricted peroxisome proliferator-activated receptor-delta deletion perturbs myocardial fatty acid oxidation and leads to cardiomyopathy. *Nat. Med.* **2004**, *10*, 1245–1250. [CrossRef] [PubMed]
102. Mello, T.; Materozzi, M.; Galli, A. PPARs and Mitochondrial Metabolism: From NAFLD to HCC. *PPAR Res.* **2016**, *2016*, 7403230. [CrossRef] [PubMed]
103. Strosznajder, A.K.; Wójtowicz, S.; Jeżyna, M.J.; Sun, G.Y.; Strosznajder, J.B. Recent Insights on the Role of PPAR-β/δ in Neuroinflammation and Neurodegeneration, and Its Potential Target for Therapy. *Neuromolecular Med.* **2021**, *23*, 86–98. [CrossRef] [PubMed]
104. Dietschy, J.M.; Turley, S.D. Thematic review series: Brain Lipids. Cholesterol metabolism in the central nervous system during early development and in the mature animal. *J. Lipid Res.* **2004**, *45*, 1375–1397. [CrossRef]
105. Villapol, S. Roles of Peroxisome Proliferator-Activated Receptor Gamma on Brain and Peripheral Inflammation. *Cell. Mol. Neurobiol.* **2018**, *38*, 121–132. [CrossRef]
106. Koo, Y.D.; Choi, J.W.; Kim, M.; Chae, S.; Ahn, B.Y.; Kim, M.; Oh, B.C.; Hwang, D.; Seol, J.H.; Kim, Y.B.; et al. SUMO-Specific Protease 2 (SENP2) Is an Important Regulator of Fatty Acid Metabolism in Skeletal Muscle. *Diabetes* **2015**, *64*, 2420–2431. [CrossRef]
107. Taheri, M.; Salamian, A.; Ghaedi, K.; Peymani, M.; Izadi, T.; Nejati, A.S.; Atefi, A.; Nematollahi, M.; Ahmadi Ghahrizjani, F.; Esmaeili, M.; et al. A ground state of PPARγ activity and expression is required for appropriate neural differentiation of hESCs. *Pharmacol. Rep. PR* **2015**, *67*, 1103–1114. [CrossRef]
108. Marion-Letellier, R.; Savoye, G.; Ghosh, S. Fatty acids, eicosanoids and PPAR gamma. *Eur. J. Pharmacol.* **2016**, *785*, 44–49. [CrossRef]
109. Li, Y.; Qi, Y.; Huang, T.H.; Yamahara, J.; Roufogalis, B.D. Pomegranate flower: A unique traditional antidiabetic medicine with dual PPAR-alpha/-gamma activator properties. *Diabetes Obes. Metab.* **2008**, *10*, 10–17. [CrossRef]
110. Ravi Kiran Ammu, V.V.V.; Garikapati, K.K.; Krishnamurthy, P.T.; Chintamaneni, P.K.; Pindiprolu, S. Possible role of PPAR-γ and COX-2 receptor modulators in the treatment of Non-Small Cell lung carcinoma. *Med. Hypotheses* **2019**, *124*, 98–100. [CrossRef]
111. Ohshima, T.; Koga, H.; Shimotohno, K. Transcriptional activity of peroxisome proliferator-activated receptor gamma is modulated by SUMO-1 modification. *J. Biol. Chem.* **2004**, *279*, 29551–29557. [CrossRef] [PubMed]
112. Diezko, R.; Suske, G. Ligand binding reduces SUMOylation of the peroxisome proliferator-activated receptor γ (PPARγ) activation function 1 (AF1) domain. *PLoS ONE* **2013**, *8*, e66947. [CrossRef] [PubMed]
113. Nadra, K.; Quignodon, L.; Sardella, C.; Joye, E.; Mucciolo, A.; Chrast, R.; Desvergne, B. PPARgamma in placental angiogenesis. *Endocrinology* **2010**, *151*, 4969–4981. [CrossRef] [PubMed]
114. Murthi, P.; Kalionis, B.; Cocquebert, M.; Rajaraman, G.; Chui, A.; Keogh, R.J.; Evain-Brion, D.; Fournier, T. Homeobox genes and down-stream transcription factor PPARγ in normal and pathological human placental development. *Placenta* **2013**, *34*, 299–309. [CrossRef] [PubMed]

115. Xu, Y.; Wang, Q.; Cook, T.J.; Knipp, G.T. Effect of placental fatty acid metabolism and regulation by peroxisome proliferator activated receptor on pregnancy and fetal outcomes. *J. Pharm. Sci.* **2007**, *96*, 2582–2606. [CrossRef]
116. Matsuo, H.; Strauss, J.F., 3rd. Peroxisome proliferators and retinoids affect JEG-3 choriocarcinoma cell function. *Endocrinology* **1994**, *135*, 1135–1145. [CrossRef]
117. Kubota, N.; Terauchi, Y.; Miki, H.; Tamemoto, H.; Yamauchi, T.; Komeda, K.; Satoh, S.; Nakano, R.; Ishii, C.; Sugiyama, T.; et al. PPAR gamma mediates high-fat diet-induced adipocyte hypertrophy and insulin resistance. *Mol. Cell* **1999**, *4*, 597–609. [CrossRef]
118. Barak, Y.; Nelson, M.C.; Ong, E.S.; Jones, Y.Z.; Ruiz-Lozano, P.; Chien, K.R.; Koder, A.; Evans, R.M. PPAR gamma is required for placental, cardiac, and adipose tissue development. *Mol. Cell* **1999**, *4*, 585–595. [CrossRef]
119. Waite, L.L.; Person, E.C.; Zhou, Y.; Lim, K.H.; Scanlan, T.S.; Taylor, R.N. Placental peroxisome proliferator-activated receptor-gamma is up-regulated by pregnancy serum. *J. Clin. Endocrinol. Metab.* **2000**, *85*, 3808–3814. [CrossRef]
120. Tarrade, A.; Schoonjans, K.; Pavan, L.; Auwerx, J.; Rochette-Egly, C.; Evain-Brion, D.; Fournier, T. PPARgamma/RXRalpha heterodimers control human trophoblast invasion. *J. Clin. Endocrinol. Metab.* **2001**, *86*, 5017–5024. [CrossRef]
121. Qiao, L.; Wattez, J.S.; Lee, S.; Guo, Z.; Schaack, J.; Hay, W.W., Jr.; Zita, M.M.; Parast, M.; Shao, J. Knockout maternal adiponectin increases fetal growth in mice: Potential role for trophoblast IGFBP-1. *Diabetologia* **2016**, *59*, 2417–2425. [CrossRef] [PubMed]
122. Tarrade, A.; Schoonjans, K.; Guibourdenche, J.; Bidart, J.M.; Vidaud, M.; Auwerx, J.; Rochette-Egly, C.; Evain-Brion, D. PPAR gamma/RXR alpha heterodimers are involved in human CG beta synthesis and human trophoblast differentiation. *Endocrinology* **2001**, *142*, 4504–4514. [CrossRef] [PubMed]
123. Holdsworth-Carson, S.J.; Lim, R.; Mitton, A.; Whitehead, C.; Rice, G.E.; Permezel, M.; Lappas, M. Peroxisome proliferator-activated receptors are altered in pathologies of the human placenta: Gestational diabetes mellitus, intrauterine growth restriction and preeclampsia. *Placenta* **2010**, *31*, 222–229. [CrossRef] [PubMed]
124. Zhao, C.; Zhang, T.; Shi, Z.; Ding, H.; Ling, X. MicroRNA-518d regulates PPARα protein expression in the placentas of females with gestational diabetes mellitus. *Mol. Med. Rep.* **2014**, *9*, 2085–2090. [CrossRef] [PubMed]
125. Nomura, Y.; Lambertini, L.; Rialdi, A.; Lee, M.; Mystal, E.Y.; Grabie, M.; Manaster, I.; Huynh, N.; Finik, J.; Davey, M.; et al. Global methylation in the placenta and umbilical cord blood from pregnancies with maternal gestational diabetes, preeclampsia, and obesity. *Reprod. Sci.* **2014**, *21*, 131–137. [CrossRef]
126. Ruchat, S.M.; Houde, A.A.; Voisin, G.; St-Pierre, J.; Perron, P.; Baillargeon, J.P.; Gaudet, D.; Hivert, M.F.; Brisson, D.; Bouchard, L. Gestational diabetes mellitus epigenetically affects genes predominantly involved in metabolic diseases. *Epigenetics* **2013**, *8*, 935–943. [CrossRef]
127. Huin, C.; Corriveau, L.; Bianchi, A.; Keller, J.M.; Collet, P.; Krémarik-Bouillaud, P.; Domenjoud, L.; Bécuwe, P.; Schohn, H.; Ménard, D.; et al. Differential expression of peroxisome proliferator-activated receptors (PPARs) in the developing human fetal digestive tract. *J. Histochem. Cytochem.* **2000**, *48*, 603–611. [CrossRef]
128. Braissant, O.; Wahli, W. Differential expression of peroxisome proliferator-activated receptor-alpha, -beta, and -gamma during rat embryonic development. *Endocrinology* **1998**, *139*, 2748–2754. [CrossRef]
129. Wieser, F.; Waite, L.; Depoix, C.; Taylor, R.N. PPAR Action in Human Placental Development and Pregnancy and Its Complications. *PPAR Res.* **2008**, *2008*, 527048. [CrossRef]
130. Jawerbaum, A.; Capobianco, E. Review: Effects of PPAR activation in the placenta and the fetus: Implications in maternal diabetes. *Placenta* **2011**, *32* (Suppl. 2), S212–S217. [CrossRef]
131. Ringseis, R.; Gutgesell, A.; Dathe, C.; Brandsch, C.; Eder, K. Feeding oxidized fat during pregnancy up-regulates expression of PPARalpha-responsive genes in the liver of rat fetuses. *Lipids Health Dis.* **2007**, *6*, 6. [CrossRef] [PubMed]
132. Lindegaard, M.L.; Nielsen, L.B. Maternal diabetes causes coordinated down-regulation of genes involved with lipid metabolism in the murine fetal heart. *Metab. Clin. Exp.* **2008**, *57*, 766–773. [CrossRef] [PubMed]
133. Martínez, N.; White, V.; Kurtz, M.; Higa, R.; Capobianco, E.; Jawerbaum, A. Activation of the nuclear receptor PPARα regulates lipid metabolism in foetal liver from diabetic rats: Implications in diabetes-induced foetal overgrowth. *Diabetes Metab. Res. Rev.* **2011**, *27*, 35–46. [CrossRef] [PubMed]
134. Devchand, P.R.; Keller, H.; Peters, J.M.; Vazquez, M.; Gonzalez, F.J.; Wahli, W. The PPARα-leukotriene B4 pathway to inflammation control. *Nature* **1996**, *384*, 39–43. [CrossRef]
135. Kersten, S.; Seydoux, J.; Peters, J.M.; Gonzalez, F.J.; Desvergne, B.; Wahli, W. Peroxisome proliferator-activated receptor α mediates the adaptive response to fasting. *J. Clin. Investig.* **1999**, *103*, 1489–1498. [CrossRef]
136. Higa, R.; González, E.; Pustovrh, M.C.; White, V.; Capobianco, E.; Martínez, N.; Jawerbaum, A. PPARdelta and its activator PGI2 are reduced in diabetic embryopathy: Involvement of PPARdelta activation in lipid metabolic and signalling pathways in rat embryo early organogenesis. *Mol. Hum. Reprod.* **2007**, *13*, 103–110. [CrossRef]
137. Barak, Y.; Liao, D.; He, W.; Ong, E.S.; Nelson, M.C.; Olefsky, J.M.; Boland, R.; Evans, R.M. Effects of peroxisome proliferator-activated receptor delta on placentation, adiposity, and colorectal cancer. *Proc. Natl. Acad. Sci. USA* **2002**, *99*, 303–308. [CrossRef]
138. Pustovrh, M.C.; Capobianco, E.; Martínez, N.; Higa, R.; White, V.; Jawerbaum, A. MMP/ TIMP balance is modulated in vitro by 15dPGJ(2) in fetuses and placentas from diabetic rats. *Eur. J. Clin. Investig.* **2009**, *39*, 1082–1090. [CrossRef]
139. Capobianco, E.; Martínez, N.; Higa, R.; White, V.; Jawerbaum, A. The effects of maternal dietary treatments with natural PPAR ligands on lipid metabolism in fetuses from control and diabetic rats. *Prostaglandins Leukot. Essent. Fat. Acids* **2008**, *79*, 191–199. [CrossRef]

140. Shimizu, N.; Chikahisa, S.; Nishi, Y.; Harada, S.; Iwaki, Y.; Fujihara, H.; Kitaoka, K.; Shiuchi, T.; Séi, H. Maternal dietary restriction alters offspring's sleep homeostasis. *PLoS ONE* **2013**, *8*, e64263. [CrossRef]
141. De Felice, M.; Melis, M.; Aroni, S.; Muntoni, A.L.; Fanni, S.; Frau, R.; Devoto, P.; Pistis, M. The PPARα agonist fenofibrate attenuates disruption of dopamine function in a maternal immune activation rat model of schizophrenia. *CNS Neurosci. Ther.* **2019**, *25*, 549–561. [CrossRef] [PubMed]
142. Nema, J.; Randhir, K.; Wadhwani, N.; Sundrani, D.; Joshi, S. Maternal vitamin D deficiency reduces docosahexaenoic acid, placental growth factor and peroxisome proliferator activated receptor gamma levels in the pup brain in a rat model of preeclampsia. *Prostaglandins Leukot. Essent. Fat. Acids* **2021**, *175*, 102364. [CrossRef] [PubMed]
143. Zhao, Q.; Wang, Q.; Wang, J.; Tang, M.; Huang, S.; Peng, K.; Han, Y.; Zhang, J.; Liu, G.; Fang, Q.; et al. Maternal immune activation-induced PPARγ-dependent dysfunction of microglia associated with neurogenic impairment and aberrant postnatal behaviors in offspring. *Neurobiol. Dis.* **2019**, *125*, 1–13. [CrossRef]
144. Liu, W.C.; Wu, C.W.; Fu, M.H.; Tain, Y.L.; Liang, C.K.; Hung, C.Y.; Chen, I.C.; Lee, Y.C.; Wu, C.Y.; Wu, K.L.H. Maternal high fructose-induced hippocampal neuroinflammation in the adult female offspring via PPARγ-NF-κB signaling. *J. Nutr. Biochem.* **2020**, *81*, 108378. [CrossRef]
145. Ke, X.; Xing, B.; Yu, B.; Yu, X.; Majnik, A.; Cohen, S.; Lane, R.; Joss-Moore, L. IUGR disrupts the PPARγ-Setd8-H4K20me(1) and Wnt signaling pathways in the juvenile rat hippocampus. *Int. J. Dev. Neurosci.* **2014**, *38*, 59–67. [CrossRef] [PubMed]
146. Rehan, V.K.; Liu, J.; Sakurai, R.; Torday, J.S. Perinatal nicotine-induced transgenerational asthma. *Am. J. Physiol. Lung Cell. Mol. Physiol.* **2013**, *305*, L501–L507. [CrossRef] [PubMed]
147. Liu, Y.; Ji, B.; Zhao, G.; Su, H.; Ge, Y.; Dai, J.; Lu, Y.; Sakurai, R.; Rehan, V.K. Protective effect of electro-acupuncture at maternal different points on perinatal nicotine exposure-induced pulmonary dysplasia in offspring based on HPA axis and signal transduction pathway. *Biochem. Biophys. Res. Commun.* **2018**, *505*, 586–592. [CrossRef]
148. Cannon, D.T.; Liu, J.; Sakurai, R.; Rossiter, H.B.; Rehan, V.K. Impaired Lung Mitochondrial Respiration Following Perinatal Nicotine Exposure in Rats. *Lung* **2016**, *194*, 325–328. [CrossRef]
149. Joss-Moore, L.A.; Wang, Y.; Baack, M.L.; Yao, J.; Norris, A.W.; Yu, X.; Callaway, C.W.; McKnight, R.A.; Albertine, K.H.; Lane, R.H. IUGR decreases PPARγ and SETD8 Expression in neonatal rat lung and these effects are ameliorated by maternal DHA supplementation. *Early Hum. Dev.* **2010**, *86*, 785–791. [CrossRef]
150. Slater-Jefferies, J.L.; Lillycrop, K.A.; Townsend, P.A.; Torrens, C.; Hoile, S.P.; Hanson, M.A.; Burdge, G.C. Feeding a protein-restricted diet during pregnancy induces altered epigenetic regulation of peroxisomal proliferator-activated receptor-α in the heart of the offspring. *J. Dev. Orig. Health Dis.* **2011**, *2*, 250–255. [CrossRef]
151. Higa, R.; Roberti, S.L.; Capobianco, E.; Fornes, D.; White, V.; Jawerbaum, A. Pro-oxidant/pro-inflammatory alterations in the offspring's heart of mild diabetic rats are regulated by maternal treatments with a mitochondrial antioxidant. *Reprod. Toxicol.* **2017**, *73*, 269–279. [CrossRef] [PubMed]
152. Vaughan, O.R.; Rosario, F.J.; Chan, J.; Cox, L.A.; Ferchaud-Roucher, V.; Zemski-Berry, K.A.; Reusch, J.E.B.; Keller, A.C.; Powell, T.L.; Jansson, T. Maternal obesity causes fetal cardiac hypertrophy and alters adult offspring myocardial metabolism in mice. *J. Physiol.* **2022**, *600*, 3169–3191. [CrossRef] [PubMed]
153. Torres Tda, S.; Aguila, M.B.; Mandarim-de-Lacerda, C.A. Rosiglitazone reverses cardiac adverse remodeling (fibrosis and vascularization) in perinatal low protein rat offspring. *Pathol. Res. Pract.* **2010**, *206*, 642–646. [CrossRef] [PubMed]
154. Chan, L.L.; Sébert, S.P.; Hyatt, M.A.; Stephenson, T.; Budge, H.; Symonds, M.E.; Gardner, D.S. Effect of maternal nutrient restriction from early to midgestation on cardiac function and metabolism after adolescent-onset obesity. *Am. J. Physiol. Regul. Integr. Comp. Physiol.* **2009**, *296*, R1455–R1463. [CrossRef]
155. Nair, A.R.; Silva, S.D., Jr.; Agbor, L.N.; Wu, J.; Nakagawa, P.; Mukohda, M.; Lu, K.T.; Sandgren, J.A.; Pierce, G.L.; Santillan, M.K.; et al. Endothelial PPARγ (Peroxisome Proliferator-Activated Receptor-γ) Protects from Angiotensin II-Induced Endothelial Dysfunction in Adult Offspring Born from Pregnancies Complicated by Hypertension. *Hypertension* **2019**, *74*, 173–183. [CrossRef] [PubMed]
156. Torres, T.S.; D'Oliveira Silva, G.; Aguila, M.B.; de Carvalho, J.J.; Mandarim-De-Lacerda, C.A. Effects of rosiglitazone (a peroxysome proliferator-activated receptor gamma agonist) on the blood pressure and aortic structure in metabolically programmed (perinatal low protein) rats. *Hypertens. Res.* **2008**, *31*, 965–975. [CrossRef] [PubMed]
157. Li, D.; Zhang, L.; Zhang, Y.; Guan, S.; Gong, X.; Wang, X. Maternal exposure to perfluorooctanoic acid (PFOA) causes liver toxicity through PPAR-α pathway and lowered histone acetylation in female offspring mice. *Environ. Sci. Pollut. Res. Int.* **2019**, *26*, 18866–18875. [CrossRef] [PubMed]
158. Peng, H.; Xu, H.; Wu, J.; Li, J.; Zhou, Y.; Ding, Z.; Siwko, S.K.; Yuan, X.; Schalinske, K.L.; Alpini, G.; et al. Maternal high-fat diet disrupted one-carbon metabolism in offspring, contributing to nonalcoholic fatty liver disease. *Liver Int.* **2021**, *41*, 1305–1319. [CrossRef]
159. Huang, S.J.; Chen, S.Q.; Lin, Y.; Yang, H.Y.; Ran, J.; Yan, F.F.; Huang, M.; Liu, X.L.; Hong, L.C.; Zhang, X.D.; et al. Maternal nicotine exposure aggravates metabolic associated fatty liver disease via PI3K/Akt signaling in adult offspring mice. *Liver Int.* **2021**, *41*, 1867–1878. [CrossRef]
160. Castaño-Moreno, E.; Castillo, V.; Peñailillo, R.; Llanos, M.N.; Valenzuela, R.; Ronco, A.M. Fatty acid and lipid metabolism in liver of pregnant mice and their offspring is influenced by unbalanced folates/vitamin B12 diets. *Prostaglandins Leukot. Essent. Fat. Acids* **2020**, *154*, 102057. [CrossRef]

161. Pang, H.; Ling, D.; Cheng, Y.; Akbar, R.; Jin, L.; Ren, J.; Wu, H.; Chen, B.; Zhou, Y.; Zhu, H.; et al. Gestational high-fat diet impaired demethylation of Pparα and induced obesity of offspring. *J. Cell. Mol. Med.* **2021**, *25*, 5404–5416. [CrossRef] [PubMed]
162. Shen, L.; Liu, Z.; Gong, J.; Zhang, L.; Wang, L.; Magdalou, J.; Chen, L.; Wang, H. Prenatal ethanol exposure programs an increased susceptibility of non-alcoholic fatty liver disease in female adult offspring rats. *Toxicol. Appl. Pharmacol.* **2014**, *274*, 263–273. [CrossRef] [PubMed]
163. Li, X.; Shi, X.; Hou, Y.; Cao, X.; Gong, L.; Wang, H.; Li, J.; Li, J.; Wu, C.; Xiao, D.; et al. Paternal hyperglycemia induces transgenerational inheritance of susceptibility to hepatic steatosis in rats involving altered methylation on Pparα promoter. *Biochim. Et Biophys. Acta. Mol. Basis Dis.* **2019**, *1865*, 147–160. [CrossRef] [PubMed]
164. Nguyen, L.T.; Chen, H.; Zaky, A.; Pollock, C.; Saad, S. SIRT1 overexpression attenuates offspring metabolic and liver disorders as a result of maternal high-fat feeding. *J. Physiol.* **2019**, *597*, 467–480. [CrossRef]
165. Long, Z.; Fan, J.; Wu, G.; Liu, X.; Wu, H.; Liu, J.; Chen, Y.; Su, S.; Cheng, X.; Xu, Z.; et al. Gestational bisphenol A exposure induces fatty liver development in male offspring mice through the inhibition of HNF1b and upregulation of PPARγ. *Cell Biol. Toxicol.* **2021**, *37*, 65–84. [CrossRef]
166. da Silva Aragão, R.; Guzmán-Quevedo, O.; Pérez-García, G.; Manhães-de-Castro, R.; Bolaños-Jiménez, F. Maternal protein restriction impairs the transcriptional metabolic flexibility of skeletal muscle in adult rat offspring. *Br. J. Nutr.* **2014**, *112*, 328–337. [CrossRef]
167. Zhou, Y.; Gu, P.; Shi, W.; Li, J.; Hao, Q.; Cao, X.; Lu, Q.; Zeng, Y. MicroRNA-29a induces insulin resistance by targeting PPARδ in skeletal muscle cells. *Int. J. Mol. Med.* **2016**, *37*, 931–938. [CrossRef]
168. Morino, K.; Petersen, K.F.; Sono, S.; Choi, C.S.; Samuel, V.T.; Lin, A.; Gallo, A.; Zhao, H.; Kashiwagi, A.; Goldberg, I.J.; et al. Regulation of mitochondrial biogenesis by lipoprotein lipase in muscle of insulin-resistant offspring of parents with type 2 diabetes. *Diabetes* **2012**, *61*, 877–887. [CrossRef]
169. Liu, J.; Zhao, H.; Yang, L.; Wang, X.; Yang, L.; Xing, Y.; Lv, X.; Ma, H.; Song, G. The role of CD36-Fabp4-PPARγ in skeletal muscle involves insulin resistance in intrauterine growth retardation mice with catch-up growth. *BMC Endocr. Disord.* **2022**, *22*, 10. [CrossRef]
170. Bayol, S.A.; Simbi, B.H.; Stickland, N.C. A maternal cafeteria diet during gestation and lactation promotes adiposity and impairs skeletal muscle development and metabolism in rat offspring at weaning. *J. Physiol.* **2005**, *567*, 951–961. [CrossRef]
171. Zolezzi, J.M.; Santos, M.J.; Bastías-Candia, S.; Pinto, C.; Godoy, J.A.; Inestrosa, N.C. PPARs in the central nervous system: Roles in neurodegeneration and neuroinflammation. *Biol. Rev. Camb. Philos. Soc.* **2017**, *92*, 2046–2069. [CrossRef] [PubMed]
172. Cuzzocrea, S.; Mazzon, E.; Di Paola, R.; Peli, A.; Bonato, A.; Britti, D.; Genovese, T.; Muià, C.; Crisafulli, C.; Caputi, A.P. The role of the peroxisome proliferator-activated receptor-alpha (PPAR-alpha) in the regulation of acute inflammation. *J. Leukoc. Biol.* **2006**, *79*, 999–1010. [CrossRef] [PubMed]
173. Esmaeili, M.A.; Yadav, S.; Gupta, R.K.; Waggoner, G.R.; Deloach, A.; Calingasan, N.Y.; Beal, M.F.; Kiaei, M. Preferential PPAR-α activation reduces neuroinflammation, and blocks neurodegeneration in vivo. *Hum. Mol. Genet.* **2016**, *25*, 317–327. [CrossRef] [PubMed]
174. Besson, V.C.; Chen, X.R.; Plotkine, M.; Marchand-Verrecchia, C. Fenofibrate, a peroxisome proliferator-activated receptor alpha agonist, exerts neuroprotective effects in traumatic brain injury. *Neurosci. Lett.* **2005**, *388*, 7–12. [CrossRef] [PubMed]
175. Roy, A.; Jana, M.; Kundu, M.; Corbett, G.T.; Rangaswamy, S.B.; Mishra, R.K.; Luan, C.H.; Gonzalez, F.J.; Pahan, K. HMG-CoA Reductase Inhibitors Bind to PPARα to Upregulate Neurotrophin Expression in the Brain and Improve Memory in Mice. *Cell Metab.* **2015**, *22*, 253–265. [CrossRef]
176. Xu, J.; Racke, M.K.; Drew, P.D. Peroxisome proliferator-activated receptor-alpha agonist fenofibrate regulates IL-12 family cytokine expression in the CNS: Relevance to multiple sclerosis. *J. Neurochem.* **2007**, *103*, 1801–1810. [CrossRef]
177. Johri, A.; Calingasan, N.Y.; Hennessey, T.M.; Sharma, A.; Yang, L.; Wille, E.; Chandra, A.; Beal, M.F. Pharmacologic activation of mitochondrial biogenesis exerts widespread beneficial effects in a transgenic mouse model of Huntington's disease. *Hum. Mol. Genet.* **2012**, *21*, 1124–1137. [CrossRef]
178. Kreisler, A.; Gelé, P.; Wiart, J.F.; Lhermitte, M.; Destée, A.; Bordet, R. Lipid-lowering drugs in the MPTP mouse model of Parkinson's disease: Fenofibrate has a neuroprotective effect, whereas bezafibrate and HMG-CoA reductase inhibitors do not. *Brain Res.* **2007**, *1135*, 77–84. [CrossRef]
179. Schnegg, C.I.; Robbins, M.E. Neuroprotective Mechanisms of PPARδ: Modulation of Oxidative Stress and Inflammatory Processes. *PPAR Res.* **2011**, *2011*, 373560. [CrossRef]
180. Chang, G.Q.; Karatayev, O.; Lukatskaya, O.; Leibowitz, S.F. Prenatal fat exposure and hypothalamic PPAR β/δ: Possible relationship to increased neurogenesis of orexigenic peptide neurons. *Peptides* **2016**, *79*, 16–26. [CrossRef]
181. Gold, P.W.; Licinio, J.; Pavlatou, M.G. Pathological parainflammation and endoplasmic reticulum stress in depression: Potential translational targets through the CNS insulin, klotho and PPAR-γ systems. *Mol. Psychiatry* **2013**, *18*, 154–165. [CrossRef] [PubMed]
182. Cramer, P.E.; Cirrito, J.R.; Wesson, D.W.; Lee, C.Y.; Karlo, J.C.; Zinn, A.E.; Casali, B.T.; Restivo, J.L.; Goebel, W.D.; James, M.J.; et al. ApoE-directed therapeutics rapidly clear β-amyloid and reverse deficits in AD mouse models. *Science* **2012**, *335*, 1503–1506. [CrossRef] [PubMed]
183. Chen, Z.; Zhong, C. Decoding Alzheimer's disease from perturbed cerebral glucose metabolism: Implications for diagnostic and therapeutic strategies. *Prog. Neurobiol.* **2013**, *108*, 21–43. [CrossRef] [PubMed]

184. P, P.; Justin, A.; Ananda Kumar, T.D.; Chinaswamy, M.; Kumar, B.R.P. Glitazones Activate PGC-1α Signaling via PPAR-γ: A Promising Strategy for Antiparkinsonism Therapeutics. *ACS Chem. Neurosci.* **2021**, *12*, 2261–2272. [CrossRef] [PubMed]
185. Landreth, G. PPARgamma agonists as new therapeutic agents for the treatment of Alzheimer's disease. *Exp. Neurol.* **2006**, *199*, 245–248. [CrossRef] [PubMed]
186. d'Angelo, M.; Castelli, V.; Catanesi, M.; Antonosante, A.; Dominguez-Benot, R.; Ippoliti, R.; Benedetti, E.; Cimini, A. PPARγ and Cognitive Performance. *Int. J. Mol. Sci.* **2019**, *20*, 5068. [CrossRef]
187. Lakatos, H.F.; Thatcher, T.H.; Kottmann, R.M.; Garcia, T.M.; Phipps, R.P.; Sime, P.J. The Role of PPARs in Lung Fibrosis. *PPAR Res.* **2007**, *2007*, 71323. [CrossRef]
188. Failla, M.; Genovese, T.; Mazzon, E.; Fruciano, M.; Fagone, E.; Gili, E.; Barera, A.; La Rosa, C.; Conte, E.; Crimi, N.; et al. 16,16-Dimethyl prostaglandin E2 efficacy on prevention and protection from bleomycin-induced lung injury and fibrosis. *Am. J. Respir. Cell Mol. Biol.* **2009**, *41*, 50–58. [CrossRef]
189. Liu, P.; Feng, Y.; Li, H.; Chen, X.; Wang, G.; Xu, S.; Li, Y.; Zhao, L. Ferrostatin-1 alleviates lipopolysaccharide-induced acute lung injury via inhibiting ferroptosis. *Cell. Mol. Biol. Lett.* **2020**, *25*, 10. [CrossRef]
190. Liu, H.; Hao, J.; Wu, C.; Liu, G.; Wang, X.; Yu, J.; Liu, Y.; Zhao, H. Eupatilin Alleviates Lipopolysaccharide-Induced Acute Lung Injury by Inhibiting Inflammation and Oxidative Stress. *Med. Sci. Monit.* **2019**, *25*, 8289–8296. [CrossRef]
191. Selle, J.; Dinger, K.; Jentgen, V.; Zanetti, D.; Will, J.; Georgomanolis, T.; Vohlen, C.; Wilke, R.; Kojonazarov, B.; Klymenko, O.; et al. Maternal and perinatal obesity induce bronchial obstruction and pulmonary hypertension via IL-6-FoxO1-axis in later life. *Nat. Commun.* **2022**, *13*, 4352. [CrossRef] [PubMed]
192. Mandras, S.; Kovacs, G.; Olschewski, H.; Broderick, M.; Nelsen, A.; Shen, E.; Champion, H. Combination Therapy in Pulmonary Arterial Hypertension-Targeting the Nitric Oxide and Prostacyclin Pathways. *J. Cardiovasc. Pharmacol. Ther.* **2021**, *26*, 453–462. [CrossRef] [PubMed]
193. Harrington, L.S.; Moreno, L.; Reed, A.; Wort, S.J.; Desvergne, B.; Garland, C.; Zhao, L.; Mitchell, J.A. The PPARbeta/delta agonist GW0742 relaxes pulmonary vessels and limits right heart hypertrophy in rats with hypoxia-induced pulmonary hypertension. *PLoS ONE* **2010**, *5*, e9526. [CrossRef]
194. Fukumoto, K.; Yano, Y.; Virgona, N.; Hagiwara, H.; Sato, H.; Senba, H.; Suzuki, K.; Asano, R.; Yamada, K.; Yano, T. Peroxisome proliferator-activated receptor delta as a molecular target to regulate lung cancer cell growth. *FEBS Lett.* **2005**, *579*, 3829–3836. [CrossRef] [PubMed]
195. Keith, R.L.; Miller, Y.E.; Hudish, T.M.; Girod, C.E.; Sotto-Santiago, S.; Franklin, W.A.; Nemenoff, R.A.; March, T.H.; Nana-Sinkam, S.P.; Geraci, M.W. Pulmonary prostacyclin synthase overexpression chemoprevents tobacco smoke lung carcinogenesis in mice. *Cancer Res.* **2004**, *64*, 5897–5904. [CrossRef]
196. Huang, T.H.; Razmovski-Naumovski, V.; Kota, B.P.; Lin, D.S.; Roufogalis, B.D. The pathophysiological function of peroxisome proliferator-activated receptor-gamma in lung-related diseases. *Respir. Res.* **2005**, *6*, 102. [CrossRef] [PubMed]
197. Cerny, L.; Torday, J.S.; Rehan, V.K. Prevention and treatment of bronchopulmonary dysplasia: Contemporary status and future outlook. *Lung* **2008**, *186*, 75–89. [CrossRef]
198. Joss-Moore, L.A.; Albertine, K.H.; Lane, R.H. Epigenetics and the developmental origins of lung disease. *Mol. Genet. Metab.* **2011**, *104*, 61–66. [CrossRef]
199. Joss-Moore, L.A.; Metcalfe, D.B.; Albertine, K.H.; McKnight, R.A.; Lane, R.H. Epigenetics and fetal adaptation to perinatal events: Diversity through fidelity. *J. Anim. Sci.* **2010**, *88*, E216–E222. [CrossRef]
200. Pattenden, S.; Antova, T.; Neuberger, M.; Nikiforov, B.; De Sario, M.; Grize, L.; Heinrich, J.; Hruba, F.; Janssen, N.; Luttmann-Gibson, H.; et al. Parental smoking and children's respiratory health: Independent effects of prenatal and postnatal exposure. *Tob. Control* **2006**, *15*, 294–301. [CrossRef]
201. Gong, M.; Liu, J.; Sakurai, R.; Corre, A.; Anthony, S.; Rehan, V.K. Perinatal nicotine exposure suppresses PPARγ epigenetically in lung alveolar interstitial fibroblasts. *Mol. Genet. Metab.* **2015**, *114*, 604–612. [CrossRef] [PubMed]
202. Krebs, M.; Sakurai, R.; Torday, J.S.; Rehan, V.K. Evidence for in vivo nicotine-induced alveolar interstitial fibroblast-to-myofibroblast transdifferentiation. *Exp. Lung Res.* **2010**, *36*, 390–398. [CrossRef] [PubMed]
203. Rehan, V.K.; Wang, Y.; Sugano, S.; Santos, J.; Patel, S.; Sakurai, R.; Boros, L.G.; Lee, W.P.; Torday, J.S. In utero nicotine exposure alters fetal rat lung alveolar type II cell proliferation, differentiation, and metabolism. *Am. J. Physiol. Lung Cell. Mol. Physiol.* **2007**, *292*, L323–L333. [CrossRef] [PubMed]
204. Montaigne, D.; Butruille, L.; Staels, B. PPAR control of metabolism and cardiovascular functions. *Nat. Rev. Cardiol.* **2021**, *18*, 809–823. [CrossRef] [PubMed]
205. Lee, W.S.; Kim, J. Peroxisome Proliferator-Activated Receptors and the Heart: Lessons from the Past and Future Directions. *PPAR Res.* **2015**, *2015*, 271983. [CrossRef]
206. Kersten, S.; Desvergne, B.; Wahli, W. Roles of PPARs in health and disease. *Nature* **2000**, *405*, 421–424. [CrossRef]
207. Watanabe, K.; Fujii, H.; Takahashi, T.; Kodama, M.; Aizawa, Y.; Ohta, Y.; Ono, T.; Hasegawa, G.; Naito, M.; Nakajima, T.; et al. Constitutive regulation of cardiac fatty acid metabolism through peroxisome proliferator-activated receptor alpha associated with age-dependent cardiac toxicity. *J. Biol. Chem.* **2000**, *275*, 22293–22299. [CrossRef]
208. Capobianco, E.; Pelesson, M.; Careaga, V.; Fornes, D.; Canosa, I.; Higa, R.; Maier, M.; Jawerbaum, A. Intrauterine programming of lipid metabolic alterations in the heart of the offspring of diabetic rats is prevented by maternal diets enriched in olive oil. *Mol. Nutr. Food Res.* **2015**, *59*, 1997–2007. [CrossRef]

209. Toyama, T.; Nakamura, H.; Harano, Y.; Yamauchi, N.; Morita, A.; Kirishima, T.; Minami, M.; Itoh, Y.; Okanoue, T. PPARalpha ligands activate antioxidant enzymes and suppress hepatic fibrosis in rats. *Biochem. Biophys. Res. Commun.* **2004**, *324*, 697–704. [CrossRef]
210. Lecarpentier, Y.; Claes, V.; Hébert, J.L. PPARs, Cardiovascular Metabolism, and Function: Near- or Far-from-Equilibrium Pathways. *PPAR Res.* **2010**, *2010*, 783273. [CrossRef]
211. Son, N.H.; Park, T.S.; Yamashita, H.; Yokoyama, M.; Huggins, L.A.; Okajima, K.; Homma, S.; Szabolcs, M.J.; Huang, L.S.; Goldberg, I.J. Cardiomyocyte expression of PPARgamma leads to cardiac dysfunction in mice. *J. Clin. Investig.* **2007**, *117*, 2791–2801. [CrossRef]
212. Mahaffey, K.W.; Hafley, G.; Dickerson, S.; Burns, S.; Tourt-Uhlig, S.; White, J.; Newby, L.K.; Komajda, M.; McMurray, J.; Bigelow, R.; et al. Results of a reevaluation of cardiovascular outcomes in the RECORD trial. *Am. Heart J.* **2013**, *166*, 240–249.e1. [CrossRef] [PubMed]
213. Chen, Y.E.; Fu, M.; Zhang, J.; Zhu, X.; Lin, Y.; Akinbami, M.A.; Song, Q. Peroxisome proliferator-activated receptors and the cardiovascular system. *Vitam. Horm.* **2003**, *66*, 157–188. [CrossRef] [PubMed]
214. Plutzky, J. The PPAR-RXR transcriptional complex in the vasculature: Energy in the balance. *Circ. Res.* **2011**, *108*, 1002–1016. [CrossRef] [PubMed]
215. Ruan, X.; Zheng, F.; Guan, Y. PPARs and the kidney in metabolic syndrome. *Am. J. Physiol. Ren. Physiol.* **2008**, *294*, F1032–F1047. [CrossRef] [PubMed]
216. Marx, N.; Sukhova, G.K.; Collins, T.; Libby, P.; Plutzky, J. PPARalpha activators inhibit cytokine-induced vascular cell adhesion molecule-1 expression in human endothelial cells. *Circulation* **1999**, *99*, 3125–3131. [CrossRef]
217. Kintscher, U.; Lyon, C.; Wakino, S.; Bruemmer, D.; Feng, X.; Goetze, S.; Graf, K.; Moustakas, A.; Staels, B.; Fleck, E.; et al. PPARalpha inhibits TGF-beta-induced beta5 integrin transcription in vascular smooth muscle cells by interacting with Smad4. *Circ. Res.* **2002**, *91*, e35–e44. [CrossRef]
218. Usuda, D.; Kanda, T. Peroxisome proliferator-activated receptors for hypertension. *World J. Cardiol.* **2014**, *6*, 744–754. [CrossRef]
219. Tordjman, K.; Bernal-Mizrachi, C.; Zemany, L.; Weng, S.; Feng, C.; Zhang, F.; Leone, T.C.; Coleman, T.; Kelly, D.P.; Semenkovich, C.F. PPARalpha deficiency reduces insulin resistance and atherosclerosis in apoE-null mice. *J. Clin. Investig.* **2001**, *107*, 1025–1034. [CrossRef]
220. Elkeles, R.S.; Diamond, J.R.; Poulter, C.; Dhanjil, S.; Nicolaides, A.N.; Mahmood, S.; Richmond, W.; Mather, H.; Sharp, P.; Feher, M.D. Cardiovascular outcomes in type 2 diabetes. A double-blind placebo-controlled study of bezafibrate: The St. Mary's, Ealing, Northwick Park Diabetes Cardiovascular Disease Prevention (SENDCAP) Study. *Diabetes Care* **1998**, *21*, 641–648. [CrossRef]
221. Zarzuelo, M.J.; Jiménez, R.; Galindo, P.; Sánchez, M.; Nieto, A.; Romero, M.; Quintela, A.M.; López-Sepúlveda, R.; Gómez-Guzmán, M.; Bailón, E.; et al. Antihypertensive effects of peroxisome proliferator-activated receptor-β activation in spontaneously hypertensive rats. *Hypertension* **2011**, *58*, 733–743. [CrossRef] [PubMed]
222. Bae, E.H.; Kim, I.J.; Ma, S.K.; Kim, S.W. Rosiglitazone prevents the progression of renal injury in DOCA-salt hypertensive rats. *Hypertens. Res.* **2010**, *33*, 255–262. [CrossRef] [PubMed]
223. Zarzuelo, M.J.; Gómez-Guzmán, M.; Jiménez, R.; Quintela, A.M.; Romero, M.; Sánchez, M.; Zarzuelo, A.; Tamargo, J.; Pérez-Vizcaíno, F.; Duarte, J. Effects of peroxisome proliferator-activated receptor-β activation in endothelin-dependent hypertension. *Cardiovasc. Res.* **2013**, *99*, 622–631. [CrossRef] [PubMed]
224. Luo, H.; Lan, C.; Fan, C.; Gong, X.; Chen, C.; Yu, C.; Wang, J.; Luo, X.; Hu, C.; Jose, P.A.; et al. Down-regulation of AMPK/PPARδ signalling promotes endoplasmic reticulum stress-induced endothelial dysfunction in adult rat offspring exposed to maternal diabetes. *Cardiovasc. Res.* **2022**, *118*, 2304–2316. [CrossRef] [PubMed]
225. Weatherford, E.T.; Itani, H.; Keen, H.L.; Sigmund, C.D. Is peroxisome proliferator-activated receptor-gamma a new "pal" of renin? *Hypertension* **2007**, *50*, 844–846. [CrossRef]
226. Duan, S.Z.; Ivashchenko, C.Y.; Whitesall, S.E.; D'Alecy, L.G.; Duquaine, D.C.; Brosius, F.C., 3rd; Gonzalez, F.J.; Vinson, C.; Pierre, M.A.; Milstone, D.S.; et al. Hypotension, lipodystrophy, and insulin resistance in generalized PPARgamma-deficient mice rescued from embryonic lethality. *J. Clin. Investig.* **2007**, *117*, 812–822. [CrossRef]
227. Tontonoz, P.; Nagy, L.; Alvarez, J.G.; Thomazy, V.A.; Evans, R.M. PPARgamma promotes monocyte/macrophage differentiation and uptake of oxidized LDL. *Cell* **1998**, *93*, 241–252. [CrossRef]
228. Bakris, G.; Viberti, G.; Weston, W.M.; Heise, M.; Porter, L.E.; Freed, M.I. Rosiglitazone reduces urinary albumin excretion in type II diabetes. *J. Hum. Hypertens.* **2003**, *17*, 7–12. [CrossRef]
229. Dormandy, J.A.; Charbonnel, B.; Eckland, D.J.; Erdmann, E.; Massi-Benedetti, M.; Moules, I.K.; Skene, A.M.; Tan, M.H.; Lefèbvre, P.J.; Murray, G.D.; et al. Secondary prevention of macrovascular events in patients with type 2 diabetes in the PROactive Study (PROspective pioglitAzone Clinical Trial In macroVascular Events): A randomised controlled trial. *Lancet* **2005**, *366*, 1279–1289. [CrossRef]
230. Saha, P.S.; Kim Sawtelle, K.R.; Bamberg, B.N.; Arrick, D.M.; Watt, M.J.; Scholl, J.L.; Zheng, H.; Mayhan, W.G. Rosiglitazone restores nitric oxide synthase-dependent reactivity of cerebral arterioles in rats exposed to prenatal alcohol. *Alcohol. Clin. Exp. Res.* **2021**, *45*, 1359–1369. [CrossRef]
231. Jones, J.R.; Barrick, C.; Kim, K.A.; Lindner, J.; Blondeau, B.; Fujimoto, Y.; Shiota, M.; Kesterson, R.A.; Kahn, B.B.; Magnuson, M.A. Deletion of PPARgamma in adipose tissues of mice protects against high fat diet-induced obesity and insulin resistance. *Proc. Natl. Acad. Sci. USA* **2005**, *102*, 6207–6212. [CrossRef] [PubMed]

232. Dubois, V.; Eeckhoute, J.; Lefebvre, P.; Staels, B. Distinct but complementary contributions of PPAR isotypes to energy homeostasis. *J. Clin. Investig.* **2017**, *127*, 1202–1214. [CrossRef] [PubMed]
233. Kersten, S. Integrated physiology and systems biology of PPARα. *Mol. Metab.* **2014**, *3*, 354–371. [CrossRef] [PubMed]
234. Patsouris, D.; Reddy, J.K.; Müller, M.; Kersten, S. Peroxisome proliferator-activated receptor alpha mediates the effects of high-fat diet on hepatic gene expression. *Endocrinology* **2006**, *147*, 1508–1516. [CrossRef]
235. Aoyama, T.; Peters, J.M.; Iritani, N.; Nakajima, T.; Furihata, K.; Hashimoto, T.; Gonzalez, F.J. Altered constitutive expression of fatty acid-metabolizing enzymes in mice lacking the peroxisome proliferator-activated receptor alpha (PPARalpha). *J. Biol. Chem.* **1998**, *273*, 5678–5684. [CrossRef]
236. Nakajima, T.; Yang, Y.; Lu, Y.; Kamijo, Y.; Yamada, Y.; Nakamura, K.; Koyama, M.; Yamaguchi, S.; Sugiyama, E.; Tanaka, N.; et al. Decreased Fatty Acid β-Oxidation Is the Main Cause of Fatty Liver Induced by Polyunsaturated Fatty Acid Deficiency in Mice. *Tohoku J. Exp. Med.* **2017**, *242*, 229–239. [CrossRef]
237. Meher, A.; Joshi, A.; Joshi, S. Differential regulation of hepatic transcription factors in the Wistar rat offspring born to dams fed folic acid, vitamin B12 deficient diets and supplemented with omega-3 fatty acids. *PLoS ONE* **2014**, *9*, e90209. [CrossRef]
238. Heinecke, F.; Mazzucco, M.B.; Fornes, D.; Roberti, S.; Jawerbaum, A.; White, V. The offspring from rats fed a fatty diet display impairments in the activation of liver peroxisome proliferator activated receptor alpha and features of fatty liver disease. *Mol. Cell. Endocrinol.* **2020**, *511*, 110818. [CrossRef]
239. Campioli, E.; Lau, M.; Papadopoulos, V. Effect of subacute and prenatal DINCH plasticizer exposure on rat dams and male offspring hepatic function: The role of PPAR-α. *Environ. Res.* **2019**, *179*, 108773. [CrossRef]
240. Ahmadian, M.; Suh, J.M.; Hah, N.; Liddle, C.; Atkins, A.R.; Downes, M.; Evans, R.M. PPARγ signaling and metabolism: The good, the bad and the future. *Nat. Med.* **2013**, *19*, 557–566. [CrossRef]
241. Lee, Y.J.; Ko, E.H.; Kim, J.E.; Kim, E.; Lee, H.; Choi, H.; Yu, J.H.; Kim, H.J.; Seong, J.K.; Kim, K.S.; et al. Nuclear receptor PPARγ-regulated monoacylglycerol O-acyltransferase 1 (MGAT1) expression is responsible for the lipid accumulation in diet-induced hepatic steatosis. *Proc. Natl. Acad. Sci. USA* **2012**, *109*, 13656–13661. [CrossRef] [PubMed]
242. Matsusue, K.; Haluzik, M.; Lambert, G.; Yim, S.H.; Gavrilova, O.; Ward, J.M.; Brewer, B., Jr.; Reitman, M.L.; Gonzalez, F.J. Liver-specific disruption of PPARgamma in leptin-deficient mice improves fatty liver but aggravates diabetic phenotypes. *J. Clin. Investig.* **2003**, *111*, 737–747. [CrossRef] [PubMed]
243. Yu, S.; Matsusue, K.; Kashireddy, P.; Cao, W.Q.; Yeldandi, V.; Yeldandi, A.V.; Rao, M.S.; Gonzalez, F.J.; Reddy, J.K. Adipocyte-specific gene expression and adipogenic steatosis in the mouse liver due to peroxisome proliferator-activated receptor γ1 (PPARγ1) overexpression. *J. Biol. Chem.* **2003**, *278*, 498–505. [CrossRef] [PubMed]
244. Gavrilova, O.; Haluzik, M.; Matsusue, K.; Cutson, J.J.; Johnson, L.; Dietz, K.R.; Nicol, C.J.; Vinson, C.; Gonzalez, F.J.; Reitman, M.L. Liver peroxisome proliferator-activated receptor gamma contributes to hepatic steatosis, triglyceride clearance, and regulation of body fat mass. *J. Biol. Chem.* **2003**, *278*, 34268–34276. [CrossRef] [PubMed]
245. Konstantinopoulos, P.A.; Vandoros, G.P.; Sotiropoulou-Bonikou, G.; Kominea, A.; Papavassiliou, A.G. NF-κB/PPARγ and/or AP-1/PPARγ 'on/off' switches and induction of CBP in colon adenocarcinomas: Correlation with COX-2 expression. *Int. J. Colorectal Dis.* **2007**, *22*, 57–68. [CrossRef] [PubMed]
246. Bae, M.A.; Rhee, S.D.; Jung, W.H.; Ahn, J.H.; Song, B.J.; Cheon, H.G. Selective inhibition of activated stellate cells and protection from carbon tetrachloride-induced liver injury in rats by a new PPARgamma agonist KR62776. *Arch. Pharmacal Res.* **2010**, *33*, 433–442. [CrossRef]
247. Magliano, D.C.; Bargut, T.C.; de Carvalho, S.N.; Aguila, M.B.; Mandarim-de-Lacerda, C.A.; Souza-Mello, V. Peroxisome proliferator-activated receptors-alpha and gamma are targets to treat offspring from maternal diet-induced obesity in mice. *PLoS ONE* **2013**, *8*, e64258. [CrossRef]
248. Bai, J.; Yao, X.; Jiang, L.; Zhang, Q.; Guan, H.; Liu, S.; Wu, W.; Qiu, T.; Gao, N.; Yang, L.; et al. Taurine protects against As_2O_3-induced autophagy in livers of rat offsprings through PPARγ pathway. *Sci. Rep.* **2016**, *6*, 27733. [CrossRef]
249. Huang, Y.; Gao, J.M.; Zhang, C.M.; Zhao, H.C.; Zhao, Y.; Li, R.; Yu, Y.; Qiao, J. Assessment of growth and metabolism characteristics in offspring of dehydroepiandrosterone-induced polycystic ovary syndrome adults. *Reproduction* **2016**, *152*, 705–714. [CrossRef]
250. Erhuma, A.; Salter, A.M.; Sculley, D.V.; Langley-Evans, S.C.; Bennett, A.J. Prenatal exposure to a low-protein diet programs disordered regulation of lipid metabolism in the aging rat. *Am. J. Physiol. Endocrinol. Metab.* **2007**, *292*, E1702–E1714. [CrossRef]
251. Huang, J.; Jia, Y.; Fu, T.; Viswakarma, N.; Bai, L.; Rao, M.S.; Zhu, Y.; Borensztajn, J.; Reddy, J.K. Sustained activation of PPARα by endogenous ligands increases hepatic fatty acid oxidation and prevents obesity in ob/ob mice. *FASEB J.* **2012**, *26*, 628–638. [CrossRef] [PubMed]
252. Tan, N.S.; Vázquez-Carrera, M.; Montagner, A.; Sng, M.K.; Guillou, H.; Wahli, W. Transcriptional control of physiological and pathological processes by the nuclear receptor PPARβ/δ. *Prog. Lipid Res.* **2016**, *64*, 98–122. [CrossRef] [PubMed]
253. Lazar, M.A. PPARγ, 10 years later. *Biochimie* **2005**, *87*, 9–13. [CrossRef] [PubMed]
254. Oak, S.; Tran, C.; Castillo, M.O.; Thamotharan, S.; Thamotharan, M.; Devaskar, S.U. Peroxisome proliferator-activated receptor-gamma agonist improves skeletal muscle insulin signaling in the pregestational intrauterine growth-restricted rat offspring. *Am. J. Physiol. Endocrinol. Metab.* **2009**, *297*, E514–E524. [CrossRef]
255. Yuan, X.; Tsujimoto, K.; Hashimoto, K.; Kawahori, K.; Hanzawa, N.; Hamaguchi, M.; Seki, T.; Nawa, M.; Ehara, T.; Kitamura, Y.; et al. Epigenetic modulation of Fgf21 in the perinatal mouse liver ameliorates diet-induced obesity in adulthood. *Nat. Commun.* **2018**, *9*, 636. [CrossRef]

256. Liu, J.; Sakurai, R.; Rehan, V.K. PPAR-γ agonist rosiglitazone reverses perinatal nicotine exposure-induced asthma in rat offspring. *Am. J. Physiol. Lung Cell. Mol. Physiol.* **2015**, *308*, L788–L796. [CrossRef]
257. Mirza, R.; Sharma, B. A selective peroxisome proliferator-activated receptor-γ agonist benefited propionic acid induced autism-like behavioral phenotypes in rats by attenuation of neuroinflammation and oxidative stress. *Chem.-Biol. Interact.* **2019**, *311*, 108758. [CrossRef]
258. Sağır, D. Dose-dependent effects of prenatal exposure of pioglitazone, the PPARγ agonist, on the hippocampus development and learning and memory performance of rat offspring. *Toxicol. Appl. Pharmacol.* **2021**, *421*, 115544. [CrossRef]

cells

MDPI

Article

Alternatively Spliced Landscape of PPARγ mRNA in Podocytes Is Distinct from Adipose Tissue

Claire Bryant [1], Amy Webb [2], Alexander S. Banks [3], Dawn Chandler [4,5], Rajgopal Govindarajan [6,7] and Shipra Agrawal [1,5,8,*]

1 Center for Clinical and Translational Research, Abigail Wexner Research Institute at Nationwide Children's Hospital, Columbus, OH 43205, USA
2 Department of Bioinformatics, The Ohio State University, Columbus, OH 43210, USA
3 Beth Israel Deaconess Medical Center, Harvard Medical School, Boston, MA 02215, USA
4 Center for Childhood Cancer and Blood Disease, Abigail Wexner Research Institute at Nationwide Children's Hospital, Columbus, OH 43205, USA
5 Department of Pediatrics, College of Medicine, The Ohio State University, Columbus, OH 43210, USA
6 Division of Pharmaceutics and Pharmacology, College of Pharmacy, The Ohio State University, Columbus, OH 43210, USA
7 Translational Therapeutics, The Ohio State University Comprehensive Cancer Center, Columbus, OH 43210, USA
8 Division of Nephrology and Hypertension, Department of Medicine, Renaissance School of Medicine, Stony Brook University, Stony Brook, NY 11794, USA
* Correspondence: shipra.agrawal@stonybrookmedicine.edu

Citation: Bryant, C.; Webb, A.; Banks, A.S.; Chandler, D.; Govindarajan, R.; Agrawal, S. Alternatively Spliced Landscape of PPARγ mRNA in Podocytes Is Distinct from Adipose Tissue. *Cells* **2022**, *11*, 3455. https://doi.org/10.3390/cells11213455

Academic Editors: Kay-Dietrich Wagner and Nicole Wagner

Received: 1 August 2022
Accepted: 21 October 2022
Published: 1 November 2022

Abstract: Podocytes are highly differentiated epithelial cells, and their structural and functional integrity is compromised in a majority of glomerular and renal diseases, leading to proteinuria, chronic kidney disease, and kidney failure. Traditional agonists (e.g., pioglitazone) and selective modulators (e.g., GQ-16) of peroxisome-proliferator-activated-receptor-γ (PPARγ) reduce proteinuria in animal models of glomerular disease and protect podocytes from injury via PPARγ activation. This indicates a pivotal role for PPARγ in maintaining glomerular function through preservation of podocytes distinct from its well-understood role in driving insulin sensitivity and adipogenesis. While its transcriptional role in activating adipokines and adipogenic genes is well-established in adipose tissue, liver and muscle, understanding of podocyte PPARγ signaling remains limited. We performed a comprehensive analysis of PPARγ mRNA variants due to alternative splicing, in human podocytes and compared with adipose tissue. We found that podocytes express the ubiquitous PPARγ Var 1 (encoding γ1) and not Var2 (encoding γ2), which is mostly restricted to adipose tissue and liver. Additionally, we detected expression at very low level of Var4, and barely detectable levels of other variants, Var3, Var11, VartORF4 and Var9, in podocytes. Furthermore, a distinct podocyte vs. adipocyte PPAR-promoter-response-element containing gene expression, enrichment and pathway signature was observed, suggesting differential regulation by podocyte specific PPARγ1 variant, distinct from the adipocyte-specific γ2 variant. In summary, podocytes and glomeruli express several PPARγ variants, including Var1 (γ1) and excluding adipocyte-specific Var2 (γ2), which may have implications in podocyte specific signaling and pathophysiology. This suggests that that new selective PPARγ modulators can be potentially developed that will be able to distinguish between the two forms, γ1 and γ2, thus forming a basis of novel targeted therapeutic avenues.

Keywords: PPARγ; podocyte; glomerular disease; proteinuria; alternative splicing

1. Introduction

Thiazolidinediones (TZDs) or peroxisome proliferator activated receptor γ (PPARγ) agonists such as pioglitazone, directly protect podocytes from injury as demonstrated in podocyte culture in vitro studies [1–5], and reduce proteinuria and glomerular injury in various animal models of glomerular disease, as reported in preclinical in vivo studies [5–13]. Moreover, these beneficial effects of PPARγ have been shown to be mediated by activation of podocyte PPARγ as demonstrated by elegant studies using podocyte specific *Pparg* knock out (KO) mouse model [7,10]. This indicates a pivotal role for PPARγ in maintaining glomerular function through preservation of podocytes [1,5–7,10]. Podocytes are highly differentiated epithelial cells in the kidneys, whose structural and functional integrity is critical for the maintenance of glomerular filtration barrier [14–16]. Accordingly, their dysfunction or loss is the initiating and progressing characteristic factor in a vast majority of renal diseases, leading to chronic kidney disease and kidney failure. Glomerular diseases characterized by high proteinuria manifest as nephrotic syndrome (NS) which is often associated with co-morbidities such as hypoalbuminemia, hypercholesterolemia, edema, and hyper-coagulopathy [17–23]. Recently, our group has further demonstrated that selective modulation of PPARγ with a partial agonist, GQ-16, provides high efficacy in reducing proteinuria, as well as NS-associated comorbidities in an experimental model of nephrotic syndrome [24]. Thus, a multitude of evidence shows that targeting the podocyte PPARγ pathway offers an attractive therapeutic strategy in NS. While PPARγ has an established role in driving many adipogenic and lipid metabolizing genes mainly in adipose tissue, liver, and muscle, the understanding of molecular pathways regulated by PPARγ in podocytes and glomeruli remains limited [5–7,9,10,25–29]. Furthermore, it is well-known that PPARγ exists in two major isoforms, γ1 and γ2, which are a result of different promoter usage as well as alternative splicing (AS) [5,30,31]. As a result, γ2 contains 30 additional amino acids at its N terminal, despite a longer mRNA sequence towards the 5′UTR of γ1. While γ1 is more widely expressed, γ2 is mostly restricted to adipose tissue and liver. Few other forms have been discovered, including some with dominant negative function [32–37]. Moreover, cell-specific expression of PPARγ variants has been demonstrated to play a differential role in downstream gene expression pattern in adipocytes (expressing the γ2 variant) vs. the macrophages (expressing the γ1 variant) [38]. However, the specific variant or isoform(s) expression of *PPARγ* and the role of its AS in podocytes and glomerular disease is unexplored. In this study, we sought out to answer this critical gap in the literature by hypothesizing that alternative splice variants of PPARγ expressed in podocytes are distinct from adipocytes. To address this hypothesis, we performed a comprehensive analysis of different AS variants of PPARγ in human podocytes and studied the expression of γ1 vs. γ2 variant in podocyte vs. adipose tissue in detail.

2. Materials and Methods

2.1. Podocyte Cell Culture and Treatments

Immortalized human podocytes were cultured in RMPI 1640 (Corning, Tewksbury, MA, USA) supplemented with 10% fetal bovine serum (FBS) (Corning, Tewksbury, MA, USA), 1% 100 X Penicillin Streptomycin L-Glutamine (PSG) (Corning, Tewksbury, MA, USA), and 1% Insulin-Transferrin-Selenium-Ethanolamine (ITS-X) (Gibco, Gaithersburg, MD, USA) [39]. Proliferating podocytes were cultured in a humidified atmosphere with 5% CO_2 at 33 °C and the media was changed twice per week and cells were passaged at ~70–80% confluence. Differentiation was induced by placing the cells at 37 °C under the same atmospheric conditions, for 14 days. Differentiated cells were treated with puromycin aminonucleoside (PAN) (Sigma-Aldrich, St. Louis, MO, USA) at 25 ug/mL or with pioglitazone (Alfa Aesar, Tewksbury, MA, USA) at 10 μM. Control cells received treatment of vehicle dimethyl sulfoxide (DMSO) (Sigma-Aldrich, St. Louis, MO, USA). Cells were harvested after 24 h and total RNA isolated.

2.2. Kidney, Liver and Adipose Tissues

This study was approved by the Institution Animal Care and Use Committee at Nationwide Children's Hospital and the experiments were performed according to their guidelines. Male Wistar rats weighing ~150–200 g (Envigo, Indianapolis, IN, USA) were purchased, acclimated for 3 days, and housed under standard conditions. They were sacrificed to collect kidneys, liver, and white adipose tissue (WAT) epididymal fat. Tissues were flash frozen in liquid nitrogen prior to RNA isolation. Glomeruli were isolated from harvested kidneys using the sequential sieving method as previously described [24] and total RNA was isolated. Commercially available human and rat kidneys total RNA was purchased from Zyagen (San Diego, CA, USA).

2.3. RNA Isolation

Rat epididymal fat and liver tissue samples were lysed in RNA extraction buffer (RLT) from the RNeasy kit (Qiagen, Germantown, MD, USA) with stainless-steel disruption beads for 4 min at 30.0 Hz using the Qiagen TissueLyser (Germantown, MD, USA). Total RNA was immediately isolated from the lysate using the RNeasy kit (Qiagen, Germantown, MD, USA), following the manufacturer's instructions. Total RNA was extracted from rat glomeruli tissue and human podocyte samples using the mirVanaTM Isolation Kit (Invitrogen, Carlsbad, CA, USA), according to the manufacturer's instructions. RNA yield and purity was checked prior to downstream applications by measuring the absorbance at 260, 280, and 230 nm with a nanodrop spectrophotometer (ThermoFisher, Waltham, MA, USA) and by calculating appropriate ratios (260/280, 260/230).

2.4. DNase Treatment, cDNA Synthesis, and PCR

Total RNA was DNase treated according to manufacturer's instructions (Zymo Research, Irvine, CA, USA). 500 ng–1 µg DNase-treated RNA was reverse transcribed using the iScript cDNA Synthesis Kit (Bio-Rad, Hercules, CA, USA), according to the manufacturer's instruction and the resulting cDNA was used for reverse transcription-polymerase chain reaction (RT-PCR) using HotStarTaq Plus Master Mix Kit (Qiagen, Germantown, MD, USA). The PCR conditions used were as follows: 95 °C for 5 min, 40 cycles [95 °C for 30 s, 50–60 °C (depending on the primer pair) for 30 s, 72 °C for 30 s], hold at 4 °C. Housekeeping gene *RPL6* was amplified only for 28 cycles to prevent over-saturation. The PCR-products were separated on a 1.5–2.5% agarose gel along with molecular weight ladder (ThermoFisher, Waltham, MA, USA), stained with 0.5 µg/mL ethidium bromide (Research Products International, Mount Prospect, IL, USA), and images were captured using the Chemidoc (Bio-Rad, Hercules, CA, USA) equipment. Densitometry was performed using ImageJ Software (National Institutes of Health, Bethesda, MD, USA) and band density was calculated by subtracting background and normalizing to *RPL6*. Quantitative RT-PCR (qRT-PCR) was performed on samples using SYBR green (Bio-Rad, Hercules, CA, USA) on the Applied Biosystems 7500 Real-Time PCR System (Waltham, MA, USA). The PCR conditions were 95 °C for 10 min, 40 X (95 °C for 15 s, 50–60 °C for 1 min) followed by a melt curve to ensure specific products. Analysis was performed using the ΔΔCt method [40] with normalization to *RPL6*.

2.5. Primer Design and Synthesis

Primers were custom designed to detect various PPARγ variants for alternatively spliced mRNA products using the Reference Sequence Database (National Center for Biotechnology Information, Bethesda, MD, USA) and confirmed for the potential binding transcripts in the annotated database using the Primer Blast program (National Center for Biotechnology Information, Bethesda, MD, USA) (Table 1, Figure 1A). For variants with exon skipping, primers were designed to span the exon-exon junction to create a specific primer pair that would exclusively amplify the desired mRNA variant.

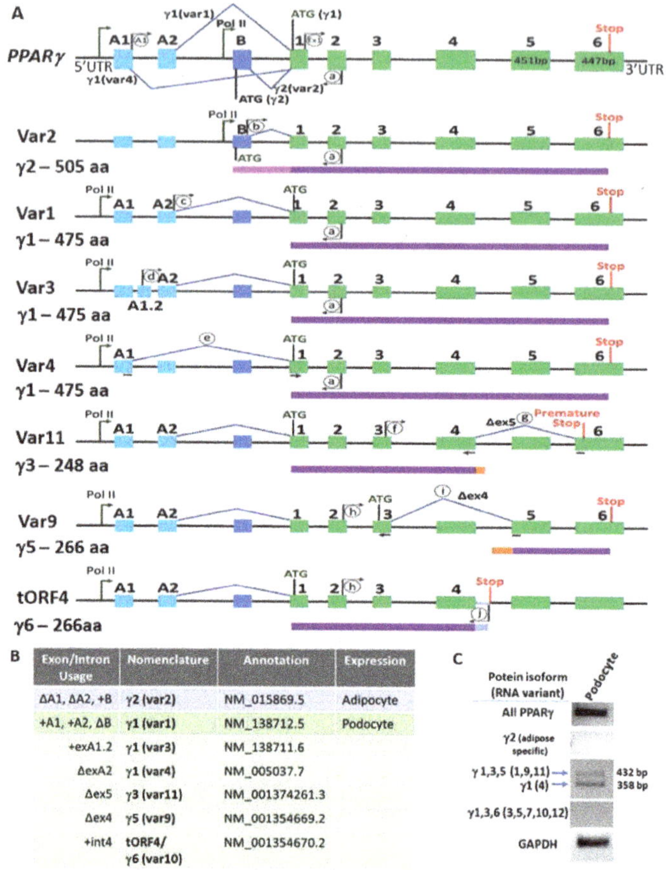

Figure 1. Alternatively Spliced Variants and Isoforms of PPARγ in Podocytes. (**A**) The schematic depicts the exon and intron usage and the transcription and translation start and stop sites for various PPARγ variants. The major isoforms of PPARγ are isoforms γ1 and γ2, which encode for proteins containing 475 and 505 amino acids, respectively. The most common human annotated variants 1 (var1: NM_138712) and 2 (var2: NM_015869) encode isoforms γ1 and γ2 (depicted in the topmost PPARγ illustration). These two variants are transcribed by different promoter usage by RNA Polymerase II (depicted in tall arrows) and alternative splicing (depicted by exon skipping). PPARγ1 variant 1 includes exons A1 and A2, a start codon (ATG) on exon 1, and γ2 variant 2 excludes exons A1 and A2, but includes exon B and a start codon in the same exon B. Thus, γ1 protein product is 30 amino acids shorter than γ2. The less known and understood variants are represented below, which include Variants 3, 4, 11, 9 and tORF4/Var 10. These are depicted with A1 and A2 usage in the schematic to emphasize those likely to be expressed in podocytes, although their adipocyte counterparts with exon B usage at the 5′ UTR also exist. A1 and A2 exons are shown in light blue, exon B in dark blue, internal exons are in green boxes, introns are black lines, spliced out exons are marked with blue lines. Primer pairs used to amplify major variants and other variants are depicted in grey arrows and circles (A1, Ex1, a–j, Table 1). For the variants that skip an exon, the primer spans the exon-exon junctions and is shown as a split grey arrow underneath the introns (c, g, i). Protein isoforms are shown below their corresponding RNA in purple and amino acid (aa) length is listed. (**B**) Summary of the protein isoforms (γ) and RNA variants (var) with detailed exon/intron usage and annotations. γ2 is found primarily in adipose tissue, while γ1 is found in podocytes. Variants 1, 3, and 4 are variants of γ1 that are likely to be found in podocytes, while variants 9, 10, and 11 are

additional less explored variants that would encode different isoforms. The variant 10 version of tORF4 is annotated that contains exon A1.2, but there is not an annotated version of tORf4 that would contain exon A1. (C) A representative gel showing the expression of several variants of PPARγ in differentiated podocytes. PCR products were generated using specific primer pairs and run on an agarose gel and stained with ethidium bromide. All PPARγ [primer pair Ex1 F, aR; 315 bp]; γ2 [primer pair bF, aR; expected product 476 bp]; variants 1, 9, 11 (432 bp PCR product) and Var 4 (358 bp product) [doublet primer pair, A1F, aR]; variants 3, 5, and 10 (415 bp product); variant 7 (341 bp product); and variant 12 (503 bp product) [primer pair dF, aR]. Primers are listed in Table 1.

Table 1. Primers Used in the Current Study.

Name	Species	Forward	Reverse
	PPARγ		
Total (All)	Homo Sapiens	ATGACCATGGTTGACACAG $_{\mathrm{Ex1}}$	GGAGTTGGAAGGCTCTTCAT $_{\mathrm{a}}$
Doublet	Homo Sapiens	CGAGGACACCGGAGAGGG $_{\mathrm{A1}}$	GGAGTTGGAAGGCTCTTCAT $_{\mathrm{a}}$
Var2 (γ2)	Homo Sapiens	TTTTAACGGATTGATCTTTTG $_{\mathrm{b}}$	GGAGTTGGAAGGCTCTTCAT $_{\mathrm{a}}$
Var1 (γ1)	Homo Sapiens	GAAAGAAGCCAACACTAAACCAC $_{\mathrm{c}}$	GGAGTTGGAAGGCTCTTCAT $_{\mathrm{a}}$
Var3 (γ1)	Homo Sapiens	TTCTGCTTAATTCCCTTTCC $_{\mathrm{d}}$	GGAGTTGGAAGGCTCTTCAT $_{\mathrm{a}}$
Var4 (γ1)	Homo Sapiens	GCCGCCGTGGCCGCAGAAA $_{\mathrm{e}}$	GGAGTTGGAAGGCTCTTCAT $_{\mathrm{a}}$
Var11 (γ3)	Homo Sapiens	CTTGCAGTGGGGATGTCTCA $_{\mathrm{f}}$	CAGCAAACCTGGGCGGTTGA $_{\mathrm{g}}$
Var9 (γ5)	Homo Sapiens	AGAGCCTTCCAACTCCCTCA $_{\mathrm{h}}$	GTCATAGATAACGAATGGCAT $_{\mathrm{i}}$
tORF4	Homo Sapiens	AGAGCCTTCCAACTCCCTCA $_{\mathrm{h}}$	AAACCCAAAACAACTTCCCG $_{\mathrm{j}}$
Var1 (γ1) *	Rattus Norvegicus	GAAAGAAGCTGTGAACCACT $_{\mathrm{Rat\ c}}$	GGAGTTTGAAGGCTCTTCAT $_{\mathrm{Rat\ a}}$
(γ2) *	Rattus Norvegicus	GCACTTCACAAGAAATTACC $_{\mathrm{Rat\ b}}$	GGAGTTTGAAGGCTCTTCAT $_{\mathrm{Rat\ a}}$
Total (All) *	Rattus Norvegicus	ATTACCATGGTTGACACAG $_{\mathrm{Rat\ Ex1}}$	GGAGTTTGAAGGCTCTTCAT $_{\mathrm{Rat\ a}}$
Podocyte Specific Genes			
NPHS1	Homo Sapiens	CGCAGGAGGAGGTGTCTTATTC	CGGGTTCCAGAGTGTCCAAG
Nphs1	Rattus Norvegicus	TGCTCTTTGCAGTTGGTGGT	TCCTGATCCTGTCCTCCGAC
SYNPO	Homo Sapiens	AGGTAGGCGTGGAGGAGG	GAGGTTCTGGTTGGGTTTGG
Synpo	Rattus Norvegicus	CCGCTTGGGTCCCCTTC	TGAACTCGTTCACCCTCTGC
WT1	Rattus Norvegicus/ Homo Sapiens	CCCTACAGCAGTGACAATTTATAC	TGCCCTTCTGTCCATTTC
Adipose Tissue Specific Genes			
CD36	Homo Sapiens	GGACTGCAGTGTAGGACTTTCC	TTCCGGTCACAGCCCATTTT
Cd36	Rattus Norvegicus	AAGTTATTGCGACATGATT	GATCCAAACACAGCATAGAT
ADIPOQ	Homo Sapiens	ACTGCAGTCTGTGGTTCTGA	ACTCCGGTTTCACCGATGTC
Adipoq	Rattus Norvegicus	TGTTCCTCTTAATCCTGCCCA	CCAACCTGCACAAGTTTCCTT
ADIPSIN	Homo Sapiens	CGCCCCGTGGTCGGAT	GACAGCTGTAGCAGCAGGAG
Adipsin	Rattus Norvegicus	AAGCTCTCCCACAATGCCTC	CATGGTACGTGCGCAGATTG
AP2	Homo Sapiens	AAAGTCAAGAGCACCATAACCTT	TGACGCCTTTCATGACGCATTCC
Ap2	Rattus Norvegicus	AAAGTGAAGAGCATCATAACCCT	TCACGCCTTTCATGACACATTCC
Housekeeping Genes			
GAPDH	Homo Sapiens	AGACACCATGGGGAAGGTGA	GACGGTGCCATGGAATTTGC
RPL6	Homo Sapiens	GCAACCCTGTCCTTGTCAGA	GCCAGCTGCTTCAGGAAAAC
Rpl6	Rattus Norvegicus	AGGCATCAGGAGGGTGAGAT	TGTGAGGGTACATGCCGTTC

* Rat nomenclature is annotated differently than human γ1 and γ2 variants. Primers $_{\mathrm{a–j,\ Ex1,\ A1}}$ are depicted in Figure 1A. Rat primers correspond to locations listed for human primers in Figure 1A.

2.6. PPAR-Promoter Responsive Element Prediction

Bulk RNASeq datasets from untreated control differentiated human immortalized podocytes (GEO GSE124622) [41] and untreated control differentiated human adipocytes (GEO GSE129153) [42] were analyzed. The average expression of control samples from the aforementioned GEO datasets were calculated and normalized expression greater than 2 were counted as detected. Genes were further cross referenced against the verified target and predicted PPAR Response Element (PPRE) genes that were downloaded from PPARgene database (http://www.ppargene.org/downloads.php (accessed on 31 July 2022)) [43] and using computational genomics approach [44]. Key podocyte and adipocyte specific genes measured in this study were also queried through the *PPAR* gene database (ppargene.org) to predict PPAR-responsive elements (PPRE) on their promoters Upon query submission, p-value and confidence level were generated if the gene was predicted as a PPAR target gene and contained putative PPREs in the 5 kb transcription start site flanking region. They were assigned into categories of confidence as: high ($p > 0.8$), medium ($0.8 \geq p > 0.6$), and low ($0.6 \geq p > 0.45$). A p value ≤ 0.45 was predicted as negative. Functional annotation of PPRE containing genes detectable exclusively in adipocytes vs. podocytes were performed using clusterProfiler (RRID:SCR_016884) [45] and plotted as many as 10 terms per cell type. The PPRE containing genes detectable in adipocyte vs. podocytes the overlapping genesets, and the associated functional terms were also compared.

2.7. Statistical Analysis

Statistical analysis was performed using GraphPad Prism Software version 8.2.0 for Windows (GraphPad Software, San Diego, CA, USA). Data are expressed as mean ± standard error of mean and compared using unpaired Student's t-test. p value significance is depicted as * $p < 0.05$, ** $p < 0.01$, **** $p < 0.0001$.

3. Results

3.1. Podocyte Specific PPARγ Splice Variants

To identify the PPARγ variants expressed in podocytes and to elucidate differences in podocytes vs. adipocytes, we examined the known variants of PPARγ by reviewing the literature, scoping out the annotated versions (RefSeq database, NCBI), and by performing variant specific PCRs using custom-designed primers. A schematic of different variant forms (Figure 1A) and their exon/intron usage information and annotation (Figure 1B) is presented. Conducting PCR utilizing variant-specific primer pairs (Figure 1A and Table 1), we could identify total PPARγ expression, specific variants corresponding to γ1 (Var1 and 4) and other forms in differentiated podocytes, but not the γ2 form (Var 2) (Figure 1C), which is highly expressed in differentiated adipocytes [46]. Variant 2 (encodes γ2, a protein of 505 amino acids) includes exon B which contains a start codon. Variant 1 (encodes γ1, a protein of 475 amino acids) utilizes exons A1 and A2 instead of B in the 5'UTR of the gene, resulting in the usage of a start codon ATG located on exon 1. Variants 3 and 4 both encode γ1 as well, but they have differing 5'UTRs. Variant 3 contains an alternative exon A1.2, and variant 4 skips exon A2. Variant 11 has a similar 5'UTR to variant 1, but it skips exon 5, resulting in a frameshift and a premature stop codon in exon 6, encoding a potential product of 248 amino acids (γ3). Variant 9 utilizes a start codon located in exon 3 and skips exon 4, encoding γ5 (266 amino acids). Variant 10 or tORF4 uses a partially retained intron after exon 4 which contains a stop codon, making a product of 266 amino acids (γ6). Variant 10 version of tORF4 is annotated and contains exon A1.2, but there is no annotated variant that uses exon A1. The possibility of expression of other variants was considered using primer pairs that would amplify specific products (Figure 1C). The doublet primer pair was able to discriminate between the cumulative expression of variants 1, 9 and 11 vs. variant 4. Cumulative expression of variants 3, 5, 7, 10 and 12 was ruled out using primer pair for Var3 (dF and aR), which would also amplify these other mentioned variants.

3.2. PPARγ Splice Variants in Podocytes vs. Adipose Tissue

PPARγ

PPARγ is a transcription factor and an established master regulator of fat cell physiology and differentiation in adipose tissue and liver [28,47]. While podocyte specific *Pparg* KO mice illustrated its role in podocytopathies [7,10], to understand the cell- and tissue-specific determinants of PPARγ function in glomeruli and in podocytes, we performed specific PCRs to detect the expression of total (all) variants and exclusively of variant 1 (encoding γ1) and variant 2 (encoding γ2) in human podocytes and whole kidney and rat tissues of glomeruli, whole kidney, liver and WAT. In accordance with the hypothesis of cell specific action of PPARγ, while expression of variant encoding the PPARγ2 isoform was observed in WAT, liver, and whole kidneys, γ2 variant was undetected in human podocytes and in the rat glomeruli (Figure 2A). As shown in Figure 1A, γ1 and γ2 use different, promoter sites and undergo AS. Expression analysis of several uncommon variants described in Figure 1 demonstrated that they are present at low and varying levels in the whole human kidney (Figure 2B). Most of these (variants 4, 3, 11, 9, and tORF4) were found to be expressed at very low to undetectable levels in podocytes (Figure 2B). Moreover, analysis of the variants encoding γ1 (Var 1 and 4) showed that their levels remained unchanged with PAN-induced injury of podocytes or treatment with PPARγ agonist, pioglitazone (Figure 2C).

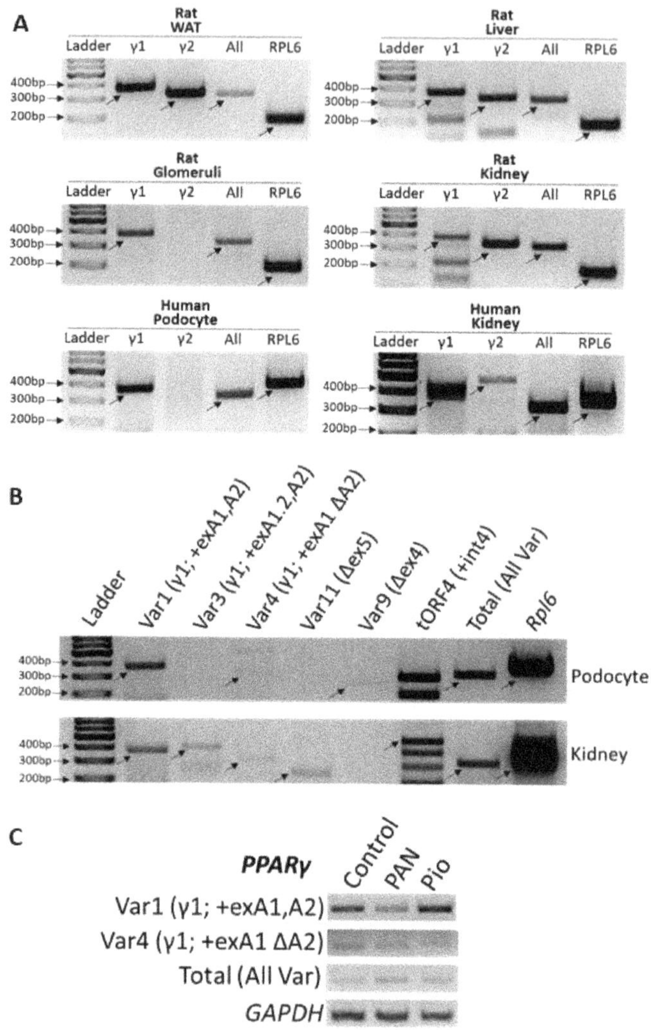

Figure 2. Expression of PPARγ Splice Variants in Podocytes vs. Adipose Tissue. (**A**) Total RNA was isolated from glomeruli, liver, and WAT samples harvested from rats, and from cultured differentiated human podocytes. Expression of the RNA variants encoding PPARγ1 and γ2 (All variants), γ1 only (Var1), and γ2 only (Var 2) isoforms was measured by RT-PCR in isolated samples as well as commercially available human and rat whole kidney RNA. Rat total (all) γ [primer pair ex1 F, ex 2 R; 315 bp], rat γ1 [primer pair exA2 F, ex2 R; 380 bp], rat γ2 [primer pair exB F, ex2 R; 329 bp], rat Rpl6 [178 bp], human total (all) γ [primer pair Ex1 F, aR; 315 bp], human γ1 [primer pair cF, aR; 387 bp], human γ2 [primer pair bF, aR; 476 bp], Human RPL6 [370 bp]. (**B**) Several PPARγ mRNA AS variants were analyzed by RT-PCR in cultured differentiated human podocytes and commercially available human whole kidney RNA using the primer pairs drawn in Figure 1A and detailed in Supplementary Table S1. Var1 [primer pair cF, aR; 387 bp], Var3 [primer pair dF, aR; 415 bp], Var4 [primer pair eF, aR; 333 bp], Var11 [primer pair fF, gR; 242 bp], Var9 [primer pair hF, iR; 254 bp], tORF4 [primer pair hF, jR; 490 bp], Total [primer pair Ex1 F, aR; 315 bp], RPL6 [370 bp] (**C**) Expression of PPARγ mRNA variants 1 [primer pair cF, aR] and 4 [primer pair eF, aR], as well as total (All) PPARγ [primer pair Ex1 F, aR] and GAPDH in puromycin aminonucleoside (PAN)-injured and pioglitazone-treated podocytes. Primers are listed in Table 1.

3.3. PPAR-Response Element (PPRE) Containing Genes in Podocyte vs. Adipocyte

PPARγ binds to its promoter response elements (PPRE) along with other transcription factors such as retinoid X receptor (RXR) or to other DNA elements in association with factors such as nuclear factor kappa B (NFκB), activator protein 1 (AP1), and in proximity to CCAAT enhancer-binding proteins (C/EBPα) [5,25,38]. We carried out promoter element prediction using a PPARgene database (ppargene.org (accessed on 31 July 2022)) [43] and cross-referenced it with bulk RNASeq GEO datasets obtained from control podocytes (GSE124622) [41] and adipocytes (GSE129153) [42]. We found that while 1669 PPRE-containing genes were detectable in both podocyte and adipocyte datasets, 453 were found to be unique to podocytes and 77 unique to adipocytes (Figure 3A(i), Supplementary Table S1). These were further classified into subsets based on the confidence level of PPRE prediction (see Methods; Figure 3A(ii)). We further analyzed a select set of PPRE containing genes known to play critical roles in podocyte pathophysiology, such as *NPHS1* (nephrin), *SYNPO* (synpatopodin), and *WT1* (wilms tumor 1) or those that have established roles in adipogenesis, such as *CD36*, *ADIPOQ*, *ADIPSIN*, and *AP2* [29,47–52] (Table 2). All the adipogenic genes depicted high level of PPRE prediction, and *NPHS1*, *SYNPO*, and *WT1* showed low to high levels of PPRE prediction. These genes which are known to play prominent roles in podocytes vs. adipocytes correlated with their expression in the GEO datasets in podocytes vs. adipocyte. In accordance, we also observed a marked disparity in the expression of these genes in podocytes vs. white adipose tissue (WAT) (Figure 3B). While the podocyte markers (*NPHS1*, *SYNPO* and *WT1*) are expressed at a higher level in podocytes vs. adipose tissue, expression of adipokines (*ADIPOQ* and *ADIPSIN*) and adipogenic genes (*CD36* and *AP2*) was found to be high in WAT and absent or modestly expressed in podocytes (Figure 3B, Table 2). Furthermore, functional enrichment analysis of PPRE-containing genes exclusive to podocyte or adipocyte generated distinct maps of biological processes, cellular components, molecular functions and pathways (Figure 3C–F). While the adipocyte gene set was rich in genes involved in temperature homeostasis, lipid catabolic process, lipase activity and regulation of lipolysis, podocyte gene set was rich in genes involved in cell–cell adhesion and membrane transporter activity. Adipocyte PPRE-containing gene products were found located in lipid droplet and collagen-containing extracellular matrix, and podocyte PPRE-containing gene products were found in cell–cell junction and tight junction. Notably, functional enrichment analysis of all PPRE-containing genes from podocytes and adipocytes (including overlapping genes) generated mostly overlapping maps of biological processes, cellular components, molecular functions and pathways (Supplementary Figure S1A–D). We further cross-referenced the podocyte and adipocyte bulk RNASeq GEO datasets with PPRE genes from another independent source wherein PPREs were predicted in the conserved elements of within 5000 bps of transcription site of human genes [44]. We found that out of 1074 PPRE genes, 478 were detectable in both podocyte and adipocyte datasets, 139 were found to be unique to podocytes and 22 unique to adipocytes (Supplementary Figure S2A, Table S2). A total of 168 genes were common between the PPRE dataset and PPRE genes described by Lemay and Hwang [44] (Supplementary Figure S2B, Table S3). A majority of these overlapping genes were detected in both podocyte and adipocyte databases (Supplementary Figure S2C).

In summary, this data suggests the existence of a PPRE-containing gene expression, enrichment and pathway signature that is distinct in podocytes vs. adipocytes. This is likely driven by podocyte specific γ1 form of PPARγ vs. the adipocyte-specific γ2 form.

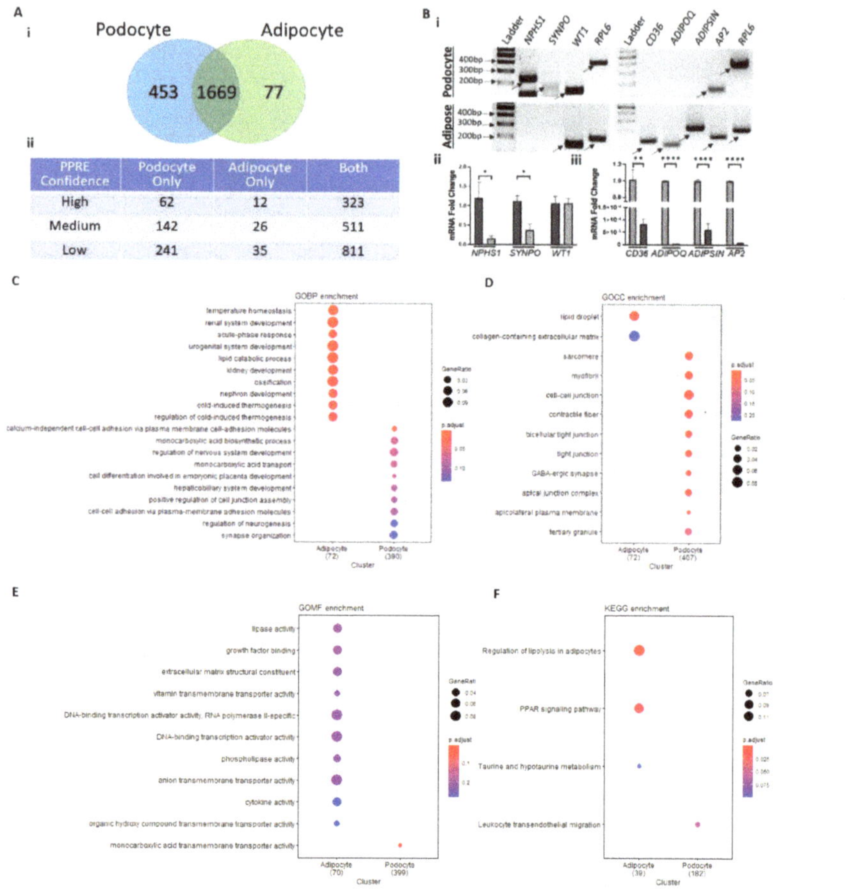

Figure 3. PPARγ-Response Element (PPRE) Containing Genes in Podocytes vs. Adipocytes. (**A**) Normalized average log counts of untreated control human differentiated podocyte and adipocyte GEO datasets (GSE124622 and GSE129153, respectively) were cross-referenced against the verified target and predicted PPRE genes from the PPARgene database. (i) 1669 PPRE-containing genes were detectable in both podocyte and adipocyte datasets, 453 unique to podocytes and 77 unique to adipocytes. (ii) PPRE-containing genes classified by their confidence level. High-confidence ($p > 0.8$), median-confidence ($0.8 \geq p > 0.6$), low-confidence category ($0.6 \geq p > 0.45$). Genes with p value < 0.45 were predicted as negative. (**B**) Total RNA was isolated from cultured differentiated human podocytes and WAT harvested from rats. Expression of the podocyte marker genes (NPHS1 [Hu: 234 bp, Rat: 138 bp], SYNPO [Hu: 182 bp, Rat: 325 bp], and WT1 [Hu/Rat: 134 bp]) and adipose tissue genes (CD36 [Hu: 304 bp, Rat: 123 bp], ADIPOQ [Hu: 347 bp, Rat: 95 bp], ADIPSIN [Hu: 301 bp, Rat: 202 bp], and AP2 [Hu: 133 bp, Rat: 133 bp]), and housekeeping gene RPL6 [Hu: 370 bp, Rat: 178 bp] were analyzed. (i) Representative gels depicting the PCR products of key genes in podocytes and adipose tissue. (ii–iii) Quantitative mRNA fold-changes measured and graphically presented using RT-PCR and real time qRT-PCR quantification of key genes in podocytes and adipose tissue. Primers are listed in Table 1. (**C**–**F**) clusterProfiles generated functional enrichment of PPRE containing genes detectable exclusively in adipocytes (77) vs. podocytes (453) and plotted as 10 terms per cell type for (**C**) biological processes, BP, (**D**) cellular components, CC, (**E**) molecular functions, MF, and (**F**) kyoto encycopedia of genes and genomes, KEGG. The color of the dot indicates the intensity of adj p value (smaller adj p value is more red) and size of the dot indicates the proportion of genes from the term that are present in the cell-specific PPRE containing genes. * $p < 0.05$, ** $p < 0.01$, **** $p < 0.0001$.

Table 2. Expression of Select PPRE Containing Genes.

Gene	Cells/Tissue Tested in the Current Study	p Value [a]	Confidence Level	PPREs Predicted [b]	Podocyte Count log [c]	Adipocyte Count log [d] (Normalized)
Podocyte Prominent Genes						
Nphs1	Podocyte/WAT	0.63696	Medium	4	2.27	–
Synpo	Podocyte/WAT	0.8715	High	5	7.89	9.03 (1.3)
Wt1	Podocyte/WAT	0.55556	Low	5	4.91	–
Adipocyte Prominent Genes						
Cd36	Podocyte/WAT	0.97886	High	3	3.77	846.26 (117.2)
Adipoq	Podocyte/WAT	0.97153	High	12	2.59	112.19 (15.5)
Adipsin	Podocyte/WAT	0.96216	High	6	4.22	991.32 (137.3)
Ap2/Fabp4	Podocyte/WAT	0.99997	High	10	–	1663.6 (230.5)
Housekeeping						
Rpl6	Podocyte/WAT	–	–	–	13.42	96.86 (13.42)

[a] Probability of PPAR target gene, higher value means a higher confidence. High-confidence ($p > 0.8$), median-confidence ($0.8 \geq p > 0.6$), low-confidence category ($0.6 \geq p > 0.45$). Genes with p value ≤ 0.45 were predicted as negative. [b] Putative PPREs in the 5Kb transcription start site (TSS) flanking region. [c] Podocyte GEO Dataset # GSE124622. [d] Adipocyte GEO Dataset # GSE129153.

4. Discussion

PPARγ agonism has a well-established beneficial role in the setting of diabetes, which has led to the generation of TZDs for the treatment of type II diabetes (Figure 4) [53,54]. Its beneficial role in kidney cells beyond its favorable systemic metabolic effects in diabetes originated from numerous preclinical studies and meta-analyses [55–59]. Earlier these effects of PPARγ in influencing non-diabetic glomerular disease was thought to be through its anti-inflammatory actions on endothelial and myeloid cells [60–62]. However, recent discoveries and advances suggest the role of podocytes in mediating these effects. Seminal studies from our group and other research teams suggested that TZDs such as pioglitazone, directly protect podocytes from injury [1–5] and reduce proteinuria and glomerular injury in various animal models of glomerular disease [5–13] (Figure 4). Moreover, studies using podocyte specific *Pparg* KO mouse demonstrated that the beneficial effects of PPARγ are mediated by activation of podocyte PPARγ, thus indicating a pivotal role for PPARγ in maintaining glomerular function through preservation of podocytes [1,5–7,10]. However, the specific variant or isoform(s) expression of *PPARγ* and the role of its AS in podocytes and glomerular disease is unexplored. We hypothesized that AS variants of PPARγ expressed in podocytes are distinct from adipocytes. We addressed this hypothesis by performing a comprehensive analysis of different AS variants of PPARγ in human podocytes and by studying the expression of γ1 vs. γ2 variant in podocyte vs. adipose tissue in detail. Our findings suggest that the podocytes mainly express the y1 form, along with minimal expression of other AS forms. They do not express the γ2 form, yet they are responsive to treatments by PPARγ agonists, implicating that new selective modulators can be potentially developed that will be able to distinguish between the two forms, γ1 and γ2, thus forming a basis of novel targeted therapeutic avenues.

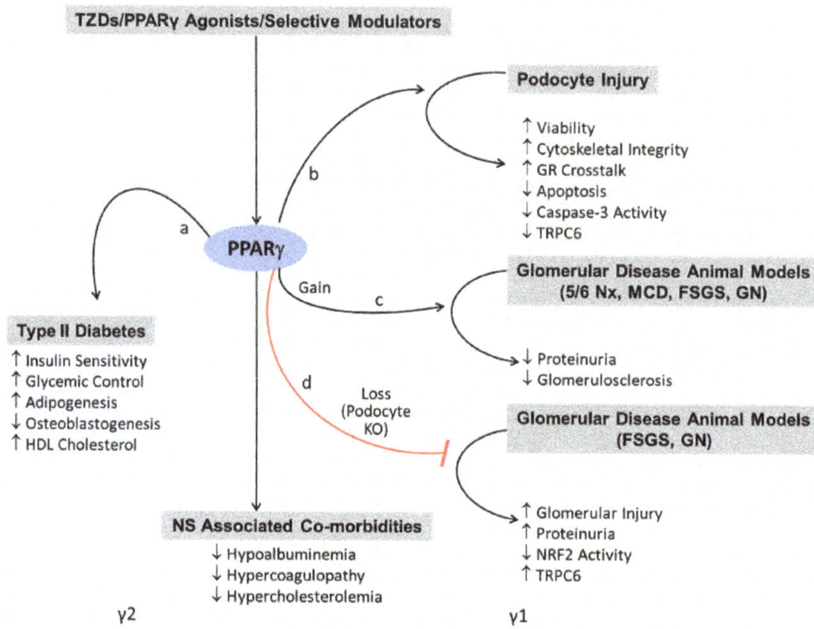

Figure 4. Illustration Describing the Roles of PPARγ1 and 2 Isoforms. Schematic of the suggested roles of PPARγ 1 and 2 forms in regulating podocyte/glomerular disease specific effects (right panel) vs. metabolic/systemic and adipogenic effects (left panel). PPARγ agonism plays a well-established role in the setting of treatment of Type II Diabetes, and treatment with its agonists, thiazolidinediones (TZDs), leads to increased ('a' pathway) insulin sensitivity and glycemic control [53,54]. However, these beneficial effects are also accompanied with adverse effects ('a' pathway), such as increased cholesterol, adipogenesis and decreased osteoblastogenesis [25]. TZDs or PPARγ agonists have now been demonstrated to reduce podocyte injury ('b' pathway) by enhancing their viability and cytoskeletal integrity and decreasing apoptosis [1–5]. A few pathways involved in this process include crosstalk with the glucocorticoid receptor (GR), decreased caspase-3 activity and decreased TRPC6 expression. Gain of PPARγ activity or its activation by TZDs ('c' pathway) has been shown to reduce proteinuria and glomerulosclerosis in various animal models of glomerular disease, such as 5/6 nephrectomy (Nx), minimal change disease (MCD), focal segmental glomerulosclerosis (FSGS) and glomerulonephritis (GN) [5–13,24]. On the other hand, loss of PPARγ in podocyte specific *Pparg* KO mouse ('d' pathway) has demonstrated exacerbation of glomerular injury and proteinuria in animal disease models of FSGS and GN [1,5–7,10]. Analysis of the podocyte vs. adipose tissue specific expression of genes downstream of PPARγ containing putative or identified PPREs (Figure 3) informs us that these downstream effects in podocytes/glomeruli are likely directed by the PPARγ1 splice variant, distinct from the adipocyte-regulatory γ2 variant.

Since their identification in 1990, PPARγ has been recognized as a nuclear receptor superfamily member, a ligand-dependent transcription factor and a master regulator of adipogenesis and metabolism, which accounts for the insulin sensitizing effects of its agonists or anti-diabetic drugs such as TZDs [63]. While its role in regulating adipogenesis and lipid metabolism in adipose tissue, liver and skeletal muscle is well-characterized, the knowledge of PPARγ-regulated signaling pathways in kidneys and in podocytes is scarce [5–10,25–29,64]. In our previous studies we speculated a cross-talk model between the PPARγ and glucocorticoid receptor and further demonstrated that while pioglitazone could provide proteinuria reducing benefits, it could also enhance the ability of low-dose steroids to reduce proteinuria [6]. This suggested the potential clinical use of these FDA-approved drugs in enabling the reduction of steroid dose, toxicity, and side effects to treat NS. We further translated these findings and demonstrated the ability of pioglitazone administration to improve clinical outcome in a child with steroid-refractory NS, which has now been extended in other NS patients as well [6,65]. Moreover, we have recently found that targeting PPARγ with a selective modulator, GQ-16, offers an even better therapeutic strategy in NS, by reducing proteinuria and NS-associated side effects with a greater efficacy while providing reduced adipogenic effects than traditional TZDs [24]. However, to understand the proteinuria-reducing podocyte protective effects of PPARγ activation, it is critical to build an understanding of podocyte specific transcriptional activity of PPARγ as the lack of such knowledge is a barrier to modulating PPARγ activity effectively. The existence of two isoforms γ1 and γ2 as a result of different promoter usage and AS further complicates the PPARγ biology [5,30,31]. As a result, γ2 contains 30 additional amino acids at its N terminal, despite a longer γ1 mRNA sequence at its 5′UTR. We found that podocytes as well as glomeruli, express the more ubiquitous form of PPARγ, the γ1 form and not the γ2 form, which is restricted to adipose tissue and liver. Interestingly a few other spliced forms of PPARγ have been discovered in the past few years, which may confer varied functions [32–36]. Approximately 95% of mammalian genes undergo AS that can result in different protein products as well as microRNA sensitivity, mRNA stability, localization, and translation efficiency [66,67]. Moreover, AS is associated with many cellular processes and diseases such as differentiation, cancer, and immunity to name a few [66,68]. Moreover, expression of PPARγ1 and 2 mRNAs and isoforms in adipose tissue, liver, and skeletal muscle is known to be regulated in animal models of obesity, diabetes, and nutrition [69]. As the expression and role of the AS of PPARγ in podocyte biology is unexplored, we performed a landscape analysis of other PPARγ forms, which have been identified mostly in adipose tissues in recent years, including some with dominant negative function [32–37]. We could detect low expression of variants 4, 3, 11, and tORF4, and variant 9 was undetectable in the whole kidney. In podocytes, Var 9 was detectable at very low levels, and these additional variants were barely detectable consistently. We have recently also observed the presence of variant 11 form due to skipping of exon 5 (delta 5) in the WAT, which was particularly reduced with both pioglitazone and GQ-16 [24]. Ligand-mediated PPARγ activation has been shown to induce skipping of exon 5 which by itself lacks ligand-dependent transactivation ability, and thus acts as a dominant negative form of PPARγ [32]. In particular, recent evidence points towards distinct roles of γ1 vs. γ2 variant of PPARγ governing specific and separate gene expression and metabolic functions at different stages in adipocytes [70], as well as between adipocytes and macrophages [38]. In the current study, analysis of the podocyte vs. adipocyte specific expression of genes downstream of PPARγ containing putative or identified PPREs informs us that these downstream effects in podocytes are likely directed by the PPARγ1 splice variant, distinct from the adipocyte-regulatory γ2 variant (Figures 3 and 4). This is also evident from differential gene expression, enrichment and signaling pathways in the podocytes vs. adipocytes. Notably, we found podocyte specific PPRE-containing genes to be functionally enriched in cell–cell junction and cell–cell adhesion, which are known to play critical roles in the maintenance of podocyte structure and function and in its normal physiology [71]. This also suggests that activation of podocyte PPARγ with agonists or selective modulators to drive these pathways is a viable therapeutic option to restore podocyte health during injury. At the same time, we acknowledge the limitations of in silico prediction of PPREs, species-specific differences that may exist in the location and functionality of these PPREs [72], and the involvement of other co-transcription factors that determine cell specificity of PPARγ [38]. Overall, previous evidence indicates that the: (1) activation of PPARγ results in reduction in podocyte injury and proteinuria [1,4–6,12,13], (2) podocyte specific PPARγ KO results in resistance to protection of proteinuria by PPARγ agonists [7,10], and (3) selective modulation of PPARγ enhances proteinuria reducing efficacy with reduced adipogenic potential [24]. Taken together with the findings in the current study suggesting that the podocytes do not express the γ2 form, yet they are responsive to treatments by PPARγ agonists, implicate that those new selective modulators can be potentially developed that will be able to distinguish between the two forms, γ1 and γ2, thus forming a basis of novel targeted therapeutic avenues.

PPARγ also undergoes post-translational modifications and its differential phosphorylation at serine (Ser) 273 is an important determinant of its effect on adipogenesis and insulin sensitivity [47,73,74]. PPARγ phosphorylation at Ser273 was demonstrated in obesity models and treatment with the PPARγ agonist dephosphorylates this residue in the adipose tissue. GQ-16, a selective modulator of PPARγ has been shown to dephosphorylate PPARγ in an in vitro assay like traditional TZDs [73,74]. Since this site of phosphorylation on Ser273 is encoded by the last codon on exon 4, it remains conserved in major variants, 1 and 2, and several other variants in the current study, except for Var9, which was only detectable in podocytes at very low levels. It should be noted that the relative position is Ser243 in the variants including A1 and A2 exons instead of Ser273 in the B exon. Moreover, our recent findings suggested that relative phosphorylated PPARγ levels (at Ser273) remained unaltered during glomerular injury and with treatments with pioglitazone and GQ-16 treatments in the glomeruli [24]. These data suggest that while PPARγ-Ser273 is critical in determining the insulin sensitizing effects of PPARγ in adipose tissue, it is unlikely to be a major mechanism of regulation in podocytes or glomeruli.

In the present study, we have observed a clear indication of the disparate roles of γ1 vs. γ2 in podocyte vs. adipocyte function and physiology and demonstrated the presence of other AS variants of PPARγ mRNA in podocytes that would share the 5′UTR with variant 1 (encoding γ1 isoform), although the roles of these other variants in podocytopathies and podocyte differentiation is still unclear. Their detection of differential expression is a challenge as they are observed to be present at too low levels in podocytes (unlike the adipocyte variants). It might be worthwhile to pursue studies in future wherein these variants are overexpressed using reporter plasmids followed by the analysis of their effects on podocyte physiology, actin cytoskeleton and function in the presence of injury causing agents, such as PAN. The demonstration of presence of these variants also warrants their detection in kidney biopsies from patients treated with PPARγ agonists for diabetic nephropathy or type II diabetes.

In summary, we have shown that podocytes and glomeruli express the variants encoding the major form of PPARγ1 isoform and do not express the variants encoding the other major form γ2. Moreover, expression of other AS forms of PPARγ has been detected at very low levels in podocytes which may have implications in its pathophysiology. Future studies directed at developing new selective modulators that can distinguish between the two forms, γ1 and γ2, to enhance beneficial effects and reduce harmful effects of PPARγ activation has the potential to provide novel targeted therapeutic avenues.

Supplementary Materials: The following supporting information can be downloaded at: https://www.mdpi.com/article/10.3390/cells11213455/s1, Figure S1: clusterProfiles generated functional enrichment of all PPRE containing genes detectable in adipocytes and/or podocytes; Figure S2: PPARγ-Response Element (PPRE) containing Genes in PPARgene dataset and Lemay and Hwang Manuscript detected in podocytes and/or adipocytes; Table S1: PPRE-driven genes from the PPARgene dataset, identified in podocyte and adipocyte datasets; Table S2: PPRE-driven genes from Lemay and Hwang manuscript, identified in podocyte and adipocyte datasets; Table S3: PPRE-driven genes in the PPARgene dataset and Lemay and Hwang manuscript

Author Contributions: Conceptualization, S.A.; Methodology, S.A., R.G., A.W., D.C., A.S.B. and C.B.; Formal Analysis, C.B., A.W. and S.A.; Investigation, C.B. and S.A.; Resources, S.A.; Data Curation, C.B., A.W. and S.A.; Writing—Original Draft Preparation, C.B. and S.A.; Writing—Review and Editing, S.A., R.G., A.W., D.C., A.S.B. and C.B.; Visualization, C.B. and S.A.; Supervision, S.A.; Project Administration, S.A.; Funding Acquisition, S.A. All authors have read and agreed to the published version of the manuscript.

Funding: This study was supported by the American Heart Association Career Development Award (CDA34110287), funds from Nationwide Children's Hospital (NCH) and Stony Brook University Renaissance School of Medicine to S.A.

Data Availability Statement: The original contributions presented in the study are included in the article material, further inquiries can be directed to the corresponding author.

Acknowledgments: The authors thank Jeffrey Kopp at the National Institute of Health and Moin Saleem at the University of Bristol for sharing the immortalized human podocyte cell line, and members of the animal resource core at the Abigail Wexner Research Institute, Nationwide Children's Hospital (NCH).

Conflicts of Interest: The authors declare no conflict of interest.

References

1. Agrawal, S.; Guess, A.J.; Benndorf, R.; Smoyer, W.E. Comparison of direct action of thiazolidinediones and glucocorticoids on renal podocytes: Protection from injury and molecular effects. *Mol. Pharmacol.* **2011**, *80*, 389–399. [CrossRef] [PubMed]
2. Miglio, G.; Rosa, A.C.; Rattazzi, L.; Grange, C.; Camussi, G.; Fantozzi, R. Protective effects of peroxisome proliferator-activated receptor agonists on human podocytes: Proposed mechanisms of action. *Br. J. Pharmacol.* **2012**, *167*, 641–653. [CrossRef] [PubMed]
3. Miglio, G.; Rosa, A.C.; Rattazzi, L.; Grange, C.; Collino, M.; Camussi, G.; Fantozzi, R. The subtypes of peroxisome proliferator-activated receptors expressed by human podocytes and their role in decreasing podocyte injury. *Br. J. Pharmacol.* **2011**, *162*, 111–125. [CrossRef] [PubMed]
4. Kanjanabuch, T.; Ma, L.J.; Chen, J.; Pozzi, A.; Guan, Y.; Mundel, P.; Fogo, A.B. PPAR-gamma agonist protects podocytes from injury. *Kidney Int.* **2007**, *71*, 1232–1239. [CrossRef]
5. Agrawal, S.; He, J.C.; Tharaux, P.L. Nuclear receptors in podocyte biology and glomerular disease. *Nat. Rev. Nephrol.* **2021**, *17*, 185–204. [CrossRef]
6. Agrawal, S.; Chanley, M.A.; Westbrook, D.; Nie, X.; Kitao, T.; Guess, A.J.; Benndorf, R.; Hidalgo, G.; Smoyer, W.E. Pioglitazone Enhances the Beneficial Effects of Glucocorticoids in Experimental Nephrotic Syndrome. *Sci. Rep.* **2016**, *6*, 24392. [CrossRef]
7. Henique, C.; Bollee, G.; Lenoir, O.; Dhaun, N.; Camus, M.; Chipont, A.; Flosseau, K.; Mandet, C.; Yamamoto, M.; Karras, A.; et al. Nuclear Factor Erythroid 2-Related Factor 2 Drives Podocyte-Specific Expression of Peroxisome Proliferator-Activated Receptor gamma Essential for Resistance to Crescentic GN. *J. Am. Soc. Nephrol.* **2016**, *27*, 172–188. [CrossRef]
8. Ma, L.J.; Marcantoni, C.; Linton, M.F.; Fazio, S.; Fogo, A.B. Peroxisome proliferator-activated receptor-gamma agonist troglitazone protects against nondiabetic glomerulosclerosis in rats. *Kidney Int.* **2001**, *59*, 1899–1910. [CrossRef]
9. Platt, C.; Coward, R.J. Peroxisome proliferator activating receptor-gamma and the podocyte. *Nephrol. Dial. Transplant.* **2016**, *32*, 423–433.
10. Sonneveld, R.; Hoenderop, J.G.; Isidori, A.M.; Henique, C.; Dijkman, H.B.; Berden, J.H.; Tharaux, P.L.; van der Vlag, J.; Nijenhuis, T. Sildenafil Prevents Podocyte Injury via PPAR-gamma-Mediated TRPC6 Inhibition. *J. Am. Soc. Nephrol.* **2017**, *28*, 1491–1505. [CrossRef]
11. Yang, H.C.; Deleuze, S.; Zuo, Y.; Potthoff, S.A.; Ma, L.J.; Fogo, A.B. The PPARgamma agonist pioglitazone ameliorates aging-related progressive renal injury. *J. Am. Soc. Nephrol.* **2009**, *20*, 2380–2388. [CrossRef] [PubMed]
12. Yang, H.C.; Ma, L.J.; Ma, J.; Fogo, A.B. Peroxisome proliferator-activated receptor-gamma agonist is protective in podocyte injury-associated sclerosis. *Kidney Int.* **2006**, *69*, 1756–1764. [CrossRef] [PubMed]
13. Zuo, Y.; Yang, H.C.; Potthoff, S.A.; Najafian, B.; Kon, V.; Ma, L.J.; Fogo, A.B. Protective effects of PPARgamma agonist in acute nephrotic syndrome. *Nephrol. Dial. Transplant.* **2012**, *27*, 174–181. [CrossRef] [PubMed]
14. Benzing, T. Molecular Design of the Kidney Filtration Barrier. *Trans. Am. Clin. Climatol. Assoc.* **2020**, *131*, 125–139.
15. Butt, L.; Unnersjo-Jess, D.; Hohne, M.; Edwards, A.; Binz-Lotter, J.; Reilly, D.; Hahnfeldt, R.; Ziegler, V.; Fremter, K.; Rinschen, M.M.; et al. A molecular mechanism explaining albuminuria in kidney disease. *Nat. Metab.* **2020**, *2*, 461–474. [CrossRef]
16. Assady, S.; Wanner, N.; Skorecki, K.L.; Huber, T.B. New Insights into Podocyte Biology in Glomerular Health and Disease. *J. Am. Soc. Nephrol.* **2017**, *28*, 1707–1715. [CrossRef]
17. Kerjaschki, D. 2015 Homer W. Smith Award: The Podocyte from Periphery to Center Stage. *J. Am. Soc. Nephrol.* **2016**, *27*, 3266–3270. [CrossRef]
18. Siddall, E.C.; Radhakrishnan, J. The pathophysiology of edema formation in the nephrotic syndrome. *Kidney Int.* **2012**, *82*, 635–642. [CrossRef]
19. Araya, C.E.; Wasserfall, C.H.; Brusko, T.M.; Mu, W.; Segal, M.S.; Johnson, R.J.; Garin, E.H. A case of unfulfilled expectations. Cytokines in idiopathic minimal lesion nephrotic syndrome. *Pediatr. Nephrol.* **2006**, *21*, 603–610.
20. Araya, C.; Diaz, L.; Wasserfall, C.; Atkinson, M.; Mu, W.; Johnson, R.; Garin, E. T regulatory cell function in idiopathic minimal lesion nephrotic syndrome. *Pediatr. Nephrol.* **2009**, *24*, 1691–1698. [CrossRef]
21. Agrawal, S.; Zaritsky, J.J.; Fornoni, A.; Smoyer, W.E. Dyslipidaemia in nephrotic syndrome: Mechanisms and treatment. *Nat. Rev. Nephrol.* **2018**, *14*, 57–70. [CrossRef] [PubMed]

22. Kendall, A.G.; Lohmann, R.C.; Dossetor, J.B. Nephrotic syndrome. A hypercoagulable state. *Arch. Intern. Med.* **1971**, *127*, 1021–1027. [CrossRef] [PubMed]
23. Alkjaersig, N.; Fletcher, A.P.; Narayanan, M.; Robson, A.M. Course and resolution of the coagulopathy in nephrotic children. *Kidney Int.* **1987**, *31*, 772–780. [CrossRef] [PubMed]
24. Bryant, C.; Rask, G.; Waller, A.P.; Webb, A.; Galdino-Pitta, M.R.; Amato, A.A.; Cianciolo, R.; Govindarajan, R.; Becknell, B.; Kerlin, B.A.; et al. Selective modulator of nuclear receptor PPARgamma with reduced adipogenic potential ameliorates experimental nephrotic syndrome. *iScience* **2022**, *25*, 104001. [CrossRef]
25. Lefterova, M.I.; Haakonsson, A.K.; Lazar, M.A.; Mandrup, S. PPARgamma and the global map of adipogenesis and beyond. *Trends Endocrinol. Metab.* **2014**, *25*, 293–302. [CrossRef]
26. Berger, J.P.; Akiyama, T.E.; Meinke, P.T. PPARs: Therapeutic targets for metabolic disease. *Trends Pharmacol. Sci.* **2005**, *26*, 244–251. [CrossRef]
27. Ahmadian, M.; Suh, J.M.; Hah, N.; Liddle, C.; Atkins, A.R.; Downes, M.; Evans, R.M. PPARgamma signaling and metabolism: The good, the bad and the future. *Nat. Med.* **2013**, *19*, 557–566. [CrossRef]
28. Evans, R.M.; Barish, G.D.; Wang, Y.X. PPARs and the complex journey to obesity. *Nat. Med.* **2004**, *10*, 355–361. [CrossRef]
29. Agrawal, S.; Guess, A.J.; Chanley, M.A.; Smoyer, W.E. Albumin-induced podocyte injury and protection are associated with regulation of COX-2. *Kidney Int.* **2014**, *86*, 1150–1160. [CrossRef]
30. Fajas, L.; Auboeuf, D.; Raspe, E.; Schoonjans, K.; Lefebvre, A.M.; Saladin, R.; Najib, J.; Laville, M.; Fruchart, J.C.; Deeb, S.; et al. The organization, promoter analysis, and expression of the human PPARgamma gene. *J. Biol. Chem.* **1997**, *272*, 18779–18789. [CrossRef]
31. Mukherjee, R.; Jow, L.; Croston, G.E.; Paterniti, J.R., Jr. Identification, characterization, and tissue distribution of human peroxisome proliferator-activated receptor (PPAR) isoforms PPARgamma2 versus PPARgamma1 and activation with retinoid X receptor agonists and antagonists. *J. Biol. Chem.* **1997**, *272*, 8071–8076. [CrossRef] [PubMed]
32. Aprile, M.; Cataldi, S.; Ambrosio, M.R.; D'Esposito, V.; Lim, K.; Dietrich, A.; Bluher, M.; Savage, D.B.; Formisano, P.; Ciccodicola, A.; et al. PPARgammaDelta5, a Naturally Occurring Dominant-Negative Splice Isoform, Impairs PPARgamma Function and Adipocyte Differentiation. *Cell Rep.* **2018**, *25*, 1577–1592.e6. [CrossRef] [PubMed]
33. Aprile, M.; Ambrosio, M.R.; D'Esposito, V.; Beguinot, F.; Formisano, P.; Costa, V.; Ciccodicola, A. PPARG in Human Adipogenesis: Differential Contribution of Canonical Transcripts and Dominant Negative Isoforms. *PPAR Res.* **2014**, *2014*, 537865. [CrossRef] [PubMed]
34. Chen, Y.; Jimenez, A.R.; Medh, J.D. Identification and regulation of novel PPAR-gamma splice variants in human THP-1 macrophages. *Biochim. Biophys. Acta* **2006**, *1759*, 32–43. [CrossRef]
35. Sabatino, L.; Casamassimi, A.; Peluso, G.; Barone, M.V.; Capaccio, D.; Migliore, C.; Bonelli, P.; Pedicini, A.; Febbraro, A.; Ciccodicola, A.; et al. A novel peroxisome proliferator-activated receptor gamma isoform with dominant negative activity generated by alternative splicing. *J. Biol. Chem.* **2005**, *280*, 26517–26525. [CrossRef]
36. Zhou, J.; Wilson, K.M.; Medh, J.D. Genetic analysis of four novel peroxisome proliferator activated receptor-gamma splice variants in monkey macrophages. *Biochem. Biophys. Res. Commun.* **2002**, *293*, 274–283. [CrossRef]
37. Mukha, A.; Kalkhoven, E.; van Mil, S.W.C. Splice variants of metabolic nuclear receptors: Relevance for metabolic disease and therapeutic targeting. *Biochim. Biophys. Acta Mol. Basis Dis.* **2021**, *1867*, 166183. [CrossRef]
38. Lefterova, M.I.; Steger, D.J.; Zhuo, D.; Qatanani, M.; Mullican, S.E.; Tuteja, G.; Manduchi, E.; Grant, G.R.; Lazar, M.A. Cell-specific determinants of peroxisome proliferator-activated receptor gamma function in adipocytes and macrophages. *Mol. Cell. Biol.* **2010**, *30*, 2078–2089. [CrossRef]
39. Saleem, M.A.; O'Hare, M.J.; Reiser, J.; Coward, R.J.; Inward, C.D.; Farren, T.; Xing, C.Y.; Ni, L.; Mathieson, P.W.; Mundel, P. A conditionally immortalized human podocyte cell line demonstrating nephrin and podocin expression. *J. Am. Soc. Nephrol.* **2002**, *13*, 630–638. [CrossRef]
40. Pfaffl, M.W. A new mathematical model for relative quantification in real-time RT-PCR. *Nucleic Acids Res.* **2001**, *29*, e45. [CrossRef]
41. Solanki, A.K.; Srivastava, P.; Rahman, B.; Lipschutz, J.H.; Nihalani, D.; Arif, E. The Use of High-Throughput Transcriptomics to Identify Pathways with Therapeutic Significance in Podocytes. *Int. J. Mol. Sci.* **2019**, *21*, 274. [CrossRef] [PubMed]
42. Whitehead, A.; Krause, F.N.; Moran, A.; MacCannell, A.D.V.; Scragg, J.L.; McNally, B.D.; Boateng, E.; Murfitt, S.A.; Virtue, S.; Wright, J.; et al. Brown and beige adipose tissue regulate systemic metabolism through a metabolite interorgan signaling axis. *Nat. Commun.* **2021**, *12*, 1905. [CrossRef] [PubMed]
43. Fang, L.; Zhang, M.; Li, Y.; Liu, Y.; Cui, Q.; Wang, N. PPARgene: A Database of Experimentally Verified and Computationally Predicted PPAR Target Genes. *PPAR Res.* **2016**, *2016*, 6042162. [CrossRef] [PubMed]
44. Lemay, D.G.; Hwang, D.H. Genome-wide identification of peroxisome proliferator response elements using integrated computational genomics. *J. Lipid Res.* **2006**, *47*, 1583–1587. [CrossRef]
45. Yu, G.; Wang, L.G.; Han, Y.; He, Q.Y. clusterProfiler: An R package for comparing biological themes among gene clusters. *OMICS* **2012**, *16*, 284–287. [CrossRef] [PubMed]
46. Tontonoz, P.; Hu, E.; Spiegelman, B.M. Stimulation of adipogenesis in fibroblasts by PPAR gamma 2, a lipid-activated transcription factor. *Cell* **1994**, *79*, 1147–1156. [CrossRef]

47. Choi, J.H.; Banks, A.S.; Estall, J.L.; Kajimura, S.; Bostrom, P.; Laznik, D.; Ruas, J.L.; Chalmers, M.J.; Kamenecka, T.M.; Bluher, M.; et al. Anti-diabetic drugs inhibit obesity-linked phosphorylation of PPARgamma by Cdk5. *Nature* **2010**, *466*, 451–456. [CrossRef]
48. Ettou, S.; Jung, Y.L.; Miyoshi, T.; Jain, D.; Hiratsuka, K.; Schumacher, V.; Taglienti, M.E.; Morizane, R.; Park, P.J.; Kreidberg, J.A. Epigenetic transcriptional reprogramming by WT1 mediates a repair response during podocyte injury. *Sci. Adv.* **2020**, *6*, eabb5460. [CrossRef]
49. Eddy, A.A.; Fogo, A.B. Plasminogen activator inhibitor-1 in chronic kidney disease: Evidence and mechanisms of action. *J. Am. Soc. Nephrol.* **2006**, *17*, 2999–3012. [CrossRef]
50. Kobayashi, N.; Ueno, T.; Ohashi, K.; Yamashita, H.; Takahashi, Y.; Sakamoto, K.; Manabe, S.; Hara, S.; Takashima, Y.; Dan, T.; et al. Podocyte injury-driven intracapillary plasminogen activator inhibitor type 1 accelerates podocyte loss via uPAR-mediated beta1-integrin endocytosis. *Am. J. Physiol. Renal Physiol.* **2015**, *308*, 614–626. [CrossRef]
51. Cheng, H.; Fan, X.; Guan, Y.; Moeckel, G.W.; Zent, R.; Harris, R.C. Distinct roles for basal and induced COX-2 in podocyte injury. *J. Am. Soc. Nephrol.* **2009**, *20*, 1953–1962. [CrossRef] [PubMed]
52. Asanuma, K.; Kim, K.; Oh, J.; Giardino, L.; Chabanis, S.; Faul, C.; Reiser, J.; Mundel, P. Synaptopodin regulates the actin-bundling activity of alpha-actinin in an isoform-specific manner. *J. Clin. Investig.* **2005**, *115*, 1188–1198. [CrossRef] [PubMed]
53. Ahsan, W. The Journey of Thiazolidinediones as Modulators of PPARs for the Management of Diabetes: A Current Perspective. *Curr. Pharm. Des.* **2019**, *25*, 2540–2554. [CrossRef] [PubMed]
54. Hofmann, C.A.; Colca, J.R. New oral thiazolidinedione antidiabetic agents act as insulin sensitizers. *Diabetes Care* **1992**, *15*, 1075–1078. [CrossRef]
55. Sarafidis, P.A.; Stafylas, P.C.; Georgianos, P.I.; Saratzis, A.N.; Lasaridis, A.N. Effect of thiazolidinediones on albuminuria and proteinuria in diabetes: A meta-analysis. *Am. J. Kidney Dis.* **2010**, *55*, 835–847. [CrossRef]
56. Buckingham, R.E.; Al-Barazanji, K.A.; Toseland, C.D.; Slaughter, M.; Connor, S.C.; West, A.; Bond, B.; Turner, N.C.; Clapham, J.C. Peroxisome proliferator-activated receptor-gamma agonist, rosiglitazone, protects against nephropathy and pancreatic islet abnormalities in Zucker fatty rats. *Diabetes* **1998**, *47*, 1326–1334. [CrossRef]
57. Cha, D.R.; Zhang, X.; Zhang, Y.; Wu, J.; Su, D.; Han, J.Y.; Fang, X.; Yu, B.; Breyer, M.D.; Guan, Y. Peroxisome proliferator activated receptor alpha/gamma dual agonist tesaglitazar attenuates diabetic nephropathy in db/db mice. *Diabetes* **2007**, *56*, 2036–2045. [CrossRef]
58. Tanimoto, M.; Fan, Q.; Gohda, T.; Shike, T.; Makita, Y.; Tomino, Y. Effect of pioglitazone on the early stage of type 2 diabetic nephropathy in KK/Ta mice. *Metabolism* **2004**, *53*, 1473–1479. [CrossRef]
59. Calkin, A.C.; Giunti, S.; Jandeleit-Dahm, K.A.; Allen, T.J.; Cooper, M.E.; Thomas, M.C. PPAR-alpha and -gamma agonists attenuate diabetic kidney disease in the apolipoprotein E knockout mouse. *Nephrol. Dial. Transplant.* **2006**, *21*, 2399–2405. [CrossRef]
60. Chinetti, G.; Griglio, S.; Antonucci, M.; Torra, I.P.; Delerive, P.; Majd, Z.; Fruchart, J.C.; Chapman, J.; Najib, J.; Staels, B. Activation of proliferator-activated receptors alpha and gamma induces apoptosis of human monocyte-derived macrophages. *J. Biol. Chem.* **1998**, *273*, 25573–25580. [CrossRef]
61. Bouhlel, M.A.; Derudas, B.; Rigamonti, E.; Dievart, R.; Brozek, J.; Haulon, S.; Zawadzki, C.; Jude, B.; Torpier, G.; Marx, N.; et al. PPARgamma activation primes human monocytes into alternative M2 macrophages with anti-inflammatory properties. *Cell Metab.* **2007**, *6*, 137–143. [CrossRef] [PubMed]
62. Straus, D.S.; Glass, C.K. Anti-inflammatory actions of PPAR ligands: New insights on cellular and molecular mechanisms. *Trends Immunol.* **2007**, *28*, 551–558. [CrossRef] [PubMed]
63. Heikkinen, S.; Auwerx, J.; Argmann, C.A. PPARgamma in human and mouse physiology. *Biochim. Biophys. Acta* **2007**, *1771*, 999–1013. [CrossRef] [PubMed]
64. Afzal, S.; Sattar, M.A.; Johns, E.J.; Eseyin, O.A. Renoprotective and haemodynamic effects of adiponectin and peroxisome proliferator-activated receptor agonist, pioglitazone, in renal vasculature of diabetic Spontaneously hypertensive rats. *PLoS ONE* **2020**, *15*, e0229803. [CrossRef] [PubMed]
65. Hunley, T.E.; Hidalgo, G.; Ng, K.H.; Shirai, Y.; Miura, K.; Beng, H.M.; Wu, Q.; Hattori, M.; Smoyer, W.E. Pioglitazone enhances proteinuria reduction in complicated pediatric nephrotic syndrome. *Pediatr. Nephrol.* **2022**, 1–12. [CrossRef]
66. Scotti, M.M.; Swanson, M.S. RNA mis-splicing in disease. *Nat. Rev. Genet.* **2016**, *17*, 19–32. [CrossRef]
67. Blijlevens, M.; Li, J.; van Beusechem, V.W. Biology of the mRNA Splicing Machinery and Its Dysregulation in Cancer Providing Therapeutic Opportunities. *Int. J. Mol. Sci.* **2021**, *22*, 5110. [CrossRef]
68. Montes, M.; Sanford, B.L.; Comiskey, D.F.; Chandler, D.S. RNA Splicing and Disease: Animal Models to Therapies. *Trends Genet.* **2019**, *35*, 68–87. [CrossRef]
69. Vidal-Puig, A.; Jimenez-Linan, M.; Lowell, B.B.; Hamann, A.; Hu, E.; Spiegelman, B.; Flier, J.S.; Moller, D.E. Regulation of PPAR gamma gene expression by nutrition and obesity in rodents. *J. Clin. Investig.* **1996**, *97*, 2553–2561. [CrossRef]
70. Hu, W.; Jiang, C.; Kim, M.; Xiao, Y.; Richter, H.J.; Guan, D.; Zhu, K.; Krusen, B.M.; Roberts, A.N.; Miller, J.; et al. Isoform-specific functions of PPARgamma in gene regulation and metabolism. *Genes Dev.* **2022**, *36*, 300–312. [CrossRef]
71. Lennon, R.; Randles, M.J.; Humphries, M.J. The importance of podocyte adhesion for a healthy glomerulus. *Front. Endocrinol.* **2014**, *5*, 160.

72. Schmidt, S.F.; Jorgensen, M.; Chen, Y.; Nielsen, R.; Sandelin, A.; Mandrup, S. Cross species comparison of C/EBPalpha and PPARgamma profiles in mouse and human adipocytes reveals interdependent retention of binding sites. *BMC Genom.* **2011**, *12*, 152. [CrossRef] [PubMed]
73. Amato, A.A.; Rajagopalan, S.; Lin, J.Z.; Carvalho, B.M.; Figueira, A.C.; Lu, J.; Ayers, S.D.; Mottin, M.; Silveira, R.L.; Souza, P.C.; et al. GQ-16, a novel peroxisome proliferator-activated receptor gamma (PPARgamma) ligand, promotes insulin sensitization without weight gain. *J. Biol. Chem.* **2012**, *287*, 28169–28179. [CrossRef] [PubMed]
74. Coelho, M.S.; de Lima, C.L.; Royer, C.; Silva, J.B.; Oliveira, F.C.; Christ, C.G.; Pereira, S.A.; Bao, S.N.; Lima, M.C.; Pitta, M.G.; et al. GQ-16, a TZD-Derived Partial PPARgamma Agonist, Induces the Expression of Thermogenesis-Related Genes in Brown Fat and Visceral White Fat and Decreases Visceral Adiposity in Obese and Hyperglycemic Mice. *PLoS ONE* **2016**, *11*, e0154310. [CrossRef]

cells

MDPI

Review

PPAR-γ Partial Agonists in Disease-Fate Decision with Special Reference to Cancer

Sangeeta Ballav, Bini Biswas, Vishal Kumar Sahu, Amit Ranjan * and Soumya Basu *

Cancer and Translational Research Centre, Dr. D. Y. Patil Biotechnology and Bioinformatics Institute, Dr. D. Y. Patil Vidyapeeth, Tathawade, Pune 411033, India
* Correspondence: bom.amit@gmail.com (A.R.); soumya.bs@gmail.com (S.B.)

Abstract: Peroxisome proliferator-activated receptor-γ (PPAR-γ) has emerged as one of the most extensively studied transcription factors since its discovery in 1990, highlighting its importance in the etiology and treatment of numerous diseases involving various types of cancer, type 2 diabetes mellitus, autoimmune, dermatological and cardiovascular disorders. Ligands are regarded as the key determinant for the tissue-specific activation of PPAR-γ. However, the mechanism governing this process is merely a contradictory debate which is yet to be systematically researched. Either these receptors get weakly activated by endogenous or natural ligands or leads to a direct over-activation process by synthetic ligands, serving as complete full agonists. Therefore, fine-tuning on the action of PPAR-γ and more subtle modulation can be a rewarding approach which might open new avenues for the treatment of several diseases. In the recent era, researchers have sought to develop safer partial PPAR-γ agonists in order to dodge the toxicity induced by full agonists, akin to a balanced activation. With a particular reference to cancer, this review concentrates on the therapeutic role of partial agonists, especially in cancer treatment. Additionally, a timely examination of their efficacy on various other disease-fate decisions has been also discussed.

Keywords: PPAR-γ; partial agonist; TZDs; disease; cancer; type 2 diabetes mellitus; autoimmune; dermatology; cardiovascular disorders

Citation: Ballav, S.; Biswas, B.; Sahu, V.K.; Ranjan, A.; Basu, S. PPAR-γ Partial Agonists in Disease-Fate Decision with Special Reference to Cancer. *Cells* **2022**, *11*, 3215. https://doi.org/10.3390/cells11203215

Academic Editors: Kay-Dietrich Wagner and Nicole Wagner

Received: 10 September 2022
Accepted: 9 October 2022
Published: 13 October 2022

1. Introduction

Disease complexities involving cancer, cardiovascular diseases (CVDs), type 2 diabetes mellitus (T2DM), autoimmune disorders (AIDs), namely, multiple sclerosis, rheumatoid arthritis and autoimmune thyroid diseases and dermatological diseases (DDs) are significantly linked to morbidity and mortality around the world. Although the survival rates for these diseases are rising, they have a noteworthy mortality rate of 25% worldwide, wherein cancer stands at 35% [1], CVDs at 32% [2], T2DM at 27% [3], AIDs at 20% [4], and DDs at 15% [5] (Figure 1). Thus, the statistics depict the severeness of these illnesses in a susceptible cohort of individuals over the last 5 years. The aforementioned diseases may coexist and share significant risk factors as well as pathophysiological mechanisms. This has opened doors to decipher new therapeutic approaches common to all the outlined disorders. Peroxisome proliferator-activated receptor (PPAR) has been an underlying transcription factor highlighting its importance in numerous diseases including cancer, CVDs, T2DM, AIDs and DDs [6–8]. Being a member of the nuclear hormone receptor (NHR) superfamily, PPAR is primarily known to serve as a regulator of glucose metabolism and homeostasis [9].

The NHR superfamily comprises a group of transcription factors (TFs) that participate in a variety of critical biological mechanisms such as cellular respiration, reproduction and inflammatory processes. The foremost member of this superfamily was cloned in 1985; however, it now contains around 48 members in mammals, specifically humans [10]. The majority of NHRs are controlled by naturally occurring lipophilic ligands such as retinoids, phospholipids, and steroids. The ability of PPAR to cause hepatic peroxisome proliferation in mice in response to xenobiotic stimuli gave rise to the family's name [11]. Research on the structure and function of the three human PPAR isotypes, namely PPAR-alpha (PPAR-α), PPAR-beta/delta (PPAR-β/δ), and PPAR-gamma (PPAR-γ), have confirmed their crucial roles in regulating carbohydrate and lipid metabolism as well as nutrition sensing. While PPAR-α and PPAR-β/δ appear to be primarily driving oxidative lipid metabolism, PPAR-γ is predominantly engaged in the cellular digestion of lipids via anabolic processes (Table 1), which also demonstrates its functional diversity in regulating glucose metabolism, homeostasis, adipocyte differentiation and cellular maintenance [9].

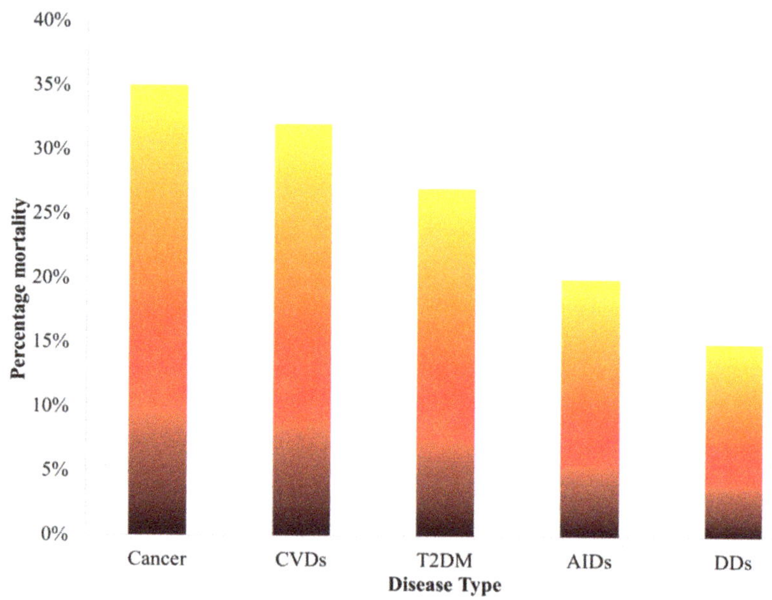

Figure 1. Statistics of percentage mortality caused by various diseases over the past five years.

PPAR-γ performs the process of transcription in the nucleus after getting bound with ligand, forming a heterodimeric complex with another NHR, retinoid-X-receptor (RXR). They are locally present in cytoplasm and translocate in nucleus after the stimulation of transcription. Later, this transcriptional complex, PPAR-γ-RXR binds to the enhancer region of PPAR response elements (PPRE) that initiates the transcription of target genes [12]. Therefore, ligands are regarded as the key determinant for tissue-specific activation of PPAR-γ protein receptor. These ligands can be categorized into endogenous, natural or synthetic. However, the mechanism that governs this process is merely a contradictory debate which is yet to be systematically researched. Either these receptors get weakly activated by endogenous or natural ligands [13] or over-activated by synthetic ligands which serve as complete full agonists [14]. Therefore, a moderate activation may serve to be a novel and intriguing therapeutic strategy which would encompass balanced PPAR-γ activity. Numerous partial agonists of PPAR-γ offer an exciting prospect for different diseases. To elucidate the therapeutic role of PPAR-γ modulators in aforementioned diseases, this review summarizes various agonists and endogenous, natural and synthetic ligands involved in the activation of the aforementioned protein receptor and their effect on different disease-fate decisions. With a particular reference to cancer, as the statistics highlights the highest mortality rate of the same, this review concentrates on the therapeutic role of various PPAR-γ agonists in the treatment of cancer. Additionally, with a special focus on partial agonists, a timely examination on their efficacy and how they can be far more effective than natural and synthetic agonists have been also discussed.

Table 1. Overview of the basic metabolic function of PPAR isotypes.

PPAR Isotype	Site of Elevated Expression	Cellular Mechanisms Initiated	Biological Function	Genes Targeted	References
PPAR-α	Heart, liver and kidney	β-oxidation of fatty acids, synthesis of lipoprotein and regulating metabolic pathways of amino acids	Complementation of metabolic reaction to fasting	Hydroxymethylglutaryl CoA synthase 2 (HMG-CoA S2)	[15]
PPAR-β/δ	Adipocytes and macrophages	Differentiation pathways of adipocytes and production of triglycerides	Differentiation pathways of adipocytes and fatty acid trapping	Fatty acid binding protein 4 (FABP4)	[16]
PPAR-γ	Adipocytes, skin and brain	β-oxidation of fatty acids	Coordination of muscle fibers and determination of its types	Acyl CoA Oxidase (AOX)	[17]

2. Structural Mechanism of PPAR-γ

Like the other nuclear receptors, there are five functional domains present in the transcription apparatus of PPAR-γ: an amino terminal transactivation domain (activation function (AF)-1) with highly conserved DNA binding domain (DBD) that conducts the binding of heterodimerized complex to the promoter regions of the target gene, a hinge region and a carboxy terminus transactivation domain (AF-2) which encompass the ligand binding domain (LBD), which is 200–300 amino acids long, with a volume of 1440 Å^3 (Figure 2). As the volume of LBD is large, it can bind and interact with a diverse group of ligands. The structural diversity of PPAR-γ sense varying ligands (weak or high affinity), and further, the receptor responds to environmental changes, such as metabolic changes or food intake. Along with this, it consists of 11–13 α-helices sandwiched in a three-layer anti-parallel helical structure and a four-stranded β-sheet that folds and becomes a large hydrophobic cavity which facilitate the binding of ligand to the receptor gene [18]. The transcriptional complex (PPAR-γ-RXR) induces a conformational change in transactivation domain at carboxyl terminus (AF-2) [19], followed by recruitment of certain co-activators with the release of co-repressors. After the shuttling of PPAR into nucleus, they associate with its various other transcription apparatus (Figure 3).

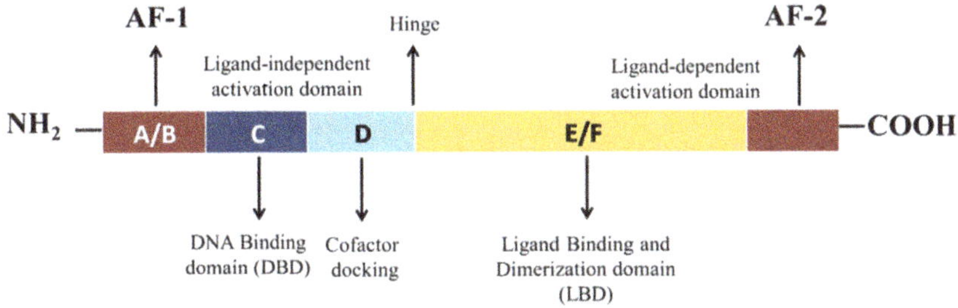

Figure 2. Schematic representation of functional domains of PPAR-γ. It comprises N-terminal AF-1 (ligand-independent activation domain) A/B domain, DNA-binding domain or the C domain) with two zinc fingers, a hinge (D domain), and C-terminal AF-2 (ligand-binding domain) E/F domain.

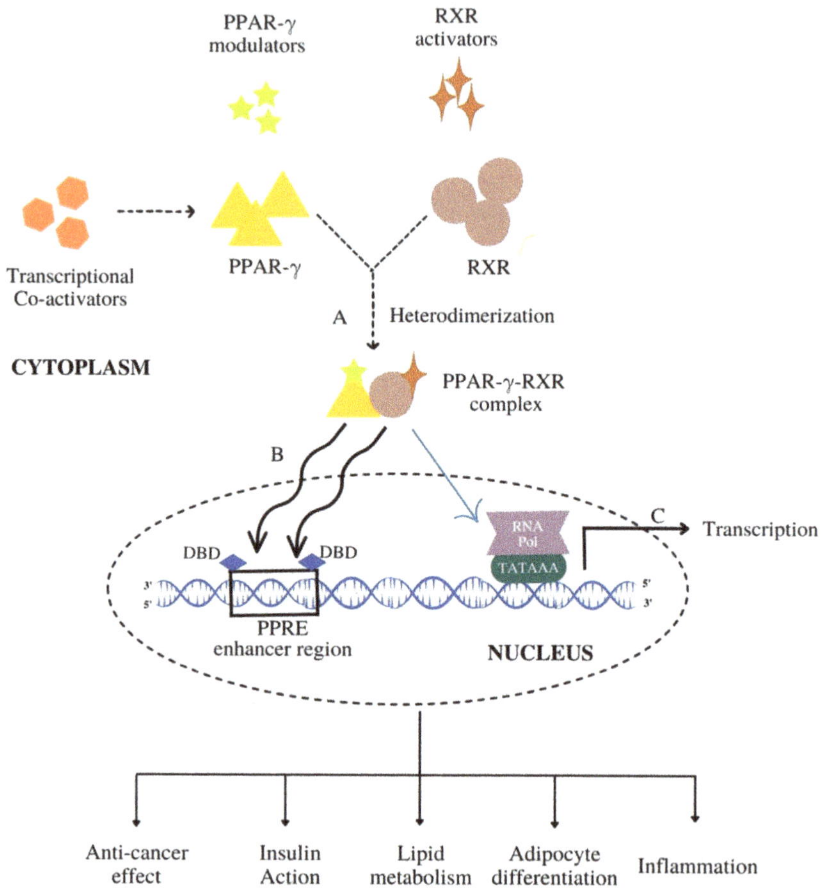

Figure 3. The mechanism of PPAR-γ activation followed by the gene transcription. (**A**) PPAR-γ upon binding with its modulators (endogenous, natural, synthetic) heterodimerize with RXR to form PPAR-γ-RXR heterodimerized complex. (**B**) This complex gets translocated inside the nucleus and binds to the DNA-binding domain (DBD) of PPAR response elements (PPRE) enhancer region. (**C**) Such a binding leads to the transcriptional activation of target genes resulting in various metabolic cascades such as insulin action, lipid metabolism and anti-cancer effects.

The conformational change induced by LBD triggers the accompanying process of co-activators (active conformations) and co-repressors (repressive conformations). Generally, a diverse group of co-activators, namely, p160/steroid receptor co-activator (SRC) family (SRC1–SRC3), cyclic-AMP (cAMP)-responsive element binding protein (CREB), CREB-binding protein (CBP) and its homologue p300, PPAR-binding protein (PBP) and PPAR-γ co-activator (PGC-1) have been adding vital impact in regulation of PPAR-γ transcriptional activation [20,21]. A typical co-activator comprises one or more LXXLL motifs (L: leucine and X: any amino acid) in the distal C-terminal region that facilitate the ligand-bound-PPAR to undergo its process. A growing body of evidence points to the organized assembly of these co-activators which is greatly influenced upon ligand binding to PPAR-γ [22]. A computational study determined the agonist mode of non-polymer ligand AZ 242 (Tesaglitazar) for PPAR-α and PPAR-γ which showed that AZ 242 activated both the receptors with the recruitment of SRC-1 co-activator. AZ 242 is a dihydrocinnamate derivative that is known to be an effective drug for abnormalities associated with glucose and lipid metabolism for insulin [23].

Along with co-activators, co-repressors also have been seen to modulate PPAR-γ transcriptional activity. Distinct co-repressors such as nuclear receptor co-repressor (NCoR) [24] and the silencing mediator of retinoid and thyroid hormone receptors (SMRT) [25] are destined to unliganded PPARs which repress the gene expression upon the recruitment of co-activators. Further, the structural transition leads to heterodimerization of the complex to achieve A PPAR-γ-RXR complex that binds strongly to PPRE [26]. This heterodimer complex recognizes the sequences located within six nucleotides of direct repeat spaced by one nucleotide (DR1) of PPRE which consists of a repeated sequence of AGGTCA (hexameric DNA consensus sequence) [27].

3. Functional Diversity of PPAR-γ

The nuclear receptor PPAR-γ gene on chromosome 3p25.2 in humans was identified as the master regulator of adipose tissue biology by Tontonoz and colleagues, in the early 1990s [28]. By using distinct promoters and alternative splicing mechanism, the human PPAR-γ gene possess nine exons that produce four PPAR-γ splice variants (PPAR-γ1-4), each code for one of two protein isoforms [29]. However, the mRNA translation of three splice variants, PPAR-γ1, γ3 and γ4, is regulated simultaneously and gets combined, forming one common isoform. Thusly, only two isoforms are considered, namely, PPAR-γ1 and γ2. PPAR-γ1 (477 amino acids) and PPAR-γ2 (505 amino acids) use two distinct promoters and alternate splicing for encoding the same gene. The absorbing difference between these two sub-isoforms lies in their amino-terminal A/B domain. The PPAR-γ1 isoform is produced by the mRNAs, PPAR-γ1, PPAR-γ3, and PPAR-γ4 [11]. PPAR-γ1 protein is widely expressed and found in skeletal muscle, liver, colon, cardiac tissue, adipose tissue, and many epithelial cell types. Additionally, PPAR-γ1 is expressed in numerous immune system cells, including T lymphocytes, dendritic cells, and monocytes and macrophages. The PPAR-γ2 isoform is produced via the translation of PPAR-γ2 mRNA transcript. Adipose tissue is the primary site of expression for PPAR-γ2, which has an extra 28 amino acids in its NH2-terminus. In addition to regulatory T cells (Tregs) and other T cell populations, PPAR-γ2 is also expressed in transitional epithelial cells of the urinary system's organs, including the bladder [30]. Total PPAR-γ expression is, however, low in non-Tregs. PPAR-γ plays a crucial function at the nexus of immunity, obesity and cancer (Figure 3). In order to understand the shared and distinctive molecular mechanism underlying the functions of PPAR-γ, we need to assess its critical biological function in detail [31].

3.1. Anti-Cancer Effect

PPAR-γ is frequently expressed in various cancer cells, including breast, lung, colon, lips, kidney, pancreatic and thyroid [32]. Several studies have revealed that PPAR-γ activation by its agonists imposes cell cycle arrest [33], apoptosis [34], angiogenesis [35], inhibition [36], and redifferentiation [37], which are the key molecular processes associated with the prevention of tumor growth and progression. The expression of angiogenesis-related proteins such as vascular endothelial growth factor (VEGF) and cyclooxygenase-2 as well as inflammatory mediators in the tumor microenvironment are all inhibited by PPAR-γ activation [38]. The stimulation of cellular differentiation by PPAR-γ activation may serve as a tumor suppressor approach. In vivo experiments suggest that PPAR-γ interacts with downstream signaling pathways of the insulin-like growth factor (IGF) axis that involves mitogen-activated protein kinase (MAPK), phosphatidylinositol 3-kinase (PI3K), and the mammalian target of rapamycin (mTOR) cascades [39–41]. According to the anti-tumor function of PPAR-γ and the anti-apoptotic function of IGF system, MAPK activation by various growth stimuli such as insulin, IGF, epidermal growth factor (EGF), and platelet-derived growth factor (PDGF) may cause serine phosphorylation of PPAR-γ. This interference between IGF system and PPAR-γ phosphorylation constitutes various metabolic actions, such as lowered insulin level in blood and increased insulin sensitivity. As a result, this leads to the inactivation of its genomic activity [42].

PPAR-γ agonists may also encourage the differentiation of cancer cells. Experimental studies in cancer cells from the thyroid, breast, and lung cancers have demonstrated that thiazolidinedione (TZD) alters the epithelial expression profile and reverses the epithelial-mesenchymal transition (EMT) process [32,43]. Additionally, various studies have shown that PPAR-γ agonists are effective in preventing the cancer stem cells (CSCs) survival, obtained from human cell lines or samples of breast, prostate, colon, bladder, and venous tissues. This evidence supports the role of PPAR-γ agonists in controlling the biology of CSCs [44,45]. As previously indicated, the IGF pathway plays a critical role in the regulation of pluripotency, EMT, and self-renewal, which is crucial for the growth and expansion of CSCs. Gianì and collaborators have recently discovered that IGF-insulin receptor (IGF-IR), which controls the capacity for self-renewal and stem cell progression, is overexpressed in human thyroid progenitor/stem cells [46]. The IGF system is crucial for controlling stem cell biology and the early stages of the carcinogenesis process, as evidenced by similar findings in cancer progenitor/stem cells from solid and hematopoietic malignancies [47]. It should be noted that the molecular mechanisms by which TZDs control differentiation and stemness processes in adipocytes and normal cells have been investigated, although in cancer cells and CSC, they are still not fully understood [32].

3.2. Insulin Action

Adipose tissue combines cerebral and peripheral metabolic signals that control energy balance as a key tissue for whole-body energy homeostasis. They expand in response to an imbalance between energy intake and expenditure, which is characterized by an increase in cell size (hypertrophy) and cell number (hyperplasia) [48]. The underlying biological mechanism by which multipotent mesenchymal precursor cells commit to the adipocyte lineage and exhibit the usual hallmarks of mature fat cells is represented by a complicated and yet poorly understood set of transcriptional events. A number of distinct transcriptional factors have recently been found to control the expression of a group of genes involved in lipid and glucose metabolism. PPAR-γ, being one of them, have been demonstrated to play crucial role in the transcriptional regulation of genes encoding proteins involved in the aforementioned activities [9]. According to Leonardini et al. the treatment strategies of metabolic diseases linked to adipose tissue expansion, T2DM presupposes the identification and thorough comprehension of the molecular processes. They regulate these disorders, as well as the creation of therapies that specifically target the contributing elements [49].

Even in the absence of obese conditions, ageing and insulin resistance are linked to escalating flaws in mitochondrial oxidation. The elevated levels of fatty acetyl-CoA and diacylglycerol produced by this mitochondrial change can affect insulin signaling in skeletal muscle and other tissues, leading to insulin resistance [50]. PPAR-γ co-activators are thought to express less in people with insulin resistance. This results in fewer muscle mitochondria and a lower ratio of type 1 oxidative muscle fibers to type 2 fibers, which eventually becomes more glycolytic and possess fewer mitochondria [51]. Both non-diabetic and diabetic whites and Mexican Americans have been found to undergo these changes. PPAR-γ is considered as the molecular target of a series of insulin-sensitizing medications known as thiazolidinediones (TZDs) and is essential for maintaining glucose homeostasis [52]. The first synthetic PPAR-γ ligand was troglitazone; however, it was discontinued from usage due to rare but substantial liver damage. Rosiglitazone and pioglitazone are the two clinically available PPAR-γ ligands that are frequently used to treat T2DM [53]. TZDs lessen peripheral insulin resistance, which is a feature of T2DM patients. The overall glycemic control is improved as a result of the increased peripheral glucose utilization, decreased hepatic glucose output, and decreased blood glucose levels [54]. In addition, PPAR-γ ligands also lower plasma fat levels and improve glucose metabolism. Both pioglitazone and rosiglitazone raise high-density lipoprotein (HDL) levels in the serum, while pioglitazone significantly lowers triglyceride levels in the blood [55]. Additionally, PPAR-γ ligands prevent the production of numerous pro-inflammatory genes in macrophages, such as matrix metalloproteinases, interleukins, and inducible nitric oxide synthase. Since the gene expression and accumulation of macrophage in adipose tissue has been found to play in the pathophysiology of obesity-related insulin resistance, these actions may also be pertinent source for obesity-related insulin resistance [56].

3.3. Lipid Metabolism

Liver is a significant organ that regulates lipid intake from the circulatory system, de novo synthesis, and distribution of the generated lipids in the form of very low-density lipoprotein (VLDL) to peripheral tissues, thus controlling total body lipid homeostasis. White adipose tissue is thought to be the main organ for storing additional lipids, even though the liver regulates the body's overall lipid balance [57]. In order to maintain and control homeostatic lipid transports under normal circumstances, the liver thus fulfills the aforementioned particular duties and is not meant to accumulate fat. However, unbalanced lipid flow in the liver causes fat buildup in pathophysiological diseases such as diabetes and obesity. Yoon Kwang Lee and collaborators confirmed in their study that it is the involvement of PPAR-γ in the hepatic fat buildup in pathologic circumstances, and hence, we discuss its critical role in lipid metabolism [58].

Overexpression of PPAR-γ1 or PPAR-γ2 strongly increases fat storage in the liver, which is expected to lead its adipogenic activity in white adipocytes [58]. Although the two PPAR-γ isoforms share the common DNA binding specificity, they exhibit different fat accumulation efficacy due to the variation in their transcription activity, wherein PPAR-γ2 has a five- to ten-fold higher transcription activity than PPAR-γ1. This is due to the additional amino acids at the N-terminus of PPAR-γ2 [59]. It has been established that PPAR-γ overexpression is brought by a high-fat diet (HFD) or other pathophysiological stresses that eventually causes lipid buildup in the liver. This is the first stage for the onset of non-alcoholic fatty liver disease (NAFLD), T2DM [60]. Given its direct involvement in the formation of lipid droplets, PPAR-γ might act as a defense against the lipotoxicity caused by free fatty acids (FFAs) and their derivatives. Under normal settings, hepatic PPAR-γ2 expression remains low, allowing the liver to transport freshly generated and/or dietary FAs in the form of VLDL to other peripheral tissues as an energy source. By storing the extra FFAs as triglycerides (TGs), hepatic PPAR-γ2 expression is up-regulated in response to HFD. This appears to protect the surrounding tissues from lipotoxicity and would be a typical procedure to counteract the continual flux of fat in the circulation [58].

3.4. Adipocyte Differentiation

Mammalian adipocytes have been identified as white, beige, and brown. Despite the fact that they all originate from separate progenitors and have fundamentally diverse morphologies and functions, the adipocytes undergo a well-coordinated differentiation process to develop into mature, fully functional cells. PPAR-γ has a well-established role in the longevity of adipocytes at various stages [31]. Non-hematopoietic pluripotent stem cells with a strong capacity for self-renewal and multi-differentiation activities are known as mesenchymal stem cells (MSCs). They develop into certain somatic cells of each hypoderm under normal differentiating process [61]. There are two distinct stages in adipocyte differentiation. Firstly, the MSCs begin to differentiate into precursor, adipocyte which is the crucial stage after receiving external stimulus. Molecular pathways underlying this particular mechanism are still unknown. Secondly, terminal differentiation involves three stages: proliferative, differentiative, and maturation phase, wherein the precursor adipocytes get transformed into mature adipocytes [61]. In the proliferative stage, the precursor adipocytes fuse and undergo contact inhibition followed by upregulation of genes, namely, CCAAT-enhancer-binding proteins (C/EBP β, C/EBP δ), Myogenic factor 5 (Myf5), before entering the cell cycle. Cell differentiation starts when the PPAR-γ binds to its respective ligands, e.g., TZD and its subtypes. Resting cells transition into the growth phase during the terminal differentiation stage, and the expression of related genes is activated by high levels of PPAR-γ. The cells finally develop into mature adipocytes after accumulating lipid droplets. The mature adipocyte stage is characterized by the presence of a single, big fat droplet inside the cell, strong expression of adipocyte marker genes such as aP2 and adiponectin, and production of cytokines that control energy balance, insulin sensitivity, and other activities [62].

3.5. Inflammation

As a member of the PPAR superfamily, PPAR-γ also controls inflammatory and immune reactions. PPAR-γ might stimulate trans-repression on pro-inflammatory genes genetically through SUMOylation or the conjugation of PPAR-γ with Small Ubiquitin-like Modifier (SUMO), a type of post-translational modification. Target genes for nuclear factor kappa B (NF-κB) are trans-repressed after the binding of a nuclear co-repressor complex, which keeps NF-κB in a repressed and promoter-bound state [63]. PPAR-γ may also control the immune system by virtue of dendritic cells (DC) and macrophages. Antigen acquisition, processing, stimulation, migration, cytokine generation, and lipid antigen presentation are all altered, which has an impact on the function of DC. While promoting the expression of anti-inflammatory mediators in macrophages, PPAR-γ suppresses the genes, namely, tumor necrosis factor alpha (TNF-α), Interleukin 1 beta (IL-1β), and Interleukin-6 (IL-6) that code for pro-inflammatory molecules [64]. Additionally, by altering cell differentiation, it prevents the maturation of pro-inflammatory wild-type "M1" macrophages by upregulating genes such as IL-4, CD163 and adiponectin that promote the development of anti-inflammatory "M2" macrophages, producing a bilateral anti-inflammatory effect (Figure 4) [64].

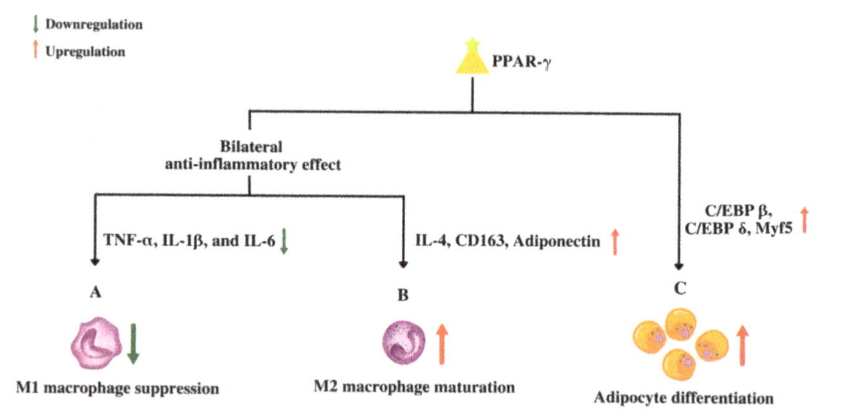

Figure 4. Role of PPAR-γ in macrophage conversion and adipocyte differentiation. (**A**) PPAR-γ suppresses the genes that code for pro-inflammatory molecules, which in-turn prevents the maturation of pro-inflammatory wild-type "M1" macrophages. (**B**) PPAR-γ promotes the maturation of anti-inflammatory "M2" macrophages by up-regulating anti-inflammatory genes leading to an overall bilateral anti-inflammatory effect. (**C**) PPAR-γ regulates adipocyte differentiation by upregulating the genes involved in proliferative stage of terminal differentiation.

By enhancing tight junction proteins, PPAR-γ has been also shown to defend the blood–brain barrier's (BBB) integrity. Through the measurement of NF-κB transcriptional activity and FITC-Dextran permeability, Talé and colleagues evaluated the impact of the PPAR-γ agonist, pioglitazone on markers of the inflammatory response and permeability changes brought on by hyperglycemia [64]. Their results indicated that high glucose-mediated activation of NF-κB is blocked by PPAR-γ agonist. Additionally, PPAR-γ agonists guard against FITC-Dextran permeability elevation that is aggravated by hyperglycemic conditions viz. T2DM. This shows that PPAR-γ has a protective function against the inflammatory and permeability changes that are caused by hyperglycemia in endothelial cells. It is noteworthy that mild PPAR-γ activators such as telmisartan, amorfrutin, and other PPAR-γ partial agonists significantly defend against PPAR-γ gene expression downregulation brought on by hyperglycemia [64].

4. PPAR-γ Activation by Various Ligands

Ligands are chiefly considered to play central importance for PPAR-γ. This is either achieved by ligand-dependent or ligand-independent activity, which is conducted by co-activators and co-repressors, respectively [65]. Figure 5 illustrates the activation of PPAR-γ receptor through both ligand dependent and independent mechanisms. The ligand-dependent mechanism involves the activation of PPAR-γ after getting bound with ligand in LBD of C-terminus (AF-2), which later leads to the process of transcription. While, the ligand-independent mechanism is mainly due to N-terminus and kinase-dependent action, wherein ligands do not participate in the activation. Mu et al. have demonstrated the importance of ligand-independent action of PPAR-γ through N-terminus. They found that the activation rate of one sub-form, PPAR-γ2 was 5-fold greater than the other sub-form, PPAR-γ1 in adipose tissue via AF-1 activation domain, which suggests that there is a distinct activation profile between the two types [66]. These mechanisms provide a base for designing drugs. Further, the ligand-dependent mechanism comprises an agonist or antagonist mode of activation. The agonist mode may be full or partial depending on the structural differences and activation profile. Full agonist activity includes binding of ligand to the LBD of PPAR-γ that provides a flexible interaction higher affinity. This is carried out by folding of 12th helix together with helices H3 and H5 that leads to formation of a hydrophobic groove. H12 of LBD ought to define the activation efficiency through its stabilization or destabilization [67]. Therefore, binding of a ligand directly to helix 12 imparts a full agonist activity.

Previous structural studies have investigated that the potent synthetic ligand of PPAR-γ; TZDs have been seen to exert full agonist activity through such process. Mainly, TZDs have an acidic head group which tends to interact with intermolecular hydrogen bonds with amino-acid residues Ser289, His323, His449 and Tyr473 [68]. In contrast, partial agonist is characterized by indirectly interacting with H12by destabilizing it and lacks the intermolecular hydrogen bonds such as that of the full agonist with the four amino acids. However, the partial agonist stabilizes H3 and β-sheet region of LBD, which defines their activation efficiency. They are considered as weak activators of PPAR-γ compared with full agonists, which shows lower transactivation potential and imparts desirable effect [69]. Agonist-mediated activation of PPAR-γ has been reported to exhibit anti-inflammatory properties, growth inhibitory effects and prevent proliferation of many human cancer cell lines [34,64]. Likewise, the antagonist activity has also been observed to destabilize H12 and stabilize H3 and β-sheet. Thereafter, they inhibit co-activators binding and transcriptional activation [70]. Therefore, PPAR-γ antagonism is exploited to combat its aberrant expression. Numerous antagonists such as BADGE, PD068235, MK886, SR-202, GW6471, LG100641 and SR13904 have been identified [71]. Thus, the position of ligand binding within LBD constitutes central importance for PPAR-γ activation. At the core of this approach, research studies are targeting PPAR-γ and aiming to build up novel drug leads.

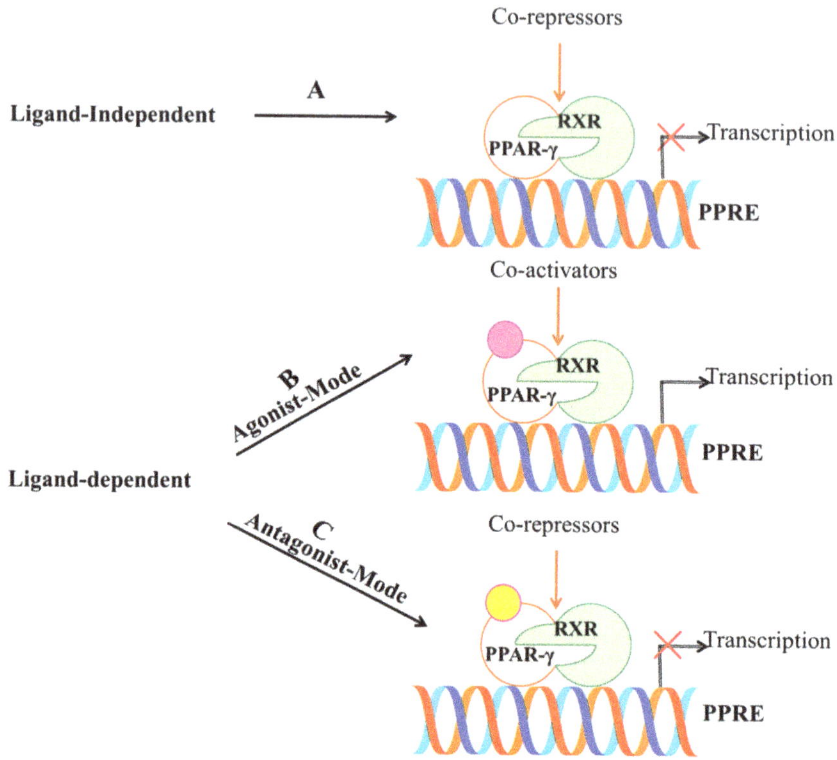

Figure 5. Schematic representation of activation of PPAR-γ receptor through ligand-dependent or -independent mechanisms. (**A**) Ligand-independent mechanism does not involve the active participation of ligand, and co-repressors bind to unliganded PPAR-γ, which repress the gene expression by chromatin remodeling. (**B**) Agonist mode of action involves the binding of ligand to LBD of PPAR-γ with the help of co-activators, which further leads to transcription of targeted genes. (**C**) Antagonist mode of action also involves the binding of ligand without leading to transactivation activity, due to the presence of co-repressors.

4.1. Endogenous Ligands

Due to the structural diversity of PPAR-γ, a variety of ligands bind to the protein receptor (Figure 6). Of note, owing to the above-mentioned mechanism of activation, even the ligands with weak efficacy can employ PPAR-γ driven effects. Endogenous and natural compounds are known as the weak agonists for PPAR-γ. The endogenous ligands constitute long chain polyunsaturated fatty acids (LC-PUFA) [72], J-series of prostaglandins [73] and arachidonic acids (15-Deoxy-Δ-12,14-Prostaglandin J2-15d-PGJ2), eicosanoids, nitroalkanes, and oxidized phospholipids [74] including docosahexanoic acid (DHA). Compelling evidence indicates that these endogenous ligands can interfere with some hallmarks of cancer via PPAR-γ activation, which promotes growth inhibition, cell apoptosis, and resisting angiogenesis [75]. Substantial literature suggests that 15d-PGJ2 exerts various anti-inflammatory effects such as inhibiting cytokine expression, nuclear factor κB (NF-κB), and IkappaB (I-κB) kinase [76]. Zhao and colleagues demonstrated that the DNA binding activity of PPAR-γ significantly increased on intrastriatal injections on the locus of striatal hematoma. The compound markedly decreased NF-κB activation, prevented neutrophil infiltration and reduced cell apoptosis [77]. Given the central role of pro-inflammatory gene expression, low-density lipoprotein of PPAR-γ such as 9-hydroxyoctadecadienoicacid (HODE) and 13-HODE have also seen to attribute PPAR-γ activation [78]. Moreover, the endogenous ligands aids in low affinity for PPAR-γ, whereas high-affinity endogenous ligands are yet not known.

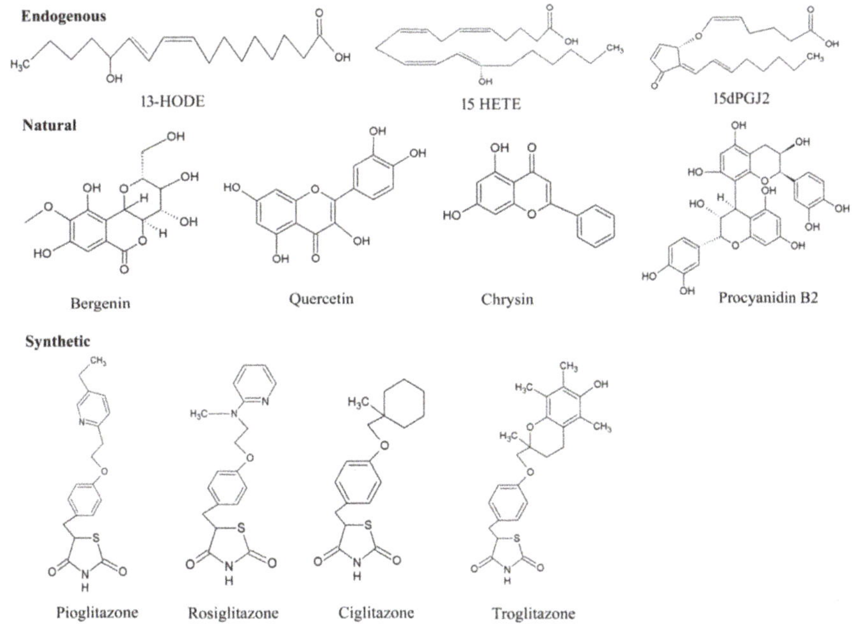

Figure 6. Chemical structures of various types PPAR-γ ligands (Endogenous, natural, and synthetic).

4.2. Natural Ligands

Certain natural products derived from plants have also appeared in the picture, which provides a promising pool of structures through their effective medicinal value for drug discovery. Traditionally, a wide range of medicinal plants have inspired the researchers to explore the mechanistic insights for PPAR-γ-activating potentiality. Vitamins, namely, tocopherol (α and γ), have been shown to serve as better modulator of PPAR-γ expression by upregulating in colon cancer cells [79]. A major group of antioxidants [80], herbs [81], fruits [82], seeds [83,84] and flowers [85] suggest protection against wide array of cancers. Some of the noteworthy examples for these natural products are extracts of *Maerua subcordata* (MS). The fruit, root, and seed extracts were revealed to up-regulate PPAR-γ expression level at 30g dry weight per liter (gDW/L), being non-cytotoxic in nature [86]. These plant natural products are known to have very fewer effects on the activation of PPAR-γ protein receptor.

According to epidemiological studies, some of the naturally occurring dietary flavonoid such as quercetin (3,5,7,3,4-pentahydroxyflavone), kaempferol (3,5,7,4tetrahydroxyflavone) and apigenin (4,5,7-trihydroxyflavone) also contribute in the activation process of PPAR-γ, which in turn are associated to the diminishing incidence of certain disorders: diabetes, coronary heart diseases and various types of cancer [87,88]. A study has implicated the efficiency of quercetin metabolites in stimulating PPAR-γ expression by inhibiting A549 cell growth with downregulation of cdk1, cyclin B and Matrix metallopeptidase 2 (MMP-2) expressions [87]. Quercetin-driven upregulation of PPAR-γ have fostered sufficient interest in generating several anti-cancer activities: promoting apoptosis, inducting cell cycle arrest and cell death in gastric, breast, lung, colon, prostate, and many other cancers [89]. Although there are only a few studies that have been able to demonstrate the efficacy of kaempferol in activating the protein receptor, PPAR-γ, it has gained attention for its anti-inflammatory and anti-cancerous effects [88]. Strikingly, Zhong et al. have illustrated the potentiality of natural flavonoid modulator, apigenin, in activating PPAR-γ by inducing apoptosis with cell cycle arrest at the G2/M phase in HCT-116, SW480, HT-29 and LoVo colon cancer cells, thereby upregulating pro-apoptotic proteins (NAG-1 and p53) and cell cycle inhibitor (p21) [90]. Unfortunately, despite such comprehensive evidence, these studies have been less convincing as the dietary flavonoids have offered fewer significant advantages in activating PPAR-γ, which pushes forth the search to uncover more novel agonists.

4.3. Synthetic Ligands

In addition to the endogenous and natural ligands, several synthetic ligands have been identified as strong agonists over the past few years. TZD were the first synthetic PPAR-γ agonists that were primarily designed to control T2DM by sensitizing insulin, which emerged as a major remedy for this disease. As a result, PPAR-γ is also known as the 'glitazone receptor' [91]. Although it comprises a diverse class of compounds, most importantly, rosiglitazone, troglitazone, ciglitazone and pioglitazone have been gaining a lot of attention towards anti-diabetic, anti-cancer, anti-microbial and anti-inflammatory activities [92–94]. Rosiglitazone represents the most potential compound amongst all TZDs, due to its superior pharmacokinetics and high affinity therapeutic value for cancer and inflammation. For instance, the anti-neoplastic effect of neuroblastoma cells (stromal (S) (SK-N-AS) and a neuroblast (N) (SH-SY5Y) phenotype) were deciphered, where Cellai et al. evaluated the efficacy of rosiglitazone for cell proliferation and invasion through the transactivation of PPAR-γ [95]. The findings addressed a significant reduction in proliferation and viability of cells by rosiglitazone, which relies on inhibiting cell adhesion and invasiveness in SK-NAS, but not in the SH-SY5Y neuroblastoma (NB) cell line. Interestingly, similar results were reported with a decrease in anti-apoptotic protein (Bcl-2) and an increase in pro-apoptotic protein (caspase 3) on human bladder cancer 5637 and T24 cell lines with the treatment of more than 10 μM concentration of rosiglitazone with PPAR-γ expression [96]. Further sensitization of breast cancer cell line, MDA-MB-231, predicted a mechanism by which rosiglitazone produces anti-tumor effects with a higher dose of 100 μM because it was required to transactivate a PPRE reporter construct for early changes in gene expression [97]. In addition, pioglitazone is under clinical trials for treatment of various diseases such as Alzheimer's disease, CVD and diabetes. Recent research showed that administration of pioglitazone-loaded Poly (lactic-co-glycolic acid) delivery system on bleomycin-induced scleroderma model mice inhibited skin fibrosis within 60 min. Simultaneously, the researchers' in vitro experiments revealed that pioglitazone reduced the migration ability and myofibroblast differentiation mediated by TGF-β in cultured fibroblasts [98]. Considerable interest has proved that along with troglitazone, ciglitazone has also been found to have an anti-cancer potential, owing to the property of chemopreventive agents. Ciglitazone induced cell cycle arrest at in the G1 phase in stomach cancer and suppressed cell growth [98]. Furthermore, it exerted a dose- and time-dependent anti-proliferative effect on A549 lung cancer cells both in vivo and in vitro, with significant upregulation of PPAR-γ expression [99]. Apart from the aforementioned TZDs, non-steroidal anti-inflammatory drugs (NSAIDs) were also discovered to exert stimulation for PPAR-γ but with low affinity. For example, indomethacin activated PPAR-γ in colorectal cancer with weak efficacy and did not result in anti-proliferative activity [100]. Other than this, sulindac sulfide, ibuprofen and diclofenacs activated PPAR-γ moderately [101]. Moreover, NSAIDs have been proven to reduce the risk of Alzheimer's disease [63].

5. PPAR-γ Agonists in Various Diseases

5.1. Cancer

Cancer, one of the most frequent diseases worldwide, are characterized by continuous cell proliferation and dysregulation of the cell cycle. This outlines the importance of chemotherapeutic agents to the modulate cell cycle and/or apoptosis [102,103]. The approach for understanding means of PPAR-γ activation has gained considerable momentum in recent years which is found to be expressed in a variety of cancer cells. A vast amount of literature points to the fact that stimulation of PPAR-γ may be a key factor in producing various anti-cancer effects (Table 2). The PPAR-γ ligand, 15d-PGJ2, resists angiogenesis, promotes apoptosis, and inhibits migration [104]. A study accounted for apoptosis in colon cancer cells by 15d-PGJ2 via PPAR-γ activation by inhibiting telomerase activity and gene expression of human telomerase reverse transcriptase (hTERT) [105]. Numerous studies illustrated that 15d-PGJ2 may serve as an anti-cancer agent in oral squamous cell carcinoma cells [106] and gastric cancer [107] by promoting cell apoptosis. In addition, the protein receptor has been readily reported to induce apoptosis and inhibited proliferation of numerous other tumor cells [36,108,109]. Recently, a study illustrated chemopreventive nature of pioglitazone in a pre-clinical mouse model of squamous lung carcinoma. The authors have observed that pioglitazone prevented lung tumor development, reduced squamous lesions, and reduced that squamous dysplasia in an N-nitroso-trischloroethylurea (NTCU)-induced mouse model [110]. In one instance, rosiglitazone was to attributed an anti-fibrotic effect and inhibitory effect on paraquat (PQ)-induced acute pulmonary fibrosis in rats, administered with intraperitoneal injection of rosiglitazone [111]. Further, rosiglitazone was also observed to exert protective role in liver cancer cells by inducing apoptosis [112].

On the contrary, pioglitazone has also marked its therapeutic importance to combat cancer-associated pathological conditions. Recently, pioglitazone has been reported to overcome the effect of doxorubicin (DOX) resistance by modulating P-glycoprotein (P-gp) in a patient-derived orthotopicxenograft (PDOX) model of osteosarcoma. P-gp is well known to have a vital role in multidrug resistance (MDR) activity, which pumps out chemotherapy agents. In lieu of its significance, the study has broadened the application of pioglitazone on MDR in osteosarcoma treatment by successful activation of PPAR-γ [113]. One of the studies examined the importance of PPAR-γ activation with ERK1/2 accumulation in lung cancer cell line, NCI-H23, via mitochondrial pathway by inducing apoptosis on treating with troglitazone [93]. Considerable interest proved that along with troglitazone, ciglitazone has encouraged the observations on anti-cancer potential, owing to the property of chemopreventive agents. In one study, ciglitazone induced cell cycle arrest at G1 phase in stomach cancer, which suppressed cell growth [114]. Further, it exerted anti-proliferative effects on A549 lung cancer cells both in vivo and in vitro, with significant upregulation of PPAR-γ expression. With regard to this point, it is imperative to note that the TZD class of drugs has reduced the incidence of various cancers; however, to date, there is no such evidence on the potential impact of TZDs for active malignant disease.

5.2. Cardiovascular Disease (CVDs)

CVDs constitute a group of disease of the heart and blood vessels. These include ischemic stroke, arrhythmia, atherosclerosis, heart attack, heart failure and heart valve problems. The present lifestyle such as unhealthy diet habits, psychological stress and physical activity has collectively underlined the increasing mortality rate [115]. Nevertheless, muscles of heart, termed as myocardium, are the most vital part that surrounds and protects the heart. It requires a massive amount of energy (ATP production) in order to maintain the cardiac structure and function. Minor changes in flux are mediated by altering the substrate concentrations and allosteric modification of enzymes involved in these metabolic pathways. However, prolonged changes in cardiac metabolism are mediated at the gene transcriptional level. PPAR-γ has been shown to regulate the cardiac metabolism transcriptionally [116]. A study has documented the inhibition of cardiac hypertrophy, a condition resulting due to congestive heart failure through PPAR-γ dependent pathway in both in vivo and in vitro methods [8]. Another study demonstrated the efficacy of PPAR-γ natural agonist, quercetin, to impede the action of AP-1 protein in cardiac hypertrophy via PPAR-γ signaling. Quercetin lowered the blood pressure level and, remarkably, minimized the left-ventricular-to-body-weight (LVW/BW) ratio in hypertensive rats, while in vitro experiments suggested the suppression of transcription activity of AP-1 (c-fos, c-jun) protein (typically involved in cardiac hypertrophy) in H9C2 cells by PPAR-γ activation by quercetin [117]. Furthermore, PPAR-γ agonists have also been extensively reviewed for causing anti-inflammatory actions in ischemic stroke [118,119]. In a cohort study conducted between 2001 and 2013, it was observed that patients admitted due to ischemic stroke were potentially shown to be more cured with the administration of pioglitazone compared to the patients without pioglitazone. The recurrent ischemic stroke was prevented by pioglitazone [120]. Presently, pioglitazone is considered as a reliable cardioprotective agent due to its ability to reduce the risk of ischemic stroke and myocardial infarction without any direct harm on the myocardium [121]. On the contrary, there are studies where PPAR-γ can protect the cells from oxidative stress in oxidative stress-induced cardiomyocyte apoptosis by increasing the expression of Bcl-2 protein. Ren et al. explained the apoptotic effects induced by roziglitazone, wherein the ligand remarkably downregulated Bcl-2 protein in oxidative stress, H_2O_2-induced cardiomyocytes [122].

5.3. Type 2 Diabetes Mellitus (T2DM)

Diabetes has contributed massively as a public health problem in a global context. Intensive studies have reported that insulin resistance plays a vital role in the development of T2DM. This insulin is fueled by obesity which arises due to imbalanced lifestyle patterns and increased consumption of high-caloric food. These factors cause a decline in the response of pancreatic-β-cells that eventually develop resistance to increased insulin secretion. At this point, glucose intolerance and elevated levels of insulin in the body lead to T2DM [123]. Thus, hyperinsulinemia is known to regulate glucose metabolism that further overcome the insulin resistance. In addition, diabetes is also associated with other metabolic syndromes such as dyslipidemias, hypertension, and polycystic ovarian syndrome (PCOS). These are the prominent risk factors that underline the causes of T2DM [124]. For the past 30 years, PPAR-γ has been thought to serve as a significant target for the treatment of insulin resistance and T2DM [52]. There is evidence that attests to the fact that the activation of PPAR-γ induces insulin sensitization. The agonists of PPAR-γ are known to indirectly normalize the glucose profile by increasing the glucose uptake stimulated by the peripheral tissues and decreasing hepatic gluconeogenesis [125]. TZD merely aids in strong stimulation of PPAR-γ and improves the pharmacological treatment of T2DM (Table 2). Currently, TZD agonists for PPAR-γ are used therapeutically. Of note, they are considered as the second-line oral drug, which is sometimes administered alone or in combination with metformin, the first-line oral drug [126]. They are often known as insulin-sensitizing agents or anti-diabetic drugs. There are in vitro studies illustrating the binding potential of several TZD ligands with PPAR-γ, which connects well with in vivo affinity as insulin sensitizers [8]. On its activation, PPAR-γ pancreatic insulin secretion is found to decrease and reduces fatty acids in blood. Most of the effects of TZDs are driven by adipocyte differentiation, which increases glucose transporters (GLUT4) and induces lipogenic genes (AP2 and CD36) [127]. Reportedly, some derivatives of TZD, MSDC-0160 and MSDC-0602 were observed to cause anti-diabetic effects via PPAR-γ, mitochondrial membranes and the pyruvate carriers (MCP1 and MCP2) [6,128]. Based on these concepts, it is clear that PPAR-γ activation by TZDs accelerates the fibroblast differentiation process into adipocyte, which enhances GLUT4 expression and increases the insulin sensibility.

The members of TZD, namely, troglitazone, pioglitazone and rosiglitazone, were once approved for the treatment of T2DM. Evidence shows that troglitazone was the first TZD to be declared as an anti-diabetic drug due to its effective regulation of glycemia. Unfortunately, it was discontinued from the market because of serious liver toxicity which was reported in 100,000 patients [129]. On the other hand, pioglitazone and rosiglitazone were also licensed for controlling hyperglycemia in T2DM, but they have been also removed from the market. Treatment of pioglitazone proved to increase body weight in an in vivo study wherein 48 volunteers with T2DM were subjected to pioglitazone (30 mg/day) or 12 weeks with placebo. The study found that pioglitazone was associated with increased expression of genes in glycerol-3-phosphate synthesis, adipocytes, c-Cbl-associated protein, tumor necrosis factor-alpha, angiopoietin-like 4, leptin, resistin, and 11-beta-hydroxysteroid dehydrogenase type 1 via the activation of PPAR-γ [130].

There are various small molecules that act as selective PPAR-γ agonists and are reported to elicit anti-diabetic effects. For example, 30 μM of F12016 has been shown to selectively activate PPAR-γ and no other isoforms. It has been further characterized for possessing glucose-lowering and insulin-sensitizing properties in diabetic KK-Ay mice [131]. The compound remarkably increased glucose uptake and obstructed phosphorylation mediated by cyclin-dependent kinase [132]. In contrast, some of the endogenous agonists such as 13-hidroxioctadecanoic acid (13-HODE) and 15-hidroxieicosatetraenoic acid (15-HODE) and prostaglandins of the A, D, and J series, which are low-density lipoproteins, are also recognized as anti-diabetic agents. In addition, natural agonists of PPAR-γ have also been reported to improve disorders related to diabetes such as glycolipid metabolism and obesity. Pan et al. confirmed the underlying mechanism, a bioactive compound, curcumin, in glycolipid metabolism. In their in vivo experiments in male C57BL/6 J obese mice for eight weeks, curcumin displayed diminished activities of body weight, serum lipid profiles and fat mass with a concomitant increase in the insulin sensitivity via activation of PPAR-γ. Meanwhile, in their in vitro experiments in 3T3-L1 adipocytes, curcumin decreased glycerol release and elevated the uptake of glucose through stimulation of PPAR-γ and C/EBP-α [133]. Altogether, PPAR-γ agonists have the ability to regulate gene expression in diabetes and its related disorders.

5.4. Autoimmune Diseases (AIDs)

AIDs is a condition that occurs in an immune system which is characterized by prolonged inflammatory reaction with the production of auto-antibodies and loss of self-tolerance or immune tolerance. AIDs are categorized into organ-specific diseases, such as rheumatoid arthritis and autoimmune thyroid diseases, and systemic diseases, such as systemic lupus erythematosus and systemic sclerosis [134]. As we know that PPAR-γ agonists exert anti-inflammatory responses, the molecules participating in immune feedback are considered as potential therapeutic targets for its treatment.Recent studies have exemplified that PPAR-γ agonists also exert a protective role in AIDs (Table 2). One group suggested the upregulation of microRNA (miR)-124 by PPAR-γ in their in vitro and in vivo work. They found that the elevation of miR-124 could attenuate the generation of pro-inflammatory cytokines and augment the expression of miR-142-3p. This was in turn observed to inhibit pro-inflammatory mediator high-mobility group box-1 (HMGB1) expression, which is normally found to be increased in AIDs [7]. Another group demonstrated the development of autoimmune kidney disease, glomerulonephritis, in mice lacking macrophage-specific PPAR-γ or RXR-α, which led to the production of auto-antibodies to nuclear antigens. The lack of PPAR-γ or RXR-α manifested a deficiency in phagocytosis, loss of immune tolerance, and clearance of apoptotic cells in the mice [135].

The activation of PPAR-γ has been also reported to persuade the macrophage polarization towards an immune-modulatory M2-like phenotype that ultimately reduces neutrophil migration [136]. Cheng et al. showed that the activation of PPAR-γ by pioglitazone diminished TNF-α-induced TGF-β, hyaluronan (HA), and HAS3 expressions substantially in the active stage patients with Graves' ophthalmopathy (GO) over normal controls [137]. It has been proven that some of the potent PPAR-γ agonists such as ziglitazone, pioglitazone and GW347845 diminished the proliferation of T-cell and production of IFN-γ, TNF-α and cytokine [138]. Studies have also attested that continuous activation of PPAR-γ can prevent Th17 differentiation in murine CD4+ T cells and human models. Further, IL-17 expression is weakened, and the release of inflammatory cytokines is decreased [139]. Moreover, it has been elucidated that PPAR-γ ligands can lead to synovial cell apoptosis. For instance, for the maintenance of rheumatoid synovitis, a significant transcription factor, NF-κB, is required, and the activation of fibroblast-like synoviocytes (FLSs) with PPAR-γ can impede the pro-inflammatory activity of NF-κB [140]. In addition, a group has established inhibition of inflammation in the lupus-prone mouse model with primary biliary cirrhosis-like cholangitis via PPAR-γ activation by 15d-PGJ2 [141]. It appears that curcumin, a bioactive compound, effectively suppresses the autoimmune response by decreasing the activity of pro-inflammatory interleukins and cytokines. Bernardo et al. discerned that curcumin enhanced the differentiation of oligodendrocyte progenitor and inhibited the arrest of maturation in them via PPAR-γ activation [142]. These studies laid the foundation for PPAR-γ agonists to be promising in various AIDs.

5.5. Inflammatory Diseases

Inflammatory diseases emerge in the central nervous system (CNS), and the devastating effects include nerve damage, inflammation of CNS, loss of vision, fatigue, pain, demyelination and impaired coordination. The critical role of PPAR-γ agonists in modulating immune responses has been established and extensively documented [63,143] (Table 2). The agonists have the ability to inhibit the activated microglia, manage inflammation and defend the neurons from various degenerative diseases of CNS such as Parkinson's disease, multiple sclerosis and Alzheimer's disease [144]. It was observed that PPAR-γ agonists have key role in suppressing the activation of macrophage or monocyte lineages [145]. A computational study screened a natural product library that revealed a total of potent 29 agonists for PPAR-γ. These agonists were further carried for in vitro analysis wherein six flavonoids were detected to stimulate transcriptional activity of PPAR-γ in THP-1 macrophages. Among these, psi-baptigenin was observed to be the most potent agonist with an EC50 of 2.9 μM [146]. Xu et al. identified an endogenous ligand, 25-hydroxycholesterol-3-sulfate (25HC3S) to activate PPAR-γ in human macrophages. 25HC3S is an oxysterol that has key role in regulating lipid homeostasis and metabolism. The authors found that this cholesterol metabolite, 25HC3S markedly elevated the levels of nuclear PPAR-γ with a decrease in NF-κB protein levels [147]. Therefore, these observations suggest that PPAR-γ agonists have the ability to inhibit various transcription factors of immune response such as Signal transducer and activator of transcription 1 (STAT-1), NF-κB and activator protein 1 (AP-1), which in turn impedes their gene expression. A pioneering work by Glass and co-workers reported the molecular mechanisms that regulate transrepression of NF-κB responsive genes. Their work showed SUMOylation of PPAR-γ upon binding of PPAR-γ ligands with SUMO1 via NCoRco-repressor. Although the process of SUMOylation seems to take place in NF-κB-activating stimuli, it tends to maintain the responsive genes of promoter region in repressed state [148,149].

In addition, the PPAR-γ endogenous agonist, 15d-PGJ2, also acts to restrict the degradation of I-κB by inhibiting I-κB kinase activation [76]. Further, 15d-PGJ2 has also been documented to inhibit the binding of NF-κB to its DNA-response elements [77]. The effects of 15d-PGJ2 on immune function were first described by Petrova et al. The authors explained the inhibitory effect of 15d-PGJ2 of LPS induction in murine BV-2 microglial cell line for NO and iNOS expression. In comparison, the potent synthetic PPAR-γ agonist failed to suppress the LPS induction [150]. In addition, Cuzzocrea et al. illustrated the potency of synthetic agonist, rosiglitazone, which elicited various anti-inflammatory effects in Carrageenan rat paw oedema model such as formation of pleural exudate, attenuation of paw oedema, mononuclear cell infiltration and histological injury. Thus, rosiglitazone led to a substantial decrease in acute inflammation in the rat [151]. PPAR-γ activation plays a key role in suppressing various gene expressions inflammatory responses. In an instance, rosiglitazone has been shown to repress the transcription of Fractalkine receptor gene via activation of PPAR-γ. Fractalkine receptors potentially regulate leukocyte adhesion and migration in immune responses to inflamed peripheral tissues. In addition, PPAR-γ activation also led to the inhibition of nuclear export of Fractalkine in endothelial cells and, thus, prevented the translocation of Fractalkine receptor [152]. From this point of view, PPAR-γ agonist offers a new angle in the pharmacologic management of various inflammatory diseases as well.

5.6. Dermatological Diseases (DDs)

Human epidermis and hair follicles (HFs) are found to be expressed by all three PPAR isoforms. Most of the prominently differentiated stratum basal keratinocytes in the epidermis contain PPAR-γ. The basal layer of hair cuticle, outer root sheath, cortex and connective tissue sheath are all shown to express PPAR-α, β/δ and γ in the HF [153]. However, only PPAR-γ and PPAR-β/δ are expressed in the inner root sheath keratinocytes. PPAR-γ expression decreases with terminal sebaceous differentiation in human sebaceous glands (SGs), with substantial expression in basal and early developed sebocytes. During puberty, PPAR-γ is the only one which is expressed more significantly in sebocytes [154]. Numerous inflammatory mediators and cytokines are produced by many different cell types, including macrophages, epithelium, smooth muscle cells, endothelium dendritic cells, and lymphocytes, which have been found to be inhibited by certain PPAR-γ ligands. By opposing the actions of transcription factors such as those in the NF-κB family, PPAR-γ directly controls the expression of pro-inflammatory genes in a ligand-dependent way [149]. Transrepression is a key mechanism that explains how PPARs can obstruct the functions of these transcription factors. Furthermore, PPAR-γ reduces the production of adhesion molecules and inhibits Langerhans cell functions. Based on its anti-inflammatory properties, PPAR-γ represents a significant research target for the comprehension and management of numerous DDs [155]. Numerous studies have also shown that TZDs have a number of additional and possibly significant effects on the structure and function of the skin, such as promoting keratinocyte differentiation, reducing inflammation, and enhancing permeability barrier cellular homeostasis, which has led to their use in the treatment of various skin pathologies [156]. However, the widespread use of TZDs has been restricted due to the drugs' potential side effects, some of which may be life-threatening. As a result, the researchers are focusing on creating new classes of partial and efficient PPAR-γ modulators that maintain the anti-inflammatory action of its agonists while minimizing their negative side effects [154].

Table 2. List of ligands activating PPAR-γ and their significant role in various diseases.

Type of PPAR-γ Ligand	Name of the Ligand	Source	Disease	Effect in Disease	References
Endogenous	13-HODE *	n-3 LC-PUFA *	Cancer	Anti-proliferative activity, cell cycle arrest (G1) and apoptosis	[157]
			Multiple sclerosis	Reduced clinical severity of allergic encephalomyelitis	[158]
	15-HETE *	n-3 LC-PUFA	Cancer	Anti-proliferative activity, cell cycle arrest (G1), and apoptosis	[159]
			CVD	Anti-platelet and anti-thrombotic effects	[160]
	15d-PGJ2*	Prostaglandin J2 derivative	Cancer	Cell cycle arrest, apoptosis and reducing ornithine decarboxylase activity	[161]
			Inflammatory disorders	Regulates expression of surface proteins, T-cell activation, and related inflammatory cytokines	[162]
			AID	Anti-inflammatory effects in primary biliary cirrhosis patients;	[141]
			Asthma	Inhibited T(H)2 type cytokine IL-5production	[163]

Table 2. *Cont.*

Type of PPAR-γ Ligand	Name of the Ligand	Source	Disease	Effect in Disease	References
Natural	Procyanidin B2	Flavonoid	Hepatic diseases	Inhibited nicotine-induced pyroptosis	[164]
	Artepillin C	*Baccharisdracunculifolia*	T2DM *	Induced adipocyte differentiation and glucose uptake	[165]
	Lectins and viscotoxins	Herbs-*Viscum album* L.	Cancer	Apoptosis, inhibition of angiogenesis	[166]
	Bergenin	Herb of *Saxifragastolonifera* Curt.	Inflammatory disorders	Alleviated disease symptoms of dextran sulfate sodium (DSS)-induced colitis	[81]
			Asthma	Prevented GLS1-dependent glutaminolysis	[167]
	Antioxidants (Ascorbic acid and phytochemicals)	Whole-apple extracts	Cancer	Inhibition of tumor-cell proliferation in prostate and breast cancer	[82]
	1,1-Bis(3′-indolyl)-1-(*p*-trifluoromethylphenyl)methane	*p*-substituted phenyl analogues	Cancer	Cell cycle arrest (G_0/G_1-S) in endometrial cancer	[168]
	Chrysin	Flavonoid	Asthma	Alleviated ovalbumin-airway hyperresponsiveness	[169]
	Quercetin	Flavonoid	Cancer	Tumor-inhibitory effects in breast cancer	[170]
			Cancer	Anti-proliferative and anti-migratory effects in lung cancer	[171]
	Cucurbitane Triterpenoid	Extract of wild bitter gourd (*Momordicacharantia*)	Cancer	Anti-proliferative effect induced apoptotic death in breast cancer cells	[172]
			Insulin resistance	Induced adipocyte differentiation and glucose uptake	[173]
			T2DM	Induced glucose uptake	[174]
	Methanolic extract *Pterocarpus marsupium*	isoflavone	T2DM	Induced glucose uptake and elevated Glut-4	[175]
Synthetic	Pioglitazone	TZD *	Neurological disease	Inhibited mTOR activation and prevented increase in IL-1β and IL-6.	[176]
			Neurological disease	Reduced hyperalgesia and astrocyte activation	[177]
			Psoriatic Arthritis Response	Inhibited angiogenesis and suppressed pro-inflammatory cytokines	[178]
			Asthma	Reduced regulator of G protein 4	[179]
			SRD	Inhibited bleomycin-induced skin fibrosis	[98]

Table 2. *Cont.*

Type of PPAR-γ Ligand	Name of the Ligand	Source	Disease	Effect in Disease	References
	Rosiglitazone	TZD	Ischemia Stroke	Limited postischemic injury in normal and diabetic hearts	[180]
			Asthma	Reduced bronchial inflammation	[181]
			Allergy	Decreased ROS generation, expression of T(H)2 cell cytokines in lungs after ovalbumin inhalation	[182]
	Ciglitazone	TZD	Cancer	Inhibitory effects on lung cancer	[183]
			Cancer	Inhibitory effects on prostate cancer	[184]
	Troglitazone	TZD	Cancer	Reduced c-Myc levels in prostate cancer	[185]
			Psoriasis	Inhibited proliferation of psoriatic human keratinocytes	[186]
	GW347845	Non-TZD	AID	Anti-inflammatory and anti-proliferative effects	[138]

* 13-HODE-13-Hydroxyoctadecadienoic acid; n-3 LC-PUFA-n-3 long chain polyunsaturated fatty acids; 15-HETE-15-hydroxyeicosatetraenoic acid; 15d-PGJ2-15-Deoxy-Δ-12,14-Prostaglandin J2; T2DM, Type 2 diabetes mellitus; TZD, Thiazolidinedione.

6. Challenges Faced and Knowledge Gaps for PPAR-γ Agonists

There has been controversy over many endogenous ligands due to their uncertain intracellular levels and interaction with PPAR-γ, which is biologically insignificant [187]. Thus, they activate PPAR-γ with relatively low affinity, i.e., they act as weak agonists. On the other hand, the natural compounds have some concerns about their limited efficiency, poor bioavailability, inadequate absorption, inappropriate solubility, and non-specificity [188]. Further, increasing evidence accounts for the adverse effects of the potent PPAR-γ agonist, TZDs. TZDs have been reported as powerful activators of PPAR-γ, ultimately showing profound anti-inflammatory, anti-cancer, and anti-diabetic activities. However, the use of TZDs have been withdrawn from market due to higher risk in cardiovascular, heart failure, weight gain, edema, sodium retention and decreased glucosuria [189,190]. Although TZDs exhibit various activities, the class of drugs has been involved for increased risk. Prolonged usage of TZDs has been reported to increase the level of PPAR-γ expression in bladder cancer [191]. According to investigations, rosiglitazone seems to be associated with a lower risk of thyroid cancer [192] and breast cancer [193], whereas it brings a higher risk of bladder cancer [194], due to which it has been banned in Europe. The first TZD to be stopped from the market of USA as well as UK was troglitazone, in the year 1999 and 2000, due to its major hepatotoxicity activities [195].

It is worth noting that a study conducted by Nissen et al. reported that rosiglitazone was shown to have a 1.4-fold increased risk of acute myocardial infarction (AMI) in the year 2007 [196]. In particular, the incidence rates of heart failure, strokes and death was observed for the patients when administered with pioglitazone for long-term treatment [94,197]. In addition, ciglitazone has been never approved for medication of its weak clinical activity, though it serves as a prototype for all TZDs. Moreover, it is also reported that an elevated level of TZDs tends to cause visceral fat accumulation, which contributes to major weight gain [190] and increased sodium, and poor water reabsorption in kidney, which leads to fluid retention, ultimately causing edema. Further consequences include fluid retention and weight gain, which lead to congestive heart failure [198]. TZDs are therefore restricted due to concerns for their adverse side-effects, which has weakened their preventive measures in cancer therapies as well as in diabetes treatment. As it acts as a full agonist for PPAR-γ, studies have shown it to express the protein receptor more than its optimized level, which leads to over-activation of PPAR-γ, eventually leading to tumor progression. Hence, proper examination is required to select the exact ligand for PPAR-γ, because natural and endogenous ligands serve as weak agonists, whereas synthetic ligands such as TZDs act as full agonist for PPAR-γ. A selective measure is necessary to re-explore the agonists which will have anti-inflammatory, anti-diabetic and anti-cancer characteristics, sans toxicity and akin to the proper activation of PPAR-γ.

7. Significance of PPAR-γ Partial Agonist

The compounds that stimulate PPAR-γ in a desired manner, i.e., lesser effect than synthetic agonists and greater effect than weak agonists, are referred to as partial agonist. One such example is selective PPAR-γ modulators (SPPAR-γMs), which were built to reduce the unwanted side-effects of synthetic agonists by optimizing the gene expression signature [199]. A natural product named amorfrutins, were observed to act as SPPAR-γM that exerted decreased gene expression in the fat storage process compared to the synthetic agonist, rosiglitazone [200]. Amorfrutin is a low nanomolar-binding2-hydroxy benzoic acid derivative of salicylic acids that promoted anti-inflammatory and insulin-sensitizing effects in diabetes mouse models. Notably, some of the synthesized derivatives of amorfrutin from a common building block library served as PPAR-γ partial agonists and promising anti-diabetic drugs [201]. Besides SPPAR-γM, telmisartan is angiotensin receptor blocker which also aids in partial activation of PPAR-γ. The compound was noticed to exert noteworthy effects against inflammatory responses, oxidative stress and EMT [202]. A group demonstrated the counteraction of TGF-β1-induced EMT in human renal proximal tubular epithelial (HK-2) cells through the partial activation of PPAR-γ [203]. Similarly, another study suggested that Treatment of telmisartan in HK2 cells inhibited EMT induction via PPAR-γ-AKT/STAT3/p38 MAPK-Snail pathway in vitro and in vivo by oxalate and calcium oxalate crystals [204]. In contrast, it has been demonstrated that flavonoids extracted from lemon balm (*Melissa officinalis*) [205] and chamomile (*Matricaria chamomilla/Matricaria recutita*) [80] flowers behave as a partial agonist for PPAR-γ by activating it with a half-maximal effective concentration (EC50) of 86 mg/mL and had 26% of maximal potency as compared to rosiglitazone (TZD). According to epidemiological studies some of the naturally occurring dietary flavonoids such as quercetin (3,5,7,3,4-pentahydroxyflavone), Kaempferol (3,5,7,4tetrahydroxyflavone) and apigenin (4,5,7-trihydroxyflavone) also contribute to the activation process of PPAR-γ which in turn are associated with a diminishing incidence of various types of cancer [87,88]. These studies certainly prove that the biological function and mechanism of PPAR-γ could improve the rationale of ligand development. Targeting various molecules to the ligand binding region of PPAR-γ can anticipate selective modulation. This would result in greater expression and a promising pharmacological approach.

8. PPAR-γ Partial Agonists Involved in Post-Transcriptional Modification and Disease-Fate Decision

It has been discovered that partial agonists are neoteric substances that have a substantial affinity for PPAR-γ and display the effects that cause insulin sensitivity, anti-cancer and anti-inflammatory characteristics, and rescue various heart disease. The beneficial effects of PPAR-γ partial agonists in various diseases are summarized in Table 3. These ligands cannot completely saturate PPAR-γ activity; nevertheless, they lessen the negative effects of TZDs. Zheng and colleagues virtually screened a library of compounds using AlphaScreen assay and found ionomycin, an antibiotic to act as partial agonist for PPAR-γ. Their basic aim was to develop alternative and better PPAR-γ ligands than TZDs, which are reported for their severe side effects. Ionomycin showed distinctly interacted with PPAR-γ LBD over the TZDs and improved hyperglycemia and insulin resistance in a mouse model of diabetes. Further in vitro and in vivo experiments showed the inhibition of PPAR-γ phosphorylation at Ser273 by cyclin-dependent kinase 5 [206]. The partial agonists also constitute SPPAR-γMs, NSAID and non-TZD partial agonist (nTZDpa) [207]. Some of these SPPAR-γMs share an indole moiety with NSAIDS. Schug and collaborators proved that the indole containing non-TZD partial agonist (nTZDpa) enhance insulin sensitivity in obese mice while reducing unfavorable effects on weight gain, adiposity, and cardiovascular hypertrophy [208]. Similarly, NSAID compounds have undergone in vitro testing as partial PPAR-γ agonists, followed by pharmacokinetic studies in rats and in insulin-resistant mice models. These are benzoyl 2-methyl indoles, often known as carboxylic acid indoles. A PPAR-γ modulator termed as SPPARM5 functioned as a partial agonist of PPAR-γ, with some reduction in the ability to promote adipose gene expression, while retaining the insulin sensitizing capabilities [209]. SPPARM5 was also examined in Zucker rats in contrast to rosiglitazone for effects on plasma and extracellular volume, heart weight, and fluid retention [207].

The PPAR-γ activation by partial agonists controls a variety of parameters, including protein expression levels, ligands, and transcriptional cofactors. These factors eventually influence the course of the disease, thus helping in determining the disease fate. Post-transcriptional modifications (PTMs) following PPAR-γ activation have the ability to modify protein shape, control protein interactions, and change the moiety between receptors and ligands, all of which affect how transcription of downstream genes is regulated [210]. The key PTMs which influence the course of development of a disease are highlighted as follows:

8.1. Phosphorylation

PPAR-γ could be phosphorylated at various locations with various stimuli, leading to various biological effects [211]. Ser273 (Ser245 in isoform 1) and Ser112 (Ser82 in isoform 1) are the primary sites for PPAR-γ phosphorylation by cyclin-dependent kinase (Cdk) and MAPK. As a result of PPAR-γ S273 being phosphorylated by Cdk5, less adipogenesis and transcriptional activity is generated [212]. TNF-α, IL-1β, and other inflammatory cytokines can be expressed as a result of PPAR-γ phosphorylation, which can also encourage the growth of foam cells and hasten the progression of atherosclerosis [213]. In response to exposure to chemicals that damage DNA, cancer cells phosphorylate the PPAR-γ Ser273 protein. Genetically or pharmacologically inhibiting this phosphorylation causes a build-up of DNA damage that leads to apoptotic cell death. Furthermore, p53 signaling is deregulated when PPAR-γ phosphorylation is inhibited, and biochemical studies demonstrate that PPAR physically interacts with p53 that is dependent on Ser273 phosphorylation [202]. These findings suggest that PPAR-γ plays a vital role in regulating the p53 response to cytotoxic therapy, which can be controlled for therapeutic benefits.

8.2. SUMOylation

The SUMOylation of PPAR-γ is termed as transrepression. According to Ying et al. cellular inflammation brought about by lipopolysaccharide is suppressed by SUMOylation of PPAR-γ by a partial agonist, which inhibits NF-κB [210]. The SUMOylation pathway is composed of the proteases namely SUMO E1, E2, and E3, which can change the way that target proteins are regulated in transcription. This mechanism is extremely important for controlling the course of the cell cycle and the tumorigenic processes [214]. Therefore, targeting SUMOylation of PPAR-γ can provide a promising solution to determine disease fate and, hence, their potential cure [215]. According to Phan et al., SUMOylation of PPAR-γ links lipid metabolism to its tumor-suppressive properties in lung cancer. They discovered that both in vitro and in vivo, PPAR-γ ligand activation significantly increased de novo lipid production as well as fatty acid beta (β)-oxidation in lung cancer [216]. More significantly, it transpires that SUMOylation of PPAR-γ was necessary for regulation of lipid metabolism. More in-depth biochemical research showed that PPAR-γ-mediated lipid production degrades nicotinamide adenine dinucleotide phosphate (NADPH), which raises the quantity of reactive oxygen species (ROS) in the mitochondria and disrupts the equilibrium of REDOX reactions in lung cancer [217]. As a result, liganded PPAR-γ SUMOylation is essential for cellular lipid metabolism as well as for inducing oxidative stress, which helps PPAR-γ act as a tumor suppressor. This study provides crucial insight into future translational and clinical research into addressing PPAR-γ control of lipid metabolism in lung cancer patients with T2DM.

8.3. Ubiquitination

Ubiquitination modification of PPAR-γ not only controls the proteasome-mediated destruction of target proteins but also functions as a "scaffold" to draw in more proteins to form signal complexes. When PPAR-γ binds to the selective ligand, it experiences substantial conformational changes. PPAR-γ is targeted for proteasomal degradation by the PPAR E3 ligases Makorin RING finger protein 1 (MKRN1) [218] and seven in absentia homolog 2 (SIAH2) [219]. Alternatively, it can encourage proteasome-dependent disassembly and bind with ubiquitination-associated enzymes, thus negatively affecting its overall transcriptional activity [215]. An E3 ubiquitin ligase, namely, neural precursor cell expressed developmentally down-regulated protein 4 (NEDD4) interacted with the hinge and LBD of PPAR-γ. Further, it underwent ubiquitination of PPAR-γ in adipocytes, as reported by Carvalho et al. [220]. The E3 ubiquitin ligase tripartite motif containing 23 (TRIM23) promotes PPAR-γ stability by inhibiting its proteasomal degradation and controlling adipocyte development. This could thus provide a potential solution to trace various diseases which involve adipogenesis dysregulation [221]. In clinical research, TZDs activators (used to treat diabetes) and the proteasome inhibitor Bortezomib (used to treat cancer) have both been used to pharmacologically control the PPAR-γ and the ubiquitin proteasome system [222]. The development of medications for the treatment of colorectal cancer may be attracted to a combination used in order to activate the transcription factor at least twice. It would be crucial to identify different cancer subtypes that, as a result of particular molecular abnormalities, may be especially vulnerable to PPAR-γ ubiquitination.

Table 3. Effect of PPAR-γ partial agonists in disease-fate decision.

PPAR-γ Partial Agonist	Type of Compound	Disease	Effect in Disease	References
SPPAR-γM5	SPPAR-γM *	T2DM	Reduced the insulin resistance index	[69]
PAR-1622 *	SPPAR-γM	T2DM	Induced adipocyte differentiation and improved hyperglycemia	[214]
PAM-1616 *	SPPAR-γM	T2DM	Improved hyperglycemia	[223]
FK614 *	SPPAR-γM	T2DM	Reduced the insulin resistance index	[224]
F12016 *	SPPAR-γM	T2DM	Insulin-sensitizing and glucose-lowering properties	[132]
KDT501 *	Chemically derived from substituted 1,3-cyclopentadione	Inflammatory disorders	Anti-inflammatory effects in monocytes/macrophages	[225]

Table 3. *Cont.*

PPAR-γ Partial Agonist	Type of Compound	Disease	Effect in Disease	References
GQ-16 *	TZD-Derived	Obesity	Reduced high fat diet-induced weight gain	[211]
		Cancer	Anti-proliferative effects in breast cancer	[212]
Telmisartan	Angiotensin type 1 receptor blocker	Inflammatory disorders	cerebroprotective effect	[213]
		T2DM	Ameliorated vascular endothelial dysfunction and protected against diabetic vascular complications	[202]

* SPPAR-γM-selective PPAR-γ modulators; PAR-1622-(S)-2-ethoxy-3(4-(5-(4-(5-(methoxymethyl)isoxazol-3-yl)phenyl)-3-methylthiophen-2-yl)methoxy)phenyl)propanoic Acid; PAM-1616-(S)-2-ethoxy-3-(4-((3-methyl-5-(4-(3-methylisoxazol-5-yl) phenyl) thiophen-2-yl) methoxy) phenyl) propanoic acid; FK614-3-(2,4-dichlorobenzyl)-2-methyl-N-(pentylsulfonyl)-3-Hbenzimidazole-5-carboxamide; F12016-2-[2-(1,2-dimethyl-1H-indol-3-yl)-2-oxo-acetylamino]-benzamide; KDT501-Potassium salt of the n-(isobutyl) congener of a tetrahydro iso α acid; GQ-16-(5Z)-5-(5-bromo-2-methoxy-benzylidene)-3-(4-methyl-benzyl)-thiazolidine-2,4-dione.

9. PPAR-γ Partial Agonists in Cancer Therapeutics

As already discussed in the above sections that inflammation and immunity are strongly regulated by PPAR-γ, suggesting being useful in cancer immunotherapy. The activation of PPAR-γ may activate many signaling pathways in anti-tumorigenic activity. According to Zhao and collaborators, myeloid-derived suppressor cells (MDSCs) infiltrate PPAR-γ attenuated mouse melanoma cells and caused generalized non-specific inflammatory reactions [214]. In addition, PPAR-γ ligand binding interaction results in a positive effect leading to prevention of tumor growth (Table 4). This positive action is accomplished via preventing the overproduction of reactive oxygen species (ROS) by MDSCs, mTOR pathway, and an additional receptor for advanced glycation end products (RAGE pathway) which work in conjunction [64,215]. Macrophages are highly diverse and plastic. In terms of malignancy, tumor-associated macrophages are particularly prevalent and pro-proliferative within tumors. They promote tumor growth and spread by suppressing the immune system and promoting angiogenesis. When activated particularly by a partial agonist, PPAR-γ may decrease the release of M1 pro-inflammatory and pro-tumor M2-cytokines without changing macrophage polarization, thus producing anti-tumor effects [215,217]. When administered as monotherapy, TZDs provide a significant clinical anti-cancer action in the majority of trials, but they also have potentially fatal adverse effects. As a result, the researchers are now trying to test partial agonists rather than the synthetic TZDs in order to sequester positive cancer therapeutic results [218].

Telmisartan is a partial PPAR-γ agonist that was noticed to exert noteworthy effects against inflammatory responses, oxidative stress and EMT [204]. Deoxyelephantopin (ESD), another PPAR-γ partial agonist, was reported to potentiate apoptosis, inhibit invasion, and abolish osteoclastogenesis by Zou et al. [69]. It was isolated from the Wild plant of *Elephantopus carolinianus* and exhibits anti-tumor, anti-inflammatory, and invasion-inhibiting activities. Their findings demonstrated that ESD-induced PPAR-γ knockdown could cause HeLa cell line death and cell cycle arrest during G2/M phase in a dose-dependent manner [219]. Gan and associates in their investigation revealed Tetrazanbigen (TNBG), a new sterol isoquinoline derivative with poor water solubility that produced mild inhibitory effects on human tumor cell lines via lipoapoptosis induction [220]. The primary goal of employing TNBG as a PPAR gamma partial agonist was to cause tumor cells to undergo lipoapoptosis. The underlying theory demonstrates how excessive lipid accumulation interferes with cancer cells ability to use lipids and how an unrestricted accumulation of lipid droplets would take up most of the cytoplasm, impair the operation of other organelles, and eventually cause differentiated cancer cells to undergo lipoapoptosis [221].

Table 4. List of PPAR-γ partial agonists and their significant role in cancer therapeutics.

PPAR-γ Partial Agonist	Type of Compound	Effect in Disease	Cell Line	Type of Cancer	References
Deoxyelephantopin	Natural	Apoptosis and cell cycle arrest (G(2)/M)	HeLa	Cervix	[69]
Halofenate	SPPAR-γM	Anti-proliferative effects	MM96L	Melanoma	[226]
Tetrazanbigen	Sterol isoquinoline derivative	Anti-proliferative effects	HepG2 and A549	Liver and lung	[220]
HydroxyCinnamic Acid Derivatives	p-coumaric acid and ferulic acid	Anti-proliferative effects	K562	Chronic Myeloid Leukemia	[222]
Telmisartan	Angiotensin II (Ang II) receptor blocker	Apoptosis and anti-proliferative effects	Caki-1,T24, LNCaP, PC3, DU-145 and NEC-8	Renal, bladder, prostate and testicular	[203]
		Anti-proliferative effects	A549	Lung	[227]

10. PPAR-γ Partial Agonists under Clinical Trials

Although various PPAR-γ partial agonists are now undergoing clinical studies, the partial agonists identified to date have not yet acquired FDA approval. INT131 (formerly AMG-131), which has advanced through Phase II clinical trials, is the most well-known and maybe most promising example [228]. Netoglitazone, Balaglitazone, Metaglidasen, and Halofenate are further PPAR-γ partial agonists that have progressed to or finished Phase II clinical trials [229–231]. When compared to rosiglitazone, partial agonists of PPAR-γ can exhibit a broad spectrum of transcriptional activation [229]. In adipocyte cell models such 3T3-L1 mouse fibroblast cells, partial agonists have demonstrated effects that cause insulin sensitization but have not been shown to cause fatty acid accumulation. Beyond their functional efficacy, the exact atomic and molecular characteristics of PPAR-γ partial agonists are still unknown.

A novel partial PPAR-γ agonist called balaglitazone was introduced by Dr. Reddy's laboratory in India as a cure for T2DM. Balaglitazone dramatically lowers HbA1c levels and has a better safety profile than full agonists, since it is a selective partial PPAR-γ agonist. As a supplement to insulin therapy, balaglitazone offers strong glycemic control, and phase trials revealed a tendency towards fewer severe side effects. However, because of unfavorable side effects such inflammation, ROS production, and altered gene expression, the investment was stopped in 2011 [231]. Murakami et al. reported Efatutazone, an oral, highly selective PPAR-γ agonist, is superior to second-generation TZDs such as pioglitazone and rosiglitazone in terms of effectiveness [232]. Efatutazone suppressed the growth of human anaplastic thyroid and pancreatic tumor cell lines in pre-clinical tumor models, as well as human colorectal and anaplastic thyroid tumor cell xenografts in nude mice. Efatutazone demonstrated good safety, tolerability, and disease control at dosages of 0.10–1.15 mg bis in die (bid) in a phase I study in patients with advanced solid tumors [233]. A recent phase I clinical trial using efatutazone plus paclitaxel in patients with advanced anaplastic thyroid carcinoma revealed similar positive safety results, disease control, and disease stability [234]. These phase I clinical studies proved that Efatutazone can be used alone or in combination with other chemotherapeutic drugs as a novel approach for the treatment of advanced metastatic tumors, as a result of its unique property as a highly selective PPAR-γ activator. Additionally, this might offer useful information about the therapeutic implications of selective PPAR-γ activation in the regulation of carcinogenesis.

Such clinical studies have thus proven that potential PPAR-γ partial agonists can be worked upon and implemented into the therapeutic lines of cancer and are expected to yield better outcomes than the synthetic ligands such as TZDs. Although none of the PPAR-γ partial agonists have yet been approved by FDA due to possible adverse effects, their efficacy and favorable outcomes against cancer have gained attention from scientists around the world who are extensively working to diminish the side-effects and treat cancer effectively.

11. Conclusions and Future Applications of PPAR-Gamma Partial Agonists for Precision Oncology

In particular, witnessing the data together tries to update with the scenario that in spite of extensive formulation of ligands for PPAR-γ, the application is limited and largely abandoned by drug companies due to several adverse effects. In light of the evidence, the partial activation of PPAR-γ stands out as an attractive approach, as the ligands would serve as promising drug leads in inducing anti-tumorigenic, anti-inflammatory and anti-diabetic effects. Through their partial agonist effect and preventive mechanism, the ligands would not only exhibit potent activators for transcription, but also be apt for co-activator recruitment with a better implication of genetic expression profiling. The partial activation approach of PPAR-γ may advocate a strategy to combat the deleterious effect of synthetic agonists, TZDs. Owing to their direct involvement in cancer therapy; the ligands may exhibit desired effects by abrogating the side effects caused by certain other agonists. These challenges clearly lie ahead to decipher how such an approach should be carried out as targets and treatments of patients with existing therapies.

Author Contributions: S.B. (Sangeeta Ballav), B.B. and V.K.S. designed the review and drafted the manuscript. S.B. (Soumya Basu) and A.R. designed and supervised the work. All authors have read and agreed to the published version of the manuscript.

Funding: This research received no external funding.

Institutional Review Board Statement: Not applicable.

Informed Consent Statement: Not applicable.

Data Availability Statement: Not applicable.

Acknowledgments: The authors are thankful to Dr. D. Y. Patil Biotechnology and Bioinformatics Institute, Dr. D. Y. Patil Vidyapeeth, Pune for the physical infrastructure. Sangeeta Ballav is thankful to Dr. D. Y. Patil Vidyapeeth, Pune for Junior Research Fellowship (DPU/291/2021). Vishal Kumar Sahu acknowledges Council of Scientific and Industrial Research (CSIR), New Delhi for Junior Research Fellowship (File No.: 09/1340(11487)/2021-EMR-I).

Conflicts of Interest: The authors declare no conflict of interest.

References

1. Siegel, R.L.; Miller, K.D.; Fuchs, H.E.; Jemal, A. Cancer statistics, 2022. *CA A Cancer J. Clin.* **2022**, *72*, 7–33. [CrossRef] [PubMed]
2. Hu, M.; Wang, X.; Tan, J.; Yang, J.; Gao, X.; Yang, Y. Causal Associations between Paternal Longevity and Risks of Cardiovascular Diseases. *J. Cardiovasc. Dev. Dis.* **2022**, *9*, 233. [CrossRef] [PubMed]
3. Knudsen, J.S.; Knudsen, S.S.; Hulman, A.; Witte, D.R.; Gregg, E.W.; Lauritzen, T.; Pedersen, L.; Sørensen, H.T.; Thomsen, R.W. Changes in type 2 diabetes incidence and mortality associated with introduction of HbA1c as diagnostic option: A Danish 24-year population-based study. *Lancet Reg. Health-Eur.* **2022**, *14*, 100291. [CrossRef] [PubMed]
4. Angum, F.; Khan, T.; Kaler, J.; Siddiqui, L.; Hussain, A. The Prevalence of Autoimmune Disorders in Women: A Narrative Review. *Cureus* **2020**, *12*, e8094. [CrossRef]
5. Quinto, R.M.; Iani, L.; Vincenzo, F.D.; Russo, F.; Porcelli, P.; Abeni, D. Does Guided Written Disclosure Reduce Distress and Improve Psychological Functioning in Patients with Skin Diseases? *Int. J. Environ. Res. Public Health* **2022**, *19*, 2943. [CrossRef]
6. Colca, J.R. The TZD insulin sensitizer clue provides a new route into diabetes drug discovery. *Expert Opin. Drug Discov.* **2015**, *10*, 1259–1270. [CrossRef]
7. Yuan, Z.; Luo, G.; Li, X.; Chen, J.; Wu, J.; Peng, Y. PPAR-γ inhibits HMGB1 expression through upregulation of miR-142-3p In Vitro and In Vivo. *Cell. Signal.* **2016**, *28*, 158–164. [CrossRef]

8. Asakawa, M.; Takano, H.; Nagai, T.; Uozumi, H.; Hasegawa, H.; Kubota, N.; Saito, T.; Masuda, Y.; Kadowaki, T.; Komuro, I. Peroxisome Proliferator-Activated Receptor Plays a Critical Role in Inhibition of Cardiac Hypertrophy In Vitro and In Vivo. *Circulation* **2002**, *105*, 1240–1246. [CrossRef]
9. Tyagi, S.; Gupta, P.; Saini, A.; Kaushal, C.; Sharma, S. The peroxisome proliferator-activated receptor: A family of nuclear receptors role in various diseases. *J. Adv. Pharm. Technol. Res.* **2011**, *2*, 236. [CrossRef]
10. Weikum, E.R.; Liu, X.; Ortlund, E.A. The nuclear receptor superfamily: A structural perspective. *Protein Sci.* **2018**, *27*, 1876–1892. [CrossRef]
11. Youssef, J.; Badr, M.Z. PPARs: History and Advances. *Methods Mol. Biol.* **2013**, *952*, 1–6. [CrossRef]
12. Decara, J.; Rivera, P.; López-Gambero, A.J.; Serrano, A.; Pavón, F.J.; Baixeras, E.; de Fonseca, F.R.; Suárez, J. Peroxisome Proliferator-Activated Receptors: Experimental Targeting for the Treatment of Inflammatory Bowel Diseases. *Front. Pharmacol.* **2020**, *11*, 730. [CrossRef]
13. Rosen, E.D.; Spiegelman, B.M. PPAR: A Nuclear Regulator of Metabolism, Differentiation, and Cell Growth. *J. Biol. Chem.* **2001**, *276*, 37731–37734. [CrossRef]
14. Capelli, D.; Cerchia, C.; Montanari, R.; Loiodice, F.; Tortorella, P.; Laghezza, A.; Cervoni, L.; Pochetti, G.; Lavecchia, A. Structural basis for PPAR partial or full activation revealed by a novel ligand binding mode. *Sci. Rep.* **2016**, *6*, 34792. [CrossRef]
15. Han, L.; Shen, W.; Bittner, S.; Kraemer, F.B.; Azhar, S. PPARs: Regulators of metabolism and as therapeutic targets in cardiovascular disease. Part I: PPAR-α. *Future Cardiol.* **2017**, *13*, 259–278. [CrossRef]
16. Rachid, T.L.; Silva-Veiga, F.M.; Graus-Nunes, F.; Bringhenti, I.; Mandarim-de-Lacerda, C.A.; Souza-Mello, V. Differential actions of PPAR-α and PPAR-β/δ on beige adipocyte formation: A study in the subcutaneous white adipose tissue of obese male mice. *PLoS ONE* **2018**, *13*, e0191365. [CrossRef]
17. Sikder, K.; Shukla, S.K.; Patel, N.; Singh, H.; Rafiq, K. High Fat Diet Up-regulates Fatty Acid Oxidation and Ketogenesis via Intervention of PPAR-γ. *Cell. Physiol. Biochem.* **2018**, *48*, 1317–1331. [CrossRef]
18. Plutzky, J. The PPAR-RXR Transcriptional Complex in the Vasculature. *Circ. Res.* **2011**, *108*, 1002–1016. [CrossRef]
19. Garcia-Bates, T.M.; Lehmann, G.M.; Simpson-Haidaris, P.J.; Bernstein, S.H.; Sime, P.J.; Phipps, R.P. Role of Peroxisome Proliferator-Activated Receptor Gamma and Its Ligands in the Treatment of Hematological Malignancies. *PPAR Res.* **2008**, *2008*, 834612. [CrossRef]
20. Michalik, L.; Desvergne, B.; Wahli, W. Peroxisome-proliferator-activated receptors and cancers: Complex stories. *Nat. Rev. Cancer* **2004**, *4*, 61–70. [CrossRef]
21. Grommes, C.; Landreth, G.L.; Heneka, M.T. Antineoplastic effects of peroxisome proliferatoractivated receptor gamma agonists. *Lancet Oncol.* **2004**, *5*, 419–429. [CrossRef]
22. Bennett, R.G. *Ligand-Independent Coactivation of Peroxisome Proliferator-Activated Receptor Gamma*; Springer: Cham, Switherland, 2021; pp. 519–535. [CrossRef]
23. Cronet, P.; Petersen, J.F.; Folmer, R.; Blomberg, N.; Sjöblom, K.; Karlsson, U.; Lindstedt, E.; Bamberg, K. Structure of the PPAR-alpha and -gamma Ligand Binding Domain in Complex with AZ 242 Ligand Selectivity and Agonist Activation in the PPAR Family. *Structure* **2001**, *9*, 699–706. [CrossRef]
24. Yu, C.; Markan, K.; Temple, K.A.; Deplewski, D.; Brady, M.J.; Cohen, R.N. The Nuclear Receptor Co-repressors NCoR and SMRT Decrease Peroxisome Proliferator-activated Receptor gamma Transcriptional Activity and Repress 3T3-L1 Adipogenesis. *J. Biol. Chem.* **2005**, *280*, 13600–13605. [CrossRef] [PubMed]
25. Emont, M.P.; Mantis, S.; Kahn, J.H.; Landeche, M.; Han, X.; Sargis, R.M.; Cohen, R.N. Silencing Mediator of Retinoid and Thyroid Hormone Receptors (SMRT) regulates glucocorticoid action in adipocytes. *Mol. Cell. Endocrinol.* **2015**, *407*, 52–56. [CrossRef] [PubMed]
26. Chandra, V.; Huang, P.; Hamuro, Y.; Raghuram, S.; Wang, Y.; Burris, T.P.; Rastinejad, F. Structure of the intact PPAR-γ–RXR-nuclear receptor complex on DNA. *Nature* **2008**, *456*, 350–356. [CrossRef]
27. Bugge, A.; Grøntved, L.; Aagaard, M.M.; Borup, R.; Mandrup, S. The PPARgamma2 A/B-Domain Plays a Gene-Specific Role in Transactivation and Cofactor Recruitment. *Mol. Endocrinol.* **2009**, *23*, 794–808. [CrossRef]
28. Tontonoz, P.; Hu, E.; Graves, R.A.; Budavari, A.I.; Spiegelman, B.M. mPPAR gamma 2: Tissue-specific regulator of an adipocyte enhancer. *Genes Dev.* **1994**, *8*, 1224–1234. [CrossRef] [PubMed]
29. Lehrke, M.; Lazar, M.A. The Many Faces of PPAR. *Cell* **2005**, *123*, 993–999. [CrossRef]
30. Cipolletta, D.; Feuerer, M.; Li, A.; Kamei, N.; Lee, J.; Shoelson, S.E.; Benoist, C.; Mathis, D. PPAR-γ is a major driver of the accumulation and phenotype of adipose tissue Treg cells. *Nature* **2012**, *486*, 549–553. [CrossRef]
31. Hernandez-Quiles, M.; Broekema, M.F.; Kalkhoven, E. PPARgamma in Metabolism, Immunity, and Cancer: Unified and Diverse Mechanisms of Action. *Front. Endocrinol.* **2021**, *12*, 624112. [CrossRef]
32. Vella, V.; Nicolosi, M.L.; Giuliano, S.; Bellomo, M.; Belfiore, A.; Malaguarnera, R. PPAR-Agonists As Antineoplastic Agents in Cancers with Dysregulated IGF Axis. *Front. Endocrinol.* **2017**, *8*, 31. [CrossRef] [PubMed]
33. Augimeri, G.; Bonofiglio, D. PPARgamma: A Potential Intrinsic and Extrinsic Molecular Target for Breast Cancer Therapy. *Biomedicines* **2021**, *9*, 543. [CrossRef] [PubMed]
34. Wagner, N.; Wagner, K. Peroxisome Proliferator-Activated Receptors and the Hallmarks of Cancer. *Cells* **2022**, *11*, 2432. [CrossRef] [PubMed]
35. Wagner, N.; Wagner, K. PPARs and Angiogenesis—Implications in Pathology. *Int. J. Mol. Sci.* **2020**, *21*, 5723. [CrossRef]

36. Basilotta, R.; Lanza, M.; Casili, G.; Chisari, G.; Munao, S.; Colarossi, L.; Cucinotta, L.; Campolo, M.; Esposito, E.; Paterniti, I. Potential Therapeutic Effects of PPAR Ligands in Glioblastoma. *Cells* **2022**, *11*, 621. [CrossRef]
37. Cheng, H.S.; Yip, Y.S.; Lim, E.K.Y.; Wahli, W.; Tan, N.S. PPARs and Tumor Microenvironment: The Emerging Roles of the Metabolic Master Regulators in Tumor Stromal–Epithelial Crosstalk and Carcinogenesis. *Cancers* **2021**, *13*, 2153. [CrossRef]
38. Yang, F.; Zhang, Z.; Xin, D.; Shi, C.; Wu, J.; Guo, Y.; Guan, Y. Peroxisome proliferator-activated receptor gamma ligands induce cell cycle arrest and apoptosis in human renal carcinoma cell lines. *Acta Pharmacol. Sin.* **2005**, *26*, 753–761. [CrossRef]
39. Bruin, J.E.; Petrik, J.J.; Hyslop, J.R.; Raha, S.; Tarnopolsky, M.A.; Gerstein, H.C.; Holloway, A.C. Rosiglitazone improves pancreatic mitochondrial function in an animal model of dysglycemia: Role of the insulin-like growth factor axis. *Endocrine* **2010**, *37*, 303–311. [CrossRef]
40. Vijayababu, M.R.; Arunkumar, A.; Kanagaraj, P.; Arunakaran, J. Effects of quercetin on insulin-like growth factors (IGFs) and their binding protein-3 (IGFBP-3) secretion and induction of apoptosis in human prostate cancer cells. *J. Carcinog.* **2006**, *5*, 10. [CrossRef]
41. Babcook, M.A.; Gupta, S. Apigenin modulates insulin-like growth factor axis: Implicationsfor prevention and therapy of prostate cancer. *Curr. Drug Targets* **2012**. *Online ahead of print*.
42. Camp, H.S.; Tafuri, S.R. Regulation of Peroxisome Proliferator-activated Receptor Activity by Mitogen-activated Protein Kinase. *J. Biol. Chem.* **1997**, *272*, 10811–10816. [CrossRef]
43. Chen, K.; Lu, P.; Beeraka, N.M.; Sukocheva, O.A.; Madhunapantula, S.V.; Liu, J.; Sinelnikov, M.Y.; Nikolenko, V.N.; Bulygin, K.V.; Mikhaleva, L.M.; et al. Mitochondrial mutations and mitoepigenetics: Focus on regulation of oxidative stress-induced responses in breast cancers. *Semin. Cancer Biol.* **2022**, *83*, 556–569. [CrossRef]
44. Wood, W.M.; Sharma, V.; Bauerle, K.T.; Pike, L.A.; Zhou, Q.; Fretwell, D.L.; Schweppe, R.E.; Haugen, B.R. PPARγ Promotes Growth and Invasion of Thyroid Cancer Cells. *PPAR Res.* **2011**, *2011*, 171765. [CrossRef]
45. Mueller, E.; Smith, M.; Sarraf, P.; Kroll, T.; Aiyer, A.; Kaufman, D.S.; Oh, W.; Demetri, G.; Figg, W.D.; Zhou, X.; et al. Effects of ligand activation of peroxisome proliferator-activated receptor gamma in human prostate cancer. *Proc. Natl. Acad. Sci. USA* **2000**, *97*, 10990–10995. [CrossRef]
46. Gianì, F.; Vella, V.; Nicolosi, M.L.; Fierabracci, A.; Lotta, S.; Malaguarnera, R.; Belfiore, A.; Vigneri, R.; Frasca, F. Thyrospheres From Normal or Malignant Thyroid Tissue Have Different Biological, Functional, and Genetic Features. *J. Clin. Endocrinol. Metab.* **2015**, *100*, E1168–E1178. [CrossRef]
47. Medyouf, H.; Gusscott, S.; Wang, H.; Tseng, J.; Wai, C.; Nemirovsky, O.; Trumpp, A.; Pflumio, F.; Carboni, J.; Gottardis, M.; et al. High-level IGF1R expression is required for leukemia-initiating cell activity in T-ALL and is supported by Notch signaling. *J. Exp. Med.* **2011**, *208*, 1809–1822. [CrossRef]
48. Longo, M.; Zatterale, F.; Naderi, J.; Parrillo, L.; Formisano, P.; Alex, G.; Raciti, G.A.; Beguinot, F.; Miele, C. Adipose Tissue Dysfunction as Determinant of Obesity-Associated Metabolic Complications. *Int. J. Mol. Sci.* **2019**, *20*, 2358. [CrossRef]
49. Leonardini, A.; Laviola, L.; Perrini, S.; Natalicchio, A.; Giorgino, F. Cross-Talk between PPARγ and Insulin Signaling and Modulation of Insulin Sensitivity. *PPAR Res.* **2009**, *2009*, 818945. [CrossRef]
50. Wondmkun, Y.T. Obesity, Insulin Resistance, and Type 2 Diabetes: Associations and Therapeutic Implications. *Diabetes Metab. Syndr. Obes. Targets Ther.* **2020**, *13*, 3611–3616. [CrossRef]
51. Phua, W.; Wong, M.; Liao, Z.; Tan, N. An aPPARent Functional Consequence in Skeletal Muscle Physiology via Peroxisome Proliferator-Activated Receptors. *Int. J. Mol. Sci.* **2018**, *19*, 1425. [CrossRef]
52. Choi, S.; Park, J.; Choi, J.H. Revisiting PPARγ as a target for the treatment of metabolic disorders. *BMB Rep.* **2014**, *47*, 599–608. [CrossRef]
53. Qaoud, M.T.; Almasri, I.; Önkol, T. Peroxisome Proliferator-Activated Receptors as Superior Targets for Treating Diabetic Disease, Design Strategies-Review Article. *Turk. J. Pharm. Sci.* **2022**, *19*, 353–370. [CrossRef]
54. Quinn, C.E.; Hamilton, P.K.; Lockhart, C.J.; McVeigh, G.E. Thiazolidinediones: Effects on insulin resistance and the cardiovascular system. *Br. J. Pharmacol.* **2008**, *153*, 636–645. [CrossRef]
55. Saltiel, A.R.; Olefsky, J.M. Thiazolidinediones in the Treatment of Insulin Resistance and Type II Diabetes. *Diabetes* **1996**, *45*, 1661–1669. [CrossRef]
56. Blaschke, F.; Takata, Y.; Caglayan, E.; Law, R.E.; Hsueh, W.A. Obesity, Peroxisome Proliferator-Activated Receptor, and Atherosclerosis in Type 2 Diabetes. *Arterioscler. Thromb. Vasc. Biol.* **2006**, *26*, 28–40. [CrossRef]
57. Alves-Bezerra, M.; Cohen, D.E. Triglyceride Metabolism in the Liver. *Compr. Physiol.* **2017**, *8*, 1–22. [CrossRef]
58. Lee, Y.K.; Park, J.E.; Lee, M.; Hardwick, J.P. Hepatic lipid homeostasis by peroxisome proliferator-activated receptor gamma 2. *Liver Res.* **2018**, *2*, 209–215. [CrossRef]
59. Gesta, S.; Tseng, Y.; Kahn, C.R. Developmental Origin of Fat: Tracking Obesity to Its Source. *Cell* **2007**, *131*, 242–256. [CrossRef]
60. He, W.; Barak, Y.; Hevener, A.; Olson, P.; Liao, D.; Le, J.; Nelson, M.; Ong, E.; Olefsky, J.M.; Evans, R.M. Adipose-specific peroxisome proliferator-activated receptor knockout causes insulin resistance in fat and liver but not in muscle. *Proc. Natl. Acad. Sci. USA* **2003**, *100*, 15712–15717. [CrossRef]
61. Wu, H.; Li, X.; Shen, C. Peroxisome Proliferator-Activated Receptor in White and Brown Adipocyte Regulation and Differentiation. *Physiol. Res.* **2020**, *67*, 759–773. [CrossRef]
62. Chu, D.; Tao, Y. Human thermogenic adipocytes: A reflection on types of adipocyte, developmental origin, and potential application. *J. Physiol. Biochem.* **2016**, *73*, 1–4. [CrossRef] [PubMed]

63. Cai, W.; Yang, T.; Liu, H.; Han, L.; Zhang, K.; Hu, X.; Zhang, X.; Yin, K.; Gao, Y.; Bennett, M.V.; et al. Peroxisome proliferator-activated receptorγ (PPARγ): A master gatekeeper in CNS injury and repair. *Prog. Neurobiol.* **2018**, *163–164*, 27–58. [CrossRef] [PubMed]
64. Chi, T.; Wang, M.; Wang, X.; Yang, K.; Xie, F.; Liao, Z.; Wei, P. PPARγ Modulators as Current and Potential Cancer Treatments. *Front. Oncol.* **2021**, *11*, 737776. [CrossRef] [PubMed]
65. Grygiel-Górniak, B. Peroxisome proliferator-activated receptors and their ligands: Nutritional and clinical implications—A review. *Nutr. J.* **2014**, *13*, 17. [CrossRef]
66. Mu, F.; Jing, Y.; Ning, B.; Huang, J.; Cui, T.; Guo, Y.; You, X.; Yan, X.; Li, H.; Wang, N. Peroxisome proliferator-activated receptor isoforms differentially regulate preadipocyte proliferation, apoptosis, and differentiation in chickens. *Poult. Sci.* **2020**, *99*, 6410–6421. [CrossRef]
67. Bruning, J.B.; Chalmers, M.J.; Prasad, S.; Busby, S.A.; Kamenecka, T.M.; He, Y.; Nettles, K.W.; Griffin, P.R. Partial Agonists Activate PPARgamma Using a Helix 12 Independent Mechanism. *Structure* **2007**, *15*, 1258–1271. [CrossRef]
68. Ahsan, W. The Journey of Thiazolidinediones as Modulators of PPARs for the Management of Diabetes: A Current Perspective. *Curr. Pharm. Des.* **2019**, *25*, 2540–2554. [CrossRef]
69. Zou, G.; Gao, Z.; Wang, J.; Zhang, Y.; Ding, H.; Huang, J.; Chen, L.; Guo, Y.; Jiang, H.; Shen, X. Deoxyelephantopin inhibits cancer cell proliferation and functions as a selective partial agonist against PPARgamma. *Biochem. Pharmacol.* **2008**, *75*, 1381–1392. [CrossRef]
70. Zhang, H.; Zhou, R.; Li, L.; Chen, J.; Chen, L.; Li, C.; Ding, H.; Yu, L.; Hu, L.; Jiang, H.; et al. Danthron Functions as a Retinoic X Receptor Antagonist by Stabilizing Tetramers of the Receptor. *J. Biol. Chem.* **2011**, *286*, 1868–1875. [CrossRef]
71. Ammazzalorso, A.; DeFilippis, B.; Giampietro, L.; Amoroso, R. Blocking the Peroxisome Proliferator-Activated Receptor (PPAR): An Overview. *ChemMedChem* **2013**, *8*, 1609–1616. [CrossRef]
72. Li, Y.; Yin, Z.; Dong, Y.; Wang, S.; Monroig, Ó.; Tocher, D.R.; You, C. Pparγ Is Involved in the Transcriptional Regulation of Liver LC-PUFA Biosynthesis by Targeting the Δ6Δ5 Fatty Acyl Desaturase Gene in the Marine Teleost Siganus canaliculatus. *Mar. Biotechnol.* **2018**, *21*, 19–29. [CrossRef]
73. Jara-Gutiérrez, Á.; Baladrón, V. The Role of Prostaglandins in Different Types of Cancer. *Cells* **2021**, *10*, 1487. [CrossRef]
74. Corona, J.C.; Duchen, M.R. PPARγ as a therapeutic target to rescue mitochondrial function in neurological disease. *Free Radic. Biol. Med.* **2016**, *100*, 153–163. [CrossRef]
75. Lathion, C.; Michalik, L.; Wahli, W. Physiological ligands of PPARs in inflammation and lipid homeostasis. *Future Lipidol.* **2006**, *1*, 191–201. [CrossRef]
76. Marcone, S.; Evans, P.; Fitzgerald, D.J. 15-Deoxy-Δ12,14-Prostaglandin J2 Modifies Components of the Proteasome and Inhibits Inflammatory Responses in Human Endothelial Cells. *Front. Immunol.* **2016**, *7*, 459. [CrossRef]
77. Zhao, X.; Zhang, Y.; Strong, R.; Grotta, J.C.; Aronowski, J. 15d-Prostaglandin J2 Activates Peroxisome Proliferator-Activated Receptor-gamma, Promotes Expression of Catalase, and Reduces Inflammation, Behavioral Dysfunction, and Neuronal Loss after Intracerebral Hemorrhage in Rats. *J. Cereb. Blood Flow Metab.* **2005**, *26*, 811–820. [CrossRef]
78. Vangaveti, V.; Shashidhar, V.; Collier, F.; Hodge, J.; Rush, C.; Malabu, U.; Baune, B.; Kennedy, R.L. 9- and 13-HODE regulate fatty acid binding protein-4 in human macrophages, but does not involve HODE/GPR132 axis in PPAR-γ regulation of FABP4. *Ther. Adv. Endocrinol. Metab.* **2018**, *9*, 137–150. [CrossRef]
79. Campbell, S.E.; Stone, W.L.; Whaley, S.G.; Qui, M.; Krishnan, K. Gamma (gamma) tocopherol up-regulates peroxisome proliferator activated receptor (PPAR) gamma (gamma) expression in SW 480 human colon cancer cell lines. *BMC Cancer* **2003**, *3*, 25. [CrossRef]
80. Afzal, S.; Sattar, M.A.; Johns, E.J.; Eseyin, O.A.; Attiq, A. Antioxidant Potential of Adiponectin and Full PPAR-γ Agonist in Correcting Streptozotocin-Induced Vascular Abnormality in Spontaneously Hypertensive Rats. *PPAR Res.* **2021**, *2021*, 6661181. [CrossRef]
81. Wang, K.; Li, Y.; Lv, Q.; Li, X.; Dai, Y.; Wei, Z. Bergenin, Acting as an Agonist of PPARgamma, Ameliorates Experimental Colitis in Mice through Improving Expression of SIRT1, and Therefore Inhibiting NF-B-Mediated Macrophage Activation. *Front. Pharmacol.* **2018**, *8*, 981. [CrossRef]
82. Reagan-Shaw, S.; Eggert, D.; Mukhtar, H.; Ahmad, N. Antiproliferative Effects of Apple Peel Extract Against Cancer Cells. *Nutr. Cancer* **2010**, *62*, 517–524. [CrossRef]
83. Pferschy-Wenzig, E.; Atanasov, A.G.; Malainer, C.; Noha, S.M.; Kunert, O.; Schuster, D.; Heiss, E.H.; Oberlies, N.H.; Wagner, H.; Bauer, R.; et al. Identification of Isosilybin A from Milk Thistle Seeds as an Agonist of Peroxisome Proliferator-Activated Receptor Gamma. *J. Nat. Prod.* **2014**, *77*, 842–847. [CrossRef]
84. Ma, S.; Huang, Y.; Zhao, Y.; Du, G.; Feng, L.; Huang, C.; Li, Y.; Guo, F. Prenylflavone derivatives from the seeds of Psoralea corylifolia exhibited PPAR-γ agonist activity. *Phytochem. Lett.* **2016**, *16*, 213–218. [CrossRef]
85. Weidner, C.; Wowro, S.J.; Rousseau, M.; Freiwald, A.; Kodelja, V.; Abdel-Aziz, H.; Kelber, O.; Sauer, S. Antidiabetic Effects of Chamomile Flowers Extract in Obese Mice through Transcriptional Stimulation of Nutrient Sensors of the Peroxisome Proliferator-Activated Receptor (PPAR) Family. *PLoS ONE* **2013**, *8*, e80335. [CrossRef]
86. Hiben, M.G.; Haan, L.d.; Spenkelink, B.; Wesseling, S.; Vervoort, J.; Rietjens, I.M.C.M. Induction of peroxisome proliferator activated receptorγ (PPARγ) mediated gene expression and inhibition of induced nitric oxide production by Maerua subcordata (Gilg) DeWolf. *BMC Complement. Med. Ther.* **2020**, *20*, 80. [CrossRef]

87. Chuang, C.; Yeh, C.; Yeh, S.; Lin, E.; Wang, L.; Wang, Y. Quercetin metabolites inhibit MMP-2 expression in A549 lung cancer cells by PPAR-γ associated mechanisms. *J. Nutr. Biochem.* **2016**, *33*, 45–53. [CrossRef]
88. Han, X.; Liu, C.; Gao, N.; Zhao, J.; Xu, J. RETRACTED: Kaempferol suppresses proliferation but increases apoptosis and autophagy by up-regulating microRNA-340 in human lung cancer cells. *Biomed. Pharmacother.* **2018**, *108*, 809–816. [CrossRef]
89. Rauf, A.; Imran, M.; Khan, I.A.; ur-Rehman, M.; Gilani, S.A.; Mehmood, Z.; Mubarak, M.S. Anticancer potential of quercetin: A comprehensive review. *Phytother. Res.* **2018**, *32*, 2109–2130. [CrossRef]
90. Zhong, Y.; Krisanapun, C.; Lee, S.; Nualsanit, T.; Sams, C.; Peungvicha, P.; Baek, S.J. Molecular targets of apigenin in colorectal cancer cells: Involvement of p21, NAG-1 and p53. *Eur. J. Cancer* **2010**, *46*, 3365–3374. [CrossRef]
91. Dumasia, R.; Eagle, K.; Kline-Rogers, E.; May, N.; Cho, L.; Mukherjee, D. Role of PPAR-γ Agonist Thiazolidinediones in Treatment of Pre-Diabetic and Diabetic Individuals: A Cardiovascular Perspective. *Curr. Drug Target-Cardiovasc. Hematol. Disord.* **2005**, *5*, 377–386. [CrossRef]
92. Wu, Y.; Sreeharsha, N.; Sharma, S.; Mishra, A.; Singh, A.K.; Gubbiyappa, S.K. Anticancer Effect of Rosiglitazone, a PPAR-γ Agonist against Diethylnitrosamine-Induced Lung Carcinogenesis. *ACS Omega* **2020**, *5*, 5334–5339. [CrossRef] [PubMed]
93. Li, M.; Lee, T.W.; Yim, A.P.; Mok, T.S.; Chen, G.G. Apoptosis induced by troglitazone is both peroxisome proliferator-activated receptor-γ and ERK-dependent in human non-small lung cancer cells. *J. Cell. Physiol.* **2006**, *209*, 428–438. [CrossRef] [PubMed]
94. Lewis, J.D.; Ferrara, A.; Peng, T.; Hedderson, M.; Bilker, W.B.; Quesenberry, C.P.; Vaughn, D.J.; Nessel, L.; Selby, J.; Strom, B.L. Risk of Bladder Cancer Among Diabetic Patients Treated With Pioglitazone. *Diabetes Care* **2011**, *34*, 916–922. [CrossRef] [PubMed]
95. Cellai, I.; Benvenuti, S.; Luciani, P.; Galli, A.; Ceni, E.; Simi, L.; Baglioni, S.; Muratori, M.; Ottanelli, B.; Serio, M.; et al. Antineoplastic effects of rosiglitazone and PPARgamma transactivation in neuroblastoma cells. *Br. J. Cancer* **2006**, *95*, 879–888. [CrossRef] [PubMed]
96. Xu, X.; Wang, J.; Jiang, H.; Meng, L.; Lang, B. Rosiglitazone induces apoptosis on human bladder cancer 5637 and T24 cell lines. *Int. J. Clin. Exp. Pathol.* **2017**, *10*, 10197–10204. [PubMed]
97. Mody, M.; Dharker, N.; Bloomston, M.; Wang, P.; Chou, F.; Glickman, T.S.; McCaffrey, T.; Yang, Z.; Pumfery, A.; Lee, D.; et al. Rosiglitazone sensitizes MDA-MB-231 breast cancer cells to anti-tumor effects of tumor necrosis factor-, CH11 and CYC202. *Endocr.-Relat. Cancer* **2007**, *14*, 305–315. [CrossRef] [PubMed]
98. Kanemaru, M.; Asai, J.; Jo, J.; Arita, T.; Kawai-Ohnishi, M.; Tsutsumi, M.; Wada, M.; Tabata, Y.; Katoh, N. Nanoparticle-mediated local delivery of pioglitazone attenuates bleomycin-induced skin fibrosis. *J. Dermatol. Sci.* **2019**, *93*, 41–49. [CrossRef]
99. Zhang, H.-L.; Zhang, Z.-X.; Xu, Y.-J. Ciglitazone inhibits growth of lung cancer cells A549 In Vitro and In Vivo: An experimental study. *Zhonghua Zhong Liu Za Zhi* **2004**, *26*, 531–534.
100. Hawcroft, G.; Gardner, S.H.; Hull, M.A. Activation of Peroxisome Proliferator-Activated Receptor gamma Does Not Explain the Antiproliferative Activity of the Nonsteroidal Anti-Inflammatory Drug Indomethacin on Human Colorectal Cancer Cells. *J. Pharmacol. Exp. Ther.* **2003**, *305*, 632–637. [CrossRef]
101. Puhl, A.C.; Milton, F.A.; Cvoro, A.; Sieglaff, D.H.; Campos, J.C.; Bernardes, A.; Filgueira, C.S.; Lindemann, J.L.; Deng, T.; Neves, F.A.; et al. Mechanisms of Peroxisome Proliferator Activated Receptor Regulation by Non-steroidal Anti-inflammatory Drugs. *Nucl. Recept. Signal.* **2015**, *13*, e004. [CrossRef]
102. Chen, K.; Zhang, J.; Beeraka, N.M.; Tang, C.; Babayeva, Y.V.; Sinelnikov, M.Y.; Zhang, X.; Zhang, J.; Liu, J.; Reshetov, I.V.; et al. Advances in the Prevention and Treatment of Obesity-Driven Effects in Breast Cancers. *Front. Oncol.* **2022**, *12*, 820968. [CrossRef]
103. Liu, Y.; Chen, C.; Wang, X.; Sun, Y.; Zhang, J.; Chen, J.; Shi, Y. An Epigenetic Role of Mitochondria in Cancer. *Cells* **2022**, *11*, 2518. [CrossRef]
104. Li, J. 15-Deoxy-Δ-12,14-Prostaglandin J2 (15d-PGJ2), an Endogenous Ligand of PPAR-γ Function and Mechanism. *PPAR Res.* **2019**, *2019*, 7242030. [CrossRef]
105. Toaldo, C.; Pizzimenti, S.; Cerbone, A.; Pettazzoni, P.; Menegatti, E.; Daniela, B.; Minelli, R.; Giglioni, B.; Dianzani, M.U.; Ferretti, C.; et al. PPAR ligands inhibit telomerase activity and hTERT expression through modulation of the Myc/Mad/Max network in colon cancer cells. *J. Cell. Mol. Med.* **2009**, *14*, 1347–1357. [CrossRef]
106. Siavash, H.; Nikitakis, N.G.; Sauk, J.J. Abrogation of IL-6-mediated JAK signalling by the cyclopentenone prostaglandin 15d-PGJ2 in oral squamous carcinoma cells. *Br. J. Cancer* **2004**, *91*, 1074–1080. [CrossRef]
107. Chen, Y. 15d-PGJ2 inhibits cell growth and induces apoptosis of MCG-803 human gastric cancer cell line. *World J. Gastroenterol.* **2003**, *9*, 2149. [CrossRef]
108. Tate, T.; Xiang, T.; Wobker, S.E.; Zhou, M.; Chen, X.; Kim, H.; Batourina, E.; Lin, C.; Kim, W.Y.; Lu, C.; et al. Pparg signaling controls bladder cancer subtype and immune exclusion. *Nat. Commun.* **2021**, *12*, 6160. [CrossRef]
109. Augimeri, G.; Giordano, C.; Gelsomino, L.; Plastina, P.; Barone, I.; Catalano, S.; Andò, S.; Bonofiglio, D. The Role of PPARγ Ligands in Breast Cancer: From Basic Research to Clinical Studies. *Cancers* **2020**, *12*, 2623. [CrossRef]
110. Dwyer-Nield, L.D.; McArthur, D.G.; Hudish, T.M.; Hudish, L.I.; Mirita, C.; Sompel, K.; Smith, A.J.; Alavi, K.; Ghosh, M.; Merrick, D.T.; et al. PPARgamma agonism inhibits progression of premalignant lesions in a murine lung squamous cell carcinoma model. *Int. J. Cancer* **2022**. *Online ahead of print*. [CrossRef]
111. Zhang, H.; You, L.; Zhao, M. Rosiglitazone attenuates paraquat-induced lung fibrosis in rats in a PPAR gamma-dependent manner. *Eur. J. Pharmacol.* **2019**, *851*, 133–143. [CrossRef]
112. Cao, L.; Chen, X.; Wang, Q.; Huang, X.; Zhen, M.; Zhang, L.; Li, W.; Bi, J. Upregulation of PTEN involved in rosiglitazone-induced apoptosis in human hepatocellular carcinoma cells. *Acta Pharmacol. Sin.* **2007**, *28*, 879–887. [CrossRef]

113. Higuchi, T.; Yamamoto, J.; Sugisawa, N.; Tashiro, Y.; Nishino, H.; Yamamoto, N.; Hayashi, K.; Kimura, H.; Miwa, S.; Igarashi, K.; et al. PPAR Agonist Pioglitazone in Combination With Cisplatinum Arrests a Chemotherapy-resistant Osteosarcoma PDOX Model. *Cancer Genom.-Proteom.* **2019**, *17*, 35–40. [CrossRef]
114. Cheon, C.W.; Kim, D.H.; Kim, D.H.; Cho, Y.H.; Kim, J.H. Effects of ciglitazone and troglitazone on the proliferation of human stomach cancer cells. *World J. Gastroenterol.* **2009**, *15*, 310. [CrossRef]
115. Adorni, R.; Zanatta, F.; D'Addario, M.; Atella, F.; Costantino, E.; Iaderosa, C.; Petarle, G.; Steca, P. Health-Related Lifestyle Profiles in Healthy Adults: Associations with Sociodemographic Indicators, Dispositional Optimism, and Sense of Coherence. *Nutrients* **2021**, *13*, 3778. [CrossRef]
116. Pasqua, T.; Rocca, C.; Giglio, A.; Angelone, T. Cardiometabolism as an Interlocking Puzzle between the Healthy and Diseased Heart: New Frontiers in Therapeutic Applications. *J. Clin. Med.* **2021**, *10*, 721. [CrossRef]
117. Yan, L.; Zhang, J.D.; Wang, B.; Lv, Y.J.; Jiang, H.; Liu, G.L.; Qiao, Y.; Ren, M.; Guo, X.F. Quercetin Inhibits Left Ventricular Hypertrophy in Spontaneously Hypertensive Rats and Inhibits Angiotensin II-Induced H9C2 Cells Hypertrophy by Enhancing PPAR- Expression and Suppressing AP-1 Activity. *PLoS ONE* **2013**, *8*, e72548. [CrossRef]
118. Giaginis, C.; Tsourouflis, G.; Theocharis, S. Peroxisome Proliferator-Activated Receptor-gamma (PPAR-gamma) Ligands: Novel Pharmacological Agents in the Treatment of Ischemia Reperfusion Injury. *Curr. Mol. Med.* **2008**, *8*, 562–579. [CrossRef]
119. Liu, J.; Wang, L. Peroxisome proliferator-activated receptor gamma agonists for preventing recurrent stroke and other vascular events in people with stroke or transient ischaemic attack. *Cochrane Database Syst. Rev.* **2019**, *2019*. [CrossRef]
120. Liu, C.; Lee, T.; Lin, Y.; Sung, P.; Wei, Y.; Li, Y. Pioglitazone and PPAR-γ modulating treatment in hypertensive and type 2 diabetic patients after ischemic stroke: A national cohort study. *Cardiovasc. Diabetol.* **2020**, *19*, 2. [CrossRef]
121. Nesti, L.; Tricò, D.; Mengozzi, A.; Natali, A. Rethinking pioglitazone as a cardioprotective agent: A new perspective on an overlooked drug. *Cardiovasc. Diabetol.* **2021**, *20*, 109. [CrossRef]
122. Ren, Y.; Sun, C.; Sun, Y.; Tan, H.; Wu, Y.; Cui, B.; Wu, Z. PPAR gamma protects cardiomyocytes against oxidative stress and apoptosis via Bcl-2 upregulation. *Vasc. Pharmacol.* **2009**, *51*, 169–174. [CrossRef] [PubMed]
123. Satin, L.S.; Butler, P.C.; Ha, J.; Sherman, A.S. Pulsatile insulin secretion, impaired glucose tolerance and type 2 diabetes. *Mol. Asp. Med.* **2015**, *42*, 61–77. [CrossRef] [PubMed]
124. Rojas, J.; Chávez, M.; Olivar, L.; Rojas, M.; Morillo, J.; Mejías, J.; Calvo, M.; Bermúdez, V. Polycystic Ovary Syndrome, Insulin Resistance, and Obesity: Navigating the Pathophysiologic Labyrinth. *Int. J. Reprod. Med.* **2014**, *2014*, 719050. [CrossRef] [PubMed]
125. Sohn, K.K.; Cruciani-Guglielmacci, C.; Kassis, N.; Clément, L.; Ouali, F.; Caüzac, M.; Lebègue, N.; Berthelot, P.; Caignard, D.; Pégorier, J.; et al. S26948, a new specific peroxisome proliferator activated receptor gamma modulator improved *in vivo* hepatic insulin sensitivity in 48 h lipid infused rats. *Eur. J. Pharmacol.* **2009**, *608*, 104–111. [CrossRef] [PubMed]
126. Holman, R. Metformin as first choice in oral diabetes treatment: The UKPDS experience. *J. Annu. Diabetol. Hotel. Dieu.* **2007**, 13–20.
127. Harvey, I.; Stephens, J.M. Artemisia scoparia promotes adipogenesis in the absence of adipogenic effectors. *Obesity* **2021**, *29*, 1309–1319. [CrossRef]
128. Colca, J.R.; Scherer, P.E. The metabolic syndrome, thiazolidinediones, and implications for intersection of chronic and inflammatory disease. *Mol. Metab.* **2022**, *55*, 101409. [CrossRef]
129. Graham, D.J.; Green, L.; Senior, J.R.; Nourjah, P. Troglitazone-induced liver failure: A case study. *Am. J. Med.* **2003**, *114*, 299–306. [CrossRef]
130. Bogacka, I.; Xie, H.; Bray, G.A.; Smith, S.R. The Effect of Pioglitazone on Peroxisome Proliferator-Activated Receptor-gamma Target Genes Related to Lipid Storage In vivo. *Diabetes Care* **2004**, *27*, 1660–1667. [CrossRef]
131. Takahashi, Y.; Fukusato, T. *Animal Models of Liver Diseases*; Academic Press: Cambridge, MA, USA, 2017; pp. 313–339. [CrossRef]
132. Liu, C.; Feng, T.; Zhu, N.; Liu, P.; Han, X.; Chen, M.; Wang, X.; Li, N.; Li, Y.; Xu, Y.; et al. Identification of a novel selective agonist of PPARγ with no promotion of adipogenesis and less inhibition of osteoblastogenesis. *Sci. Rep.* **2015**, *5*, 9530. [CrossRef]
133. Pan, Y.; Zhao, D.; Yu, N.; An, T.; Miao, J.; Mo, F.; Gu, Y.; Zhang, D.; Gao, S.; Jiang, G. Curcumin improves glycolipid metabolism through regulating peroxisome proliferator activated receptor signalling pathway in high-fat diet-induced obese mice and 3T3-L1 adipocytes. *R. Soc. Open Sci.* **2017**, *4*, 170917. [CrossRef]
134. Liu, Y.; Wang, J.; Luo, S.; Zhan, Y.; Lu, Q. The roles of PPARγ and its agonists in autoimmune diseases: A comprehensive review. *J. Autoimmun.* **2020**, *113*, 102510. [CrossRef]
135. Rőszer, T.; Menéndez-Gutiérrez, M.P.; Lefterova, M.I.; Alameda, D.; Núñez, V.; Lazar, M.A.; Fischer, T.; Ricote, M. Autoimmune Kidney Disease and Impaired Engulfment of Apoptotic Cells in Mice with Macrophage Peroxisome Proliferator-Activated Receptor gamma or Retinoid X Receptor Deficiency. *J. Immunol.* **2010**, *186*, 621–631. [CrossRef]
136. Hucke, S.; Floßdorf, J.; Grützke, B.; Dunay, I.R.; Frenzel, K.; Jungverdorben, J.; Linnartz, B.; Mack, M.; Peitz, M.; Brüstle, O.; et al. Licensing of myeloid cells promotes central nervous system autoimmunity and is controlled by peroxisome proliferator-activated receptor-γ. *Brain* **2012**, *135*, 1586–1605. [CrossRef]
137. Cheng, A.M.S.; Yin, H.Y.; Chen, A.; Liu, Y.; Chuang, M.; He, H.; Tighe, S.; Sheha, H.; Liao, S. Celecoxib and Pioglitazone as Potential Therapeutics for Regulating TGF—Induced Hyaluronan in Dysthyroid Myopathy. *Investig. Opthalmol. Vis. Sci.* **2016**, *57*, 1951. [CrossRef]
138. Schmidt, S.; Moric, E.; Schmidt, M.; Sastre, M.; Feinstein, D.L.; Heneka, M.T. Anti-inflammatory and antiproliferative actions of PPAR- agonists on T lymphocytes derived from MS patients. *J. Leukoc. Biol.* **2003**, *75*, 478–485. [CrossRef]

139. Vallée, A.; Lecarpentier, Y.; Guillevin, R.; Vallée, J. Demyelination in Multiple Sclerosis: Reprogramming Energy Metabolism and Potential PPAR Agonist Treatment Approaches. *Int. J. Mol. Sci.* **2018**, *19*, 1212. [CrossRef]
140. Li, X.; Sun, Y.; Bao, J.; Chen, X.; Li, Y.; Yang, Y.; Zhang, L.; Huang, C.; Wu, B.; Meng, X.; et al. Functional role of PPAR-γ on the proliferation and migration of fibroblast-like synoviocytes in rheumatoid arthritis. *Sci. Rep.* **2017**, *7*, 12671. [CrossRef]
141. Nozaki, Y.; Harada, K.; Sanzen, T.; Nakanuma, Y. PPAR ligand attenuates portal inflammation in the MRL-lpr mouse: A new strategy to restrain cholangiopathy in primary biliary cirrhosis. *Med. Mol. Morphol.* **2013**, *46*, 153–159. [CrossRef]
142. Bernardo, A.; Plumitallo, C.; Nuccio, C.D.; Visentin, S.; Minghetti, L. Curcumin promotes oligodendrocyte differentiation and their protection against TNF-α through the activation of the nuclear receptor PPAR-γ. *Sci. Rep.* **2021**, *11*, 4952. [CrossRef]
143. Christofides, A.; Konstantinidou, E.; Jani, C.; Boussiotis, V.A. The role of peroxisome proliferator-activated receptors (PPAR) in immune responses. *Metabolism* **2021**, *114*, 154338. [CrossRef]
144. Khan, M.A.; Alam, Q.; Haque, A.; Ashafaq, M.; Khan, M.J.; Ashraf, G.M.; Ahmad, M. Current Progress on Peroxisome Proliferator-activated Receptor Gamma Agonist as an Emerging Therapeutic Approach for the Treatment of Alzheimer's Disease: An Update. *Curr. Neuropharmacol.* **2019**, *17*, 232–246. [CrossRef]
145. Toobian, D.; Ghosh, P.; Katkar, G.D. Parsing the Role of PPARs in Macrophage Processes. *Front. Immunol.* **2021**, *12*, 783780. [CrossRef]
146. Salam, N.K.; Huang, T.H.; Kota, B.P.; Kim, M.S.; Li, Y.; Hibbs, D.E. Novel PPAR-gamma Agonists Identified from a Natural Product Library: A Virtual Screening, Induced-Fit Docking and Biological Assay Study. *Chem. Biol. Drug Des.* **2007**, *71*, 57–70. [CrossRef]
147. Xu, L.; Shen, S.; Ma, Y.; Kim, J.K.; Rodriguez-Agudo, D.; Heuman, D.M.; Hylemon, P.B.; Pandak, W.M.; Ren, S. 25-Hydroxycholesterol-3-sulfate attenuates inflammatory response via PPAR-γ signaling in human THP-1 macrophages. *Am. J. Physiol.-Endocrinol. Metab.* **2012**, *302*, E788–E799. [CrossRef]
148. Li, M.; Pascual, G.; Glass, C.K. Peroxisome Proliferator-Activated Receptor-γ Dependent Repression of the Inducible Nitric Oxide Synthase Gene. *Mol. Cell. Biol.* **2000**, *20*, 4699–4707. [CrossRef] [PubMed]
149. Ricote, M.; Glass, C.K. PPARs and molecular mechanisms of transrepression. *Biochim. Biophys. Acta BBA-Mol. Cell Biol. Lipids* **2007**, *1771*, 926–935. [CrossRef] [PubMed]
150. Petrova, T.V.; Akama, K.T.; Eldik, L.J.V. Cyclopentenone prostaglandins suppress activation of microglia: Down-regulation of inducible nitric-oxide synthase by 15-deoxy-Δ. *Proc. Natl. Acad. Sci. USA* **1999**, *96*, 4668–4673. [CrossRef] [PubMed]
151. Cuzzocrea, S.; Pisano, B.; Dugo, L.; Ianaro, A.; Maffia, P.; Patel, N.S.; Paola, R.D.; Ialenti, A.; Genovese, T.; Chatterjee, P.K.; et al. Rosiglitazone, a ligand of the peroxisome proliferator-activated receptor-γ, reduces acute inflammation. *Eur. J. Pharmacol.* **2004**, *483*, 79–93. [CrossRef] [PubMed]
152. Wan, Y.; Evans, R.M. Rosiglitazone activation of PPARgamma suppresses fractalkine signaling. *J. Mol. Endocrinol.* **2009**, *44*, 135–142. [CrossRef] [PubMed]
153. Yessoufou, A.; Wahli, W. Multifaceted roles of peroxisome proliferator-activated receptors (PPARs) at the cellular and whole organism levels. *Swiss Med. Wkly.* **2010**, *140*, w13071. [CrossRef]
154. Ramot, Y.; Mastrofrancesco, A.; Camera, E.; Desreumaux, P.; Paus, R.; Picardo, M. The role of PPARγ-mediated signalling in skin biology and pathology: New targets and opportunities for clinical dermatology. *Exp. Dermatol.* **2015**, *24*, 245–251. [CrossRef]
155. Dubrac, S.; Stoitzner, P.; Pirkebner, D.; Elentner, A.; Schoonjans, K.; Auwerx, J.; Sael, S.; Hengster, P.; Fritsch, P.; Romani, N.; et al. Peroxisome Proliferator-Activated Receptor-γ Activation Inhibits Langerhans Cell Function. *J. Immunol.* **2007**, *178*, 4362–4372. [CrossRef]
156. Sugiyama, H.; Nonaka, T.; Kishimoto, T.; Komoriya, K.; Tsuji, K.; Nakahata, T. Peroxisome proliferator-activated receptors are expressed in human cultured mast cells: A possible role of these receptors in negative regulation of mast cell activation. *Eur. J. Immunol.* **2000**, *30*, 3363–3370. [CrossRef]
157. Umeno, A.; Sakashita, M.; Sugino, S.; Murotomi, K.; Okuzawa, T.; Morita, N.; Tomii, K.; Tsuchiya, Y.; Yamasaki, K.; Horie, M.; et al. Comprehensive analysis of PPARγ agonist activities of stereo-, regio-, and enantio-isomers of hydroxyoctadecadienoic acids. *Biosci. Rep.* **2020**, *40*, BSR20193767. [CrossRef]
158. Emerson, M.R.; LeVine, S.M. Experimental allergic encephalomyelitis is exacerbated in mice deficient for 12/15-lipoxygenase or 5-lipoxygenase. *Brain Res.* **2004**, *1021*, 140–145. [CrossRef]
159. Shankaranarayanan, P.; Nigam, S. IL-4 Induces Apoptosis in A549 Lung Adenocarcinoma Cells: Evidence for the Pivotal Role of 15-Hydroxyeicosatetraenoic Acid Binding to Activated Peroxisome Proliferator-Activated Receptor Transcription Factor. *J. Immunol.* **2003**, *170*, 887–894. [CrossRef]
160. Yamaguchi, A.; Tourdot, B.E.; Yeung, J.; Holman, T.; Holinstat, M.A. The 15-lipoxygenase-derived Oxylipins 15-HETrE And 15-HETE Inhibit Platelet Activation In Part Through Activation of PPARs. *Arterioscler. Thromb. Vasc. Biol.* **2021**, *41*, AP119. [CrossRef]
161. Li, Y.; Atkinson, K.; Zhang, T. Combination of chemotherapy and cancer stem cell targeting agents: Preclinical and clinical studies. *Cancer Lett.* **2017**, *396*, 103–109. [CrossRef]
162. Illés, P.; Grycová, A.; Krasulová, K.; Dvořák, Z. Effects of Flavored Nonalcoholic Beverages on Transcriptional Activities of Nuclear and Steroid Hormone Receptors: Proof of Concept for Novel Reporter Cell Line PAZ-PPARg. *J. Agric. Food Chem.* **2018**, *66*, 12066–12078. [CrossRef]

163. Mueller, C.; Weaver, V.; Heuvel, J.P.V.; August, A.; Cantorna, M.T. Peroxisome proliferator-activated receptor gamma ligands attenuate immunological symptoms of experimental allergic asthma. *Arch. Biochem. Biophys.* **2003**, *418*, 186–196. [CrossRef] [PubMed]
164. Liu, J.; Yao, Q.; Xie, X.; Cui, Q.; Jiang, T.; Zhao, Z.; Du, X.; Lai, B.; Xiao, L.; Wang, N. Procyanidin B2 Attenuates Nicotine-Induced Hepatocyte Pyroptosis through a PPAR-Dependent Mechanism. *Nutrients* **2022**, *14*, 1756. [CrossRef] [PubMed]
165. Choi, S.; Cha, B.; Iida, K.; Lee, Y.; Yonezawa, T.; Teruya, T.; Nagai, K.; Woo, J. Artepillin C, as a PPARγ ligand, enhances adipocyte differentiation and glucose uptake in 3T3-L1 cells. *Biochem. Pharmacol.* **2011**, *81*, 925–933. [CrossRef] [PubMed]
166. Delebinski, C.I.; Twardziok, M.; Kleinsimon, S.; Hoff, F.; Mulsow, K.; Rolff, J.; Jäger, S.; Eggert, A.; Seifert, G. A Natural Combination Extract of Viscum album L. Containing Both Triterpene Acids and Lectins Is Highly Effective against AML *In vivo*. *PLoS ONE* **2015**, *10*, e0133892. [CrossRef]
167. Yang, L.; Zheng, Y.; Miao, Y.; Yan, W.; Geng, Y.; Dai, Y.; Wei, Z. Bergenin, a PPARγ agonist, inhibits Th17 differentiation and subsequent neutrophilic asthma by preventing GLS1-dependent glutaminolysis. *Acta Pharmacol. Sin.* **2021**, *43*, 963–976. [CrossRef]
168. Hong, J.; Samudio, I.; Chintharlapalli, S.; Safe, S. 1,1-bis(3′-indolyl)-1-(p-substituted phenyl)methanes decrease mitochondrial membrane potential and induce apoptosis in endometrial and other cancer cell lines. *Mol. Carcinog.* **2008**, *47*, 492–507. [CrossRef]
169. Yao, J.; Jiang, M.; Zhang, Y.; Liu, X.; Du, Q.; Feng, G. Chrysin alleviates allergic inflammation and airway remodeling in a murine model of chronic asthma. *Int. Immunopharmacol.* **2016**, *32*, 24–31. [CrossRef]
170. Wu, Q.; Needs, P.W.; Lu, Y.; Kroon, P.A.; Ren, D.; Yang, X. Different antitumor effects of quercetin, quercetin-3′-sulfate and quercetin-3-glucuronide in human breast cancer MCF-7 cells. *Food Funct.* **2018**, *9*, 1736–1746. [CrossRef]
171. Ballav, S.; Lokhande, K.B.; Dabhi, I.; Inje, S.; Ranjan, A.; Swamy, K.V.; Basu, S. Designing novel quercetin derivatives as matrix metalloproteinase-9 inhibitors in colon carcinoma: An *In vitro* and *in silico* approach. *J. Dent. Res. Rev.* **2020**, *7*, 30–35. [CrossRef]
172. Weng, J.; Bai, L.; Chiu, C.; Hu, J.; Chiu, S.; Wu, C. Cucurbitane Triterpenoid from *Momordica charantia* Induces Apoptosis and Autophagy in Breast Cancer Cells, in Part, through Peroxisome Proliferator-Activated Receptori-γ Activation. *Evid.-Based Complement. Altern. Med.* **2013**, *2013*, 935675. [CrossRef]
173. Noruddin, N.A.A.; Hamzah, M.F.; Rosman, Z.; Salin, N.H.; Shu-Chien, A.C.; Muhammad, T.S.T. Natural Compound 3,7,25-trihydroxycucurbita-5,23(E)-dien-19-al from Momordica charantia Acts as PPAR Ligand. *Molecules* **2021**, *26*, 2682. [CrossRef]
174. Kumar, R.; Balaji, S.; Uma, T.; Sehgal, P. Fruit extracts of *Momordica charantia* potentiate glucose uptake and up-regulate Glut-4, PPAR and PI3K. *J. Ethnopharmacol.* **2009**, *126*, 533–537. [CrossRef]
175. An, R.; Pathmanathan, K.; Shankernarayanan, N.; Vishwakarma, R.A.; Balakrishnan, A. Upregulation of Glut-4 and PPARgamma by an isoflavone from *Pterocarpus marsupium* on L6 myotubes: A possible mechanism of action. *J. Ethnopharmacol.* **2005**, *97*, 253–260. [CrossRef]
176. San, Y.; Liu, Y.; Zhang, Y.; Shi, P.; Zhu, Y. Peroxisome proliferator-activated receptor-γ agonist inhibits the mammalian target of rapamycin signaling pathway and has a protective effect in a rat model of status epilepticus. *Mol. Med. Rep.* **2015**, *12*, 1877–1883. [CrossRef]
177. Griggs, R.B.; Donahue, R.R.; Morgenweck, J.; Grace, P.M.; Sutton, A.; Watkins, L.R.; Taylor, B.K. Pioglitazone rapidly reduces neuropathic pain through astrocyte and nongenomic PPAR mechanisms. *Pain* **2015**, *156*, 469–482. [CrossRef]
178. Bongartz, T. Treatment of active psoriatic arthritis with the PPAR ligand pioglitazone: An open-label pilot study. *Rheumatology* **2005**, *44*, 126–129. [CrossRef]
179. Meng, X.; Sun, X.; Zhang, Y.; Shi, H.; Deng, W.; Liu, Y.; Wang, G.; Fang, P.; Yang, S. PPAR Agonist PGZ Attenuates OVA-Induced Airway Inflammation and Airway Remodeling via RGS4 Signaling in Mouse Model. *Inflammation* **2018**, *41*, 2079–2089. [CrossRef]
180. Khandoudi, N.; Delerive, P.; Berrebi-Bertrand, I.; Buckingham, R.E.; Staels, B.; Bril, A. Rosiglitazone, a Peroxisome Proliferator-Activated Receptor-, Inhibits the Jun NH2-Terminal Kinase/Activating Protein 1 Pathway and Protects the Heart From Ischemia/Reperfusion Injury. *Diabetes* **2002**, *51*, 1507–1514. [CrossRef]
181. Lee, K.S.; Park, S.J.; Hwang, P.H.; Yi, H.K.; Song, C.H.; Chai, O.H.; Kim, J.; Lee, M.K.; Lee, Y.C. PPAR-gamma modulates allergic inflammation through up-regulation of PTEN. *FASEB J.* **2005**, *19*, 1033–1035. [CrossRef]
182. Lee, K.S.; Kim, S.R.; Park, S.J.; Park, H.S.; Min, K.H.; Jin, S.M.; Lee, M.K.; Kim, U.H.; Lee, Y.C. Peroxisome proliferator activated receptor-γ modulates reactive oxygen species generation and activation of nuclear factor-B and hypoxia-inducible factor 1 in allergic airway disease of mice. *J. Allergy Clin. Immunol.* **2006**, *118*, 120–127. [CrossRef]
183. Wanguang, Z.; Huilan, Z.; Lihua, X. Influence of ciglitazone on A549 cells growth *in vitro* and *in vivo* and mechanism. *J. Huazhong Univ. Sci. Technol. Med. Sci.* **2006**, *26*, 36–39. [CrossRef]
184. Moss, P.E.; Lyles, B.E.; Stewart, L.V. The PPARγ ligand ciglitazone regulates androgen receptor activation differently in androgen-dependent versus androgen-independent human prostate cancer cells. *Exp. Cell Res.* **2010**, *316*, 3478–3488. [CrossRef]
185. Akinyeke, T.O.; Stewart, L.V. Troglitazone suppresses c-Myc levels in human prostate cancer cells via a PPAR-γ independent mechanism. *Cancer Biol. Ther.* **2011**, *11*, 1046–1058. [CrossRef]
186. Ellis, C.N.; Varani, J.; Fisher, G.J.; Zeigler, M.E.; Pershadsingh, H.A.; Benson, S.C.; Chi, Y.; Kurtz, T.W. Troglitazone Improves Psoriasis and Normalizes Models of Proliferative Skin Disease. *Arch. Dermatol.* **2000**, *136*, 609–616. [CrossRef]
187. Shang, J.; Brust, R.; Mosure, S.A.; Bass, J.; Munoz-Tello, P.; Lin, H.; Hughes, T.S.; Tang, M.; Ge, Q.; Kamenekca, T.M.; et al. Cooperative cobinding of synthetic and natural ligands to the nuclear receptor PPARγ. *eLife* **2018**, *7*, e43320. [CrossRef]
188. Thilakarathna, S.; Rupasinghe, H. Flavonoid Bioavailability and Attempts for Bioavailability Enhancement. *Nutrients* **2013**, *5*, 3367–3387. [CrossRef]

189. Horita, S.; Nakamura, M.; Satoh, N.; Suzuki, M.; Seki, G. Thiazolidinediones and Edema: Recent Advances in the Pathogenesis of Thiazolidinediones-Induced Renal Sodium Retention. *PPAR Res.* **2015**, *2015*, 646423. [CrossRef]
190. Guan, Y.; Hao, C.; Cha, D.R.; Rao, R.; Lu, W.; Kohan, D.E.; Magnuson, M.A.; Redha, R.; Zhang, Y.; Breyer, M.D. Thiazolidinediones expand body fluid volume through PPARgamma stimulation of ENaC-mediated renal salt absorption. *Nat. Med.* **2005**, *11*, 861–866. [CrossRef]
191. Filipova, E.; Uzunova, K.; Kalinov, K.; Vekov, T. Pioglitazone and the Risk of Bladder Cancer: A Meta-Analysis. *Diabetes Ther.* **2017**, *8*, 705–726. [CrossRef]
192. Tseng, C. Rosiglitazone may reduce thyroid cancer risk in patients with type 2 diabetes. *Ann. Med.* **2013**, *45*, 539–544. [CrossRef]
193. Tseng, C. Rosiglitazone reduces breast cancer risk in Taiwanese female patients with type 2 diabetes mellitus. *Oncotarget* **2016**, *8*, 3042–3048. [CrossRef] [PubMed]
194. Han, E.; Jang, S.; Kim, G.; Lee, Y.; Choe, E.Y.; Nam, C.M.; Kang, E.S. Rosiglitazone Use and the Risk of Bladder Cancer in Patients With Type 2 Diabetes. *Medicine* **2016**, *95*, e2786. [CrossRef] [PubMed]
195. Krentz, A.J.; Bailey, C.J. Oral antidiabetic agents: Current role in type 2 diabetes mellitus. *Drugs* **2005**, *65*, 385–411. [CrossRef] [PubMed]
196. Nissen, S.E.; Wolski, K. Effect of Rosiglitazone on the Risk of Myocardial Infarction and Death from Cardiovascular Causes. *N. Engl. J. Med.* **2007**, *356*, 2457–2471. [CrossRef]
197. Dormandy, J.; Bhattacharya, M.; van Troostenburg de Bruyn, A.R.; PROactive Investigators. Safety and Tolerability of Pioglitazone in High-Risk Patients with Type 2 Diabetes. *Drug Saf.* **2009**, *32*, 187–202. [CrossRef]
198. Nesto, R.W.; Bell, D.; Bonow, R.O.; Fonseca, V.; Grundy, S.M.; Horton, E.S.; Winter, M.L.; Porte, D.; Semenkovich, C.F.; Smith, S.; et al. Thiazolidinedione Use, Fluid Retention, and Congestive Heart Failure. *Circulation* **2003**, *108*, 2941–2948. [CrossRef]
199. Doshi, L.S.; Brahma, M.K.; Bahirat, U.A.; Dixit, A.V.; Nemmani, K.V. Discovery and development of selective PPAR gamma modulators as safe and effective antidiabetic agents. *Expert Opin. Investig. Drugs* **2010**, *19*, 489–512. [CrossRef]
200. Weidner, C.; de Groot, J.C.; Prasad, A.; Freiwald, A.; Quedenau, C.; Kliem, M.; Witzke, A.; Kodelja, V.; Han, C.; Giegold, S.; et al. Amorfrutins are potent antidiabetic dietary natural products. *Proc. Natl. Acad. Sci. USA* **2012**, *109*, 7257–7262. [CrossRef]
201. Aidhen, I.S.; Mukkamala, R.; Weidner, C.; Sauer, S. A Common Building Block for the Syntheses of Amorfrutin and Cajaninstilbene Acid Libraries toward Efficient Binding with Peroxisome Proliferator-Activated Receptors. *Org. Lett.* **2014**, *17*, 194–197. [CrossRef]
202. Ayza, M.A.; Zewdie, K.A.; Tesfaye, B.A.; Gebrekirstos, S.T.; Berhe, D.F. Anti-Diabetic Effect of Telmisartan Through its Partial PPAR-gamma Agonistic Activity. *Diabetes Metab. Syndr. Obes. Targets Ther.* **2020**, *13*, 3627–3635. [CrossRef]
203. Matsuyama, M.; Funao, K.; Kuratsukuri, K.; Tanaka, T.; Kawahito, Y.; Sano, H.; Chargui, J.; Touraine, J.; Yoshimura, N.; Yoshimura, R. Telmisartan inhibits human urological cancer cell growth through early apoptosis. *Exp. Ther. Med.* **2010**, *1*, 301–306. [CrossRef]
204. Liu, Y.; Chen, S.; Liu, J.; Jin, Y.; Yu, S.; An, R. Telmisartan inhibits oxalate and calcium oxalate crystal-induced epithelial-mesenchymal transformation via PPAR-AKT/STAT3/p38 MAPK-Snail pathway. *Life Sci.* **2020**, *241*, 117108. [CrossRef]
205. Weidner, C.; Wowro, S.J.; Freiwald, A.; Kodelja, V.; Abdel-Aziz, H.; Kelber, O.; Sauer, S. Lemon balm extract causes potent antihyperglycemic and antihyperlipidemic effects in insulin-resistant obese mice. *Mol. Nutr. Food Res.* **2013**, *58*, 903–907. [CrossRef]
206. Zheng, W.; Feng, X.; Qiu, L.; Pan, Z.; Wang, R.; Lin, S.; Hou, D.; Jin, L.; Li, Y. Identification of the antibiotic ionomycin as an unexpected peroxisome proliferator-activated receptorγ (PPARγ) ligand with a unique binding mode and effective glucose-lowering activity in a mouse model of diabetes. *Diabetologia* **2012**, *56*, 401–411. [CrossRef]
207. Villacorta, L.; Schopfer, F.J.; Zhang, J.; Freeman, B.A.; Chen, Y.E. PPARγ and its ligands: Therapeutic implications in cardiovascular disease. *Clin. Sci.* **2009**, *116*, 205–218. [CrossRef]
208. Schug, T.T.; Berry, D.C.; Shaw, N.S.; Travis, S.N.; Noy, N. Opposing Effects of Retinoic Acid on Cell Growth Result from Alternate Activation of Two Different Nuclear Receptors. *Cell* **2007**, *129*, 723–733. [CrossRef]
209. Bray, G.A. The Zucker-fatty rat: A review. *Fed. Proc.* **1977**, *36*, 148–153.
210. Brunmeir, R.; Xu, F. Functional Regulation of PPARs through Post-Translational Modifications. *Int. J. Mol. Sci.* **2018**, *19*, 1738. [CrossRef]
211. Coelho, M.S.; de Lima, C.L.; Royer, C.; Silva, J.B.; Oliveira, F.C.B.; Christ, C.G.; Pereira, S.A.; Bao, S.N.; Lima, M.C.A.; Pitta, M.G.R.; et al. GQ-16, a TZD-Derived Partial PPARgamma Agonist, Induces the Expression of Thermogenesis-Related Genes in Brown Fat and Visceral White Fat and Decreases Visceral Adiposity in Obese and Hyperglycemic Mice. *PLoS ONE* **2016**, *11*, e0154310. [CrossRef]
212. Ferreira, A.P.B.; Coelho, M.S.; Amato, A.A.; Neves, F.d.R.; Rodrigues, I.C.; Royer, C. Effect of PPAR Partial Agonist, GQ-16, on Viability of Breast Cancer Cells in Culture. *FASEB J.* **2017**, *31*, 876.5. [CrossRef]
213. Haraguchi, T.; Iwasaki, K.; Takasaki, K.; Uchida, K.; Naito, T.; Nogami, A.; Kubota, K.; Shindo, T.; Uchida, N.; Katsurabayashi, S.; et al. Telmisartan, a partial agonist of peroxisome proliferator-activated receptor gamma, improves impairment of spatial memory and hippocampal apoptosis in rats treated with repeated cerebral ischemia. *Brain Res.* **2010**, *1353*, 125–132. [CrossRef]
214. Zhao, T.; Du, H.; Blum, J.S.; Yan, C. Critical role of PPARγ in myeloid-derived suppressor cell-stimulated cancer cell proliferation and metastasis. *Oncotarget* **2015**, *7*, 1529–1543. [CrossRef]
215. Riehl, A.; Németh, J.; Angel, P.; Hess, J. The receptor RAGE: Bridging inflammation and cancer. *Cell Commun. Signal.* **2009**, *7*, 12. [CrossRef]

216. Phan, A.N.; Vo, V.T.; Hua, T.N.; Kim, M.; Jo, S.; Choi, J.; Kim, H.; Son, J.; Suh, Y.; Jeong, Y. PPAR sumoylation-mediated lipid accumulation in lung cancer. *Oncotarget* **2017**, *8*, 82491–82505. [CrossRef]
217. Gionfriddo, G.; Plastina, P.; Augimeri, G.; Catalano, S.; Giordano, C.; Barone, I.; Morelli, C.; Giordano, F.; Gelsomino, L.; Sisci, D.; et al. Modulating Tumor-Associated Macrophage Polarization by Synthetic and Natural PPAR Ligands as a Potential Target in Breast Cancer. *Cells* **2020**, *9*, 174. [CrossRef]
218. Heudobler, D.; Rechenmacher, M.; Lüke, F.; Vogelhuber, M.; Pukrop, T.; Herr, W.; Ghibelli, L.; Gerner, C.; Reichle, A. Peroxisome Proliferator-Activated Receptors (PPAR)γ Agonists as Master Modulators of Tumor Tissue. *Int. J. Mol. Sci.* **2018**, *19*, 3540. [CrossRef]
219. Huang, C.; Lo, C.; Chiu, C.; Shyur, L. Deoxyelephantopin, a novel multifunctional agent, suppresses mammary tumor growth and lung metastasis and doubles survival time in mice. *Br. J. Pharmacol.* **2010**, *159*, 856–871. [CrossRef]
220. Gan, L.; Gan, Z.; Dan, Y.; Li, Y.; Zhang, P.; Chen, S.; Ye, Z.; Pan, T.; Wan, C.; Hu, X.; et al. Tetrazanbigen Derivatives as Peroxisome Proliferator-Activated Receptor Gamma (PPARγ) Partial Agonists:Design, Synthesis, Structure–Activity Relationship, and Anticancer Activities. *J. Med. Chem.* **2021**, *64*, 1018–1036. [CrossRef] [PubMed]
221. Gbelcová, H.; Švéda, M.; Laubertová, L.; Varga, I.; Vítek, L.; Kolář, M.; Strnad, H.; Zelenka, J.; Böhmer, D.; Ruml, T. The effect of simvastatin on lipid droplets accumulation in human embryonic kidney cells and pancreatic cancer cells. *Lipids Health Dis.* **2013**, *12*, 126. [CrossRef] [PubMed]
222. Joshi, H.; Marulkar, K.; Gota, V.; Ramaa, C.S. Hydroxy Cinnamic Acid Derivatives as Partial PPARγ Agonists: *In silico* Studies, Synthesis and Biological Characterization Against Chronic Myeloid Leukemia Cell Line (K562). *Anti-Cancer Agents Med. Chem.* **2017**, *17*, 524–541. [CrossRef] [PubMed]
223. Kim, M.; Chae, Y.N.; Choi, S.; Moon, H.S.; Son, M.; Bae, M.; Choi, H.; Hur, Y.; Kim, E.; Park, Y.H.; et al. PAM-1616, a selective peroxisome proliferator-activated receptor modulator with preserved anti-diabetic efficacy and reduced adverse effects. *Eur. J. Pharmacol.* **2011**, *650*, 673–681. [CrossRef]
224. Fujimura, T.; Kimura, C.; Oe, T.; Takata, Y.; Sakuma, H.; Aramori, I.; Mutoh, S. A Selective Peroxisome Proliferator-Activated Receptor Modulator with Distinct Fat Cell Regulation Properties. *J. Pharmacol. Exp. Ther.* **2006**, *318*, 863–871. [CrossRef]
225. Konda, V.R.; Desai, A.; Darland, G.; Grayson, N.; Bland, J.S. KDT501, a Derivative from Hops, Normalizes Glucose Metabolism and Body Weight in Rodent Models of Diabetes. *PLoS ONE* **2014**, *9*, e87848. [CrossRef]
226. Smith, A.G.; Beaumont, K.A.; Smit, D.J.; Thurber, A.E.; Cook, A.L.; Boyle, G.M.; Parsons, P.G.; Sturm, R.A.; Muscat, G.E. PPAR agonists attenuate proliferation and modulate Wnt/-catenin signalling in melanoma cells. *Int. J. Biochem. Cell Biol.* **2009**, *41*, 844–852. [CrossRef]
227. Li, J.; Chen, L.; Yu, P.; Liu, B.; Zhu, J.; Yang, Y. Telmisartan Exerts Anti-Tumor Effects by Activating Peroxisome Proliferator-Activated Receptor- in Human Lung Adenocarcinoma A549 Cells. *Molecules* **2014**, *19*, 2862–2876. [CrossRef]
228. Higgins, L.S.; Mantzoros, C.S. The Development of INT131 as a Selective PPARgamma Modulator: Approach to a Safer Insulin Sensitizer. *PPAR Res.* **2008**, *2008*, 936906. [CrossRef]
229. Kroker, A.J.; Bruning, J.B. Review of the Structural and Dynamic Mechanisms of PPARγPartial Agonism. *PPAR Res.* **2015**, *2015*, 816856. [CrossRef]
230. Laghezza, A.; Montanari, R.; Lavecchia, A.; Piemontese, L.; Pochetti, G.; Iacobazzi, V.; Infantino, V.; Capelli, D.; DeBellis, M.; Liantonio, A.; et al. On the Metabolically Active Form of Metaglidasen: Improved Synthesis and Investigation of Its Peculiar Activity on Peroxisome Proliferator-Activated Receptors and Skeletal Muscles. *ChemMedChem* **2015**, *10*, 555–565. [CrossRef]
231. Hong, F.; Xu, P.; Zhai, Y. The Opportunities and Challenges of Peroxisome Proliferator-Activated Receptors Ligands in Clinical Drug Discovery and Development. *Int. J. Mol. Sci.* **2018**, *19*, 2189. [CrossRef]
232. Chen, L.; Bush, C.R.; Necela, B.M.; Su, W.; Yanagisawa, M.; Anastasiadis, P.Z.; Fields, A.P.; Thompson, E.A. RS5444, a novel PPARgamma agonist, regulates aspects of the differentiated phenotype in nontransformed intestinal epithelial cells. *Mol. Cell. Endocrinol.* **2006**, *251*, 17–32. [CrossRef]
233. Pishvaian, M.J.; Marshall, J.L.; Wagner, A.J.; Hwang, J.J.; Malik, S.; Cotarla, I.; Deeken, J.F.; He, A.R.; Daniel, H.; Halim, A.; et al. A phase 1 study of efatutazone, an oral peroxisome proliferator-activated receptor gamma agonist, administered to patients with advanced malignancies. *Cancer* **2012**, *118*, 5403–5413. [CrossRef]
234. Smallridge, R.C.; Copland, J.A.; Brose, M.S.; Wadsworth, J.T.; Houvras, Y.; Menefee, M.E.; Bible, K.C.; Shah, M.H.; Gramza, A.W.; Klopper, J.P.; et al. Efatutazone, an Oral PPAR-gamma Agonist, in Combination With Paclitaxel in Anaplastic Thyroid Cancer: Results of a Multicenter Phase 1 Trial. *J. Clin. Endocrinol. Metab.* **2013**, *98*, 2392–2400. [CrossRef]

cells

MDPI

Article

Activation of PPARα Ameliorates Cardiac Fibrosis in Dsg2-Deficient Arrhythmogenic Cardiomyopathy

Zirui Qiu [1,†], Yawen Zhao [1,†], Tian Tao [1], Wenying Guo [1], Ruonan Liu [1], Jingmin Huang [1] and Geyang Xu [1,2,*]

1 Department of Physiology, School of Medicine, Jinan University, 601 Huangpu Avenue West, Tianhe District, Guangzhou 510632, China
2 Center for Clinical Epidemiology and Methodology (CCEM), Guangdong Second Provincial General Hospital, Guangzhou 510317, China
* Correspondence: xugeyangliang@163.com; Tel.: +86-20-85-22-02-60; Fax: +86-20-85-22-13-43
† These authors contributed equally to this work.

Highlights:

- Cardiac-specific Dsg2 deletion induces excessive cardiac fibrosis in mice.
- Fenofibrate alleviates cardiac fibrosis in CS-Dsg2$^{-/-}$ mice.
- Cardiac-specific activation of PPARα ameliorates cardiac fibrosis in CS-Dsg2$^{-/-}$ mice.
- The inhibitory effect of PPARα on cardiac fibrosis is mediated by STAT3 and TGF-β /SMAD3 signaling.
- PPARα is a promising target for the intervention of ACM by ameliorating cardiac fibrosis.

Abstract: Background: Arrhythmogenic cardiomyopathy (ACM) is a genetic heart muscle disease characterized by progressive fibro-fatty replacement of cardiac myocytes. Up to now, the existing therapeutic modalities for ACM are mostly palliative. About 50% of ACM is caused by mutations in genes encoding desmosomal proteins including Desmoglein-2 (Dsg2). In the current study, the cardiac fibrosis of ACM and its underlying mechanism were investigated by using a cardiac-specific knockout of Dsg2 mouse model. Methods: Cardiac-specific *Dsg2* knockout (CS-Dsg2$^{-/-}$) mice and wild-type (WT) mice were respectively used as the animal model of ACM and controls. The myocardial collagen volume fraction was determined by histological analysis. The expression levels of fibrotic markers such as α-SMA and Collagen I as well as signal transducers such as STAT3, SMAD3, and PPARα were measured by Western blot and quantitative real-time PCR. Results: Increased cardiac fibrosis was observed in CS-Dsg2$^{-/-}$ mice according to Masson staining. PPARα deficiency and hyperactivation of STAT3 and SMAD3 were observed in the myocardium of CS-Dsg2$^{-/-}$ mice. The biomarkers of fibrosis such as α-SMA and Collagen I were upregulated after gene silencing of Dsg2 in HL-1 cells. Furthermore, STAT3 gene silencing by Stat3 siRNA inhibited the expression of fibrotic markers. The activation of PPARα by fenofibrate or AAV9-Pparα improved the cardiac fibrosis and decreased the phosphorylation of STAT3, SMAD3, and AKT in CS-Dsg2$^{-/-}$ mice. Conclusions: Activation of PPARα alleviates the cardiac fibrosis in ACM.

Keywords: arrhythmogenic cardiomyopathy; desmoglein-2; cardiac fibrosis; PPARα; fenofibrate; STAT3

Citation: Qiu, Z.; Zhao, Y.; Tao, T.; Guo, W.; Liu, R.; Huang, J.; Xu, G. Activation of PPARα Ameliorates Cardiac Fibrosis in Dsg2-Deficient Arrhythmogenic Cardiomyopathy. *Cells* **2022**, *11*, 3184. https://doi.org/10.3390/cells11203184

Academic Editors: Kay-Dietrich Wagner and Nicole Wagner

Received: 16 August 2022
Accepted: 28 September 2022
Published: 11 October 2022

1. Introduction

Arrhythmogenic cardiomyopathy (ACM) is a fatal heart disease characterized by cardiac dysfunction, heart failure, and life-threatening ventricular arrhythmias [1,2]. The population prevalence of ACM has been estimated between 1:1000 and 1:5000 [2]. Studies have shown that ACM causes 10% to 15% of sudden cardiac death (SCD) cases, especially among young people and athletes [2]. Pathological features of ACM include loss of myocytes and progressive fibro-fatty replacement. These pathological features tend to occur in the right ventricle (RV), with left ventricular (LV) or bilateral ventricular involvement [2,3].

Previous studies found that one or more mutations in genes encoding desmosomal proteins led to about 50% of ACM cases [4], including desmoglein2 (DSG2) [5], desmocollin2 [6], plakoglobin [7], desmoplakin [8], and plakophilin-2 [9]. DSG2 is a major cadherin of the cardiac desmosome; it was reported that mutations in the Dsg2 gene are associated with severe lethal heart muscle diseases such as ACM [10]. Till now, the main purpose of existing treatment for ACM is to prevent SCD [11].

Necrotic and apoptotic cardiomyocytes are replaced by fibrosis during the ACM disease's progress [12]. Although cardiac fibrosis plays a critical role in enhancing cardiac structural stability, it results in cardiac structural remodeling and impaired cardiac function, finally increasing the risk of potentially lethal cardiac arrhythmias [13]. Thus, improvement of cardiac fibrosis might be beneficial to avoid further deterioration of the ACM. The signal transducer and activator of transcription 3 (STAT3) is hyperactivated in fibrotic diseases and STAT3 inhibitors are currently used in the treatments of fibrotic diseases, especially in cardiac fibrosis [14]. Transforming growth factor-β (TGF-β) is a central mediator in hypertrophic and fibrotic of the heart. Canonical and non-canonical pathways for TGF-β-induced fibrosis in the heart are known [15]. Furthermore, interaction of Stat3 and TGF-β/Smad3 signaling is regarded as playing a critical role in cardiac fibrotic processes [16]. Recent studies illustrated the antagonistic effects and bidirectional regulation between STATs and peroxisome proliferator-activated receptors (PPARs) and suggested a potential cross-talk between STAT and PPAR pathways [17–19]. PPARs are the nuclear receptor superfamily of ligand-activated transcription factors. As the predominant PPAR isoform in the heart, peroxisome proliferator-activated receptor α (PPARα) modulates cardiac metabolism substrate conversion in cardiac hypertrophy, cardiac hypoxia, and diabetic heart [20]. PPARα gene deletion contributes to cardiac hypertrophy and deterioration of cardiac function [21]. Previous studies illustrated that PPARα activation alleviated cardiac fibrosis and reversed cardiac dysfunction [22] and PPARα could inhibit the TGF-β-induced profibrotic pathway in cardiac fibrosis [23,24]. Fenofibrate alleviated myocardial inflammation and collagen deposition in Ang II-infused rats [25]. Recently, we reported that activation of PPARα reduced the cardiac lipid accumulation and restored cardiac function in ACM mice [26]. Although PPARα plays a critical role in lipid accumulation in ACM, the effects of PPARα on cardiac fibrosis in ACM is still unclear. We hypothesized that the PPARα-STAT3/SMAD pathway is critical to cardiac fibrosis in ACM mice. In our current study, we found that PPARα was downregulated in the hearts of cardiac-specific Dsg2 knockout mice; restoring the activity of PPARα by using fenofibrate (a PPARα agonist) or AAV9-Pparα improved cardiac fibrosis via the PPARα-STAT3/SMAD pathway in cardiac-specific Dsg2 deletion mice. Our findings suggest that PPARα is a potential therapeutic target of cardiac fibrosis in ACM.

2. Materials and Methods

2.1. Materials

Fenofibrate was purchased from Sigma-Aldrich (St. Louis, MO, USA). Rabbit anti-Phospho-stat3 (Tyr705), rabbit anti-Phospho-SMAD3 (Ser423/425), rabbit anti-SMAD, rabbit anti-Phospho-AKT (Ser473), rabbit anti-AKT, rabbit anti-α-SMA, rabbit anti-Collagen I antibodies, mouse anti-stat3, and mouse monoclonal anti-β-actin were purchased from Cell Signaling Technology (Beverly, MA, USA). Rabbit anti-DSG2, rabbit anti-PPARα, rabbit anti-GAPDH antibodies were from Abcam Inc. (Cambridge, MA, USA).

2.2. Animals and Treatments

Cardiac-specific dsg2 gene knockout (CS-Dsg2$^{-/-}$) on C57-based genetic backgrounds were successfully constructed by mating DSG2 flox with CKMM cre [26]. Mice were housed in standard plastic rodent cages and maintained in a regulated environment (24 °C, 12 h light and 12 h dark cycles with lights on at 7:00 a.m.). All mice used in this study were 8–12 weeks old.

In order to activate PPARα in vivo, PPARα agonist fenofibrate (150 mg/kg body weight) was administrated by daily oral gavage for 28 days. To overexpress PPARα in the heart, male CS-Dsg2$^{-/-}$ mice were tail-vein infused with adeno-associated virus carrying PPARα (AAV9-cTnT-*Pparα*, 5×10^{11} vg per mouse). Adeno-associated virus carrying GFP (AAV9-cTnT-GFP, 5×10^{11} vg per mouse) was used as control. The mice were sacrificed 28 days after AAV injections [26].

2.3. Histological Analysis

Hearts were harvested from mice, fixed overnight in 4% paraformaldehyde, embedded in paraffin, and then, sectioned serially. Masson's trichrome staining was performed to evaluate collagen deposition using a kit following manufacturer's instructions (G1006-20 ML, Servicebio, Wuhan, China). The collagen volume fraction (CVF) was determined by Image J software as an index of cardiac fibrosis. The ratio of myocardial collagen area to the total myocardial area was used to calculate the collagen volume fraction.

2.4. Cell Culture and Treatment

The murine atrial cardiac myocyte cell line HL-1 was maintained in 10% fetal bovine serum (FBS) in Dulbecco's modified Eagle's medium at 37 °C in an atmosphere of 5% CO_2. For transient transfection, cells were plated at optimal densities and grown for 24 h. Cells were then transfected with Dsg2 siRNA (MBS828119, MyBioSource, San Diego, CA, USA) or Stat3 siRNA (6354, Cell Signaling Technology) using lipofectamine reagent according to the manufacturer's instructions.

2.5. Western Blot Analysis

The tissues and cells were homogenized in the lysis buffer. After protein quantification, 40 μg of protein was loaded onto SDS-PAGE gels. Then, protein extracts were electrophoresed, blotted, and then, incubated with primary antibodies. The antibodies were detected using 1:10,000 horseradish peroxidase-conjugated donkey anti-rabbit IgG and donkey anti-mouse IgG (Jackson ImmunoResearch, West Grove, PA, USA). Western blotting luminol reagent was used to visualize bands. The band intensities were quantitated by Image J software.

2.6. RNA Extraction, Quantitative Real-Time PCR

For gene expression analysis, RNA was isolated from mouse tissues and cells by using Trizol (Takara, Kusatsu, Shiga) and reverse-transcribed into cDNAs using the first-strand synthesis system for RT-PCR kit (Takara). SYBR green-based real-time PCR was performed using the Bio-Rad IQ5 PCR system (Bio-Rad, Foster City, CA, USA). Sequences for the primer pairs used in this study are shown in Table 1.

Table 1. List and sequences of primers used in RT-PCR experiments.

	Upstream Primer (5′-3′)	Downstream Primer (5′-3′)	Accession Number(s)
α-SMA	CCCTGAAGAGCATCCGACAC	TGCTGTTATAGGTGGTTTCGTG	NM_007392.3
Collagen I	TGTTCAGCTTTGTGGACCTC	GGACCCTTAGGCCATTGTGT	NM_007742.4
Dsg2	CGCACCAGGAAAGTACCAG	CCACAGTGGCATATCAACAGC	NM_007883.3
PPARα	AGAGCCCCATCTGTCCTCTC	ACTGGTAGTCTGCAAAACCAAA	XM_006520624.3
TGF-β	AGCCCTGGATACCAAC-TATTGCTTCAGCTCCACAG	AGGGGCGGGGCGGGGCGGGGCTTCAGCTGC	NM_011577.2
β-actin	CCACAGCTGAGAGGGAAATC	AAGGAAGGCTGGAAAAGAGC	NM_007393.5

2.7. Statistical Analysis

Data are expressed as mean ± SEM. Statistical significance was analyzed with a student's t-test. Differences were considered statistically significant with p values < 0.05.

3. Results

3.1. Cardiac-Specific Dsg2 Gene Deletion Provokes Cardiac Fibrosis

The ACM mouse model was generated by crossing $Dsg2^{fl\text{-}neo/+}$ mice with Ckmm-Cre mice which resulted in cardiac-specific Dsg2 deletion (CS-$Dsg2^{-/-}$). Increased cardiac fibrosis was observed in CS-$Dsg2^{-/-}$ mice according to Masson staining (Figure 1A). Several studies indicated that the activation of STAT3 contributes to cardiac fibrosis [27–29]. In our study, increased phosphorylation levels of STAT3 at Tyr705, SMAD3 at Ser423/425, and AKT at Ser473, and decreased expression levels of PPARα were observed in the LV, interventricular septum (IVS), and RV of CS-$Dsg2^{-/-}$ mice (Figure 1B). We next investigated the expression of TGF-β, α-smooth muscle actin (α-SMA), and collagen type I (Collagen I) in CS-$Dsg2^{-/-}$ mice. Expression levels of TGF-β, α-SMA, and Collagen I in LV, IVS, and RV of CS-$Dsg2^{-/-}$ mice were higher than that of littermate controls (Figure 1B,C).

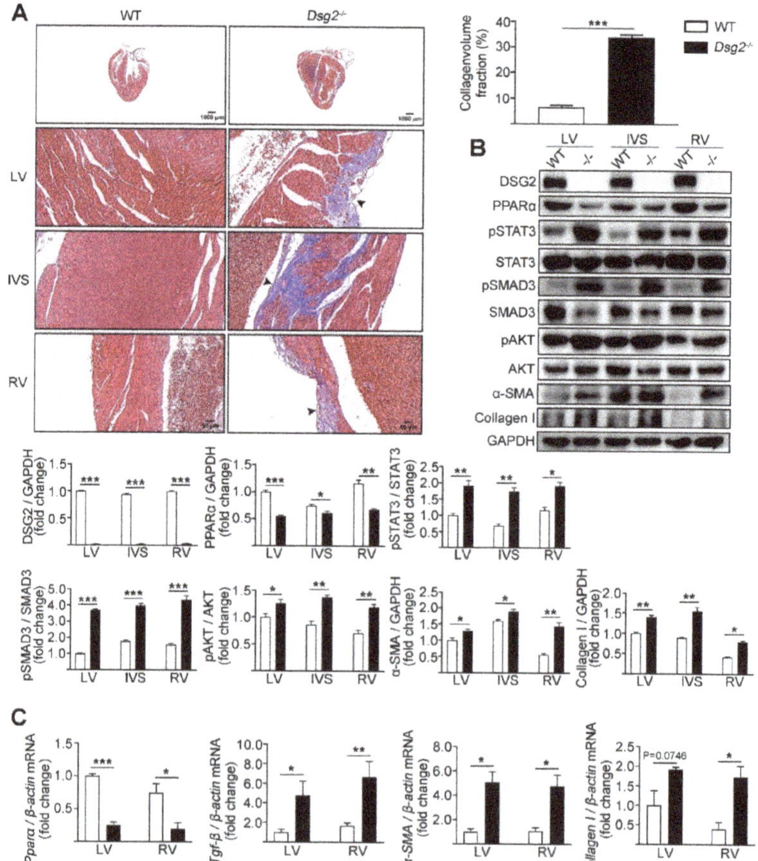

Figure 1. Cardiac-specific Dsg2 knockout induced cardiac fibrosis. (**A**) Masson staining of heart sections in WT and CS-$Dsg2^{-/-}$ (−/−) mice. Arrow shows cardiac fibrosis. Collagen volume fraction in the hearts of WT and CS-$Dsg2^{-/-}$ mice was assessed. (**B**) Representative Western blots from mouse left ventricular (LV), interventricular septum (IVS), and right ventricle (RV). DSG2, PPARα, pSTAT3, pSMAD3, pAKT, α-SMA, and Collagen I were detected using specific antibodies. STAT3, SMAD3, AKT, and GAPDH were used as loading controls. (**C**) Results of quantitative PCR analysis of PPARα, TGF-β, α-SMA, and Collagen I mRNA levels in mouse LV and RV are expressed as fold change of control using β-actin as loading control. Results are expressed as mean values ± SEM. n = 6. * $p < 0.05$, ** $p < 0.01$, *** $p < 0.001$ vs. WT.

The effects of Dsg2 on STAT3 activity and fibrosis were assessed in the cardiac myocyte cell line HL-1. To silence the expression of Dsg2 and Stat3, HL-1 cells were transfected with Dsg2 siRNA and Stat3 siRNA. The knockdown efficiency of Dsg2 siRNA was 66% and 51% for Stat3 siRNA. Consistent to the in vivo study, knockdown of Dsg2 in the HL-1 cells led to an increase in the phosphorylation of STAT3 (Tyr705) and the expression levels of α-SMA and Collagen I (Figure 2A,B). Furthermore, knockdown of Stat3 in the HL-1 cells decreased the expression levels of α-SMA and Collagen I (Figure 2C,D).

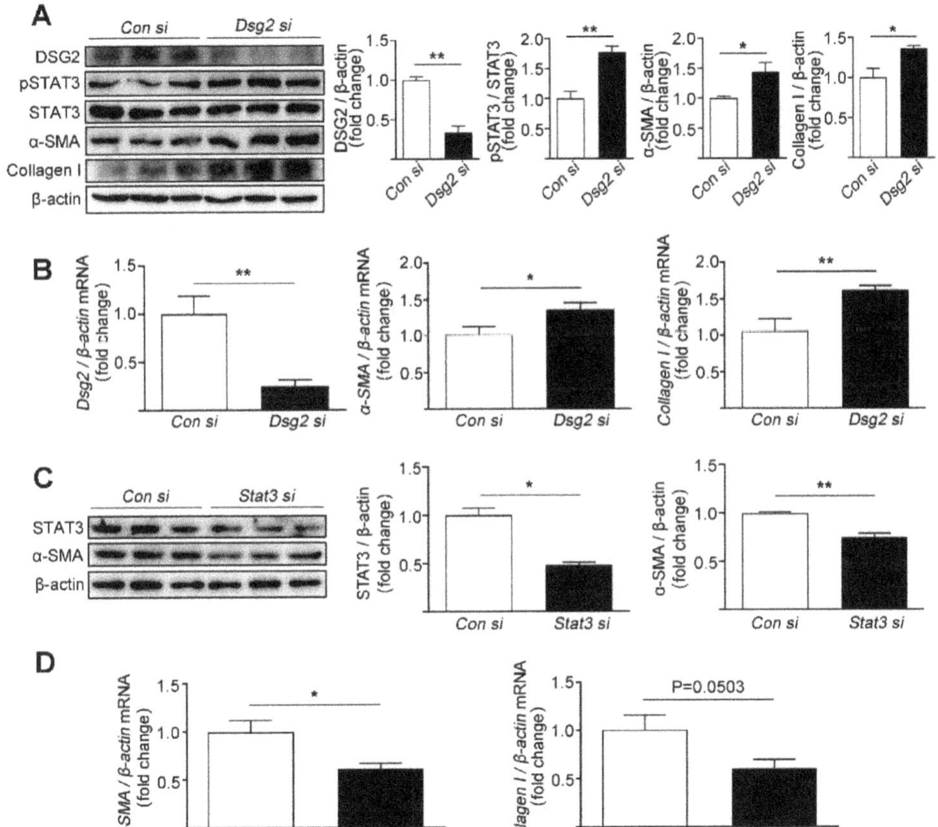

Figure 2. Effects of Dsg2 siRNA and Stat3 siRNA on the expression levels of fibrotic markers in HL-1 cells. (**A**,**B**) HL-1 cells were transfected with control siRNA or Dsg2 siRNA. (**A**) Representative Western blots for DSG2, pSTAT3, α-SMA, and Collagen I were detected using specific antibodies. STAT3 and β-actin were used as loading controls. (**B**) Results of quantitative PCR analysis of Dsg2, α-SMA, and Collagen I mRNA levels in HL-1 cells treated with control or Dsg2 siRNA are expressed as fold change of control using β-actin as loading control. Results are expressed as mean values ± SEM. n = 3. * $p < 0.05$, ** $p < 0.01$ vs. control. (**C**,**D**) HL-1 cells were transfected with control siRNA or Stat3 siRNA. (**C**) Representative Western blots for STAT3 and α-SMA were detected using specific antibodies. β-actin were used as loading controls. (**D**) Results of quantitative PCR analysis of α-SMA and Collagen I mRNA levels in HL-1 cells treated with control or Stat3 siRNA are expressed as fold change of control using β-actin as loading control. Results are expressed as mean values ± SEM. n = 3. * $p < 0.05$, ** $p < 0.01$, vs. control.

3.2. Fenofibrate Alleviated Cardiac Fibrosis in CS-Dsg2$^{-/-}$ Mice

Fenofibrate, a PPARα agonist, affords myocardial protection apart from its lipid lowering effects [30]. We next assessed the effect of fenofibrate on cardiac fibrosis in CS-Dsg2$^{-/-}$ mice. Interestingly, a significant improvement in cardiac fibrosis was observed in CS-Dsg2$^{-/-}$ mice after being treated with fenofibrate (150 mg/kg/day, for 4 weeks) (Figure 3A). Fenofibrate decreased the phosphorylation levels of STAT3, SMAD3, and AKT as well as the expression levels of TGF-β, α-SMA, and Collagen I in CS-Dsg2$^{-/-}$ mice (Figure 3B,C).

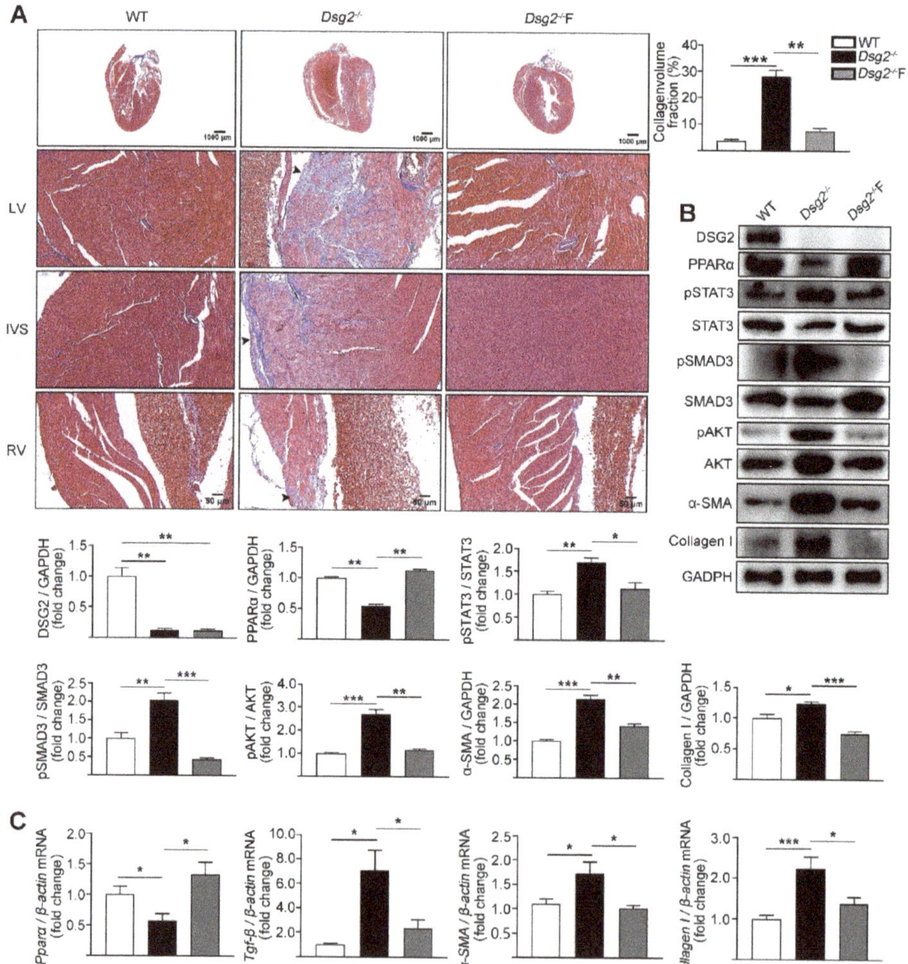

Figure 3. Fenofibrate alleviated cardiac fibrosis in CS-Dsg2$^{-/-}$ mice. (**A**) Masson staining of heart sections in WT, CS-Dsg2$^{-/-}$ mice, and CS-Dsg2$^{-/-}$ mice treated with fenofibrate (Dsg2$^{-/-}$F). Collagen volume fraction in the hearts of WT, CS-Dsg2$^{-/-}$, and Dsg2$^{-/-}$F mice were assessed. (**B**) Representative Western blots from ventricles of WT, CS-Dsg2$^{-/-}$, and Dsg2$^{-/-}$F mice. DSG2, PPARα, pSTAT3, pSMAD3, pAKT, α-SMA, and Collagen I were detected using specific antibodies. STAT3, SMAD3, AKT, and GAPDH were used as loading controls. (**C**) Results of quantitative PCR analysis of PPARα, TGF-β, α-SMA, and Collagen I mRNA levels in mouse ventricles are expressed as fold change of control using β-actin as loading control. Results are expressed as mean values ± SEM. n = 6. * $p < 0.05$, ** $p < 0.01$, *** $p < 0.001$ vs. control.

3.3. Cardiac-Specific Activation of PPARα Alleviated Cardiac Fibrosis in CS-Dsg2$^{-/-}$ Mice

To further confirm that PPARα activation in the heart could improve cardiac fibrosis in Dsg2$^{-/-}$ mice, cardiac-specific activation of PPARα was performed by tail-vein infusion of AAV9. AAV9-cTnT promoter-Pparα significantly reduced cardiac fibrosis in CS-Dsg2$^{-/-}$ mice (Figure 4A). Simultaneously, the levels of phosphorylated STAT3, phosphorylated SMAD3, phosphorylated AKT, and the expression levels of TGF-β, α-SMA, and Collagen I were decreased after cardiac-specific activation of PPARα (Figure 4B,C).

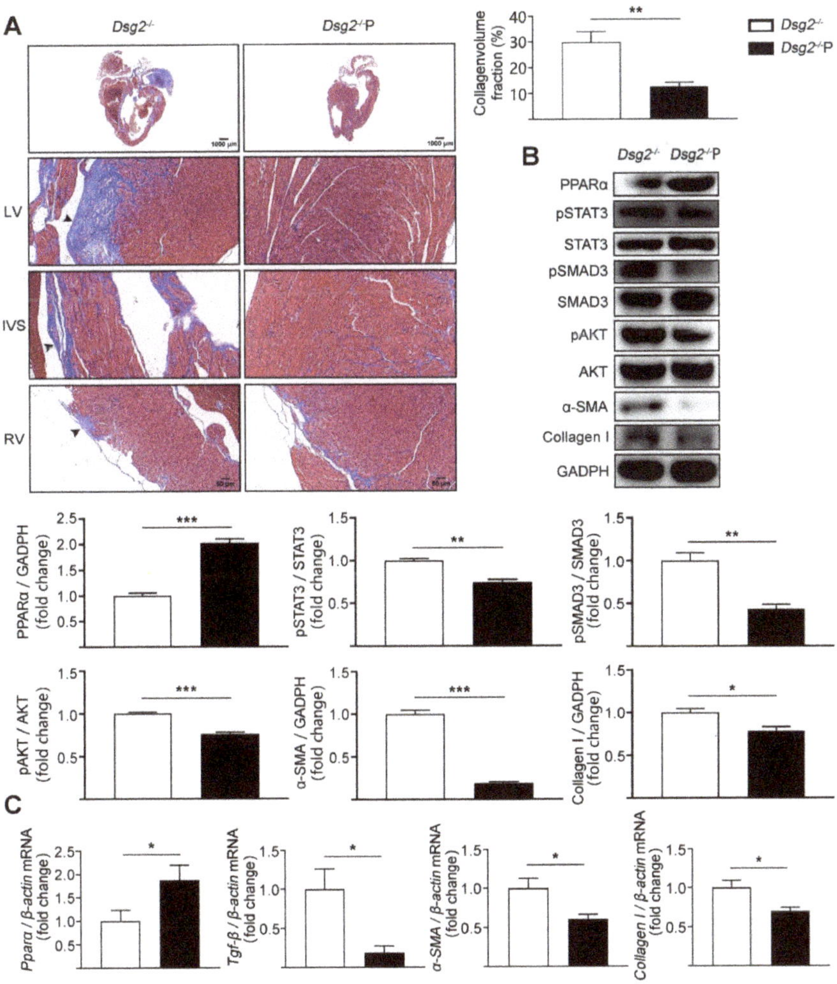

Figure 4. AAV9-Pparα alleviated cardiac fibrosis in CS-Dsg2$^{-/-}$ mice. (**A**) Masson staining of heart sections in CS-Dsg2$^{-/-}$ mice and CS-Dsg2$^{-/-}$ mice received AAV9-Pparα (Dsg2$^{-/-}$P). Collagen volume fraction in the hearts of CS-Dsg2$^{-/-}$ and Dsg2$^{-/-}$P mice were assessed. (**B**) Representative Western blots from ventricles of CS-Dsg2$^{-/-}$ and Dsg2$^{-/-}$P mice. PPARα, pSTAT3, pSMAD3, pAKT, α-SMA, and Collagen I were detected using specific antibodies. STAT3, SMAD3, AKT, and GAPDH were used as loading controls. (**C**) Results of quantitative PCR analysis of PPARα, TGF-β, α-SMA, and Collagen I mRNA levels in mouse ventricles are expressed as fold change of control using β-actin as loading control. Results are expressed as mean values ± SEM. n = 6. * $p < 0.05$, ** $p < 0.01$, *** $p < 0.001$ vs. CS-Dsg2$^{-/-}$.

4. Discussion

ACM is characterized by progressive replacement of cardiomyocytes by fibro-fatty tissue, cardiac dysfunction, ventricular arrhythmias, and heart failure. Mutation of desmoglein-2 (Dsg2) is one of the major causes of ACM and has been shown to lead to a loss of adhesive function [31]. Dsg2 mutation carriers display more severe heart muscle disease, which is associated with biventricular involvement and rapid evolution to end-stage heart failure [32]. In our previous study, we generated an ACM mouse model by cardiac-specific knockout of the DSG2 gene and discovered that downregulation of PPARα contributed to the impairment of fatty acid oxidation and, thus, to lipid accumulation in the DSG2 deletion-induced ACM [26]. However, whether downregulation of PPARα also contributes to the fibrosis in ACM was unsolved. In the present study, we uncovered a previously unrecognized role of PPARα in cardiac fibrosis in Dsg2-deficient ACM mice. Cardiac-specific Dsg2 knockout contributed to a severe cardiac fibrosis. Decreased expression of PPARα and the increased phosphorylation of STAT3 and SMAD3 were observed in this model. Moreover, activation of PPARα, either by fenofibrate or AAV9-Pparα, decreased the activity of STAT3 and SMAD3 and improved cardiac fibrosis in Dsg2 deletion-induced ACM.

Cardiac fibrosis is defined as excessive deposition of extracellular matrix (ECM) proteins by cardiac fibroblasts (CFs). CFs are transformed into myofibroblasts when they respond to stress and pathological stimuli [33]. Myocardial fibrosis reduces tissue compliance and accelerates the progression to heart failure [34]. In our study, histological analysis showed excessive deposition of collagen in the hearts of CS-Dsg2$^{-/-}$ mice when compared to WT mice. Furthermore, fibrotic markers such as TGF-β, α-SMA, and Collagen I were activated after cardiac-specific Dsg2 deletion. Cardiac fibrosis has been implicated in the progression of ACM. Increased cardiac fibrosis has been associated with altered cardiac conduction, resulting in conduction slowing, blockage, and re-entry [35]. Recent evidence indicates that the fibrosis state preceded the development of cardiac dysfunction in cardiomyopathies [36]. Thus, identification of druggable targets that can alleviate cardiac fibrosis might be beneficial to the treatment of ACM.

As a ligand-activated transcription factor which is highly expressed in cardiomyocytes, the role of PPARα in the heart is complex and vital. PPARα involvement in the regulation of inflammation [37], hypertrophy [38], energy metabolism [39], ischemia/reperfusion injury [40], and cardiac fibrosis [25] in hearts has been established in recent studies. Fenofibrate is a member of the fibrate family of PPARα receptor agonists and has regulating efficacy of inflammation and extracellular matrix remodeling of the heart [25,41]. As a PPARα agonist, fenofibrate has been widely used for hyperlipidemia in clinics and can also promote fatty acid oxidation in the mitochondria and improve myocardial energy metabolism [42]. Fenofibrate alleviated myocardial inflammation and fibrosis in diabetic mice via PPARα receptor [43]. Our previous study showed that PPARα was downregulated in the heart of the Dsg2 deletion ACM model and reactivation of PPARα significantly alleviated the lipid accumulation and improved cardiac function in CS-Dsg2$^{-/-}$ mice [26]. In the current study, we demonstrated that downregulation of PPARα also contributed to the cardiac fibrosis in the Dsg2 deletion-induced ACM model. Moreover, reactivation of PPARα either by tail-vein injection of AAV9-Pparα or oral treatment of fenofibrate improved the cardiac fibrosis in CS-Dsg2$^{-/-}$ mice. Our results suggested that PPARα is a promising therapeutic target for ACM intervention which not only alleviates lipid accumulation but also improves cardiac fibrosis.

Although cardiac fibrosis is one of the pathological characteristics of ACM, the mechanism of how mutations of desmosomal proteins lead to fibrosis is elusive. TGF-β is a core mediator in cardiac fibrosis. Canonical (SMAD-dependent) and non-canonical (SMAD-independent) pathways for TGF-β-induced fibrosis in the heart are documented [44]. In the canonical pathway, TGF-β activates SMAD2/3 signaling, which in turn regulates the expressions of collagen and α-SMA in myofibroblasts [45]. Non-canonical pathways involve STAT, MAPK, and PI3K pathways [46–48]. Our study showed that cardiac-specific Dsg2 deletion led to enhanced phosphorylation of SMAD3, STAT3, and AKT, suggesting that both canonical and non-canonical TGF-β pathways are activated in Dsg2 deletion-induced ACM. Among these pathways, STAT3 is reported to be critical to cardiac fibrosis and hypertrophy and activated in the hearts of mouse models of cardiac hypertrophy and heart failure [49]. Several studies have demonstrated that STAT3 maintains ECM homeostasis by regulating collagen synthesis and secretion in CFs [50]. Continuous STAT3 activation (tyrosine 705 residue phosphorylation) was regarded as a poor indicator in cardiac hypertrophy and heart failure [51]. Our study showed that knockdown of Dsg2 by Dsg2 siRNA induced the activation of fibrotic markers and STAT3 in HL-1 cells, while Stat3 siRNA did the reverse. These results suggested that activation of STAT3 contributes to cardiac fibrosis in cardiac-specific Dsg2 deletion mice. Furthermore, reactivation of PPARα either by AAV9-PPARα or fenofibrate decreased the phosphorylation of SMAD3, STAT3, and AKT in the Dsg2 deletion-induced ACM model, implying that PPARα modulated these pathways and deficiency in PPARα contributed to the activation of them and, thus, to cardiac fibrosis. Although the mechanistic link between PPARα and the STAT3 and TGF-β /SMAD3 pathways remains unclear, potential cross-talk between PPARα and STAT3 and TGF-β /SMAD3 pathways were reported in recent studies [17]. Chang H et al. reported that activation of PPARα ameliorates autoimmune myocarditis by suppressing Th17 cell differentiation through reducing phosphorylated STAT3 [52]. Gervois et al. demonstrated that fenofibrate treatment decreased the phosphorylation of STAT3 in livers [53]. Bansal T et al. reported that activation of PPARα improves cardiac fibrosis by inhibiting non-canonical TGF-β signaling [24]. Sekiguchi K et al. demonstrated that TGF-β signaling pathways directly inhibit PPARα activity in cardiac myocytes [54]. These studies suggest a role of PPARα in modulating STAT3 and TGF-β /SMAD3 pathways.

Current treatments for ACM lack effective treatment to improve or reverse cardiac fibrosis. In the present study, we established that activation of PPARα by fenofibrate or AAV9-Pparα improved the cardiac fibrosis in Dsg2 deletion-induced ACM. At the same time, activation of PPARα provided a cardioprotective effect through reducing the phosphorylation of STAT3 and SMAD3. These results indicated that the inhibitory effect of PPARα on cardiac fibrosis is mediated by a downregulation of STAT3 and TGF-β /SMAD3, and PPARα may be a significant target of ACM treatment. PPARα agonist fenofibrate may be a potential drug against cardiac fibrosis in ACM. In conclusion, our study generated an ACM model by cardiac-specific Dsg2 knockout and suggested that activation of PPARα ameliorates the excessive cardiac fibrosis in ACM.

Supplementary Materials: The following supporting information can be downloaded at: https://www.mdpi.com/article/10.3390/cells11203184/s1, Figure S1: The expression of PPARα in the skeletal muscle of cardiac specific DSG2 null mice received with AAV9-cTnT-GFP ($DSG2^{-/-}$) or AAV9-cTnT-Pparα ($DSG2^{-/-}$ P).

Author Contributions: G.X. designed research; Z.Q., Y.Z., T.T., W.G., R.L. and J.H. performed research; Z.Q., Y.Z. and G.X. analyzed data; G.X., Z.Q. and Y.Z. wrote and edited the paper. All authors contributed to the discussion and revised the article, and all approved the final versions of the manuscript. G.X. was responsible for the integrity of the work as a whole. All authors have read and agreed to the published version of the manuscript.

Funding: This work was supported by grants from the National Natural Science Foundation of China (82170818, 81770794) and the Fundamental Research Funds for the Central Universities (21620423).

Institutional Review Board Statement: Animal studies were conducted in accordance with the Guidelines on Laboratory Animals of Jinan University and were reviewed and approved by the Animal Care and Use Committee of Jinan University (No:20191128-09).

Informed Consent Statement: Not applicable.

Data Availability Statement: All data relevant to the study are included in the article or uploaded as supplemental information.

Acknowledgments: We thank Yubi Lin from Guangdong Medical University for providing the CS-$Dsg2^{-/-}$ mice.

Conflicts of Interest: On behalf of all authors, the corresponding author states that there is no conflict of interest.

Abbreviations

ACM	arrhythmogenic cardiomyopathy
CFs	cardiac fibroblasts
Collagen I	collagen type I
DSG2	Desmoglein-2
LV	left ventricular
PPARα	peroxisome proliferator-activated receptor α
RV	right ventricle
STAT3	signal transducer and activator of transcription 3
SMAD	mothers against decapentaplegic homolog 3
SCD	sudden cardiac death
TGF-β	transforming growth factor–β
α-SMA	alpha-smooth muscle actin

References

1. Thiene, G.; Nava, A.; Corrado, D.; Rossi, L.; Pennelli, N. Right ventricular cardiomyopathy and sudden death in young people. *N. Engl. J. Med.* **1988**, *318*, 129–133. [CrossRef] [PubMed]
2. Corrado, D.; Basso, C.; Judge, D.P. Arrhythmogenic Cardiomyopathy. *Circ. Res.* **2017**, *121*, 784–802. [CrossRef] [PubMed]
3. Bueno-Beti, C.; Asimaki, A. Histopathological Features and Protein Markers of Arrhythmogenic Cardiomyopathy. *Front. Cardiovasc. Med.* **2021**, *8*, 746321. [CrossRef] [PubMed]
4. Austin, K.M.; Trembley, M.A.; Chandler, S.F.; Sanders, S.P.; Saffitz, J.E.; Abrams, D.J.; Pu, W.T. Molecular mechanisms of arrhythmogenic cardiomyopathy. *Nat. Rev. Cardiol.* **2019**, *16*, 519–537. [CrossRef]
5. Awad, M.M.; Dalal, D.; Cho, E.; Amat-Alarcon, N.; James, C.; Tichnell, C.; Tucker, A.; Russell, S.D.; Bluemke, D.A.; Dietz, H.C.; et al. DSG2 mutations contribute to arrhythmogenic right ventricular dysplasia/cardiomyopathy. *Am. J. Hum. Genet.* **2006**, *79*, 136–142. [CrossRef]
6. Heuser, A.; Plovie, E.R.; Ellinor, P.T.; Grossmann, K.S.; Shin, J.T.; Wichter, T.; Basson, C.T.; Lerman, B.B.; Sasse-Klaassen, S.; Thierfelder, L.; et al. Mutant desmocollin-2 causes arrhythmogenic right ventricular cardiomyopathy. *Am. J. Hum. Genet.* **2006**, *79*, 1081–1088. [CrossRef]
7. McKoy, G.; Protonotarios, N.; Crosby, A.; Tsatsopoulou, A.; Anastasakis, A.; Coonar, A.; Norman, M.; Baboonian, C.; Jeffery, S.; McKenna, W.J. Identification of a deletion in plakoglobin in arrhythmogenic right ventricular cardiomyopathy with palmoplantar keratoderma and woolly hair (Naxos disease). *Lancet* **2000**, *355*, 2119–2124. [CrossRef]
8. Norgett, E.E.; Hatsell, S.J.; Carvajal-Huerta, L.; Cabezas, J.C.; Common, J.; Purkis, P.E.; Whittock, N.; Leigh, I.M.; Stevens, H.P.; Kelsell, D.P. Recessive mutation in desmoplakin disrupts desmoplakin-intermediate filament interactions and causes dilated cardiomyopathy, woolly hair and keratoderma. *Hum. Mol. Genet.* **2000**, *9*, 2761–2766. [CrossRef]
9. Gerull, B.; Heuser, A.; Wichter, T.; Paul, M.; Basson, C.T.; McDermott, D.A.; Lerman, B.B.; Markowitz, S.M.; Ellinor, P.T.; MacRae, C.A.; et al. Mutations in the desmosomal protein plakophilin-2 are common in arrhythmogenic right ventricular cardiomyopathy. *Nat. Genet.* **2004**, *36*, 1162–1164. [CrossRef]
10. Pilichou, K.; Nava, A.; Basso, C.; Beffagna, G.; Bauce, B.; Lorenzon, A.; Frigo, G.; Vettori, A.; Valente, M.; Towbin, J.; et al. Mutations in desmoglein-2 gene are associated with arrhythmogenic right ventricular cardiomyopathy. *Circulation* **2006**, *113*, 1171–1179. [CrossRef]
11. Rigato, I.; Corrado, D.; Basso, C.; Zorzi, A.; Pilichou, K.; Bauce, B.; Thiene, G. Pharmacotherapy and other therapeutic modalities for managing Arrhythmogenic Right Ventricular Cardiomyopathy. *Cardiovasc. Drugs Ther.* **2015**, *29*, 171–177. [CrossRef]
12. de Jong, S.; van Veen, T.A.; van Rijen, H.V.; de Bakker, J.M. Fibrosis and cardiac arrhythmias. *J. Cardiovasc. Pharmacol.* **2011**, *57*, 630–638. [CrossRef] [PubMed]
13. Nguyen, T.P.; Qu, Z.; Weiss, J.N. Cardiac fibrosis and arrhythmogenesis: The road to repair is paved with perils. *J. Mol. Cell. Cardiol.* **2014**, *70*, 83–91. [CrossRef] [PubMed]
14. Chakraborty, D.; Šumová, B.; Mallano, T.; Chen, C.W.; Distler, A.; Bergmann, C.; Ludolph, I.; Horch, R.E.; Gelse, K.; Ramming, A.; et al. Activation of STAT3 integrates common profibrotic pathways to promote fibroblast activation and tissue fibrosis. *Nat. Commun.* **2017**, *8*, 1130. [CrossRef]
15. Dobaczewski, M.; Chen, W.; Frangogiannis, N.G. Transforming growth factor (TGF)-β signaling in cardiac remodeling. *J. Mol. Cell. Cardiol.* **2011**, *51*, 600–606. [CrossRef] [PubMed]
16. Su, S.A.; Yang, D.; Wu, Y.; Xie, Y.; Zhu, W.; Cai, Z.; Shen, J.; Fu, Z.; Wang, Y.; Jia, L.; et al. EphrinB2 Regulates Cardiac Fibrosis Through Modulating the Interaction of Stat3 and TGF-β/Smad3 Signaling. *Circ. Res.* **2017**, *121*, 617–627. [CrossRef]
17. Zhou, Y.C.; Waxman, D.J. Cross-talk between janus kinase-signal transducer and activator of transcription (JAK-STAT) and peroxisome proliferator-activated receptor-alpha (PPARalpha) signaling pathways. Growth hormone inhibition of pparalpha transcriptional activity mediated by stat5b. *J. Biol. Chem.* **1999**, *274*, 2672–2681.

18. Shipley, J.M.; Waxman, D.J. Down-regulation of STAT5b transcriptional activity by ligand-activated peroxisome proliferator-activated receptor (PPAR) alpha and PPARgamma. *Mol. Pharmacol.* **2003**, *64*, 355–364. [CrossRef]
19. Zhou, Y.C.; Waxman, D.J. STAT5b down-regulates peroxisome proliferator-activated receptor alpha transcription by inhibition of ligand-independent activation function region-1 trans-activation domain. *J. Biol. Chem.* **1999**, *274*, 29874–29882. [CrossRef]
20. Finck, B.N.; Kelly, D.P. Peroxisome proliferator-activated receptor alpha (PPARalpha) signaling in the gene regulatory control of energy metabolism in the normal and diseased heart. *J. Mol. Cell. Cardiol.* **2002**, *34*, 1249–1257. [CrossRef]
21. Loichot, C.; Jesel, L.; Tesse, A.; Tabernero, A.; Schoonjans, K.; Roul, G.; Carpusca, I.; Auwerx, J.; Andriantsitohaina, R. Deletion of peroxisome proliferator-activated receptor-alpha induces an alteration of cardiac functions. *Am. J. Physiol. Heart Circ. Physiol.* **2006**, *291*, H161–H166. [CrossRef] [PubMed]
22. Ogata, T.; Miyauchi, T.; Sakai, S.; Takanashi, M.; Irukayama-Tomobe, Y.; Yamaguchi, I. Myocardial fibrosis and diastolic dysfunction in deoxycorticosterone acetate-salt hypertensive rats is ameliorated by the peroxisome proliferator-activated receptor-alpha activator fenofibrate, partly by suppressing inflammatory responses associated with the nuclear factor-kappa-B pathway. *J. Am. Coll. Cardiol.* **2004**, *43*, 1481–1488. [PubMed]
23. Zhang, Y.; Ji, H.; Qiao, O.; Li, Z.; Pecoraro, L.; Zhang, X.; Han, X.; Wang, W.; Zhang, X.; Man, S.; et al. Nanoparticle conjugation of ginsenoside Rb3 inhibits myocardial fibrosis by regulating PPARα pathway. *Biomed. Pharmacother.* **2021**, *139*, 111630. [CrossRef]
24. Bansal, T.; Chatterjee, E.; Singh, J.; Ray, A.; Kundu, B.; Thankamani, V.; Sengupta, S.; Sarkar, S. Arjunolic acid, a peroxisome proliferator-activated receptor α agonist, regresses cardiac fibrosis by inhibiting non-canonical TGF-β signaling. *J. Biol. Chem.* **2017**, *292*, 16440–16462. [CrossRef] [PubMed]
25. Diep, Q.N.; Benkirane, K.; Amiri, F.; Cohn, J.S.; Endemann, D.; Schiffrin, E.L. PPAR alpha activator fenofibrate inhibits myocardial inflammation and fibrosis in angiotensin II-infused rats. *J. Mol. Cell. Cardiol.* **2004**, *36*, 295–304. [CrossRef]
26. Lin, Y.; Liu, R.; Huang, Y.; Yang, Z.; Xian, J.; Huang, J.; Qiu, Z.; Lin, X.; Zhang, M.; Chen, H.; et al. Reactivation of PPARα alleviates myocardial lipid accumulation and cardiac dysfunction by improving fatty acid β-oxidation in Dsg2-deficient arrhythmogenic cardiomyopathy. In *Acta Pharmaceutica Sinica B*; Elsevier: Amsterdam, The Netherlands, 2022. [CrossRef]
27. Bao, Q.; Zhang, B.; Suo, Y.; Liu, C.; Yang, Q.; Zhang, K.; Yuan, M.; Yuan, M.; Zhang, Y.; Li, G. Intermittent hypoxia mediated by TSP1 dependent on STAT3 induces cardiac fibroblast activation and cardiac fibrosis. *eLife* **2020**, *9*, e49923. [CrossRef]
28. Cao, W.; Shi, P.; Ge, J.J. miR-21 enhances cardiac fibrotic remodeling and fibroblast proliferation via CADM1/STAT3 pathway. *BMC Cardiovasc. Disord.* **2017**, *17*, 88. [CrossRef]
29. Chen, X.; Su, J.; Feng, J.; Cheng, L.; Li, Q.; Qiu, C.; Zheng, Q. TRIM72 contributes to cardiac fibrosis via regulating STAT3/Notch-1 signaling. *J. Cell. Physiol.* **2019**, *234*, 17749–17756. [CrossRef]
30. Balakumar, P.; Rohilla, A.; Mahadevan, N. Pleiotropic actions of fenofibrate on the heart. *Pharmacol. Res.* **2011**, *63*, 8–12. [CrossRef]
31. Kant, S.; Holthöfer, B.; Magin, T.M.; Krusche, C.A.; Leube, R.E. Desmoglein 2-Dependent Arrhythmogenic Cardiomyopathy Is Caused by a Loss of Adhesive Function. *Circ. Cardiovasc. Genet.* **2015**, *8*, 553–563. [CrossRef]
32. Hermida, A.; Fressart, V.; Hidden-Lucet, F.; Donal, E.; Probst, V.; Deharo, J.C.; Chevalier, P.; Klug, D.; Mansencal, N.; Delacretaz, E.; et al. High risk of heart failure associated with desmoglein-2 mutations compared to plakophilin-2 mutations in arrhythmogenic right ventricular cardiomyopathy/dysplasia. *Eur. J. Heart Fail.* **2019**, *21*, 792–800. [CrossRef] [PubMed]
33. Souders, C.A.; Bowers, S.L.; Baudino, T.A. Cardiac fibroblast: The renaissance cell. *Circ. Res.* **2009**, *105*, 1164–1176. [CrossRef] [PubMed]
34. Travers, J.G.; Kamal, F.A.; Robbins, J.; Yutzey, K.E.; Blaxall, B.C. Cardiac Fibrosis: The Fibroblast Awakens. *Circ. Res.* **2016**, *118*, 1021–1040. [CrossRef] [PubMed]
35. Morita, N.; Mandel, W.J.; Kobayashi, Y.; Karagueuzian, H.S. Cardiac fibrosis as a determinant of ventricular tachyarrhythmias. *J. Arrhythm.* **2014**, *30*, 389–394. [CrossRef] [PubMed]
36. Ho, C.Y.; López, B.; Coelho-Filho, O.R.; Lakdawala, N.K.; Cirino, A.L.; Jarolim, P.; Kwong, R.; González, A.; Colan, S.D.; Seidman, J.G.; et al. Myocardial fibrosis as an early manifestation of hypertrophic cardiomyopathy. *N. Engl. J. Med.* **2010**, *363*, 552–563. [CrossRef]
37. Smeets, P.J.; Planavila, A.; van der Vusse, G.J.; van Bilsen, M. Peroxisome proliferator-activated receptors and inflammation: Take it to heart. *Acta Physiol.* **2007**, *191*, 171–188. [CrossRef]
38. Smeets, P.J.; Teunissen, B.E.; Willemsen, P.H.; van Nieuwenhoven, F.A.; Brouns, A.E.; Janssen, B.J.; Cleutjens, J.P.; Staels, B.; van der Vusse, G.J.; van Bilsen, M. Cardiac hypertrophy is enhanced in PPAR alpha−/− mice in response to chronic pressure overload. *Cardiovasc. Res.* **2008**, *78*, 79–89. [CrossRef]
39. Finck, B.N. The PPAR regulatory system in cardiac physiology and disease. *Cardiovasc. Res.* **2007**, *73*, 269–277. [CrossRef]
40. Li, L.X.; Yin, L.H.; Gao, M.; Xu, L.N.; Qi, Y.; Peng, J.Y. MiR-23a-5p exacerbates intestinal ischemia-reperfusion injury by promoting oxidative stress via targeting PPAR alpha. *Biochem. Pharmacol.* **2020**, *180*, 114194. [CrossRef]
41. Lockyer, P.; Schisler, J.C.; Patterson, C.; Willis, M.S. Minireview: Won't get fooled again: The nonmetabolic roles of peroxisome proliferator-activated receptors (PPARs) in the heart. *Mol. Endocrinol.* **2010**, *24*, 1111–1119. [CrossRef]
42. Li, P.; Luo, S.; Pan, C.; Cheng, X. Modulation of fatty acid metabolism is involved in the alleviation of isoproterenol-induced rat heart failure by fenofibrate. *Mol. Med. Rep.* **2015**, *12*, 7899–7906. [CrossRef] [PubMed]
43. Zhang, J.; Cheng, Y.; Gu, J.; Wang, S.; Zhou, S.; Wang, Y.; Tan, Y.; Feng, W.; Fu, Y.; Mellen, N.; et al. Fenofibrate increases cardiac autophagy via FGF21/SIRT1 and prevents fibrosis and inflammation in the hearts of Type 1 diabetic mice. *Clin. Sci.* **2016**, *130*, 625–641. [CrossRef] [PubMed]

44. Meng, X.M.; Nikolic-Paterson, D.J.; Lan, H.Y. TGF-β: The master regulator of fibrosis. *Nat. Rev. Nephrol.* **2016**, *12*, 325–338. [CrossRef] [PubMed]
45. Khalil, H.; Kanisicak, O.; Prasad, V.; Correll, R.N.; Fu, X.; Schips, T.; Vagnozzi, R.J.; Liu, R.; Huynh, T.; Lee, S.J.; et al. Fibroblast-specific TGF-β-Smad2/3 signaling underlies cardiac fibrosis. *J. Clin. Investig.* **2017**, *127*, 3770–3783. [CrossRef]
46. Liu, S.; Zhang, C.; Wang, B.; Zhang, H.; Qin, G.; Li, C.; Cao, L.; Gao, Q.; Ping, Y.; Zhang, K.; et al. Regulatory T cells promote glioma cell stemness through TGF-β-NF-κB-IL6-STAT3 signaling. *Cancer Immunol. Immunother.* **2021**, *70*, 2601–2616. [CrossRef]
47. Wang, L.; Ma, R.; Flavell, R.A.; Choi, M.E. Requirement of mitogen-activated protein kinase kinase 3 (MKK3) for activation of p38alpha and p38delta MAPK isoforms by TGF-beta 1 in murine mesangial cells. *J. Biol. Chem.* **2002**, *277*, 47257–47262. [CrossRef] [PubMed]
48. Torrealba, N.; Vera, R.; Fraile, B.; Martínez-Onsurbe, P.; Paniagua, R.; Royuela, M. TGF-β/PI3K/AKT/mTOR/NF-kB pathway. Clinicopathological features in prostate cancer. *Aging Male* **2020**, *23*, 801–811. [CrossRef]
49. Zhuang, L.; Jia, K.; Chen, C.; Li, Z.; Zhao, J.; Hu, J.; Zhang, H.; Fan, Q.; Huang, C.; Xie, H.; et al. DYRK1B-STAT3 Drives Cardiac Hypertrophy and Heart Failure by Impairing Mitochondrial Bioenergetics. *Circulation* **2022**, *145*, 829–846. [CrossRef]
50. Patel, N.J.; Nassal, D.M.; Gratz, D.; Hund, T.J. Emerging therapeutic targets for cardiac arrhythmias: Role of STAT3 in regulating cardiac fibroblast function. *Expert Opin. Ther. Targets* **2021**, *25*, 63–73. [CrossRef]
51. Boengler, K.; Hilfiker-Kleiner, D.; Drexler, H.; Heusch, G.; Schulz, R. The myocardial JAK/STAT pathway: From protection to failure. *Pharmacol. Ther.* **2008**, *120*, 172–185. [CrossRef]
52. Chang, H.; Zhao, F.; Xie, X.; Liao, Y.; Song, Y.; Liu, C.; Wu, Y.; Wang, Y.; Liu, D.; Wang, Y.; et al. PPARα suppresses Th17 cell differentiation through IL-6/STAT3/RORγt pathway in experimental autoimmune myocarditis. *Exp. Cell Res.* **2019**, *375*, 22–30. [CrossRef] [PubMed]
53. Gervois, P.; Kleemann, R.; Pilon, A.; Percevault, F.; Koenig, W.; Staels, B.; Kooistra, T. Global suppression of IL-6-induced acute phase response gene expression after chronic in vivo treatment with the peroxisome proliferator-activated receptor-alpha activator fenofibrate. *J. Biol. Chem.* **2004**, *279*, 16154–16160. [CrossRef] [PubMed]
54. Sekiguchi, K.; Tian, Q.; Ishiyama, M.; Burchfield, J.; Gao, F.; Mann, D.L.; Barger, P.M. Inhibition of PPAR-alpha activity in mice with cardiac-restricted expression of tumor necrosis factor: Potential role of TGF-beta/Smad3. *Am. J. Physiol. Heart Circ. Physiol.* **2007**, *292*, 1443–1451. [CrossRef] [PubMed]

cells

Article

Suppression of Hepatic PPARα in Primary Biliary Cholangitis Is Modulated by miR-155

Monika Adamowicz [1,*], Agnieszka Kempinska-Podhorodecka [1], Joanna Abramczyk [1], Jesus M. Banales [2,3], Piotr Milkiewicz [4,5] and Malgorzata Milkiewicz [1]

1 Department of Medical Biology, Pomeranian Medical University in Szczecin, 70-111 Szczecin, Poland
2 Department of Liver and Gastrointestinal Diseases, Biodonostia Health Research Institute, Donostia University Hospital, University of the Basque Country (UPV/EHU), CIBERehd, Ikerbasque, 20014 San Sebastian, Spain
3 Department of Biochemistry and Genetics, School of Sciences, University of Navarra, 31009 Pamplona, Spain
4 Liver and Internal Medicine Unit, Medical University of Warsaw, 02-097 Warsaw, Poland
5 Translational Medicine Group, Pomeranian Medical University, 70-111 Szczecin, Poland
* Correspondence: monikadamowicz@gmail.com; Tel.: +48-91-466-1867

Abstract: Background: PPARα is a ligand-activated transcription factor that shows protective effects against metabolic disorders, inflammation and apoptosis. Primary biliary cholangitis and primary sclerosing cholangitis result in the intrahepatic accumulation of bile acids that leads to liver dysfunction and damage. Small, non-coding RNAs such as miR-155 and miR-21 are associated with silencing PPARα. Methods: The expression of miR-155, miR-21 and PPARα were evaluated using real-time PCR on liver tissue, as well as on human hepatocytes (HepG2) or cholangiocytes (NHCs) following exposure to lipopolysaccharide (LPS), glycodeoxycholic acid (GCDCA), lithocholic acid (LCA) and/or ursodeoxycholic acid (UDCA). Results: A reduction of PPARα in primary biliary cholangitis (PBC) livers was associated with miR-21 and miR-155 upregulation. Experimental overexpression of either miR-155 or miR-21 inhibited PPARα in hepatocytes, whereas, in cholangiocytes, only miR-21 suppressed PPARα. Both GCDCA and LCA induced the cell type-specific upregulation of miR-155 or miR-21. In HepG2, LPS-induced miR-155 expression was blocked by a cotreatment with UDCA and was associated with PPARα upregulation. In NHC cells, the expression of miR-21 was induced by LPS but did not affect PPARα expression. Conclusions: Hepatic PPARα expression is reduced in PBC livers as a likely result of miR-155 overexpression. UDCA effectively reduced both baseline and LPS-induced miR-155 expression, thus preventing the suppression of PPARα.

Keywords: miRNA; PPARα; liver; primary biliary cholangitis

Citation: Adamowicz, M.; Kempinska-Podhorodecka, A.; Abramczyk, J.; Banales, J.M.; Milkiewicz, P.; Milkiewicz, M. Suppression of Hepatic PPARα in Primary Biliary Cholangitis Is Modulated by miR-155. *Cells* **2022**, *11*, 2880. https://doi.org/10.3390/cells11182880

Academic Editors: Kay-Dietrich Wagner and Nicole Wagner

Received: 21 July 2022
Accepted: 10 September 2022
Published: 15 September 2022

1. Introduction

Primary biliary cholangitis (PBC) is a slow, progressive, chronic liver disease that predominantly affects middle-aged women [1]. While the aetiology of PBC has not been established, it is believed that cholangiocyte secretory failure and/or autoimmunity against intrahepatic bile ducts is linked to the presence of auto-reactive T-lymphocytes and raised plasma concentrations of specific anti-mitochondrial antibodies (AMA) [1]. Disrupted bile acid metabolism in the entero-hepatic circulation, enhanced oxidative stress, and induced inflammation cytokines causes cholestatic liver damage, which ultimately leads to liver fibrosis and cirrhosis. Another chronic cholestatic condition is primary sclerosing cholangitis (PSC), which frequently is associated with inflammatory bowel disease. Ursodeoxycholic acid (UDCA) is the first-line treatment for patients with PBC or PSC and significantly delays the progression of liver disease in the majority of cases [2].

Peroxisome proliferator-activated receptor alpha (PPARα) belongs to the superfamily of nuclear receptors (PPARs) that are ligand-activated transcription factors. PPARα regulates gene expression by binding with its heterodimeric partner, retinoid X receptor, to specific PPAR-response elements. PPARα is primarily expressed in tissues with fatty acid oxidation activity, including the liver, and regulates the expression of multiple genes involved in lipid metabolism and energy homeostasis. It is also involved in protecting against inflammation and cell apoptosis [3]. PPARα has an important role in both the inhibition of excessive inflammatory responses and in the development of innate host defences [4]. PPARα also protects against hyperglycaemia-induced endothelial inflammation and the retinal cell apoptosis pathway via blocking of the nuclear factor-kB pathway [5]. The pivotal role of PPARα in the maintenance of self-tolerance and immune homeostasis is mediated via iTregs induced by PPARα-dependent Foxp3 expression [6]. PPAR agonists (fibric acid derivatives) contribute to a range of actions, including cholesterol and bile acid (BA) homeostasis, and hinder the proinflammatory response. Both bezafibrate, an agonist of all three isoforms of human PPARs, and fenofibrate, a PPARα–selective agonist, lower serum liver biochemical markers in patients with PBC [7]. The beneficial effect of fibrates in PBC is explained by its anti-cholestatic function, as they have the ability to inhibit bile acid uptake and synthesis, as well as reduce the toxicity of bile through the translocation of phosphatidylcholine into bile [8,9]. In PBC patients who do not respond satisfactorily to UDCA treatment, the addition of bezafibrate has led to the reduction of both fibrosis and the inflammatory response [10]. Although PPARα agonists are given consideration in the treatment of various cholestatic liver disorders, knowledge of hepatic expression of PPARα at different stages of the disease is scarce.

MicroRNAs (miRs) are naturally occurring, highly conserved families of short, non-coding RNAs that regulate gene expression either via the inhibition of transcription or by repressing mRNA translation. A number of human diseases, including cancer, metabolic disorders, immune dysfunction and liver diseases, are associated with abnormal miRNA profiles [11]. A single miRNA can target numerous transcripts; therefore, the dysregulated expression of miRs can modify multiple target proteins. Some miRs, such as miR-155 or miR-21, are able to suppress the expression of different PPAR isoforms in distinct tissues [12,13]. MiR-155 was one of the earliest to be identified as a modulator of both the immune response and autoimmune development. Moreover, it appears to be the most relevant miRNA involved in several liver diseases [13–15]. In contrast, miR-21 is most abundantly expressed in hematopoietic cells, and its main role is in resolving inflammation and the suppression of proinflammatory responses [16,17]. Its absence gives rise to vascular inflammation and plaque formation.

In view of the critical role of PPARα signalling in the regulation of the immune response and the physiological relevance of miR-155 or miR-21 in the regulation of the PPARα gene, we evaluated the expression of these factors in liver tissue. Primary normal human cholangiocyte and hepatocyte cell lines were used to investigate: (i) whether PPARα expression is modulated by miR-155 or miR-21, (ii) the effect of toxic bile acids and LPS stimulation on miR-15 and miR-21 expression and (iii) the effect of UDCA treatment on PPARα expression.

2. Materials and Methods

2.1. Liver Tissue

Liver tissue specimens were obtained either during routine percutaneous liver biopsies from patients with early-stage (F0-F2) esPBC ($n = 18$) or were collected from explanted livers of patients with advanced (F4) PBC ($n = 24$) or primary sclerosing cholangitis (PSC) ($n = 18$) who underwent liver transplantation. Control liver samples ($n = 16$) were comprised of large-margin liver resections of colorectal metastases that showed no pathologist-identified microscopic changes indicative of liver disease. The samples were collected in the Hepatology and Internal Medicine Unit of the Medical University of Warsaw. Each patient gave informed consent prior to participating in the study. Table 1 lists the patient demographic details.

Table 1. Demographic and laboratory features of all analysed subjects.

	Control (n = 16)	esPBC (n = 18)	PBC (n = 24)	PSC (n = 18)
Gender (Female/Male)	7/9	18/0	22/2	6/12
Age (years)	50 (25–60)	55 (28–64)	57 (36–69)	33 (20–57)
Bilirubin (mgL/dL, NR: 0.1–1.1)	0.5 (0.2–1.0)	0.6 (0.3–7.8)	4.4 (0.6–21)	2.7 (0.4–32.2)
ALP (IU/L, NR: 40–120)	24 (40–118)	178 (47–456)	400 (119–1373)	387 (114–2181)
AST (IU/L, NR: 5–35)	23 (9–34)	40 (13–182)	104 (51–295)	99 (24–500)

Median and range values (in parentheses). Abbreviations: esPBC, early-stage primary biliary cholangitis; PBC, primary biliary cholangitis; PSC, primary sclerosing cholangitis; ALP, alkaline phosphatase; AST, aspartate aminotransferase and NR, normal range.

2.2. Cell Culture and Treatments

Primary normal human cholangiocytes (NHC), as well as the human hepatocarcinoma cell line (HepG2, American Type Culture Collection), were used for the in vitro studies [18,19]. NHC cells were established, characterised and cultured, as previously described [20–22]. For all analyses, NHCs and HepG2s were seeded in 6-well plates (3 × 105 cells/well) and allowed to attach overnight. Cells were transfected with commercially available miRNA Mimics for miR-155 and miR-21 (mirVana® miRNA mimic hsa-miR-155; ID: MC28440; hsa-miR-21 ID: 477975_mir Ambion, Austin, TX, USA). Transient transfection was performed using Lipofectamine RNAiMAX (Invitrogen, Carlsbad, CA, USA) according to the manufacturer's protocol. Vehicle-treated cells (Lipofectamine) were used as the control group. Forty-eight hours after transfection, the HepG2 and NHC cells were lysed and frozen as pellets for further analysis. For all experiments, UDCA (*U5127-1G*, Sigma-Aldrich, St. Louis, MO, USA) was dissolved as a 100 mM stock solution in EtOH. HepG2 and NHC cells were incubated with UDCA alone (50–200 µM) or two hours prior to 24-h stimulation with a lipopolysaccharide from *Escherichia coli* 0111:B4 (LPS, 5 µg/mL L4391-1MG SIGMA). To investigate the effect of bile acids, HepG2 and NHC cells were exposed to 500 µM of GCDCA and 150 µM GCDCA (ID: 24895023, Sigma-Aldrich, St. Louis, MO, USA), respectively. The effect of lithocholic acid (LCA) at a dose of 100 µM (LCA, Sigma-Aldrich, St. Louis, MO, USA) was tested independently in both cell types for 24 h. All experiments were repeated at least three times, and the untreated cells were used as a negative control. Cells were stored at −80 °C until molecular analyses were performed.

2.3. MicroRNA and mRNA Extraction and Quantification

Total RNA was extracted using the RNeasy Kit (Qiagen, Hilden, Germany) and subjected to reverse transcription using either the TaqMan Advanced miRNA cDNA Synthesis Kit (Applied Biosystems, Thermo Fisher Scientific, Waltham, MA, USA) for the quantitative analysis of microRNA or SuperScript IV RT (Invitrogen, Carlsbad, CA, USA) for further gene expression analysis according to the manufacturer's protocol. The expression of miR-155, miR-21 and the reference miRNA miR-16-5p were measured using TaqMan® Advanced miRNA Assays (Assays ID 002623_mir, 477975_mir and 477860_mir, respectively) and TaqMan® Fast Advanced Master Mix (Applied Biosystems, Waltham, MA, USA). The quantitative analyses of the change in expression of specific target genes were measured using the 7500 Fast Real-Time PCR System (Applied Biosystems, Foster City, CA, USA) using human TaqMan Gene Expression Assays for PPARα (Hs00947539_m1), PDCD4 (Hs00377253_m1), PTEN (Hs02621230_s1), IL-6 (Hs001741131-m1), IL-1B (Hs01555410-m1) and 18S RNA (Hs99999901_s1). Relative amounts of transcripts in comparison to controls were determined using the $2^{-\Delta\Delta Ct}$ formula [23].

2.4. Immunoblot Analysis

Proteins were extracted from liver tissue samples by homogenisation with lysis buffer (RIPA buffer) supplemented with a protease inhibitor cocktail (Roche, Basel, Switzerland) and phosphatase inhibitors (PhosSTOP EASYpack; Roche, Basel, Switzerland). Proteins were electrophoresed on SDS-polyacrylamide gels and then blotted onto a polyvinylidene difluoride PVDF membrane (Thermo Scientific, Rockford, IL, USA) under semi-dry transfer conditions (Thermo Scientific, Rockford, IL, USA). After blocking with 5% non-fat dried milk, membranes were probed overnight at 4 °C using the primary antibodies: anti-PPARα (H-2): SC-398394, Santa Cruz Biotechnology, Inc.) followed by incubation with peroxidase-conjugated secondary anti-mouse (1:1000) antibodies (GE Healthcare, code: NA9310). Protein loading was normalised to anti-GAPDH (1:5000, sc-25,778 + HRP; Santa Cruz). Bands were visualised through a chemiluminescence detection system (Chemiluminescent HRP Substrate, Millipore, MA, USA) and quantified using the MicroChemi 2.0 System and GelQuant software (Maale HaHamisha, Jerusalem, Israel).

2.5. Immunohistochemistry

Immunohistochemical analyses of liver sections were performed using the ImmPRESS Universal Reagent kit (Vector Laboratories, Burlingame, CA, USA, #SP-2001). The deparaffinisation of the tissue sections were followed by antigen unmasking with antigen retrieval buffer (citrate-based solution, pH 6.0; 95 °C for 20 min). After blocking with ready-to-use normal horse serum (2.5%), samples were incubated with primary antibodies against PPARα (sc-398394, Santa Cruz Biotechnology, Inc. Oregon, USA) for 90 min in room temperature. After washing, samples were left for 30 min with ImmPRESS reagent and then dyed with a substrate/chromogen mixture (ImmPACT™ DAB). After washing, samples were counterstained with haematoxylin and mounted (Aqueous Permanent Medium, Dako, Denmark). A Zeiss Axio Imager Z2 optical microscope equipped with the Zen Pro 2011 acquisition program was used to acquire the images.

2.6. Statistical Analysis

StatView software version 5.0 (SAS Institute, Cary, NC, USA) was used for the statistical analyses. Statistical differences between groups were analysed using the Student's t-test and multiple groups' comparisons were performed with one-way analysis of variance (ANOVA). All graphs were generated using GraphPad Prism version 7 software (GraphPad Software, San Diego, CA, USA). Data are expressed as the mean ± SEM. Results were considered statistically significant when p-values were less than 0.05.

3. Results

First, human hepatic samples obtained from patients with PBC or PSC during liver transplantation were examined for PPARα expression. The analysis showed a marked reduction both at the mRNA (Figure 1a) and protein (Figure 1b) levels in PBC livers. The 50% reduction in mRNA expression was significant in comparison to the controls ($p = 0.01$), early-stage esPBC ($p = 0.01$) and in comparison to another cholestatic liver disease, PSC ($p = 0.001$; Figure 1a). A histological evaluation of PPARα in the control (Figure 1c) and PBS livers (Figure 1d) demonstrated the expression of this protein both in hepatocytes and cholangiocytes within bile ducts. However, in contrast to PBC livers, in the control tissue, a strong nuclear localisation of PPARα within hepatocytes was observed.

Knowing that both miR-155 and miR-21 target PPARα, we estimated the levels of these miRs in the liver tissue. The observed phenomenon of the reduction of PPARα in cirrhotic PBC livers was associated with a substantial induction of miR-155 (3.5-fold increase vs. controls, $p = 0.004$; Figure 2a) and miR-21 (50-fold increase vs. controls, $p = 0.0001$, and $p = 0.01$ vs. PSC; Figure 2b). There was a negative correlation between PPARα and miR-21 ($r = -0.45$, $p = 0.01$), and the expression of miR-21 positively correlated with the inflammatory cytokines, both IL-6 ($r = 0.56$, $p = 0.003$) and IL-1b ($r = 0.46$, $p = 0.027$).

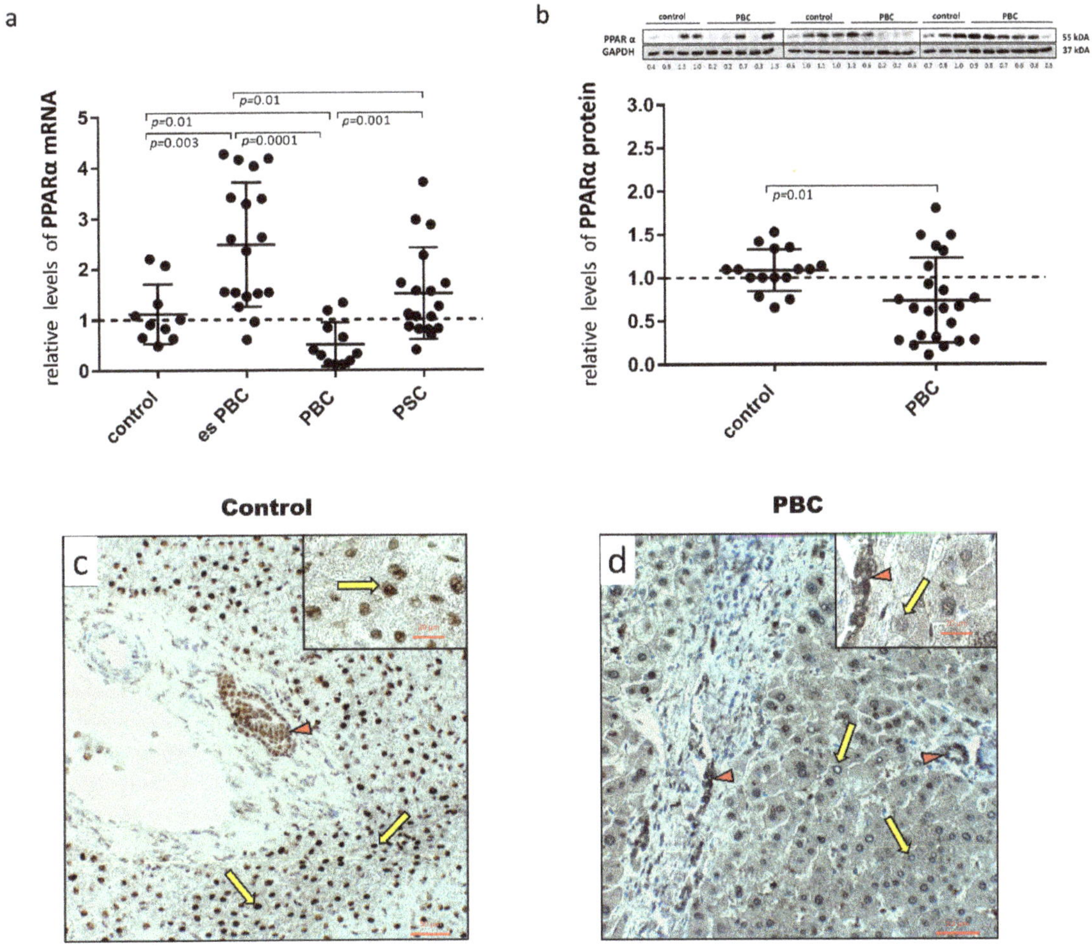

Figure 1. Presence of PPARα in liver tissue samples obtained from the controls, the early-stages of primary biliary cholangitis (esPBC), cirrhotic PBC and primary sclerosing cholangitis (PSC) patients. PPARα mRNA expression was suppressed in cirrhotic PBC (**a**), and Western blot analysis confirmed lower levels of PPARα at the protein level (**b**). Levels of gene expression were normalised to the endogenous reference, 18S RNA and the levels of each protein were normalised to GAPDH. Dots illustrate each patient, and the data are presented as mean plus interquartile range (IQR). Statistical analysis was performed using ANOVA or a Student's *t*-test. Immunohistochemical staining clearly showed a dominant nuclear localisation of the PPARα protein in the control tissue (**c**) in contrast to liver tissue from patients with PBC (**d**). Both hepatocytes (yellow arrows) and cholangiocytes (red arrow heads) were positive for PPARα. Original magnification 200× or 400× (inserts).

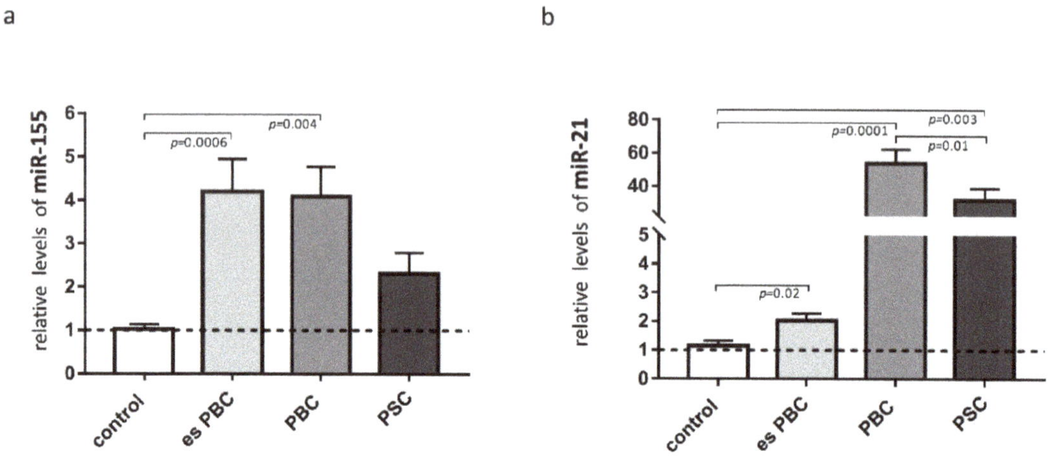

Figure 2. Expression of miR-155 and miR-21 in liver tissue. In patients with primary biliary cholangitis (PBC), both miR-155 (**a**) and miR-21 (**b**) expression were increased in comparison to the healthy controls. In livers of PSC patients, only miR-21 was substantially induced. MiR-16 served as the reference for loading. Bars indicate the mean ± SEM. Statistical analysis was performed using ANOVA or a Student's *t*-test.

To further investigate the specific role of these miRNAs in the aforementioned liver diseases, we transfected HepG2 and NHC cells with either miR-155 or miR-21 mimics. The experimental overexpression of miR-155 reduced the PPARα mRNA levels in HepG2 cells (0.6 ± 0.01 vs. 1.0 ± in the controls $p = 0.02$; Figure 3a) but not in NHC cells (Figure 3a). Similarly, the overexpression of miR-21 inhibited PPARα in HepG2 (0.8 ± 0.05. vs. 1.0 ± in the controls $p = 0.01$ Figure 3b) but not in NHC cells (Figure 3b).

Inflammation contributes to the pathogenesis of PBC; therefore, to investigate the effect of activated inflammatory response on the miRs expression, we exposed HEPG2 and NHC cells to lipopolysaccharide (LPS), which activates Toll-like receptor 4 (TLR4). The incubation of HepG2 cells with LPS led to the induction of miR-155 expression (40-fold, $p = 0.001$ vs. controls, Figure 4a), which was blocked by the UDCA cotreatment ($p = 0.04$ vs. LPS). Moreover, UDCA alone suppressed the baseline expression of miR-155 ($p = 0.008$ vs. controls) in HepG2. In contrast, in NHC cells, miR-155 expression did not change after LPS exposure, but, similarly to HepG2, UDCA substantially reduced both the baseline expression of miR-155 (20% reduction, $p = 0.03$ vs. nontreated control cells) and after LPS exposure (80% reduction, $p = 0.001$ vs. nontreated cells; Figure 4a). The expression of miR-21 was induced by LPS only in NHC cells (1.4-fold increase, $p = 0.0001$ vs. nontreated cells) and was further enhanced by the UDCA cotreatment (4-fold increase, $p = 0.04$ vs. nontreated cells; Figure 4b). *PPARα* gene expression was enhanced by UDCA in both LPS-stimulated ($p = 0.0001$ vs. nontreated cells; Figure 4c), and non-LPS-stimulated HepG2 cells ($p = 0.002$ vs. nontreated cells; Figure 4c).

The development of cholestatic liver diseases such as PBC is negatively impacted not only by inflammation but also by chronic exposure to toxic bile acid. There have been few studies on the effect of bile acids on miR profiles. It was reported that, in primary human hepatocytes, chenodeoxycholic acid affected the expression of different miRs; however, neither miR-155 or miR-21 were evaluated in the study [24]. Our study showed that both glycochenodeoxycholic acid (GCDCA) and lithocholic acid (LCA) induced miR-21 expression in HepG2 cells (1.7 ± 0.15, $p = 0.004$ and 1.3 ± 0.06, $p = 0.01$, respectively, Figure 5a,b), whereas, in NHC cells, these bile acids upregulated miR-155 (2.13 ± 0.39, $p = 0.05$ and 2.253 ± 0.66, $p = 0.05$, respectively, Figure 5a,b). Moreover, in NHC cells, the expression of miR-21 was stimulated by LCA exposure (Figure 5b). Interestingly, one of these bile acids species, namely GCDCA, decreased the PPARα level but only in the HepG2 cell line (0.84 ± 0.03, p = 0.002, Figure 5a).

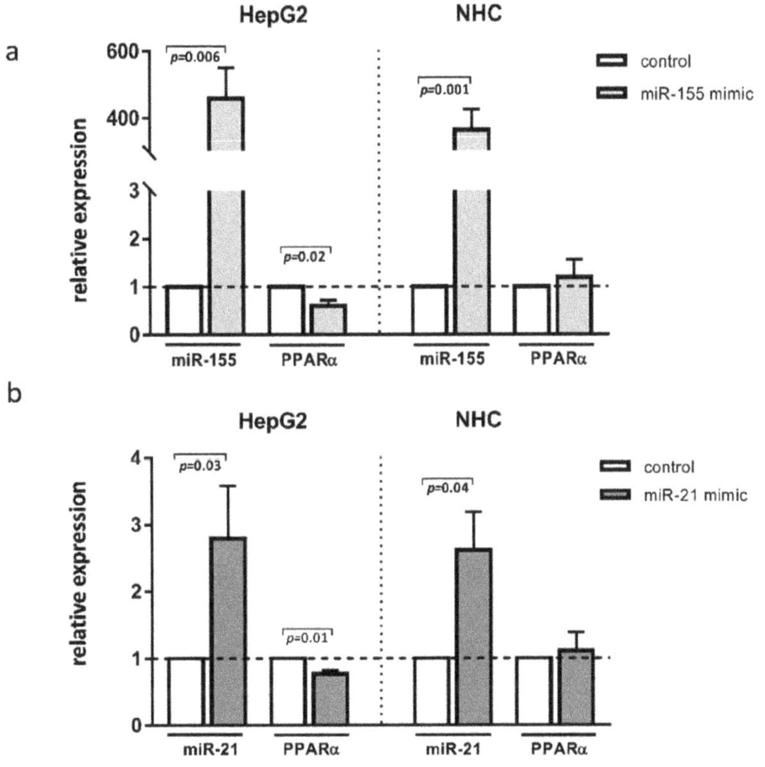

Figure 3. PPARα, expression after miR-155 or miR-21 activation. Human hepatocarcinoma (HepG2) and normal human cholangiocyte (NHC) cells were transfected with miR-155 mimic (**a**) or miR-21 mimic (**b**). Increased levels of these miRNAs were confirmed in both cell lines. Overexpression of both miR-155 and miR-21 led to the strong downregulation of PPARα in HepG2 but not in NHC cells. Each experiment was repeated at least three times. Levels of gene expression were normalised to the reference miR-16 for miRNA or 18S RNA for other genes. Bars indicate the mean ± SEM. Student's t-test was used for the quantitative data analysis.

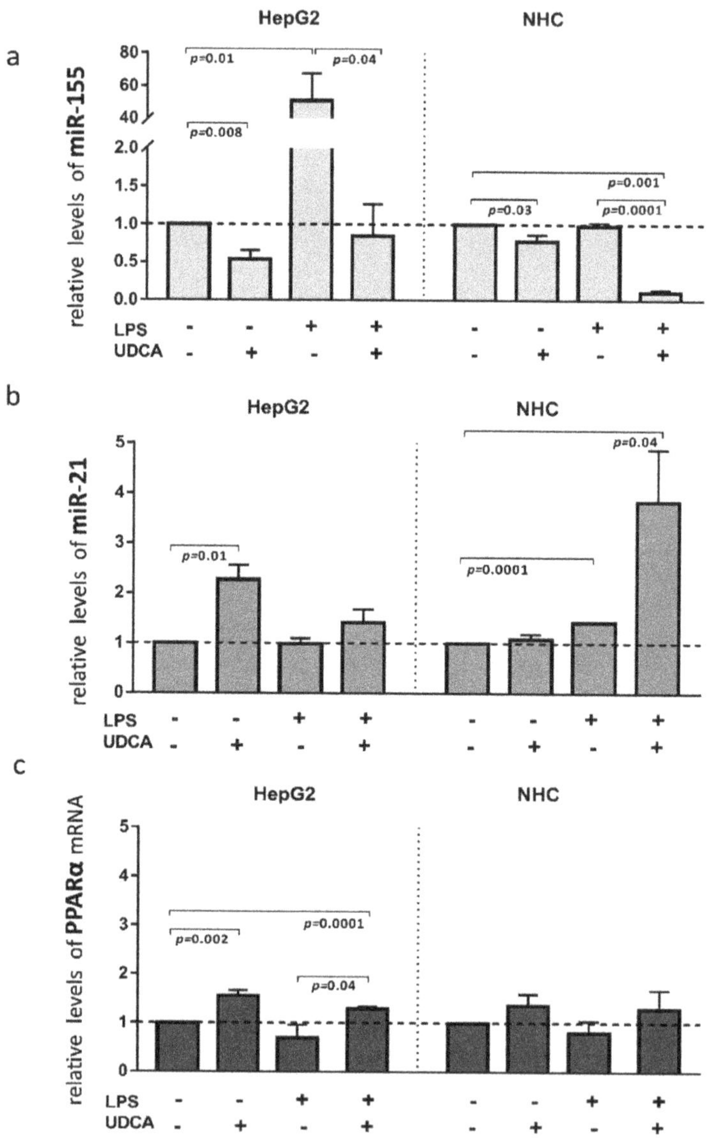

Figure 4. The effect of lipopolysaccharide (LPS) and/or ursodeoxycholic acid (UDCA) exposure in human hepatocarcinoma (HepG2) and normal human cholangiocyte (NHC) cell lines. LPS stimuli enhanced miR-155 (**a**) in HepG2 cells and miR-21 (**b**) in NHC cells. MiR-155 was enhanced after the incubation of HepG2 cells with LPS, whereas, in NHC, the expression of miR-21 was induced by LPS stimuli. UDCA reduced both the baseline and LPS-induced miR-155 expression in HepG2 cells, which was accompanied by the upregulation of PPARα. In NHC cells, UDCA enhanced the expression of miR-21 but did not affect PPARα expression (**c**). Bars indicate the mean ± SEM. The statistical analysis was performed using ANOVA.

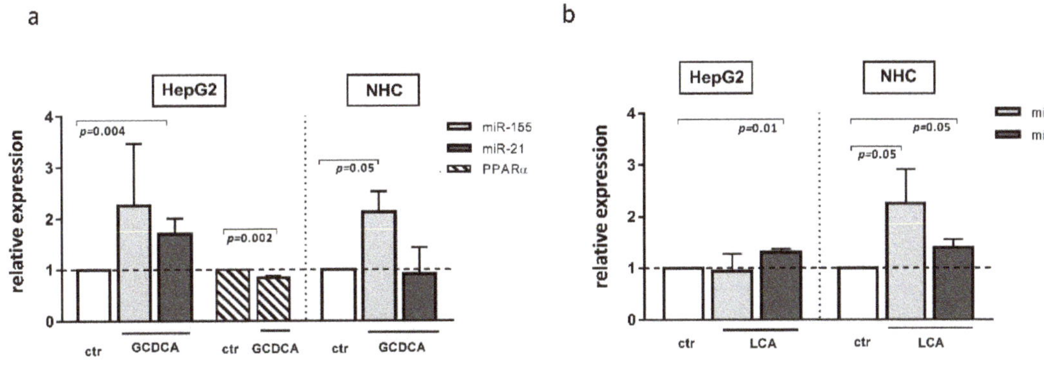

Figure 5. The effect of glycochenodeoxycholic acid (GCDCA) and lithocholic acid (LCA) in human hepatocarcinoma (HepG2) and normal human cholangiocytes (NHC) cells. In HepG2 cells, both GCDCA (**a**) and LCA (**b**) induced miR-21. In NHC cells, both bile acids induced miR-155, whereas miR-21 was enhanced only after LCA stimulation. The PPARα level was reduced in HepG2 after GCDCA treatment (**a**). Each experiment was repeated at least three times. MiR-16 was used as an endogenous reference for miRNA or 18S RNA for other genes. Bars indicate the mean ± SEM. A Student's *t*-test were used for a quantitative data analysis.

4. Discussion

This study provides new insight into the regulation of PPARα in cholestatic livers. In cirrhotic PBC livers, a substantial reduction of PPARα expression was associated with the upregulation of both miR-21 and miR-155. A cell-based analysis demonstrated that the experimental overexpression of either miR-155 or miR-21 inhibited *PPARα* mRNA in hepatocytes, whereas, in cholangiocytes, only the overexpression of miR-21 led to *PPARα* downregulation. The factors responsible for the induction of these miRNAs appeared to be cell type-specific when the HepG2 and NHC cell lines were compared. Moreover, a new biological function of UDCA as a modulator of miR-21 and miR-155 in those cells was found.

We observed a substantial reduction of both the mRNA and protein levels of PPARα in cirrhotic PBC livers. In contrast, in another cholestatic condition such as PSC, the hepatic expression of this nuclear receptor was comparable to the control values. Moreover, this study showed that the inhibition of PPAR was present only in advanced phases and not in the early stages of the disease. This is in line with reports on altered hepatic PPARα expression in liver diseases. In patients with Wilson's disease, PPARα expression was found to be altered in proportion to the progression of liver injury, i.e., it was enhanced in patients with mild liver impairment but reduced in patients with moderate or intense liver damage [25]. Similarly, in subjects with non-alcoholic steatohepatitis (NASH), hepatic PPARα expression declined with the development of NASH features and was negatively correlated with the severity of steatosis, hepatocyte ballooning or fibrosis [26]. In the context of cholestatic liver diseases, our study represents a novel report, as there is a lack of information on the hepatic expression of this nuclear expression under conditions of sustained cholestasis.

PPARα plays a crucial role in bile acid homeostasis via the regulation of bile acid biosynthesis, transport and secretion [27]. Furthermore, it has been demonstrated that fenofibrate-activated PPARα signalling eliminates oxidative stress and attenuates cholestatic liver injury [28]. In patients with PBC, therapies based on PPARα agonists are well-tolerated and allied with a significant decrease in the alkaline phosphatase (ALP) levels and anti-inflammatory markers [29,30]. Moreover, a prospective, long-term, longitudinal study showed a potentially beneficial effect of bezafibrate in combination with UDCA in patients who had an inadequate response to UDCA [10]. A key factor influencing the effectiveness of fibrate-based therapy is the adequate expression of PPARα; however, our study showed a substantial reduction of PPARα expression in cirrhotic PBC livers. Therefore, understanding the molecular mechanism responsible for the hepatic reduction of PPARα is of particular importance.

This study focused on two miRNAs that are known to modulate the *PPARα* gene, i.e., miR-155 and miR-21. We showed that, in cirrhotic PBC livers, the expressions of both miRNAs were substantially upregulated, which is in contrast to cirrhotic PSC, where only miR-21 was increased. MiR-155 modifies proinflammatory responses that affect not only immune cells but also hepatic parenchymal cells, including hepatocytes. MiR-155 is known to exert pleiotropic functions depending on the aetiology and disease context. However, miR-155 expression in cholestatic diseases has not been described to date, although there are reports from other hepatobiliary diseases. The serum level of miR-155 is increased in patients with alcoholic cirrhosis [13] and in non-alcoholic fatty liver disease (NAFLD) [31]. In patients with biliary atresia or PBC, an inverse correlation between miR-155 and the suppressor of cytokine signalling 1 (SOCS1) has been reported [32,33]. In addition, the antiviral treatment of hepatitis C (HCV) patients normalised the level of miR-155 in peripheral monocytes in contrast to non-responders [14].

Since the specific suppression of PPARα in PBC livers was accompanied by a substantial upregulation of miR-155 and miR-21, we conducted functional studies in hepatocyte and cholangiocyte cells. Of note, the experimental overexpression of either miR-155 or miR-21 suppressed the PPARα levels in HepG2 cells but not in NHC cells. The downregulation of PPARα by miR-21 has been described in numerous pathologic processes. For example, miR-21 directly inhibits PPARα translation [12,34], which promotes the expression of vascular cell adhesion molecule-1 (VCAM-1) and favours the adhesion of inflammatory cells [35] or leads to retinal microvascular dysfunction [36]. In the context of liver pathology, PPARα was demonstrated to be a direct target of miR-155 or miR-21 in mouse biliary, hepatic and inflammatory cells in a mouse model of alcohol-induced steatohepatitis, NASH and in the development of hepatocellular carcinoma (HCC) [13,34,35,37].

Further, to investigate the factors involved in the hepatic upregulation of these miRNAs, we induced inflammatory responses by LPS or exposed the cells to toxic bile acids. The factors involved in the hepatic upregulation of the miRNAs were cell-specific. Consistent with previous studies [38,39], we found that miR-155 expression was induced by the inflammatory response activated by LPS exposure but only in HepG2 cells. In NHC cells, the induction of this miRNA was only observed in response to toxic bile acids (both GCDA and LCA). Comparatively little is known about the effects of bile acids on the cellular microRNAome. In accordance with our observations, chenodeoxycholic acid did not affect miR-155 expression in primary human hepatocytes [24]; however, the acidic, bile-induced upregulation of miR-155 was noted in human hypopharyngeal primary cells [40]. Here, we present the first report that prolonged exposure to bile acids may induce miR-155 expression in normal human cholangiocytes.

Interestingly, LPS-induced miR-155 expression in HepG2 was overridden by UDCA treatment. Moreover, UDCA effectively reduced the baseline expression of miR-155 in both HepG2 and NHC cells. Given the relevance of miR-155 dysregulation in the proper homeostasis of the immune response and macrophage polarisation [41], this is a noteworthy observation that confirms the positive role of UDCA in modulating miR-155 expression. Previously, UDCA was shown to effectively decrease both miR-34 in primary rat hepatocytes and miR-122 in human serum [42,43].

Considerable evidence has highlighted miR-21 as one of the key switches that controls the magnitude of inflammation [16,44]. However, its presence is not entirely attributed to a proinflammatory or an immunosuppressive condition. Recently, miR-21 has been proposed as a negative modulator of Toll-Like Receptor 4 (TLR4) signalling by targeting PTEN and PDCD4, which resulted in the elevated production of IL-10 [44,45]. Moreover, miR-21 influences the fine balance between Th1 and Th2 responses, and elevated miR-21 expression leads to a reduction of IL-6 secretion and the induction of IL-10 production in macrophages [46–48]. Similarly, a negative regulation of the TNF-a levels by miR-21 has been reported [16,44,48]. Thus, miR-21 dysregulation that has been observed in a number of inflammatory diseases promotes an anti-inflammatory, immunosuppressive environment. The absence of miR-21 in hematopoietic cells also enhances vascular inflammation and atherosclerosis [17]. Interestingly, we found a substantially induced expression of miR-21 in the liver tissue of both PBC and PSC patients. miR-21 was previously implicated in the development of fibrosis; however, in this study, a small increase was observed in the early stages of PBC (F0-F2), followed by a substantial increase in cirrhotic PBC. There was no correlation between the levels of miR-21 and the stages of fibrosis, which is in agreement with previous reports [49,50]. Even though miR-21 ablation has been shown to protect from fibrosis and acute oxidative stress in the livers of mice with bile duct ligation, it was an acute model of cholestasis, which did not completely mimic the sustained cholestasis that occurs in PBC [51]. Interestingly, the authors noticed that miR-21−/− mice displayed an increased hepatic TLR4 expression, which was attributed to the anti-inflammatory function of miR-21 [51].

Our current study demonstrated that LPS, and two distinct bile acid species (LCA and GCDA) induced miR-21 expression; however, the responses were cell-dependent. Consequently, the upregulation of miR-21 was observed in NHC cells after LPS or LCA incubation. This was in contrast to HepG2 cells, where GCDA and LCA exposure triggered miR-21 induction. Previously, LPS was reported to induce miR-21 in a number of cell lines, including human biliary epithelial cells and hepatic stellate cells [52,53]. In contrast to our study, an inhibition of miR-21 by another bile acid, cytotoxic deoxycholic acid (DCA) in a dose-dependent manner was reported in primary rat hepatocytes [54]. Interestingly, we noticed a further upregulation of miR-21 by UDCA in NHC cells incubated with LPS, which is in line with the observation that UDCA is a strong inducer of miR-21 in regenerating rat livers and cultured HepG2 cells [54,55]. The induction of miR-21 in murine macrophages via a treatment with LPS was associated with silencing its target genes, PTEN and PDCD4, which are powerful inhibitors of the AP-1 transcription factor [44]. In this study, the forced overexpression of miR-21 in HepG2 cells decreased PDCD4 and PTEN mRNA expressions (Figure S1). Bile acids are strong modulators of AP-1 activity, and the increase of miR-21 expression stimulated by UDCA was shown to inhibit the activation of AP-1 and thus favour a pro-proliferative environment [56].

Cumulatively, hepatic PPARα expression is substantially reduced in PBC livers, potentially as a result of enhanced miR-155 expression. Furthermore, the increased miR-21 expression in PBC and PSC livers may be implicated in resolving inflammation. UDCA effectively reduced both baseline and lipopolysaccharide-induced miR-155 expression, which prevented the suppression of PPAR.

Supplementary Materials: The following supporting information can be downloaded at: https://www.mdpi.com/article/10.3390/cells11182880/s1, Figure S1: Programmed cell death 4 (PDCD4) and phosphatase and tensin homologue (PTEN) expression after miR-21 activation.

Author Contributions: Conceptualisation, M.M.; methodology, M.A., A.K.-P., J.A. and M.M.; validation, M.M.; formal analysis, M.A. and J.A.; investigation, M.M., A.K.-P. and P.M.; resources, M.M. and P.M.; data curation, M.A. and J.A.; writing—original draft preparation, M.M.; writing—review and editing, all authors; visualisation, M.A. and J.A.; supervision M.M., P.M. and J.M.B.; project administration, M.M. and funding acquisition, M.M. All authors have read and agreed to the published version of the manuscript.

Funding: This work was supported by Program of Polish Minister of Science and Higher Education "Regional Initiative of Excellence" in 2019–2022, project no, 002/RID/2018/19.

Institutional Review Board Statement: The study was conducted in accordance with the Declaration of Helsinki and approved by the Ethics Committee of Pomeranian Medical University (BN-001/43/06).

Informed Consent Statement: Not applicable.

Data Availability Statement: Not applicable.

Conflicts of Interest: The authors declare no conflict of interest.

References

1. Lleo, A.; Wang, G.Q.; Gershwin, M.E.; Hirschfield, G. Primary biliary cholangitis. *Lancet* **2020**, *396*, 1915–1926. [CrossRef]
2. Corpechot, C.; Abenavoli, L.; Rabahi, N.; Chretien, Y.; Andreani, T.; Johanet, C.; Chazouilleres, O.; Poupon, R. Biochemical response to ursodeoxycholic acid and long-term prognosis in primary biliary cirrhosis. *Hepatology* **2008**, *48*, 871–877. [CrossRef] [PubMed]
3. Ghonem, N.S.; Assis, D.N.; Boyer, J.L. Fibrates and cholestasis. *Hepatology* **2015**, *62*, 635–643. [CrossRef]
4. Christofides, A.; Konstantinidou, E.; Jani, C.; Boussiotis, V.A. The role of peroxisome proliferator-activated receptors (PPAR) in immune responses. *Metabolism* **2021**, *114*, 154338. [CrossRef]
5. Hu, Y.; Chen, Y.; Ding, L.; He, X.; Takahashi, Y.; Gao, Y.; Shen, W.; Cheng, R.; Chen, Q.; Qi, X.; et al. Pathogenic role of diabetes-induced PPAR-+- down-regulation in microvascular dysfunction. *Proc. Natl. Acad. Sci. USA* **2013**, *110*, 15401–15406. [CrossRef]
6. Lei, J.; Hasegawa, H.; Matsumoto, T.; Yasukawa, M. Peroxisome proliferator-activated receptor alpha and gamma agonists together with TGF-beta convert human CD4+. *J. Immunol.* **2010**, *185*, 7186–7198. [CrossRef]
7. Corpechot, C.; Chazouilleres, O.; Rousseau, A. Bezafibrate in primary biliary cholangitis. *N. Engl. J. Med.* **2018**, *379*, 985.
8. Grigorian, A.Y.; Mardini, H.E.; Corpechot, C.; Poupon, R.; Levy, C. Fenofibrate is effective adjunctive therapy in the treatment of primary biliary cirrhosis: A meta-analysis. *Clin. Res. Hepatol. Gastroenterol.* **2015**, *39*, 296–306. [CrossRef]
9. Gallucci, G.M.; Trottier, J.; Hemme, C.; Assis, D.N.; Boyer, J.L.; Barbier, O.; Ghonem, N.S. Adjunct fenofibrate up-regulates bile acid glucuronidation and improves treatment response for patients with cholestasis. *Hepatol. Commun.* **2021**, *5*, 2035–2051. [CrossRef]
10. Sorda, J.A.; Gonzalez, B.E.; Barreyro, F.J.; Avagnina, A.; Carballo, P.; Paes de Lima, A.; Daruich, J. Bezafibrate therapy in primary biliary cholangitis refractory to ursodeoxycholic acid: A longitudinal study of paired liver biopsies at 5 years of follow up. *Aliment. Pharmacol. Ther.* **2021**, *54*, 1202–1212. [CrossRef]
11. Szabo, G.; Bala, S. MicroRNAs in liver disease. *Nat. Rev. Gastroenterol. Hepatol.* **2013**, *10*, 542–552. [CrossRef] [PubMed]
12. Kida, K.; Nakajima, M.; Mohri, T.; Oda, Y.; Takagi, S.; Fukami, T.; Yokoi, T. PPARalpha is regulated by miR-21 and miR-27b in human liver. *Pharm. Res.* **2011**, *28*, 2467–2476. [CrossRef] [PubMed]
13. Bala, S.; Csak, T.; Saha, B.; Zatsiorsky, J.; Kodys, K.; Catalano, D.; Satishchandran, A.; Szabo, G. The pro-inflammatory effects of miR-155 promote liver fibrosis and alcohol-induced steatohepatitis. *J. Hepatol.* **2016**, *64*, 1378–1387. [CrossRef] [PubMed]
14. Bala, S.; Tilahun, Y.; Taha, O.; Alao, H.; Kodys, K.; Catalano, D.; Szabo, G. Increased microRNA-155 expression in the serum and peripheral monocytes in chronic HCV infection. *J. Transl. Med.* **2012**, *10*, 151. [CrossRef] [PubMed]
15. Hartmann, P.; Tacke, F. Tiny RNA with great effects: miR-155 in alcoholic liver disease. *J. Hepatol.* **2016**, *64*, 1214–1216. [CrossRef] [PubMed]
16. Sheedy, F.J. Turning 21: Induction of miR-21 as a key switch in the inflammatory response. *Front. Immunol.* **2015**, *6*, 19. [CrossRef]
17. Canfran-Duque, A.; Rotllan, N.; Zhang, X.; Fernandez-Fuertes, M.; Ramirez-Hidalgo, C.; Araldi, E.; Daimiel, L.; Busto, R.; Fernandez-Hernando, C.; Suarez, Y. Macrophage deficiency of miR-21 promotes apoptosis, plaque necrosis, and vascular inflammation during atherogenesis. *EMBO Mol. Med.* **2017**, *9*, 1244–1262. [CrossRef]
18. Weerachayaphorn, J.; Amaya, M.J.; Spirli, C.; Chansela, P.; Mitchell-Richards, K.A.; Ananthanarayanan, M.; Nathanson, M.H. Nuclear factor, erythroid 2-like 2 regulates expression of type 3 inositol 1,4,5-trisphosphate receptor and calcium signaling in cholangiocytes. *Gastroenterology* **2015**, *149*, 211–222. [CrossRef]
19. Kilanczyk, E.; Ruminkiewicz, D.; Banales, J.M.; Milkiewicz, P.; Milkiewicz, M. DHEA protects human cholangiocytes and hepatocytes against apoptosis and oxidative stress. *Cells* **2022**, *11*, 1038. [CrossRef]
20. Urribarri, A.D.; Munoz-Garrido, P.; Perugorria, M.J.; Erice, O.; Merino-Azpitarte, M.; Arbelaiz, A.; Lozano, E.; Hijona, E.; Jimenez-Aguero, R.; Fernandez-Barrena, M.G.; et al. Inhibition of metalloprotease hyperactivity in cystic cholangiocytes halts the development of polycystic liver diseases. *Gut* **2014**, *63*, 1658–1667. [CrossRef]
21. Banales, J.M.; Sáez, E.; Uriz, M.; Sarvide, S.; Urribarri, A.D.; Splinter, P.; Bogert, P.S.T.; Bujanda, L.; Prieto, J.; Medina, J.F.; et al. Upregulation of mir-506 leads to decreased AE2 expression in biliary epithelium of patients with primary biliary cirrhosis. *Hepatology* **2012**, *56*, 687–697. [CrossRef]
22. De Urturi, D.S.; Buque, X.; Porteiro, B.; Folgueira, C.; Mora, A.; Delgado, T.C.; Prieto-Fernandez, E.; Olaizola, P.; Gomez-Santos, B.; Apodaka-Biguri, M.; et al. Methionine adenosyltransferase 1a antisense oligonucleotides activate the liver-brown adipose tissue axis preventing obesity and associated hepatosteatosis. *Nat. Commun.* **2022**, *13*, 1096. [CrossRef]
23. Applied Biosystems. Guide to performing relative quantitation of gene expression using real-time quantitative PCR. *Rev. Biol.* **2008**, *6*, 4371095.
24. Krattinger, R.; Bostrom, A.; Lee, S.M.L.; Thasler, W.E.; Schioth, H.B.; Kullak-Ublick, G.A.; Mwinyi, J. Chenodeoxycholic acid significantly impacts the expression of miRNAs and genes involved in lipid, bile acid and drug metabolism in human hepatocytes. *Life Sci.* **2016**, *156*, 47–56. [CrossRef]

25. Nagasaka, H.; Miida, T.; Inui, A.; Inoue, I.; Tsukahara, H.; Komatsu, H.; Hiejima, E.; Fujisawa, T.; Yorifuji, T.; Hiranao, K.; et al. Fatty liver and anti-oxidant enzyme activities along with peroxisome proliferator-activated receptors gamma and alpha expressions in the liver of Wilson's disease. *Mol Genet. Metab.* **2012**, *107*, 542–547. [CrossRef]
26. Francque, S.; Verrijken, A.; Caron, S.; Prawitt, J.; Paumelle, R.; Derudas, B.; Lefebvre, P.; Taskinen, M.R.; Van, H.W.; Mertens, I.; et al. PPARalpha gene expression correlates with severity and histological treatment response in patients with non-alcoholic steatohepatitis. *J. Hepatol.* **2015**, *63*, 164–173. [CrossRef]
27. Li, F.; Patterson, A.D.; Krausz, K.W.; Tanaka, N.; Gonzalez, F.J. Metabolomics reveals an essential role for peroxisome proliferator-activated receptor alpha in bile acid homeostasis. *J. Lipid Res.* **2012**, *53*, 1625–1635. [CrossRef]
28. Zhao, Q.; Yang, R.; Wang, J.; Hu, D.D.; Li, F. PPARalpha activation protects against cholestatic liver injury. *Sci. Rep.* **2017**, *7*, 9967. [CrossRef]
29. Carrion, A.F.; Lindor, K.D.; Levy, C. Safety of fibrates in cholestatic liver diseases. *Liver Int.* **2021**, *41*, 1335–1343. [CrossRef]
30. Schattenberg, J.M.; Pares, A.; Kowdley, K.V.; Heneghan, M.A.; Caldwell, S.; Pratt, D.; Bonder, A.; Hirschfield, G.M.; Levy, C.; Vierling, J.; et al. A randomized placebo-controlled trial of elafibranor in patients with primary biliary cholangitis and incomplete response to UDCA. *J. Hepatol.* **2021**, *74*, 1344–1354. [CrossRef]
31. Longchamps, R.J.; Abey, S.K.; Martino, A.C.; Henderson, W.A. Letter: Gender-associated cell-free microRNA profiles in non-alcoholic fatty liver disease. *Aliment. Pharmacol. Ther.* **2014**, *39*, 997–998. [CrossRef]
32. Kempinska-Podhorodecka, A.; Milkiewicz, M.; Wasik, U.; Ligocka, J.; Zawadzki, M.; Krawczyk, M.; Milkiewicz, P. Decreased expression of vitamin D receptor affects an immune response in primary biliary cholangitis via the VDR-miRNA155-SOCS1 pathway. *Int. J. Mol. Sci.* **2017**, *18*, 289. [CrossRef]
33. Zhao, R.; Dong, R.; Yang, Y.; Wang, Y.; Ma, J.; Wang, J.; Li, H.; Zheng, S. MicroRNA-155 modulates bile duct inflammation by targeting the suppressor of cytokine signaling 1 in biliary atresia. *Pediatr. Res.* **2017**, *82*, 1007–1016. [CrossRef]
34. Loyer, X.; Paradis, V.; Henique, C.; Vion, A.C.; Colnot, N.; Guerin, C.L.; Devue, C.; On, S.; Scetbun, J.; Romain, M.; et al. Liver microRNA-21 is overexpressed in non-alcoholic steatohepatitis and contributes to the disease in experimental models by inhibiting PPARalpha expression. *Gut* **2016**, *65*, 1882–1894. [CrossRef]
35. Zhou, J.; Wang, K.C.; Wu, W.; Subramaniam, S.; Shyy, J.Y.; Chiu, J.J.; Li, J.Y.; Chien, S. MicroRNA-21 targets peroxisome proliferators-activated receptor-alpha in an autoregulatory loop to modulate flow-induced endothelial inflammation. *Proc. Natl. Acad. Sci. USA* **2011**, *108*, 10355–10360. [CrossRef]
36. Chen, Q.; Qiu, F.; Zhou, K.; Matlock, H.G.; Takahashi, Y.; Rajala, R.V.S.; Yang, Y.; Moran, E.; Ma, J.X. Pathogenic role of microRNA-21 in diabetic retinopathy through downregulation of PPARalpha. *Diabetes* **2017**, *66*, 1671–1682. [CrossRef]
37. Koenig, A.B.; Barajas, J.M.; Guerrero, M.J.; Ghoshal, K. A comprehensive analysis of argonaute-CLIP data identifies novel, conserved and species-specific targets of miR-21 in human liver and hepatocellular carcinoma. *Int. J. Mol. Sci.* **2018**, *19*, 851. [CrossRef]
38. Bala, S.; Marcos, M.; Kodys, K.; Csak, T.; Catalano, D.; Mandrekar, P.; Szabo, G. Up-regulation of microRNA-155 in macrophages contributes to increased tumor necrosis factor {alpha} (TNF{alpha}) production via increased mRNA half-life in alcoholic liver disease. *J. Biol. Chem.* **2011**, *286*, 1436–1444. [CrossRef]
39. Csak, T.; Bala, S.; Lippai, D.; Kodys, K.; Catalano, D.; Iracheta-Vellve, A.; Szabo, G. MicroRNA-155 deficiency attenuates liver steatosis and fibrosis without reducing inflammation in a mouse model of steatohepatitis. *PLoS ONE* **2015**, *10*, e0129251. [CrossRef]
40. Doukas, S.G.; Vageli, D.P.; Sasaki, C.T. NF-kappaB inhibition reverses acidic bile-induced miR-21, miR-155, miR-192, miR-34a, miR-375 and miR-451a deregulations in human hypopharyngeal cells. *J. Cell. Mol. Med.* **2018**, *22*, 2922–2934. [CrossRef]
41. Mashima, R. Physiological roles of miR-155. *Immunology* **2015**, *145*, 323–333. [CrossRef] [PubMed]
42. Kim, D.J.; Yoon, S.; Ji, S.C.; Yang, J.; Kim, Y.K.; Lee, S.; Yu, K.S.; Jang, I.J.; Chung, J.Y.; Cho, J.Y. Ursodeoxycholic acid improves liver function via phenylalanine/tyrosine pathway and microbiome remodelling in patients with liver dysfunction. *Sci. Rep.* **2018**, *8*, 11874. [CrossRef] [PubMed]
43. Castro, R.E.; Ferreira, D.M.; Afonso, M.B.; Borralho, P.M.; Machado, M.V.; Cortez-Pinto, H.; Rodrigues, C.M. miR-34a/SIRT1/p53 is suppressed by ursodeoxycholic acid in the rat liver and activated by disease severity in human non-alcoholic fatty liver disease. *J. Hepatol.* **2013**, *58*, 119–125. [CrossRef]
44. Sheedy, F.J.; Palsson-McDermott, E.; Hennessy, E.J.; Martin, C.; O'Leary, J.J.; Ruan, Q.; Johnson, D.S.; Chen, Y.; O'Neill, L.A. Negative regulation of TLR4 via targeting of the proinflammatory tumor suppressor PDCD4 by the microRNA miR-21. *Nat. Immunol.* **2010**, *11*, 141–147. [CrossRef]
45. Merline, R.; Moreth, K.; Beckmann, J.; Nastase, M.V.; Zeng-Brouwers, J.; Tralhao, J.G.; Lemarchand, P.; Pfeilschifter, J.; Schaefer, R.M.; Iozzo, R.V.; et al. Signaling by the matrix proteoglycan decorin controls inflammation and cancer through PDCD4 and MicroRNA-21. *Sci. Signal.* **2011**, *4*, ra75. [CrossRef]
46. Caescu, C.I.; Guo, X.; Tesfa, L.; Bhagat, T.D.; Verma, A.; Zheng, D.; Stanley, E.R. Colony stimulating factor-1 receptor signaling networks inhibit mouse macrophage inflammatory responses by induction of microRNA-21. *Blood* **2015**, *125*, e1–e13. [CrossRef]
47. Barnett, R.E.; Conklin, D.J.; Ryan, L.; Keskey, R.C.; Ramjee, V.; Sepulveda, E.A.; Srivastava, S.; Bhatnagar, A.; Cheadle, W.G. Anti-inflammatory effects of miR-21 in the macrophage response to peritonitis. *J. Leukoc. Biol.* **2016**, *99*, 361–371. [CrossRef]
48. Zhu, W.D.; Xu, J.; Zhang, M.; Zhu, T.M.; Zhang, Y.H.; Sun, K. MicroRNA-21 inhibits lipopolysaccharide-induced acute lung injury by targeting nuclear factor-kappaB. *Exp. Ther. Med.* **2018**, *16*, 4616–4622.

49. Caviglia, J.M.; Yan, J.; Jang, M.K.; Gwak, G.Y.; Affo, S.; Yu, L.; Olinga, P.; Friedman, R.A.; Chen, X.; Schwabe, R.F. MicroRNA-21 and dicer are dispensable for hepatic stellate cell activation and the development of liver fibrosis. *Hepatology* **2018**, *67*, 2414–2429. [CrossRef]
50. Wasik, U.; Kempinska-Podhorodecka, A.; Bogdanos, D.P.; Milkiewicz, P.; Milkiewicz, M. Enhanced expression of miR-21 and miR-150 is a feature of anti-mitochondrial antibody-negative primary biliary cholangitis. *Mol. Med.* **2020**, *26*, 8. [CrossRef]
51. Afonso, M.B.; Rodrigues, P.M.; Simao, A.L.; Gaspar, M.M.; Carvalho, T.; Borralho, P.; Banales, J.M.; Castro, R.E.; Rodrigues, C.M.P. miRNA-21 ablation protects against liver injury and necroptosis in cholestasis. *Cell Death Differ.* **2018**, *25*, 857–872. [CrossRef] [PubMed]
52. Zhou, R.; Hu, G.; Gong, A.Y.; Chen, X.M. Binding of NF-kappaB p65 subunit to the promoter elements is involved in LPS-induced transactivation of miRNA genes in human biliary epithelial cells. *Nucleic Acids Res.* **2010**, *38*, 3222–3232. [CrossRef] [PubMed]
53. Wu, N.; McDaniel, K.; Zhou, T.; Ramos-Lorenzo, S.; Wu, C.; Huang, L.; Chen, D.; Annable, T.; Francis, H.; Glaser, S.; et al. Knockout of microRNA-21 attenuates alcoholic hepatitis through the VHL/NF-kappaB signaling pathway in hepatic stellate cells. *Am. J. Physiol Gastrointest. Liver Physiol.* **2018**, *315*, G385–G398. [CrossRef]
54. Castro, R.E.; Ferreira, D.M.; Zhang, X.; Borralho, P.M.; Sarver, A.L.; Zeng, Y.; Steer, C.J.; Kren, B.T.; Rodrigues, C.M. Identification of microRNAs during rat liver regeneration after partial hepatectomy and modulation by ursodeoxycholic acid. *Am. J. Physiol Gastrointest. Liver Physiol.* **2010**, *299*, G887–G897. [CrossRef]
55. Rodrigues, P.M.; Afonso, M.B.; Simao, A.L.; Borralho, P.M.; Rodrigues, C.M.P.; Castro, R.E. Inhibition of NF-kappaB by deoxycholic acid induces miR-21/PDCD4-dependent hepatocellular apoptosis. *Sci. Rep.* **2015**, *5*, 17528. [CrossRef]
56. Shah, S.A.; Volkov, Y.; Arfin, Q.; Abdel-Latif, M.M.; Kelleher, D. Ursodeoxycholic acid inhibits interleukin beta 1 and deoxycholic acid-induced activation of NF-kappaB and AP-1 in human colon cancer cells. *Int. J. Cancer* **2006**, *118*, 532–539. [CrossRef] [PubMed]

cells

MDPI

Article

Rosiglitazone Ameliorates Cardiac and Skeletal Muscle Dysfunction by Correction of Energetics in Huntington's Disease

Marta Tomczyk [1,2,*], Alicja Braczko [1], Paulina Mierzejewska [1], Magdalena Podlacha [2], Oliwia Krol [1], Patrycja Jablonska [1], Agata Jedrzejewska [1], Karolina Pierzynowska [1,2], Grzegorz Wegrzyn [2], Ewa M. Slominska [1] and Ryszard T. Smolenski [1,*]

1 Department of Biochemistry, Medical University of Gdansk, 80-211 Gdansk, Poland
2 Department of Molecular Biology, University of Gdansk, 80-308 Gdansk, Poland
* Correspondence: marta.tomczyk@gumed.edu.pl (M.T.); rt.smolenski@gumed.edu.pl (R.T.S.)

Abstract: Huntington's disease (HD) is a rare neurodegenerative disease that is accompanied by skeletal muscle atrophy and cardiomyopathy. Tissues affected by HD (central nervous system [CNS], skeletal muscle, and heart) are known to suffer from deteriorated cellular energy metabolism that manifests already at presymptomatic stages. This work aimed to test the effects of peroxisome proliferator-activated receptor (PPAR)-γ agonist—rosiglitazone on grip strength and heart function in an experimental HD model—on R6/1 mice and to address the mechanisms. We noted that rosiglitazone treatment lead to improvement of R6/1 mice grip strength and cardiac mechanical function. It was accompanied by an enhancement of the total adenine nucleotides pool, increased glucose oxidation, changes in mitochondrial number (indicated as increased citric synthase activity), and reduction in mitochondrial complex I activity. These metabolic changes were supported by increased total antioxidant status in HD mice injected with rosiglitazone. Correction of energy deficits with rosiglitazone was further indicated by decreased accumulation of nucleotide catabolites in HD mice serum. Thus, rosiglitazone treatment may not only delay neurodegeneration but also may ameliorate cardio- and myopathy linked to HD by improvement of cellular energetics.

Keywords: Huntington's disease; myopathy; cardiomyopathy; rosiglitazone; molecular mechanisms; therapy; energy metabolism

Citation: Tomczyk, M.; Braczko, A.; Mierzejewska, P.; Podlacha, M.; Krol, O.; Jablonska, P.; Jedrzejewska, A.; Pierzynowska, K.; Wegrzyn, G.; Slominska, E.M.; et al. Rosiglitazone Ameliorates Cardiac and Skeletal Muscle Dysfunction by Correction of Energetics in Huntington's Disease. *Cells* **2022**, *11*, 2662. https://doi.org/10.3390/cells11172662

Academic Editors: Kay-Dietrich Wagner and Nicole Wagner

Received: 2 August 2022
Accepted: 24 August 2022
Published: 27 August 2022

1. Introduction

Huntington's disease (HD) is a rare neurodegenerative disease that is known to primarily affect the central nervous system. The genetic cause of HD is the occurrence of multiple repeats of the CAG nucleotide sequence within the huntingtin gene (*HTT*) localized on chromosome 4, which results in the elongation of the polyglutamine stretch in the HTT protein [1]. The elongation of the polyglutamine stretch in exon 1 *HTT* leads to the formation of insoluble huntingtin aggregates, which are observed in both the early and advanced stages of the disease [2,3]. Aggregates of the mutated form of HTT (mHTT) have been identified not only in the brain but also outside the central nervous system (CNS), e.g., in skeletal muscle [3–5]. Interestingly, mHTT is absent in the HD-affected heart [6,7]. It has been shown that HD patients suffer from reduced (by about 50%) muscular strength compared to healthy controls [8]. Besides skeletal muscle pathology, multiple epidemiological studies have shown that heart failure is the second leading cause of death in HD patients [9,10]. Similar findings were observed in preclinical HD models [11]. HD mice models were characterized by skeletal muscle atrophy and altered ultrastructure of transverse tubules in skeletal muscle fibers [12,13]. mHTT formation in skeletal muscle leads to defects, such as myofiber size reduction or type switching [12,14–17]. HD animal models reaffirmed cardiac pathological events noted in HD patients, such as variations in the heart rate and cardiac remodeling [7,18,19]. Moreover, heart contractile dysfunctions, which might be a part of dilated cardiomyopathy were noted [7]. Thus, nowadays HD is considered as a multisystem disorder [11,20].

HD-affected CNS and non-CNS tissues were characterized by defects in energy metabolism [21]. The striatum mitochondrial oxidative metabolism investigation underlined the selective defect of glycolysis in early and clinical symptoms in HD patients [22]. In a few independent studies of the striatum of mHTT knock-in mice, HD patients' postmortem brains, and lymphoblasts, the adenosine-5′-triphosphate (ATP)/ adenosine-5′-diphosphate (ADP) ratio was reduced as a consequence of mHTT aggregation [23–25]. A decreased ATP/ADP ratio was found also in mHTT-containing striatal cells, which were linked to increased Ca^{2+} influx through N-methyl-D-aspartate (NMDA) receptors. Interestingly, the disrupted ATP/ADP ratio was normalized by blocking Ca^{2+} influx [26]. Deteriorations in energy metabolism occur also in HD-affected skeletal muscle [27]. It has been noted that the skeletal muscles of HD patients are characterized by dysfunction of oxidative metabolism [28]. Moreover, muscle ATP/phosphocreatine and inorganic phosphate levels were significantly reduced in both symptomatic and presymptomatic HD subjects [29]. Previously, we have noted that R6/2, as well as HdhQ150, two well-established HD mice models, exhibited decreased ATP, ADP, and adenosine-5′-monophosphate (AMP) concentrations in three different skeletal muscle tissues—extensor digitorum longus, tibialis anterior, and soleus. Moreover, a significant reduction of phosphocreatine (PCr) and creatine (Cr) levels and the PCr/Cr were noted [17]. Similar changes were observed in HD mice models' hearts. We highlighted decreased concentrations of ATP and phosphocreatine as well as diminished ATP/ADP ratios [30].

Interestingly, as mentioned above, energy metabolism deterioration manifests not only in the advanced stages of the disease but also in the presymptomatic. It could be suggested that energy deficit is likely to be an early phenomenon in the cascade of events leading to HD pathogenesis. Moreover, impaired bioenergetics in HD likely represent downstream effects of both an mHTT toxic gain-of-function and an HTT loss-of-function [21]. Thus, therapeutic strategies include compounds that directly correct disrupted ATP levels in affected HD CNS as well as non-CNS tissues might be an interesting therapeutic target. Nevertheless, compounds such as the coenzyme Q10 or creatinine were widely tested and even investigated in clinical trials, but the results were not promising [31].

An alternative might be the application of peroxisome proliferator-activated receptors (PPARs) agonists, which have already undergone preclinical studies for the treatment of CNS, cardiovascular as well as skeletal muscle diseases. PPARs belong to the group of nuclear receptors that activate or repress target genes as heterodimers with retinoic X receptors (RxR). PPARs family included: PPAR alpha (PPARα), PPAR beta/delta (PPARβ/δ), and PPAR gamma (PPARγ) [32]. Different types of cells exhibited various expressions of PPARs; thus, the outcome of its activation might be different in various tissues [33]. In 2016, the PPAR delta receptor agonist KD3010 was tested in the HD N171-82Q mouse model. Study revealed improved motor function, reducing the progression of the neurodegenerative process, and longer survival of treated animals [34]. Nevertheless, this study was focused mainly on the evaluation of CNS function improvement. Thus, our work for the first time highlighted the effect of PPAR agonist treatment on HD mouse model grip strength, cardiac function, and HD-affected skeletal muscle and heart metabolism.

2. Materials and Methods

2.1. Animal Maintenance and Treatment

All experiments were conducted following the *Guide for the Care and Use of the Laboratory Animals* published by the European Parliament, Directive 2010/63/EU, and were approved by the local bioethical committee for the Medical University of Gdansk. Animals were maintained on a 12:12 h light-dark cycle at 25 °C, 30–40% humidity, and were provided with free access to water and a standard chow diet (Morawski, Kcynia, Poland). R6/1 (n = 30) aged 21 weeks old and C57BL/6J (n = 11) as WT mice were used in the study. R6/1 mice (n = 12) were treated daily for six weeks with 10 mg/kg of rosiglitazone

(Sigma-Aldrich, St. Louis, MO, USA) (dissolved in 0.09% DMSO) or 0.09% DMSO (Sigma-Aldrich, St. Louis, MO, USA) administered intraperitoneally [35].

2.1.1. Forelimb Grip Strength Measurement

Forelimb grip strength was measured by a grip strength meter (GSM Grip strength meter, Ugo Basile, Gemonio VA, Italy) as described earlier [36]. Briefly, the animal was held on the apparatus so that only the forelimb paws grasped the specially designed mouse flat mesh assembly. Then, the mouse was pulled back until its grip was broken, which was recorded from a digital display. The maximum values were used for analysis. Forelimb and maximal muscle strength were obtained as values of GF (gram-force) and normalized to body weights as "g/g mouse body weight."

2.1.2. Echocardiography

Echocardiographic examination was performed with a high-resolution ultrasound system (Vevo 1100, VisualSonics Inc, Toronto, Ontario, Canada) [37]. Mice were anesthetized with ketamine (Biowet Pulawy, Pulawy, Poland) (100 mg/kg) and xylazine (Biowet Pulawy, Pulawy, Poland) (10 mg/kg) intraperitoneally (i.p)., then their chest hair was removed and mice were placed on a heating pad. The probe (70 MHz) was placed over the anterior chest wall and directed to the ascending aorta in 2D mode. Then the mode was switched to Doppler flow velocity. Stroke Volume (SV), Cardiac Output (CO), Left Ventricular mass (LVmass), Ejection Fraction (EF), and Fractional shortening (FS) were recorded.

2.1.3. Analysis of Cardiac and Skeletal Muscle Glucose Usage

Analysis of cardiac and skeletal muscle glucose usage was performed within the method described before [30,38]. D-glucose-1,6-$^{13}C_2$ (Sigma-Aldrich, St. Louis, MO, USA) was administrated in the subcutaneous injection of a 1.8 mg/g body weight dose. Moreover, blood samples were collected from the tail vein before and after 30, 60, and 90 min of $^{13}C_2$ glucose administration. Next, after animal anesthesia, heart and skeletal muscle were rapidly excised (after 90 min), and freeze clamped (after animal intubation and under artificial ventilation).

Hearts were placed for 24 h in a freeze dryer (Modulyo, Thermo Electron Corporation, Waltham, MA, USA) at −55 °C, and then were extracted with 0.4 M perchloric acid (Sigma-Aldrich, St. Louis, MO, USA) in a 1:25 ratio, followed by neutralization with 2 M KOH (Sigma-Aldrich, St. Louis, MO, USA). Supernatants (obtained from centrifugation at 4 °C, 14,000 RPM/min for 10 min) were analyzed by LC/MS.

Blood extraction was performed using ice-cooled acetone (Sigma-Aldrich, St. Louis, MO, USA) in a 1:3 ratio. Next, samples were placed in ice for 15 min and centrifuged at 4 °C, 14,000 RPM/min for 10 min. This was followed by drying in a vacuum concentrator (JW Electronic, Warsaw, Poland) and sediments were dissolved in high-purity water (Nanopure—ultrapure water system, Barnstead, Thermo, Waltham, MA, USA) and analyzed with LC/MS.

The heart extracts were analyzed by LC/MS using a TSQ-Vantage triple quadrupole mass detector (Thermo, Waltham, MA, USA), linked to a Surveyor chromatography system (Thermo, Waltham, MA, USA) in positive heated electrospray ionization with fragmentation mode (Tandem MS), monitoring ^{13}C isotopic enrichment of fragments containing C3 of alanine or C4 of glutamate. The ^{13}C glucose enrichment in blood was measured using liquid chromatography-mass spectrometry—an LCQ-Deca XP mass detector (Thermo Finnigan, San Jose, CA, USA). Fragments containing ^{12}C and ^{13}C glucose were detected in negative electrospray ionization with the selected ion monitoring (SIM) mode for ^{12}C glucose m/z 178.00–179.40 and m/z 179.00–180.40 for D-glucose-1,6-$^{13}C_2$.

2.1.4. Mice Tissues and Serum Collection

Tissues and serum for further analysis were collected after mice anesthesia with a ketamine/xylazine mixture (Biowet Pulawy, Pulawy, Poland) (50 mg/kg + 5 mg/kg)

and artificial ventilation. Blood was collected from inferior vena cava (IVC). For serum collection, blood was centrifuged at 2000 RPM for 4 min. Mice heart and skeletal muscle were also isolated.

2.2. Measurement of Total Adenine Nucleotides Pool, Phosphocreatine and Creatine, and Nicotinamide Dinucleotides

Hearts and skeletal muscle (soleus) were prepared and analyzed with the high-pressure liquid chromatography (HPLC) method as previously described [30].

2.3. Investigation of Cardiac and Skeletal Muscle Mitochondrial Chain Complexes Activities

Mitochondria were isolated from the soleus muscle and heart, and prepared based on the previously described procedure [36,39]. Analysis was performed by Seahorse Metabolic Flux Analyzer (Agilent Technologies, Santa Clara, CA, USA). For electron flow experiments, isolated mitochondria were diluted in cold MAS buffer (enriched with 10 mM pyruvate (Sigma-Aldrich, St. Louis, MO, USA)2 mM malate (Sigma-Aldrich, St. Louis, MO, USA), and 4 μM FCCP (Sigma-Aldrich, St. Louis, MO, USA). A mitochondrial suspension of 25 μL was placed into Seahorse plate wells and centrifuged at 2000× g for 15 min at 4 °C. The concentration of mitochondrial protein was 6 μg per well. After centrifugation, 180 μL of prewarmed MAS buffer supplemented with pyruvate, malate, and FCCP was added to each well, and the plate was then placed into a non-CO_2 incubator for 8 min. The Seahorse cartridge was filled with the following reagents: 2 μM Rotenone (Sigma-Aldrich, St. Louis, MO, USA), 2 mM succinate (Sigma-Aldrich, St. Louis, MO, USA), 4 μM Antimycin (Sigma-Aldrich, St. Louis, MO, USA), and a mix of 10 mM ascorbate (Sigma-Aldrich, St. Louis, MO, USA) and 100 μM TMPD (Sigma-Aldrich, St. Louis, MO, USA).

2.4. Evaluation of Cardiac and Skeletal Muscle Citric Synthase Activity

Citric synthase activity (in soleus muscle and heart) was measured within the assay kit (Sigma-Aldrich, St. Louis, MO, USA). The activity of the enzyme is measured by following the color of 5-thio-2-nitrobenzoic acid (TNB), which is generated from 5,5′-Dithiobis-(2-nitrobenzoic acid) (DTNB) present in the reaction of citrate synthesis, and caused by the deacetylation of Acetyl-CoA. Citric synthase activity was presented as μmol/mL/min.

2.5. Measurement of Nucleotides Catabolites in Serum

Mice serum was extracted with 1.3 M perchloric acid (Sigma-Aldrich, Burlington, MA, USA) (1:1 ratio). Levels of nucleotides were measured by a reverse-phase high-pressure liquid chromatography (RP-HPLC) method using the liquid chromatography (LC) system (Agilent Technologies 1100 series, Agilent Technologies Inc., Santa Clara, CA, USA), as described previously [15,30]. Results are presented as μmol/L.

2.6. Analysis of Total Plasma Antioxidant Status

The total antioxidant status (TAOS) in plasma was measured by the 2,2′-azino-bis(3-ethylbenzothiazoline-6-sulphonic acid (ABTS) assay, which was based on the capacity of plasma to scavenge the ABTS+ radical [40]. Briefly, the relative inhibition of ABTS+ formation, after the plasma addition, is proportional to the antioxidant capacity of the sample. For the measurement, plasma was diluted with 180 μL phosphate buffer (0.076 M NaH_2PO_4 (POCH, Gliwice, Poland)+ 0.23 M Na_2HPO_4 (Sigma-Aldrich, Burlington, MA, USA) in pure water), and then it was incubated for 10 min at room temperature in a 96-well plate with a 5 μL reaction mixture containing 7 mM ABTS (Sigma-Aldrich, Burlington, MA, USA) and 2.45 mM potassium persulfate (Sigma-Aldrich, Burlington, MA, USA) (in phosphate buffer: 0.22 M NaH_2PO_4 (POCH, Gliwice, Poland) + 0.37 M Na_2HPO_4 (Sigma-Aldrich, Burlington, MA, USA)) solved in pure water. The absorbance in the test and control samples (saline instead of plasma) was read at 630 nm. Results expressed as percentage inhibition of the reaction were calculated as follows: TAOS [%] = 100 × (Ac − At)/Ac, where Ac is the absorbance of the control sample absorbance, and At is the test sample absorbance.

2.7. Investigation of Serum-Free Fatty Acids and Blood Glucose Levels

The free fatty acids (FFA) concentration in serum was measured using a commercial colorimetric assay kit (Wako NEFA C test kit; Wako Chemicals, Neuss, Germany). Serum was collected after 24 h starvation. Random blood glucose levels were measured with a glucometer (Accu check Active, Roche Diabetes Care, F. Hoffmann-La Roche Ltd., Basel, Switzerland). Blood drop was collected from the tail vein of non-starved mice.

2.8. Statistical Analysis

Statistical significance was evaluated using Student's t-test for comparatives of two groups. A value of $p < 0.05$ was used to denote statistical significance, and the results are expressed as mean ± SEM. All statistics were carried out using GraphPad Prism 5.00 (GraphPad Software, San Diego, CA, USA).

3. Results

3.1. Rosiglitazone Improved Grip Strength and Cardiac Function in an HD Mouse Model

Previous experimental research that investigated the cardiac and skeletal muscle function in Huntington's disease (HD) examined those mainly other than R6/1 HD mice models (R6/2, HdhQ150, or N171-82Q). Thus, to ensure that the investigated HD mouse model exhibited any changes in skeletal muscle and cardiac functionality, we assessed the forelimb grip strength and as well as cardiac function parameters (stroke volume, ejection fraction, fractional shortening, cardiac output, and left ventricular mass) in R6/1 in comparison to healthy controls. Similar to other HD mice models, the R6/1 mice model also exhibited a reduction of forelimb grip strength as well as normalized grip strength (Supplement Figure S1). Furthermore, significant reduction in ejection fraction, fractional shortening, cardiac output as well as left ventricular mass relative to wild-type (WT) were noted (Supplement Figure S2). That results suggested the presence of serious depletion of grip strength and cardiac function.

One of the main goals of our study was to investigate the influence of rosiglitazone on HD mouse model skeletal muscle functionality. Thus, we measured the forelimb grip strength and normalized grip strength (force normalized for mouse body weight) in R6/1 and R6/1 mice treated with rosiglitazone. We found no changes in body weight in peroxisome proliferator-activated receptor (PPAR) agonists treated mice in comparison to non-treated HD mice, while maximum, as well as normalized grip strength evaluation, indicated higher values of these parameters in HD treated with rosiglitazone (Figure 1A–C).

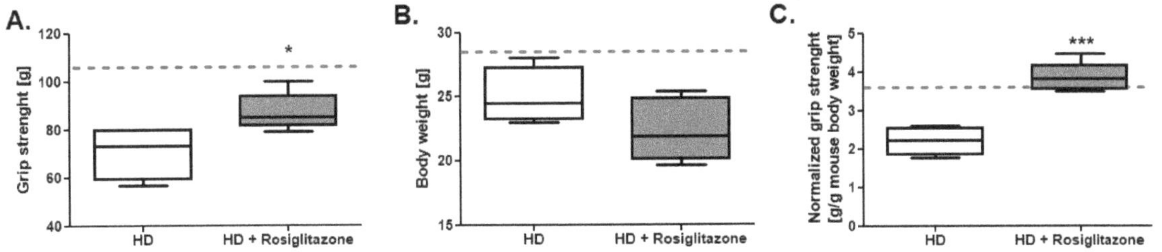

Figure 1. Improved grip strength in R6/1 mice treated with rosiglitazone. (**A**) Maximum forelimb grip strength. (**B**) Bodyweight (**C**) Normalized forelimb grip strength (maximum forelimb grip strength/g of body weight) in R6/1 (HD) and R6/1 with rosiglitazone treated mice (HD + Rosiglitazone). Results presented as mean ± SEM, n = 5–6, * $p < 0.05$, *** $p < 0.001$. The red dotted lines present the mean value of the investigated parameter in control, C57BL/6J mice (adapted from supplementary material data).

We examined also the R6/1 mice's heart function after rosiglitazone treatment (representative echocardiograms in Supplement Figure S3). We noted tendencies in the improvement of stroke volume (SV), cardiac output, and ejection fraction in HD mice treated with rosiglitazone and no changes in left ventricular mass in comparison to HD control (treated with 0.09% DMSO) (Figure 2A–D). Interestingly, we found the statistically confirmed improvement of fractional shortening in the HD mouse model injected with investigated PPAR agonist (Figure 2E).

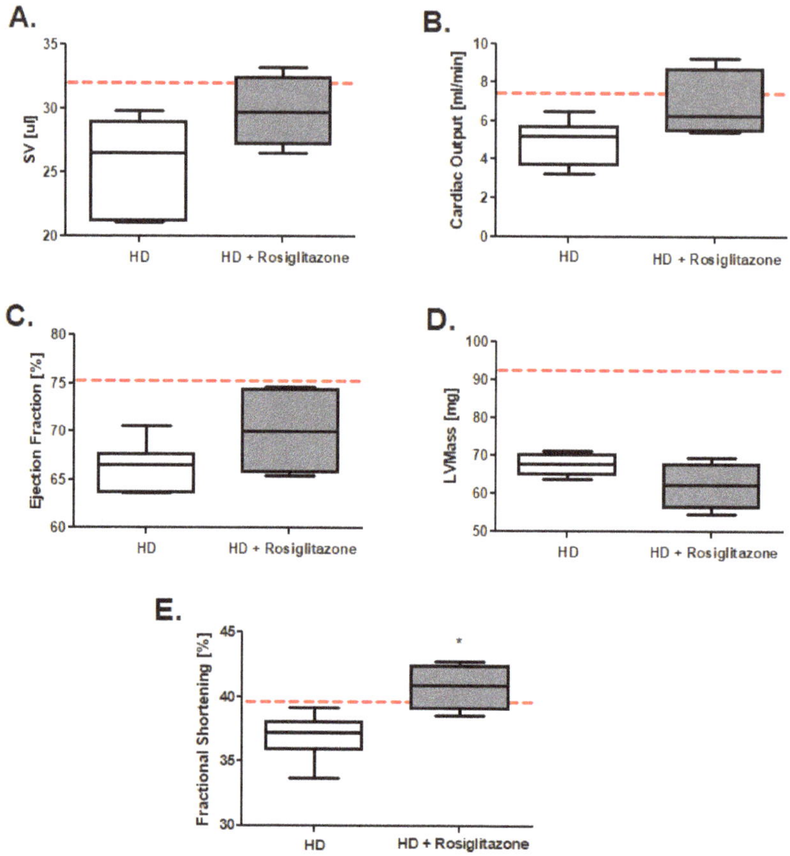

Figure 2. Huntington's disease (HD) mouse model heart function improvement after rosiglitazone treatment (**A**) Stroke volume (SV), (**B**) Cardiac output, (**C**) Ejection fraction, (**D**) Left ventricle mass, (**E**) Fractional shortening in R6/1 (HD) and R6/1 with rosiglitazone treated mice (HD + Rosiglitazone). Results presented, n = 4–6, * $p < 0.05$. The red dotted lines present the mean value of the investigated parameter in control, C57BL/6J mice (adapted from supplementary material data).

3.2. *Rosiglitazone Enhanced Skeletal Muscle and Cardiac Glucose Usage in an HD Mouse Model*

To unravel the source of the noticed skeletal muscle as well as cardiac functionality improvement, we evaluated the in vivo glycolytic and oxidative metabolism of labeled ^{13}C glucose. Metabolite tracking (4-^{13}C glutamate and 3-^{13}C alanine) after ^{13}C glucose administration was previously extensively studied by our group [38]. Theoretical assumptions, supported by experimental studies, indicate that after ^{13}C glucose administration, the heart accumulates 3-^{13}C pyruvate in proportion to the fraction of glycolytic substrate, supplied by exogenous glucose relative to alternative unlabeled substrate sources (e.g., endogenous glycogen) and 4-^{13}C α-ketoglutarate in proportion to the fraction of tricarboxylic acid (TCA) cycle carbon flux supported by flux through pyruvate dehydrogenase (PDH), relative to other acetyl-CoA sources (e.g., free fatty acids (FFA)). It has to be mentioned that 3-^{13}C pyruvate, as well as 4-^{13}C α-ketoglutarate, were present in small quantities in the heart, but occur in isotopic equilibrium with tracked 3-^{13}C alanine and 4-^{13}C glutamate [41]. Thus, the measurement of myocardial or skeletal muscle 3-^{13}C alanine/^{12}C alanine (^{13}C alanine enrichment) and 4-^{13}C glutamate/^{12}C glutamate (^{13}C glutamate enrichment) in steady-state ^{13}C glucose enrichment in the blood allows for the estimation of the contribution of circulating glucose to myocardial glycolytic and oxidative flux.

We observed no changes in ^{13}C alanine enrichment in skeletal muscle as well as in heart to ^{13}C glucose enrichment in the mouse blood ratio (Figure 3A,D). On the other hand, we noted an increased ^{13}C glutamate/^{13}C glucose ratio and ^{13}C glutamate/^{13}C alanine ratio in skeletal muscle and heart in the R6/1 mice model treated with rosiglitazone relative to HD treated with 0.09% DMSO (control), which indicates enhanced glucose oxidation as well as its overall use in cardiac and skeletal muscle metabolism (Figure 3 B,C,E,F).

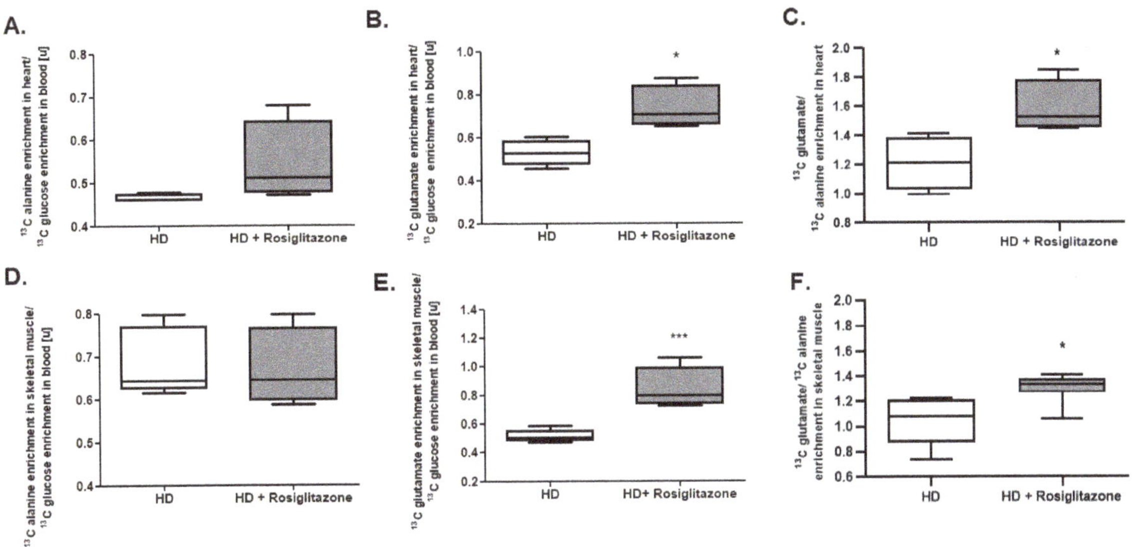

Figure 3. Increased glucose usage in cardiac and skeletal muscle metabolism in Huntington's disease (HD) mouse model treated with rosiglitazone. (**A**) ^{13}C alanine enrichment in heart/^{13}C glucose enrichment ratio in the blood, (**B**) ^{13}C glutamate enrichment in heart/^{13}C glucose enrichment ratio in the blood, (**C**) ^{13}C glutamate/^{13}C alanine enrichment ratio in the heart, (**D**) ^{13}C alanine enrichment in skeletal muscle/^{13}C glucose enrichment ratio in the blood, (**E**) ^{13}C glutamate enrichment in skeletal muscle/^{13}C glucose enrichment ratio in the blood, (**F**) ^{13}C glutamate/^{13}C alanine ratio enrichment in the skeletal muscle of R6/1 (HD) and R6/1 with rosiglitazone treated mice (HD + Rosiglitazone). Data presented; n = 5; * $p < 0.05$, *** $p < 0.001$. Due to the methodological inability to compare the obtained values with previous experiments, values from control experiments (wild-type mice) are not shown. Nevertheless, the comparison of glucose usage in cardiac and skeletal muscle metabolism between control and HD mice models was already published in our two previous studies [17,30].

Additionally, we assessed the serum-free fatty acids (FFA) as well as blood glucose concentration (Supplementary Figure S4). We found that HD mice treated with PPAR agonist were characterized by reduced blood glucose levels in comparison to non-treated mice. In the case of FFA, we found that HD mice injected with rosiglitazone do not exhibit statistically significant changes in this parameter relative to HD non-treated mice.

3.3. Rosiglitazone Improved Cardiac and Skeletal Muscle Adenine Nucleotides Pool

Further analysis tested whether rosiglitazone may affect overall skeletal muscle and cardiac energy metabolism. While accurate quantitative analysis of unstable metabolites such as adenosine-5′-triphosphate (ATP) and phosphocreatine was not possible due to limitations of the tissue collection procedure we were able to collect data on total pools of relevant metabolites. We noted that total cardiac and skeletal muscle adenine nucleotides pools were elevated in rosiglitazone-injected mice (Figure 4A,D). Similar increases were found in total creatine pools (Figure 4B,E). On the other hand, there were no changes in the total cardiac and skeletal muscle NAD^+ and NADH pool (Figure 4C,F).

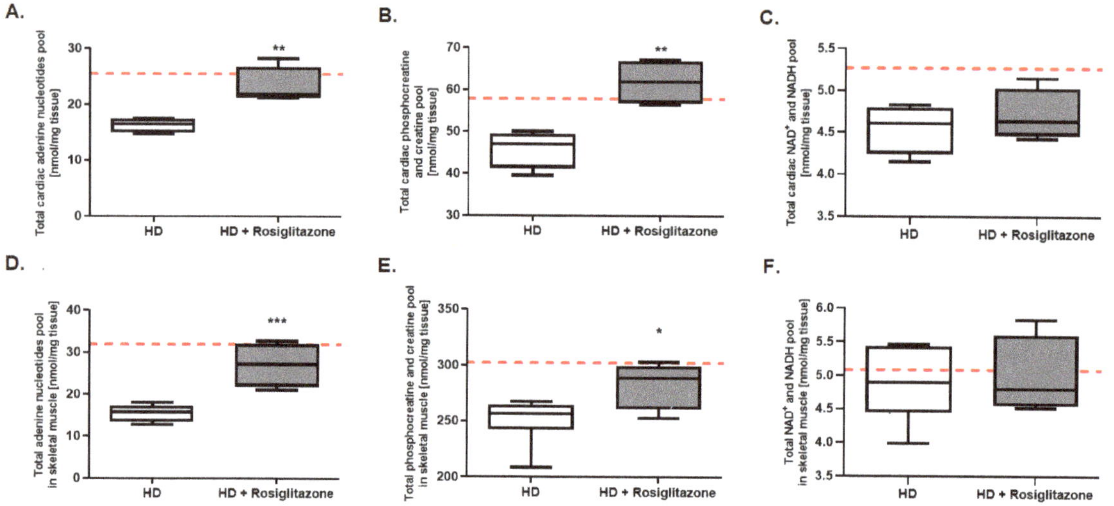

Figure 4. Enhanced total adenine nucleotides and total phosphocreatine creatine pools in hearts and skeletal muscles of Huntington's disease (HD) mouse model treated with rosiglitazone. Total adenine nucleotides pool in hearts (**A**) and skeletal muscle (**D**), total phosphocreatine and creatine pool in hearts (**B**) and skeletal muscle (**E**), and total NAD+ and NADH pools in hearts (**C**) and skeletal muscle (**F**) in R6/1 (HD) and R6/1 with rosiglitazone treated mice (HD + Rosiglitazone). Results presented as mean ± SEM, n = 4–6, * $p < 0.05$, ** $p < 0.01$, *** $p < 0.001$. The red dotted lines present the mean value of the investigated parameter in control, C57BL/6J mice (adapted from our previous works [17,42]).

3.4. Rosiglitazone Led to Cardiac and Skeletal Muscle Mitochondria Functionality Changes

The initial step of this part of our research was the examination of oxygen consumption rate (OCR) in coupled and un-coupled cardiac and soleus mitochondria of R6/1 (HD) as well as control WT mice. Moreover, we assessed the activities of mitochondrial complex I (with pyruvate, malate, and FCCP), II (after rotenone and succinate addition), and complex IV (after addition of TMPD and ascorbate) in isolated mitochondria.

We did not observe differences in OCR between coupled cardiac and soleus muscle mitochondria isolated from HD and WT mice at any measurement points (Supplementary Figure S5A,C). The level of respiration in isolated cardiac mitochondria in state 2, state 3 ADP, state 4o, and state 3u were similar in both strains (Supplementary Figure S5B). On the other hand, examination of mitochondrial respiration states in soleus muscle highlighted the reduced state 3u (controlled exclusively by substrate oxidation, its reduction might suggest dysfunction in respiratory chain components, substrates translocases, or dehydrogenases) (Supplementary Figure S5D). Interestingly, evaluation of R6/1 mice cardiac and soleus muscle mitochondrial OCR and complexes activities in an uncoupled state revealed complex I hyper-activation in comparison to WT (Supplementary Figures S6A,B and S7A,B). No changes in cardiac and skeletal muscle mitochondrial complex IV respiration were found between HD and control mice (Supplementary Figures S6D and S7D). Nevertheless, some difference was found between those investigated tissues. Indeed, R6/1 mice heart exhibited increased mitochondrial complex II respiration (Supplementary Figure S6A,C), while soleus muscle exhibited opposite reduced tendency (Supplementary Figure S7A,C).

Next, we evaluated the activity of citric synthase (CS, the indicator of mitochondria amount) and reactive oxygen species levels (ROS, the indicator of oxidative stress) in the hearts and soleus of HD and control mice. We found no changes in CS activity and increased ROS levels in R6/1 mice's hearts and skeletal muscle in comparison to control mice (Supplementary Figure S8A–D). Results ensure the presence of mitochondrial functionality changes in HD-affected heart and skeletal muscle in investigated HD mice model.

It is well known that rosiglitazone may improve mitochondrial number and metabolism. Thus, we aimed to evaluate the mitochondrial oxidative chain complexes' respiration and activities in hearts and mitochondria-rich, red soleus muscle, isolated from HD mice treated with rosiglitazone as well as non-treated HD mice. There were no changes in complex II and complex IV respiration, while reduced complex I respiration in hearts and soleus of R6/1 treated with PPAR agonist (HD + Rosiglitazone) relative to R6/1 (HD) were noted (Figure 5A,B,D,E).

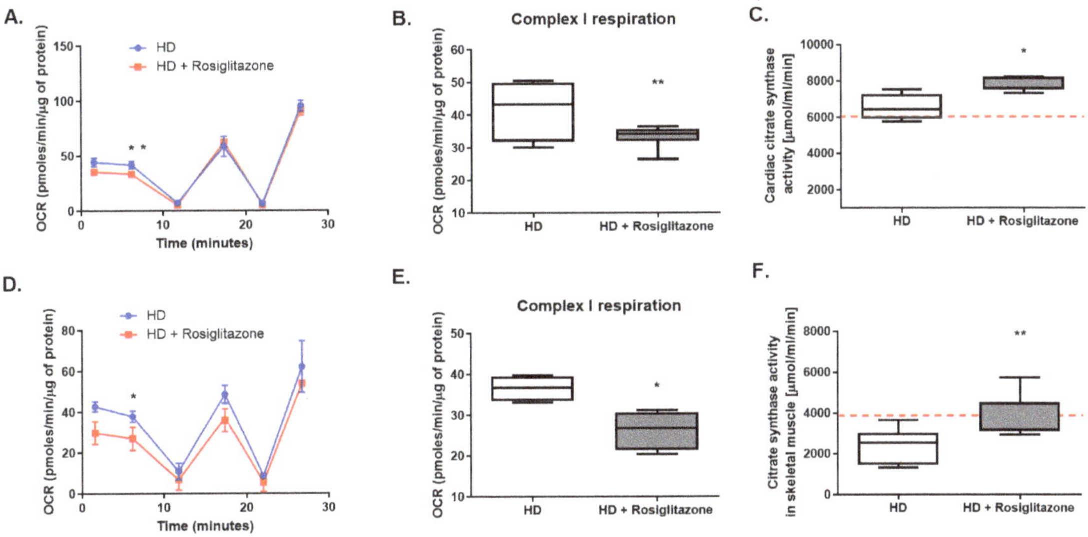

Figure 5. Diminished complex I respiration in mitochondria as well as increased cardiac and skeletal muscle synthase activity in R6/1 mice treated with rosiglitazone. OCR data from isolated cardiac (**A**) and soleus muscle (**D**), complex I respiration in cardiac (**B**) and soleus muscle (**E**) mitochondria, and cardiac (**C**) and skeletal muscle (**F**) citrate synthase activity in R6/1 (HD) and R6/1 with rosiglitazone treated mice (HD + Rosiglitazone). Data presented as mean ± SEM, n = 4–6, * $p < 0.05$, ** $p < 0.01$. The red dotted lines present the mean value of the investigated parameter in control, C57BL/6J mice (adapted from supplementary materials). Due to the methodological inability to compare the obtained values with values from other experiments conducted on Seahorse metabolic analyzer, values from control experiments (wild-type mice) are not shown. Nevertheless, the comparison of the OCR values between control and HD mice models is present in the supplementary materials.

To investigate the effect of rosiglitazone on mitochondrial biogenesis in HD-affected tissues (heart and skeletal muscle), we assessed the activity of citric synthase. Interestingly, we noted elevated activity of this enzyme in investigated tissues isolated from HD mice treated with rosiglitazone in comparison to non-treated HD mice (Figure 5C,F).

3.5. Rosiglitazone Abolished Changes in Energy Deficits Biomarkers and Improved Total Antioxidant Status in HD Mouse Model Serum

Improvement of energy metabolism may lead to changes in nucleotides' catabolite profile in serum. Thus, the next step of our research was the evaluation of the concentration of uric acid, hypoxanthine, and inosine in the serum of R6/1 mice treated with rosiglitazone. We found reduced levels of all investigated nucleotide catabolites in HD mice injected with PPAR agonist in comparison to non-treated mice (Figure 6A–C).

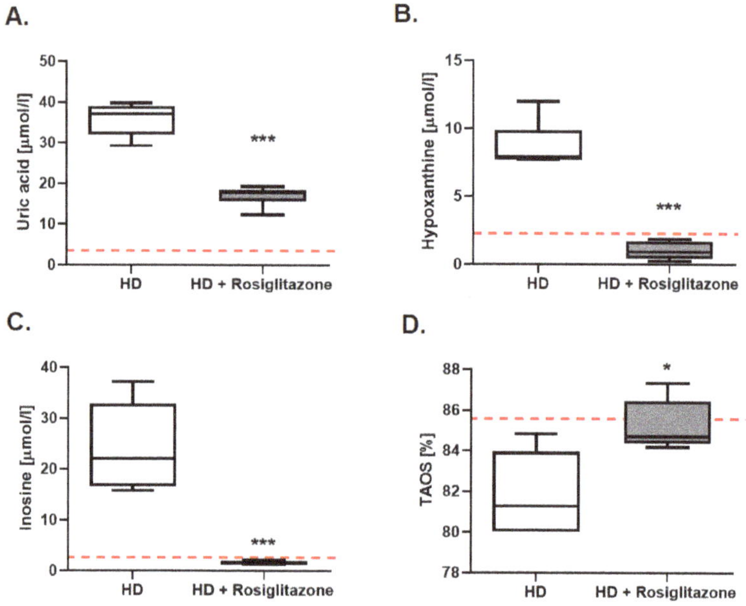

Figure 6. The reduced serum concentration of purine catabolites and improved total plasma antioxidant status (TAOS) in R6/1 mice treated with rosiglitazone. Serum uric acid (**A**), hypoxanthine (**B**), inosine (**C**) concentration, and total plasma antioxidant status (**D**) in R6/1 (HD) and R6/1 with rosiglitazone treated mice (HD + Rosiglitazone). Data presented as mean ± SEM, n = 5–6, * $p < 0.05$, *** $p < 0.001$. The red dotted lines present the mean value of the investigated parameter in control, C57BL/6J mice (adapted from supplementary materials and our previous work [30]).

Due to the earlier noted increase of ROS in HD mice hearts and skeletal muscles, we also investigated total plasma antioxidant status (TAOS) in R6/1 and wild-type littermates. We established a reduced value of this parameter in HD mice (Supplementary Figure S9). Following these findings, we aimed to assess TAOS in HD mice treated with rosiglitazone in comparison to the non-treated group, and noted enhancement of its value in mice injected with PPAR agonist (Figure 6D).

4. Discussion

This study highlighted that the rosiglitazone treatment improves grip strength and cardiac function in Huntington's disease (HD) mouse model R6/1 (Figure 7). These functional changes were accompanied by enhanced total adenine nucleotide and total creatine pools, increased glucose oxidation, changes in mitochondrial number (indicated as increased citric synthase activity), and reduction of mitochondrial complex I activity. Correction of energy deficits with rosiglitazone abolished, as noted in our previous research, the accumulation of nucleotide catabolites in HD mice serum [30]. Moreover, enhancement of energy metabolism and changes in mitochondrial complex I lead to improvement of oxidative balance highlighted as an increased total antioxidant status in HD mice injected with rosiglitazone.

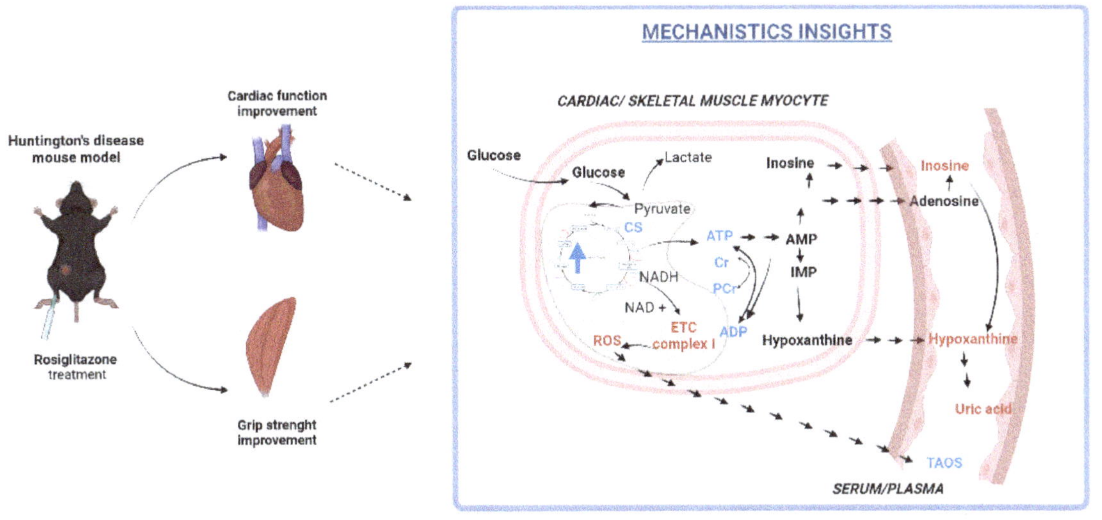

Figure 7. A model depicting the mechanism by which rosiglitazone may lead to improvement of cardiac and skeletal muscle functionality in the R6/1 (Huntington's disease) mouse model. Rosiglitazone led to cardiac and skeletal muscle: 1. Mitochondrial number enhancement (measured by citric synthase activity), 2. Increased glucose oxidation and its use in overall metabolism (measured by ^{13}C glutamate/^{13}C glucose as well as ^{13}C glutamate/^{13}C alanine enrichment) and 3. Changes in mitochondrial functionality (diminished mitochondrial complex I activity). Those factors may lead to improvement of cardiac and skeletal muscle energy metabolism (enhancement of total adenine nucleotides and total phosphocreatine and creatine levels) and may contribute to diminished serum concentration of adenine nucleotides catabolites. Changes in mitochondria functionality may also contribute to improvement of total plasma antioxidant status.Red color: downregulation, blue color: upregulation. Created with Bioreder.com on 7 April 2022.

Rosiglitazone is a synthetic agonist of peroxisome proliferator-activated receptor (PPAR)-γ, which is commonly used to reverse insulin resistance in patients with type II diabetes [43]. Interestingly, it has also been tested as a neuroprotective agent in HD, where it significantly attenuated toxicity induced by mutant huntingtin (mHTT) in striatal cells [44]. Rosiglitazone treatment also significantly reduced mHTT aggregates in the mHTT expressing neuroblastoma cell line [45]. Moreover, administration of rosiglitazone significantly improved motor function, rescued brain-derived neurotrophic factor (BDNF) deficiency in the cerebral cortex, and prevented loss of orexin-A-immunopositive neurons in the hypothalamus of N171-82Q HD mice [35]. Similar results were also noted in the HD rat model (injected with quinolinic acid) treated with rosiglitazone [46]. However, our study for the first time indicated that PPAR-γ agonist treatment of the HD mouse model also improved skeletal muscle as well as heart functionality. Although PPAR-γ expression in skeletal muscle and the heart is relatively small, there are studies indicating that these receptors may play an important role in its metabolism and function [47,48]. Treating fibromyalgia (multisystem failure process involving the immune, musculoskeletal, and central nervous system) in rats with PPAR-γ agonist, pioglitazone resulted in a significant improvement of skeletal muscle functions, reduced fatigability, and rapid recovery from fatigue [49], which is consistent with our results. Blocking the PPAR-γ pathway, though administration of GW9662, counteracted pioglitazone's protective effects [49]. Interestingly, experimental studies have also shown a reduction in the level of peroxisome proliferator-activated receptor-gamma coactivator (PGC-1α), one of the proteins activated by PPAR-γ in skeletal muscles of HD mice models as well as HD patients [50]. Additionally, the pharmacological activation of this co-activator led to the increased expression of skeletal muscle fiber proteins that suggested an important role of the PPAR pathway in the development of HD-related myopathy [51]. Moreover, rosiglitazone treatment seems to have an impact also on the cardiovascular system [43]. Consistently with our observation, rosiglitazone administration led to improvement of cardiac function (contractile dysfunction and the protection of myocardial injury during ischemic/reperfusion) in different animal models [52–54]. Nevertheless, opposite findings have also been reported. Growing evidence has demonstrated adverse effects of rosiglitazone, including increased risk of acute myocardial infarction and heart failure, which was one of the causes of its withdrawal from EU countries [55,56]. Besides its controversy, rosiglitazone seems to be an interesting therapeutic tool in HD due to unique metabolic alterations reported in this disease.

Indeed, our previous research indicated that HD cardio- and myopathy were linked with deficits in energy metabolism [17,30]. We noted that HD mice models exhibited decreased glucose oxidation in skeletal muscle and heart. This change may reduce production capacity for adenosine-5′-triphosphate (ATP), especially if combined with reduced oxygen supply. Glucose is a better substrate for energy supply ensuring cell survival. Its oxidation generates more ATP molecules concerning oxygen consumption than other metabolic fuels. Thus, increasing glucose oxidation in HD-affected skeletal muscle and heart may result in an improvement of its function. PPAR agonists could induce such an effect. Indeed, How et al. noted that rosiglitazone treatment diminished cardiac fatty acids and increased cardiac glucose oxidation in diabetic mice [57]. Rosiglitazone enhanced also glucose oxidation, and thus its overall use in metabolism in our study. Similar to our results, it was accompanied by reduced blood glucose levels caused by increased whole-body glucose uptake and its reduced hepatic release [57,58]. One may conclude that rosiglitazone treatment may help cardiac and skeletal muscle function in HD mice by treating impaired glucose homeostasis that may develop at later stages of the disease instead of directly targeting primary HD defects [59]. Although R6/1 displayed some signs of impaired glucose tolerance (including abnormal glucose handling and higher glucose plasma and insulin levels in the glucose challenge), it did not manifest as diabetes due to normal peripheral insulin sensitivity. Moreover, the fasting plasma glucose levels were similar to the values in wild-type mice, as described by different investigators; thus, this experimental model is recognized as a nondiabetic HD mouse model [59–61]. Interestingly, the glucose concentration reduction after rosiglitazone was also observed in non-diabetic individuals [58]. This suggests that rosiglitazone treatment could be beneficial for skeletal muscle and cardiac metabolism in HD in a way that is independent of diabetic status.

As mentioned earlier, PPAR-γ is not highly expressed in cardiac and skeletal muscle tissue, and so the effect of rosiglitazone on substrate metabolism might be most likely indirect to the changes in cardiac substrate supply, rather than a direct effect of PPARγ on cardiac or skeletal muscle. Nevertheless, our study indicated that this metabolic shift resulted in improvement of total adenine nucleotides as well as total phosphocreatine and creatine pool in R6/1 mice model skeletal muscle and heart. Moreover, energy deficits improved via rosiglitazone abolished also the massive elevation in nucleotides catabolites (inosine, hypoxanthine, and uric acid) concentration in HD mouse serum observed in our previous research [30]. Indeed, we found an increase in inosine, hypoxanthine, xanthine, and uric acid in the sera of the HD murine models. More importantly, we found also that hypoxanthine levels were elevated also in the plasma of symptomatic HD patients and correlated with HD progression parameters [30]. We suggested earlier that catabolites that we observed in sera were released by the affected heart and/or skeletal muscle tissues, which is in the line with the current study.

Changes in energy metabolism in HD-related myopathies were linked with disruption in mitochondrial structure [50,62,63]. It has been reported that rosiglitazone treatment increases mitochondrial biogenesis in the brain and others mice tissues [44,64,65]. Indeed, we also found the increased activity of citrate synthase, a commonly used marker of mitochondrial abundance, in HD mouse model skeletal muscle and hearts treated with rosiglitazone [66]. Mitochondrial dysfunction in HD was linked also to its functionality changes including disruption of mitochondrial metabolism, calcium overload, and thus oxidative stress induction [21,67–70]. It has been shown that mHTT intracellular aggregation leads to increased reactive oxygen species (ROS) production [71]. ROS play also an important role in skeletal muscle atrophy and heart failure [72–74]. Abnormalities of superoxide dismutase activity and glutathione peroxidase, antioxidant enzymes involved in ROS breakdown, were also found in the HD-affected cardiac mitochondria [75]. Interestingly, we noted also an increased ROS level in R6/1 mice's hearts and skeletal muscle.

It is well known that mitochondrial complex I and III, but especially complex I, are considered to be the main sites of ROS production [76]. However, data that highlighted mitochondrial complexes activity in HD-affected skeletal muscle and heart are ambiguous. Indeed, no alterations of the mitochondrial electron transport chain activity were found in the skeletal muscle of 12-weeks-old R6/2 mice or skeletal muscle of 15-months-old HdhQ111 knock-in mice [77,78]. On the other hand, other analyses, performed on muscles of R6/2 mice at late stages of disease progression, reported a significant reduction of the activity of the complex IV or both, complex I and IV [6,79]. Our current study underlined the opposite tendency. We found that mitochondria of the soleus of the R6/1 mice model, despite reduced complex II and no changes in IV activity, exhibited elevated complex I activity. A similar tendency in complex I respiration we found in the mitochondria isolated from R6/1 mice hearts. Moreover, in contrast to mitochondria isolated from skeletal muscle the improvement of complex II respiration was noted. While Kojer et al. highlighted no changes in mitochondrial oxidative chain complexes in R6/2 mice hearts [6]. We suggested that such complex I hyperactivation may be the cause of observed increased ROS production and thus HD cardiac and skeletal muscle mitochondria dysfunction. Nevertheless, more mechanistic studies on isolated mitochondria or cardiac and muscle cells are needed to clarify the origin of this activation.

There is a study indicating that rosiglitazone treatment rapidly decreases the activities of mitochondrial respiratory complex I and III without modifying complex II in liver mitochondria, and in this way diminished intracellular ROS production [80]. Rabol et al. showed that skeletal muscle mitochondria isolated from patients with diabetes type 2 supplemented with rosiglitazone for 12 weeks also exerted an inhibitory effect on complex I [81]. It is in the line with our data, highlighting that rosiglitazone treatment of HD mouse model leads to reduction of complex I activity in mitochondria isolated from skeletal muscle and heart. That may suppress the ROS overproduction and improve cardiac and skeletal muscle functionality. Indeed we also found the improvement of total plasma antioxidant status (TAOS), one of the oxidative balance indicators in rosiglitazone-treated mice, which suggested its correction [82]. It is known that the evaluation of this parameter was applied to reveal oxidative balance instead of measuring different oxidant and antioxidant molecules individually [83,84]. We noted also the reduction in TAOS in R6/1 relative to control mice. It supports the paradigms mentioned above regarding the presence of oxidative stress in HD mouse models. Interestingly, recently Yildiz et al. also underlined a significant reduction in TAOS in patients with Alzheimer's disease and other CNS diseases [85]. Unfortunately, no data were available from HD patients.

5. Conclusions and Perspectives

Several earlier studies highlighted the usefulness of anti-diabetic drugs (primarily peroxisome proliferator-activated receptors (PPARs) agonists, like rosiglitazone) in the treatment of patients with other neurodegenerations such as Alzheimer's disease [86]. Hervas et al. (2017) noted that administration of metformin—another anti-diabetic drug that stimulates adenosine monophosphate-activated protein kinase (AMPK)—was associated with better cognitive function and motor function improvement in Huntington's disease (HD) patients [87]. Although, usage of an AMPK activator in HD might be beneficial only in the first stages of the disease where AMPK is downregulated, not in advanced stages where AMPK is already upregulated in HD-affected tissues (brain as well as skeletal muscles and heart) [30,87–89]. PPAR agonists could be more effective since PPAR and its signaling are downregulated in all HD stages [90].

Despite numerous basic studies highlighting the role of PPAR agonists in HD, more clinical studies are needed to clarify its possible usage in HD treatment. These studies should especially identify whether non-diabetic or diabetic HD patients would benefit more, and identify side effects. Our research conducted on non-diabetic R6/1 HD mouse models indicated that even in non-diabetic HD patients, such treatment could be considered.

6. Limitations

It has to be mentioned that grip strength evaluated in this study does not selectively identify a change in skeletal muscle function. Besides skeletal muscle functionality, the central nervous system (CNS), motor neurons, and neuromuscular transmission also contribute to overall grip strength. Thus, enhancement of this parameter after peroxisome proliferator-activated receptors (PPARs) agonist treatment might be related to the improvement of the CNS and its signaling (peripheral nerve functionality enhancement), while cardiac function recovery after rosiglitazone therapy might be linked with the improvement of the heart–brain axis. Nonetheless, dysfunction of peripheral cells was noted also in vitro studies, suggesting that the pathogenesis of Huntington's disease (HD) in the heart and skeletal muscle may be independent of the CNS [1,91].

It is well known that skeletal muscle can continue to generate force at low adenosine-5′-triphosphate (ATP) levels. Indeed, it has been shown that maximum skeletal muscle force does not decrease until the concentration of ATP is less than 20 μM [92]. Moreover, skeletal muscle force was shown to increase by 10% in 0.5–1 mM of ATP concentration [93,94]. Changes in ATP levels might also affect muscle function by altering the SR Ca2+ ATPase (SERCA pump) and excitation-contraction coupling [95]. Interestingly, there is work showing altered excitation-contraction coupling in HD-affected muscle [96,97].

Furthermore, maximal skeletal muscle force occurs when blood flow is blocked to the muscle due to the contraction. These adaptations gave the ability of a skeletal muscle to continue its function in extreme metabolic conditions. Thus, as observed in our research, improved skeletal muscle glucose metabolism and total adenine nucleotides pool after rosiglitazone treatment may not result in improved skeletal muscle contractility parameters. On the other hand, our previous research indicated that the skeletal muscle of HD mice models was characterized by progressive impairment of the contractile and significant loss of motor units, accompanied by diminished skeletal muscle glucose oxidation and deterioration in energy metabolism [17]. Thus, one may conclude that improvement of these parameters in HD-affected skeletal muscle may have a positive impact on its functionality; however, more experimental studies are needed to clarify this assumption.

Despite these limitations, our study for the first time highlighted that the treatment of the HD mouse model with PPAR-γ agonists like rosiglitazone induces alterations in skeletal muscle and heart metabolism that may contribute to enhanced grip strength and improvement of cardiac function. Thus, these molecules might be an interesting therapeutic tool to treat not only neurodegeneration but also cardiomyopathy and myopathy linked to HD.

Supplementary Materials: The following supporting information can be downloaded at: https://www.mdpi.com/article/10.3390/cells11172662/s1, Figure S1: Reduced grip strength in R6/1 mice in comparison to healthy controls (C57BL/6J), Figure S2: Deterioration of heart function in R6/1 mice model, Figure S3: Representative echocardiograms from R6/1 mice model (A.) and R6/1 mice model treated with rosiglitazone (B.), Figure S4: Blood glucose and serum-free fatty acids concentration in R6/1 mice (HD) as well as HD with Rosiglitazone (HD+ Rosiglitazone), Figure S5: Representative Seahorse XF assays on coupled cardiac and skeletal muscle mitochondria isolated from HD and control mice, Figure S6: Representative Seahorse XF assays on isolated cardiac mitochondria of HD and control mice, Figure S7: Representative Seahorse XF assays on isolated soleus muscle mitochondria of HD and control mice., Figure S8: Mitochondrial functionality parameters (citrate synthase and ROS) in hearts and skeletal muscle of HD and control mice, Figure S9: Total plasma antioxidant status in C57Bl (control) and R6/1 (HD) mice.

Author Contributions: Conceptualization, M.T. and R.T.S.; Formal analysis, M.T., A.B., P.M., M.P., O.K., P.J., A.J. and K.P.; Funding acquisition, M.T. and R.T.S.; Investigation, M.T., A.B., P.M., M.P., P.J., A.J. and K.P.; Methodology, M.T., A.B., P.M., O.K., P.J. and A.J.; Supervision, G.W., E.M.S. and R.T.S.; Writing—original draft, M.T. and R.T.S. All authors have read and agreed to the published version of the manuscript.

Funding: This research was supported by the National Science Centre in Poland (number 2016/23/B/NZ4/03877 and number 2015/17/N/NZ4/02841).

Institutional Review Board Statement: All experimental protocols used in this study were approved by the local commission for animal experiments in Bydgoszcz, Poland (Resolution 43/2015).

Informed Consent Statement: Not applicable.

Data Availability Statement: Not applicable.

Conflicts of Interest: The authors declare no conflict of interest.

References

1. Walker, F.O. Huntington's Disease. *Lancet* **2007**, *369*, 218–228. [CrossRef]
2. Davies, S.W.; Turmaine, M.; Cozens, B.A.; DiFiglia, M.; Sharp, A.H.; Ross, C.A.; Scherzinger, E.; Wanker, E.E.; Mangiarini, L.; Bates, G.P. Formation of Neuronal Intranuclear Inclusions Underlies the Neurological Dysfunction in Mice Transgenic for the HD Mutation. *Cell* **1997**, *90*, 537–548. [CrossRef]
3. Moffitt, H.; McPhail, G.D.; Woodman, B.; Hobbs, C.; Bates, G.P. Formation of Polyglutamine Inclusions in a Wide Range of Non-CNS Tissues in the HdhQ150 Knock-in Mouse Model of Huntington's Disease. *PLoS ONE* **2009**, *4*, e8025. [CrossRef] [PubMed]
4. Li, S.H.; Schilling, G.; Young, W.S., 3rd; Li, X.J.; Margolis, R.L.; Stine, O.C.; Wagster, M.V.; Abbott, M.H.; Franz, M.L.; Ranen, N.G.; et al. Huntington's Disease Gene (IT15) Is Widely Expressed in Human and Rat Tissues. *Neuron* **1993**, *11*, 985–993. [CrossRef]
5. Li, S.H.; Li, X.J. Huntingtin-Protein Interactions and the Pathogenesis of Huntington's Disease. *Trends Genet.* **2004**, *20*, 146–154. [CrossRef]
6. Kojer, K.; Hering, T.; Bazenet, C.; Weiss, A.; Herrmann, F.; Taanman, J.W.; Orth, M. Huntingtin Aggregates and Mitochondrial Pathology in Skeletal Muscle but Not Heart of Late-Stage R6/2 Mice. *J. Huntingtons. Dis.* **2019**, *8*, 145–159. [CrossRef]
7. Mielcarek, M.; Inuabasi, L.; Bondulich, M.K.; Muller, T.; Osborne, G.F.; Franklin, S.A.; Smith, D.L.; Neueder, A.; Rosinski, J.; Rattray, I.; et al. Dysfunction of the CNS-Heart Axis in Mouse Models of Huntington's Disease. *PLoS Genet* **2014**, *10*, e1004550. [CrossRef]
8. Busse, M.E.; Hughes, G.; Wiles, C.M.; Rosser, A.E. Use of Hand-Held Dynamometry in the Evaluation of Lower Limb Muscle Strength in People with Huntington's Disease. *J. Neurol.* **2008**, *255*, 1534–1540. [CrossRef]
9. Lanska, D.J.; Lavine, L.; Lanska, M.J.; Schoenberg, B.S. Huntington's Disease Mortality in the United States. *Neurology* **1988**, *38*, 769–772. [CrossRef]
10. Chiu, E.; Alexander, L. Causes of Death in Huntington's Disease. *Med. J. Aust.* **1982**, *1*, 153. [CrossRef]
11. van der Burg, J.M.; Bjorkqvist, M.; Brundin, P. Beyond the Brain: Widespread Pathology in Huntington's Disease. *Lancet Neurol.* **2009**, *8*, 765–774. [CrossRef]
12. Ribchester, R.R.; Thomson, D.; Wood, N.I.; Hinks, T.; Gillingwater, T.H.; Wishart, T.M.; Court, F.A.; Morton, A.J. Progressive Abnormalities in Skeletal Muscle and Neuromuscular Junctions of Transgenic Mice Expressing the Huntington's Disease Mutation. *Eur. J. Neurosci.* **2004**, *20*, 3092–3114. [CrossRef] [PubMed]
13. Romer, S.H.; Metzger, S.; Peraza, K.; Wright, M.C.; Jobe, D.S.; Song, L.S.; Rich, M.M.; Foy, B.D.; Talmadge, R.J.; Voss, A.A. A Mouse Model of Huntington's Disease Shows Altered Ultrastructure of Transverse Tubules in Skeletal Muscle Fibers. *J. Gen. Physiol.* **2021**, *153*, e202012637. [CrossRef] [PubMed]
14. Valadão, P.A.C.; de Aragão, B.C.; Andrade, J.N.; Magalhães-Gomes, M.P.S.; Foureaux, G.; Joviano-Santos, J.V.; Nogueira, J.C.; Machado, T.C.G.; de Jesus, I.C.G.; Nogueira, J.M.; et al. Abnormalities in the Motor Unit of a Fast-Twitch Lower Limb Skeletal Muscle in Huntington's Disease. *ASN Neuro* **2019**, *11*, 1759091419886212. [CrossRef] [PubMed]
15. Valadão, P.A.C.; de Aragão, B.C.; Andrade, J.N.; Magalhães-Gomes, M.P.S.; Foureaux, G.; Joviano-Santos, J.V.; Nogueira, J.C.; Ribeiro, F.M.; Tapia, J.C.; Guatimosim, C. Muscle Atrophy Is Associated with Cervical Spinal Motoneuron Loss in BACHD Mouse Model for Huntington's Disease. *Eur. J. Neurosci.* **2017**, *45*, 785–796. [CrossRef]
16. Strand, A.D.; Aragaki, A.K.; Shaw, D.; Bird, T.; Holton, J.; Turner, C.; Tapscott, S.J.; Tabrizi, S.J.; Schapira, A.H.; Kooperberg, C.; et al. Gene Expression in Huntington's Disease Skeletal Muscle: A Potential Biomarker. *Hum. Mol. Genet.* **2005**, *14*, 1863–1876. [CrossRef]

17. Mielcarek, M.; Toczek, M.; Smeets, C.J.; Franklin, S.A.; Bondulich, M.K.; Jolinon, N.; Muller, T.; Ahmed, M.; Dick, J.R.; Piotrowska, I.; et al. HDAC4-Myogenin Axis As an Important Marker of HD-Related Skeletal Muscle Atrophy. *PLoS Genet.* **2015**, *11*, e1005021. [CrossRef]
18. Wood, N.I.; Sawiak, S.J.; Buonincontri, G.; Niu, Y.; Kane, A.D.; Carpenter, T.A.; Giussani, D.A.; Morton, A.J. Direct Evidence of Progressive Cardiac Dysfunction in a Transgenic Mouse Model of Huntington's Disease. *J. Huntingt. Dis.* **2012**, *1*, 57–64. [CrossRef]
19. Kiriazis, H.; Jennings, N.L.; Davern, P.; Lambert, G.; Su, Y.; Pang, T.; Du, X.; La Greca, L.; Head, G.A.; Hannan, A.J.; et al. Neurocardiac Dysregulation and Neurogenic Arrhythmias in a Transgenic Mouse Model of Huntington's Disease. *J. Physiol.* **2012**, *590*, 5845–5860. [CrossRef]
20. Mielcarek, M. Huntington's Disease Is a Multi-System Disorder. *Rare Dis.* **2015**, *3*, e1058464. [CrossRef]
21. Mochel, F.; Haller, R.G. Energy Deficit in Huntington Disease: Why It Matters. *Journal of Clinical Investigation* **2011**, *121*, 493–499. [CrossRef] [PubMed]
22. Jodeiri Farshbaf, M.; Ghaedi, K. Huntington's Disease and Mitochondria. *Neurotox Res* **2017**, *32*, 518–529. [CrossRef] [PubMed]
23. Gines, S.; Seong, I.S.; Fossale, E.; Ivanova, E.; Trettel, F.; Gusella, J.F.; Wheeler, V.C.; Persichetti, F.; MacDonald, M.E. Specific Progressive CAMP Reduction Implicates Energy Deficit in Presymptomatic Huntington's Disease Knock-in Mice. *Hum. Mol. Genet.* **2003**, *12*, 497–508. [CrossRef]
24. Mochel, F.; Durant, B.; Meng, X.; O'Callaghan, J.; Yu, H.; Brouillet, E.; Wheeler, V.C.; Humbert, S.; Schiffmann, R.; Durr, A. Early Alterations of Brain Cellular Energy Homeostasis in Huntington Disease Models. *J. Biol. Chem.* **2012**, *287*, 1361–1370. [CrossRef]
25. Seong, I.S.; Ivanova, E.; Lee, J.M.; Choo, Y.S.; Fossale, E.; Anderson, M.; Gusella, J.F.; Laramie, J.M.; Myers, R.H.; Lesort, M.; et al. HD CAG Repeat Implicates a Dominant Property of Huntingtin in Mitochondrial Energy Metabolism. *Hum. Mol. Genet.* **2005**, *14*, 2871–2880. [CrossRef] [PubMed]
26. Cui, L.; Jeong, H.; Borovecki, F.; Parkhurst, C.N.; Tanese, N.; Krainc, D. Transcriptional Repression of PGC-1alpha by Mutant Huntingtin Leads to Mitochondrial Dysfunction and Neurodegeneration. *Cell* **2006**, *127*, 59–69. [CrossRef] [PubMed]
27. Zielonka, D.; Piotrowska, I.; Marcinkowski, J.T.; Mielcarek, M. Skeletal Muscle Pathology in Huntington's Disease. *Front. Physiol.* **2014**, *5*, 380. [CrossRef]
28. Saft, C.; Zange, J.; Andrich, J.; Muller, K.; Lindenberg, K.; Landwehrmeyer, B.; Vorgerd, M.; Kraus, P.H.; Przuntek, H.; Schols, L. Mitochondrial Impairment in Patients and Asymptomatic Mutation Carriers of Huntington's Disease. *Mov. Disord.* **2005**, *20*, 674–679. [CrossRef]
29. Lodi, R.; Schapira, A.H.V.; Manners, D.; Styles, P.; Wood, N.W.; Taylor, D.J.; Warner, T.T. Abnormal in Vivo Skeletal Muscle Energy Metabolism in Huntington's Disease and Dentatorubropallidoluysian Atrophy. *Ann. Neurol.* **2000**, *48*, 72–76. [CrossRef]
30. Toczek, M.; Zielonka, D.; Zukowska, P.; Marcinkowski, J.T.; Slominska, E.; Isalan, M.; Smolenski, R.T.; Mielcarek, M. An Impaired Metabolism of Nucleotides Underpins a Novel Mechanism of Cardiac Remodeling Leading to Huntington's Disease Related Cardiomyopathy. *Biochim. Biophys. Acta-Mol. Basis Dis.* **2016**, *1862*, 2147–2157. [CrossRef]
31. Duan, W.; Jiang, M.; Jin, J. Metabolism in HD: Still a Relevant Mechanism? *Mov. Disord.* **2014**, *29*, 1366–1374. [CrossRef] [PubMed]
32. Wagner, N.; Wagner, K.-D. Peroxisome Proliferator-Activated Receptors and the Hallmarks of Cancer. *Cells* **2022**, *11*, 2432. [CrossRef] [PubMed]
33. Wagner, K.D.; Wagner, N. Peroxisome Proliferator-Activated Receptor Beta/Delta (PPARβ/δ) Acts as Regulator of Metabolism Linked to Multiple Cellular Functions. *Pharmacol. Ther.* **2010**, *125*, 423–435. [CrossRef] [PubMed]
34. Dickey, A.S.; Pineda, V.V.; Tsunemi, T.; Liu, P.P.; Miranda, H.C.; Gilmore-Hall, S.K.; Lomas, N.; Sampat, K.R.; Buttgereit, A.; Torres, M.J.M.; et al. PPAR-δ Is Repressed in Huntington's Disease, Is Required for Normal Neuronal Function and Can Be Targeted Therapeutically. *Nat. Med.* **2016**, *22*, 37–45. [CrossRef]
35. Jin, J.; Albertz, J.; Guo, Z.; Peng, Q.; Rudow, G.; Troncoso, J.C.; Ross, C.A.; Duan, W. Neuroprotective Effects of PPAR-γ Agonist Rosiglitazone in N171-82Q Mouse Model of Huntington's Disease. *J. Neurochem.* **2013**, *125*, 410–419. [CrossRef]
36. Tomczyk, M.; Braczko, A.; Jablonska, P.; Mika, A.; Przyborowski, K.; Jedrzejewska, A.; Krol, O.; Kus, F.; Sledzinski, T.; Chlopicki, S.; et al. Enhanced Muscle Strength in Dyslipidemic Mice and Its Relation to Increased Capacity for Fatty Acid Oxidation. *Int. J. Mol. Sci.* **2021**, *22*, 12251. [CrossRef]
37. Zukowska, P.; Kutryb-Zajac, B.; Jasztal, A.; Toczek, M.; Zabielska, M.; Borkowski, T.; Khalpey, Z.; Smolenski, R.T.; Slominska, E.M. Deletion of CD73 in Mice Leads to Aortic Valve Dysfunction. *Biochim. Biophys. Acta Mol. Basis Dis.* **2017**, *1863*, 1464–1472. [CrossRef]
38. Tomczyk, M.; Olkowicz, M.; Slominska, E.M.; Smolenski, R.T. High Throughput Procedure for Comparative Analysis of in Vivo Cardiac Glucose or Amino Acids Use in Cardiovascular Pathologies and Pharmacological Treatments. *Metabolites* **2021**, *11*, 497. [CrossRef]
39. Rogers, G.W.; Brand, M.D.; Petrosyan, S.; Ashok, D.; Elorza, A.A.; Ferrick, D.A.; Murphy, A.N. High Throughput Microplate Respiratory Measurements Using Minimal Quantities of Isolated Mitochondria. *PLoS ONE* **2011**, *6*, e21746. [CrossRef]
40. Cichońska, D.; Król, O.; Słomińska, E.M.; Kochańska, B.; Świetlik, D.; Ochocińska, J.; Kusiak, A. Influence of Electronic Cigarettes on Antioxidant Capacity and Nucleotide Metabolites in Saliva. *Toxics* **2021**, *9*, 263. [CrossRef]
41. McNulty, P.H.; Cline, G.W.; Whiting, J.M.; Shulman, G.I. Regulation of Myocardial [^{13}C]Glucose Metabolism in Conscious Rats. *Am. J. Physiol.-Heart Circ. Physiol.* **2000**, *279*, H375–H381. [CrossRef] [PubMed]

42. Olkowicz, M.; Tomczyk, M.; Debski, J.; Tyrankiewicz, U.; Przyborowski, K.; Borkowski, T.; Zabielska-Kaczorowska, M.; Szupryczynska, N.; Kochan, Z.; Smeda, M.; et al. Enhanced Cardiac Hypoxic Injury in Atherogenic Dyslipidaemia Results from Alterations in the Energy Metabolism Pattern. *Metabolism.* **2021**, *114*, 154400. [CrossRef]
43. Palee, S. PPARγ Activator, Rosiglitazone: Is It Beneficial or Harmful to the Cardiovascular System? *World J. Cardiol.* **2011**, *3*, 144. [CrossRef] [PubMed]
44. Quintanilla, R.A.; Jin, Y.N.; Fuenzalida, K.; Bronfman, M.; Johnson, G.V.W. Rosiglitazone Treatment Prevents Mitochondrial Dysfunction in Mutant Huntingtin-Expressing Cells. *J. Biol. Chem.* **2008**, *283*, 25628–25637. [CrossRef]
45. Chiang, M.C.; Cheng, Y.C.; Nicol, C.J.; Lin, K.H.; Yen, C.H.; Chen, S.J.; Huang, R.N. Rosiglitazone Activation of PPARγ-Dependent Signaling Is Neuroprotective in Mutant Huntingtin Expressing Cells. *Exp. Cell Res.* **2015**, *338*, 183–193. [CrossRef] [PubMed]
46. Mishra, J.; Chaudhary, T.; Kumar, A. Rosiglitazone Synergizes the Neuroprotective Effects of Valproic Acid against Quinolinic Acid-Induced Neurotoxicity in Rats: Targeting PPARγ and HDAC Pathways. *Neurotox. Res.* **2014**, *26*, 130–151. [CrossRef] [PubMed]
47. Han, L.; Shen, W.J.; Bittner, S.; Kraemer, F.B.; Azhar, S. PPARs: Regulators of Metabolism and as Therapeutic Targets in Cardiovascular Disease. Part II: PPAR-β/δ and PPAR-γ. *Future Cardiol.* **2017**, *13*, 279–296. [CrossRef]
48. Janani, C.; Ranjitha Kumari, B.D. PPAR Gamma Gene—A Review. *Diabetes Metab. Syndr. Clin. Res. Rev.* **2015**, *9*, 46–50. [CrossRef]
49. Hassan, F.E.; Sakr, H.I.; Mohie, P.M.; Suliman, H.S.; Mohamed, A.S.; Attia, M.H.; Eid, D.M. Pioglitazone Improves Skeletal Muscle Functions in Reserpine-Induced Fibromyalgia Rat Model. *Ann. Med.* **2021**, *53*, 1033–1041. [CrossRef]
50. Chaturvedi, R.K.; Adhihetty, P.; Shukla, S.; Hennessy, T.; Calingasan, N.; Yang, L.; Starkov, A.; Kiaei, M.; Cannella, M.; Sassone, J.; et al. Impaired PGC-1α Function in Muscle in Huntington's Disease. *Hum. Mol. Genet.* **2009**, *18*, 3048–3065. [CrossRef]
51. Johri, A.; Calingasan, N.Y.; Hennessey, T.M.; Sharma, A.; Yang, L.; Wille, E.; Chandra, A.; Beal, M.F. Pharmacologic Activation of Mitochondrial Biogenesis Exerts Widespread Beneficial Effects in a Transgenic Mouse Model of Huntington's Disease. *Hum. Mol. Genet.* **2012**, *21*, 1124–1137. [CrossRef] [PubMed]
52. Yue, T.L.; Chen, J.; Bao, W.; Narayanan, P.K.; Bril, A.; Jiang, W.; Lysko, P.G.; Gu, J.L.; Boyce, R.; Zimmerman, D.M.; et al. In Vivo Myocardial Protection from Ischemia/Reperfusion Injury by the Peroxisome Proliferator-Activated Receptor-γ Agonist Rosiglitazone. *Circulation* **2001**, *104*, 2588–2594. [CrossRef] [PubMed]
53. Khandoudi, N.; Delerive, P.; Berrebi-Bertrand, I.; Buckingham, R.E.; Staels, B.; Bril, A. Rosiglitazone, a Peroxisome Proliferator-Activated Receptor-γ, Inhibits the Jun NH2-Terminal Kinase/Activating Protein 1 Pathway and Protects the Heart from Ischemia/Reperfusion Injury. *Diabetes* **2002**, *51*, 1507–1514. [CrossRef] [PubMed]
54. Sidell, R.J.; Cole, M.A.; Draper, N.J.; Desrois, M.; Buckingham, R.E.; Clarke, K. Thiazolidinedione Treatment Normalizes Insulin Resistance and Ischemic Injury in the Zucker Fatty Rat Heart. *Diabetes* **2002**, *51*, 1110–1117. [CrossRef]
55. Gersh, B.J. Effect of Rosiglitazone on the Risk of Myocardial Infarction and Death from Cardiovascular Causes. *Yearb. Cardiol.* **2008**, *2008*, 2457–2471. [CrossRef]
56. Home, P.D.; Pocock, S.J.; Beck-Nielsen, H.; Gomis, R.; Hanefeld, M.; Jones, N.P.; Komajda, M.; McMurray, J.J.V. Rosiglitazone Evaluated for Cardiovascular Outcomes—An Interim Analysis. *N. Engl. J. Med.* **2007**, *357*, 28–38. [CrossRef]
57. How, O.J.; Larsen, T.S.; Hafstad, A.D.; Khalid, A.; Myhre, E.S.P.; Murray, A.J.; Boardman, N.T.; Cole, M.; Clarke, K.; Severson, D.L.; et al. Rosiglitazone Treatment Improves Cardiac Efficiency in Hearts from Diabetic Mice. *Arch. Physiol. Biochem.* **2007**, *113*, 211–220. [CrossRef]
58. Kim, S.H.; Abbasi, F.; Chu, J.W.; McLaughlin, T.L.; Lamendola, C.; Polonsky, K.S.; Reaven, G.M. Rosiglitazone Reduces Glucose-Stimulated Insulin Secretion Rate and Increases Insulin Clearance in Nondiabetic, Insulin-Resistant Individuals. *Diabetes* **2005**, *54*, 2447–2452. [CrossRef]
59. Montojo, M.T.; Aganzo, M.; González, N. Huntington's Disease and Diabetes: Chronological Sequence of Its Association. *J. Huntingt. Dis.* **2017**, *6*, 179–188. [CrossRef]
60. Hansson, O.; Petersén, Å.; Leist, M.; Nicotera, P.; Castilho, R.F.; Brundin, P. Transgenic Mice Expressing a Huntington's Disease Mutation Are Resistant to Quinolinic Acid-Induced Striatal Excitotoxicity. *Proc. Natl. Acad. Sci. USA* **1999**, *96*, 8727–8732. [CrossRef]
61. Van Dellen, A.; Blakemore, C.; Deacon, R.; York, D.; Hannan, A.J. Delaying the Onset of Huntington's in Mice. *Nature* **2000**, *404*, 721–722. [CrossRef] [PubMed]
62. Mihm, M.J.; Amann, D.M.; Schanbacher, B.L.; Altschuld, R.A.; Bauer, J.A.; Hoyt, K.R. Cardiac Dysfunction in the R6/2 Mouse Model of Huntington's Disease. *Neurobiol. Dis.* **2007**, *25*, 297–308. [CrossRef] [PubMed]
63. Bozzi, M.; Sciandra, F. Molecular Mechanisms Underlying Muscle Wasting in Huntington's Disease. *Int. J. Mol. Sci.* **2020**, *21*, 8314. [CrossRef] [PubMed]
64. Strum, J.C.; Shehee, R.; Virley, D.; Richardson, J.; Mattie, M.; Selley, P.; Ghosh, S.; Nock, C.; Saunders, A.; Roses, A. Rosiglitazone Induces Mitochondrial Biogenesis in Mouse Brain. *J. Alzheimer's Dis.* **2007**, *11*, 45–51. [CrossRef]
65. Wilson-Fritch, L.; Nicoloro, S.; Chouinard, M.; Lazar, M.A.; Chui, P.C.; Leszyk, J.; Straubhaar, J.; Czech, M.P.; Corvera, S. Mitochondrial Remodeling in Adipose Tissue Associated with Obesity and Treatment with Rosiglitazone. *J. Clin. Investig.* **2004**, *114*, 1281–1289. [CrossRef]
66. Scaini, G.; Rochi, N.; Benedet, J.; Ferreira, G.K.; Teodorak, B.P.; Comim, C.M.; de Constantino, L.S.; Vuolo, F.; Constantino, L.C.; Quevedo, J.; et al. Inhibition of Brain Citrate Synthase Activity in an Animal Model of Sepsis. *Rev. Bras. Ter. Intensiva* **2011**, *23*, 158–163. [CrossRef]

67. Intihar, T.A.; Martinez, E.A.; Gomez-Pastor, R. Mitochondrial Dysfunction in Huntington's Disease; Interplay between HSF1, P53 and PGC-1α Transcription Factors. *Front. Cell. Neurosci.* **2019**, *13*, 103. [CrossRef]
68. Ismailoglu, I.; Chen, Q.; Popowski, M.; Yang, L.; Gross, S.S.; Brivanlou, A.H. Huntingtin Protein Is Essential for Mitochondrial Metabolism, Bioenergetics and Structure in Murine Embryonic Stem Cells. *Dev. Biol.* **2014**, *391*, 230–240. [CrossRef]
69. Kumar, A.; Ratan, R.R. Oxidative Stress and Huntington's Disease: The Good, the Bad, and the Ugly. *J. Huntingt. Dis.* **2016**, *5*, 217–237. [CrossRef]
70. Ayala-Peña, S. Role of Oxidative DNA Damage in Mitochondrial Dysfunction and Huntington's Disease Pathogenesis. *Free. Radic. Biol. Med.* **2013**, *62*, 102–110. [CrossRef]
71. Hands, S.; Sajjad, M.U.; Newton, M.J.; Wyttenbach, A. In Vitro and in Vivo Aggregation of a Fragment of Huntingtin Protein Directly Causes Free Radical Production. *J. Biol. Chem.* **2011**, *286*, 44512–44520. [CrossRef] [PubMed]
72. Zuo, L.; Best, T.M.; Roberts, W.J.; Diaz, P.T.; Wagner, P.D. Characterization of Reactive Oxygen Species in Diaphragm. *Acta Physiol.* **2015**, *213*, 700–710. [CrossRef] [PubMed]
73. Zuo, L.; Pannell, B.K. Redox Characterization of Functioning Skeletal Muscle. *Front. Physiol.* **2015**, *6*, 338. [CrossRef] [PubMed]
74. He, F.; Zuo, L. Redox Roles of Reactive Oxygen Species in Cardiovascular Diseases. *Int. J. Mol. Sci.* **2015**, *16*, 27770–27780. [CrossRef]
75. Joviano-Santos, J.V.; Santos-Miranda, A.; Botelho, A.F.M.; de Jesus, I.C.G.; Andrade, J.N.; de Oliveira Barreto, T.; Magalhães-Gomes, M.P.S.; Valadão, P.A.C.; dos Cruz, J.S.; Melo, M.M.; et al. Increased Oxidative Stress and CaMKII Activity Contribute to Electro-Mechanical Defects in Cardiomyocytes from a Murine Model of Huntington's Disease. *FEBS J.* **2019**, *286*, 110–123. [CrossRef]
76. Zhao, R.Z.; Jiang, S.; Zhang, L.; Yu, Z. Bin. Mitochondrial Electron Transport Chain, ROS Generation and Uncoupling (Review). *Int. J. Mol. Med.* **2019**, *44*, 3–15. [CrossRef]
77. Tabrizi, S.J.; Workman, J.; Hart, P.E.; Mangiarini, L.; Mahal, A.; Bates, G.; Cooper, J.M.; Schapira, A.H.V. Mitochondrial Dysfunction and Free Radical Damage in the Huntington R6/2 Transgenic Mouse. *Ann. Neurol.* **2000**, *47*, 80–86. [CrossRef]
78. Buck, E.; Zügel, M.; Schumann, U.; Merz, T.; Gumpp, A.M.; Witting, A.; Steinacker, J.M.; Landwehrmeyer, G.B.; Weydt, P.; Calzia, E.; et al. High-Resolution Respirometry of Fine-Needle Muscle Biopsies in Pre-Manifest Huntington's Disease Expansion Mutation Carriers Shows Normal Mitochondrial Respiratory Function. *PLoS ONE* **2017**, *12*, e0175248. [CrossRef]
79. Gizatullina, Z.Z.; Lindenberg, K.S.; Harjes, P.; Chen, Y.; Kosinski, C.M.; Landwehrmeyer, B.G.; Ludolph, A.C.; Striggow, F.; Zierz, S.; Gellerich, F.N. Low Stability of Huntington Muscle Mitochondria against Ca^{2+} in R6/2 Mice. *Ann. Neurol.* **2006**, *59*, 407–411. [CrossRef]
80. Sanz, M.N.; Sánchez-Martín, C.; Detaille, D.; Vial, G.; Rigoulet, M.; El-Mir, M.Y.; Rodríguez-Villanueva, G. Acute Mitochondrial Actions of Glitazones on the Liver: A Crucial Parameter for Their Antidiabetic Properties. *Cell. Physiol. Biochem.* **2011**, *28*, 899–910. [CrossRef]
81. Rabøl, R.; Boushel, R.; Almdal, T.; Hansen, C.N.; Ploug, T.; Haugaard, S.B.; Prats, C.; Madsbad, S.; Dela, F. Opposite Effects of Pioglitazone and Rosiglitazone on Mitochondrial Respiration in Skeletal Muscle of Patients with Type 2 Diabetes. *Diabetes, Obes. Metab.* **2010**, *12*, 806–814. [CrossRef]
82. Tugrul, S.; Koçyiğit, A.; Doğan, R.; Eren, S.B.; Senturk, E.; Ozturan, O.; Ozar, O.F. Total Antioxidant Status and Oxidative Stress in Recurrent Aphthous Stomatitis. *Int. J. Dermatol.* **2016**, *55*, e130–e135. [CrossRef]
83. Erel, O. A New Automated Colorimetric Method for Measuring Total Oxidant Status. *Clin. Biochem.* **2005**, *38*, 1103–1111. [CrossRef] [PubMed]
84. Erel, O. A Novel Automated Direct Measurement Method for Total Antioxidant Capacity Using a New Generation, More Stable ABTS Radical Cation. *Clin. Biochem.* **2004**, *37*, 277–285. [CrossRef] [PubMed]
85. Yildiz, Z.; Eren, N.; Orcun, A.; Münevver Gokyigit, F.; Turgay, F.; Gündogdu Celebi, L. Serum Apelin-13 Levels and Total Oxidant/Antioxidant Status of Patients with Alzheimer's Disease. *Aging Med.* **2021**, *4*, 201–205. [CrossRef] [PubMed]
86. Rizvi, S.M.D.; Shaikh, S.; Waseem, S.M.A.; Shakil, S.; Abuzenadah, A.M.; Biswas, D.; Tabrez, S.; Ashraf, G.M.; Kamal, M.A. Role of Anti-Diabetic Drugs as Therapeutic Agents in Alzheimer's Disease. *EXCLI J.* **2015**, *14*, 684. [CrossRef]
87. Hervás, D.; Fornés-Ferrer, V.; Gómez-Escribano, A.P.; Sequedo, M.D.; Peiró, C.; Millán, J.M.; Vázquez-Manrique, R.P. Metformin Intake Associates with Better Cognitive Function in Patients with Huntington's Disease. *PLoS ONE* **2017**, *12*, e0179283. [CrossRef]
88. Mielcarek, M.; Smolenski, R.T.; Isalan, M. Transcriptional Signature of an Altered Purine Metabolism in the Skeletal Muscle of a Huntington's Disease Mouse Model. *Front. Physiol.* **2017**, *8*, 127. [CrossRef]
89. Short, B. AMPK Amplifies Huntington's Disease. *J. Cell Biol.* **2011**, *194*, 2247–2262. [CrossRef]
90. Chiang, M.C.; Chen, C.M.; Lee, M.R.; Chen, H.W.; Chen, H.M.; Wu, Y.S.; Hung, C.H.; Kang, J.J.; Chang, C.P.; Chang, C.; et al. Modulation of Energy Deficiency in Huntington's Disease via Activation of the Peroxisome Proliferator-Activated Receptor Gamma. *Hum. Mol. Genet.* **2010**, *19*, 4043–4058. [CrossRef]
91. Tomczyk, M.; Glaser, T.; Slominska, E.M.; Ulrich, H.; Smolenski, R.T. Purine Nucleotides Metabolism and Signaling in Huntington's Disease: Search for a Target for Novel Therapies. *Int. J. Mol. Sci.* **2021**, *22*, 6545. [CrossRef] [PubMed]
92. Cooke, R.; Bialek, W. Contraction of Glycerinated Muscle Fibers as a Function of the ATP Concentration. *Biophys. J.* **1979**, *28*, 241–258. [CrossRef]
93. Dutka, T.L.; Mollica, J.P.; Lamb, G.D. Differential Effects of Peroxynitrite on Contractile Protein Properties in Fast- and Slow-Twitch Skeletal Muscle Fibers of Rat. *J. Appl. Physiol.* **2011**, *110*, 705–716. [CrossRef] [PubMed]

94. Godt, R.E.; Nosek, T.M. Changes of Intracellular Milieu with Fatigue or Hypoxia Depress Contraction of Skinned Rabbit Skeletal and Cardiac Muscle. *J. Physiol.* **1989**, *412*, 155–180. [CrossRef]
95. Allen, D.G.; Lamb, G.D.; Westerblad, H. Skeletal Muscle Fatigue: Cellular Mechanisms. *Physiol. Rev.* **2008**, *88*, 287–332. [CrossRef]
96. Braubach, P.; Orynbayev, M.; Andronache, Z.; Hering, T.; Landwehrmeyer, G.B.; Lindenberg, K.S.; Melzer, W. Altered Ca^{2+} Signaling in Skeletal Muscle Fibers of the R6/2 Mouse, a Model of Huntington's Disease. *J. Gen. Physiol.* **2014**, *144*, 393–413. [CrossRef]
97. Dridi, H.; Liu, X.; Yuan, Q.; Reiken, S.; Yehya, M.; Sittenfeld, L.; Apostolou, P.; Buron, J.; Sicard, P.; Matecki, S.; et al. Role of Defective Calcium Regulation in Cardiorespiratory Dysfunction in Huntington's Disease. *JCI Insight* **2020**, *5*, e140614. [CrossRef]

cells

Review

Peroxisome Proliferator-Activated Receptors and the Hallmarks of Cancer

Nicole Wagner * and Kay-Dietrich Wagner *

CNRS, INSERM, iBV, Université Côte d'Azur, 06107 Nice, France
* Correspondence: nwagner@unice.fr (N.W.); kwagner@unice.fr (K.-D.W.); Tel.: +33-489-153-713 (K.-D.W.)

Abstract: Peroxisome proliferator-activated receptors (PPARs) function as nuclear transcription factors upon the binding of physiological or pharmacological ligands and heterodimerization with retinoic X receptors. Physiological ligands include fatty acids and fatty-acid-derived compounds with low specificity for the different PPAR subtypes (alpha, beta/delta, and gamma). For each of the PPAR subtypes, specific pharmacological agonists and antagonists, as well as pan-agonists, are available. In agreement with their natural ligands, PPARs are mainly focused on as targets for the treatment of metabolic syndrome and its associated complications. Nevertheless, many publications are available that implicate PPARs in malignancies. In several instances, they are controversial for very similar models. Thus, to better predict the potential use of PPAR modulators for personalized medicine in therapies against malignancies, it seems necessary and timely to review the three PPARs in relation to the didactic concept of cancer hallmark capabilities. We previously described the functions of PPAR beta/delta with respect to the cancer hallmarks and reviewed the implications of all PPARs in angiogenesis. Thus, the current review updates our knowledge on PPAR beta and the hallmarks of cancer and extends the concept to PPAR alpha and PPAR gamma.

Keywords: PPAR; cell proliferation; angiogenesis; cellular metabolism; immune surveillance; metastasis; resistance to cell death; tumor growth suppressors

Citation: Wagner, N.; Wagner, K.-D. Peroxisome Proliferator-Activated Receptors and the Hallmarks of Cancer. *Cells* **2022**, *11*, 2432. https://doi.org/10.3390/cells11152432

Academic Editor: Gérard Lizard

Received: 13 July 2022
Accepted: 4 August 2022
Published: 5 August 2022

1. Introduction

In addition to receptors for steroid and thyroid hormones, vitamin D and retinoids, and several orphan receptors, peroxisome proliferator-activated receptors (PPARs) belong to the group of nuclear receptors [1,2]. Although peroxisome proliferation in response to hypolipidemic fibrate drugs (PPAR alpha agonist) was described already in 1970s [3,4], it took nearly 20 years for PPAR alpha (PPARα), PPAR beta/delta (PPARβ/δ), and PPAR gamma (PPARγ) to be identified [5–7]. On the molecular level, PPARs activate/repress target genes as heterodimers with retinoic X receptors (RxR), which exist in three different isoforms. Liver X receptor α (LxRα) and retinoic acid receptors (RAR)s also form heterodimers with RxR. Thus, depending on the level of expression of the different receptors, the outcome of PPAR activation might differ between cell types (reviewed in [1]). In addition to the classical PPAR/RxR transcriptional complexes [8], PPARs might also interact with glucocorticoid receptors, photoreceptor-specific nuclear receptors, and estrogen-related receptors, which could additionally modify the responses of PPAR activation [9]. As a general PPAR response element, a direct repeat of the sequence AGGTCA, spaced by a single nucleotide, has been originally identified (DR1); in fact for PPAR alpha only [10]. Binding exclusively to this element would not explain the specificity of the identified PPAR alpha, beta/delta, and gamma target genes. Furthermore, thousands of these elements are found in the genome, mostly far away from the gene promoter regions. Experimental evidence suggests a higher heterogeneity of binding elements for PPARs [1,11]. The ligand-dependent and ligand-independent effects, posttranscriptional modifications, co-activators, and co-repressors of PPARs have been extensively reviewed [1,12,13].

Endogenous ligands for PPARs include unsaturated fatty acids, eicosanoids, prostaglandins, and prostacyclins [1,14]. Synthetic activators and inhibitors for all PPARs are available. Until now, only PPARα agonists (e.g., fibrates) have been in clinical use for lipid lowering, the prevention of atherosclerosis, and cardiovascular disease [15,16], while PPARγ agonists (e.g., thiazolidinediones) lower glucose by increasing insulin sensitivity, mainly in skeletal muscle and adipose tissue [17]. In addition to these "classical" applications for the treatment of metabolism-related diseases and metabolic syndrome, PPARs might be involved in a variety of diseases [18] and PPAR modulators might become interesting candidates for neurodegenerative disorders [19], addiction [20], psychiatric disorders [21,22], hepatic and kidney diseases [12,23–25], and autoimmune and inflammatory diseases [16,26–29]. Importantly, PPARs are also critically involved in cancer. The expression of PPARs has been detected in various cancer types and cancer cell lines, but PPARs also play important roles in the tumor stroma, i.e., cancer-associated fibroblasts, mesenchymal cells, endothelial cells, and macrophages (reviewed in [30]). In addition to cancer cell growth, angiogenesis, and the antitumor immune response play an important role in cancer progression and metastasis [31]. Here, we will use the didactic concept of the "Hallmarks of Cancer" by Hanahan and Weinberg [32–37] to delineate the functions of the different PPARs in cancer hallmark capabilities. We already used this concept for PPARβ/δ [18,38]. Thus, here, we will describe PPARα and PPARγ functions with respect to the hallmarks of cancer and updates for PPARβ/δ.

2. PPARs and Cell Proliferation

2.1. PPARα

PPARα expression has been demonstrated in human breast cancer cell lines, which showed increased proliferation upon PPARα activation [39] (Table 1). Leptin and glucose treatment stimulated breast cancer proliferation, which was accompanied by an upregulation of PPARα, suggesting the involvement of PPARα in this process [40]. Similarly, arachidonic acid (AA) has been found to promote breast cancer cell proliferation through the activation of PPARα [41]. However, contrasting results were obtained by another group [42]. The PPAR agonist fenofibrate reduced the proliferation of triple-negative breast cancer cells [43]. Similar results were obtained with clofibrate in inflammatory breast cancer cell lines [44]. Different outcomes on breast cancer cell proliferation may be explained by the different types of breast cancer cell lines used, but also by the different concentrations of fibrates. Tauber and colleagues reported stimulation of the proliferation of MCF-7 breast cancer cells with low fibrate concentrations, and suppression with high doses [45]. Dose-dependent effects of fibrates on cell proliferation have also been reported for human liver cancer cells [46]. The sustained activation of PPARα leads to liver tumorigenesis in rodents. However, in a PPARα humanized model, sustained PPARα activation very rarely provoked liver cancers, which suggests that structural differences between human and mouse PPARα are responsible for the differential susceptibility to peroxisome proliferator-induced hepatocarcinogenesis [47]. In an excellent study, Tanaka and colleagues provided evidence that the hepatitis C virus (HCV) core protein induces heterogeneous activation of PPARα in transgenic mice. The stabilization of PPARα through interaction with the Hepatitis C virus (HCV) core protein and an increase in non-esterified fatty acids, serving as endogenous PPARα ligands, were suggested to contribute to the age-dependent and multicentric hepatocarcinogenesis mediated by the core protein [48]. Interestingly, the hepatocyte restricted the constitutive activation of the PPARα-induced proliferation of hepatocytes, but not carcinogenesis, indicating that the PPARα activation of other cell types than hepatocytes is responsible for the carcinogenic effect of PPARα activation [49]. The existence of an alternatively spliced transcript variant (PPARA-tr) in humans, but not in rodents, with a deficient ligand-binding domain that is unable to bind to peroxisome proliferator-responsive DNA elements (PPREs) could partially explain the species differences in hepatocarcinogenesis [50,51]. A later study suggested a higher susceptibility of PPARα-knockout mice to diethylnitrosamine (DEN)-induced hepatocellular carcinoma (HCC) [52]. However, Kaipainen and colleagues evidenced a tumor-suppressive phenotype in PPARα-deficient mice. The absence of PPARα switches tumor-associated inflammation into tumor-suppressive inflammatory infiltrates, which inhibit tumor angiogenesis and tumor progression independently of the cellular tumor type [53]. Later, PPARα deficiency was also proposed to impair regulatory T-cell functions, leading to the inhibition of melanoma growth [54]. These studies confirm the importance of the molecular properties of stromal host cells for cancer progression, which also explains the differential outcomes of analyses in pure in vitro studies, leading to potential false therapeutic deductions. The PPARα agonist fenofibrate, for example, decreased endometrial cancer cell proliferation in vitro but failed to improve outcomes in vivo [55]. Yokoyama and co-workers reported an inhibition of proliferation in ovarian cancer cell lines in vitro, as well as a reduction in ovarian cancer cell tumor growth in vivo via the activation of PPARα with clofibrate [56]. PPARα is expressed in medulloblastoma cells, and PPARα activation with fenofibrate inhibited cell proliferation in medulloblastoma cell lines [57]. Similar results were proposed using fenofibrate treatment in a glioblastoma cell line [58] and neuroblastoma cells [59]. However, the overexpression of PPARα in glioma stem cells (GSCs) has been observed. GSCs are responsible for tumor initiation, treatment resistance, and recurrence. The knockdown (KD) of PPARα reduced the proliferative and tumor-forming capacities of GSCs, and xenografts failed to establish viable intracranial tumors [60]. PPARα was found to induce carnitine palmitoyltransferase 1C (CPT1C) in a breast and a pancreatic cancer cell line, leading to the activation of cell proliferation [61]. Using syngenic implantation of B16 melanoma, LLC1 lung carcinoma, and SKOV-3 ovarian cancer xenograft models, the efficiency of the tumor growth-inhibiting properties of the PPARα antagonist NXT629 has been demonstrated [62]. Li and colleagues showed that the level of PPARα and its activity were increased in 4-(methylnitrosamino)-l-(3-pyridyl)-lbutanone (NNK)-induced mouse-lung tumors. An increase in PPARα occurred before the formation of lung tumors, indicating that the molecular changes play a role in lung carcinogenesis [63]. In contrast, in two lung cancer cell lines, fenofibrate reduced cell proliferation [64]. PPARα activation in vivo using Wy-14,643 or bezafibrate reduced non-small-cell lung cancer (NSCLC) growth through the inhibition of a proangiogenic epoxygenase. Epoxygenases oxidize arachidonic acid to epoxyeicosatrienoic acids (EET), pro-angiogenic lipids which support tumor growth [65]. Although PPARα activation by Wy-14,643 did not alter proliferation of cancer cell lines in vitro, it reduced tumorigenesis in vivo through the inhibition of angiogenesis [66]. The PPARα agonist fenofibrate has further been demonstrated to suppress B cell lymphoma in mice through the modulation of lipid metabolism. B cell tumors trigger systemic lipid mobilization from white adipose tissue to the liver and increase very-low-density lipoprotein (VLDL)/low-density lipoprotein (LDL) release from the liver to promote tumor growth. B cell lymphoma cells express extremely low levels of PPARα; therefore, fenofibrate did not increase lipid utilization in the tumors but enhanced the clearance of lipids and blocked hepatic lipid release, leading to reduced tumor growth [67]. Fenofibrate has also been proposed to suppress colon cancer cell proliferation in vitro and in in vivo xenograft models through epigenetic modifications involving the inhibition of DNA Methyltransferase 1 (DNMT1) [68]. To summarize, given the highly controversial results regarding the tumor-suppressing or -promoting effects of therapeutic PPARα modulation, especially activation, this intervention seems to be inadequate in the context of cancer. To the best of our knowledge, no clinical trials for the use of PPARα agonists in cancer therapy exist. One trial with the PPARα antagonist TPST-1120 as a monotherapy, and in combination with Nivolumab, Docetaxel or Cetuximab, in subjects with advanced cancers (NCT03829436) is ongoing.

Table 1. Effects of PPARα on cell proliferation and tumor growth.

Model	Intervention	Outcome	References
In vitro			
MCF-7, MDA-MB-231 breast cancer cell lines	Clofibrate, Wy-14,643	Proliferation⇧	[39]
MCF-7 breast cancer cell line	Leptin, glucose	Proliferation⇧	[40]
MDA-MB-231, MCF-7, BT-474 breast cancer cell lines	AA	Proliferation⇧	[41]
MDA-MB-231, MCF-7 breast cancer cell line	AA	Proliferation⇩	[42]
Triple-negative breast cancer cell lines	Fenofibrate	Proliferation⇩	[43]
SUM149PT and SUM1315MO2 inflammatory breast cancer cell lines	Clofibrate	Proliferation⇩	[44]
Ishikawa endometrial cancer cells	Fenofibrate	Proliferation⇩, tumor growth≈	[55]
BsB8 mouse medulloblastoma cells, human D384, and Daoy medulloblastoma cells	Fenofibrate	Proliferation⇩	[57]
U87 glioblastoma cell line	Fenofibrate	Proliferation⇩	[58]
Neuroblastoma cell line	Fenofibrate	Proliferation⇩	[59]
MDA-MB-231 breast, Panc-1 pancreatic cancer cell line	GW6471 (antagonist), Wy-14,643	Proliferation⇩ upon antagonist, proliferation⇧ upon agonist	[61]
A549 and SK-MES-1 lung cancer cell lines	Fenofibrate	Proliferation⇩	[64]
In vivo			
Mouse xenograft models	Fenofibrate	Tumor growth⇩	[43]
Wildtype mice	Wy-14,643	Liver tumorigenesis⇧	[47]
Hepatitis C virus transgenic mice with activated PPARα		Liver tumorigenesis⇧	[48]
Transgenic mice with PPARα activation in hepatocytes	Hepatocytic overexpression	Proliferation⇧	[49]
PPARα-knockout mice	Diethylnitrosamine-induced hepatocarcinoma	Liver tumorigenesis⇧	[52]
PPARα-knockout mice	Syngenic MEF/RS tumors, LLC1 lung cancer, B16 melanoma	Tumor growth⇩	[53]
PPARα-knockout mice	B16 melanoma	Tumor growth⇩	[69]
Ovcar-3 and Diss ovarian cancer cell lines, implanted tumors in nude mice	Clofibrate	Proliferation⇩, tumor growth⇩	[56]
PPARα knockdown in glioma stem cells, xenograft models	PPARα knockdown	Proliferation⇩, tumor growth⇩	[60]
Wildtype mice with LLC1 lung, B16 melanoma, or SKOV-3 ovarian cancer	NXT969 antagonist	Tumor growth⇩	[62]
KRasLA2 mouse model of spontaneous primary NSCLC, orthotopic lung cancer cell injection	Wy-14,643, bezafibrate	Tumor growth⇩	[65]
Wildtype and PPARα-knockout mice injected with Bcr/Abl-transformed B cells	Fenofibrate	Tumor growth⇩	[67]
HCT-116 colon cancer cell line, Xenograft model	Fenofibrate	Proliferation⇩, tumor growth⇩	[68]

⇧ Indicates increase, ⇩ indicates decrease.

2.2. *PPARβ/δ*

PPARβ/δ expression has been reported in a variety of cancer tissues and cell lines. The effects of PPARβ/δ on cell proliferation and tumor growth are highly controversial, and have been reviewed recently; summarizing tables are provided [38]. Many studies focused on colon cancer. The discrepancy between the observed effects of PPARβ/δ activation can only lead to the conclusion that any therapeutical use of PPARβ/δ modulation has to be avoided. Most studies report a colon cancer-enhancing effect of PPARβ/δ. Examination of PPARβ/δ in human multistage carcinogenesis of the colorectum revealed that its expression increased from normal mucosa to adenomatous polyps to colorectal cancer. The most elevated PPARβ/δ levels were observed in colon cancer cells with a highly malignant morphology [70]. PPARβ/δ expression in human colon cancer tissues was associated with poor prognosis and a higher metastatic risk [71]. An opposite report has been published for human and mouse colon cancer samples; however, no histomorphological detection analysis of PPARβ/δ has been performed to allow for the correlation of PPARβ/δ with expression in malignant cancer cells [72]. It has been demonstrated that PPARβ/δ mediates mitogenic vascular endothelial growth factor (VEGF) release in colon cancer [73–75], although one report also claimed that a loss of PPARβ/δ would enhance vascular endothelial growth factor (VEGF) release [76]. PPARβ/δ has been shown to promote [73,77–82] or to inhibit [76,83,84] colon cancer in vivo. In line with a pro-tumorigenic role, PPARβ/δ activation via a high-fat diet (HFD) or PPARβ/δ agonist treatment allowed stem and progenitor cells to initiate tumorigenesis in the setting of a loss of the adenomatous polyposis coli (APC) tumor-suppressor gene [85]. PPARβ/δ-mediated epithelial hyperproliferation, which increases the risk for gastric adenocarcinoma, was further found to be induced by Helicobacter pylori infection [86]. Regarding breast cancer, most studies suggest a pro-tumorigenic function of PPARβ/δ. Only two in vitro studies from the same group using the same breast cancer cell line suggest a reduction in cell proliferation upon PPARβ/δ activation [87,88]. The same group published two very similar studies, one using neuroblastoma cell lines, and the other testicular embryonal carcinoma cells, in which PPARβ/δ overexpression and/or activation had beneficial tumor-cell proliferation- or growth-inhibiting effects [89,90]. In contrast, by applying a variety of different molecular tools as either overexpression or knockout models, or conducting pharmacological activation or inhibition of PPARβ/δ, it has been shown, in vivo, that PPARβ/δ favors mammary tumorigenesis [91–94]. 3-phosphoinositide-dependent kinase-1 (DK1) favors these tumorigenic properties of PPARβ/δ in breast cancer [92,93]. Fatty-acid-binding protein 5 (FABP5), which shuttles ligands from the cytosol to PPARβ/δ, underlines the importance of endogenous PPARβ/δ ligands for cancer growth, as knockout of FABP5 was sufficient to reduce mammary tumorigenesis [95]. In line with this, FABP5 has been shown to convert the strong anticarcinogenic properties of retinoic acid (RA) into tumor-promoting functions as it delivers RA to the mitogenic and anti-apoptotic PPARβ/δ receptor [96]. Similar to the effects observed in mammary carcinomas, activation of the FABP5/PPARβ/δ pathway was shown to promote cell survival, proliferation, and anchorage-independent growth in prostate cancer cells [97]. The oncogenic redirection of transforming growth factor (TGF)-β1 signaling via the activation of PPARβ/δ was also identified to promote prostate cancer growth [98]. One study, however, suggested the inhibition of prostate cancer growth by PPARβ/δ through a noncanonical and ligand-independent pathway [99]. The activation of PPARβ/δ has been proposed to inhibit liver tumorigenesis in hepatitis B transgenic mice [100]; however, in different human hepatocellular carcinoma cell lines, the activation of PPARβ/δ enhanced the growth of these cancer cells through the activation of cyclooxygenase (COX)-2 [101]. PPARβ/δ activation has been shown to inhibit melanoma skin cancer cell proliferation through repression of the Wilms tumor suppressor (WT)1 [102], which favors human melanoma progression [103]. PPARβ/δ-knockout animals were more susceptible to skin carcinogenesis as their wildtype counterparts and PPARβ/δ agonists inhibited keratinocyte proliferation [104], as well as proliferation in a human squamous-cell carcinoma cell line [105]. In line with these finding, the authors proposed a protective effect of PPARβ/δ activation, coupled with the inhibition of COX-2 activity, to increase the efficacy of chemoprevention in skin tumorigenesis [106,107]. However, a later report from this group showed that PPARβ/ δ is not involved in the suppression of skin carcinogenesis by non-steroidal anti-inflammatory drugs (NSAID) which inhibit COX-2 [108]. In contrast to an inhibitory function of PPARβ/ δ in the tumorigenesis of non-melanoma skin cancers, one study clearly evidenced the pro-tumorigenic role of PPARβ/δ involving the direct activation of proto-oncogene tyrosine-protein kinase Src, which promotes the development of ultraviolet (UV)-induced skin cancer in mice [109]. An elegant study focused on the importance of fibroblast PPARβ/ δ expression in non-melanoma skin tumorigenesis. Although the chemically induced skin tumors of animals with the conditional deletion of PPARβ/ δ in fibroblasts showed increased proliferation, the tumor burden was smaller and the tumor onset delayed; this indicates the role of fibroblast PPARβ/δ in epithelial–mesenchymal communication, which further influences tumor growth [110]. Regarding lung cancer, high expression of PPARβ/δ limited to cancer cells has been demonstrated in human cancer samples. In lung cancer cell lines, the activation of PPARβ/δ stimulated proliferation and inhibited apoptosis [111,112]. Nicotine increases PPARβ/δ expression in lung carcinoma cells, which contributes to increased proliferation [113]. In contrast, one study using the activation of PPARβ/δ in two lung cancer cell lines in vitro did not find differences for proliferation upon stimulation of PPARβ/δ [114]. In transgenic mice lacking one or both PPARβ/δ alleles, the growth of RAF-induced lung adenomas was decreased [115]. Although cell proliferation in mouse LLC1 lung cancer cells was decreased upon activation of PPARβ/δ, LLC1 tumor growth in vivo was enhanced in mice with conditional vascular overexpression of PPARβ/δ, underlining the importance of crosstalk between the tumor stroma and cancer cells for tumor growth [11]. One study reported that PPARβ/δ activation promoted apoptosis and reduced the tumor growth of nasopharyngeal carcinoma cells [116]. PPARβ/δ was found to be highly expressed in liposarcoma compared to benign lipoma, and PPARβ/δ activation increased liposarcoma cell proliferation, which was mediated via the direct transcriptional repression of leptin by PPARβ/δ [117]. Additionally, in thyroid tumors, PPARβ/δ was increased and correlated with the expression of the proliferation marker Ki67. PPARβ/δ activation increased the cell proliferation of thyroid cells [118]. PPARβ/δ was highly expressed in epithelial ovarian cancer cell lines and the inhibition of PPARβ/δ reduced their proliferation and tumor growth in vivo. Interestingly, aspirin, a NSAID that preferentially inhibits COX-1, compromised PPARβ/δ function and cell growth by inhibiting extracellular signal-regulated kinases 1/2 [119]. PPARβ/δ promoted the survival and proliferation of chronic lymphocytic leukemia cells [120] and changed the outcome of signaling from cytokines such as interferons (IFNs) [121]. A detailed table on the effects of PPARβ/δ on cell proliferation and tumor growth can be found in [38]. In conclusion, most studies identified PPARβ/δ as a tumor-promoting factor which increases cell proliferation and cancer growth. Although some studies report the inhibition of cancer cell proliferation upon PPARβ/δ activation, the therapeutic modulation of PPARβ/δ appears dangerous. Consequently, no cancer-related clinical trials are reported.

2.3. *PPARγ*

PPARγ expression is found in a variety of cancer tissues and cell lines. The activation of PPARγ by different agonists increased the frequency and size of colon tumors in C57BL/6J-APCMin/+ mice [122,123] (Table 2). However, in human colon cancer cell lines, PPARγ inhibited tumor-cell proliferation [124–127]. Prostate cancers were found to overexpress PPARγ. The PPARγ agonist troglitazone inhibited the proliferation of PC-3 prostate cancer cells in vitro and in xenograft models in vivo [128], which was confirmed by others in later studies [129,130]. Similarly, growth inhibition via PPARγ activation has been described for liposarcoma [131], gastric cancer [132,133], bladder carcinoma [130,134], renal cell carcinoma [130], neuroblastoma [135,136], glioblastoma [137,138], melanoma [139–142], NSCLC [143,144], adrenocortical cancer [145,146], hepatocellular carcinoma [147], endometrial carcinoma [148], ovarian cancer [149,150], multiple myeloma [151], B cell lymphoma [152], mesothelioma [153], and esophageal squamous-cell carcinoma [154]. Most of these studies used cancer cell lines and PPARγ agonist treatment in vitro. Exciting results for therapeutic effects of PPARγ activation have been obtained in chronic myeloid leukemia (CML). With standard therapies, mainly tyrosine kinase inhibitors (TKIs), only 10% of patients achieve a complete molecular response/remission (CMR). This is mainly due to a pool of quiescent CML leukemia stem cells (LSCs), which are not completely eradicated by TKIs. Prost and colleagues demonstrated that thiazolidinediones target this pool of LSCs through the decreased transcription of signal transducer and activator of transcription (STAT) 5, leading to sustained CMR in a small group of patients [155]. A proof-of-concept study including 24 patients yielded positive outcomes with a combined therapy of pioglitazone and imatinib (TKI) [156]. A phase 2 trial is ongoing (EudraCT 2009-011675-79). PPARγ has been identified as a critical modifier in thyroid carcinogenesis using transgenic animals harboring a knock-in dominant-negative mutant thyroid hormone receptor beta (TRbetaPV/PV mouse), which spontaneously develop follicular thyroid carcinoma. TRbetaPV/PV mice were crossed with PPARγ +/− mice, and it was shown that thyroid carcinogenesis progressed faster in animals with PPARγ haplo-insufficiency. Reduced PPARγ led to the activation of the nuclear factor-kappaB signaling pathway, resulting in the repression of apoptosis. Furthermore, the treatment of TRbetaPV/PV mice with rosiglitazone delayed the progression of thyroid carcinogenesis by decreasing cell proliferation [157]. Wu and colleagues showed that the inhibition of PPARγ via the overexpression of dominant negative PPARγ (dnPPARγ) in the myeloid cell lineage provokes systemic inflammation and an increase in myeloid-derived suppressor cells (MDSC), which led to immunosuppression and the appearance of multiple cancers [158]. In breast cancer [159,160] and uterine leiomyomas [161], the growth-inhibiting effect of PPARγ activation was attributed to the inhibition of estrogen-receptor signaling. This seems to be partially mediated through the repression of leptin's stimulatory effects on estrogen signaling by PPARγ [162]. However, later, it was shown that the PPARγ agonist prostaglandin 15-deoxy-$\Delta^{12,14}$-PGJ2 (15d-PGJ2) inhibits the transcriptional activity of estrogen receptor alpha via PPARγ-independent covalent modification of its DNA-binding domain [163]. Methylene-substituted diindolylmethanes (C-DIMs) are PPARγ-activating agents. They reduce the proliferation of breast cancer cell lines. However, the decrease in cell growth was not inhibited by PPARγ antagonists, indicating that the observed effect might be PPARγ-independent [164]. An elegant study used transgenic mice prone to mammary-gland cancer crossed with mice expressing a constitutively active form of PPARγ in the mammary gland. The resulting PyV/VpPPARγ females developed tumors with accelerated kinetics. Even before reaching maturity at around 30 days of age, female mice displayed palpable tumor masses. These results indicate that once an initiating event has taken place, increased PPARγ signaling exacerbates mammary-gland tumor development [165]; this is similar to the observed situation of accelerated colon cancer formation in APCMin/+ mice treated with thiazolidinediones described before [122,123]. Avena and colleagues focused on the importance of the tumor stroma for cancer growth. They demonstrated that the overexpression of PPARγ in breast cancer cells reduced tumor growth in a xenograft model and demonstrated increased autophagy in the tumor cells. However, when breast cancer cells were co-injected with PPARγ-overexpressing fibroblasts, tumor growth was significantly increased. Stromal cells with overexpression of PPARγ displayed metabolic features of cancer-associated fibroblasts, with increased autophagy, glycolysis, and senescence; this supports a catabolic pro-inflammatory microenvironment that metabolically enhances cancer growth. The activation of an autophagic program, therefore, have pro- or antitumorigenic effects, depending on the cellular context [166]. The mammary secretory-epithelial-cell-specific knockout of PPARγ enhanced tumor growth in a 7,12-dimethylbenz[a]anthracene (DMBA)-induced breast cancer model [167]. A small clinical trial in patients with early-stage breast cancer did not evidence differences in breast tumor-cell proliferation upon treatment with rosiglitazone, administered between the time of diagnostic biopsy and definitive surgery [168]. PPARγ ligands did not prevent chemically or UV-induced skin tumors, although they significantly inhibited basal-level keratinocyte proliferation [169].

It is important to note that the anti-cancer effects of thiazolidinediones (rosiglitazone, pioglitazone, and troglitazone) might be independent of PPARγ activation, as it has been demonstrated that they are mediated by translation inhibition [170]. In osteosarcoma cell lines, troglitazone enhanced proliferation in one study [171], and inhibited proliferation in another [172]. Srivastava and colleagues demonstrated, in a lung cancer model, that treatment with the PPARγ agonist pioglitazone triggers a metabolic switch that inhibits pyruvate oxidation and reduces glutathione levels. These metabolic changes increase reactive oxygen species (ROS) levels, which leads to the rapid hypophosphorylation of the retinoblastoma protein (RB) and cell-cycle arrest [173]. In a very recent study, Musicant and colleagues demonstrated that the inhibition of PPARγ might be beneficial in mucoepidermoid carcinoma (MEC), a salivary-gland cancer that is driven primarily by a transcriptional coactivator fusion composed of cyclic AMP-regulated transcriptional coactivator 1 (CRTC1) and mastermind-like 2 (MAML2). The chimeric CRTC1/MAML2 (C1/M2) oncoprotein induces transcriptional activation of the non-canonical peroxisome proliferator-activated receptor gamma coactivator-1 alpha (PGC-1α) splice variant PGC-1α4, which regulates PPARγ-mediated insulin-like growth factor (IGF) 1 expression. The inhibition of PPARγ by inverse agonists inhibits MEC cell proliferation and tumor growth in xenograft models [174]. Besides the clinical trials already mentioned, one trial (NCT00408434) of efatutazone in patients with advanced solid malignancies and no curative therapeutic options reported evidence of disease control [175]. In other clinical trials investigating the effects of efatutazone in combination with carboplatin/paclitazel in NSCLC (NCT01199055), or in combination with erlotinib (NCT01199068), partial responses were around 40%. However, in a clinical trial for liposarcoma (NCT02249949), efatutazone resulted in neither complete nor partial responses. The development of efatutazone has been discontinued. Clinical trials for pioglitazone in the treatment of leukoplakia in head and neck cancer (NCT00099021) resulted in partial responses of 70%, and in another trial for oral leukoplakia (NCT00951379), partial responses of 46% were achieved. Over twenty years ago, a very small clinical trial in three patients with liposarcoma treated with troglitazone already provided some evidence for adipocytic differentiation and decreased proliferation [176]. However, no results are available for later trials with a higher number of patients enrolled (NCT00003058 and NCT00004180). A table with detailed information regarding clinical trials using PPARγ agonists for cancer treatment is given in [177]. Although a large body of evidence suggests that PPARγ functions as a tumor suppressor, the role of PPARγ in tumorigenesis remains controversial. The predominant use of in vitro cell culture studies is limited in its elucidation of the biological relevance of PPARγ in cancer, as complex gene–gene and gene–environment interactions are not considered. It can be concluded that the role of PPARγ in cancer depends on the specific cancer type, the tumor stage, and the tumor environment, which implies that the therapeutical modulation of PPARγ must be considered with caution.

Table 2. Effects of PPARγ on cell proliferation and tumor growth.

Model	Intervention	Outcome	References
In vitro			
Colon cancer cell lines	BRL 49653 activator	Proliferation⇩	[124]
Colon cancer cell lines	Troglitazone	Proliferation⇩	[126]
Liposarcoma cell lines	Pioglitazone	Proliferation⇩	[131]
Gastric cancer cell lines	Troglitazone, pioglitazone	Proliferation⇩	[132]
Gastric cancer cell lines	Troglitazone, 15d-PGJ2	Proliferation⇩	[133]
LA-N-5 nb neuroblastoma cell line	15d-PGJ2, GW1929	Proliferation⇩	[135]
SK-N-AS, SH-SY5Y neuroblastoma cell lines	Rosiglitazone	Proliferation⇩	[136]
U87MG, T98G glioblastoma cell lines	15d-PGJ2	Proliferation⇩	[137]
U87, U251 glioblastoma cell lines	Rosiglitazone	Proliferation⇩	[138]
A375 melanoma cell line	15d-PGJ2, ciglitazone	Proliferation⇩	[142]
Different melanoma cell lines	Multiple thiazolidinediones	Proliferation⇩	[140]
A375 melanoma cell line, xenograft model	Ciglitazone	Proliferation⇩	[141]
H1792 and H1838 NSCLC lines	Rosiglitazone	Proliferation⇩	[144]
H295R adrenocortical cancer cell line	Rosiglitazone, pioglitazone	Proliferation⇩	[145,146]
MCF-7 breast cancer cell line	15d-PGJ2	Proliferation⇩	[159]
MCF-7 breast cancer cell line	15d-PGJ2, ciglitazone	Proliferation⇩	[160]

Table 2. *Cont.*

Model	Intervention	Outcome	References
In vitro			
MDA-MB-231, MDA-MB-453 breast cancer cell lines	C-DIM	Proliferation⇩	[164]
MDA-MB-231 breast cancer cells	Overexpression of PPARγ	Tumor growth⇩	[166]
MDA-MB-231 breast cancer cells	+ PPARγ-overexpressing fibroblasts	Tumor growth⇧	[166]
Leiomyoma cell line	Ciglitazone, troglitazone	Proliferation⇩	[161]
Ishikawa, Sawano, RL95-2 endometrial carcinoma cell lines	15d-PGJ2	Proliferation⇩	[148]
SKOV3 ovarian cancer cell line	C-DIM	Proliferation⇩	[149]
A2780, OVCAR3, OVCAR5, OVCAR8, OVCAR432, SKOV3, IGROV1 ovarian cancer cell lines	Ciglitazone, PPAR-γ antagonist GW9662	Proliferation⇩(agonist), proliferation⇧ (antagonist)	[150]
RPMI 8226 multiple-myeloma cell line	Overexpression of PPAR-γ	Proliferation⇩	[151]
B-cell lymphoma cell line	Silencing, overexpression of PPAR-γ	Proliferation⇧ (silencing), proliferation⇩ (overexpression)	[152]
G292, MG63, SAOS and U2OS osteosarcoma cell lines	Troglitazone	Proliferation⇧	[171]
143B, MNNG/HOS, MG-63, and TE-85 osteosarcoma cell lines	Troglitazone, ciglitazone	Proliferation⇩	[172]
H292, H3118, HMC1, HMC3A, HMC3B mucoepidermoid carcinoma cell lines	SR10221, SR2595, T0070907 inverse agonists	Proliferation⇩, tumor growth⇩	[174]
In vivo			
C57BL/6J-APC^{Min}/+ mice	BRL-49,653, troglitazone	Tumor growth⇧	[122]
C57BL/6J-APC^{Min}/+ mice	Troglitazone	Tumor growth⇧	[123]
Colon cancer cell lines, xenograft mouse model	Troglitazone	Proliferation⇩, tumor growth⇩	[125]
SW480 colon cancer cell line, xenograft model	C-DIM	Proliferation⇩, tumor growth⇩	[127]
A549 NSCLC line, xenograft models	Troglitazone, pioglitazone	Proliferation⇩, tumor growth⇩	[143]
NCI-H2347, NCI-H1993 lung adenocarcinoma cell lines, xenograft models	Pioglitazone	Proliferation⇩, tumor growth⇩	[173]
Huh7 and Hep3B hepatocellular cancer cell lines, xenograft models	Troglitazone	Proliferation⇩, tumor growth⇩	[147]
Dominant-negative mutant thyroid hormone receptor beta (TRbetaPV/PV mice)	Rosiglitazone	Tumor growth⇩	[157]
MMTV-VpPPARγ animals	Breeding with MMTV-PyV strain	Tumor growth⇧	[165]
MSE cell-specific PPARγ knockout (PPARγ-MSE KO)	7,12-dimethylbenz[a]anthracene (DMBA)-induced breast tumorigenesis	Tumor growth⇧	[167]
Thirty-eight patients with early-stage breast cancer	Rosiglitazone	Proliferation≈, tumor growth≈	[168]
UV and chemically induced skin carcinogenesis	Rosiglitazone, troglitazone	Tumor growth≈	[169]
CML LSCs Three patients with CML	Pioglitazone in combination with imatinib	Proliferation⇩, CMR≈5 years	[155]
EHMES-10, MSTO-211H mesothelioma cell lines, xenograft models	Troglitazone	Proliferation⇩, tumor growth⇩	[153]
Esophageal squamous-cell carcinoma line, xenograft model	Efatutazone; troglitazone	Proliferation⇩, tumor growth⇩; proliferation≈, tumor growth≈	[154]
Overexpression of dn PPAR-γ in myeloid lineage cells		Tumor growth⇧	[158]

⇧ Indicates increase, ⇩ indicates decrease.

The major effects of PPARα, PPARβ/δ, and PPARγ on proliferation are depicted in Figure 1.

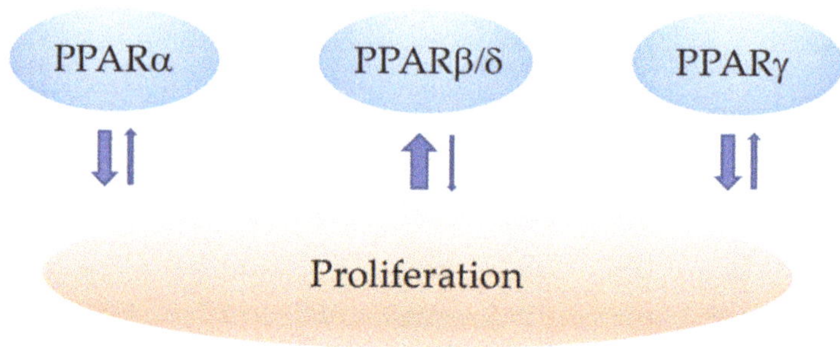

Figure 1. Schematic illustration of the influence of PPARα, PPARβ/δ, and PPARγ on cancer-cell proliferation. ↓ indicates inhibition and ↑ an increase in cell growth and proliferation. The width of the arrows corresponds to the number of studies reporting similar effects. Note that for a certain cancer type, the situation might be different (see the main text for details).

3. PPARs and Cell Death

3.1. PPARα

The PPARα activator fenofibrate has been shown to induce apoptosis in a human hepatocellular carcinoma cell line through an increase in reactive oxygen species (ROS) [178]. As another molecular mechanism of PPARα-dependent apoptosis, it has been proposed that PPARα serves as an E3 ubiquitin ligase to induce Bcl2 ubiquitination and degradation, leading to apoptosis [179]. Additionally in endometrial cancer [180], breast cancer [181], glioblastoma [182], colon cancer [68,183], ovarian cancer [56], medulloblastoma [57], neuroblastoma [59], pancreatic cancer [184], and NSCLC [185], the activation of PPARα induced apoptosis. These studies were mainly performed using a cancer cell line in in vitro assays. Conjugated linoleic acids induced apoptosis in a variety of human cancer cell lines, which was accompanied by a strong increase in PPARα [186]. The synergistic pro-apoptotic anticancer activity of clioquinol (5-chloro-7-iodo-8-hydroxyquinoline) and docosahexaenoic acid (DHA) in human cancer cells has also been suggested to be mediated by PPARα signaling [187]. Zang and colleagues reported that the dual PPARα/γ agonist TZD18 provoked apoptosis in human leukemia, glioblastoma, and breast cancer cell lines through the induction of the endoplasmic reticulum stress response [188]. Later, the same observations were made in gastric cancer cell lines [189]. However, it is not clear if these actions were mediated through combined PPARα/γ signaling or solely through PPARα or PPARγ signaling. Crowe and colleagues evidenced that combined therapy using PPAR and RXR ligands for breast cancer treatment resulted in growth inhibition. This was due to apoptosis when PPARα ligands were used. In contrast, PPARγ agonists provoked decreased growth characterized by S-phase inhibition [181]. In mantle-cell lymphoma (MCL), a type of aggressive B cell non-Hodgkin's lymphoma, which is frequently resistant to conventional chemotherapies, fenofibrate efficiently induced apoptosis through the downregulation of tumor necrosis factor (TNF) α. The addition of recombinant TNFα partially rescued fenofibrate-induced apoptosis, whereas the PPARα antagonist GW6471 did not affect the fenofibrate effects. Therefore, it might be possible that fenofibrate induced apoptosis through other mechanisms than the activation of PPARα [190]. In retinoblastoma cells, apoptosis was induced by fatty acid synthase, which led to the downregulation of PPARα; however, the relationship between these molecular events has not been investigated [191]. Similarly, in hepatic carcinoma cells, apoptosis was induced by the flavonoid quercetin, which downregulated PPARα expression [192]. The cause–effect relationship remains to be elucidated. Fenofibrate was found to induce apoptosis in triple-negative breast cancer cell lines, which involved the activation of the nuclear factor 'kappa-light-chain-enhancer' of activated B-cell (NF-κB) pathways, as the effect could be almost totally blocked by an NF-κB-specific inhibitor. The induction of apoptosis by fenofibrate was, however, independent of PPARα expression status, as the PPARα antagonist GW6471 did not change apoptosis induction by fenofibrate [43]. In contrast, the induction of apoptosis in hepatocellular carcinoma cells via the overexpression of PPARα was dependent on NF-κB signaling, as PPARα was found to directly interact with IκBα (nuclear factor kappa-light-polypeptide-gene-enhancer in B-cells inhibitor alpha) [52]. In contrast to most studies suggesting a pro-apoptotic function of PPARα activation, Li and coworkers reported that the PPARα inhibitor MT886 induced apoptosis in hepatocarcinoma cell lines, and the agonist fenofibrate significantly increased proliferation, the expression of cell-cycle-related protein (CyclinD1, CDK2), and cell-proliferation-related proteins (PCNA) [46]. Similarly, Abu Aboud and colleagues demonstrated enhanced apoptosis in renal-cell carcinoma upon PPARα inhibition in vitro [193] and in vivo through a decrease in enhanced fatty-acid oxidation and oxidative phosphorylation, and further cancer-cell-specific glycolysis inhibition [194]. The induction of apoptosis via PPARα inhibition has also been described in head and neck paragangliomas (HNPGLs); in one case, the authors described the inhibition of the PI3K/GSK3β/β-catenin signaling pathway as the underlying molecular mechanism [195]. In conclusion, most of the studies suggest that PPARα activation induces apoptosis in cancer cells. However, given that a substantial number of research works also propose the opposite, and advise the use of PPARα inhibition to provoke apoptosis in tumor cells, no clear recommendation for therapeutic PPARα modulation in cancer treatment can be postulated.

3.2. *PPARβ/δ*

The function of PPARβ/δ in cancer-cell death was reviewed in detail in [38]. Most studies support the cell-death-preventing role of PPARβ/δ in tumor cells. In 1999, it was already demonstrated that PPARβ/δ was overexpressed in colorectal cancers (CRC) with adenomatous polyposis coli (APC)/β-catenin mutations, leading to the prevention of apoptosis in colon cancer cells. NSAIDs could compensate for this defect by suppressing PPARβ/δ and promoting apoptosis [196]. Cyclooxygenase-derived prostaglandin E_2 (PGE_2), which is overexpressed in most CRCs, was further found to indirectly transactivate PPARβ/δ to inhibit colon cancer-cell apoptosis [197]. Interestingly, it has been demonstrated that fibroblasts isolated from the mucosa of hereditary non-polyposis colorectal cancer (HNPCC) patients produced 50 times more PGE_2 than normal fibroblasts. Stromal overproduction of PGE_2 in HNPCC patients is likely to prevent the apoptosis of neoplastic lesions through the activation of PPARβ/δ, thereby facilitating progression into a malignant state [198]. Studies using HCT116 colon cancer cells confirmed that treatment with the PPARβ/δ agonist GW501516 diminished serum-withdrawal-induced apoptosis, which was not the case in PPARβ/δ-deficient HCT116 cells; this indicates the specificity of the apoptosis-preventing effect for PPARβ/δ [77]. Other mechanisms for the PPARβ/δ-mediated prevention of apoptosis in colon cancer have been suggested, such as the activation of the 14-3-3ε protein [199], or survivin [200] expression by PPARβ/δ. In contrast to these studies, one report suggested a pro-apoptotic function of PPARβ/δ in colon carcinoma. GW0742 agonist treatment induced apoptosis in wildtype, but not in PPARβ/δ-knockout animals with chemically induced colon carcinoma. Apoptosis was quantified via TdT-mediated dUTP-biotin nick-end labeling (TUNEL) staining of colon sections and subsequent cell counting; however, as no images were provided, it is difficult to assume TUNEL-specific positivity for cancer cells [83]. A study from the same group using different human colon cancer cell lines treated with hydrogen peroxide to induce apoptosis, different concentrations of the PPARβ/δ agonist GW0742, and NSAIDs could not find evidence for a decrease in apoptosis upon PPARβ/δ activation [72]. Conjugated linoleic acids (CLAs) were found to reduce proliferation in different human cancer cell lines. In cancer cell lines in which the inhibition of cell proliferation was correlated with apoptosis induction, PPARβ/δ expression became strongly downregulated [186]. PPARβ/δ activation decreased human and mouse melanoma cell proliferation; however, no changes in apoptosis could be observed [102]. The activation of PPARβ/δ has been shown to inhibit cisplatin-induced apoptosis in human lung cancer cell lines [111], and the knockout of PPARβ/δ induced apoptosis in lung cancer cells [112]. In mouse LLC1 lung cancer cells, the modulation of PPARβ/δ activity did not influence apoptosis [11]. The inhibition of PPARβ/δ sensitized neuroblastoma cells to retinoic acid-induced cell death [201]. In contrast, in prostate cancer cell lines, ginsenoside Rh2- [202] and telmisartan- [203] induced apoptosis were hampered by the inhibition of PPARβ/δ. In line with a pro-apoptotic function of PPARβ/δ, enhanced apoptosis in a bladder carcinoma cell line [204] as well as in nasopharyngeal tumor cells [116] and liver cancer cells [205] was reported upon PPARβ/δ activation.

3.3. *PPARγ*

Over twenty years ago, Padilla and colleagues already described that 15d-PGJ2 that binds to PPARγ exerts cytotoxicity in malignant B-cell lymphoma via apoptosis induction. Additionally, thiazolidinedione PPARγ agonists negatively affected B-lineage cells, indicating a specific PPARγ function of counteracting the stimulatory effects of prostaglandin E_2 (PGE_2) [206,207]. Later, the inhibition of NFκB was shown to be the major mechanism of 15d-PGJ_2-induced apoptosis in aggressive B-cell malignancies. These effects were mimicked by the proteasome inhibitor MG-132, but not by troglitazone, suggesting that 15d-PGJ2-induced apoptosis is independent of PPARγ [208]. In multiple myeloma, the overexpression of PPARγ induced apoptosis through the inhibition of Interleukin-6 production [151]. Similarly, in acute myeloid leukemia (AML), the forced expression of PPARγ regulated the induction of apoptosis via caspase-8 activation [209]. The activation of PPARγ by 15d-PGJ2 has also been demonstrated to inhibit tyrosine phosphorylation of epidermal growth factor receptors ErbB-2 and ErbB-3 in a breast cancer cell line, leading to a dramatic increase in apoptosis [159]. A later study, however, showed that while 15d-PGJ2 activates PPRE-mediated transcription, PPARγ is not required for 15d-PGJ_2-induced apoptosis in breast cancer cells. As other possible mechanisms of apoptosis induction by 15d-PGJ2, the inhibition of NFκB-mediated survival pathways, the inhibition of transcriptional activation of COX-2, and the inhibition of the ubiquitin proteosome were proposed [210]. The PPARγ-independent induction of apoptosis by 15d-PGJ_2 has also been demonstrated in prostate and bladder carcinoma cells [211]. Additionally, 15d-PGJ_2 induced apoptosis in pancreatic cancer cells through the downregulation of human telomerase reverse transcriptase (hTERT) [212]. Thiazolidinediones sensitize breast cancer cells to tumor necrosis factor-related apoptosis-inducing ligand (TRAIL) therapy by reducing cyclin D3 levels, but not other D-type cyclins [213]. Later, combined treatment with TRAIL and PPARγ ligands, especially 15d-PGJ2, was proposed to overcome chemoresistance in ovarian cancers for successful apoptosis induction [214]. The simultaneous activation of PPARγ and RXR has been suggested to promote apoptosis, implicating the upregulation of p53 in breast cancer cell lines [215]. NSAIDs, considered in cancer prevention due to their inhibitory effect on cyclooxygenases (COX), have recently been proposed to exert their antineoplastic activity through the activation of PPARγ, which induces proline dehydrogenase/proline oxidase (PRODH/POX)-dependent apoptosis in breast cancer cells [216]. In many other studies PPARγ agonists induced apoptosis in bladder cancer [217], gastric carcinoma [133,218], lung cancer [219], esophageal adenocarcinoma [220], pancreatic cancer [221], hepatocellular carcinoma [222], neuroblastoma [223], melanoma [141,142], glioblastoma [224], leukemia [225], leiomyoma [226], mesothelioma [153], and colon carcinoma [227]. Nevertheless, it is not always clear if apoptosis induction is mediated via PPARγ activation. In colon carcinoma, increased PPARβ/δ expression and/or activation of PPARβ/δ antagonized the ability of PPARγ to induce cell death. The activation of PPARγ was found to decrease survivin expression and increase caspase-3 activity, whereas the activation of PPARβ/δ counteracted these effects [200]. A highly interesting study investigated the role of PPARγ coactivator-1 alpha (PGC-1α) in the induction of apoptosis in human epithelial ovarian cancer cells. The overexpression of PGC-1α in human epithelial ovarian cancer cells induced cell apoptosis through the coordinated regulation of Bcl-2 and Bax expression. The suppression of PPARγ expression via siRNA or PPARγ antagonist treatment inhibited PGC-1α-induced apoptosis, suggesting that PPARγ is required for apoptosis induction by PGC-1α [211]. Alternative promoter and mRNA splicing give rise to several PPARγ mRNA and protein isoforms, reviewed in [228]. Kim and coworkers identified a novel splice variant of human PPARγ 1 (hPPAR γ1) that exhibits dominant-negative activity in human tumor-derived cell lines and investigated the function of a truncated splice variant of hPPARγ 1 (hPPARγ1(tr)) in lung cancer. The overexpression of hPPARγ1(tr) rendered cancer cells more resistant to chemotherapeutic drug- and chemical-induced cell death [229]. PPARγ mediated apoptosis induction by n-3 polyunsaturated fatty acids (n-3 PUFA) in a breast cancer cell line, which might explain the beneficial effects of diets enriched in n-3 PUFA [230]. Like the results described above for breast cancer, in colon cancer, the anti-apoptotic activity of the PPARγ agonist troglitazone was also found to be independent of PPARγ. Instead of apoptosis induction through PPARγ, the activation of early growth response-1 (Egr-1) transcription factor was identified as the underlying molecular mechanism [231]. This has also been described for the apoptotic action of C-DIMs, PPARγ agonists, which decreased colon cancer cell survival through the PPARγ-independent activation of early growth response protein (Egr) 1 [127]. In contrast, Telmisartan, an angiotensin II receptor blocker (ARB), was found to inhibit cancer cell proliferation and induce apoptosis through the activation of PPARγ [232–234]. In contrast to these pro-apoptotic actions of PPARγ agonists, the PPARγ agonist troglitazone increased cell proliferation and inhibited staurosporine-induced apoptosis in several osteosarcoma cell lines through Akt activation [171]. Later, studies from the Kilgore lab provided evidence that the unreflected therapeutical use of PPARγ ligands in patients predisposed to or already diagnosed with cancer, especially breast cancer, could be dangerous. They identified Myc-associated zinc finger protein (MAZ) as a transcriptional mediator of PPARγ1 expression. The down-regulation of PPARγ1 expression led to reduced cellular proliferation and the induction of apoptosis in breast cancer cells [235]. Interestingly, it has been demonstrated that PPARγ ligands can have distinct activities. One relates to the ability of ligands to act as canonical agonists of the nuclear receptor on peroxisome proliferator response elements, which leads to adipogenesis. The second relates to the allosteric inhibition of phosphorylation of the Ser273 residue of PPARγ. PPARγ is phosphorylated in response to DNA damage, and the inhibition of phosphorylation by novel noncanonical ligands can sensitize cancer cells to DNA-damaging agents. They might represent a safer approach in cancer therapies as the established canonical agonists, which are used less and less frequently due to reported severe side effects or contradictory therapeutical outcomes [236]. A good study by Schaefer and colleagues using hepatocellular carcinoma cells demonstrated that PPARγ antagonists prevented adhesion to the extracellular matrix followed by caspase-dependent apoptosis (anoikis). They found that PPARγ inhibitor T0070907 was significantly more efficient in causing cancer-cell death than the activators troglitazone and rosiglitazone, which had no effect on cell adhesion and caused cell death at much higher concentrations [237]. Later studies confirmed this mechanism of anoikis induction by PPARγ antagonists in squamous-cell carcinoma [178,238]. Some reports evidenced autophagy induction in cancer cells upon PPARγ activation [239–241]. Autophagy can either suppress or promote tumor growth [242], and deducing that the induction of autophagy in cancers via PPARγ modulation might be beneficial is, consequently, erroneous. The difficulty in categorizing PPARγ activation in cancer therapy as beneficial or disadvantageous is also well-illustrated in a study from Baron and colleagues, who investigated the effects of ciglitazone in two different colon cancer cell lines: HT29 and SW480 cells. Ciglitazone induced apoptosis in HT29 cells, but stimulated SW480 cell proliferation. The authors concluded that the differential responses for growth regulation result from cell-specific protein synthesis and differences in protein regulation [243]. Based on the outcomes of all these studies, it is therefore impossible to recommend PPARγ modulation to induce cancer-cell death.

The major effects of PPARα, PPARβ/δ, and PPARγ on cell death and the underlying molecular mechanisms are summarized in Figure 2.

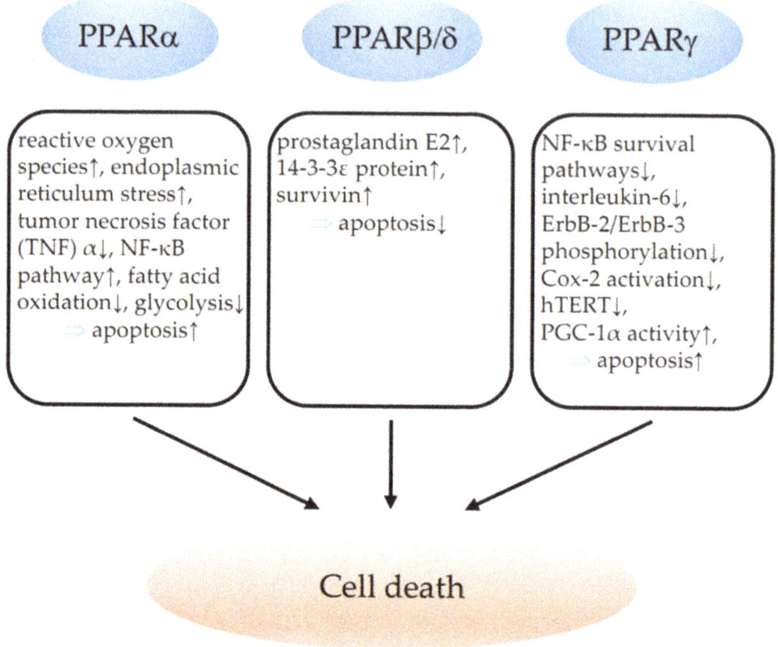

Figure 2. Illustration of the influence of PPARα, PPARβ/δ, and PPARγ on cancer-cell death. ↓ indicates inhibition and ↑ indicates an increase. ⇒: leads to; TNFα: tumor necrosis factor alpha; NF-κB: nuclear factor kappa-light-chain-enhancer of activated B cells; Cox-2: cyclooxygenase-2; hTERT: telomerase reverse transcriptase human; PGC-1α: peroxisome proliferator-activated receptor gamma coactivator-1 alpha.

4. PPARs and Angiogenesis

4.1. *PPARα*

The activation of PPARα is, in general, considered to suppress tumor angiogenesis, which has been reviewed in detail in [18,177]. One study investigated the expression of PPARα in human non-melanoma skin cancer and found less expression of PPARα in squamous-cell carcinoma and actinic lesions than in normal human skin samples; however, no correlation with vascular densities could be established [244]. A single study using syngenic tumor inoculation experiments in PPARα knockout mice observed a reduction in tumor vascularization and proposed that PPARα might favor tumor angiogenesis [53]. However, the same research group, as well as the great majority of other laboratories, could establish an anti-angiogenic effect of PPARα activation, implying different PPARα agonists in a variety of animal tumor models [56,65,245–248].

4.2. *PPARβ/δ*

In contrast to PPARα, PPARβ/δ clearly favors tumor angiogenesis. Studies of human cancer samples revealed that the overexpression of PPARβ/δ in malignant squamous-cell carcinoma [244], pancreatic, prostate, breast cancer, and endometrial adenocarcinoma [249], as well as in colon carcinoma [250], was correlated with higher vessel densities and enhanced tumor progression. Using PPARβ/δ-knockout models, several research groups reported diminished or disturbed tumor-vessel formation and impaired tumor growth upon the induction of different cancer types [249,251,252], confirming the supporting role of PPARβ/δ for tumor angiogenesis and progression. Pharmacological PPARβ/δ activation induced Il-8 and VEGF expression in endothelial cells [253,254]. Enhanced Il-8 expression caused tumor angiogenesis and metastasis formation [252]. Using mice with inducible vascular-specific overexpression of PPARβ/δ [255], it has been demonstrated that the overexpression of PPARβ/δ solely in endothelial cells is sufficient to promote tumor angiogenesis, progression, and metastasis formation. The increased tumor angiogenesis in this model is related to enhanced endothelial Vegf receptor 1, 2, and 3; platelet-derived growth factor receptor beta (Pdgfrβ); platelet-derived growth factor subunit B (Pdgfb); and tyrosinkinase KIT (c-kit) expression [11]. This clearly indicates the danger of the potential therapeutic use of PPARβ/δ agonists, which have been further shown to promote tumor vascularization, growth, and metastasis occurrence [11]. Further detailed information on the angiogenesis-promoting effects of PPARβ/δ can be found in several recent review articles [18,38,177].

4.3. PPARγ

PPARγ activation has mostly been considered to inhibit tumor angiogenesis (reviewed in detail in [177]). Although no correlation could be found between PPARγ expression and vascular density in skin squamous-cell carcinoma [244], PPARγ was found to be less expressed in highly vascularized high grade glioma than in low grade glioma [256]. Most in vivo [69,257–264] studies using different PPARγ agonists observed an inhibition of tumor angiogenesis upon PPARγ activation. In line with these findings, and suggesting that PPARγ activation inhibits tumor vascularization, the deletion of PPARγ in the mammary epithelium of an in vivo model of basal breast cancer lead to increased tumor vessel formation [265]. However, a recent study revealed that activated PPARγ promotes tumor vascularization and growth in breast cancer. Conformational changes in PPARγ induced by ligand activation provoked enhanced angiogenesis and faster tumor growth of mammary tumor cells [266]. A recent study further demonstrated that PPARγ agonists can enhance a pro-tumorigenic secretome in cancer cells, leading to increased tumor angiogenesis and progression [267].

In conclusion, although PPARα and PPARγ seem to decrease tumor angiogenesis, caution should be taken regarding the therapeutical use of any PPAR agonist in the setting of susceptibility to cancer. The example of PPARβ/δ agonists which had been in clinical trials for the treatment of hyperlipidemia and cardiovascular diseases at the beginning of 2000 and turned out to provoke cancers in mice and rats after prolonged treatment, which put a stop to phase 4 trials [268], clearly illustrates the necessity of considering the therapeutic modulation of any PPAR with great care. Regrettably, in clinical studies investigating the use of PPAR modulation in cancer, the effects on tumor vascularization have not been evaluated (reviewed in [177]). A schematic summary of the role of PPARs in tumor angiogenesis is provided in Figure 3.

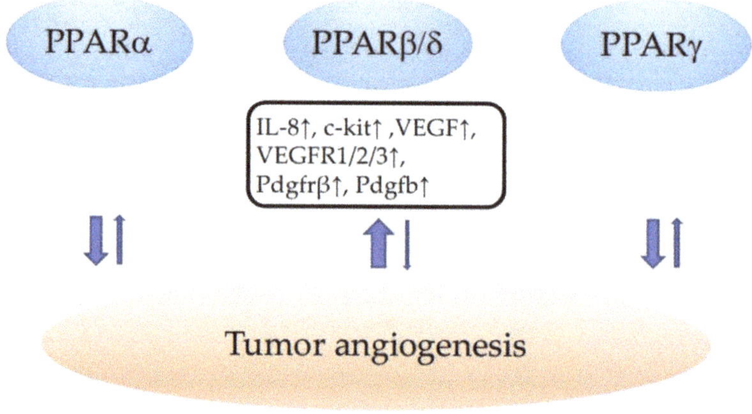

Figure 3. Summary of the influence of PPARα, PPARβ/δ, and PPARγ on tumor angiogenesis. ↓ indicates inhibition and ↑ an increase in angiogenesis. The width of the arrows corresponds to the number of studies reporting similar effects. IL-8: interleukin-8; c-kit: tyrosine-protein kinase Kit; VEGF: vascular endothelial growth factor; VEGFR1/2/3: vascular endothelial growth factor receptors 1/2/3; Pdgfrβ: platelet-derived growth factor receptor beta; Pdgfb: platelet-derived growth factor beta.

5. PPARs and Tumor Suppressors

5.1. PPARα

In addition to the positive regulation of growth-promoting signals, cancer progression is also characterized by the escape of tumor-suppressor action [32]. P53 has been shown recently to transcriptionally inhibit PPARα expression, which has been related to telomere dysfunction and aging, but a potential role in carcinogenesis remained unexplored [269]. On the contrary, PPARα binds to the p300 promoter, which results in the activation of the gene followed by the acetylation and stabilization of p53 in hepatocellular carcinoma [270]. The peroxisomal enzyme Acyl-CoA oxidase 2 (ACOX2) has been postulated as a tumor suppressor in hepatocellular carcinoma via the positive regulation of PPARα. Besides the upregulation of PPARα in hepatoma cell lines with ACOX2 overexpression, no mechanistic link between the two proteins has been explored [271]. Tribbles homolog 3 (TRIB3) has been identified as an oncoprotein in acute myeloid leukemia via the inhibition of apoptosis and autophagy. Mechanistically, this is due to the protein–protein interaction of TRIB3 with PPARα favoring the ubiquitination and degradation of PPARα; on the contrary, the pharmacological activation of PPARα promotes apoptosis and autophagy of leukemia cells [272]. PPARα expression was low in mouse and human colon cancers. The deletion of PPARα in mice reduced the expression of the retinoblastoma protein, resulting in increased expression of the methyltransferases DNMT1 and PRMT6 and, consequently, DNA and histone methylation and lower expression of the tumor suppressors p21 and p27 [273]. P21 seems to act upstream of PPARα under fasting conditions [274]. The tumor suppressor P63 represses PPARα in human keratinocytes [275]. The exact molecular regulation and consequences for tumor growth remained, in both reports, unexplored. PPARα transcriptionally activates the cell-cycle regulator p16Ink4a via a PPAR-response element and an SP1-binding site, and inhibits smooth-muscle cell proliferation, which is relevant to the prevention of intimal hyperplasia in cardiovascular disease [276]. Given the importance of p16Ink4a for cancer [277], potential relevance to tumor growth is likely. Fenofibrate treatment induced the expression of the thioredoxin-binding protein (TXNIP) tumor suppressor in neuroblastoma cells and induced apoptosis. As the inhibition of PPARα did not modify these results, it is likely that fenofibrate had a PPARα-independent effect [59] as was also shown in hepatocellular carcinoma cells [278]. N-Acetyl-Cysteine (NAC) has been described as a PPARα agonist, which inhibits the proliferation of non-small-cell lung carcinoma cells through the induction of p53 and the inhibition of p65, collaboratively reducing PDK1 promoter activity and expression [279]. PPARα activation supports the binding of HIF-1α to the von Hippel–Lindau tumor suppressor, thereby inducing HIF-1α degradation through the ubiquitin–proteasome pathway. Consequently, less Vegf is produced from cancer cells, and angiogenesis and tumor growth might be reduced [280].

Menin, the product of the MEN1 (multiple endocrine neoplasia type 1) tumor-suppressor gene was shown to physically interact with the PPARα protein to control the expression of genes involved in fatty-acid oxidation. The authors investigated a model of hepatic steatosis. Whether this interaction is relevant for tumorigenesis was not analyzed [281].

5.2. *PPARβ/δ*

We have reviewed the knowledge of PPARβ/δ and tumor suppressors before [38]. Recently, it has been shown that pancreatic intraepithelial neoplasias, which mostly harbor oncogenic KRAS mutations, are characterized by the upregulation of PPARβ/δ. PPARβ/δ stimulation via a high-fat diet, or when a specific agonist promotes tumor progression to pancreatic ductal adenocarcinoma [282]. Mechanistically, this is due to the activation of the CCL2/CCR2 axis in pancreatic epithelial cells, which induces an immunosuppressive tumor microenvironment [283]. The increased expression and activity of PPARβ/δ in K-Ras-transformed intestinal epithelial cells has already been described [284]. In hepatocellular carcinoma, SIRT4 acts as a tumor suppressor via the inhibition of PPARβ/δ-induced fatty-acid oxidation and the polarization of macrophages to a pro-inflammatory M1 phenotype [285]. The overexpression of PPARβ/δ in melanoma compared to normal skin has been reported in humans, mice, and horses [102,286]. The expression of PPARβ/δ was inversely correlated with the Wilms tumor suppressor WT1 [286], which is mostly considered as an oncogene [31,103,287–293]. PPARβ/δ activation inhibits melanoma-cell proliferation via the direct repression of WT1 [102], while WT1 stimulates melanoma-cell proliferation [103].

In smooth-muscle cells, the PPARβ/δ agonist L-165041 repressed the phosphorylation of the retinoblastoma protein pRB, and consequently, inhibited proliferation [294]. Whether a similar mechanism is acting in cancer cells is unknown. PPARβ/δ activation with GW0742 reduced SOX2 expression in neuroblastoma cell lines and induced cell differentiation, independently of the p53 status of the cells. Nevertheless, the authors concluded that PPARβ/δ induces neuroblastoma cell differentiation through the SOX2- and p53-dependent pathways [89].

The adenomatous polyposis coli (APC) tumor suppressor is frequently mutated in colon cancer and mouse models, and APC mutations are widely used in colon cancer research. Early reports showed that APC indirectly inhibits PPARβ/δ expression in colon cancer via the suppression of β-catenin/Tcf-4-mediated transcription [196]. The treatment of APCmin mice with the PPARβ/δ agonist GW501516 resulted in an increase in the number and size of intestinal polyps [77]. APC and axin tumor-suppressor-inactivating and β-catenin/Tcf-activating mutations are frequent in different types of cancers. Nearly 50% of ovarian endometrioid adenocarcinomas showed mutations with the dysregulation of β-catenin, which results in the upregulation of PPARβ/δ, MMP-7, Cyclin D1, Connexin 43, and ITF2 [295]. The overexpression of the tumor suppressor called transducer of ErbB-2.1 (Tob1) in gastric cancer cell lines reduced the expression and transcriptional activity of β-catenin, and consequently, of PPARβ/δ [296], supporting the regulation of PPARβ/δ by β-catenin in different cancer types. In breast cancer cells, PPARβ/δ activity seems to be tightly regulated via fatty-acid-binding protein 5 (FABP5). FABP5 binds natural ligands for PPARβ/δ and shuttles them to this nuclear receptor as a pre-requisite for activation. FABP5 expression is positively regulated via EGFR/ERK/phophatidylinositol-3-kinase signaling and activation of the transcription factor NF-kappaB, which is pro-tumorigenic in breast cancer, while Krüppel-like factor KLF2 inhibits FABP5 expression, reducing PPARβ/δ activity, and consequently, is tumor-suppressive [297].

5.3. PPARγ

The tumor suppressor Cyld has been proposed as a transcriptional target gene of PPARγ in mammary epithelial cells. Troglitazone stimulated Cyld mRNA expression and the activity of luciferase reporter/promoter constructs. Thereby, Cyld could act as a mediator of PPARγ-dependent anti-inflammatory and anti-proliferative activity in mammary epithelial cells [298]. The Wnt7a/Frizzled9/Gα16 pathway activates PPARγ to inhibit cell proliferation in non-small-cell lung cancer [299]. The retinoblastoma tumor-suppressor (Rb) protein interacts with E2F to suppress PPARγ expression. Consequently, in mice with compound loss of p53 and pRb, the tumor spectrum shifted from osteosarcoma (bone tumor) to hibernomas (brown-fat tumor), supporting the involvement of PPARγ in the cell-fate switch from bone- to adipose-tissue tumors [300]. The retinoic acid-producing enzyme aldehyde dehydrogenase 1a1 acts as a tumor suppressor in splenic B-cell subpopulations by regulating retinoic acid receptor alpha, zinc finger protein Zfp423, and PPARγ. The regulation of PPARγ was specific only to an IgG1(+)/CD19(+) cell population [301]. In hepatocellular carcinoma cells, PPARγ activation using rosiglitazone, or its overexpression, induced Cited2, which was associated with reduced cell growth and the induction of p15, p21, and p27. Chromatin immunoprecipitation confirmed that the binding of PPARγ to the Cited2 promoter sequence was direct [302]. Additionally, in bladder cancer cells, troglitazone increased the expression of p21 and p16Ink4a [217]. CCAAT/enhancer-binding protein-alpha (C/EBP-alpha) overexpression induced PPARγ expression, and secondary PPARγ directly activated p53 and induced apoptosis in rat hepatic stellate cells [303]. As C/EBP-alpha activating mutations are found in acute myeloid leukemia patients [304], this regulatory pathway might be relevant for cancer. In breast cancers, C/EBP-alpha shows low expression compared to its normal nuclear expression in ductal cells. Additionally, in this case, the overexpression of C/EBP-alpha was associated with increased PPARγ and p21 expression [305].

Estrogen receptor alpha (ERα) interacts physically with PPARγ, and both proteins compete for the chance to bind to PPREs. While PPARγ activates transcription from this element, ERα represses transactivation. Thus, both proteins differentially modulate the proliferation of breast cancer cell lines in vitro [306]. The relationships between the different PPARs and tumor suppressors are schematically summarized in Figure 4.

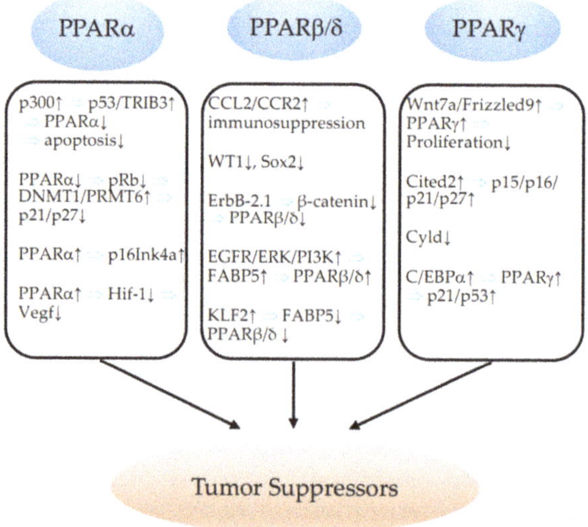

Figure 4. Illustration of the relationships between PPARα, PPARβ/δ, and PPARγ and tumor suppressors.

↓ indicates inhibition and ↑ indicates an increase. ⇒: leads to; p300: P300 transcriptional co-activator protein; p53: tumor protein p53; Trib3: Tribbles homolog 3; pRb: phosphorylated retinoblastoma protein; DNMT1: DNA (cytosine-5)-methyltransferase 1; PRMT6: protein arginine N-methyltransferase 6; p16Ink4a/21/27: tumor suppressors p16Ink4a, p21, p27; Hif-1: hypoxia-inducible factor-1; Vegf: vascular endothelial growth factor; CCL2: monocyte chemotactic protein-1; CCR2: receptor for monocyte chemoattractant protein-1; WT1: Wilms tumor 1 protein; SOX2: SRY-box transcription factor 2; ErbB-2.1: Erb-B2 receptor tyrosine kinase 2; EGFR: epidermal growth factor receptor; ERK: extracellular signal-regulated kinase; PI3K: phosphoinositide 3-kinase; FABP5: fatty-acid-binding protein 5; KLF2: Krüppel-like Factor 2; Wnt7a: Wnt family member 7A; Cited2: Cbp/p300-interacting transactivator 2; Cyld: cyld lysine 63 deubiquitinase; C/EBPa: CCAAT/enhancer-binding protein alpha.

6. PPARs in Invasion and Metastasis

6.1. PPARα

PPARα ligands were shown to inhibit the phorbol-ester-induced upregulation of Cox-2 and VEGF expression, both implicated in metastasis promotion, in a colon cancer cell line [307]. Similarly, PPARα ligands inhibited the transforming growth factor (TGF) α-induced expression of matrix metalloproteinase 9 (MMP 9), also strongly implicated in metastasis advancement [308]. Fenofibrate reduced the metastatic potential of melanoma cells in vitro and in vivo, implicating the downregulation of Akt phosphorylation [309,310]. The ligand activation of PPARα inhibited the formation of proangiogenic epoxyeicosatrienoic acids (EET) by the cytochrome P450 arachidonic acid epoxygenases (Cyp2c), and thereby reduced NSCLC growth and metastatic progression in vivo [65,247]. Acyl-CoA oxidase 2 (ACOX2) has been proposed to inhibit tumor progression and the metastasis of HCC trough a PPARα-dependent pathway [271]. In contrast, an elegant in vitro and in vivo study evidenced that PPARα favored metastasis. PPARα is required for the generation of immunosuppressive regulatory B cells, designated tBregs from B cells, which is induced by metabolites of the 5-lipoxygenase pathway. A deficiency of PPARα in B cells blocked the generation of tBregs, and thus, abrogated lung metastasis in mice with established breast cancer [311]. The metastasis of tumors to lymph nodes predicts disease progression and influences therapeutic schemes. Comparative metabolomic and transcriptomic analyses of primary tumors which had metastasized to lymph nodes demonstrated that metastasizing tumor cells undergo a metabolic shift towards fatty-acid oxidation (FAO). Most upregulated gene sets in the metastatic lymph node tumors were related to aspects of lipid biology, fatty-acid metabolism, and PPARα signaling pathways. The authors demonstrated that the activation of the transcriptional coactivator yes-associated protein (Yap) in lymph node metastatic tumors induced the upregulation of genes implicated in FAO. The inducible knockdown of Yap or of the inhibition of FAO suppressed lymph node metastasis [312]. Chen and coworkers reported that mitochondrial 3-hydroxy-3-methylglutaryl-CoA synthase (HMGCS2) enhanced the motility and metastasis formation of CRC and oral squamous-cell carcinoma (OSCC) cells in vitro and in vivo. This oncogenic function was found to be mediated through the direct binding of HMGCS2 to PPARα, which, in turn, led to the transcriptional activation of the proto-oncogene tyrosine-protein kinase Src, a target of PPARα. *HMGCS2* mRNA expression was further found to be associated with poor clinical prognoses and outcomes in patients [313]. It is highly interesting that the plasticizer di(2-ethylhexyl) phthalate (DEHP) and its hydrolysate mono(2-ethylhexyl) phthalate (MEHP) are major toxicants from plastics; nevertheless, a potential carcinogenic effect has not been investigated. Leng and colleagues demonstrated that MEHP treatment promoted the phosphorylation of Akt and the degradation of IκB-α, thus activating NF-κB and enhancing NF-κB nuclear translocation, which enhanced metastasis formation of ovarian cancer xenografts. The inhibition of PPARα by the antagonist GW6471 abrogated metastasis in vivo, indicating that the MEHP promotion of metastasis is mediated in a PPARα-dependent manner through the PI3K/Akt/NF-κB pathway [314]. In conclusion, PPARα favored metastasis in many model systems, also through its wider implication in metabolic and immunological processes. PPARα modulation is therefore, nowadays, not considered as a safe therapeutic option in the setting of cancer.

6.2. PPARβ/δ

The role of PPARβ/δ for the invasion and metastasis of cancers has recently been thoroughly reviewed in [38]. In a very detailed study, Abdollahi and colleagues demonstrated that PPARβ/δ expression levels were correlated with a higher malignant grade and distant metastasis formation in cancer patients with prostate, breast, and endometrial adenocarcinoma [249]. Additionally, in colorectal cancer, high expression of PPARβ/δ coincided with a high risk of developing distant liver metastases [71]. In contrast, in vitro studies using the PPARβ/δ agonist GW501516 in pancreatic [315] or breast cancer cells [316] reported decreased invasion capabilities of the tumor cells upon PPARβ/δ activation. A metastasis-inhibiting role of PPARβ/δ has been proposed by Lim and coworkers, who reported that treatment with the PPARβ/δ antagonist for 10 h increased melanoma cell migration and invasion. This antagonist had, so far, not been used in other studies, and the results were not confirmed by employing well-established antagonists such as GSK0660 or GSK3787 [317]. One group observed the downregulation of N-Cadherin upon PPARβ/δ agonist activation in a bladder cancer cell line, which has been suggested to diminish metastatic potential [318]. Most of the studies, however, confirm the invasion- and metastasis-promoting effects of PPARβ/δ, which were first suggested via analyses of PPARβ/δ expression in published large-scale microarray data from cancer patients [71,249]. A study by Zuo and colleagues identified several pro-metastatic genes as PPARβ/δ targets through the analysis of transcriptome profiling of HCT116 colon cancer cells, with or without the genetic deletion of PPARβ/δ. Using several experimental in vivo models (syngenic and orthotopic tumor inductions, different tumor-cell types), the authors showed that PPARβ/δ knockdown in cancer cells inhibited metastasis formation. The treatment of mice with the PPARβ/δ agonist GW0742 enhanced metastasis formation. It was further demonstrated that high expression of PPARβ/δ in cancer cells is the most important factor for metastasis formation as heterozygous PPARβ/δ mice developed fewer metastases than their wildtype littermates; however, the most important metastasis inhibition was observed when PPARβ/δ was deleted in cancer cells used for syngenic tumor induction. High PPARβ/δ expression in cancer cells additionally promoted tumor angiogenesis through increases in VEGF and IL-8. Finally, analyses of independent datasets from cancer patients (liposarcoma, colon, breast, and lung cancer) demonstrated that PPARβ/δ expression in cancer cells strongly influenced metastasis-free survival [252]. Our group confirmed the pro-metastatic effects of PPARβ/δ activation in vivo. PPARβ/δ agonist GW0742-treated animals with syngenic induced LLC1 tumors had significantly increased spontaneous lung and liver metastasis formation compared to controls injected with a vehicle. We further evidenced that the conditional inducible overexpression of PPARβ/δ in vascular cells was sufficient to promote metastasis formation [11]. High-fat diets are associated with carcinogenesis [319]; however, the underlying mechanisms are not well-understood. A recent study demonstrated the implication of PPARβ/δ in the pro-metastatic effects of dietary fats in colorectal cancer. The authors showed, first, that the activation of PPARβ/δ by GW501516 induced the expansion of colonic cancer stem cells (CSC) and boosted metastasis formation in vivo through the induction of the self-renewal regulatory factor Nanog. The activation of PPARβ/δ increased, whereas the knockout of PPARβ/δ decreased Nanog expression, and knockdown of Nanog abolished the metastasis-promoting effects of PPARβ/δ. Finally, the authors demonstrated that a high-fat diet mimicked the effects of PPARβ/δ activation by inducing Nanog, accelerating tumor formation, and increasing liver metastasis development. The knockout of PPARβ/δ inhibited the high-fat-diet-induced effects on tumorigenesis and progression [320]. Although few studies reported decreased metastasis-related events upon PPARβ/δ activation in vitro, the role of PPARβ/δ on metastasis remains to be defined in representative in vivo models, which unequivocally confirms the pro-metastatic functions of PPARβ/δ.

6.3. PPARγ

Thiazolidinediones were found to inhibit the synthesis of matrix metalloproteinases (MMPs) and adhesion to the extracellular matrix (ECM) proteins of colon cancer cell lines [321], and to abolish lymph node and lung metastases in colon cancer xenograft models [322]. Similarly, linoleic acids have been reported to inhibit colon cancer metastasis through PPARγ activation [323]. Later, the downregulation of the chemokine receptor CXCR4 was further attributed to the metastasis-preventing effects of PPARγ in colon [324,325] as well as in breast cancer [326]. In line with these findings, low levels of PPARγ in colon cancers of patients were correlated with enhanced metastatic potential [327]. NSAIDs were reported to have beneficial effects on colon metastasis inhibition through their suppression of cancer stem cells, mediated through the suppression of Cox-2 and the activation of PPARγ [328]. Mammary tumors were found to metastasize less upon PPARγ activation due to decreased MMP production [329]. 15d-PGJ2 has further been shown to inhibit osteolytic breast cancer bone metastasis [330]. Additionally, NSCLC cells overexpressing PPARγ exhibited decreased metastatic potential [331]. A good study showed that the activation of PPARγ inhibited transforming growth factor β (TGF-β)-induced epithelial mesenchymal transition (EMT) in lung cancer cells. PPARγ-antagonized TGF-β–caused a loss of E-cadherin expression and inhibited the induction of mesenchymal markers and MMPs, thus preventing migration, invasion, and metastasis formation [332]. Rosiglitazone was found to suppress metastatic potential in gastric cancer, and the enhanced activity of PPARγ resulted in increased direct transcriptional activation of cellular adhesion molecule 3, which inhibits the migration and invasion of gastric cancer cells [333,334]. Modulation of the plasminogen activator system has been proposed to be one metastasis inhibiting mechanism of PPARγ activation in pancreatic cancer [335]. In hepatocellular carcinoma (HCC), low PPARγ expression was correlated with more advanced TNM (tumor, node, metastasis) stages [335], and PPARγ activation decreased the invasive and metastatic potential of cancer cells in vitro and in vivo through the downregulation of MMP9 and 13, and the upregulation of the extracellular matrix-regulator tissue inhibitors of metalloproteinase (TIMP) 3, E-cadherin, and spleen tyrosine kinase [336]. The high expression of Micro RNA 130b (miR-130b) in HCC was correlated with enhanced metastasis and the downregulation of PPARγ. Lowering miR-130b resulted in increased PPARγ expression and suppressed EMT in HCC cells [337]. An elegant study determined that PPARγ is required for the peroxisome proliferator-activated receptor-gamma coactivator-1α (PGC1α)-mediated inhibition of HCC metastasis. PGC1α inhibits the aerobic glycolysis of cancer cells through PPARγ-dependent inhibition of the WNT/β-catenin pathway [338]. However, an in vitro study suggested that PPARγ antagonists inhibited metastasis through the cleavage of vimentin in hepatocellular carcinoma [339]. Like the situation in HCC, microRNA 27b (miR-27b) has been suggested to downregulate PPARγ, and thereby, to promote the invasion of cervical carcinoma [340]. In squamous-cell carcinoma, the inhibition of PPARγ was proposed to decrease cell adhesion through the downregulation of integrin alpha 5 [238]. Later, doubts regarding the suggested beneficial effects of PPARγ activation for metastasis inhibition in lung cancer arose. Ahn and coworkers identified mitogen-activated protein kinase kinase 4 (MAP2K4) as a tumor suppressor in lung adenocarcinoma. MAP2K4 inhibited lung cancer cell invasion through the repression of PPARγ. MAP2K4 deficiency increased PPARγ expression and promoted cancer cell invasion, which could be reversed via PPARγ inhibition [341]. PPARγ agonist activation in orthotopic and spontaneous murine lung cancer models significantly increased metastasis formation through its upregulated expression in macrophages, which contributed to tumor progression and metastasis through increased arginase 1 expression. The inducible conditional knockout of PPARγ solely in macrophages reconstituted the beneficial roles of PPARγ ligand activation in lung cancer cell growth and metastasis inhibition [342]. The increased production of transforming growth factor β 1 (TGFβ1) in macrophages upon stimulation of PPARγ has been proposed as the underlying mechanism for the promotion of invasion and metastasis in this context [343]. Similarly, bone marrow adipocytes promote bone metastasis formation in prostate cancer, which is, in part, mediated through the PPARγ-induced activation of fatty-acid-binding protein 4 (Fabp4) [344]. Liliane Michaliks' group further showed that the PPARγ agonist rosiglitazone activates a tumorigenic secretion program of cytokines, chemokines, and pro-angiogenic factors in melanoma cells, leading to a tumor progression- and metastasis-favoring microenvironment [267]. This, again, suggests that PPARγ may have anti-tumorigenic effects on cancer cells, but pro-tumorigenic effects on cells of the microenvironment, as was already described in the context of breast cancer [166]. The situation might be even more complex as truncated isoforms of PPARγ might further fuel the metastasis-promoting actions of tumor stromal cells. Niu and colleagues demonstrated that caspase-1 cleaves PPARγ, leading to a truncated isoform which translocates to mitochondria, resulting in the inhibition of medium-chain acyl-CoA dehydrogenase (MCAD) and fatty-acid oxidation. Thus, the differentiation of tumor- and metastasis-promoting macrophages is enhanced by the accumulation of lipid droplets [345]. Tumor-associated macrophages can be divided in two subgroups: M1 macrophages, which are pro-inflammatory cells involved in killing tumor cells, and M2 macrophages, which mediate tumor progression and metastasis. Shu and colleagues revealed the important role of integrin $\beta 3$ in macrophage M2 polarization. The inhibition of integrin β3 blocked M2 polarization only in the setting of high PPARγ expression and activity, which indicates that the action of integrin β3 depends on PPARγ [346]. An excellent study unveiled the mechanism by which PPARγ facilitates brain metastasis formation from primary cancers: astrocytes, brain glial cells, have a high content of polyunsaturated fatty acids, which function as donors of PPARγ activation in invading cancer cells, thus enhancing proliferation and metastatic outgrowth to the brain. PPARγ expression was significantly higher in brain metastatic lesions than in the primary tumors of breast cancer and melanoma patients. PPARγ antagonist treatment reduced melanoma or breast cancer brain metastasis burden in animals. This further adds to the complexity regarding the role of PPARγ in cancer, which depends on the stage of cancer development. PPARγ might inhibit early primary cancer growth, but fuels advanced-stage metastatic formation [347]. The situation also becomes more complicated, as in several different tumor types such as prostate [348–350], bladder [351], pancreatic cancer [352], and myxoid liposarcoma [353], high levels of PPARγ expression in tumor cells are correlated with enhanced metastasis formation; this also indicates that a general beneficial effect of PPARγ expression in tumor cells on metastasis inhibition cannot be concluded. The major effects of PPARs for invasion and metastasis formation are illustrated in Figure 5.

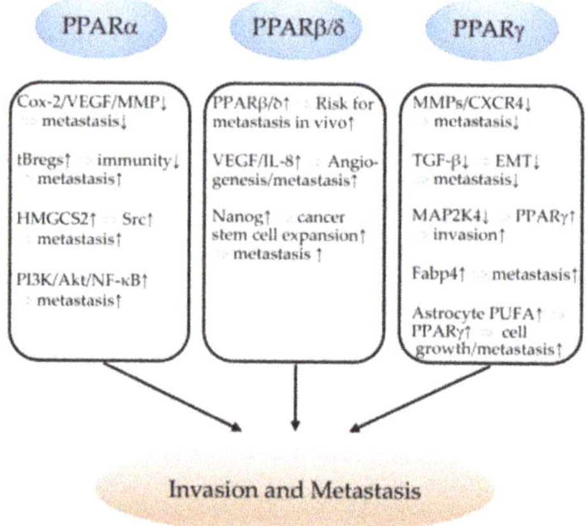

Figure 5. Schematic representation of the effects of PPARα, PPARβ/δ, and PPARγ on invasion and metastasis formation. ↓ indicates inhibition and ↑ indicates an increase. ⇒: leads to; Cox-2: cyclooxygenase-2; VEGF: vascular endothelial growth factor; MMP: matrix metalloproteinase; tBregs:

immunosuppressive regulatory B cells; HMGCS2: 3-hydroxy-3-methylglutaryl-CoA synthase 2 (mitochondrial); Src: proto-oncogene tyrosine-protein kinase Src; PI3K: phosphatidylinositol 3-kinase; Akt: AKT serine/threonine kinase; NF-κB: nuclear factor kappa-light-chain-enhancer of activated B cells; IL-8: interleukin 8; CXCR4: C-X-C chemokine receptor type 4; TGF-β: transforming growth factor beta; EMT: epithelial–mesenchymal transition; MAP2K4: dual-specificity mitogen-activated protein kinase kinase 4; Fabp4: fatty-acid-binding protein 4; PUFA: polyunsaturated fatty acid.

7. PPARs and Replicative Immortality

7.1. PPARα

Stem cells in the intestinal epithelium lose their self-renewal capacity with aging due to decreased Wnt signaling. Mechanistically, high mTORC1 activity inhibits PPARα. In turn, Notum, a Wnt inhibitor, becomes activated via a lack of PPARα, and stem cell self- is inhibited [354]. Whether this mechanism also operates in cancer stem cells remains to be determined. High PPARα expression has been described in glioma stem cells compared to fetal neuronal stem cells. The inhibition of PPARα expression induced the downregulation of stem cell markers c-Myc, Sox2, and nestin, and induced senescence. In contrast to control cells with intact PPARα expression, knockdown cells did not form tumors in vivo, suggesting PPARα inhibition as a potential target for the inhibition of glioblastoma growth [60]. In line with this, the positive transcriptional regulation of CPT1C by PPARα was shown to inhibit senescence in different cancer cell lines in vitro [61]. Whether the shortened lifespan, hepatocarcinogenesis, and age-related lesions in the heart, kidney, and liver of PPARα-knockout mice reported earlier [355] are due to modifications in senescence remains unexplored; however, it seems more likely that alterations in apoptotic pathways are responsible for these phenotypes [356].

7.2. PPARβ/δ

The role of PPARβ/δ in replicative immortality, senescence, and cancer stemness was reviewed recently [38]. The pharmacological activation of PPARβ/δ inhibited senescence in human vascular smooth-muscle cells, coronary artery endothelial cells, keratinocytes, and cardiomyocytes [357–360]. On the contrary, higher PPARβ/δ expression was correlated with increased senescence in benign neurofibromas and colon adenomas [361], and senescence, in this case, was correlated with endoplasmic reticulum stress [362], which seems unusual. In endothelial cells, the lipid peroxidation product 4-HNE activated PPARβ/δ, resulting in the induction of thioredoxin-interacting protein (TXNIP) expression and senescence [363].

PPARβ/δ activation keeps neuronal and colonic cancer stem cells in an proliferative, undifferentiated state via the induction of Sox2 and Nanog [320,364], which, in the case of colon cancer, contributes to metastasis formation in response to fatty-acid intake [320]. PPARβ/δ is expressed in gastric progenitor cells where it upregulates Ccl20 and Cxcl1, contributing to chronic inflammation and malignant transformation [80]. Furthermore, PPARβ/δ contributes to stemness through protein–protein binding with β-catenin and the transcriptional activation of low-density lipoprotein receptor-related protein 5 (LRP5), which acts as a Wnt co-receptor [365]. Whether this is the case in cancer stem cells is an open question. In general, it is currently difficult to conclude whether the PPARβ/δ-dependent induction/inhibition of senescence might promote or delay cancer progression, as senescence, on one hand, is a gatekeeper to prevent cancer, but on the other hand, it might also contribute to the initiation and progression of a second tumor [366–369].

7.3. PPARγ

Recently, it was shown that the Fanconi anemia protein FANCD2 and Hairy Enhancer Split 1 (HES1) collaborate in the transcriptional repression of PPARγ to keep hematopoietic stem cells in a quiescent state and to avoid stem cell exhaustion, as well as hematological malignancies [370]. PPARγ is also required for enhanced glucose-stimulated insulin secretion in senescent pancreatic beta cells with aging [371]. Whether this affects cancer metabolism and growth is currently undetermined. PPARγ has different effects in stroma and cancer cells. PPARγ overexpression reduced breast cancer cell growth in xenograft models, and was associated with increased autophagy and the inhibition of angiogenesis; meanwhile, overexpression in stromal cells enhanced tumor growth, which has been related to the increased expression of autophagic markers, the production of lactate, cell hypertrophy, mitochondrial dysfunction, and senescence, as illustrated by higher p16/p21 expression and beta galactosidase [166]. In cell-culture models, PPARγ inhibits the expression of silent information regulator type 1 (SIRT1), a molecule known to delay senescence, which is in agreement with the senescence-promoting effects of PPARγ described above [372]. In human fibroblasts, PPARγ transcriptionally activates p16 and induces senescence [373]. In human colon cancer samples, a significant correlation between PPARγ and the expression of pRb, cyclin D1, p16, and p21 was found; however, surprisingly, PPARγ expression did not correlate with the stage, grade of differentiation, metastasis, tumor proliferative capacity, or patient survival [374]. Additionally, the opposite effect, involving the pioglitazone-induced induction of proliferation via the inhibition of P16 expression in adipocyte progenitors, has been described [375]. Pioglitazone treatment in mice activated telomerase and inhibited p16 expression and senescence in vascular cells [376]. The effects of PPARs on replicative immortality and senescence are summarized in Figure 6.

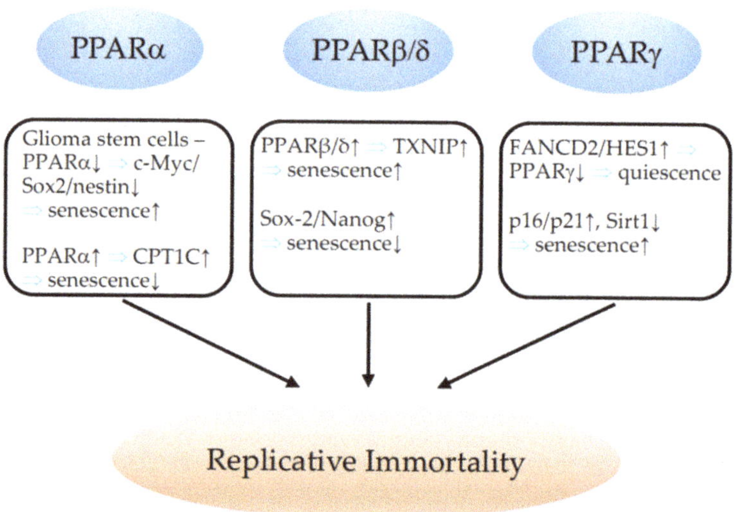

Figure 6. Summary of the effects of PPARα, PPARβ/δ, and PPARγ on senescence and replicative immortality. ↓ indicates inhibition and ↑ indicates an increase. ⇒: leads to; c-Myc: MYC proto-oncogene; Sox2: SRY (sex-determining region Y)-box 2; CPT1C: carnitine palmitoyltransferase 1C; TXNIP: thioredoxin-interacting protein; FANCD2: Fanconi anemia, complementation group D2; HES1: hes family bHLH transcription factor 1; Sirt1: sirtuin 1.

8. PPARs and Tumor Metabolism

PPARs are important mediators of lipid and glucose metabolism [1,377]. Glucose and fatty acids serve to sustain cancer-cell proliferation and fatty-acid function as signaling molecules and membrane components of cancer, as well as immune cells [32,378]. A major metabolic anomaly in cancers, i.e., the dependence on aerobic glycolysis for energy production, was described by Otto Warburg nearly 100 years ago [379]. Furthermore, as a general characteristic of cancer metabolism, the rapid growth of tumors results in hypoxia and the stabilization of hypoxia-inducible transcription factors (Hif) [380,381], which induce or repress the expression of downstream target genes, with relevance to cancer growth, e.g., VEGF [382], WT1 [383], PPARα [384], glucose transporters, and many others (reviewed in [385]). As the expression of different PPARs varies between cancer types, here, we will summarize the knowledge on PPARs in the metabolic regulation of distinct tumors.

8.1. PPARα

The hepatocarcinogenic effects of peroxisome proliferators in mice were already described in the 1970s [386]. PPARα activation induces the key genes of fatty-acid metabolism, which results in the increased generation of reactive oxygen species [387] and favors carcinogenesis. This predisposing role is modified by antioxidant defense mechanisms, age, and nutritional status (reviewed in [388]). Furthermore, interactions between different cell types modify the response to PPAR modulators.

Fibrates also favor oxidative metabolism in cytotoxic T cells. Fenofibrate reduced glucose's utilization of cancer cells and stromal cells and shifted their metabolisms to fatty-acid use [389]. The glucose in the tumor environment was available for CD8 T cells and tumor infiltrating lymphocytes, which enhanced the success of tumor vaccination in a mouse model [390]. A potential use of PPAR ligands for the metabolic reprogramming of T cells in cancer immunotherapy has been described and reviewed before [391,392]. In a recent study, it was shown that the addition of fibrates to immune checkpoint inhibitors in patients with non-small-cell lung cancer increases overall survival, which was not the case in patients receiving chemotherapy [393]. Whether this effect is due to shifts in metabolism or involves other cancer hallmark capabilities is unknown. Nevertheless, it is an exciting finding linking PPAR research to clinical application.

Further crosstalk exists between adipocytes and tumor cells. Obese or diabetic patients are at increased risk of breast cancer [394,395]. The co-culture of adipocytes and breast cancer cell induces the expression of genes involved in inflammation and lipid metabolism (IL1, PLIN2, ANGPTL4). ANGPTL4 is a downstream target of PPARα. Consequently, the pharmacological inhibition of PPARα reduced ANGPTL4 expression, which is involved in adipose-tissue-induced β-oxidation, proliferation, and the invasion of breast cancer cells [396]. High glucose activated PPARα and PPARγ expression in breast cancer cell cultures [40]. Sirt6 activated PPARα expression, promoted beta-oxidation and mediated the PPARα-dependent inhibition of SREBP-dependent cholesterol and triglyceride synthesis in the livers of mice [397]. Whether this pathway is relevant for tumorigenesis remains to be determined. Activating mutations in the beta-catenin gene are frequently found in hepatocellular carcinomas. Beta-catenin acts as an activator of PPARα, which stimulates fatty-acid oxidation as the major metabolic pathway of beta-catenin-dependent hepatocellular carcinoma. Consequently, a knockout of PPARα and the inhibition of fatty-acid oxidation using the CPT-1 inhibitor etomoxir reduced hepatocellular carcinoma progression [398].

Aldehyde dehydrogenase (ALDH7A1) acts upstream of PPARα by providing metabolites which act as ligands for this receptor. The knockdown of ALDH7A1 increased cell migration and invasion. Low levels of the aldehyde dehydrogenase protein were correlated with poor clinical outcome in hepatic and kidney cancer patients [399]. The PPARα agonist Wy14,643 reduced Glucose transporter 1 (Glut1) expression, glucose transport, and the proliferation of different cell lines, suggesting anti-tumorigenic action in this model [400,401]. In contrast, PPARα is highly expressed in glioblastoma and glioma stem cells, and its inhibition results in the downregulation of key regulators of fatty-acid oxygenation, ACOX1 and CPT1A, and reduced tumor growth in mice [60]. Surprisingly, the inhibition of aerobic glycolysis, mitochondrial damage, and reduced glioblastoma growth in mice in response to fenofibrate treatment has also been described [402]. The PPARα antagonist GW6471 attenuated enhanced fatty-acid oxidation and oxidative phosphorylation, blocked enhanced glycolysis, and reduced tumor growth in a renal-cell carcinoma model in nude mice [194].

8.2. PPARβ/δ

PPARβ/δ function in cancer and metabolic alterations were previously investigated in colon cancer. The first publications were already controversial (reviewed in [38,403]).

PPARβ/δ activation stimulates calcineurin expression [404], which induces Hif-1 stabilization [405]. Hypoxia, in turn, stimulates the transcriptional activation of PPARβ/δ in colon cancer cells via association with p300. PPARβ/δ deficiency in colon cancer cells reduces hypoxia-induced VEGF and IL6 expression, which links PPARβ/δ to tumor angiogenesis and immune response in colon cancer [406].

A mouse model of PPARβ/δ overexpression in gastric progenitor cells demonstrated the development of invasive gastric tumors in aging animals. Metabolic profiling revealed that these tumors do not require glycolysis but fatty-acid oxidation for tumor progression [407]. Additionally, a high-fat diet has been shown to induce fatty-acid oxidation depending on PPARβ/δ, which is associated with intestinal stem cell activation and enhanced tumorigenesis [408], as well as colorectal metastasis formation via the activation of Nanog in colonic cancer stem cells [320]. Epidemiological studies suggest a positive association of saturated fatty acids with colon cancer risk, while an inverse association exists for polyunsaturated fatty acids [409]. However, experimental studies showed that saturated long-chain fatty acids (SLCFAs) might inhibit the proliferation of some cancer cell lines, while unsaturated long-chain fatty acids (ULCFAs) induce cancer cell growth [410,411]. These differences could be related to the expression of fatty-acid-binding protein 5 (FABP5), retinoic acid receptors (RAR), and PPARβ/δ. Both SLCFAs and ULCFAs bind to FABP5, which dislodges retinoic acid and endogenous PPAR ligands from this transport protein. Depending on the presence of RARs, retinoic acid will bind to this receptor and activate it. SLCFAs reduce PPARβ/δ activity, while ULCFA/FABP5 complexes translocate to the nucleus where ULCFAs act as ligands for PPARβ/δ [412]. Consequently, a lack of FABP5 has been shown to inhibit mammary tumorigenesis [95]. These data are in general agreement with a pro-tumorigenic effect of PPARβ/δ, but point also to the complexity of different ligands, PPAR, RXR, and fatty-acid-binding protein expression in each individual tumor sample.

As an epigenetic mechanism, N1-methyladenosine methylation in tRNA via TRMT6/TRMT61A enhances PPARβ/δ translation, which augments cholesterol synthesis and Hedgehog signaling in liver cancer stem cells to support hepatic carcinogenesis [413]. The PPARβ/δ agonist GW501516 induced the expression of Glut1 (glucose transporter 1) and SLC1A5 (solute carrier family 1 member 5), which favors glucose and glutamine influx, thereby enhancing the proliferation of different cancer cell lines in vitro [78,414]. These effects were reversed by metformin. The molecular mechanisms were not investigated. In hepatocellular carcinoma resistant against the tyrosine kinase inhibitor sorafenib, a higher glutamine metabolism and reductive glutamine carboxylation dependent on PPARβ/δ were reported. The inhibition of PPARβ/δ reversed these metabolic alterations and sensitized the tumors to sorafenib, suggesting that sorafenib resistance in these tumors depends on PPARβ/δ-dependent metabolic alterations and might be treated with PPARβ/δ antagonists [415].

8.3. PPARγ

The role of PPARγ in metabolism and cancer has been reviewed before [228,416,417]. Part of the beneficial effects in cancer might simply be attributed to the reduction in tumor cachexia, which was associated with better survival in animal models [418,419]. The complex interactions between stroma and cancer cells are underlined by the observation that PPARγ activation in cancer cells reduces tumor growth, while overexpression in stromal cells enhances breast cancer growth in mice. In this model, cancer cells induce autophagy, glycolysis, and senescence in stromal cells, while stromal cells generate L-lactate, ketones, glutamine, amino acids, and fatty acids that are used by cancer cells to enhance their tumorigenic potential [166].

New data showed that interleukin-4 (IL-4) induces the expression of hematopoietic prostaglandin D2 synthase, thereby enhancing the endogenous levels of prostaglandin D2 and its metabolites. They act via PPARγ to reduce the severity of acute myeloid leukemia (AML) in mouse models and patient cells, suggesting IL-4 as a potential additional therapeutic option for AML [420]. Ubiquitin-specific protease 22 (USP22) stabilizes PPARγ due to deubiquitination, which increases acetyl-CoA carboxylase (ACC) and ATP citrate lyase (ACLY) expression and induces de novo lipogenesis as a risk factor for hepatocellular carcinoma (HCC). Consequently, PPARγ inhibition might reduce HCC progression [421]. In prostate cancer cells, PPARγ stimulates AKT serine/threonine kinase 3 (AKT3) expression, which favors PGC1α localization to the nucleus, mitochondrial biogenesis, and elevates ATP levels, ultimately leading to tumor-cell proliferation and metastasis via an enhanced epithelial–mesenchymal transition [350]. N-3 polyunsaturated fatty acids stimulated Syndecan 1 expression via PPARγ activation in prostate epithelium and prostate cancer cells [422]. The authors suggested chemo preventive properties of n-3 fatty acids in prostate cancer via this pathway, which was not proven experimentally. Additionally, in metastatic brain tumors, PPARγ is activated and contributes to metastatic spreading of the tumor cells due to the generation of lipid-derived endogenous PPAR activators from surrounding astrocytes [347].

Acyl-coenzyme-A-binding protein (ACBP) is a direct downstream effector of PPARγ that induces lipogenesis [423]. The long non-coding RNA MALAT1 acts upstream of PPARγ and might directly activate the PPARγ promoter to induce adipogenesis. Low expression of MALAT1 in cancer patients is associated with tumor cachexia and poor survival [424]. The esophageal adenocarcinoma-specific master regulator transcription factors (MRTFs) ELF3, KLF5, GATA6, and EHF activate PPARγ. PPARγ, in turn, enhances the synthesis of fatty acids, phospholipids, and sphingolipids and, in a positive feedback loop, induces MRTF expression, suggesting a pro-cancerogenic function in esophageal adenocarcinoma [425]. In metastatic prostate cancer, the situation seems comparable. PPARγ promotes the growth of this cancer type via the activation of lipid signaling pathways, i.e., the upregulation of fatty-acid synthase, acetyl-CoA carboxylase, and ATP citrate lyase. The inhibition of PPARγ reduces lipid synthesis and tumor growth [348]. Furthermore, PPARγ promotes prostate cancer growth via the induction of VEGF expression [426].

Hypoxia induces the stabilization of Hif-1α, which suppresses PPARγ in non-small-cell lung cancer (NSCLC). This is associated with uncoupling protein 2 (UCP2) downregulation, which results in the production of reactive oxygen species, upregulation of the ABC transporter protein ABCG2, elevated glucose uptake, and reduced oxygen consumption. These mechanisms might contribute to chemoresistance in NSCLC [427]. Whether PPARγ agonists sensitize NSCLCs to chemotherapy and are of therapeutic benefit, or whether other Hif-1α-dependent signaling pathways might interfere in this tumor type, could be relatively easily answered from researchers' long clinical experience with the use of PPARγ agonists. Earlier studies found that PPARγ inhibits the growth and invasiveness of NSCLCs and other cell lines via the inhibition of Cox-2 expression [428] and the reduction in prostaglandin E(2) production [429,430].

A clinical trial of at least phase 2 in CML patients showed some beneficial effects of the addition of pioglitazone [156]. The PPARγ agonist pioglitazone was found to induce a metabolic switch that inhibits pyruvate oxidation, reduces glutathione levels, and increases reactive oxygen species (ROS) levels, inducing the hypo-phosphorylation of the retinoblastoma protein (RB) and cell-cycle arrest [173]. In a prostate cancer cell-derived tumor spheroid culture system, pioglitazone lowered the pH, decreased oxygen consumption, and increased lactate secretion. Other glitazones had similar effects [431]. Troglitazone and ciglitazone inhibited aerobic glycolysis, induced SIRT1 expression and endoplasmic reticulum stress in cancer cells, and induced autophagy and apoptosis independently of PPARγ [432]. Thus, it remains difficult to conclude specific PPARγ effects in cancer metabolism from studies using thiazolidinediones. The major effects of PPARs on tumor metabolism and the functional consequences are summarized in Figure 7.

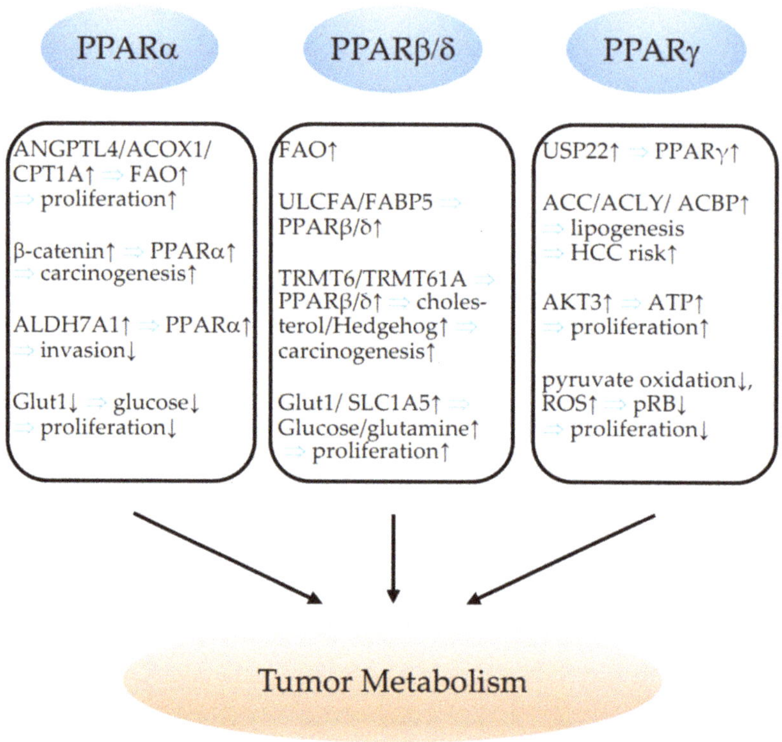

Figure 7. Summary of the effects of PPARα, PPARβ/δ, and PPARγ on tumor metabolism. ↓ indicates inhibition and ↑ indicates an increase. ⇒: leads to; ANGPTL4: angiopoietin-like 4; ACOX1: acyl-coenzyme A oxidase 1; CPT1A: carnitine palmitoyltransferase 1A; FAO: fatty-acid oxidation; ALDH7A1: aldehyde dehydrogenase 7 family member A1; Glut1: glucose transporter 1; ULCFA: unsaturated long-chain fatty acids; FABP5: fatty-acid-binding protein 5; TRMT6: tRNA methyltransferase 6 non-catalytic subunit; TRMT61A: tRNA methyltransferase 61A; SLC1A5: solute carrier family 1 member 5; USP22: ubiquitin-specific peptidase 22; ACC: acetyl-CoA carboxylase; ACLY: ATP citrate lyase; ACBP: acyl-CoA binding protein; HCC: hepatocellular carcinoma; AKT3: AKT serine/threonine kinase 3; ATP: adenosine triphosphate; ROS: reactive oxygen species; pRB: phosphorylated retinoblastoma protein.

9. PPARs and Cancer Immunity

9.1. PPARα

Over twenty years ago, a regulatory function of PPARα in inflammatory processes was already proposed. PPARα-null mice displayed a prolonged inflammatory response to stimulation with leukotriene B4, an activating ligand for PPARα [433]. PPARα has further been shown to be the predominant PPAR expressed by T and B lymphocytes. Following T-cell activation, PPARα was downregulated, whereas PPARγ expression increased [434]. PPARα is also already expressed in monocytes and upregulated during their maturation into macrophages. PPARα agonists induce the apoptosis of activated, but not of un-activated macrophages [435]. PPARα plays a major role in the immunomodulation caused by peroxisome proliferators (PPs). The group of J. W. DePierre demonstrated that several PPs, including perfluorooctanoic acid (PFOA), di(2-ethylhexyl)phthalate (DEHP), Wy-14 643, and nafenopin caused dramatic thymic and splenic atrophy in wildtype mice, with decreases in both, B- and T-cell populations, with the greatest reduction in the immature CD4⁺CD8⁺ population [436]. In contrast to wildtype animals, the authors did not observe these immunomodulatory effects of PPs in PPARα-knockout animals, identifying PPARα as the crucial regulator of PP-induced immunomodulation [437]. PPARα activation further decreases early B-cell development within the bone marrow [438]. The ability of PPs to suppress adaptive immunity in rodents may contribute to the development of hepatocarcinogenesis (reviewed in [439]) in response to these same substances. Using PPARα-deficient mice fed a high-fat diet, PPARα has been shown to protect against obesity-induced liver inflammation via the downregulation of inflammatory genes and the attenuation of adipose-tissue inflammation, partially through the prevention of fat accumulation in the liver [440]. Similarly, in a human-like hyperlipidemic mouse model (APOE2 knock-in mice) fed a western-type high-fat diet, fenofibrate treatment decreased hepatic macrophage accumulation, abolished steatosis, and reduced the expression of inflammatory genes [441]. Similarly, beneficial effects have been reported for PPARα activation in inflammatory bowel disease [442–444]. Michalik and colleagues evidenced the implication of PPARα in skin wound healing. They showed that PPARα is mainly involved in the initial inflammatory phase after injury, which precedes normal wound repair. PPARα-deficient mice exhibited a significant delay in the early-phase healing process, characterized by the impaired recruitment of neutrophils and monocytes/macrophages to the wound bed. This uncontrolled inflammation accounts for the transient delay of healing observed in PPARα-deficient animals [445]. The feeding of PPARα agonists to aged mice restored the cellular redox balance, evidenced by a lowering of tissue lipid peroxidation, an elimination of constitutively active NF-κB, and a loss of spontaneous inflammatory cytokine production [446]. PPARα further directly represses pro-inflammatory genes such as STAT, activator protein-1 (AP-1), NF-κB, and nuclear factor of activated T cells (NFAT) and activates anti-inflammatory components such as interleukin-1 receptor antagonist (IL1-Ra), Vanin-1, and mannose-binding lectin (MBL), as reviewed in [447,448]. PPARα further functions as a natural suppressor of the enzyme 11-β hydroxysteroid dehydrogenase 1 (HSD11B1), a widely expressed enzyme that converts biologically inactive cortisone to the functional glucocorticoid cortisol, known to exert multiple immunomodulatory effects [449]. In contrast to the suggested anti-inflammatory role of PPARα, Hill and colleagues observed, in a mouse model of endotoxemia, higher TNFα levels in animals treated with PPARα agonists [450]. Most studies have suggested a role for PPARα in the downregulation of endothelial cell (EC) inflammatory responses. PPARα agonists limited chronic inflammation mediated by VCAM-1 and monocytes without affecting acute inflammation mediated by E-selectin and neutrophil binding [451]. The PPARα agonist fenofibrate inhibits VCAM-1 transcription, in part, by inhibiting NF-κB [452]. The repression of NF-κB via PPARα activation was also identified as the mechanism for the inhibition of interleukin-6 and for the prostaglandin production and expression of COX-2 in human aortic smooth-muscle cells [453]. Lee and colleagues demonstrated a pro-inflammatory role of PPARα in the mediation of the activation of endothelial cells to produce monocyte chemotactic activity in response to oxidized phospholipids and lipoproteins [454]. Based on in vivo and in vitro studies, PPARα appears to have predominantly anti-inflammatory effects, although, in some studies, the pro-inflammatory consequences of PPARα activation have been demonstrated. Inflammation can either support or inhibit cancer growth. An outstanding report evidenced that PPARα-expressing granulocytes, mainly neutrophils, are required for tumor growth. PPARα deficiency in the host suppressed tumor growth via the induction of a plain inflammation capable of suppressing tumor angiogenesis, mainly through increased production of thrombospondin (TSP)-1 [53]. PPARα deficiency has further been demonstrated to inhibit tumor growth by impairing regulatory T-cell (Treg) functions and by supporting a pro-inflammatory Th1 T-cell phenotype [54]. These findings clearly support the negative impact of PPARα on the immune environment in the setting of cancer. However, from a metabolic point of view, PPARα activation could also be beneficial in reducing tumor growth. Tumor-infiltrating lymphocytes (TILs) suffer from the metabolic stress of hypoxia and hypoglycemia in the tumor environment. To preserve their effector functions, it has been demonstrated that they are able to enhance PPARα signaling and fatty-acid (FA) catabolism. Fenofibrate treatment further improved TILs' ability to reduce tumor growth via the promotion of FA catabolism [455]. Nevertheless, a recent study evidenced that PPARα drives dendritic-cell immune dysfunction in cancer. Dendritic cells are key players in the initiation, programming, and regulation of anti-tumor responses. Fatty-acid-carrying tumor-derived exosomes (TDEs) activate PPARα, which, in turn, leads to excess lipid-droplet biogenesis and enhanced FAO, provoking a metabolic shift to mitochondrial oxidative phosphorylation and dendritic-cell immune dysfunction. The inhibition of PPARα reversed the TDE-induced immune dysfunction of dendritic cells and increased immunotherapy effectiveness [456]. Cancer development and its response to therapy are regulated by inflammation. PPARα is clearly involved in both chronic inflammation, facilitating tumor progression and treatment resistance, and acute inflammatory reactions, often leading to anti-tumor immune responses. Due to its plethora of immunomodulatory and metabolic effects, PPARα might either promote or suppress tumor progression, provoking opposing effects on therapeutic outcomes.

9.2. *PPARβ/δ*

The function of PPARβ/δ in immunomodulation has been extensively reviewed in [457] and [38]. The first attestations to a possible implication of PPARβ/δ in immune processes resulted from observations following skin injury. PPARβ/δ-deficient animals displayed a greater hyperplastic response in skin after O-tetradecanoylphorbol-13-acetate (TPA) treatment than wildtype controls and did not respond to NSAID sulindac treatment in contrast to their wildtype counterparts [458]. Tan and colleagues showed that the pro-inflammatory mediators TNF-α, interferon (IFN)-γ, and tissue plasminogen activator (TPA) upregulate PPARβ/δ expression in primary keratinocytes isolated from wildtype mice. The increase in PPARβ/δ strongly accelerated the differentiation of keratinocytes and increased their resistance to apoptotic signals, which was abolished in PPARβ/δ-deficient mice [459]. PPARβ/δ immune functions have frequently been studied in the setting of atherosclerosis. PPARβ/δ, highly expressed in endothelial cells [460], inhibits endothelial-cell inflammatory responses which lead to leukocyte recruitment [461–464]. In macrophages, PPARβ/δ controls inflammation through its association with the transcriptional co-repressor B-cell lymphoma (BCL)-6 which blocks the anti-inflammatory actions of BCL-6 and increases levels of inflammatory mediators such as methyl-accepting chemotaxis proteins (MCP)-1 and 3, and IL-1β. Following ligand binding to PPARβ/δ, BCL-6 is released and can repress inflammation [463]. The PPARβ/δ agonist GW0742 was shown to inhibit COX-2 and inducible nitric oxide synthase (iNOS) in macrophages [465]. PPARβ/δ has further been implicated in the switch of pro-inflammatory M1 macrophages to the anti-inflammatory M2 phenotype [466,467]. The PPARβ/δ agonist GW0742 strongly induced arginase I expression in macrophages, which impacted the balance of Th1/Th2 responses [468]. It is highly interesting that PPARβ/δ functions as a transcriptional basis for the detection and the discarding of apoptotic cells by macrophages, thus ensuring the timely and effective clearance of dying cells and increased anti-inflammatory cytokine production [469]. Adhikary and colleagues investigated the PPARβ/δ-regulated signaling network in human monocyte-derived macrophages. PPARβ/δ agonists inhibited the expression of multiple pro-inflammatory mediators and induced an anti-inflammatory phenotype. Of note, the authors also identified the immune stimulatory effects of PPARβ/δ agonists, which were reflected functionally by enhanced macrophage survival under hypoxic stress and stimulated $CD8^+$ T-cell activation upon PPARβ/δ activation [470]. In ovarian cancer, tumor-associated ascites contains high concentrations of polyunsaturated fatty acids (PUFAs), which function as potent PPARβ/δ agonists in macrophages. They accumulate in lipid droplets in tumor-associated macrophages (TAMs), providing a reservoir of PPARβ/δ ligands, and induce the upregulation of PPARβ/δ target genes associated with immune regulation and tumor progression, such as CD300A, mitogen-activated protein kinase (MAP3K) 8 and angiopoietin-like 4 (ANGPTL4) [471]. Little is known about their expression and function in other immune cell types. PPARβ/δ expression has been described in lymphocytes [472] and has been suggested to stimulate T-cell proliferation and to inhibit INF-induced apoptosis [473]. Recently, the PPARβ/δ agonist GW501516 has been shown to enhance the efficacy of adoptive cell therapy by enhancing the expression of carnitine palmitoyl transferase 1A (CPT1A), the rate-limiting enzyme of FAO, in activated $CD8^+$ T cells. Activated T cells produced more IFN and T-bet, which prevent cell exhaustion [474]. PPARβ/δ is further implicated in monocyte-to-dendritic cell maturation. Interestingly, PPARβ/δ agonists and naturally occurring ligands such as fatty acids drive the maturation of dendritic cells with an atypical phenotype, characterized by reduced expression of IL-10 and IL-12, and reduced stimulatory effects on leucocytes [475]. Mast cells, able to rapidly respond to modifications in their environment, favor tumor progression through the induction of angiogenesis and tissue remodeling (reviewed in [476]). Recently, it has been demonstrated that PPARβ/δ might be involved in mast-cell maturation and contribute to inflammatory responses in mast cells; however, the consequences of PPARβ/δ in mast cells in the context of cancer have not been studied [477]. Natural-killer (NK) cells have major functions in anti-tumor immunity, and obesity has been shown to reduce NK cell cytotoxic effector functions. Lipids induce metabolic defects, causing NK cell failure, leading to a loss of anticancer functions. NK cells express PPARα and PPARβ/δ, and agonists for both PPARs induce a dysfunctional NK cell phenotype; this mimics the NK cell phenotype in obesity, which is unable to exert anti-tumor functions [478]. In general, PPARβ/δ appears to be anti-inflammatory. However, the few studies investigating PPARβ/δ immune function in cancer describe pro-tumorigenic consequences such as the stimulation of tumor-promoting TAMs [471], and the inhibition of the cytotoxic anti-cancer effects of NK cells [478].

9.3. PPARγ

PPARγ agonists mediate a direct inhibitory role in T-cell immune responses. They negatively regulate T-cell activation by inhibiting the nuclear factor of activated T cells (NFAT) and subsequent IL-2 production [479,480]. Consequently, the limitation of T-cell activation by PPARγ activation improves inflammatory diseases [481–483]. PPARγ activation has also been demonstrated to decrease T-cell proliferation through the induction of apoptosis [484]; however, other studies have shown that PPARγ agonists attenuate apoptosis induced by cytokine or serum withdrawal. Survival promotion was attributed to PPARγ actions in cellular metabolic activities and the maintenance of T-cell mitochondrial membrane potential [485,486]. PPARγ further mediates T-cell differentiation. IL-17-secreting T helper cells (Th17) play a crucial role in autoimmune diseases. Their differentiation is induced by TGF beta/IL-6. PPARγ acts as a negative regulator of Th17 differentiation through inhibition of TGF beta/IL-6 signaling, and was not found to influence the differentiation of Th1, Th2, or regulatory T cells [487]. A recent elegant study employing a mouse model of atopic dermatitis evidenced that obesity exacerbated inflammatory responses through the conversion of a Th2-driven inflammatory disease to a worsened Th17-driven disease status. PPARγ expression was decreased in Th2 cells from obese animals compared to their lean counterparts. Using conditional deletion of PPARγ in T cells, the authors demonstrated the necessity of PPARγ to prevent uncontrolled Th17-mediated inflammation by redirecting the T helper cells towards a Th2 inflammatory response. Consequently, PPARγ agonists could reduce Th17-aggravated inflammation [283]. Interestingly, in colon cancer patients, the hierarchical clustering of a correlation matrix revealed that patients with high expression of the Th17 cluster had a poor prognosis. In contrast, no prediction of prognosis was associated with Th2 or Treg clusters, and enhanced Th1 clusters corresponded to better outcomes [488]. PPARγ agonists also inhibited allogeneic human memory T-cell responses in a model of human artery grafts in immunodeficient mice [489]. PPARγ is further involved in Treg homeostasis, as PPARγ deficiency led to reduced Treg recruitment in a colitis model [490], whereas PPARγ activation increased the induction of Tregs [491]. In general, PPARγ-expressing Tregs are considered to suppress adipose-tissue inflammation in obesity [492,493]. PPARγ activation in group 2 innate lymphoid cells (ILC2s) sustains type 2 cytokine production. Crucial to the pathogenesis of many allergic and fibrotic diseases, these cytokines can also promote tumorigenesis and cancer growth. Consequently, PPARγ deletion, specifically in ILC2s, reduced tumor growth in a mouse colorectal cancer model [494]. PPARγ expression in dendritic cells (DCs) was reported over twenty years ago. PPARγ activators were shown to inhibit the production of dendritic-cell IL-12, a strong Th1 pro-inflammatory inductor, thereby modulating the polarization of immune responses [495]. PPARγ activation provoked CD1d glycoprotein expression in DCs, leading to the selective induction of invariant natural-killer T-cell (iNKT cell) proliferation [496]. iNKT cells represent a distinct population of T lymphocytes, which have features of both conventional T cells and natural-killer (NK) cells and are considered important mediators of immune responses and tumor surveillance. PPARγ further enhances the anti-tumor efficacy of iNKT cells by assuring cholesterol synthesis and IFN-γ production in tumor-infiltrating iNKT cells [497]. A claudin-low subtype of bladder cancers has recently been described. They show an imbalance in decreased PPARγ expression and the resulting enhanced NF-κB signaling, and high cytokine and chemokine expression. These tumors are characterized by an enrichment of immune gene signatures but a simultaneous expression of immune-checkpoint molecules, which demonstrates that despite their high immune infiltration, they are also actively immunosuppressed [498]. However, increased PPARγ expression in bladder cancer through its suppression of NF-κB leads to the phenotype of immune cold tumors, which do not respond to immunotherapies and are characterized by low immune-cell trafficking, impaired T-cell activation, an abundance of myeloid-derived suppressor cells, and Tregs that release immunosuppressive cytokines [499]. Accordingly, in a different subtype of bladder cancer, muscle-invasive bladder cancer, recurrent mutations in RXRα lead to an imbalance of the PPARγ/RXRα heterodimer, and focal amplification of PPARγ. PPARγ overexpression impairs $CD8^+$ T cell infiltration, possibly through NF-κB inhibition, and confers resistance to immunotherapies [500]. The important roles of PPARγ in affecting the immunophenotype of DCs, as well as how PPARγ-regulated processes could be employed in the design of tumor vaccination strategies, are further reviewed in [501]. Immune tolerance of local DCs is believed to induce immune evasion and to contribute to the resistance of cancers to immunotherapies. In contrast to the anti-tumorigenic function of PPARγ in DCs proposed by many studies, Zhao and colleagues identified a paracrine Wnt5a-β-catenin-PPAR-γ signaling pathway driving FAO in DCs by which melanomas escape from immunotherapies. FAO promotes Treg-cell development and suppresses T-effector-cell activation. The blockade of FAO enhanced the effectiveness of anti PD 1 immunotherapy and slowed melanoma tumor progression [502]. DCs isolated from patients with advanced breast cancer expressed high levels of the adiponectin receptors AdipoR1 and AdipoR2. Using a different pathway to AdipoR1, AdipoR2 modified the inflammatory processes by activating the PPARγ pathway through the induction of COX 2. This leads to a blockade of NF-κB activation in DCs, and thereby attenuates their ability to stimulate antigen-specific T-cell responses [503]. High levels of Glutathione peroxidase 4 (GPX4), which inhibits ferroptosis, a lipid peroxidation-mediated cell death in tumor cells, are associated with poor prognosis in cancer patients. The inhibition of GPX4 with the compound RSL3 was shown to enhance the anticancer effect of cisplatin [504]. However, therapy-enhanced ferroptosis in dendritic cells severely impaired their anti-tumor functions that should produce cytokines, promote MCH expression, and activate T cells. It has been shown that PPARγ is responsible for RSL3-induced ferroptosis, which leads to the obstruction of DC maturation, as PPARγ knockdown was sufficient to restore anti-tumor activity in RSL3 treated dendritic cells [505]. Furthermore, PPARγ agonists impair innate immunity NK cell functions through inhibition of cytolytic NK activity [506]. The early identification of high PPARγ expression in the spleen [507] led many research groups to investigate its function in monocytes/macrophages. PPARγ has a fundamental role in lipid metabolism and is consequently highly expressed in foam cells, which are cholesterol-carrying macrophages in atherosclerotic lesions [508,509]. Following exposure to oxidized low-density lipoprotein, PPARγ is induced in monocytes and leads to the transcriptional induction of the immunotolerant state marker CD36, participating in atherosclerotic arterial lesion formation through its interaction with oxidized low-density lipoprotein (oxLDL), which triggers signaling cascades for inflammatory responses [508]. A series of studies investigated the anti-inflammatory effects of PPARγ thiazolidindione ligands, which were found to inhibit the inflammatory cytokines TNFα, IL-6, IL-1β [510], iNOS, MMP9, and scavenger receptor A (SR-A) [511]. PPARγ activation, therefore, mostly suppresses the immunoreactive state of a macrophage. However, non-thiazolidindione agonists of PPARγ failed to induce anti-inflammatory responses [512], and PPARγ-deficient embryonic stem cells could be differentiated into the monocytic lineage, suggesting PPARγ-independent effects of thiazolidindiones and 15d-PGJ2 on inflammation [513,514]. Nevertheless, PPARγ is important for defining the lineage of tissue-resident macrophages through transcriptional modulation in regulating the differentiation of pre-macrophages and alveolar macrophages, Kupffer cells, adipose-associated macrophages, and intestinal macrophages (reviewed in [27]); moreover, its activation primes primary monocytes for M2 differentiation, resulting in more pronounced anti-inflammatory activity in M1 macrophages [515]. In the setting of cancer, PPARγ activation was shown to reverse the MDSC and M2 macrophage-mediated suppression of the cytotoxic T lymphocyte (CTL) anti-tumor responses [516]. The deletion of PPARγ in macrophages further exacerbated mammary-tumor development in a mouse model. Mechanistically, PPARγ was found to suppress Gpr132 protein in macrophages, which is pro-inflammatory and tumorigenic [517]. The expression of PPARγ in macrophages favors an anti-inflammatory TAM phenotype. Macrophages exposed to breast cancer cell media achieved a TAM-like phenotype with features from both M1 and M2 polarization. The further addition of rosiglitazone to the breast cancer-conditioned medium reduced the secretion of M1 pro-inflammatory and pro-tumor M2-cytokines [518]. Similarly, the conditioned medium from macrophages exposed to apoptotic lung cancer cells inhibited the EMT, migration, and invasion of cancer cells. Apoptotic 344SQ activated PPARγ in macrophages, inducing enhanced phosphatase and tensin homolog on chromosome ten (PTEN) expression, which antagonized pro-tumorigenic phosphoinositide 3-kinase (PI3K) signaling [519]. However, PPARγ agonists were shown to drive the macrophage phenotype versus the M2 form in a model of a pathogen-induced macrophage challenge. This shift was accompanied by the enhanced production of TGFβ and arginase 1 and enhanced phagocytic activity [520]. Consequently, PPARγ activation in macrophages has been shown to fuel lung cancer progression and metastasis, especially through increased arginase 1 [342] and TGFβ1 [343] expression. Similarly, in a breast cancer model, PPARγ was found to induce M2 polarization through the induction of integrin β3 [346]. The cleavage of PPARγ by caspase-1 has been shown to enhance tumor promotion through the induction of TAMs. Truncated PPARγ translocates to mitochondria and interacts with medium-chain acyl-CoA dehydrogenase (MCAD), thereby inhibiting MCAD and FAO, which leads to lipid-droplet accretion and TAM differentiation. Caspase-1 deficiency significantly impaired tumor growth, underlining the importance of this pathway for tumor promotion by TAMs [345]. Highly interestingly, Moreira and colleagues demonstrated that CLAs, which are frequently used in dietary supplementation and known to activate PPARγ, have efficient anti-inflammatory effects that prevent colitis, but worsen colorectal cancer formation. CLAs induce macrophage- and T-cell-producing TGF-β via PPARγ activation, which enhances colorectal cancer progression. The macrophage-specific deletion of PPARγ abrogated pro-tumorigenic CLA effects in colon cancer [521]. In contrast to its overall anti-tumoral role in cancer cells, PPARγ governs major immuno-metabolic switches and alternative activation in immune cells, especially macrophages, thereby facilitating tumor initiation, progression, and metastasis. The PPAR functions and molecular mechanisms in cancer immunity are summarized in Figure 8.

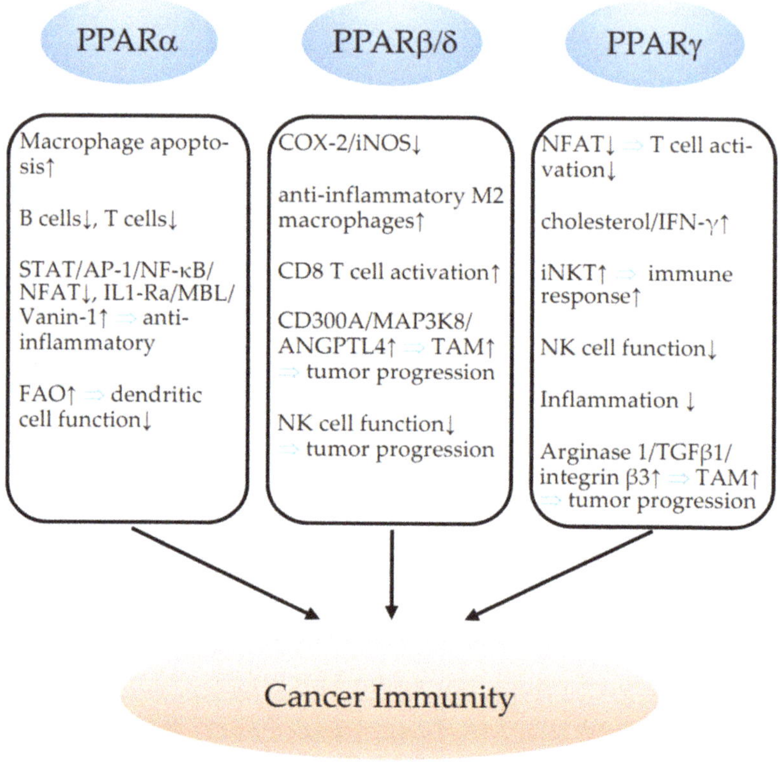

Figure 8. Effects of PPARα, PPARβ/δ, and PPARγ on cancer immunity. ↓ indicates inhibition and ↑ indicates an increase. ⇒: leads to; STAT: signal transducer and activator of transcription; AP-1: activator protein-1; NF-κB: nuclear factor kappa-light-polypeptide-gene-enhancer in B cells; NFAT: nuclear factor of activated T cells; IL1-Ra: interleukin 1 receptor antagonist; MBL: mannose-binding lectin; FAO: fatty-acid oxidation; COX-2: cyclooxygenase-2; iNOS: nitric oxide synthase 2 inducible; CD: cluster of differentiation; MAP3K8: mitogen-activated protein kinase kinase kinase 8; ANGPTL4: angiopoietin-like 4; TAM: tumor-associated macrophages; NK: natural-killer cell; IFN-γ: interferon gamma; iNKT: invariant natural-killer T cell; TGFβ1: transforming growth factor beta 1; integrin β3: integrin subunit beta 3.

10. Conclusions

Given the multiple diverse functions of PPARs in the cancer hallmarks, it is currently difficult to judge whether specific agonists or antagonists might have beneficial effects for cancer treatment. The effects in different cancer types and in each cancer type on stromal and tumor cells are divergent. Thus, with the advancement of personalized medicine, these differences should be considered for treatment decisions. In addition, research on dual- and pan-PPAR modulators might open new therapeutic strategies. The use and analysis of existing large databases, e.g., the National Veterans Health Administration (VHA) database including cancer patients with the coincidental administration of PPAR agonists, might give additional insights into the clinical role of PPAR modulation in cancer.

Author Contributions: Conceptualization, N.W. and K.-D.W.; formal analysis, N.W. and K.-D.W.; writing—original draft preparation, N.W. and K.-D.W.; writing—review and editing, N.W. and K.-D.W.; visualization, N.W.; funding acquisition, N.W. and K.-D.W. All authors have read and agreed to the published version of the manuscript.

Funding: This research was funded by the Fondation pour la Recherche Medicale, grant number FRM DPC20170139474 (K.-D.W.); the Fondation ARC pour la recherche sur le cancer, grant number n°PJA 20161204650 (N.W.); Gemluc (N.W.); Plan Cancer INSERM (K.-D.W.); the Fondation ARC pour la recherche sur le cancer, grant number ARCPJA2021060004010 (K.-D.W.); and the Agence Nationale de la Recherche, grant R19125AA "Senage" (K.-D.W.).

Institutional Review Board Statement: Not applicable.

Informed Consent Statement: Not applicable.

Conflicts of Interest: The authors declare no conflict of interest.

References

1. Wagner, K.D.; Wagner, N. Peroxisome proliferator-activated receptor beta/delta (PPARβ/δ) acts as regulator of metabolism linked to multiple cellular functions. *Pharmacol. Ther.* **2010**, *125*, 423–435. [CrossRef]
2. Miyachi, H. Structural Biology-Based Exploration of Subtype-Selective Agonists for Peroxisome Proliferator-Activated Receptors. *Int. J. Mol. Sci.* **2021**, *22*, 9223. [CrossRef] [PubMed]
3. Moody, D.E.; Reddy, J.K. Increase in hepatic carnitine acetyltransferase activity associated with peroxisomal (microbody) proliferation induced by the hypolipidemic drugs clofibrate, nafenopin, and methyl clofenapate. *Res. Commun. Chem. Pathol. Pharmacol.* **1974**, *9*, 501–510. [PubMed]
4. De Duve, C. Evolution of the peroxisome. *Ann. N. Y. Acad. Sci.* **1969**, *168*, 369–381. [CrossRef] [PubMed]
5. Issemann, I.; Green, S. Activation of a member of the steroid hormone receptor superfamily by peroxisome proliferators. *Nature* **1990**, *347*, 645–650. [CrossRef] [PubMed]
6. Göttlicher, M.; Widmark, E.; Li, Q.; Gustafsson, J.A. Fatty acids activate a chimera of the clofibric acid-activated receptor and the glucocorticoid receptor. *Proc. Natl. Acad. Sci. USA* **1992**, *89*, 4653–4657. [CrossRef] [PubMed]
7. Dreyer, C.; Krey, G.; Keller, H.; Givel, F.; Helftenbein, G.; Wahli, W. Control of the peroxisomal β-oxidation pathway by a novel family of nuclear hormone receptors. *Cell* **1992**, *68*, 879–887. [CrossRef]
8. Plutzky, J. The PPAR-RXR transcriptional complex in the vasculature: Energy in the balance. *Circ. Res.* **2011**, *108*, 1002–1016. [CrossRef]
9. De Bosscher, K.; Desmet, S.J.; Clarisse, D.; Estébanez-Perpiña, E.; Brunsveld, L. Nuclear receptor crosstalk—Defining the mechanisms for therapeutic innovation. *Nat. Rev. Endocrinol.* **2020**, *16*, 363–377. [CrossRef] [PubMed]
10. Palmer, C.N.; Hsu, M.H.; Griffin, H.J.; Johnson, E.F. Novel sequence determinants in peroxisome proliferator signaling. *J. Biol. Chem.* **1995**, *270*, 16114–16121. [CrossRef] [PubMed]
11. Wagner, K.D.; Du, S.; Martin, L.; Leccia, N.; Michiels, J.F.; Wagner, N. Vascular PPARβ/δ Promotes Tumor Angiogenesis and Progression. *Cells* **2019**, *8*, 1623. [CrossRef]
12. Fougerat, A.; Montagner, A.; Loiseau, N.; Guillou, H.; Wahli, W. Peroxisome Proliferator-Activated Receptors and Their Novel Ligands as Candidates for the Treatment of Non-Alcoholic Fatty Liver Disease. *Cells* **2020**, *9*, 1638. [CrossRef] [PubMed]
13. Feige, J.N.; Auwerx, J. Transcriptional coregulators in the control of energy homeostasis. *Trends Cell Biol.* **2007**, *17*, 292–301. [CrossRef] [PubMed]
14. Grygiel-Górniak, B. Peroxisome proliferator-activated receptors and their ligands: Nutritional and clinical implications—A review. *Nutr. J.* **2014**, *13*, 17. [CrossRef]
15. Montaigne, D.; Butruille, L.; Staels, B. PPAR control of metabolism and cardiovascular functions. *Nat. Rev. Cardiol.* **2021**, *18*, 809–823. [CrossRef]
16. Wagner, K.D.; Wagner, N. PPARs and Myocardial Infarction. *Int. J. Mol. Sci.* **2020**, *21*, 9436. [CrossRef] [PubMed]
17. Marx, N.; Davies, M.J.; Grant, P.J.; Mathieu, C.; Petrie, J.R.; Cosentino, F.; Buse, J.B. Guideline recommendations and the positioning of newer drugs in type 2 diabetes care. *Lancet Diabetes Endocrinol.* **2021**, *9*, 46–52. [CrossRef]
18. Wagner, N.; Wagner, K.D. The Role of PPARs in Disease. *Cells* **2020**, *9*, 2367. [CrossRef] [PubMed]
19. Sáez-Orellana, F.; Octave, J.N.; Pierrot, N. Alzheimer's Disease, a Lipid Story: Involvement of Peroxisome Proliferator-Activated Receptor α. *Cells* **2020**, *9*, 1215. [CrossRef] [PubMed]
20. Matheson, J.; Le Foll, B. Therapeutic Potential of Peroxisome Proliferator-Activated Receptor (PPAR) Agonists in Substance Use Disorders: A Synthesis of Preclinical and Human Evidence. *Cells* **2020**, *9*, 1996. [CrossRef]
21. Quiroga, C.; Barberena, J.J.; Alcaraz-Silva, J.; Machado, S.; Imperatori, C.; Yadollahpour, A.; Budde, H.; Yamamoto, T.; Telles-Correia, D.; Murillo-Rodríguez, E. The Role of Peroxisome Proliferator-Activated Receptor in Addiction: A Novel Drug Target. *Curr. Top. Med. Chem.* **2021**, *21*, 964–975. [CrossRef] [PubMed]
22. Elias, E.; Zhang, A.Y.; Manners, M.T. Novel Pharmacological Approaches to the Treatment of Depression. *Life* **2022**, *12*, 196. [CrossRef] [PubMed]
23. Luan, Z.L.; Zhang, C.; Ming, W.H.; Huang, Y.Z.; Guan, Y.F.; Zhang, X.Y. Nuclear receptors in renal health and disease. *EBioMedicine* **2022**, *76*, 103855. [CrossRef]

24. Mantovani, A.; Byrne, C.D.; Targher, G. Efficacy of peroxisome proliferator-activated receptor agonists, glucagon-like peptide-1 receptor agonists, or sodium-glucose cotransporter-2 inhibitors for treatment of non-alcoholic fatty liver disease: A systematic review. *Lancet Gastroenterol. Hepatol.* **2022**, *7*, 367–378. [CrossRef]
25. Kökény, G.; Calvier, L.; Hansmann, G. PPARγ and TGFβ-Major Regulators of Metabolism, Inflammation, and Fibrosis in the Lungs and Kidneys. *Int. J. Mol. Sci.* **2021**, *22*, 10431. [CrossRef]
26. Liu, Y.; Wang, J.; Luo, S.; Zhan, Y.; Lu, Q. The roles of PPARγ and its agonists in autoimmune diseases: A comprehensive review. *J. Autoimmun.* **2020**, *113*, 102510. [CrossRef] [PubMed]
27. Toobian, D.; Ghosh, P.; Katkar, G.D. Parsing the Role of PPARs in Macrophage Processes. *Front. Immunol.* **2021**, *12*, 783780. [CrossRef] [PubMed]
28. Gerussi, A.; Lucà, M.; Cristoferi, L.; Ronca, V.; Mancuso, C.; Milani, C.; D'Amato, D.; O'Donnell, S.E.; Carbone, M.; Invernizzi, P. New Therapeutic Targets in Autoimmune Cholangiopathies. *Front. Med.* **2020**, *7*, 117. [CrossRef] [PubMed]
29. Rzemieniec, J.; Castiglioni, L.; Gelosa, P.; Muluhie, M.; Mercuriali, B.; Sironi, L. Nuclear Receptors in Myocardial and Cerebral Ischemia-Mechanisms of Action and Therapeutic Strategies. *Int. J. Mol. Sci.* **2021**, *22*, 12326. [CrossRef] [PubMed]
30. Cheng, H.S.; Yip, Y.S.; Lim, E.K.Y.; Wahli, W.; Tan, N.S. PPARs and Tumor Microenvironment: The Emerging Roles of the Metabolic Master Regulators in Tumor Stromal-Epithelial Crosstalk and Carcinogenesis. *Cancers* **2021**, *13*, 2153. [CrossRef] [PubMed]
31. Wagner, K.D.; Cherfils-Vicini, J.; Hosen, N.; Hohenstein, P.; Gilson, E.; Hastie, N.D.; Michiels, J.F.; Wagner, N. The Wilms' tumour suppressor Wt1 is a major regulator of tumour angiogenesis and progression. *Nat. Commun.* **2014**, *5*, 5852. [CrossRef] [PubMed]
32. Hanahan, D.; Weinberg, R.A. Hallmarks of cancer: The next generation. *Cell* **2011**, *144*, 646–674. [CrossRef]
33. Erez, N.; Truitt, M.; Olson, P.; Arron, S.T.; Hanahan, D. Cancer-Associated Fibroblasts Are Activated in Incipient Neoplasia to Orchestrate Tumor-Promoting Inflammation in an NF-κB-Dependent Manner. *Cancer Cell* **2010**, *17*, 135–147. [CrossRef] [PubMed]
34. Nozawa, H.; Chiu, C.; Hanahan, D. Infiltrating neutrophils mediate the initial angiogenic switch in a mouse model of multistage carcinogenesis. *Proc. Natl. Acad. Sci. USA* **2006**, *103*, 12493–12498. [CrossRef] [PubMed]
35. Bergers, G.; Brekken, R.; McMahon, G.; Vu, T.H.; Itoh, T.; Tamaki, K.; Tanzawa, K.; Thorpe, P.; Itohara, S.; Werb, Z.; et al. Matrix metalloproteinase-9 triggers the angiogenic switch during carcinogenesis. *Nat. Cell Biol.* **2000**, *2*, 737–744. [CrossRef] [PubMed]
36. Hanahan, D.; Weinberg, R.A. The hallmarks of cancer. *Cell* **2000**, *100*, 57–70. [CrossRef]
37. Hanahan, D.; Folkman, J. Patterns and emerging mechanisms of the angiogenic switch during tumorigenesis. *Cell* **1996**, *86*, 353–364. [CrossRef]
38. Wagner, N.; Wagner, K.D. PPAR Beta/Delta and the Hallmarks of Cancer. *Cells* **2020**, *9*, 1133. [CrossRef] [PubMed]
39. Suchanek, K.M.; May, F.J.; Robinson, J.A.; Lee, W.J.; Holman, N.A.; Monteith, G.R.; Roberts-Thomson, S.J. Peroxisome proliferator-activated receptor α in the human breast cancer cell lines MCF-7 and MDA-MB-231. *Mol. Carcinog.* **2002**, *34*, 165–171. [CrossRef] [PubMed]
40. Okumura, M.; Yamamoto, M.; Sakuma, H.; Kojima, T.; Maruyama, T.; Jamali, M.; Cooper, D.R.; Yasuda, K. Leptin and high glucose stimulate cell proliferation in MCF-7 human breast cancer cells: Reciprocal involvement of PKC-α and PPAR expression. *Biochim. Biophys. Acta* **2002**, *1592*, 107–116. [CrossRef]
41. Chang, N.W.; Wu, C.T.; Chen, D.R.; Yeh, C.Y.; Lin, C. High levels of arachidonic acid and peroxisome proliferator-activated receptor-alpha in breast cancer tissues are associated with promoting cancer cell proliferation. *J. Nutr. Biochem.* **2013**, *24*, 274–281. [CrossRef]
42. Bocca, C.; Bozzo, F.; Martinasso, G.; Canuto, R.A.; Miglietta, A. Involvement of PPARα in the growth inhibitory effect of arachidonic acid on breast cancer cells. *Br. J. Nutr.* **2008**, *100*, 739–750. [CrossRef] [PubMed]
43. Li, T.; Zhang, Q.; Zhang, J.; Yang, G.; Shao, Z.; Luo, J.; Fan, M.; Ni, C.; Wu, Z.; Hu, X. Fenofibrate induces apoptosis of triple-negative breast cancer cells via activation of NF-κB pathway. *BMC Cancer* **2014**, *14*, 96. [CrossRef] [PubMed]
44. Chandran, K.; Goswami, S.; Sharma-Walia, N. Implications of a peroxisome proliferator-activated receptor alpha (PPARα) ligand clofibrate in breast cancer. *Oncotarget* **2016**, *7*, 15577–15599. [CrossRef] [PubMed]
45. Tauber, Z.; Koleckova, M.; Cizkova, K. Peroxisome proliferator-activated receptor α(PPAR α)-cytochrome P450 epoxygenases-soluble epoxide hydrolase axis in ER + PR + HER2− breast cancer. *Med. Mol. Morphol.* **2020**, *53*, 141–148. [CrossRef] [PubMed]
46. Li, B.; Jiang, H.Y.; Wang, Z.H.; Ma, Y.C.; Bao, Y.N.; Jin, Y. Effect of fenofibrate on proliferation of SMMC-7721 cells via regulating cell cycle. *Hum. Exp. Toxicol.* **2021**, *40*, 1208–1221. [CrossRef]
47. Morimura, K.; Cheung, C.; Ward, J.M.; Reddy, J.K.; Gonzalez, F.J. Differential susceptibility of mice humanized for peroxisome proliferator-activated receptor α to Wy-14,643-induced liver tumorigenesis. *Carcinogenesis* **2006**, *27*, 1074–1080. [CrossRef] [PubMed]
48. Tanaka, N.; Moriya, K.; Kiyosawa, K.; Koike, K.; Aoyama, T. Hepatitis C virus core protein induces spontaneous and persistent activation of peroxisome proliferator-activated receptor α in transgenic mice: Implications for HCV-associated hepatocarcinogenesis. *Int. J. Cancer* **2008**, *122*, 124–131. [CrossRef]
49. Yang, Q.; Ito, S.; Gonzalez, F.J. Hepatocyte-restricted constitutive activation of PPARα induces hepatoproliferation but not hepatocarcinogenesis. *Carcinogenesis* **2007**, *28*, 1171–1177. [CrossRef]
50. Gervois, P.; Torra, I.P.; Chinetti, G.; Grötzinger, T.; Dubois, G.; Fruchart, J.C.; Fruchart-Najib, J.; Leitersdorf, E.; Staels, B. A truncated human peroxisome proliferator-activated receptor α splice variant with dominant negative activity. *Mol. Endocrinol.* **1999**, *13*, 1535–1549. [CrossRef] [PubMed]

51. Thomas, M.; Bayha, C.; Klein, K.; Müller, S.; Weiss, T.S.; Schwab, M.; Zanger, U.M. The truncated splice variant of peroxisome proliferator-activated receptor alpha, PPARα-tr, autonomously regulates proliferative and pro-inflammatory genes. *BMC Cancer* **2015**, *15*, 488. [CrossRef] [PubMed]
52. Zhang, N.; Chu, E.S.; Zhang, J.; Li, X.; Liang, Q.; Chen, J.; Chen, M.; Teoh, N.; Farrell, G.; Sung, J.J.; et al. Peroxisome proliferator activated receptor alpha inhibits hepatocarcinogenesis through mediating NF-κB signaling pathway. *Oncotarget* **2014**, *5*, 8330–8340. [CrossRef]
53. Kaipainen, A.; Kieran, M.W.; Huang, S.; Butterfield, C.; Bielenberg, D.; Mostoslavsky, G.; Mulligan, R.; Folkman, J.; Panigrahy, D. PPARα deficiency in inflammatory cells suppresses tumor growth. *PLoS ONE* **2007**, *2*, e260. [CrossRef] [PubMed]
54. Hichami, A.; Yessoufou, A.; Ghiringhelli, F.; Salvadori, F.; Moutairou, K.; Zwetyenga, N.; Khan, N.A. Peroxisome proliferator-activated receptor alpha deficiency impairs regulatory T cell functions: Possible application in the inhibition of melanoma tumor growth in mice. *Biochimie* **2016**, *131*, 1–10. [CrossRef] [PubMed]
55. Saidi, S.A.; Holland, C.M.; Charnock-Jones, D.S.; Smith, S.K. In vitro and in vivo effects of the PPAR-alpha agonists fenofibrate and retinoic acid in endometrial cancer. *Mol. Cancer* **2006**, *5*, 13. [CrossRef] [PubMed]
56. Yokoyama, Y.; Xin, B.; Shigeto, T.; Umemoto, M.; Kasai-Sakamoto, A.; Futagami, M.; Tsuchida, S.; Al-Mulla, F.; Mizunuma, H. Clofibric acid, a peroxisome proliferator-activated receptor α ligand, inhibits growth of human ovarian cancer. *Mol. Cancer Ther.* **2007**, *6*, 1379–1386. [CrossRef] [PubMed]
57. Urbanska, K.; Pannizzo, P.; Grabacka, M.; Croul, S.; Del Valle, L.; Khalili, K.; Reiss, K. Activation of PPARα inhibits IGF-I-mediated growth and survival responses in medulloblastoma cell lines. *Int. J. Cancer* **2008**, *123*, 1015–1024. [CrossRef] [PubMed]
58. Han, D.F.; Zhang, J.X.; Wei, W.J.; Tao, T.; Hu, Q.; Wang, Y.Y.; Wang, X.F.; Liu, N.; You, Y.P. Fenofibrate induces G0/G1 phase arrest by modulating the PPARα/FoxO1/p27 kip pathway in human glioblastoma cells. *Tumour Biol.* **2015**, *36*, 3823–3829. [CrossRef]
59. Su, C.; Shi, A.; Cao, G.; Tao, T.; Chen, R.; Hu, Z.; Shen, Z.; Tao, H.; Cao, B.; Hu, D.; et al. Fenofibrate suppressed proliferation and migration of human neuroblastoma cells via oxidative stress dependent of TXNIP upregulation. *Biochem. Biophys. Res. Commun.* **2015**, *460*, 983–988. [CrossRef] [PubMed]
60. Haynes, H.R.; Scott, H.L.; Killick-Cole, C.L.; Shaw, G.; Brend, T.; Hares, K.M.; Redondo, J.; Kemp, K.C.; Ballesteros, L.S.; Herman, A.; et al. shRNA-mediated PPARα knockdown in human glioma stem cells reduces in vitro proliferation and inhibits orthotopic xenograft tumour growth. *J. Pathol.* **2019**, *247*, 422–434. [CrossRef] [PubMed]
61. Chen, Y.; Wang, Y.; Huang, Y.; Zeng, H.; Hu, B.; Guan, L.; Zhang, H.; Yu, A.M.; Johnson, C.H.; Gonzalez, F.J.; et al. PPARα regulates tumor cell proliferation and senescence via a novel target gene carnitine palmitoyltransferase 1C. *Carcinogenesis* **2017**, *38*, 474–483. [CrossRef] [PubMed]
62. Stebbins, K.J.; Broadhead, A.R.; Cabrera, G.; Correa, L.D.; Messmer, D.; Bundey, R.; Baccei, C.; Bravo, Y.; Chen, A.; Stock, N.S.; et al. In vitro and in vivo pharmacology of NXT629, a novel and selective PPARα antagonist. *Eur. J. Pharmacol.* **2017**, *809*, 130–140. [CrossRef] [PubMed]
63. Li, M.Y.; Yuan, H.; Ma, L.T.; Kong, A.W.; Hsin, M.K.; Yip, J.H.; Underwood, M.J.; Chen, G.G. Roles of peroxisome proliferator-activated receptor-α and -γ in the development of non-small cell lung cancer. *Am. J. Respir. Cell Mol. Biol.* **2010**, *43*, 674–683. [CrossRef]
64. Liang, H.; Kowalczyk, P.; Junco, J.J.; Klug-De Santiago, H.L.; Malik, G.; Wei, S.J.; Slaga, T.J. Differential effects on lung cancer cell proliferation by agonists of glucocorticoid and PPARα receptors. *Mol. Carcinog.* **2014**, *53*, 753–763. [CrossRef] [PubMed]
65. Skrypnyk, N.; Chen, X.; Hu, W.; Su, Y.; Mont, S.; Yang, S.; Gangadhariah, M.; Wei, S.; Falck, J.R.; Jat, J.L.; et al. PPARα activation can help prevent and treat non-small cell lung cancer. *Cancer Res.* **2014**, *74*, 621–631. [CrossRef] [PubMed]
66. Pozzi, A.; Ibanez, M.R.; Gatica, A.E.; Yang, S.; Wei, S.; Mei, S.; Falck, J.R.; Capdevila, J.H. Peroxisomal proliferator-activated receptor-α-dependent inhibition of endothelial cell proliferation and tumorigenesis. *J. Biol. Chem.* **2007**, *282*, 17685–17695. [CrossRef] [PubMed]
67. Huang, J.; Das, S.K.; Jha, P.; Al Zoughbi, W.; Schauer, S.; Claudel, T.; Sexl, V.; Vesely, P.; Birner-Gruenberger, R.; Kratky, D.; et al. The PPARα agonist fenofibrate suppresses B-cell lymphoma in mice by modulating lipid metabolism. *Biochim. Biophys. Acta* **2013**, *1831*, 1555–1565. [CrossRef]
68. Kong, R.; Wang, N.; Han, W.; Bao, W.; Lu, J. Fenofibrate Exerts Antitumor Effects in Colon Cancer via Regulation of DNMT1 and CDKN2A. *PPAR Res.* **2021**, *2021*, 6663782. [CrossRef] [PubMed]
69. Berger, H.; Végran, F.; Chikh, M.; Gilardi, F.; Ladoire, S.; Bugaut, H.; Mignot, G.; Chalmin, F.; Bruchard, M.; Derangère, V.; et al. SOCS3 transactivation by PPARγ prevents IL-17-driven cancer growth. *Cancer Res.* **2013**, *73*, 3578–3590. [CrossRef] [PubMed]
70. Takayama, O.; Yamamoto, H.; Damdinsuren, B.; Sugita, Y.; Ngan, C.Y.; Xu, X.; Tsujino, T.; Takemasa, I.; Ikeda, M.; Sekimoto, M.; et al. Expression of PPARδ in multistage carcinogenesis of the colorectum: Implications of malignant cancer morphology. *Br. J. Cancer* **2006**, *95*, 889–895. [CrossRef]
71. Yoshinaga, M.; Taki, K.; Somada, S.; Sakiyama, Y.; Kubo, N.; Kaku, T.; Tsuruta, S.; Kusumoto, T.; Sakai, H.; Nakamura, K.; et al. The expression of both peroxisome proliferator-activated receptor delta and cyclooxygenase-2 in tissues is associated with poor prognosis in colorectal cancer patients. *Dig. Dis. Sci.* **2011**, *56*, 1194–1200. [CrossRef] [PubMed]
72. Foreman, J.E.; Chang, W.C.; Palkar, P.S.; Zhu, B.; Borland, M.G.; Williams, J.L.; Kramer, L.R.; Clapper, M.L.; Gonzalez, F.J.; Peters, J.M. Functional characterization of peroxisome proliferator-activated receptor-β/δ expression in colon cancer. *Mol. Carcinog.* **2011**, *50*, 884–900. [CrossRef] [PubMed]

73. Wang, D.; Wang, H.; Guo, Y.; Ning, W.; Katkuri, S.; Wahli, W.; Desvergne, B.; Dey, S.K.; DuBois, R.N. Crosstalk between peroxisome proliferator-activated receptor δ and VEGF stimulates cancer progression. *Proc. Natl. Acad. Sci. USA* **2006**, *103*, 19069–19074. [CrossRef] [PubMed]
74. Röhrl, C.; Kaindl, U.; Koneczny, I.; Hudec, X.; Baron, D.M.; König, J.S.; Marian, B. Peroxisome-proliferator-activated receptors γ and β/δ mediate vascular endothelial growth factor production in colorectal tumor cells. *J. Cancer Res. Clin. Oncol.* **2011**, *137*, 29–39. [CrossRef]
75. Zuo, X.; Peng, Z.; Moussalli, M.J.; Morris, J.S.; Broaddus, R.R.; Fischer, S.M.; Shureiqi, I. Targeted genetic disruption of peroxisome proliferator-activated receptor-δ and colonic tumorigenesis. *J. Natl. Cancer Inst.* **2009**, *101*, 762–767. [CrossRef] [PubMed]
76. Yang, L.; Zhou, J.; Ma, Q.; Wang, C.; Chen, K.; Meng, W.; Yu, Y.; Zhou, Z.; Sun, X. Knockdown of PPAR δ gene promotes the growth of colon cancer and reduces the sensitivity to bevacizumab in nude mice model. *PLoS ONE* **2013**, *8*, e60715. [CrossRef]
77. Gupta, R.A.; Wang, D.; Katkuri, S.; Wang, H.; Dey, S.K.; DuBois, R.N. Activation of nuclear hormone receptor peroxisome proliferator-activated receptor-δ accelerates intestinal adenoma growth. *Nat. Med.* **2004**, *10*, 245–247. [CrossRef]
78. Ding, J.; Gou, Q.; Jin, J.; Shi, J.; Liu, Q.; Hou, Y. Metformin inhibits PPARδ agonist-mediated tumor growth by reducing Glut1 and SLC1A5 expressions of cancer cells. *Eur. J. Pharmacol.* **2019**, *857*, 172425. [CrossRef] [PubMed]
79. Liu, Y.; Deguchi, Y.; Tian, R.; Wei, D.; Wu, L.; Chen, W.; Xu, W.; Xu, M.; Liu, F.; Gao, S.; et al. Pleiotropic Effects of PPARD Accelerate Colorectal Tumorigenesis, Progression, and Invasion. *Cancer Res.* **2019**, *79*, 954–969. [CrossRef] [PubMed]
80. Zuo, X.; Deguchi, Y.; Xu, W.; Liu, Y.; Li, H.S.; Wei, D.; Tian, R.; Chen, W.; Xu, M.; Yang, Y.; et al. PPARD and Interferon Gamma Promote Transformation of Gastric Progenitor Cells and Tumorigenesis in Mice. *Gastroenterology* **2019**, *157*, 163–178. [CrossRef]
81. Zhou, D.; Jin, J.; Liu, Q.; Shi, J.; Hou, Y. PPARδ agonist enhances colitis-associated colorectal cancer. *Eur. J. Pharmacol.* **2019**, *842*, 248–254. [CrossRef] [PubMed]
82. Zuo, X.; Xu, M.; Yu, J.; Wu, Y.; Moussalli, M.J.; Manyam, G.C.; Lee, S.I.; Liang, S.; Gagea, M.; Morris, J.S.; et al. Potentiation of colon cancer susceptibility in mice by colonic epithelial PPAR-δ/β overexpression. *J. Natl. Cancer Inst.* **2014**, *106*, dju052. [CrossRef] [PubMed]
83. Marin, H.E.; Peraza, M.A.; Billin, A.N.; Willson, T.M.; Ward, J.M.; Kennett, M.J.; Gonzalez, F.J.; Peters, J.M. Ligand activation of peroxisome proliferator-activated receptor β inhibits colon carcinogenesis. *Cancer Res.* **2006**, *66*, 4394–4401. [CrossRef] [PubMed]
84. Harman, F.S.; Nicol, C.J.; Marin, H.E.; Ward, J.M.; Gonzalez, F.J.; Peters, J.M. Peroxisome proliferator-activated receptor-δ attenuates colon carcinogenesis. *Nat. Med.* **2004**, *10*, 481–483. [CrossRef] [PubMed]
85. Beyaz, S.; Mana, M.D.; Roper, J.; Kedrin, D.; Saadatpour, A.; Hong, S.J.; Bauer-Rowe, K.E.; Xifaras, M.E.; Akkad, A.; Arias, E.; et al. High-fat diet enhances stemness and tumorigenicity of intestinal progenitors. *Nature* **2016**, *531*, 53–58. [CrossRef]
86. Nagy, T.A.; Wroblewski, L.E.; Wang, D.; Piazuelo, M.B.; Delgado, A.; Romero-Gallo, J.; Noto, J.; Israel, D.A.; Ogden, S.R.; Correa, P.; et al. β-Catenin and p120 mediate PPARδ-dependent proliferation induced by Helicobacter pylori in human and rodent epithelia. *Gastroenterology* **2011**, *141*, 553–564. [CrossRef]
87. Girroir, E.E.; Hollingshead, H.E.; Billin, A.N.; Willson, T.M.; Robertson, G.P.; Sharma, A.K.; Amin, S.; Gonzalez, F.J.; Peters, J.M. Peroxisome proliferator-activated receptor-β/δ (PPARβ/δ) ligands inhibit growth of UACC903 and MCF7 human cancer cell lines. *Toxicology* **2008**, *243*, 236–243. [CrossRef]
88. Yao, P.L.; Morales, J.L.; Zhu, B.; Kang, B.H.; Gonzalez, F.J.; Peters, J.M. Activation of peroxisome proliferator-activated receptor-β/δ (PPAR-β/δ) inhibits human breast cancer cell line tumorigenicity. *Mol. Cancer Ther.* **2014**, *13*, 1008–1017. [CrossRef]
89. Yao, P.L.; Chen, L.; Dobrzański, T.P.; Zhu, B.; Kang, B.H.; Müller, R.; Gonzalez, F.J.; Peters, J.M. Peroxisome proliferator-activated receptor-β/δ inhibits human neuroblastoma cell tumorigenesis by inducing p53- and SOX2-mediated cell differentiation. *Mol. Carcinog.* **2017**, *56*, 1472–1483. [CrossRef]
90. Yao, P.L.; Chen, L.P.; Dobrzański, T.P.; Phillips, D.A.; Zhu, B.; Kang, B.H.; Gonzalez, F.J.; Peters, J.M. Inhibition of testicular embryonal carcinoma cell tumorigenicity by peroxisome proliferator-activated receptor-β/δ- and retinoic acid receptor-dependent mechanisms. *Oncotarget* **2015**, *6*, 36319–36337. [CrossRef]
91. Yuan, H.; Lu, J.; Xiao, J.; Upadhyay, G.; Umans, R.; Kallakury, B.; Yin, Y.; Fant, M.E.; Kopelovich, L.; Glazer, R.I. PPARδ induces estrogen receptor-positive mammary neoplasia through an inflammatory and metabolic phenotype linked to mTOR activation. *Cancer Res.* **2013**, *73*, 4349–4361. [CrossRef] [PubMed]
92. Pollock, C.B.; Yin, Y.; Yuan, H.; Zeng, X.; King, S.; Li, X.; Kopelovich, L.; Albanese, C.; Glazer, R.I. PPARδ activation acts cooperatively with 3-phosphoinositide-dependent protein kinase-1 to enhance mammary tumorigenesis. *PLoS ONE* **2011**, *6*, e16215. [CrossRef] [PubMed]
93. Yin, Y.; Russell, R.G.; Dettin, L.E.; Bai, R.; Wei, Z.L.; Kozikowski, A.P.; Kopelovich, L.; Kopleovich, L.; Glazer, R.I. Peroxisome proliferator-activated receptor δ and γ agonists differentially alter tumor differentiation and progression during mammary carcinogenesis. *Cancer Res.* **2005**, *65*, 3950–3957. [CrossRef] [PubMed]
94. Ghosh, M.; Ai, Y.; Narko, K.; Wang, Z.; Peters, J.M.; Hla, T. PPARδ is pro-tumorigenic in a mouse model of COX-2-induced mammary cancer. *Prostaglandins Other Lipid Mediat.* **2009**, *88*, 97–100. [CrossRef]
95. Levi, L.; Lobo, G.; Doud, M.K.; von Lintig, J.; Seachrist, D.; Tochtrop, G.P.; Noy, N. Genetic ablation of the fatty acid-binding protein FABP5 suppresses HER2-induced mammary tumorigenesis. *Cancer Res.* **2013**, *73*, 4770–4780. [CrossRef] [PubMed]
96. Schug, T.T.; Berry, D.C.; Toshkov, I.A.; Cheng, L.; Nikitin, A.Y.; Noy, N. Overcoming retinoic acid-resistance of mammary carcinomas by diverting retinoic acid from PPARβ/δ to RAR. *Proc. Natl. Acad. Sci. USA* **2008**, *105*, 7546–7551. [CrossRef] [PubMed]

97. Morgan, E.; Kannan-Thulasiraman, P.; Noy, N. Involvement of Fatty Acid Binding Protein 5 and PPARβ/δ in Prostate Cancer Cell Growth. *PPAR Res.* **2010**, *2010*, 234629. [CrossRef]
98. Her, N.G.; Jeong, S.I.; Cho, K.; Ha, T.K.; Han, J.; Ko, K.P.; Park, S.K.; Lee, J.H.; Lee, M.G.; Ryu, B.K.; et al. PPARδ promotes oncogenic redirection of TGF-β1 signaling through the activation of the ABCA1-Cav1 pathway. *Cell Cycle* **2013**, *12*, 1521–1535. [CrossRef]
99. Martín-Martín, N.; Zabala-Letona, A.; Fernández-Ruiz, S.; Arreal, L.; Camacho, L.; Castillo-Martin, M.; Cortazar, A.R.; Torrano, V.; Astobiza, I.; Zúñiga-García, P.; et al. PPARδ Elicits Ligand-Independent Repression of Trefoil Factor Family to Limit Prostate Cancer Growth. *Cancer Res.* **2018**, *78*, 399–409. [CrossRef]
100. Balandaram, G.; Kramer, L.R.; Kang, B.H.; Murray, I.A.; Perdew, G.H.; Gonzalez, F.J.; Peters, J.M. Ligand activation of peroxisome proliferator-activated receptor-β/δ suppresses liver tumorigenesis in hepatitis B transgenic mice. *Toxicology* **2016**, *363–364*, 1–9. [CrossRef]
101. Xu, L.; Han, C.; Lim, K.; Wu, T. Cross-talk between peroxisome proliferator-activated receptor δ and cytosolic phospholipase A(2)α/cyclooxygenase-2/prostaglandin E(2) signaling pathways in human hepatocellular carcinoma cells. *Cancer Res.* **2006**, *66*, 11859–11868. [CrossRef] [PubMed]
102. Michiels, J.F.; Perrin, C.; Leccia, N.; Massi, D.; Grimaldi, P.; Wagner, N. PPARβ activation inhibits melanoma cell proliferation involving repression of the Wilms' tumour suppressor WT1. *Pflugers Arch.* **2010**, *459*, 689–703. [CrossRef] [PubMed]
103. Wagner, N.; Panelos, J.; Massi, D.; Wagner, K.D. The Wilms' tumor suppressor WT1 is associated with melanoma proliferation. *Pflugers Arch.* **2008**, *455*, 839–847. [CrossRef] [PubMed]
104. Kim, D.J.; Bility, M.T.; Billin, A.N.; Willson, T.M.; Gonzalez, F.J.; Peters, J.M. PPARβ/δ selectively induces differentiation and inhibits cell proliferation. *Cell Death Differ.* **2006**, *13*, 53–60. [CrossRef]
105. Borland, M.G.; Kehres, E.M.; Lee, C.; Wagner, A.L.; Shannon, B.E.; Albrecht, P.P.; Zhu, B.; Gonzalez, F.J.; Peters, J.M. Inhibition of tumorigenesis by peroxisome proliferator-activated receptor (PPAR)-dependent cell cycle blocks in human skin carcinoma cells. *Toxicology* **2018**, *404–405*, 25–32. [CrossRef]
106. Zhu, B.; Bai, R.; Kennett, M.J.; Kang, B.H.; Gonzalez, F.J.; Peters, J.M. Chemoprevention of chemically induced skin tumorigenesis by ligand activation of peroxisome proliferator-activated receptor-β/δ and inhibition of cyclooxygenase 2. *Mol. Cancer Ther.* **2010**, *9*, 3267–3277. [CrossRef]
107. Bility, M.T.; Zhu, B.; Kang, B.H.; Gonzalez, F.J.; Peters, J.M. Ligand activation of peroxisome proliferator-activated receptor-β/δ and inhibition of cyclooxygenase-2 enhances inhibition of skin tumorigenesis. *Toxicol. Sci.* **2010**, *113*, 27–36. [CrossRef]
108. Kim, D.J.; Prabhu, K.S.; Gonzalez, F.J.; Peters, J.M. Inhibition of chemically induced skin carcinogenesis by sulindac is independent of peroxisome proliferator-activated receptor-β/δ (PPARβ/δ). *Carcinogenesis* **2006**, *27*, 1105–1112. [CrossRef]
109. Montagner, A.; Delgado, M.B.; Tallichet-Blanc, C.; Chan, J.S.; Sng, M.K.; Mottaz, H.; Degueurce, G.; Lippi, Y.; Moret, C.; Baruchet, M.; et al. Src is activated by the nuclear receptor peroxisome proliferator-activated receptor β/δ in ultraviolet radiation-induced skin cancer. *EMBO Mol. Med.* **2014**, *6*, 80–98. [CrossRef]
110. Tan, M.W.Y.; Sng, M.K.; Cheng, H.S.; Low, Z.S.; Leong, B.J.J.; Chua, D.; Tan, E.H.P.; Chan, J.S.K.; Yip, Y.S.; Lee, Y.H.; et al. Deficiency in fibroblast PPARβ/δ reduces nonmelanoma skin cancers in mice. *Cell Death Differ.* **2020**, *27*, 2668–2680. [CrossRef]
111. Pedchenko, T.V.; Gonzalez, A.L.; Wang, D.; DuBois, R.N.; Massion, P.P. Peroxisome proliferator—Activated receptor β/δ expression and activation in lung cancer. *Am. J. Respir. Cell Mol. Biol.* **2008**, *39*, 689–696. [CrossRef]
112. Genini, D.; Garcia-Escudero, R.; Carbone, G.M.; Catapano, C.V. Transcriptional and Non-Transcriptional Functions of PPARβ/δ in Non-Small Cell Lung Cancer. *PLoS ONE* **2012**, *7*, e46009. [CrossRef] [PubMed]
113. Sun, X.; Ritzenthaler, J.D.; Zhong, X.; Zheng, Y.; Roman, J.; Han, S. Nicotine stimulates PPARβ/δ expression in human lung carcinoma cells through activation of PI3K/mTOR and suppression of AP-2α. *Cancer Res.* **2009**, *69*, 6445–6453. [CrossRef] [PubMed]
114. He, P.; Borland, M.G.; Zhu, B.; Sharma, A.K.; Amin, S.; El-Bayoumy, K.; Gonzalez, F.J.; Peters, J.M. Effect of ligand activation of peroxisome proliferator-activated receptor-β/δ (PPARβ/δ) in human lung cancer cell lines. *Toxicology* **2008**, *254*, 112–117. [CrossRef]
115. Müller-Brüsselbach, S.; Ebrahimsade, S.; Jäkel, J.; Eckhardt, J.; Rapp, U.R.; Peters, J.M.; Moll, R.; Müller, R. Growth of transgenic RAF-induced lung adenomas is increased in mice with a disrupted PPARβ/δ gene. *Int. J. Oncol.* **2007**, *31*, 607–611. [CrossRef]
116. Gu, L.; Shi, Y.; Xu, W.; Ji, Y. PPARβ/δ Agonist GW501516 Inhibits Tumorigenesis and Promotes Apoptosis of the Undifferentiated Nasopharyngeal Carcinoma C666-1 Cells by Regulating miR-206. *Oncol. Res.* **2019**, *27*, 923–933. [CrossRef]
117. Wagner, K.D.; Benchetrit, M.; Bianchini, L.; Michiels, J.F.; Wagner, N. Peroxisome proliferator-activated receptor β/δ (PPARβ/δ) is highly expressed in liposarcoma and promotes migration and proliferation. *J. Pathol.* **2011**, *224*, 575–588. [CrossRef] [PubMed]
118. Zeng, L.; Geng, Y.; Tretiakova, M.; Yu, X.; Sicinski, P.; Kroll, T.G. Peroxisome proliferator-activated receptor-δ induces cell proliferation by a cyclin E1-dependent mechanism and is up-regulated in thyroid tumors. *Cancer Res.* **2008**, *68*, 6578–6586. [CrossRef]
119. Daikoku, T.; Tranguch, S.; Chakrabarty, A.; Wang, D.; Khabele, D.; Orsulic, S.; Morrow, J.D.; Dubois, R.N.; Dey, S.K. Extracellular signal-regulated kinase is a target of cyclooxygenase-1-peroxisome proliferator-activated receptor-δ signaling in epithelial ovarian cancer. *Cancer Res.* **2007**, *67*, 5285–5292. [CrossRef]
120. Li, Y.J.; Sun, L.; Shi, Y.; Wang, G.; Wang, X.; Dunn, S.E.; Iorio, C.; Screaton, R.A.; Spaner, D.E. PPAR-delta promotes survival of chronic lymphocytic leukemia cells in energetically unfavorable conditions. *Leukemia* **2017**, *31*, 1905–1914. [CrossRef]

121. Sun, L.; Shi, Y.; Wang, G.; Wang, X.; Zeng, S.; Dunn, S.E.; Fairn, G.D.; Li, Y.J.; Spaner, D.E. PPAR-delta modulates membrane cholesterol and cytokine signaling in malignant B cells. *Leukemia* **2018**, *32*, 184–193. [CrossRef] [PubMed]
122. Lefebvre, A.M.; Chen, I.; Desreumaux, P.; Najib, J.; Fruchart, J.C.; Geboes, K.; Briggs, M.; Heyman, R.; Auwerx, J. Activation of the peroxisome proliferator-activated receptor γ promotes the development of colon tumors in C57BL/6J-APCMin/+ mice. *Nat. Med.* **1998**, *4*, 1053–1057. [CrossRef]
123. Saez, E.; Tontonoz, P.; Nelson, M.C.; Alvarez, J.G.; Ming, U.T.; Baird, S.M.; Thomazy, V.A.; Evans, R.M. Activators of the nuclear receptor PPARγ enhance colon polyp formation. *Nat. Med.* **1998**, *4*, 1058–1061. [CrossRef] [PubMed]
124. Brockman, J.A.; Gupta, R.A.; Dubois, R.N. Activation of PPARγ leads to inhibition of anchorage-independent growth of human colorectal cancer cells. *Gastroenterology* **1998**, *115*, 1049–1055. [CrossRef]
125. Sarraf, P.; Mueller, E.; Jones, D.; King, F.J.; DeAngelo, D.J.; Partridge, J.B.; Holden, S.A.; Chen, L.B.; Singer, S.; Fletcher, C.; et al. Differentiation and reversal of malignant changes in colon cancer through PPARγ. *Nat. Med.* **1998**, *4*, 1046–1052. [CrossRef] [PubMed]
126. Kitamura, S.; Miyazaki, Y.; Shinomura, Y.; Kondo, S.; Kanayama, S.; Matsuzawa, Y. Peroxisome proliferator-activated receptor γ induces growth arrest and differentiation markers of human colon cancer cells. *Jpn. J. Cancer Res.* **1999**, *90*, 75–80. [CrossRef] [PubMed]
127. Chintharlapalli, S.; Papineni, S.; Safe, S. 1,1-Bis(3′-indolyl)-1-(p-substituted phenyl)methanes inhibit colon cancer cell and tumor growth through PPARγ-dependent and PPARγ-independent pathways. *Mol. Cancer Ther.* **2006**, *5*, 1362–1370. [CrossRef] [PubMed]
128. Kubota, T.; Koshizuka, K.; Williamson, E.A.; Asou, H.; Said, J.W.; Holden, S.; Miyoshi, I.; Koeffler, H.P. Ligand for peroxisome proliferator-activated receptor γ (troglitazone) has potent antitumor effect against human prostate cancer both in vitro and in vivo. *Cancer Res.* **1998**, *58*, 3344–3352.
129. Shappell, S.B.; Gupta, R.A.; Manning, S.; Whitehead, R.; Boeglin, W.E.; Schneider, C.; Case, T.; Price, J.; Jack, G.S.; Wheeler, T.M.; et al. 15S-Hydroxyeicosatetraenoic acid activates peroxisome proliferator-activated receptor γ and inhibits proliferation in PC3 prostate carcinoma cells. *Cancer Res.* **2001**, *61*, 497–503. [PubMed]
130. Yoshimura, R.; Matsuyama, M.; Hase, T.; Tsuchida, K.; Kuratsukuri, K.; Kawahito, Y.; Sano, H.; Segawa, Y.; Nakatani, T. The effect of peroxisome proliferator-activated receptor-γ ligand on urological cancer cells. *Int. J. Mol. Med.* **2003**, *12*, 861–865. [CrossRef] [PubMed]
131. Tontonoz, P.; Singer, S.; Forman, B.M.; Sarraf, P.; Fletcher, J.A.; Fletcher, C.D.; Brun, R.P.; Mueller, E.; Altiok, S.; Oppenheim, H.; et al. Terminal differentiation of human liposarcoma cells induced by ligands for peroxisome proliferator-activated receptor γ and the retinoid X receptor. *Proc. Natl. Acad. Sci. USA* **1997**, *94*, 237–241. [CrossRef] [PubMed]
132. Takahashi, N.; Okumura, T.; Motomura, W.; Fujimoto, Y.; Kawabata, I.; Kohgo, Y. Activation of PPARγ inhibits cell growth and induces apoptosis in human gastric cancer cells. *FEBS Lett.* **1999**, *455*, 135–139. [CrossRef]
133. Sato, H.; Ishihara, S.; Kawashima, K.; Moriyama, N.; Suetsugu, H.; Kazumori, H.; Okuyama, T.; Rumi, M.A.; Fukuda, R.; Nagasue, N.; et al. Expression of peroxisome proliferator-activated receptor (PPAR)γ in gastric cancer and inhibitory effects of PPARγ agonists. *Br. J. Cancer* **2000**, *83*, 1394–1400. [CrossRef]
134. Kassouf, W.; Chintharlapalli, S.; Abdelrahim, M.; Nelkin, G.; Safe, S.; Kamat, A.M. Inhibition of bladder tumor growth by 1,1-bis(3′-indolyl)-1-(p-substitutedphenyl)methanes: A new class of peroxisome proliferator-activated receptor γ agonists. *Cancer Res.* **2006**, *66*, 412–418. [CrossRef]
135. Han, S.W.; Greene, M.E.; Pitts, J.; Wada, R.K.; Sidell, N. Novel expression and function of peroxisome proliferator-activated receptor γ (PPARγ) in human neuroblastoma cells. *Clin. Cancer Res.* **2001**, *7*, 98–104. [PubMed]
136. Cellai, I.; Benvenuti, S.; Luciani, P.; Galli, A.; Ceni, E.; Simi, L.; Baglioni, S.; Muratori, M.; Ottanelli, B.; Serio, M.; et al. Antineoplastic effects of rosiglitazone and PPARγ transactivation in neuroblastoma cells. *Br. J. Cancer* **2006**, *95*, 879–888. [CrossRef] [PubMed]
137. Chearwae, W.; Bright, J.J. PPARγ agonists inhibit growth and expansion of CD133+ brain tumour stem cells. *Br. J. Cancer* **2008**, *99*, 2044–2053. [CrossRef] [PubMed]
138. Wang, P.; Yu, J.; Yin, Q.; Li, W.; Ren, X.; Hao, X. Rosiglitazone suppresses glioma cell growth and cell cycle by blocking the transforming growth factor-delta mediated pathway. *Neurochem. Res.* **2012**, *37*, 2076–2084. [CrossRef] [PubMed]
139. Jozkowicz, A.; Dulak, J.; Piatkowska, E.; Placha, W.; Dembinska-Kiec, A. Ligands of peroxisome proliferator-activated receptor-γ increase the generation of vascular endothelial growth factor in vascular smooth muscle cells and in macrophages. *Acta Biochim. Pol.* **2000**, *47*, 1147–1157. [CrossRef] [PubMed]
140. Freudlsperger, C.; Moll, I.; Schumacher, U.; Thies, A. Anti-proliferative effect of peroxisome proliferator-activated receptor γ agonists on human malignant melanoma cells in vitro. *Anticancer Drugs* **2006**, *17*, 325–332. [CrossRef]
141. Botton, T.; Puissant, A.; Bahadoran, P.; Annicotte, J.S.; Fajas, L.; Ortonne, J.P.; Gozzerino, G.; Zamoum, T.; Tartare-Deckert, S.; Bertolotto, C.; et al. In vitro and in vivo anti-melanoma effects of ciglitazone. *J. Investig. Dermatol.* **2009**, *129*, 1208–1218. [CrossRef]
142. Placha, W.; Gil, D.; Dembińska-Kieć, A.; Laidler, P. The effect of PPARγ ligands on the proliferation and apoptosis of human melanoma cells. *Melanoma Res.* **2003**, *13*, 447–456. [CrossRef] [PubMed]
143. Keshamouni, V.G.; Reddy, R.C.; Arenberg, D.A.; Joel, B.; Thannickal, V.J.; Kalemkerian, G.P.; Standiford, T.J. Peroxisome proliferator-activated receptor-γ activation inhibits tumor progression in non-small-cell lung cancer. *Oncogene* **2004**, *23*, 100–108. [CrossRef] [PubMed]

144. Han, S.; Roman, J. Rosiglitazone suppresses human lung carcinoma cell growth through PPARγ-dependent and PPARγ-independent signal pathways. *Mol. Cancer Ther.* **2006**, *5*, 430–437. [CrossRef] [PubMed]
145. Ferruzzi, P.; Ceni, E.; Tarocchi, M.; Grappone, C.; Milani, S.; Galli, A.; Fiorelli, G.; Serio, M.; Mannelli, M. Thiazolidinediones inhibit growth and invasiveness of the human adrenocortical cancer cell line H295R. *J. Clin. Endocrinol. Metab.* **2005**, *90*, 1332–1339. [CrossRef]
146. Betz, M.J.; Shapiro, I.; Fassnacht, M.; Hahner, S.; Reincke, M.; Beuschlein, F.; Network, G.a.A.A. Peroxisome proliferator-activated receptor-γ agonists suppress adrenocortical tumor cell proliferation and induce differentiation. *J. Clin. Endocrinol. Metab.* **2005**, *90*, 3886–3896. [CrossRef]
147. Yu, J.; Qiao, L.; Zimmermann, L.; Ebert, M.P.; Zhang, H.; Lin, W.; Röcken, C.; Malfertheiner, P.; Farrell, G.C. Troglitazone inhibits tumor growth in hepatocellular carcinoma in vitro and in vivo. *Hepatology* **2006**, *43*, 134–143. [CrossRef] [PubMed]
148. Ota, K.; Ito, K.; Suzuki, T.; Saito, S.; Tamura, M.; Hayashi, S.; Okamura, K.; Sasano, H.; Yaegashi, N. Peroxisome proliferator-activated receptor γ and growth inhibition by its ligands in uterine endometrial carcinoma. *Clin. Cancer Res.* **2006**, *12*, 4200–4208. [CrossRef] [PubMed]
149. Lei, P.; Abdelrahim, M.; Safe, S. 1,1-Bis(3′-indolyl)-1-(p-substituted phenyl)methanes inhibit ovarian cancer cell growth through peroxisome proliferator-activated receptor-dependent and independent pathways. *Mol. Cancer Ther.* **2006**, *5*, 2324–2336. [CrossRef]
150. Vignati, S.; Albertini, V.; Rinaldi, A.; Kwee, I.; Riva, C.; Oldrini, R.; Capella, C.; Bertoni, F.; Carbone, G.M.; Catapano, C.V. Cellular and molecular consequences of peroxisome proliferator-activated receptor-γ activation in ovarian cancer cells. *Neoplasia* **2006**, *8*, 851–861. [CrossRef] [PubMed]
151. Garcia-Bates, T.M.; Bernstein, S.H.; Phipps, R.P. Peroxisome proliferator-activated receptor γ overexpression suppresses growth and induces apoptosis in human multiple myeloma cells. *Clin. Cancer Res.* **2008**, *14*, 6414–6425. [CrossRef] [PubMed]
152. Garcia-Bates, T.M.; Peslak, S.A.; Baglole, C.J.; Maggirwar, S.B.; Bernstein, S.H.; Phipps, R.P. Peroxisome proliferator-activated receptor gamma overexpression and knockdown: Impact on human B cell lymphoma proliferation and survival. *Cancer Immunol. Immunother.* **2009**, *58*, 1071–1083. [CrossRef] [PubMed]
153. Hamaguchi, N.; Hamada, H.; Miyoshi, S.; Irifune, K.; Ito, R.; Miyazaki, T.; Higaki, J. In vitro and in vivo therapeutic efficacy of the PPAR-γ agonist troglitazone in combination with cisplatin against malignant pleural mesothelioma cell growth. *Cancer Sci.* **2010**, *101*, 1955–1964. [CrossRef] [PubMed]
154. Sawayama, H.; Ishimoto, T.; Watanabe, M.; Yoshida, N.; Sugihara, H.; Kurashige, J.; Hirashima, K.; Iwatsuki, M.; Baba, Y.; Oki, E.; et al. Small molecule agonists of PPAR-γ exert therapeutic effects in esophageal cancer. *Cancer Res.* **2014**, *74*, 575–585. [CrossRef] [PubMed]
155. Prost, S.; Relouzat, F.; Spentchian, M.; Ouzegdouh, Y.; Saliba, J.; Massonnet, G.; Beressi, J.P.; Verhoeyen, E.; Raggueneau, V.; Maneglier, B.; et al. Erosion of the chronic myeloid leukaemia stem cell pool by PPARγ agonists. *Nature* **2015**, *525*, 380–383. [CrossRef] [PubMed]
156. Rousselot, P.; Prost, S.; Guilhot, J.; Roy, L.; Etienne, G.; Legros, L.; Charbonnier, A.; Coiteux, V.; Cony-Makhoul, P.; Huguet, F.; et al. Pioglitazone together with imatinib in chronic myeloid leukemia: A proof of concept study. *Cancer* **2017**, *123*, 1791–1799. [CrossRef] [PubMed]
157. Kato, Y.; Ying, H.; Zhao, L.; Furuya, F.; Araki, O.; Willingham, M.C.; Cheng, S.Y. PPARγ insufficiency promotes follicular thyroid carcinogenesis via activation of the nuclear factor-κB signaling pathway. *Oncogene* **2006**, *25*, 2736–2747. [CrossRef]
158. Wu, L.; Yan, C.; Czader, M.; Foreman, O.; Blum, J.S.; Kapur, R.; Du, H. Inhibition of PPARγ in myeloid-lineage cells induces systemic inflammation, immunosuppression, and tumorigenesis. *Blood* **2012**, *119*, 115–126. [CrossRef]
159. Pignatelli, M.; Cortés-Canteli, M.; Lai, C.; Santos, A.; Perez-Castillo, A. The peroxisome proliferator-activated receptor γ is an inhibitor of ErbBs activity in human breast cancer cells. *J. Cell Sci.* **2001**, *114*, 4117–4126. [CrossRef]
160. Qin, C.; Burghardt, R.; Smith, R.; Wormke, M.; Stewart, J.; Safe, S. Peroxisome proliferator-activated receptor γ agonists induce proteasome-dependent degradation of cyclin D1 and estrogen receptor α in MCF-7 breast cancer cells. *Cancer Res.* **2003**, *63*, 958–964.
161. Houston, K.D.; Copland, J.A.; Broaddus, R.R.; Gottardis, M.M.; Fischer, S.M.; Walker, C.L. Inhibition of proliferation and estrogen receptor signaling by peroxisome proliferator-activated receptor γ ligands in uterine leiomyoma. *Cancer Res.* **2003**, *63*, 1221–1227. [PubMed]
162. Catalano, S.; Mauro, L.; Bonofiglio, D.; Pellegrino, M.; Qi, H.; Rizza, P.; Vizza, D.; Bossi, G.; Andò, S. In vivo and in vitro evidence that PPARγ ligands are antagonists of leptin signaling in breast cancer. *Am. J. Pathol.* **2011**, *179*, 1030–1040. [CrossRef] [PubMed]
163. Kim, H.J.; Kim, J.Y.; Meng, Z.; Wang, L.H.; Liu, F.; Conrads, T.P.; Burke, T.R.; Veenstra, T.D.; Farrar, W.L. 15-deoxy-$\Delta^{12,14}$-prostaglandin J_2 inhibits transcriptional activity of estrogen receptor-α via covalent modification of DNA-binding domain. *Cancer Res.* **2007**, *67*, 2595–2602. [CrossRef] [PubMed]
164. Vanderlaag, K.; Su, Y.; Frankel, A.E.; Grage, H.; Smith, R.; Khan, S.; Safe, S. 1,1-Bis(3′-indolyl)-1-(p-substituted phenyl)methanes inhibit proliferation of estrogen receptor-negative breast cancer cells by activation of multiple pathways. *Breast Cancer Res. Treat.* **2008**, *109*, 273–283. [CrossRef] [PubMed]
165. Saez, E.; Rosenfeld, J.; Livolsi, A.; Olson, P.; Lombardo, E.; Nelson, M.; Banayo, E.; Cardiff, R.D.; Izpisua-Belmonte, J.C.; Evans, R.M. PPAR γ signaling exacerbates mammary gland tumor development. *Genes Dev.* **2004**, *18*, 528–540. [CrossRef] [PubMed]

166. Avena, P.; Anselmo, W.; Whitaker-Menezes, D.; Wang, C.; Pestell, R.G.; Lamb, R.S.; Hulit, J.; Casaburi, I.; Andò, S.; Martinez-Outschoorn, U.E.; et al. Compartment-specific activation of PPARγ governs breast cancer tumor growth, via metabolic reprogramming and symbiosis. *Cell Cycle* **2013**, *12*, 1360–1370. [CrossRef]
167. Apostoli, A.J.; Skelhorne-Gross, G.E.; Rubino, R.E.; Peterson, N.T.; Di Lena, M.A.; Schneider, M.M.; SenGupta, S.K.; Nicol, C.J. Loss of PPARγ expression in mammary secretory epithelial cells creates a pro-breast tumorigenic environment. *Int. J. Cancer* **2014**, *134*, 1055–1066. [CrossRef]
168. Yee, L.D.; Williams, N.; Wen, P.; Young, D.C.; Lester, J.; Johnson, M.V.; Farrar, W.B.; Walker, M.J.; Povoski, S.P.; Suster, S.; et al. Pilot study of rosiglitazone therapy in women with breast cancer: Effects of short-term therapy on tumor tissue and serum markers. *Clin. Cancer Res.* **2007**, *13*, 246–252. [CrossRef]
169. He, G.; Muga, S.; Thuillier, P.; Lubet, R.A.; Fischer, S.M. The effect of PPARγ ligands on UV- or chemically-induced carcinogenesis in mouse skin. *Mol. Carcinog.* **2005**, *43*, 198–206. [CrossRef]
170. Palakurthi, S.S.; Aktas, H.; Grubissich, L.M.; Mortensen, R.M.; Halperin, J.A. Anticancer effects of thiazolidinediones are independent of peroxisome proliferator-activated receptor γ and mediated by inhibition of translation initiation. *Cancer Res.* **2001**, *61*, 6213–6218.
171. Lucarelli, E.; Sangiorgi, L.; Maini, V.; Lattanzi, G.; Marmiroli, S.; Reggiani, M.; Mordenti, M.; Alessandra Gobbi, G.; Scrimieri, F.; Zambon Bertoja, A.; et al. Troglitazione affects survival of human osteosarcoma cells. *Int. J. Cancer* **2002**, *98*, 344–351. [CrossRef] [PubMed]
172. Haydon, R.C.; Zhou, L.; Feng, T.; Breyer, B.; Cheng, H.; Jiang, W.; Ishikawa, A.; Peabody, T.; Montag, A.; Simon, M.A.; et al. Nuclear receptor agonists as potential differentiation therapy agents for human osteosarcoma. *Clin. Cancer Res.* **2002**, *8*, 1288–1294. [PubMed]
173. Srivastava, N.; Kollipara, R.K.; Singh, D.K.; Sudderth, J.; Hu, Z.; Nguyen, H.; Wang, S.; Humphries, C.G.; Carstens, R.; Huffman, K.E.; et al. Inhibition of cancer cell proliferation by PPARγ is mediated by a metabolic switch that increases reactive oxygen species levels. *Cell Metab.* **2014**, *20*, 650–661. [CrossRef] [PubMed]
174. Musicant, A.M.; Parag-Sharma, K.; Gong, W.; Sengupta, M.; Chatterjee, A.; Henry, E.C.; Tsai, Y.H.; Hayward, M.C.; Sheth, S.; Betancourt, R.; et al. CRTC1/MAML2 directs a PGC-1α-IGF-1 circuit that confers vulnerability to PPARγ inhibition. *Cell Rep.* **2021**, *34*, 108768. [CrossRef]
175. Pishvaian, M.J.; Marshall, J.L.; Wagner, A.J.; Hwang, J.J.; Malik, S.; Cotarla, I.; Deeken, J.F.; He, A.R.; Daniel, H.; Halim, A.B.; et al. A phase 1 study of efatutazone, an oral peroxisome proliferator-activated receptor gamma agonist, administered to patients with advanced malignancies. *Cancer* **2012**, *118*, 5403–5413. [CrossRef]
176. Demetri, G.D.; Fletcher, C.D.; Mueller, E.; Sarraf, P.; Naujoks, R.; Campbell, N.; Spiegelman, B.M.; Singer, S. Induction of solid tumor differentiation by the peroxisome proliferator-activated receptor-γ ligand troglitazone in patients with liposarcoma. *Proc. Natl. Acad. Sci. USA* **1999**, *96*, 3951–3956. [CrossRef]
177. Wagner, N.; Wagner, K.D. PPARs and Angiogenesis-Implications in Pathology. *Int. J. Mol. Sci.* **2020**, *21*, 5723. [CrossRef]
178. Jiao, H.L.; Zhao, B.L. Cytotoxic effect of peroxisome proliferator fenofibrate on human HepG2 hepatoma cell line and relevant mechanisms. *Toxicol. Appl. Pharmacol.* **2002**, *185*, 172–179. [CrossRef]
179. Gao, J.; Liu, Q.; Xu, Y.; Gong, X.; Zhang, R.; Zhou, C.; Su, Z.; Jin, J.; Shi, H.; Shi, J.; et al. PPARα induces cell apoptosis by destructing Bcl2. *Oncotarget* **2015**, *6*, 44635–44642. [CrossRef]
180. Holland, C.M.; Saidi, S.A.; Evans, A.L.; Sharkey, A.M.; Latimer, J.A.; Crawford, R.A.; Charnock-Jones, D.S.; Print, C.G.; Smith, S.K. Transcriptome analysis of endometrial cancer identifies peroxisome proliferator-activated receptors as potential therapeutic targets. *Mol. Cancer Ther.* **2004**, *3*, 993–1001. [CrossRef]
181. Crowe, D.L.; Chandraratna, R.A. A retinoid X receptor (RXR)-selective retinoid reveals that RXR-α is potentially a therapeutic target in breast cancer cell lines, and that it potentiates antiproliferative and apoptotic responses to peroxisome proliferator-activated receptor ligands. *Breast Cancer Res.* **2004**, *6*, R546–R555. [CrossRef] [PubMed]
182. Strakova, N.; Ehrmann, J.; Bartos, J.; Malikova, J.; Dolezel, J.; Kolar, Z. Peroxisome proliferator-activated receptors (PPAR) agonists affect cell viability, apoptosis and expression of cell cycle related proteins in cell lines of glial brain tumors. *Neoplasma* **2005**, *52*, 126–136. [PubMed]
183. Martinasso, G.; Oraldi, M.; Trombetta, A.; Maggiora, M.; Bertetto, O.; Canuto, R.A.; Muzio, G. Involvement of PPARs in Cell Proliferation and Apoptosis in Human Colon Cancer Specimens and in Normal and Cancer Cell Lines. *PPAR Res.* **2007**, *2007*, 93416. [CrossRef]
184. Xue, J.; Zhu, W.; Song, J.; Jiao, Y.; Luo, J.; Yu, C.; Zhou, J.; Wu, J.; Chen, M.; Ding, W.Q.; et al. Activation of PPARα by clofibrate sensitizes pancreatic cancer cells to radiation through the Wnt/β-catenin pathway. *Oncogene* **2018**, *37*, 953–962. [CrossRef] [PubMed]
185. Wang, M.S.; Han, Q.S.; Jia, Ż.R.; Chen, C.S.; Qiao, C.; Liu, Q.Q.; Zhang, Y.M.; Wang, K.W.; Wang, J.; Xiao, K.; et al. PPARα agonist fenofibrate relieves acquired resistance to gefitinib in non-small cell lung cancer by promoting apoptosis via PPARα/AMPK/AKT/FoxO1 pathway. *Acta Pharmacol. Sin.* **2022**, *43*, 167–176. [CrossRef]
186. Maggiora, M.; Bologna, M.; Cerù, M.P.; Possati, L.; Angelucci, A.; Cimini, A.; Miglietta, A.; Bozzo, F.; Margiotta, C.; Muzio, G.; et al. An overview of the effect of linoleic and conjugated-linoleic acids on the growth of several human tumor cell lines. *Int. J. Cancer* **2004**, *112*, 909–919. [CrossRef]

187. Tuller, E.R.; Brock, A.L.; Yu, H.; Lou, J.R.; Benbrook, D.M.; Ding, W.Q. PPARα signaling mediates the synergistic cytotoxicity of clioquinol and docosahexaenoic acid in human cancer cells. *Biochem. Pharmacol.* **2009**, *77*, 1480–1486. [CrossRef]
188. Zang, C.; Liu, H.; Bertz, J.; Possinger, K.; Koeffler, H.P.; Elstner, E.; Eucker, J. Induction of endoplasmic reticulum stress response by TZD18, a novel dual ligand for peroxisome proliferator-activated receptor α/γ, in human breast cancer cells. *Mol. Cancer Ther.* **2009**, *8*, 2296–2307. [CrossRef]
189. Ma, Y.; Wang, B.; Li, L.; Wang, F.; Xia, X. The administration of peroxisome proliferator-activated receptors α/γ agonist TZD18 inhibits cell growth and induces apoptosis in human gastric cancer cell lines. *J. Cancer Res. Ther.* **2019**, *15*, 120–125. [CrossRef]
190. Zak, Z.; Gelebart, P.; Lai, R. Fenofibrate induces effective apoptosis in mantle cell lymphoma by inhibiting the TNFα/NF-κB signaling axis. *Leukemia* **2010**, *24*, 1476–1486. [CrossRef]
191. Deepa, P.R.; Vandhana, S.; Krishnakumar, S. Fatty acid synthase inhibition induces differential expression of genes involved in apoptosis and cell proliferation in ocular cancer cells. *Nutr. Cancer* **2013**, *65*, 311–316. [CrossRef] [PubMed]
192. Casella, M.L.; Parody, J.P.; Ceballos, M.P.; Quiroga, A.D.; Ronco, M.T.; Francés, D.E.; Monti, J.A.; Pisani, G.B.; Carnovale, C.E.; Carrillo, M.C.; et al. Quercetin prevents liver carcinogenesis by inducing cell cycle arrest, decreasing cell proliferation and enhancing apoptosis. *Mol. Nutr. Food Res.* **2014**, *58*, 289–300. [CrossRef] [PubMed]
193. Abu Aboud, O.; Wettersten, H.I.; Weiss, R.H. Inhibition of PPARα induces cell cycle arrest and apoptosis, and synergizes with glycolysis inhibition in kidney cancer cells. *PLoS ONE* **2013**, *8*, e71115. [CrossRef]
194. Abu Aboud, O.; Donohoe, D.; Bultman, S.; Fitch, M.; Riiff, T.; Hellerstein, M.; Weiss, R.H. PPARα inhibition modulates multiple reprogrammed metabolic pathways in kidney cancer and attenuates tumor growth. *Am. J. Physiol. Cell Physiol.* **2015**, *308*, C890–C898. [CrossRef] [PubMed]
195. Florio, R.; De Lellis, L.; di Giacomo, V.; Di Marcantonio, M.C.; Cristiano, L.; Basile, M.; Verginelli, F.; Verzilli, D.; Ammazzalorso, A.; Prasad, S.C.; et al. Effects of PPARα inhibition in head and neck paraganglioma cells. *PLoS ONE* **2017**, *12*, e0178995. [CrossRef]
196. He, T.C.; Chan, T.A.; Vogelstein, B.; Kinzler, K.W. PPARδ is an APC-regulated target of nonsteroidal anti-inflammatory drugs. *Cell* **1999**, *99*, 335–345. [CrossRef]
197. Wang, D.; Wang, H.; Shi, Q.; Katkuri, S.; Walhi, W.; Desvergne, B.; Das, S.K.; Dey, S.K.; DuBois, R.N. Prostaglandin E(2) promotes colorectal adenoma growth via transactivation of the nuclear peroxisome proliferator-activated receptor δ. *Cancer Cell* **2004**, *6*, 285–295. [CrossRef] [PubMed]
198. Cutler, N.S.; Graves-Deal, R.; LaFleur, B.J.; Gao, Z.; Boman, B.M.; Whitehead, R.H.; Terry, E.; Morrow, J.D.; Coffey, R.J. Stromal production of prostacyclin confers an antiapoptotic effect to colonic epithelial cells. *Cancer Res.* **2003**, *63*, 1748–1751. [PubMed]
199. Liou, J.Y.; Lee, S.; Ghelani, D.; Matijevic-Aleksic, N.; Wu, K.K. Protection of endothelial survival by peroxisome proliferator-activated receptor-δ mediated 14-3-3 upregulation. *Arterioscler. Thromb. Vasc. Biol.* **2006**, *26*, 1481–1487. [CrossRef] [PubMed]
200. Wang, D.; Ning, W.; Xie, D.; Guo, L.; DuBois, R.N. Peroxisome proliferator-activated receptor δ confers resistance to peroxisome proliferator-activated receptor γ-induced apoptosis in colorectal cancer cells. *Oncogene* **2012**, *31*, 1013–1023. [CrossRef] [PubMed]
201. Bell, E.; Ponthan, F.; Whitworth, C.; Westermann, F.; Thomas, H.; Redfern, C.P. Cell survival signalling through PPARδ and arachidonic acid metabolites in neuroblastoma. *PLoS ONE* **2013**, *8*, e68859. [CrossRef] [PubMed]
202. Tong-Lin Wu, T.; Tong, Y.C.; Chen, I.H.; Niu, H.S.; Li, Y.; Cheng, J.T. Induction of apoptosis in prostate cancer by ginsenoside Rh2. *Oncotarget* **2018**, *9*, 11109–11118. [CrossRef] [PubMed]
203. Wu, T.T.; Niu, H.S.; Chen, L.J.; Cheng, J.T.; Tong, Y.C. Increase of human prostate cancer cell (DU145) apoptosis by telmisartan through PPAR-delta pathway. *Eur. J. Pharmacol.* **2016**, *775*, 35–42. [CrossRef] [PubMed]
204. Péchery, A.; Fauconnet, S.; Bittard, H.; Lascombe, I. Apoptotic effect of the selective PPARβ/δ agonist GW501516 in invasive bladder cancer cells. *Tumour Biol.* **2016**, *37*, 14789–14802. [CrossRef] [PubMed]
205. Shen, B.; Li, A.; Wan, Y.Y.; Shen, G.; Zhu, J.; Nie, Y. Lack of PPAR. *Biomed Res. Int.* **2020**, *2020*, 9563851. [CrossRef] [PubMed]
206. Padilla, J.; Kaur, K.; Harris, S.G.; Phipps, R.P. PPAR-γ-mediated regulation of normal and malignant B lineage cells. *Ann. N. Y. Acad. Sci.* **2000**, *905*, 97–109. [CrossRef] [PubMed]
207. Padilla, J.; Kaur, K.; Cao, H.J.; Smith, T.J.; Phipps, R.P. Peroxisome proliferator activator receptor-γ agonists and 15-deoxy-$\Delta^{12,1412,14}$-PGJ_2 induce apoptosis in normal and malignant B-lineage cells. *J. Immunol.* **2000**, *165*, 6941–6948. [CrossRef]
208. Piva, R.; Gianferretti, P.; Ciucci, A.; Taulli, R.; Belardo, G.; Santoro, M.G. 15-Deoxy-$\Delta^{12,14}$-prostaglandin J_2 induces apoptosis in human malignant B cells: An effect associated with inhibition of NF-κB activity and down-regulation of antiapoptotic proteins. *Blood* **2005**, *105*, 1750–1758. [CrossRef]
209. Tsao, T.; Kornblau, S.; Safe, S.; Watt, J.C.; Ruvolo, V.; Chen, W.; Qiu, Y.; Coombes, K.R.; Ju, Z.; Abdelrahim, M.; et al. Role of peroxisome proliferator-activated receptor-γ and its coactivator DRIP205 in cellular responses to CDDO (RTA-401) in acute myelogenous leukemia. *Cancer Res.* **2010**, *70*, 4949–4960. [CrossRef] [PubMed]
210. Clay, C.E.; Monjazeb, A.; Thorburn, J.; Chilton, F.H.; High, K.P. 15-Deoxy-$\Delta^{12,14}$-prostaglandin J_2-induced apoptosis does not require PPARγ in breast cancer cells. *J. Lipid Res.* **2002**, *43*, 1818–1828. [CrossRef] [PubMed]
211. Chaffer, C.L.; Thomas, D.M.; Thompson, E.W.; Williams, E.D. PPARγ-independent induction of growth arrest and apoptosis in prostate and bladder carcinoma. *BMC Cancer* **2006**, *6*, 53. [CrossRef] [PubMed]
212. Kondoh, K.; Tsuji, N.; Asanuma, K.; Kobayashi, D.; Watanabe, N. Inhibition of estrogen receptor β-mediated human telomerase reverse transcriptase gene transcription via the suppression of mitogen-activated protein kinase signaling plays an important role in 15-deoxy-$\Delta^{12,14}$-prostaglandin J_2-induced apoptosis in cancer cells. *Exp. Cell Res.* **2007**, *313*, 3486–3496. [CrossRef] [PubMed]

213. Lu, M.; Kwan, T.; Yu, C.; Chen, F.; Freedman, B.; Schafer, J.M.; Lee, E.J.; Jameson, J.L.; Jordan, V.C.; Cryns, V.L. Peroxisome proliferator-activated receptor γ agonists promote TRAIL-induced apoptosis by reducing survivin levels via cyclin D3 repression and cell cycle arrest. *J. Biol. Chem.* **2005**, *280*, 6742–6751. [CrossRef] [PubMed]
214. Bräutigam, K.; Biernath-Wüpping, J.; Bauerschlag, D.O.; von Kaisenberg, C.S.; Jonat, W.; Maass, N.; Arnold, N.; Meinhold-Heerlein, I. Combined treatment with TRAIL and PPARγ ligands overcomes chemoresistance of ovarian cancer cell lines. *J. Cancer Res. Clin. Oncol.* **2011**, *137*, 875–886. [CrossRef] [PubMed]
215. Bonofiglio, D.; Cione, E.; Qi, H.; Pingitore, A.; Perri, M.; Catalano, S.; Vizza, D.; Panno, M.L.; Genchi, G.; Fuqua, S.A.; et al. Combined low doses of PPARγ and RXR ligands trigger an intrinsic apoptotic pathway in human breast cancer cells. *Am. J. Pathol.* **2009**, *175*, 1270–1280. [CrossRef] [PubMed]
216. Kazberuk, A.; Chalecka, M.; Palka, J.; Surazynski, A. Nonsteroidal Anti-Inflammatory Drugs as PPARγ Agonists Can Induce PRODH/POX-Dependent Apoptosis in Breast Cancer Cells: New Alternative Pathway in NSAID-Induced Apoptosis. *Int. J. Mol. Sci.* **2022**, *23*, 1510. [CrossRef]
217. Guan, Y.F.; Zhang, Y.H.; Breyer, R.M.; Davis, L.; Breyer, M.D. Expression of peroxisome proliferator-activated receptor γ (PPARγ) in human transitional bladder cancer and its role in inducing cell death. *Neoplasia* **1999**, *1*, 330–339. [CrossRef]
218. Lu, J.; Imamura, K.; Nomura, S.; Mafune, K.; Nakajima, A.; Kadowaki, T.; Kubota, N.; Terauchi, Y.; Ishii, G.; Ochiai, A.; et al. Chemopreventive effect of peroxisome proliferator-activated receptor γ on gastric carcinogenesis in mice. *Cancer Res.* **2005**, *65*, 4769–4774. [CrossRef]
219. Tsubouchi, Y.; Sano, H.; Kawahito, Y.; Mukai, S.; Yamada, R.; Kohno, M.; Inoue, K.; Hla, T.; Kondo, M. Inhibition of human lung cancer cell growth by the peroxisome proliferator-activated receptor-γ agonists through induction of apoptosis. *Biochem. Biophys. Res. Commun.* **2000**, *270*, 400–405. [CrossRef]
220. Takashima, T.; Fujiwara, Y.; Higuchi, K.; Arakawa, T.; Yano, Y.; Hasuma, T.; Otani, S. PPAR-γ ligands inhibit growth of human esophageal adenocarcinoma cells through induction of apoptosis, cell cycle arrest and reduction of ornithine decarboxylase activity. *Int. J. Oncol.* **2001**, *19*, 465–471. [CrossRef]
221. Eibl, G.; Wente, M.N.; Reber, H.A.; Hines, O.J. Peroxisome proliferator-activated receptor γ induces pancreatic cancer cell apoptosis. *Biochem. Biophys. Res. Commun.* **2001**, *287*, 522–529. [CrossRef] [PubMed]
222. Li, M.Y.; Deng, H.; Zhao, J.M.; Dai, D.; Tan, X.Y. PPARγ pathway activation results in apoptosis and COX-2 inhibition in HepG2 cells. *World J. Gastroenterol.* **2003**, *9*, 1220–1226. [CrossRef] [PubMed]
223. Kim, E.J.; Park, K.S.; Chung, S.Y.; Sheen, Y.Y.; Moon, D.C.; Song, Y.S.; Kim, K.S.; Song, S.; Yun, Y.P.; Lee, M.K.; et al. Peroxisome proliferator-activated receptor-γ activator 15-deoxy-$\Delta^{12,14}$-prostaglandin J_2 inhibits neuroblastoma cell growth through induction of apoptosis: Association with extracellular signal-regulated kinase signal pathway. *J. Pharmacol. Exp. Ther.* **2003**, *307*, 505–517. [CrossRef] [PubMed]
224. Strakova, N.; Ehrmann, J.; Dzubak, P.; Bouchal, J.; Kolar, Z. The synthetic ligand of peroxisome proliferator-activated receptor-γ ciglitazone affects human glioblastoma cell lines. *J. Pharmacol. Exp. Ther.* **2004**, *309*, 1239–1247. [CrossRef]
225. Konopleva, M.; Elstner, E.; McQueen, T.J.; Tsao, T.; Sudarikov, A.; Hu, W.; Schober, W.D.; Wang, R.Y.; Chism, D.; Kornblau, S.M.; et al. Peroxisome proliferator-activated receptor γ and retinoid X receptor ligands are potent inducers of differentiation and apoptosis in leukemias. *Mol. Cancer Ther.* **2004**, *3*, 1249–1262. [CrossRef]
226. Nam, D.H.; Ramachandran, S.; Song, D.K.; Kwon, K.Y.; Jeon, D.S.; Shin, S.J.; Kwon, S.H.; Cha, S.D.; Bae, I.; Cho, C.H. Growth inhibition and apoptosis induced in human leiomyoma cells by treatment with the PPAR gamma ligand ciglitizone. *Mol. Hum. Reprod.* **2007**, *13*, 829–836. [CrossRef]
227. Shimada, T.; Kojima, K.; Yoshiura, K.; Hiraishi, H.; Terano, A. Characteristics of the peroxisome proliferator activated receptor γ (PPARγ) ligand induced apoptosis in colon cancer cells. *Gut* **2002**, *50*, 658–664. [CrossRef]
228. Hernandez-Quiles, M.; Broekema, M.F.; Kalkhoven, E. PPARgamma in Metabolism, Immunity, and Cancer: Unified and Diverse Mechanisms of Action. *Front. Endocrinol.* **2021**, *12*, 624112. [CrossRef]
229. Kim, H.J.; Hwang, J.Y.; Choi, W.S.; Lee, J.H.; Chang, K.C.; Nishinaka, T.; Yabe-Nishimura, C.; Seo, H.G. Expression of a peroxisome proliferator-activated receptor γ 1 splice variant that was identified in human lung cancers suppresses cell death induced by cisplatin and oxidative stress. *Clin. Cancer Res.* **2007**, *13*, 2577–2583. [CrossRef]
230. Sun, H.; Berquin, I.M.; Owens, R.T.; O'Flaherty, J.T.; Edwards, I.J. Peroxisome proliferator-activated receptor γ-mediated up-regulation of syndecan-1 by n-3 fatty acids promotes apoptosis of human breast cancer cells. *Cancer Res.* **2008**, *68*, 2912–2919. [CrossRef]
231. Baek, S.J.; Wilson, L.C.; Hsi, L.C.; Eling, T.E. Troglitazone, a peroxisome proliferator-activated receptor γ (PPAR γ) ligand, selectively induces the early growth response-1 gene independently of PPAR γ. A novel mechanism for its anti-tumorigenic activity. *J. Biol. Chem.* **2003**, *278*, 5845–5853. [CrossRef] [PubMed]
232. Funao, K.; Matsuyama, M.; Kawahito, Y.; Sano, H.; Chargui, J.; Touraine, J.L.; Nakatani, T.; Yoshimura, R. Telmisartan is a potent target for prevention and treatment in human prostate cancer. *Oncol. Rep.* **2008**, *20*, 295–300. [PubMed]
233. Funao, K.; Matsuyama, M.; Kawahito, Y.; Sano, H.; Chargui, J.; Touraine, J.L.; Nakatani, T.; Yoshimura, R. Telmisartan as a peroxisome proliferator-activated receptor-γ ligand is a new target in the treatment of human renal cell carcinoma. *Mol. Med. Rep.* **2009**, *2*, 193–198. [CrossRef]

234. Matsuyama, M.; Funao, K.; Kuratsukuri, K.; Tanaka, T.; Kawahito, Y.; Sano, H.; Chargui, J.; Touraine, J.L.; Yoshimura, N.; Yoshimura, R. Telmisartan inhibits human urological cancer cell growth through early apoptosis. *Exp. Ther. Med.* **2010**, *1*, 301–306. [CrossRef]
235. Zaytseva, Y.Y.; Wang, X.; Southard, R.C.; Wallis, N.K.; Kilgore, M.W. Down-regulation of PPARgamma1 suppresses cell growth and induces apoptosis in MCF-7 breast cancer cells. *Mol. Cancer* **2008**, *7*, 90. [CrossRef] [PubMed]
236. Khandekar, M.J.; Banks, A.S.; Laznik-Bogoslavski, D.; White, J.P.; Choi, J.H.; Kazak, L.; Lo, J.C.; Cohen, P.; Wong, K.K.; Kamenecka, T.M.; et al. Noncanonical agonist PPARγ ligands modulate the response to DNA damage and sensitize cancer cells to cytotoxic chemotherapy. *Proc. Natl. Acad. Sci. USA* **2018**, *115*, 561–566. [CrossRef] [PubMed]
237. Schaefer, K.L.; Wada, K.; Takahashi, H.; Matsuhashi, N.; Ohnishi, S.; Wolfe, M.M.; Turner, J.R.; Nakajima, A.; Borkan, S.C.; Saubermann, L.J. Peroxisome proliferator-activated receptor γ inhibition prevents adhesion to the extracellular matrix and induces anoikis in hepatocellular carcinoma cells. *Cancer Res.* **2005**, *65*, 2251–2259. [CrossRef]
238. Masuda, T.; Wada, K.; Nakajima, A.; Okura, M.; Kudo, C.; Kadowaki, T.; Kogo, M.; Kamisaki, Y. Critical role of peroxisome proliferator-activated receptor γ on anoikis and invasion of squamous cell carcinoma. *Clin. Cancer Res.* **2005**, *11*, 4012–4021. [CrossRef]
239. Cerquetti, L.; Sampaoli, C.; Amendola, D.; Bucci, B.; Masuelli, L.; Marchese, R.; Misiti, S.; De Venanzi, A.; Poggi, M.; Toscano, V.; et al. Rosiglitazone induces autophagy in H295R and cell cycle deregulation in SW13 adrenocortical cancer cells. *Exp. Cell Res.* **2011**, *317*, 1397–1410. [CrossRef]
240. Rovito, D.; Giordano, C.; Vizza, D.; Plastina, P.; Barone, I.; Casaburi, I.; Lanzino, M.; De Amicis, F.; Sisci, D.; Mauro, L.; et al. Omega-3 PUFA ethanolamides DHEA and EPEA induce autophagy through PPARγ activation in MCF-7 breast cancer cells. *J. Cell. Physiol.* **2013**, *228*, 1314–1322. [CrossRef]
241. To, K.K.W.; Wu, W.K.K.; Loong, H.H.F. PPARgamma agonists sensitize PTEN-deficient resistant lung cancer cells to EGFR tyrosine kinase inhibitors by inducing autophagy. *Eur. J. Pharmacol.* **2018**, *823*, 19–26. [CrossRef] [PubMed]
242. Yun, C.W.; Lee, S.H. The Roles of Autophagy in Cancer. *Int. J. Mol. Sci.* **2018**, *19*, 3466. [CrossRef] [PubMed]
243. Baron, D.M.; Kaindl, U.; Haudek-Prinz, V.J.; Bayer, E.; Röhrl, C.; Gerner, C.; Marian, B. Autonomous inhibition of apoptosis correlates with responsiveness of colon carcinoma cell lines to ciglitazone. *PLoS ONE* **2014**, *9*, e114158. [CrossRef] [PubMed]
244. Nijsten, T.; Geluyckens, E.; Colpaert, C.; Lambert, J. Peroxisome proliferator-activated receptors in squamous cell carcinoma and its precursors. *J. Cutan. Pathol.* **2005**, *32*, 340–347. [CrossRef] [PubMed]
245. Panigrahy, D.; Kaipainen, A.; Huang, S.; Butterfield, C.E.; Barnés, C.M.; Fannon, M.; Laforme, A.M.; Chaponis, D.M.; Folkman, J.; Kieran, M.W. PPARα agonist fenofibrate suppresses tumor growth through direct and indirect angiogenesis inhibition. *Proc. Natl. Acad. Sci. USA* **2008**, *105*, 985–990. [CrossRef] [PubMed]
246. Pozzi, A.; Popescu, V.; Yang, S.; Mei, S.; Shi, M.; Puolitaival, S.M.; Caprioli, R.M.; Capdevila, J.H. The anti-tumorigenic properties of peroxisomal proliferator-activated receptor α are arachidonic acid epoxygenase-mediated. *J. Biol. Chem.* **2010**, *285*, 12840–12850. [CrossRef]
247. Wu, L.; Wang, W.; Dai, M.; Li, H.; Chen, C.; Wang, D. PPARα ligand, AVE8134, and cyclooxygenase inhibitor therapy synergistically suppress lung cancer growth and metastasis. *BMC Cancer* **2019**, *19*, 1166. [CrossRef]
248. Garrido-Urbani, S.; Jemelin, S.; Deffert, C.; Carnesecchi, S.; Basset, O.; Szyndralewiez, C.; Heitz, F.; Page, P.; Montet, X.; Michalik, L.; et al. Targeting vascular NADPH oxidase 1 blocks tumor angiogenesis through a PPARα mediated mechanism. *PLoS ONE* **2011**, *6*, e14665. [CrossRef]
249. Abdollahi, A.; Schwager, C.; Kleeff, J.; Esposito, I.; Domhan, S.; Peschke, P.; Hauser, K.; Hahnfeldt, P.; Hlatky, L.; Debus, J.; et al. Transcriptional network governing the angiogenic switch in human pancreatic cancer. *Proc. Natl. Acad. Sci. USA* **2007**, *104*, 12890–12895. [CrossRef] [PubMed]
250. Yoshinaga, M.; Kitamura, Y.; Chaen, T.; Yamashita, S.; Tsuruta, S.; Hisano, T.; Ikeda, Y.; Sakai, H.; Nakamura, K.; Takayanagi, R.; et al. The simultaneous expression of peroxisome proliferator-activated receptor Delta and cyclooxygenase-2 may enhance angiogenesis and tumor venous invasion in tissues of colorectal cancers. *Dig. Dis. Sci.* **2009**, *54*, 1108–1114. [CrossRef]
251. Müller-Brüsselbach, S.; Kömhoff, M.; Rieck, M.; Meissner, W.; Kaddatz, K.; Adamkiewicz, J.; Keil, B.; Klose, K.J.; Moll, R.; Burdick, A.D.; et al. Deregulation of tumor angiogenesis and blockade of tumor growth in PPARβ-deficient mice. *EMBO J.* **2007**, *26*, 3686–3698. [CrossRef] [PubMed]
252. Zuo, X.; Xu, W.; Xu, M.; Tian, R.; Moussalli, M.J.; Mao, F.; Zheng, X.; Wang, J.; Morris, J.S.; Gagea, M.; et al. Metastasis regulation by PPARD expression in cancer cells. *JCI Insight* **2017**, *2*, e91419. [CrossRef] [PubMed]
253. Meissner, M.; Hrgovic, I.; Doll, M.; Naidenow, J.; Reichenbach, G.; Hailemariam-Jahn, T.; Michailidou, D.; Gille, J.; Kaufmann, R. Peroxisome proliferator-activated receptor δ activators induce IL-8 expression in nonstimulated endothelial cells in a transcriptional and posttranscriptional manner. *J. Biol. Chem.* **2010**, *285*, 33797–33804. [CrossRef] [PubMed]
254. Piqueras, L.; Reynolds, A.R.; Hodivala-Dilke, K.M.; Alfranca, A.; Redondo, J.M.; Hatae, T.; Tanabe, T.; Warner, T.D.; Bishop-Bailey, D. Activation of PPARβ/δ induces endothelial cell proliferation and angiogenesis. *Arterioscler. Thromb. Vasc. Biol.* **2007**, *27*, 63–69. [CrossRef]
255. Wagner, K.D.; Vukolic, A.; Baudouy, D.; Michiels, J.F.; Wagner, N. Inducible Conditional Vascular-Specific Overexpression of Peroxisome Proliferator-Activated Receptor Beta/Delta Leads to Rapid Cardiac Hypertrophy. *PPAR Res.* **2016**, *2016*, 7631085. [CrossRef]

256. Zhang, J.; Yang, W.; Zhao, D.; Han, Y.; Liu, B.; Zhao, H.; Wang, H.; Zhang, Q.; Xu, G. Correlation between TSP-1, TGF-β and PPAR-γ expression levels and glioma microvascular density. *Oncol. Lett.* **2014**, *7*, 95–100. [CrossRef]
257. Panigrahy, D.; Singer, S.; Shen, L.Q.; Butterfield, C.E.; Freedman, D.A.; Chen, E.J.; Moses, M.A.; Kilroy, S.; Duensing, S.; Fletcher, C.; et al. PPARγ ligands inhibit primary tumor growth and metastasis by inhibiting angiogenesis. *J. Clin. Investig.* **2002**, *110*, 923–932. [CrossRef]
258. Huang, H.; Campbell, S.C.; Bedford, D.F.; Nelius, T.; Veliceasa, D.; Shroff, E.H.; Henkin, J.; Schneider, A.; Bouck, N.; Volpert, O.V. Peroxisome proliferator-activated receptor γ ligands improve the antitumor efficacy of thrombospondin peptide ABT510. *Mol. Cancer Res.* **2004**, *2*, 541–550. [CrossRef]
259. Keshamouni, V.G.; Arenberg, D.A.; Reddy, R.C.; Newstead, M.J.; Anthwal, S.; Standiford, T.J. PPAR-γ activation inhibits angiogenesis by blocking ELR+CXC chemokine production in non-small cell lung cancer. *Neoplasia* **2005**, *7*, 294–301. [CrossRef]
260. Copland, J.A.; Marlow, L.A.; Kurakata, S.; Fujiwara, K.; Wong, A.K.; Kreinest, P.A.; Williams, S.F.; Haugen, B.R.; Klopper, J.P.; Smallridge, R.C. Novel high-affinity PPARγ agonist alone and in combination with paclitaxel inhibits human anaplastic thyroid carcinoma tumor growth via p21WAF1/CIP1. *Oncogene* **2006**, *25*, 2304–2317. [CrossRef]
261. Xin, B.; Yokoyama, Y.; Shigeto, T.; Futagami, M.; Mizunuma, H. Inhibitory effect of meloxicam, a selective cyclooxygenase-2 inhibitor, and ciglitazone, a peroxisome proliferator-activated receptor gamma ligand, on the growth of human ovarian cancers. *Cancer* **2007**, *110*, 791–800. [CrossRef] [PubMed]
262. Yokoyama, Y.; Xin, B.; Shigeto, T.; Mizunuma, H. Combination of ciglitazone, a peroxisome proliferator-activated receptor γ ligand, and cisplatin enhances the inhibition of growth of human ovarian cancers. *J. Cancer Res. Clin. Oncol.* **2011**, *137*, 1219–1228. [CrossRef] [PubMed]
263. Dong, Y.W.; Wang, X.P.; Wu, K. Suppression of pancreatic carcinoma growth by activating peroxisome proliferator-activated receptor γ involves angiogenesis inhibition. *World J. Gastroenterol.* **2009**, *15*, 441–448. [CrossRef] [PubMed]
264. Huang, G.; Yin, L.; Lan, J.; Tong, R.; Li, M.; Na, F.; Mo, X.; Chen, C.; Xue, J.; Lu, Y. Synergy between peroxisome proliferator-activated receptor γ agonist and radiotherapy in cancer. *Cancer Sci.* **2018**, *109*, 2243–2255. [CrossRef]
265. Kramer, K.; Wu, J.; Crowe, D.L. Tumor suppressor control of the cancer stem cell niche. *Oncogene* **2016**, *35*, 4165–4178. [CrossRef]
266. Tian, L.; Zhou, J.; Casimiro, M.C.; Liang, B.; Ojeifo, J.O.; Wang, M.; Hyslop, T.; Wang, C.; Pestell, R.G. Activating peroxisome proliferator-activated receptor γ mutant promotes tumor growth in vivo by enhancing angiogenesis. *Cancer Res.* **2009**, *69*, 9236–9244. [CrossRef]
267. Pich, C.; Meylan, P.; Mastelic-Gavillet, B.; Nguyen, T.N.; Loyon, R.; Trang, B.K.; Moser, H.; Moret, C.; Goepfert, C.; Hafner, J.; et al. Induction of Paracrine Signaling in Metastatic Melanoma Cells by PPARγ Agonist Rosiglitazone Activates Stromal Cells and Enhances Tumor Growth. *Cancer Res.* **2018**, *78*, 6447–6461. [CrossRef]
268. Mitchell, J.A.; Bishop-Bailey, D. PPARβ/δ a potential target in pulmonary hypertension blighted by cancer risk. *Pulm. Circ.* **2019**, *9*, 1–2. [CrossRef]
269. Cai, Y.; Liu, H.; Song, E.; Wang, L.; Xu, J.; He, Y.; Zhang, D.; Zhang, L.; Cheng, K.K.; Jin, L.; et al. Deficiency of telomere-associated repressor activator protein 1 precipitates cardiac aging in mice. *Theranostics* **2021**, *11*, 4710–4727. [CrossRef]
270. Di Leo, L.; Vegliante, R.; Ciccarone, F.; Salvatori, I.; Scimeca, M.; Bonanno, E.; Sagnotta, A.; Grazi, G.L.; Aquilano, K.; Ciriolo, M.R. Forcing ATGL expression in hepatocarcinoma cells imposes glycolytic rewiring through PPAR-α/p300-mediated acetylation of p53. *Oncogene* **2019**, *38*, 1860–1875. [CrossRef] [PubMed]
271. Zhang, Q.; Zhang, Y.; Sun, S.; Wang, K.; Qian, J.; Cui, Z.; Tao, T.; Zhou, J. ACOX2 is a prognostic marker and impedes the progression of hepatocellular carcinoma via PPARα pathway. *Cell Death Dis.* **2021**, *12*, 15. [CrossRef] [PubMed]
272. Luo, X.; Zhong, L.; Yu, L.; Xiong, L.; Dan, W.; Li, J.; Ye, J.; Chu, X.; Liu, C.; Liu, B. TRIB3 destabilizes tumor suppressor PPARα expression through ubiquitin-mediated proteasome degradation in acute myeloid leukemia. *Life Sci.* **2020**, *257*, 118021. [CrossRef] [PubMed]
273. Luo, Y.; Xie, C.; Brocker, C.N.; Fan, J.; Wu, X.; Feng, L.; Wang, Q.; Zhao, J.; Lu, D.; Tandon, M.; et al. Intestinal PPARα Protects Against Colon Carcinogenesis via Regulation of Methyltransferases DNMT1 and PRMT6. *Gastroenterology* **2019**, *157*, 744–759.e744. [CrossRef]
274. Lopez-Guadamillas, E.; Fernandez-Marcos, P.J.; Pantoja, C.; Muñoz-Martin, M.; Martínez, D.; Gómez-López, G.; Campos-Olivas, R.; Valverde, A.M.; Serrano, M. p21^{Cip1} plays a critical role in the physiological adaptation to fasting through activation of PPARα. *Sci. Rep.* **2016**, *6*, 34542. [CrossRef] [PubMed]
275. Pozzi, S.; Boergesen, M.; Sinha, S.; Mandrup, S.; Mantovani, R. Peroxisome proliferator-activated receptor-α is a functional target of p63 in adult human keratinocytes. *J. Investig. Dermatol.* **2009**, *129*, 2376–2385. [CrossRef]
276. Gizard, F.; Amant, C.; Barbier, O.; Bellosta, S.; Robillard, R.; Percevault, F.; Sevestre, H.; Krimpenfort, P.; Corsini, A.; Rochette, J.; et al. PPAR α inhibits vascular smooth muscle cell proliferation underlying intimal hyperplasia by inducing the tumor suppressor p16INK4a. *J. Clin. Investig.* **2005**, *115*, 3228–3238. [CrossRef] [PubMed]
277. Sherr, C.J. Cancer cell cycles. *Science* **1996**, *274*, 1672–1677. [CrossRef] [PubMed]
278. Yamasaki, D.; Kawabe, N.; Nakamura, H.; Tachibana, K.; Ishimoto, K.; Tanaka, T.; Aburatani, H.; Sakai, J.; Hamakubo, T.; Kodama, T.; et al. Fenofibrate suppresses growth of the human hepatocellular carcinoma cell via PPARα-independent mechanisms. *Eur. J. Cell Biol.* **2011**, *90*, 657–664. [CrossRef]

279. Hann, S.S.; Zheng, F.; Zhao, S. Targeting 3-phosphoinositide-dependent protein kinase 1 by N-acetyl-cysteine through activation of peroxisome proliferators activated receptor alpha in human lung cancer cells, the role of p53 and p65. *J. Exp. Clin. Cancer Res.* **2013**, *32*, 43. [CrossRef]
280. Zhou, J.; Zhang, S.; Xue, J.; Avery, J.; Wu, J.; Lind, S.E.; Ding, W.Q. Activation of peroxisome proliferator-activated receptor α (PPARα) suppresses hypoxia-inducible factor-1α (HIF-1α) signaling in cancer cells. *J. Biol. Chem.* **2012**, *287*, 35161–35169. [CrossRef]
281. Cheng, P.; Yang, S.S.; Hu, X.G.; Zhou, X.Y.; Zhang, Y.J.; Jin, G.; Zhou, Y.Q. Menin prevents liver steatosis through co-activation of peroxisome proliferator-activated receptor alpha. *FEBS Lett.* **2011**, *585*, 3403–3408. [CrossRef] [PubMed]
282. Liu, Y.; Deguchi, Y.; Wei, D.; Liu, F.; Moussalli, M.J.; Deguchi, E.; Li, D.; Wang, H.; Valentin, L.A.; Colby, J.K.; et al. Rapid acceleration of KRAS-mutant pancreatic carcinogenesis via remodeling of tumor immune microenvironment by PPARδ. *Nat. Commun.* **2022**, *13*, 2665. [CrossRef] [PubMed]
283. Bapat, S.P.; Whitty, C.; Mowery, C.T.; Liang, Y.; Yoo, A.; Jiang, Z.; Peters, M.C.; Zhang, L.J.; Vogel, I.; Zhou, C.; et al. Obesity alters pathology and treatment response in inflammatory disease. *Nature* **2022**, *604*, 337–342. [CrossRef] [PubMed]
284. Shao, J.; Sheng, H.; DuBois, R.N. Peroxisome proliferator-activated receptors modulate K-Ras-mediated transformation of intestinal epithelial cells. *Cancer Res.* **2002**, *62*, 3282–3288.
285. Li, Z.; Li, H.; Zhao, Z.B.; Zhu, W.; Feng, P.P.; Zhu, X.W.; Gong, J.P. SIRT4 silencing in tumor-associated macrophages promotes HCC development via PPARδ signalling-mediated alternative activation of macrophages. *J. Exp. Clin. Cancer Res.* **2019**, *38*, 469. [CrossRef]
286. Rangel-Sánchez, I.Y.; Salas-Treviño, D.; Soto-Domínguez, A.; Garza-Rodríguez, O.I.; Saucedo-Cárdenas, O.; Zapata-Benavides, P.; Zarate-Ramos, J.J.; Cedillo-Rosales, S.; Zamora-Ávila, D.E. Expression of the Wilms' tumour gene and its association with PPARβ/δ in healthy skin and melanoma of horses. *Acta Vet. Hung.* **2021**, *68*, 374–379. [CrossRef]
287. Wagner, K.D.; El Maï, M.; Ladomery, M.; Belali, T.; Leccia, N.; Michiels, J.F.; Wagner, N. Altered VEGF Splicing Isoform Balance in Tumor Endothelium Involves Activation of Splicing Factors Srpk1 and Srsf1 by the Wilms' Tumor Suppressor Wt1. *Cells* **2019**, *8*, 41. [CrossRef]
288. El Maï, M.; Wagner, K.D.; Michiels, J.F.; Ambrosetti, D.; Borderie, A.; Destree, S.; Renault, V.; Djerbi, N.; Giraud-Panis, M.J.; Gilson, E.; et al. The Telomeric Protein TRF2 Regulates Angiogenesis by Binding and Activating the PDGFRβ Promoter. *Cell Rep.* **2014**, *9*, 1047–1060. [CrossRef]
289. Wagner, N.; Michiels, J.F.; Schedl, A.; Wagner, K.D. The Wilms' tumour suppressor WT1 is involved in endothelial cell proliferation and migration: Expression in tumour vessels in vivo. *Oncogene* **2008**, *27*, 3662–3672. [CrossRef] [PubMed]
290. Hanada, S.; Tsuruta, T.; Haraguchi, K.; Okamoto, M.; Sugiyama, H.; Koido, S. Long-term survival of pancreatic cancer patients treated with multimodal therapy combined with WT1-targeted dendritic cell vaccines. *Hum. Vaccin. Immunother.* **2019**, *15*, 397–406. [CrossRef]
291. Sugiyama, H. WT1 (Wilms' tumor gene 1): Biology and cancer immunotherapy. *Jpn. J. Clin. Oncol.* **2010**, *40*, 377–387. [CrossRef] [PubMed]
292. Oka, Y.; Tsuboi, A.; Oji, Y.; Kawase, I.; Sugiyama, H. WT1 peptide vaccine for the treatment of cancer. *Curr. Opin. Immunol.* **2008**, *20*, 211–220. [CrossRef] [PubMed]
293. Oji, Y.; Miyoshi, S.; Maeda, H.; Hayashi, S.; Tamaki, H.; Nakatsuka, S.; Yao, M.; Takahashi, E.; Nakano, Y.; Hirabayashi, H.; et al. Overexpression of the Wilms' tumor gene WT1 in de novo lung cancers. *Int. J. Cancer* **2002**, *100*, 297–303. [CrossRef]
294. Lim, H.J.; Lee, S.; Park, J.H.; Lee, K.S.; Choi, H.E.; Chung, K.S.; Lee, H.H.; Park, H.Y. PPAR δ agonist L-165041 inhibits rat vascular smooth muscle cell proliferation and migration via inhibition of cell cycle. *Atherosclerosis* **2009**, *202*, 446–454. [CrossRef]
295. Zhai, Y.; Wu, R.; Schwartz, D.R.; Darrah, D.; Reed, H.; Kolligs, F.T.; Nieman, M.T.; Fearon, E.R.; Cho, K.R. Role of β-catenin/T-cell factor-regulated genes in ovarian endometrioid adenocarcinomas. *Am. J. Pathol.* **2002**, *160*, 1229–1238. [CrossRef]
296. Kundu, J.; Wahab, S.M.; Kundu, J.K.; Choi, Y.L.; Erkin, O.C.; Lee, H.S.; Park, S.G.; Shin, Y.K. Tob1 induces apoptosis and inhibits proliferation, migration and invasion of gastric cancer cells by activating Smad4 and inhibiting β-catenin signaling. *Int. J. Oncol.* **2012**, *41*, 839–848. [CrossRef] [PubMed]
297. Kannan-Thulasiraman, P.; Seachrist, D.D.; Mahabeleshwar, G.H.; Jain, M.K.; Noy, N. Fatty acid-binding protein 5 and PPARβ/δ are critical mediators of epidermal growth factor receptor-induced carcinoma cell growth. *J. Biol. Chem.* **2010**, *285*, 19106–19115. [CrossRef] [PubMed]
298. Pseftogas, A.; Gonidas, C.; Mosialos, G. Activation of peroxisome proliferator-activated receptor gamma in mammary epithelial cells upregulates the expression of tumor suppressor Cyld to mediate growth inhibition and anti-inflammatory effects. *Int. J. Biochem. Cell Biol.* **2017**, *82*, 49–56. [CrossRef]
299. Avasarala, S.; Bikkavilli, R.K.; Van Scoyk, M.; Zhang, W.; Lapite, A.; Hostetter, L.; Byers, J.T.; Heasley, L.E.; Sohn, J.W.; Winn, R.A. Heterotrimeric G-protein, Gα16, is a critical downstream effector of non-canonical Wnt signaling and a potent inhibitor of transformed cell growth in non small cell lung cancer. *PLoS ONE* **2013**, *8*, e76895. [CrossRef]
300. Calo, E.; Quintero-Estades, J.A.; Danielian, P.S.; Nedelcu, S.; Berman, S.D.; Lees, J.A. Rb regulates fate choice and lineage commitment in vivo. *Nature* **2010**, *466*, 1110–1114. [CrossRef]
301. Yasmeen, R.; Meyers, J.M.; Alvarez, C.E.; Thomas, J.L.; Bonnegarde-Bernard, A.; Alder, H.; Papenfuss, T.L.; Benson, D.M.; Boyaka, P.N.; Ziouzenkova, O. Aldehyde dehydrogenase-1a1 induces oncogene suppressor genes in B cell populations. *Biochim. Biophys. Acta* **2013**, *1833*, 3218–3227. [CrossRef] [PubMed]

302. Cheung, K.F.; Zhao, J.; Hao, Y.; Li, X.; Lowe, A.W.; Cheng, A.S.; Sung, J.J.; Yu, J. CITED2 is a novel direct effector of peroxisome proliferator-activated receptor γ in suppressing hepatocellular carcinoma cell growth. *Cancer* **2013**, *119*, 1217–1226. [CrossRef] [PubMed]
303. Wang, X.; Huang, G.; Mei, S.; Qian, J.; Ji, J.; Zhang, J. Over-expression of C/EBP-α induces apoptosis in cultured rat hepatic stellate cells depending on p53 and peroxisome proliferator-activated receptor-γ. *Biochem. Biophys. Res. Commun.* **2009**, *380*, 286–291. [CrossRef] [PubMed]
304. Schmidt, L.; Heyes, E.; Grebien, F. Gain-of-Function Effects of N-Terminal CEBPA Mutations in Acute Myeloid Leukemia. *Bioessays* **2020**, *42*, e1900178. [CrossRef]
305. Gery, S.; Tanosaki, S.; Bose, S.; Bose, N.; Vadgama, J.; Koeffler, H.P. Down-regulation and growth inhibitory role of C/EBPα in breast cancer. *Clin. Cancer Res.* **2005**, *11*, 3184–3190. [CrossRef] [PubMed]
306. Bonofiglio, D.; Gabriele, S.; Aquila, S.; Catalano, S.; Gentile, M.; Middea, E.; Giordano, F.; Andò, S. Estrogen receptor α binds to peroxisome proliferator-activated receptor response element and negatively interferes with peroxisome proliferator-activated receptor γ signaling in breast cancer cells. *Clin. Cancer Res.* **2005**, *11*, 6139–6147. [CrossRef] [PubMed]
307. Grau, R.; Punzón, C.; Fresno, M.; Iñiguez, M.A. Peroxisome-proliferator-activated receptor α agonists inhibit cyclo-oxygenase 2 and vascular endothelial growth factor transcriptional activation in human colorectal carcinoma cells via inhibition of activator protein-1. *Biochem. J.* **2006**, *395*, 81–88. [CrossRef] [PubMed]
308. Meissner, M.; Berlinski, B.; Gille, J.; Doll, M.; Kaufmann, R. Peroxisome proliferator activated receptor-α agonists suppress transforming growth factor-α-induced matrix metalloproteinase-9 expression in human keratinocytes. *Clin. Exp. Dermatol.* **2011**, *36*, 911–914. [CrossRef] [PubMed]
309. Grabacka, M.; Placha, W.; Plonka, P.M.; Pajak, S.; Urbanska, K.; Laidler, P.; Slominski, A. Inhibition of melanoma metastases by fenofibrate. *Arch. Dermatol. Res.* **2004**, *296*, 54–58. [CrossRef]
310. Grabacka, M.; Plonka, P.M.; Urbanska, K.; Reiss, K. Peroxisome proliferator-activated receptor α activation decreases metastatic potential of melanoma cells in vitro via down-regulation of Akt. *Clin. Cancer Res.* **2006**, *12*, 3028–3036. [CrossRef]
311. Wejksza, K.; Lee-Chang, C.; Bodogai, M.; Bonzo, J.; Gonzalez, F.J.; Lehrmann, E.; Becker, K.; Biragyn, A. Cancer-produced metabolites of 5-lipoxygenase induce tumor-evoked regulatory B cells via peroxisome proliferator-activated receptor α. *J. Immunol.* **2013**, *190*, 2575–2584. [CrossRef] [PubMed]
312. Lee, C.K.; Jeong, S.H.; Jang, C.; Bae, H.; Kim, Y.H.; Park, I.; Kim, S.K.; Koh, G.Y. Tumor metastasis to lymph nodes requires YAP-dependent metabolic adaptation. *Science* **2019**, *363*, 644–649. [CrossRef]
313. Chen, S.W.; Chou, C.T.; Chang, C.C.; Li, Y.J.; Chen, S.T.; Lin, I.C.; Kok, S.H.; Cheng, S.J.; Lee, J.J.; Wu, T.S.; et al. HMGCS2 enhances invasion and metastasis via direct interaction with PPARα to activate Src signaling in colorectal cancer and oral cancer. *Oncotarget* **2017**, *8*, 22460–22476. [CrossRef] [PubMed]
314. Leng, J.; Li, H.; Niu, Y.; Chen, K.; Yuan, X.; Chen, H.; Fu, Z.; Zhang, L.; Wang, F.; Chen, C.; et al. Low-dose mono(2-ethylhexyl) phthalate promotes ovarian cancer development through PPARα-dependent PI3K/Akt/NF-κB pathway. *Sci. Total Environ.* **2021**, *790*, 147990. [CrossRef] [PubMed]
315. Coleman, J.D.; Thompson, J.T.; Smith, R.W.; Prokopczyk, B.; Vanden Heuvel, J.P. Role of Peroxisome Proliferator-Activated Receptor β/δ and B-Cell Lymphoma-6 in Regulation of Genes Involved in Metastasis and Migration in Pancreatic Cancer Cells. *PPAR Res.* **2013**, *2013*, 121956. [CrossRef]
316. Ham, S.A.; Yoo, T.; Lee, W.J.; Hwang, J.S.; Hur, J.; Paek, K.S.; Lim, D.S.; Han, S.G.; Lee, C.H.; Seo, H.G. ADAMTS1-mediated targeting of TSP-1 by PPARδ suppresses migration and invasion of breast cancer cells. *Oncotarget* **2017**, *8*, 94091–94103. [CrossRef]
317. Lim, J.C.W.; Kwan, Y.P.; Tan, M.S.; Teo, M.H.Y.; Chiba, S.; Wahli, W.; Wang, X. The Role of PPARβ/δ in Melanoma Metastasis. *Int. J. Mol. Sci.* **2018**, *19*, 2860. [CrossRef]
318. Elie-Caille, C.; Lascombe, I.; Péchery, A.; Bittard, H.; Fauconnet, S. Molecular and nanoscale evaluation of N-cadherin expression in invasive bladder cancer cells under control conditions or GW501516 exposure. *Mol. Cell. Biochem.* **2020**, *471*, 113–127. [CrossRef]
319. Woutersen, R.A.; Appel, M.J.; van Garderen-Hoetmer, A.; Wijnands, M.V. Dietary fat and carcinogenesis. *Mutat. Res.* **1999**, *443*, 111–127. [CrossRef]
320. Wang, D.; Fu, L.; Wei, J.; Xiong, Y.; DuBois, R.N. PPARδ Mediates the Effect of Dietary Fat in Promoting Colorectal Cancer Metastasis. *Cancer Res.* **2019**, *79*, 4480–4490. [CrossRef]
321. Sunami, E.; Tsuno, N.H.; Kitayama, J.; Saito, S.; Osada, T.; Yamaguchi, H.; Tomozawa, S.; Tsuruo, T.; Shibata, Y.; Nagawa, H. Decreased synthesis of matrix metalloproteinase-7 and adhesion to the extracellular matrix proteins of human colon cancer cells treated with troglitazone. *Surg. Today* **2002**, *32*, 343–350. [CrossRef] [PubMed]
322. Yoshizumi, T.; Ohta, T.; Ninomiya, I.; Terada, I.; Fushida, S.; Fujimura, T.; Nishimura, G.; Shimizu, K.; Yi, S.; Miwa, K. Thiazolidinedione, a peroxisome proliferator-activated receptor-γ ligand, inhibits growth and metastasis of HT-29 human colon cancer cells through differentiation-promoting effects. *Int. J. Oncol.* **2004**, *25*, 631–639. [CrossRef] [PubMed]
323. Sasaki, T.; Fujii, K.; Yoshida, K.; Shimura, H.; Sasahira, T.; Ohmori, H.; Kuniyasu, H. Peritoneal metastasis inhibition by linoleic acid with activation of PPARγ in human gastrointestinal cancer cells. *Virchows Arch.* **2006**, *448*, 422–427. [CrossRef] [PubMed]
324. Richard, C.L.; Blay, J. Thiazolidinedione drugs down-regulate CXCR4 expression on human colorectal cancer cells in a peroxisome proliferator activated receptor γ-dependent manner. *Int. J. Oncol.* **2007**, *30*, 1215–1222. [CrossRef]
325. Richard, C.L.; Lowthers, E.L.; Blay, J. 15-Deoxy-$\Delta^{12,14}$-prostaglandin J_2 down-regulates CXCR4 on carcinoma cells through PPARγ- and NFkappaB-mediated pathways. *Exp. Cell Res.* **2007**, *313*, 3446–3458. [CrossRef]

326. Rovito, D.; Gionfriddo, G.; Barone, I.; Giordano, C.; Grande, F.; De Amicis, F.; Lanzino, M.; Catalano, S.; Andò, S.; Bonofiglio, D. Ligand-activated PPARγ downregulates CXCR4 gene expression through a novel identified PPAR response element and inhibits breast cancer progression. *Oncotarget* **2016**, *7*, 65109–65124. [CrossRef]
327. Pancione, M.; Forte, N.; Sabatino, L.; Tomaselli, E.; Parente, D.; Febbraro, A.; Colantuoni, V. Reduced β-catenin and peroxisome proliferator-activated receptor-γ expression levels are associated with colorectal cancer metastatic progression: Correlation with tumor-associated macrophages, cyclooxygenase 2, and patient outcome. *Hum. Pathol.* **2009**, *40*, 714–725. [CrossRef]
328. Moon, C.M.; Kwon, J.H.; Kim, J.S.; Oh, S.H.; Jin Lee, K.; Park, J.J.; Pil Hong, S.; Cheon, J.H.; Kim, T.I.; Kim, W.H. Nonsteroidal anti-inflammatory drugs suppress cancer stem cells via inhibiting PTGS2 (cyclooxygenase 2) and NOTCH/HES1 and activating PPARG in colorectal cancer. *Int. J. Cancer* **2014**, *134*, 519–529. [CrossRef]
329. Magenta, G.; Borenstein, X.; Rolando, R.; Jasnis, M.A. Rosiglitazone inhibits metastasis development of a murine mammary tumor cell line LMM3. *BMC Cancer* **2008**, *8*, 47. [CrossRef]
330. Kim, K.R.; Kim, H.J.; Lee, S.K.; Ma, G.T.; Park, K.K.; Chung, W.Y. 15-deoxy-δ12,14-prostaglandin J_2 inhibits osteolytic breast cancer bone metastasis and estrogen deficiency-induced bone loss. *PLoS ONE* **2015**, *10*, e0122764. [CrossRef]
331. Bren-Mattison, Y.; Van Putten, V.; Chan, D.; Winn, R.; Geraci, M.W.; Nemenoff, R.A. Peroxisome proliferator-activated receptor-γ (PPAR(γ)) inhibits tumorigenesis by reversing the undifferentiated phenotype of metastatic non-small-cell lung cancer cells (NSCLC). *Oncogene* **2005**, *24*, 1412–1422. [CrossRef]
332. Reka, A.K.; Kurapati, H.; Narala, V.R.; Bommer, G.; Chen, J.; Standiford, T.J.; Keshamouni, V.G. Peroxisome proliferator-activated receptor-γ activation inhibits tumor metastasis by antagonizing Smad3-mediated epithelial-mesenchymal transition. *Mol. Cancer Ther.* **2010**, *9*, 3221–3232. [CrossRef] [PubMed]
333. Chen, Q.Y.; Huang, X.B.; Zhao, Y.J.; Wang, H.G.; Wang, J.B.; Liu, L.C.; Wang, L.Q.; Zhong, Q.; Xie, J.W.; Lin, J.X.; et al. The peroxisome proliferator-activated receptor agonist rosiglitazone specifically represses tumour metastatic potential in chromatin inaccessibility-mediated FABP4-deficient gastric cancer. *Theranostics* **2022**, *12*, 1904–1920. [CrossRef] [PubMed]
334. Sawai, H.; Liu, J.; Reber, H.A.; Hines, O.J.; Eibl, G. Activation of peroxisome proliferator-activated receptor-γ decreases pancreatic cancer cell invasion through modulation of the plasminogen activator system. *Mol. Cancer Res.* **2006**, *4*, 159–167. [CrossRef] [PubMed]
335. Hsu, H.T.; Sung, M.T.; Lee, C.C.; Kuo, Y.J.; Chi, C.W.; Lee, H.C.; Hsia, C.Y. Peroxisome Proliferator-Activated Receptor γ Expression Is Inversely Associated with Macroscopic Vascular Invasion in Human Hepatocellular Carcinoma. *Int. J. Mol. Sci.* **2016**, *17*, 1226. [CrossRef]
336. Shen, B.; Chu, E.S.; Zhao, G.; Man, K.; Wu, C.W.; Cheng, J.T.; Li, G.; Nie, Y.; Lo, C.M.; Teoh, N.; et al. PPARγ inhibits hepatocellular carcinoma metastases in vitro and in mice. *Br. J. Cancer* **2012**, *106*, 1486–1494. [CrossRef]
337. Tu, K.; Zheng, X.; Dou, C.; Li, C.; Yang, W.; Yao, Y.; Liu, Q. MicroRNA-130b promotes cell aggressiveness by inhibiting peroxisome proliferator-activated receptor γ in human hepatocellular carcinoma. *Int. J. Mol. Sci.* **2014**, *15*, 20486–20499. [CrossRef]
338. Zuo, Q.; He, J.; Zhang, S.; Wang, H.; Jin, G.; Jin, H.; Cheng, Z.; Tao, X.; Yu, C.; Li, B.; et al. PPARγ Coactivator-1α Suppresses Metastasis of Hepatocellular Carcinoma by Inhibiting Warburg Effect by PPARγ-Dependent WNT/β-Catenin/Pyruvate Dehydrogenase Kinase Isozyme 1 Axis. *Hepatology* **2021**, *73*, 644–660. [CrossRef]
339. Kim, K.R.; Choi, H.N.; Lee, H.J.; Baek, H.A.; Park, H.S.; Jang, K.Y.; Chung, M.J.; Moon, W.S. A peroxisome proliferator-activated receptor γ antagonist induces vimentin cleavage and inhibits invasion in high-grade hepatocellular carcinoma. *Oncol. Rep.* **2007**, *18*, 825–832. [CrossRef]
340. Zhang, S.; Liu, F.; Mao, X.; Huang, J.; Yang, J.; Yin, X.; Wu, L.; Zheng, L.; Wang, Q. Elevation of miR-27b by HPV16 E7 inhibits PPARγ expression and promotes proliferation and invasion in cervical carcinoma cells. *Int. J. Oncol.* **2015**, *47*, 1759–1766. [CrossRef]
341. Ahn, Y.H.; Yang, Y.; Gibbons, D.L.; Creighton, C.J.; Yang, F.; Wistuba, I.I.; Lin, W.; Thilaganathan, N.; Alvarez, C.A.; Roybal, J.; et al. Map2k4 functions as a tumor suppressor in lung adenocarcinoma and inhibits tumor cell invasion by decreasing peroxisome proliferator-activated receptor γ2 expression. *Mol. Cell. Biol.* **2011**, *31*, 4270–4285. [CrossRef] [PubMed]
342. Li, H.; Sorenson, A.L.; Poczobutt, J.; Amin, J.; Joyal, T.; Sullivan, T.; Crossno, J.T.; Weiser-Evans, M.C.; Nemenoff, R.A. Activation of PPARγ in myeloid cells promotes lung cancer progression and metastasis. *PLoS ONE* **2011**, *6*, e28133. [CrossRef] [PubMed]
343. Sippel, T.R.; Johnson, A.M.; Li, H.Y.; Hanson, D.; Nguyen, T.T.; Bullock, B.L.; Poczobutt, J.M.; Kwak, J.W.; Kleczko, E.K.; Weiser-Evans, M.C.; et al. Activation of PPARγ in Myeloid Cells Promotes Progression of Epithelial Lung Tumors through TGFβ1. *Mol. Cancer Res.* **2019**, *17*, 1748–1758. [CrossRef] [PubMed]
344. Herroon, M.K.; Rajagurubandara, E.; Hardaway, A.L.; Powell, K.; Turchick, A.; Feldmann, D.; Podgorski, I. Bone marrow adipocytes promote tumor growth in bone via FABP4-dependent mechanisms. *Oncotarget* **2013**, *4*, 2108–2123. [CrossRef]
345. Niu, Z.; Shi, Q.; Zhang, W.; Shu, Y.; Yang, N.; Chen, B.; Wang, Q.; Zhao, X.; Chen, J.; Cheng, N.; et al. Caspase-1 cleaves PPARγ for potentiating the pro-tumor action of TAMs. *Nat. Commun.* **2017**, *8*, 766. [CrossRef]
346. Shu, Y.; Qin, M.; Song, Y.; Tang, Q.; Huang, Y.; Shen, P.; Lu, Y. M2 polarization of tumor-associated macrophages is dependent on integrin β3 via peroxisome proliferator-activated receptor-γ up-regulation in breast cancer. *Immunology* **2020**, *160*, 345–356. [CrossRef]
347. Zou, Y.; Watters, A.; Cheng, N.; Perry, C.E.; Xu, K.; Alicea, G.M.; Parris, J.L.D.; Baraban, E.; Ray, P.; Nayak, A.; et al. Polyunsaturated Fatty Acids from Astrocytes Activate PPARγ Signaling in Cancer Cells to Promote Brain Metastasis. *Cancer Discov.* **2019**, *9*, 1720–1735. [CrossRef]

348. Ahmad, I.; Mui, E.; Galbraith, L.; Patel, R.; Tan, E.H.; Salji, M.; Rust, A.G.; Repiscak, P.; Hedley, A.; Markert, E.; et al. Sleeping Beauty screen reveals Pparg activation in metastatic prostate cancer. *Proc. Natl. Acad. Sci. USA* **2016**, *113*, 8290–8295. [CrossRef]
349. Liu, R.Z.; Choi, W.S.; Jain, S.; Dinakaran, D.; Xu, X.; Han, W.H.; Yang, X.H.; Glubrecht, D.D.; Moore, R.B.; Lemieux, H.; et al. The FABP12/PPARγ pathway promotes metastatic transformation by inducing epithelial-to-mesenchymal transition and lipid-derived energy production in prostate cancer cells. *Mol. Oncol.* **2020**, *14*, 3100–3120. [CrossRef]
350. Galbraith, L.C.A.; Mui, E.; Nixon, C.; Hedley, A.; Strachan, D.; MacKay, G.; Sumpton, D.; Sansom, O.J.; Leung, H.Y.; Ahmad, I. PPAR-gamma induced AKT3 expression increases levels of mitochondrial biogenesis driving prostate cancer. *Oncogene* **2021**, *40*, 2355–2366. [CrossRef]
351. Yang, D.R.; Lin, S.J.; Ding, X.F.; Miyamoto, H.; Messing, E.; Li, L.Q.; Wang, N.; Chang, C. Higher expression of peroxisome proliferator-activated receptor γ or its activation by agonist thiazolidinedione-rosiglitazone promotes bladder cancer cell migration and invasion. *Urology* **2013**, *81*, e1101–e1106. [CrossRef] [PubMed]
352. Zhang, Y.; Luo, H.Y.; Liu, G.L.; Wang, D.S.; Wang, Z.Q.; Zeng, Z.L.; Xu, R.H. Prognostic significance and therapeutic implications of peroxisome proliferator-activated receptor γ overexpression in human pancreatic carcinoma. *Int. J. Oncol.* **2015**, *46*, 175–184. [CrossRef] [PubMed]
353. Takeuchi, A.; Yamamoto, N.; Shirai, T.; Hayashi, K.; Miwa, S.; Munesue, S.; Yamamoto, Y.; Tsuchiya, H. Clinical relevance of peroxisome proliferator-activated receptor-gamma expression in myxoid liposarcoma. *BMC Cancer* **2016**, *16*, 442. [CrossRef] [PubMed]
354. Pentinmikko, N.; Iqbal, S.; Mana, M.; Andersson, S.; Cognetta, A.B.; Suciu, R.M.; Roper, J.; Luopajärvi, K.; Markelin, E.; Gopalakrishnan, S.; et al. Notum produced by Paneth cells attenuates regeneration of aged intestinal epithelium. *Nature* **2019**, *571*, 398–402. [CrossRef]
355. Howroyd, P.; Swanson, C.; Dunn, C.; Cattley, R.C.; Corton, J.C. Decreased longevity and enhancement of age-dependent lesions in mice lacking the nuclear receptor peroxisome proliferator-activated receptor α (PPARα). *Toxicol. Pathol.* **2004**, *32*, 591–599. [CrossRef]
356. Youssef, J.; Badr, M. Enhanced hepatocarcinogenicity due to agonists of peroxisome proliferator-activated receptors in senescent rats: Role of peroxisome proliferation, cell proliferation, and apoptosis. *Sci. World J.* **2002**, *2*, 1491–1500. [CrossRef]
357. Kim, H.J.; Ham, S.A.; Kim, M.Y.; Hwang, J.S.; Lee, H.; Kang, E.S.; Yoo, T.; Woo, I.S.; Yabe-Nishimura, C.; Paek, K.S.; et al. PPARδ coordinates angiotensin II-induced senescence in vascular smooth muscle cells through PTEN-mediated inhibition of superoxide generation. *J. Biol. Chem.* **2011**, *286*, 44585–44593. [CrossRef]
358. Kim, M.Y.; Kang, E.S.; Ham, S.A.; Hwang, J.S.; Yoo, T.S.; Lee, H.; Paek, K.S.; Park, C.; Lee, H.T.; Kim, J.H.; et al. The PPARδ-mediated inhibition of angiotensin II-induced premature senescence in human endothelial cells is SIRT1-dependent. *Biochem. Pharmacol.* **2012**, *84*, 1627–1634. [CrossRef]
359. Ham, S.A.; Hwang, J.S.; Yoo, T.; Lee, H.; Kang, E.S.; Park, C.; Oh, J.W.; Lee, H.T.; Min, G.; Kim, J.H.; et al. Ligand-activated PPARδ inhibits UVB-induced senescence of human keratinocytes via PTEN-mediated inhibition of superoxide production. *Biochem. J.* **2012**, *444*, 27–38. [CrossRef]
360. Altieri, P.; Spallarossa, P.; Barisione, C.; Garibaldi, S.; Garuti, A.; Fabbi, P.; Ghigliotti, G.; Brunelli, C. Inhibition of doxorubicin-induced senescence by PPARδ activation agonists in cardiac muscle cells: Cooperation between PPARδ and Bcl6. *PLoS ONE* **2012**, *7*, e46126. [CrossRef]
361. Zhu, B.; Ferry, C.H.; Blazanin, N.; Bility, M.T.; Khozoie, C.; Kang, B.H.; Glick, A.B.; Gonzalez, F.J.; Peters, J.M. PPARβ/δ promotes HRAS-induced senescence and tumor suppression by potentiating p-ERK and repressing p-AKT signaling. *Oncogene* **2014**, *33*, 5348–5359. [CrossRef] [PubMed]
362. Zhu, B.; Ferry, C.H.; Markell, L.K.; Blazanin, N.; Glick, A.B.; Gonzalez, F.J.; Peters, J.M. The nuclear receptor peroxisome proliferator-activated receptor-β/δ (PPARβ/δ) promotes oncogene-induced cellular senescence through repression of endoplasmic reticulum stress. *J. Biol. Chem.* **2014**, *289*, 20102–20119. [CrossRef] [PubMed]
363. Riahi, Y.; Kaiser, N.; Cohen, G.; Abd-Elrahman, I.; Blum, G.; Shapira, O.M.; Koler, T.; Simionescu, M.; Sima, A.V.; Zarkovic, N.; et al. Foam cell-derived 4-hydroxynonenal induces endothelial cell senescence in a TXNIP-dependent manner. *J. Cell Mol. Med.* **2015**, *19*, 1887–1899. [CrossRef] [PubMed]
364. Bernal, C.; Araya, C.; Palma, V.; Bronfman, M. PPARβ/δ and PPARγ maintain undifferentiated phenotypes of mouse adult neural precursor cells from the subventricular zone. *Front. Cell. Neurosci.* **2015**, *9*, 78. [CrossRef]
365. Scholtysek, C.; Katzenbeisser, J.; Fu, H.; Uderhardt, S.; Ipseiz, N.; Stoll, C.; Zaiss, M.M.; Stock, M.; Donhauser, L.; Böhm, C.; et al. PPARβ/δ governs Wnt signaling and bone turnover. *Nat. Med.* **2013**, *19*, 608–613. [CrossRef]
366. Lecot, P.; Alimirah, F.; Desprez, P.Y.; Campisi, J.; Wiley, C. Context-dependent effects of cellular senescence in cancer development. *Br. J. Cancer* **2016**, *114*, 1180–1184. [CrossRef]
367. Milanovic, M.; Fan, D.N.Y.; Belenki, D.; Däbritz, J.H.M.; Zhao, Z.; Yu, Y.; Dörr, J.R.; Dimitrova, L.; Lenze, D.; Monteiro Barbosa, I.A.; et al. Senescence-associated reprogramming promotes cancer stemness. *Nature* **2018**, *553*, 96–100. [CrossRef]
368. Alimirah, F.; Pulido, T.; Valdovinos, A.; Alptekin, S.; Chang, E.; Jones, E.; Diaz, D.A.; Flores, J.; Velarde, M.C.; Demaria, M.; et al. Cellular Senescence Promotes Skin Carcinogenesis through p38MAPK and p44/42MAPK Signaling. *Cancer Res.* **2020**, *80*, 3606–3619. [CrossRef]

369. Abdul-Aziz, A.M.; Sun, Y.; Hellmich, C.; Marlein, C.R.; Mistry, J.; Forde, E.; Piddock, R.E.; Shafat, M.S.; Morfakis, A.; Mehta, T.; et al. Acute myeloid leukemia induces protumoral p16INK4a-driven senescence in the bone marrow microenvironment. *Blood* **2019**, *133*, 446–456. [CrossRef]
370. Wu, L.; Li, X.; Lin, Q.; Chowdhury, F.; Mazumder, M.H.; Du, W. FANCD2 and HES1 suppress inflammation-induced PPARγ to prevent haematopoietic stem cell exhaustion. *Br. J. Haematol.* **2021**, *192*, 652–663. [CrossRef]
371. Helman, A.; Klochendler, A.; Azazmeh, N.; Gabai, Y.; Horwitz, E.; Anzi, S.; Swisa, A.; Condiotti, R.; Granit, R.Z.; Nevo, Y.; et al. p16(Ink4a)-induced senescence of pancreatic beta cells enhances insulin secretion. *Nat. Med.* **2016**, *22*, 412–420. [CrossRef] [PubMed]
372. Han, L.; Zhou, R.; Niu, J.; McNutt, M.A.; Wang, P.; Tong, T. SIRT1 is regulated by a PPAR{γ}-SIRT1 negative feedback loop associated with senescence. *Nucleic Acids Res.* **2010**, *38*, 7458–7471. [CrossRef] [PubMed]
373. Gan, Q.; Huang, J.; Zhou, R.; Niu, J.; Zhu, X.; Wang, J.; Zhang, Z.; Tong, T. PPAR{γ} accelerates cellular senescence by inducing p16INK4{α} expression in human diploid fibroblasts. *J. Cell Sci.* **2008**, *121*, 2235–2245. [CrossRef] [PubMed]
374. Theocharis, S.; Giaginis, C.; Parasi, A.; Margeli, A.; Kakisis, J.; Agapitos, E.; Kouraklis, G. Expression of peroxisome proliferator-activated receptor-γ in colon cancer: Correlation with histopathological parameters, cell cycle-related molecules, and patients' survival. *Dig. Dis. Sci.* **2007**, *52*, 2305–2311. [CrossRef] [PubMed]
375. Hasan, A.U.; Ohmori, K.; Hashimoto, T.; Kamitori, K.; Hirata, Y.; Ishihara, Y.; Okamoto, N.; Noma, T.; Kosaka, H.; Tokuda, M.; et al. Pioglitazone promotes preadipocyte proliferation by downregulating p16(Ink4a). *Biochem. Biophys. Res. Commun.* **2011**, *411*, 375–380. [CrossRef]
376. Werner, C.; Gensch, C.; Pöss, J.; Haendeler, J.; Böhm, M.; Laufs, U. Pioglitazone activates aortic telomerase and prevents stress-induced endothelial apoptosis. *Atherosclerosis* **2011**, *216*, 23–34. [CrossRef]
377. Mello, T.; Materozzi, M.; Galli, A. PPARs and Mitochondrial Metabolism: From NAFLD to HCC. *PPAR Res.* **2016**, *2016*, 7403230. [CrossRef]
378. Vegliante, R.; Di Leo, L.; Ciccarone, F.; Ciriolo, M.R. Hints on ATGL implications in cancer: Beyond bioenergetic clues. *Cell Death Dis.* **2018**, *9*, 316. [CrossRef]
379. Warburg, O.; Wind, F.; Negelein, E. The metabolism of tumors in the body. *J. Gen. Physiol.* **1927**, *8*, 519–530. [CrossRef]
380. Maxwell, P.H.; Dachs, G.U.; Gleadle, J.M.; Nicholls, L.G.; Harris, A.L.; Stratford, I.J.; Hankinson, O.; Pugh, C.W.; Ratcliffe, P.J. Hypoxia-inducible factor-1 modulates gene expression in solid tumors and influences both angiogenesis and tumor growth. *Proc. Natl. Acad. Sci. USA* **1997**, *94*, 8104–8109. [CrossRef]
381. Wang, G.L.; Semenza, G.L. General involvement of hypoxia-inducible factor 1 in transcriptional response to hypoxia. *Proc. Natl. Acad. Sci. USA* **1993**, *90*, 4304–4308. [CrossRef] [PubMed]
382. Forsythe, J.A.; Jiang, B.H.; Iyer, N.V.; Agani, F.; Leung, S.W.; Koos, R.D.; Semenza, G.L. Activation of vascular endothelial growth factor gene transcription by hypoxia-inducible factor 1. *Mol. Cell. Biol.* **1996**, *16*, 4604–4613. [CrossRef] [PubMed]
383. Wagner, K.D.; Wagner, N.; Wellmann, S.; Schley, G.; Bondke, A.; Theres, H.; Scholz, H. Oxygen-regulated expression of the Wilms' tumor suppressor Wt1 involves hypoxia-inducible factor-1 (HIF-1). *FASEB J.* **2003**, *17*, 1364–1366. [CrossRef] [PubMed]
384. Narravula, S.; Colgan, S.P. Hypoxia-inducible factor 1-mediated inhibition of peroxisome proliferator-activated receptor α expression during hypoxia. *J. Immunol.* **2001**, *166*, 7543–7548. [CrossRef] [PubMed]
385. Balamurugan, K. HIF-1 at the crossroads of hypoxia, inflammation, and cancer. *Int. J. Cancer* **2016**, *138*, 1058–1066. [CrossRef]
386. Reddy, J.K.; Rao, S.; Moody, D.E. Hepatocellular carcinomas in acatalasemic mice treated with nafenopin, a hypolipidemic peroxisome proliferator. *Cancer Res.* **1976**, *36*, 1211–1217.
387. Drukala, J.; Urbanska, K.; Wilk, A.; Grabacka, M.; Wybieralska, E.; Del Valle, L.; Madeja, Z.; Reiss, K. ROS accumulation and IGF-IR inhibition contribute to fenofibrate/PPARα -mediated inhibition of glioma cell motility in vitro. *Mol. Cancer* **2010**, *9*, 159. [CrossRef]
388. Misra, P.; Reddy, J.K. Peroxisome proliferator-activated receptor-α activation and excess energy burning in hepatocarcinogenesis. *Biochimie* **2014**, *98*, 63–74. [CrossRef]
389. Wilk, A.; Wyczechowska, D.; Zapata, A.; Dean, M.; Mullinax, J.; Marrero, L.; Parsons, C.; Peruzzi, F.; Culicchia, F.; Ochoa, A.; et al. Molecular mechanisms of fenofibrate-induced metabolic catastrophe and glioblastoma cell death. *Mol. Cell. Biol.* **2015**, *35*, 182–198. [CrossRef]
390. Chekaoui, A.; Ertl, H.C.J. PPARα Agonist Fenofibrate Enhances Cancer Vaccine Efficacy. *Cancer Res.* **2021**, *81*, 4431–4440. [CrossRef]
391. Bahrambeigi, S.; Molaparast, M.; Sohrabi, F.; Seifi, L.; Faraji, A.; Fani, S.; Shafiei-Irannejad, V. Targeting PPAR ligands as possible approaches for metabolic reprogramming of T cells in cancer immunotherapy. *Immunol. Lett.* **2020**, *220*, 32–37. [CrossRef] [PubMed]
392. Chowdhury, P.S.; Chamoto, K.; Kumar, A.; Honjo, T. PPAR-Induced Fatty Acid Oxidation in T Cells Increases the Number of Tumor-Reactive CD8. *Cancer Immunol. Res.* **2018**, *6*, 1375–1387. [CrossRef] [PubMed]
393. Stokes, W.A.; Behera, M.; Jiang, R.; Gutman, D.A.; Huang, Z.; Burns, A.; Sebastian, N.T.; Sukhatme, V.; Lowe, M.C.; Ramalingam, S.S.; et al. Impact of concomitant fibrates on immunotherapy outcomes for advanced non-small cell lung cancer. *Cancer Med.* **2022**, *00*, 1–10. [CrossRef] [PubMed]
394. Haakinson, D.J.; Leeds, S.G.; Dueck, A.C.; Gray, R.J.; Wasif, N.; Stucky, C.C.; Northfelt, D.W.; Apsey, H.A.; Pockaj, B. The impact of obesity on breast cancer: A retrospective review. *Ann. Surg. Oncol.* **2012**, *19*, 3012–3018. [CrossRef]

395. La Vecchia, C.; Negri, E.; Franceschi, S.; D'Avanzo, B.; Boyle, P. A case-control study of diabetes mellitus and cancer risk. *Br. J. Cancer* **1994**, *70*, 950–953. [CrossRef]
396. Blücher, C.; Iberl, S.; Schwagarus, N.; Müller, S.; Liebisch, G.; Höring, M.; Hidrobo, M.S.; Ecker, J.; Spindler, N.; Dietrich, A.; et al. Secreted Factors from Adipose Tissue Reprogram Tumor Lipid Metabolism and Induce Motility by Modulating PPARα/ANGPTL4 and FAK. *Mol. Cancer Res.* **2020**, *18*, 1849–1862. [CrossRef]
397. Naiman, S.; Huynh, F.K.; Gil, R.; Glick, Y.; Shahar, Y.; Touitou, N.; Nahum, L.; Avivi, M.Y.; Roichman, A.; Kanfi, Y.; et al. SIRT6 Promotes Hepatic Beta-Oxidation via Activation of PPARα. *Cell Rep.* **2019**, *29*, 4127–4143.e4128. [CrossRef]
398. Senni, N.; Savall, M.; Cabrerizo Granados, D.; Alves-Guerra, M.C.; Sartor, C.; Lagoutte, I.; Gougelet, A.; Terris, B.; Gilgenkrantz, H.; Perret, C.; et al. β-catenin-activated hepatocellular carcinomas are addicted to fatty acids. *Gut* **2019**, *68*, 322–334. [CrossRef]
399. Andrejeva, D.; Kugler, J.M.; Nguyen, H.T.; Malmendal, A.; Holm, M.L.; Toft, B.G.; Loya, A.C.; Cohen, S.M. Metabolic control of PPAR activity by aldehyde dehydrogenase regulates invasive cell behavior and predicts survival in hepatocellular and renal clear cell carcinoma. *BMC Cancer* **2018**, *18*, 1180. [CrossRef]
400. Gou, Q.; Dong, C.; Jin, J.; Liu, Q.; Lu, W.; Shi, J.; Hou, Y. PPARα agonist alleviates tumor growth and chemo-resistance associated with the inhibition of glucose metabolic pathway. *Eur. J. Pharmacol.* **2019**, *863*, 172664. [CrossRef]
401. You, M.; Jin, J.; Liu, Q.; Xu, Q.; Shi, J.; Hou, Y. PPARα Promotes Cancer Cell Glut1 Transcription Repression. *J. Cell. Biochem.* **2017**, *118*, 1556–1562. [CrossRef] [PubMed]
402. Han, D.; Wei, W.; Chen, X.; Zhang, Y.; Wang, Y.; Zhang, J.; Wang, X.; Yu, T.; Hu, Q.; Liu, N.; et al. NF-κB/RelA-PKM2 mediates inhibition of glycolysis by fenofibrate in glioblastoma cells. *Oncotarget* **2015**, *6*, 26119–26128. [CrossRef] [PubMed]
403. Xu, M.; Zuo, X.; Shureiqi, I. Targeting peroxisome proliferator-activated receptor-β/δ in colon cancer: How to aim? *Biochem. Pharmacol.* **2013**, *85*, 607–611. [CrossRef] [PubMed]
404. Wagner, N.; Jehl-Piétri, C.; Lopez, P.; Murdaca, J.; Giordano, C.; Schwartz, C.; Gounon, P.; Hatem, S.N.; Grimaldi, P.; Wagner, K.D. Peroxisome proliferator-activated receptor β stimulation induces rapid cardiac growth and angiogenesis via direct activation of calcineurin. *Cardiovasc. Res.* **2009**, *83*, 61–71. [CrossRef]
405. Liu, Y.V.; Hubbi, M.E.; Pan, F.; McDonald, K.R.; Mansharamani, M.; Cole, R.N.; Liu, J.O.; Semenza, G.L. Calcineurin promotes hypoxia-inducible factor 1α expression by dephosphorylating RACK1 and blocking RACK1 dimerization. *J. Biol. Chem.* **2007**, *282*, 37064–37073. [CrossRef] [PubMed]
406. Jeong, E.; Koo, J.E.; Yeon, S.H.; Kwak, M.K.; Hwang, D.H.; Lee, J.Y. PPARδ deficiency disrupts hypoxia-mediated tumorigenic potential of colon cancer cells. *Mol. Carcinog.* **2014**, *53*, 926–937. [CrossRef]
407. Pudakalakatti, S.; Titus, M.; Enriquez, J.S.; Ramachandran, S.; Zacharias, N.M.; Shureiqi, I.; Liu, Y.; Yao, J.C.; Zuo, X.; Bhattacharya, P.K. Identifying the Metabolic Signatures of PPARD-Overexpressing Gastric Tumors. *Int. J. Mol. Sci.* **2022**, *23*, 1645. [CrossRef]
408. Mana, M.D.; Hussey, A.M.; Tzouanas, C.N.; Imada, S.; Barrera Millan, Y.; Bahceci, D.; Saiz, D.R.; Webb, A.T.; Lewis, C.A.; Carmeliet, P.; et al. High-fat diet-activated fatty acid oxidation mediates intestinal stemness and tumorigenicity. *Cell Rep.* **2021**, *35*, 109212. [CrossRef]
409. Hodge, A.M.; Williamson, E.J.; Bassett, J.K.; MacInnis, R.J.; Giles, G.G.; English, D.R. Dietary and biomarker estimates of fatty acids and risk of colorectal cancer. *Int. J. Cancer* **2015**, *137*, 1224–1234. [CrossRef] [PubMed]
410. Li, C.; Zhao, X.; Toline, E.C.; Siegal, G.P.; Evans, L.M.; Ibrahim-Hashim, A.; Desmond, R.A.; Hardy, R.W. Prevention of carcinogenesis and inhibition of breast cancer tumor burden by dietary stearate. *Carcinogenesis* **2011**, *32*, 1251–1258. [CrossRef] [PubMed]
411. Hardy, S.; Langelier, Y.; Prentki, M. Oleate activates phosphatidylinositol 3-kinase and promotes proliferation and reduces apoptosis of MDA-MB-231 breast cancer cells, whereas palmitate has opposite effects. *Cancer Res.* **2000**, *60*, 6353–6358. [PubMed]
412. Levi, L.; Wang, Z.; Doud, M.K.; Hazen, S.L.; Noy, N. Saturated fatty acids regulate retinoic acid signalling and suppress tumorigenesis by targeting fatty acid-binding protein 5. *Nat. Commun.* **2015**, *6*, 8794. [CrossRef] [PubMed]
413. Wang, Y.; Wang, J.; Li, X.; Xiong, X.; Zhou, Z.; Zhu, X.; Gu, Y.; Dominissini, D.; He, L.; Tian, Y.; et al. N^1-methyladenosine methylation in tRNA drives liver tumourigenesis by regulating cholesterol metabolism. *Nat. Commun.* **2021**, *12*, 6314. [CrossRef]
414. Zhang, W.; Xu, Y.; Xu, Q.; Shi, H.; Shi, J.; Hou, Y. PPARδ promotes tumor progression via activation of Glut1 and SLC1-A5 transcription. *Carcinogenesis* **2017**, *38*, 748–755. [CrossRef]
415. Kim, M.J.; Choi, Y.K.; Park, S.Y.; Jang, S.Y.; Lee, J.Y.; Ham, H.J.; Kim, B.G.; Jeon, H.J.; Kim, J.H.; Kim, J.G.; et al. PPARδ Reprograms Glutamine Metabolism in Sorafenib-Resistant HCC. *Mol. Cancer Res.* **2017**, *15*, 1230–1242. [CrossRef]
416. Kim, J.H.; Song, J.; Park, K.W. The multifaceted factor peroxisome proliferator-activated receptor γ (PPARγ) in metabolism, immunity, and cancer. *Arch. Pharm. Res.* **2015**, *38*, 302–312. [CrossRef]
417. Sakharkar, M.K.; Shashni, B.; Sharma, K.; Dhillon, S.K.; Ranjekar, P.R.; Sakharkar, K.R. Therapeutic implications of targeting energy metabolism in breast cancer. *PPAR Res.* **2013**, *2013*, 109285. [CrossRef]
418. Asp, M.L.; Tian, M.; Kliewer, K.L.; Belury, M.A. Rosiglitazone delayed weight loss and anorexia while attenuating adipose depletion in mice with cancer cachexia. *Cancer Biol. Ther.* **2011**, *12*, 957–965. [CrossRef] [PubMed]
419. Beluzi, M.; Peres, S.B.; Henriques, F.S.; Sertié, R.A.; Franco, F.O.; Santos, K.B.; Knobl, P.; Andreotti, S.; Shida, C.S.; Neves, R.X.; et al. Pioglitazone treatment increases survival and prevents body weight loss in tumor-bearing animals: Possible anti-cachectic effect. *PLoS ONE* **2015**, *10*, e0122660. [CrossRef]
420. Qian, F.; Arner, B.E.; Kelly, K.M.; Annageldiyev, C.; Sharma, A.; Claxton, D.F.; Paulson, R.F.; Prabhu, K.S. Interleukin-4 treatment reduces leukemia burden in acute myeloid leukemia. *FASEB J.* **2022**, *36*, e22328. [CrossRef]

421. Ning, Z.; Guo, X.; Liu, X.; Lu, C.; Wang, A.; Wang, X.; Wang, W.; Chen, H.; Qin, W.; Zhou, L.; et al. USP22 regulates lipidome accumulation by stabilizing PPARγ in hepatocellular carcinoma. *Nat. Commun.* **2022**, *13*, 2187. [CrossRef] [PubMed]
422. Edwards, I.J.; Sun, H.; Hu, Y.; Berquin, I.M.; O'Flaherty, J.T.; Cline, J.M.; Rudel, L.L.; Chen, Y.Q. In vivo and in vitro regulation of syndecan 1 in prostate cells by n-3 polyunsaturated fatty acids. *J. Biol. Chem.* **2008**, *283*, 18441–18449. [CrossRef] [PubMed]
423. Anagnostopoulos, G.; Motiño, O.; Li, S.; Carbonnier, V.; Chen, H.; Sica, V.; Durand, S.; Bourgin, M.; Aprahamian, F.; Nirmalathasan, N.; et al. An obesogenic feedforward loop involving PPARγ, acyl-CoA binding protein and GABA. *Cell Death Dis.* **2022**, *13*, 356. [CrossRef] [PubMed]
424. Han, J.; Shen, L.; Zhan, Z.; Liu, Y.; Zhang, C.; Guo, R.; Luo, Y.; Xie, Z.; Feng, Y.; Wu, G. The long noncoding RNA MALAT1 modulates adipose loss in cancer-associated cachexia by suppressing adipogenesis through PPAR-γ. *Nutr. Metab.* **2021**, *18*, 27. [CrossRef] [PubMed]
425. Ma, S.; Zhou, B.; Yang, Q.; Pan, Y.; Yang, W.; Freedland, S.J.; Ding, L.W.; Freeman, M.R.; Breunig, J.J.; Bhowmick, N.A.; et al. A Transcriptional Regulatory Loop of Master Regulator Transcription Factors, PPARG, and Fatty Acid Synthesis Promotes Esophageal Adenocarcinoma. *Cancer Res.* **2021**, *81*, 1216–1229. [CrossRef] [PubMed]
426. Forootan, F.S.; Forootan, S.S.; Gou, X.; Yang, J.; Liu, B.; Chen, D.; Al Fayi, M.S.; Al-Jameel, W.; Rudland, P.S.; Hussain, S.A.; et al. Fatty acid activated PPARγ promotes tumorigenicity of prostate cancer cells by up regulating VEGF via PPAR responsive elements of the promoter. *Oncotarget* **2016**, *7*, 9322–9339. [CrossRef] [PubMed]
427. Wang, M.; Li, G.; Yang, Z.; Wang, L.; Zhang, L.; Wang, T.; Zhang, Y.; Zhang, S.; Han, Y.; Jia, L. Uncoupling protein 2 downregulation by hypoxia through repression of peroxisome proliferator-activated receptor γ promotes chemoresistance of non-small cell lung cancer. *Oncotarget* **2017**, *8*, 8083–8094. [CrossRef]
428. Han, S.; Inoue, H.; Flowers, L.C.; Sidell, N. Control of COX-2 gene expression through peroxisome proliferator-activated receptor γ in human cervical cancer cells. *Clin. Cancer Res.* **2003**, *9*, 4627–4635.
429. Bren-Mattison, Y.; Meyer, A.M.; Van Putten, V.; Li, H.; Kuhn, K.; Stearman, R.; Weiser-Evans, M.; Winn, R.A.; Heasley, L.E.; Nemenoff, R.A. Antitumorigenic effects of peroxisome proliferator-activated receptor-γ in non-small-cell lung cancer cells are mediated by suppression of cyclooxygenase-2 via inhibition of nuclear factor-κB. *Mol. Pharmacol.* **2008**, *73*, 709–717. [CrossRef]
430. Hazra, S.; Dubinett, S.M. Ciglitazone mediates COX-2 dependent suppression of PGE2 in human non-small cell lung cancer cells. *Prostaglandins Leukot. Essent. Fat. Acids* **2007**, *77*, 51–58. [CrossRef]
431. Gottfried, E.; Rogenhofer, S.; Waibel, H.; Kunz-Schughart, L.A.; Reichle, A.; Wehrstein, M.; Peuker, A.; Peter, K.; Hartmannsgruber, G.; Andreesen, R.; et al. Pioglitazone modulates tumor cell metabolism and proliferation in multicellular tumor spheroids. *Cancer Chemother. Pharmacol.* **2011**, *67*, 117–126. [CrossRef] [PubMed]
432. Wei, S.; Kulp, S.K.; Chen, C.S. Energy restriction as an antitumor target of thiazolidinediones. *J. Biol. Chem.* **2010**, *285*, 9780–9791. [CrossRef] [PubMed]
433. Devchand, P.R.; Keller, H.; Peters, J.M.; Vazquez, M.; Gonzalez, F.J.; Wahli, W. The PPARα-leukotriene B4 pathway to inflammation control. *Nature* **1996**, *384*, 39–43. [CrossRef] [PubMed]
434. Jones, D.C.; Ding, X.; Daynes, R.A. Nuclear receptor peroxisome proliferator-activated receptor α (PPARα) is expressed in resting murine lymphocytes. The PPARα in T and B lymphocytes is both transactivation and transrepression competent. *J. Biol. Chem.* **2002**, *277*, 6838–6845. [CrossRef] [PubMed]
435. Chinetti, G.; Griglio, S.; Antonucci, M.; Torra, I.P.; Delerive, P.; Majd, Z.; Fruchart, J.C.; Chapman, J.; Najib, J.; Staels, B. Activation of proliferator-activated receptors α and γ induces apoptosis of human monocyte-derived macrophages. *J. Biol. Chem.* **1998**, *273*, 25573–25580. [CrossRef]
436. Yang, Q.; Xie, Y.; Depierre, J.W. Effects of peroxisome proliferators on the thymus and spleen of mice. *Clin. Exp. Immunol.* **2000**, *122*, 219–226. [CrossRef]
437. Yang, Q.; Xie, Y.; Alexson, S.E.; Nelson, B.D.; DePierre, J.W. Involvement of the peroxisome proliferator-activated receptor α in the immunomodulation caused by peroxisome proliferators in mice. *Biochem. Pharmacol.* **2002**, *63*, 1893–1900. [CrossRef]
438. Yang, Q.; Gonzalez, F.J. Peroxisome proliferator-activated receptor α regulates B lymphocyte development via an indirect pathway in mice. *Biochem. Pharmacol.* **2004**, *68*, 2143–2150. [CrossRef]
439. Roberts, R.A. Peroxisome proliferators: Mechanisms of adverse effects in rodents and molecular basis for species differences. *Arch. Toxicol.* **1999**, *73*, 413–418. [CrossRef]
440. Stienstra, R.; Mandard, S.; Patsouris, D.; Maass, C.; Kersten, S.; Müller, M. Peroxisome proliferator-activated receptor α protects against obesity-induced hepatic inflammation. *Endocrinology* **2007**, *148*, 2753–2763. [CrossRef]
441. Shiri-Sverdlov, R.; Wouters, K.; van Gorp, P.J.; Gijbels, M.J.; Noel, B.; Buffat, L.; Staels, B.; Maeda, N.; van Bilsen, M.; Hofker, M.H. Early diet-induced non-alcoholic steatohepatitis in APOE2 knock-in mice and its prevention by fibrates. *J. Hepatol.* **2006**, *44*, 732–741. [CrossRef] [PubMed]
442. Riccardi, L.; Mazzon, E.; Bruscoli, S.; Esposito, E.; Crisafulli, C.; Di Paola, R.; Caminiti, R.; Riccardi, C.; Cuzzocrea, S. Peroxisome proliferator-activated receptor-α modulates the anti-inflammatory effect of glucocorticoids in a model of inflammatory bowel disease in mice. *Shock* **2009**, *31*, 308–316. [CrossRef] [PubMed]
443. Azuma, Y.T.; Nishiyama, K.; Matsuo, Y.; Kuwamura, M.; Morioka, A.; Nakajima, H.; Takeuchi, T. PPARα contributes to colonic protection in mice with DSS-induced colitis. *Int. Immunopharmacol.* **2010**, *10*, 1261–1267. [CrossRef] [PubMed]

444. Manoharan, I.; Suryawanshi, A.; Hong, Y.; Ranganathan, P.; Shanmugam, A.; Ahmad, S.; Swafford, D.; Manicassamy, B.; Ramesh, G.; Koni, P.A.; et al. Homeostatic PPARα Signaling Limits Inflammatory Responses to Commensal Microbiota in the Intestine. *J. Immunol.* **2016**, *196*, 4739–4749. [CrossRef] [PubMed]
445. Michalik, L.; Desvergne, B.; Tan, N.S.; Basu-Modak, S.; Escher, P.; Rieusset, J.; Peters, J.M.; Kaya, G.; Gonzalez, F.J.; Zakany, J.; et al. Impaired skin wound healing in peroxisome proliferator-activated receptor (PPAR)α and PPARβ mutant mice. *J. Cell. Biol.* **2001**, *154*, 799–814. [CrossRef] [PubMed]
446. Poynter, M.E.; Daynes, R.A. Peroxisome proliferator-activated receptor α activation modulates cellular redox status, represses nuclear factor-kappaB signaling, and reduces inflammatory cytokine production in aging. *J. Biol. Chem.* **1998**, *273*, 32833–32841. [CrossRef]
447. Mandard, S.; Patsouris, D. Nuclear control of the inflammatory response in mammals by peroxisome proliferator-activated receptors. *PPAR Res.* **2013**, *2013*, 613864. [CrossRef]
448. Daynes, R.A.; Jones, D.C. Emerging roles of PPARs in inflammation and immunity. *Nat. Rev. Immunol.* **2002**, *2*, 748–759. [CrossRef]
449. Hermanowski-Vosatka, A.; Gerhold, D.; Mundt, S.S.; Loving, V.A.; Lu, M.; Chen, Y.; Elbrecht, A.; Wu, M.; Doebber, T.; Kelly, L.; et al. PPARα agonists reduce 11β-hydroxysteroid dehydrogenase type 1 in the liver. *Biochem. Biophys. Res. Commun.* **2000**, *279*, 330–336. [CrossRef]
450. Hill, M.R.; Clarke, S.; Rodgers, K.; Thornhill, B.; Peters, J.M.; Gonzalez, F.J.; Gimble, J.M. Effect of peroxisome proliferator-activated receptor alpha activators on tumor necrosis factor expression in mice during endotoxemia. *Infect. Immun.* **1999**, *67*, 3488–3493. [CrossRef]
451. Jackson, S.M.; Parhami, F.; Xi, X.P.; Berliner, J.A.; Hsueh, W.A.; Law, R.E.; Demer, L.L. Peroxisome proliferator-activated receptor activators target human endothelial cells to inhibit leukocyte-endothelial cell interaction. *Arterioscler. Thromb. Vasc. Biol.* **1999**, *19*, 2094–2104. [CrossRef] [PubMed]
452. Marx, N.; Sukhova, G.K.; Collins, T.; Libby, P.; Plutzky, J. PPARα activators inhibit cytokine-induced vascular cell adhesion molecule-1 expression in human endothelial cells. *Circulation* **1999**, *99*, 3125–3131. [CrossRef] [PubMed]
453. Staels, B.; Koenig, W.; Habib, A.; Merval, R.; Lebret, M.; Torra, I.P.; Delerive, P.; Fadel, A.; Chinetti, G.; Fruchart, J.C.; et al. Activation of human aortic smooth-muscle cells is inhibited by PPARα but not by PPARγ activators. *Nature* **1998**, *393*, 790–793. [CrossRef] [PubMed]
454. Lee, H.; Shi, W.; Tontonoz, P.; Wang, S.; Subbanagounder, G.; Hedrick, C.C.; Hama, S.; Borromeo, C.; Evans, R.M.; Berliner, J.A.; et al. Role for peroxisome proliferator-activated receptor α in oxidized phospholipid-induced synthesis of monocyte chemotactic protein-1 and interleukin-8 by endothelial cells. *Circ. Res.* **2000**, *87*, 516–521. [CrossRef]
455. Zhang, Y.; Kurupati, R.; Liu, L.; Zhou, X.Y.; Zhang, G.; Hudaihed, A.; Filisio, F.; Giles-Davis, W.; Xu, X.; Karakousis, G.C.; et al. Enhancing CD8. *Cancer Cell* **2017**, *32*, 377–391.e379. [CrossRef]
456. Yin, X.; Zeng, W.; Wu, B.; Wang, L.; Wang, Z.; Tian, H.; Jiang, Y.; Clay, R.; Wei, X.; Qin, Y.; et al. PPARα Inhibition Overcomes Tumor-Derived Exosomal Lipid-Induced Dendritic Cell Dysfunction. *Cell Rep.* **2020**, *33*, 108278. [CrossRef]
457. Bishop-Bailey, D.; Bystrom, J. Emerging roles of peroxisome proliferator-activated receptor-β/δ in inflammation. *Pharmacol. Ther.* **2009**, *124*, 141–150. [CrossRef]
458. Peters, J.M.; Lee, S.S.; Li, W.; Ward, J.M.; Gavrilova, O.; Everett, C.; Reitman, M.L.; Hudson, L.D.; Gonzalez, F.J. Growth, adipose, brain, and skin alterations resulting from targeted disruption of the mouse peroxisome proliferator-activated receptor β(δ). *Mol. Cell. Biol.* **2000**, *20*, 5119–5128. [CrossRef]
459. Tan, N.S.; Michalik, L.; Noy, N.; Yasmin, R.; Pacot, C.; Heim, M.; Flühmann, B.; Desvergne, B.; Wahli, W. Critical roles of PPAR β/δ in keratinocyte response to inflammation. *Genes Dev.* **2001**, *15*, 3263–3277. [CrossRef]
460. Bishop-Bailey, D.; Hla, T. Endothelial cell apoptosis induced by the peroxisome proliferator-activated receptor (PPAR) ligand 15-deoxy-$\Delta^{12,14}$-prostaglandin J_2. *J. Biol. Chem.* **1999**, *274*, 17042–17048. [CrossRef]
461. Rival, Y.; Benéteau, N.; Taillandier, T.; Pezet, M.; Dupont-Passelaigue, E.; Patoiseau, J.F.; Junquéro, D.; Colpaert, F.C.; Delhon, A. PPARα and PPARδ activators inhibit cytokine-induced nuclear translocation of NF-κB and expression of VCAM-1 in EAhy926 endothelial cells. *Eur. J. Pharmacol.* **2002**, *435*, 143–151. [CrossRef]
462. Fan, Y.; Wang, Y.; Tang, Z.; Zhang, H.; Qin, X.; Zhu, Y.; Guan, Y.; Wang, X.; Staels, B.; Chien, S.; et al. Suppression of pro-inflammatory adhesion molecules by PPAR-δ in human vascular endothelial cells. *Arterioscler. Thromb. Vasc. Biol.* **2008**, *28*, 315–321. [CrossRef] [PubMed]
463. Lee, C.H.; Chawla, A.; Urbiztondo, N.; Liao, D.; Boisvert, W.A.; Evans, R.M.; Curtiss, L.K. Transcriptional repression of atherogenic inflammation: Modulation by PPARδ. *Science* **2003**, *302*, 453–457. [CrossRef] [PubMed]
464. Piqueras, L.; Sanz, M.J.; Perretti, M.; Morcillo, E.; Norling, L.; Mitchell, J.A.; Li, Y.; Bishop-Bailey, D. Activation of PPARβ/δ inhibits leukocyte recruitment, cell adhesion molecule expression, and chemokine release. *J. Leukoc. Biol.* **2009**, *86*, 115–122. [CrossRef]
465. Welch, J.S.; Ricote, M.; Akiyama, T.E.; Gonzalez, F.J.; Glass, C.K. PPARγ and PPARδ negatively regulate specific subsets of lipopolysaccharide and IFN-γ target genes in macrophages. *Proc. Natl. Acad. Sci. USA* **2003**, *100*, 6712–6717. [CrossRef] [PubMed]
466. Kang, K.; Reilly, S.M.; Karabacak, V.; Gangl, M.R.; Fitzgerald, K.; Hatano, B.; Lee, C.H. Adipocyte-derived Th2 cytokines and myeloid PPARδ regulate macrophage polarization and insulin sensitivity. *Cell Metab.* **2008**, *7*, 485–495. [CrossRef]

467. Odegaard, J.I.; Ricardo-Gonzalez, R.R.; Red Eagle, A.; Vats, D.; Morel, C.R.; Goforth, M.H.; Subramanian, V.; Mukundan, L.; Ferrante, A.W.; Chawla, A. Alternative M2 activation of Kupffer cells by PPARδ ameliorates obesity-induced insulin resistance. *Cell Metab.* **2008**, *7*, 496–507. [CrossRef]
468. Gallardo-Soler, A.; Gómez-Nieto, C.; Campo, M.L.; Marathe, C.; Tontonoz, P.; Castrillo, A.; Corraliza, I. Arginase I induction by modified lipoproteins in macrophages: A peroxisome proliferator-activated receptor-γ/δ-mediated effect that links lipid metabolism and immunity. *Mol. Endocrinol.* **2008**, *22*, 1394–1402. [CrossRef]
469. Mukundan, L.; Odegaard, J.I.; Morel, C.R.; Heredia, J.E.; Mwangi, J.W.; Ricardo-Gonzalez, R.R.; Goh, Y.P.; Eagle, A.R.; Dunn, S.E.; Awakuni, J.U.; et al. PPAR-δ senses and orchestrates clearance of apoptotic cells to promote tolerance. *Nat. Med.* **2009**, *15*, 1266–1272. [CrossRef]
470. Adhikary, T.; Wortmann, A.; Schumann, T.; Finkernagel, F.; Lieber, S.; Roth, K.; Toth, P.M.; Diederich, W.E.; Nist, A.; Stiewe, T.; et al. The transcriptional PPARβ/δ network in human macrophages defines a unique agonist-induced activation state. *Nucleic Acids Res.* **2015**, *43*, 5033–5051. [CrossRef]
471. Schumann, T.; Adhikary, T.; Wortmann, A.; Finkernagel, F.; Lieber, S.; Schnitzer, E.; Legrand, N.; Schober, Y.; Nockher, W.A.; Toth, P.M.; et al. Deregulation of PPARβ/δ target genes in tumor-associated macrophages by fatty acid ligands in the ovarian cancer microenvironment. *Oncotarget* **2015**, *6*, 13416–13433. [CrossRef] [PubMed]
472. Schote, A.B.; Turner, J.D.; Schiltz, J.; Muller, C.P. Nuclear receptors in human immune cells: Expression and correlations. *Mol. Immunol.* **2007**, *44*, 1436–1445. [CrossRef] [PubMed]
473. al Yacoub, N.; Romanowska, M.; Krauss, S.; Schweiger, S.; Foerster, J. PPARδ is a type 1 IFN target gene and inhibits apoptosis in T cells. *J. Investig. Dermatol.* **2008**, *128*, 1940–1949. [CrossRef] [PubMed]
474. Saibil, S.D.; St Paul, M.; Laister, R.C.; Garcia-Batres, C.R.; Israni-Winger, K.; Elford, A.R.; Grimshaw, N.; Robert-Tissot, C.; Roy, D.G.; Jones, R.G.; et al. Activation of Peroxisome Proliferator-Activated Receptors α and δ Synergizes with Inflammatory Signals to Enhance Adoptive Cell Therapy. *Cancer Res.* **2019**, *79*, 445–451. [CrossRef]
475. Jakobsen, M.A.; Petersen, R.K.; Kristiansen, K.; Lange, M.; Lillevang, S.T. Peroxisome proliferator-activated receptor α, δ, γ1 and γ2 expressions are present in human monocyte-derived dendritic cells and modulate dendritic cell maturation by addition of subtype-specific ligands. *Scand. J. Immunol.* **2006**, *63*, 330–337. [CrossRef] [PubMed]
476. Maciel, T.T.; Moura, I.C.; Hermine, O. The role of mast cells in cancers. *F1000Prime Rep.* **2015**, *7*, 09. [CrossRef]
477. Yao, P.L.; Morales, J.L.; Gonzalez, F.J.; Peters, J.M. Peroxisome proliferator-activated receptor-β/δ modulates mast cell phenotype. *Immunology* **2017**, *150*, 456–467. [CrossRef]
478. Michelet, X.; Dyck, L.; Hogan, A.; Loftus, R.M.; Duquette, D.; Wei, K.; Beyaz, S.; Tavakkoli, A.; Foley, C.; Donnelly, R.; et al. Metabolic reprogramming of natural killer cells in obesity limits antitumor responses. *Nat. Immunol.* **2018**, *19*, 1330–1340. [CrossRef]
479. Yang, X.Y.; Wang, L.H.; Chen, T.; Hodge, D.R.; Resau, J.H.; DaSilva, L.; Farrar, W.L. Activation of human T lymphocytes is inhibited by peroxisome proliferator-activated receptor γ (PPARγ) agonists. PPARγ co-association with transcription factor NFAT. *J. Biol. Chem.* **2000**, *275*, 4541–4544. [CrossRef]
480. Clark, R.B.; Bishop-Bailey, D.; Estrada-Hernandez, T.; Hla, T.; Puddington, L.; Padula, S.J. The nuclear receptor PPAR γ and immunoregulation: PPAR γ mediates inhibition of helper T cell responses. *J. Immunol.* **2000**, *164*, 1364–1371. [CrossRef]
481. Desreumaux, P.; Dubuquoy, L.; Nutten, S.; Peuchmaur, M.; Englaro, W.; Schoonjans, K.; Derijard, B.; Desvergne, B.; Wahli, W.; Chambon, P.; et al. Attenuation of colon inflammation through activators of the retinoid X receptor (RXR)/peroxisome proliferator-activated receptor γ (PPARγ) heterodimer. A basis for new therapeutic strategies. *J. Exp. Med.* **2001**, *193*, 827–838. [CrossRef] [PubMed]
482. Su, C.G.; Wen, X.; Bailey, S.T.; Jiang, W.; Rangwala, S.M.; Keilbaugh, S.A.; Flanigan, A.; Murthy, S.; Lazar, M.A.; Wu, G.D. A novel therapy for colitis utilizing PPAR-γ ligands to inhibit the epithelial inflammatory response. *J. Clin. Investig.* **1999**, *104*, 383–389. [CrossRef] [PubMed]
483. Hontecillas, R.; Bassaganya-Riera, J. Peroxisome proliferator-activated receptor γ is required for regulatory CD4+ T cell-mediated protection against colitis. *J. Immunol.* **2007**, *178*, 2940–2949. [CrossRef] [PubMed]
484. Harris, S.G.; Phipps, R.P. The nuclear receptor PPAR γ is expressed by mouse T lymphocytes and PPAR γ agonists induce apoptosis. *Eur. J. Immunol.* **2001**, *31*, 1098–1105. [CrossRef]
485. Wang, Y.L.; Frauwirth, K.A.; Rangwala, S.M.; Lazar, M.A.; Thompson, C.B. Thiazolidinedione activation of peroxisome proliferator-activated receptor γ can enhance mitochondrial potential and promote cell survival. *J. Biol. Chem.* **2002**, *277*, 31781–31788. [CrossRef]
486. Jo, S.H.; Yang, C.; Miao, Q.; Marzec, M.; Wasik, M.A.; Lu, P.; Wang, Y.L. Peroxisome proliferator-activated receptor γ promotes lymphocyte survival through its actions on cellular metabolic activities. *J. Immunol.* **2006**, *177*, 3737–3745. [CrossRef] [PubMed]
487. Klotz, L.; Burgdorf, S.; Dani, I.; Saijo, K.; Flossdorf, J.; Hucke, S.; Alferink, J.; Nowak, N.; Novak, N.; Beyer, M.; et al. The nuclear receptor PPAR γ selectively inhibits Th17 differentiation in a T cell-intrinsic fashion and suppresses CNS autoimmunity. *J. Exp. Med.* **2009**, *206*, 2079–2089. [CrossRef]
488. Tosolini, M.; Kirilovsky, A.; Mlecnik, B.; Fredriksen, T.; Mauger, S.; Bindea, G.; Berger, A.; Bruneval, P.; Fridman, W.H.; Pagès, F.; et al. Clinical impact of different classes of infiltrating T cytotoxic and helper cells (Th1, th2, treg, th17) in patients with colorectal cancer. *Cancer Res.* **2011**, *71*, 1263–1271. [CrossRef]

489. Tobiasova, Z.; Zhang, L.; Yi, T.; Qin, L.; Manes, T.D.; Kulkarni, S.; Lorber, M.I.; Rodriguez, F.C.; Choi, J.M.; Tellides, G.; et al. Peroxisome proliferator-activated receptor-γ agonists prevent in vivo remodeling of human artery induced by alloreactive T cells. *Circulation* **2011**, *124*, 196–205. [CrossRef]
490. Guri, A.J.; Mohapatra, S.K.; Horne, W.T.; Hontecillas, R.; Bassaganya-Riera, J. The role of T cell PPAR γ in mice with experimental inflammatory bowel disease. *BMC Gastroenterol.* **2010**, *10*, 60. [CrossRef]
491. Wohlfert, E.A.; Nichols, F.C.; Nevius, E.; Clark, R.B. Peroxisome proliferator-activated receptor γ (PPARγ) and immunoregulation: Enhancement of regulatory T cells through PPARγ-dependent and -independent mechanisms. *J. Immunol.* **2007**, *178*, 4129–4135. [CrossRef] [PubMed]
492. Feuerer, M.; Hill, J.A.; Mathis, D.; Benoist, C. Foxp3+ regulatory T cells: Differentiation, specification, subphenotypes. *Nat. Immunol.* **2009**, *10*, 689–695. [CrossRef] [PubMed]
493. Cipolletta, D.; Feuerer, M.; Li, A.; Kamei, N.; Lee, J.; Shoelson, S.E.; Benoist, C.; Mathis, D. PPAR-γ is a major driver of the accumulation and phenotype of adipose tissue Treg cells. *Nature* **2012**, *486*, 549–553. [CrossRef] [PubMed]
494. Ercolano, G.; Gomez-Cadena, A.; Dumauthioz, N.; Vanoni, G.; Kreutzfeldt, M.; Wyss, T.; Michalik, L.; Loyon, R.; Ianaro, A.; Ho, P.C.; et al. PPARɣ drives IL-33-dependent ILC2 pro-tumoral functions. *Nat. Commun.* **2021**, *12*, 2538. [CrossRef] [PubMed]
495. Faveeuw, C.; Fougeray, S.; Angeli, V.; Fontaine, J.; Chinetti, G.; Gosset, P.; Delerive, P.; Maliszewski, C.; Capron, M.; Staels, B.; et al. Peroxisome proliferator-activated receptor γ activators inhibit interleukin-12 production in murine dendritic cells. *FEBS Lett.* **2000**, *486*, 261–266. [CrossRef]
496. Szatmari, I.; Gogolak, P.; Im, J.S.; Dezso, B.; Rajnavolgyi, E.; Nagy, L. Activation of PPARγ specifies a dendritic cell subtype capable of enhanced induction of iNKT cell expansion. *Immunity* **2004**, *21*, 95–106. [CrossRef]
497. Fu, S.; He, K.; Tian, C.; Sun, H.; Zhu, C.; Bai, S.; Liu, J.; Wu, Q.; Xie, D.; Yue, T.; et al. Impaired lipid biosynthesis hinders anti-tumor efficacy of intratumoral iNKT cells. *Nat. Commun.* **2020**, *11*, 438. [CrossRef]
498. Kardos, J.; Chai, S.; Mose, L.E.; Selitsky, S.R.; Krishnan, B.; Saito, R.; Iglesia, M.D.; Milowsky, M.I.; Parker, J.S.; Kim, W.Y.; et al. Claudin-low bladder tumors are immune infiltrated and actively immune suppressed. *JCI Insight* **2016**, *1*, e85902. [CrossRef]
499. Tate, T.; Xiang, T.; Wobker, S.E.; Zhou, M.; Chen, X.; Kim, H.; Batourina, E.; Lin, C.S.; Kim, W.Y.; Lu, C.; et al. Pparg signaling controls bladder cancer subtype and immune exclusion. *Nat. Commun.* **2021**, *12*, 6160. [CrossRef]
500. Korpal, M.; Puyang, X.; Jeremy Wu, Z.; Seiler, R.; Furman, C.; Oo, H.Z.; Seiler, M.; Irwin, S.; Subramanian, V.; Julie Joshi, J.; et al. Evasion of immunosurveillance by genomic alterations of PPARγ/RXRα in bladder cancer. *Nat. Commun.* **2017**, *8*, 103. [CrossRef]
501. Gyöngyösi, A.; Nagy, L. Potential Therapeutic Use of PPARγ-Programed Human Monocyte-Derived Dendritic Cells in Cancer Vaccination Therapy. *PPAR Res.* **2008**, *2008*, 473804. [CrossRef] [PubMed]
502. Zhao, F.; Xiao, C.; Evans, K.S.; Theivanthiran, T.; DeVito, N.; Holtzhausen, A.; Liu, J.; Liu, X.; Boczkowski, D.; Nair, S.; et al. Paracrine Wnt5a-β-Catenin Signaling Triggers a Metabolic Program that Drives Dendritic Cell Tolerization. *Immunity* **2018**, *48*, 147–160.e147. [CrossRef] [PubMed]
503. Tan, P.H.; Tyrrell, H.E.; Gao, L.; Xu, D.; Quan, J.; Gill, D.; Rai, L.; Ding, Y.; Plant, G.; Chen, Y.; et al. Adiponectin receptor signaling on dendritic cells blunts antitumor immunity. *Cancer Res.* **2014**, *74*, 5711–5722. [CrossRef] [PubMed]
504. Zhang, X.; Sui, S.; Wang, L.; Li, H.; Zhang, L.; Xu, S.; Zheng, X. Inhibition of tumor propellant glutathione peroxidase 4 induces ferroptosis in cancer cells and enhances anticancer effect of cisplatin. *J. Cell. Physiol.* **2020**, *235*, 3425–3437. [CrossRef] [PubMed]
505. Han, L.; Bai, L.; Qu, C.; Dai, E.; Liu, J.; Kang, R.; Zhou, D.; Tang, D.; Zhao, Y. PPARG-mediated ferroptosis in dendritic cells limits antitumor immunity. *Biochem. Biophys. Res. Commun.* **2021**, *576*, 33–39. [CrossRef] [PubMed]
506. Zhang, X.; Rodriguez-Galán, M.C.; Subleski, J.J.; Ortaldo, J.R.; Hodge, D.L.; Wang, J.M.; Shimozato, O.; Reynolds, D.A.; Young, H.A. Peroxisome proliferator-activated receptor-γ and its ligands attenuate biologic functions of human natural killer cells. *Blood* **2004**, *104*, 3276–3284. [CrossRef]
507. Kliewer, S.A.; Forman, B.M.; Blumberg, B.; Ong, E.S.; Borgmeyer, U.; Mangelsdorf, D.J.; Umesono, K.; Evans, R.M. Differential expression and activation of a family of murine peroxisome proliferator-activated receptors. *Proc. Natl. Acad. Sci. USA* **1994**, *91*, 7355–7359. [CrossRef] [PubMed]
508. Tontonoz, P.; Nagy, L.; Alvarez, J.G.; Thomazy, V.A.; Evans, R.M. PPARγ promotes monocyte/macrophage differentiation and uptake of oxidized LDL. *Cell* **1998**, *93*, 241–252. [CrossRef]
509. Ricote, M.; Huang, J.; Fajas, L.; Li, A.; Welch, J.; Najib, J.; Witztum, J.L.; Auwerx, J.; Palinski, W.; Glass, C.K. Expression of the peroxisome proliferator-activated receptor γ (PPARγ) in human atherosclerosis and regulation in macrophages by colony stimulating factors and oxidized low density lipoprotein. *Proc. Natl. Acad. Sci. USA* **1998**, *95*, 7614–7619. [CrossRef]
510. Jiang, C.; Ting, A.T.; Seed, B. PPAR-γ agonists inhibit production of monocyte inflammatory cytokines. *Nature* **1998**, *391*, 82–86. [CrossRef]
511. Ricote, M.; Li, A.C.; Willson, T.M.; Kelly, C.J.; Glass, C.K. The peroxisome proliferator-activated receptor-γ is a negative regulator of macrophage activation. *Nature* **1998**, *391*, 79–82. [CrossRef] [PubMed]
512. Thieringer, R.; Fenyk-Melody, J.E.; Le Grand, C.B.; Shelton, B.A.; Detmers, P.A.; Somers, E.P.; Carbin, L.; Moller, D.E.; Wright, S.D.; Berger, J. Activation of peroxisome proliferator-activated receptor γ does not inhibit IL-6 or TNF-α responses of macrophages to lipopolysaccharide in vitro or in vivo. *J. Immunol.* **2000**, *164*, 1046–1054. [CrossRef] [PubMed]
513. Chawla, A.; Barak, Y.; Nagy, L.; Liao, D.; Tontonoz, P.; Evans, R.M. PPAR-γ dependent and independent effects on macrophage-gene expression in lipid metabolism and inflammation. *Nat. Med.* **2001**, *7*, 48–52. [CrossRef] [PubMed]

514. Moore, K.J.; Rosen, E.D.; Fitzgerald, M.L.; Randow, F.; Andersson, L.P.; Altshuler, D.; Milstone, D.S.; Mortensen, R.M.; Spiegelman, B.M.; Freeman, M.W. The role of PPAR-γ in macrophage differentiation and cholesterol uptake. *Nat. Med.* **2001**, *7*, 41–47. [CrossRef] [PubMed]
515. Bouhlel, M.A.; Derudas, B.; Rigamonti, E.; Dièvart, R.; Brozek, J.; Haulon, S.; Zawadzki, C.; Jude, B.; Torpier, G.; Marx, N.; et al. PPARγ activation primes human monocytes into alternative M2 macrophages with anti-inflammatory properties. *Cell Metab.* **2007**, *6*, 137–143. [CrossRef] [PubMed]
516. Van Ginderachter, J.A.; Meerschaut, S.; Liu, Y.; Brys, L.; De Groeve, K.; Hassanzadeh Ghassabeh, G.; Raes, G.; De Baetselier, P. Peroxisome proliferator-activated receptor γ (PPARγ) ligands reverse CTL suppression by alternatively activated (M2) macrophages in cancer. *Blood* **2006**, *108*, 525–535. [CrossRef] [PubMed]
517. Cheng, W.Y.; Huynh, H.; Chen, P.; Peña-Llopis, S.; Wan, Y. Macrophage PPARγ inhibits Gpr132 to mediate the anti-tumor effects of rosiglitazone. *Elife* **2016**, *5*, e18501. [CrossRef] [PubMed]
518. Gionfriddo, G.; Plastina, P.; Augimeri, G.; Catalano, S.; Giordano, C.; Barone, I.; Morelli, C.; Giordano, F.; Gelsomino, L.; Sisci, D.; et al. Modulating Tumor-Associated Macrophage Polarization by Synthetic and Natural PPARγ Ligands as a Potential Target in Breast Cancer. *Cells* **2020**, *9*, 174. [CrossRef]
519. Kim, Y.B.; Ahn, Y.H.; Jung, J.H.; Lee, Y.J.; Lee, J.H.; Kang, J.L. Programming of macrophages by UV-irradiated apoptotic cancer cells inhibits cancer progression and lung metastasis. *Cell Mol. Immunol.* **2019**, *16*, 851–867. [CrossRef]
520. Penas, F.; Mirkin, G.A.; Vera, M.; Cevey, Á.; González, C.D.; Gómez, M.I.; Sales, M.E.; Goren, N.B. Treatment in vitro with PPARα and PPARγ ligands drives M1-to-M2 polarization of macrophages from T. cruzi-infected mice. *Biochim. Biophys. Acta* **2015**, *1852*, 893–904. [CrossRef]
521. Moreira, T.G.; Horta, L.S.; Gomes-Santos, A.C.; Oliveira, R.P.; Queiroz, N.M.G.P.; Mangani, D.; Daniel, B.; Vieira, A.T.; Liu, S.; Rodrigues, A.M.; et al. CLA-supplemented diet accelerates experimental colorectal cancer by inducing TGF-β-producing macrophages and T cells. *Mucosal Immunol.* **2019**, *12*, 188–199. [CrossRef] [PubMed]

cells

MDPI

Article

Transcriptomics Reveals Discordant Lipid Metabolism Effects between In Vitro Models Exposed to Elafibranor and Liver Samples of NAFLD Patients after Bariatric Surgery

Joost Boeckmans *,†, Alexandra Gatzios †, Anja Heymans, Matthias Rombaut, Vera Rogiers, Joery De Kock, Tamara Vanhaecke ‡ and Robim M. Rodrigues *,‡

Department of In Vitro Toxicology and Dermato-Cosmetology, Faculty of Medicine and Pharmacy, Vrije Universiteit Brussel, 1090 Brussels, Belgium; alexandra.gatzios@vub.be (A.G.); anja.heymans@vub.be (A.H.); matthias.rombaut@vub.be (M.R.); vera.rogiers@vub.be (V.R.); joery.de.kock@vub.be (J.D.K.); tamara.vanhaecke@vub.be (T.V.)

* Correspondence: joost.boeckmans@vub.be (J.B.); robim.marcelino.rodrigues@vub.be (R.M.R.); Tel.: +32-(0)-2-477-45-19 (R.M.R.)

† These authors contributed equally to this work as first authors.

‡ These authors contributed equally to this work as senior authors.

Abstract: Background and aims: Non-alcoholic steatohepatitis (NASH) is a life-threatening stage of non-alcoholic fatty liver disease (NAFLD) for which no drugs have been approved. We have previously shown that human-derived hepatic in vitro models can be used to mimic key cellular mechanisms involved in the progression of NASH. In the present study, we first characterize the transcriptome of multiple in vitro NASH models. Subsequently, we investigate how elafibranor, which is a peroxisome proliferator-activated receptor (PPAR)-α/δ agonist that has recently failed a phase 3 clinical trial as a potential anti-NASH compound, modulates the transcriptome of these models. Finally, we compare the elafibranor-induced gene expression modulation to transcriptome data of patients with improved/resolved NAFLD/NASH upon bariatric surgery, which is the only proven clinical NASH therapy. Methods: Human whole genome microarrays were used for the transcriptomics evaluation of hepatic in vitro models. Comparison to publicly available clinical datasets was conducted using multiple bioinformatic application tools. Results: Primary human hepatocytes (PHH), HepaRG, and human skin stem cell-derived hepatic progenitors (hSKP-HPC) exposed to NASH-inducing triggers exhibit up to 35% overlap with datasets of liver samples from NASH patients. Exposure of the in vitro NASH models to elafibranor partially reversed the transcriptional modulations, predicting an inhibition of toll-like receptor (TLR)-2/4/9-mediated inflammatory responses, NFκB-signaling, hepatic fibrosis, and leukocyte migration. These transcriptomic changes were also observed in the datasets of liver samples of patients with resolved NASH. Peroxisome Proliferator Activated Receptor Alpha (PPARA), PPARG Coactivator 1 Alpha (PPARGC1A), and Sirtuin 1 (SIRT1) were identified as the major common upstream regulators upon exposure to elafibranor. Analysis of the downstream mechanistic networks further revealed that angiopoietin Like 4 (ANGPTL4), pyruvate dehydrogenase kinase 4 (PDK4), and perilipin 2 (PLIN2), which are involved in the promotion of hepatic lipid accumulation, were also commonly upregulated by elafibranor in all in vitro NASH models. Contrarily, these genes were not upregulated in liver samples of patients with resolved NASH. Conclusion: Transcriptomics comparison between in vitro NASH models exposed to elafibranor and clinical datasets of NAFLD patients after bariatric surgery reveals commonly modulated anti-inflammatory responses, but discordant modulations of key factors in lipid metabolism. This discordant adverse effect of elafibranor deserves further investigation when assessing PPAR-α/δ agonism as a potential anti-NASH therapy.

Keywords: NASH; NAFLD; elafibranor; in vitro; hSKP-HPC; HepaRG; hepatocytes; bariatric surgery; transcriptomics

Citation: Boeckmans, J.; Gatzios, A.; Heymans, A.; Rombaut, M.; Rogiers, V.; De Kock, J.; Vanhaecke, T.; Rodrigues, R.M. Transcriptomics Reveals Discordant Lipid Metabolism Effects between In Vitro Models Exposed to Elafibranor and Liver Samples of NAFLD Patients after Bariatric Surgery. *Cells* **2022**, *11*, 893. https://doi.org/10.3390/cells11050893

Academic Editors: Kay-Dietrich Wagner and Nicole Wagner

Received: 1 February 2022
Accepted: 2 March 2022
Published: 4 March 2022

1. Introduction

Non-alcoholic fatty liver disease (NAFLD) is an umbrella term that covers a range of liver pathologies, starting from uncomplicated hepatic steatosis to more severe disease stadia, including non-alcoholic steatohepatitis (NASH), fibrosis, cirrhosis, and ultimately hepatocellular carcinoma [1]. NASH is characterized by hepatic fat accumulation and inflammation and affects approximately 5 percent of the global population [1,2]. It is considered as the tipping point to the most severe stages of NAFLD and also carries high risk of (extra)hepatic complications [1,2]. Despite the urgent medical need, no drugs have been approved yet to treat NASH [3]. Weight loss through lifestyle intervention and bariatric surgery are still the only options to treat NASH [4,5]. Multiple phase 2 and 3 clinical studies are ongoing, but recently several phase 3 trials failed to meet their primary endpoints. This indicates that more performant and predictive preclinical models should be employed during the investigation of novel anti-NASH therapeutics. We recently showed that human hepatic in vitro models, including primary human hepatocytes (PHH), human skin precursor-derived hepatic progenitor cells (hSKP-HPC), and HepaRG and HepG2 cell lines, can model key NASH-specific cellular mechanisms and capture potential anti-NASH properties of novel compounds such as peroxisome proliferator-activated receptor (PPAR) agonists, a drug class that is under clinical evaluation for anti-NASH treatment [6,7]. PPARs have a regulatory role in hepatic lipid metabolism and inflammation and multiple dual- and pan-PPAR agonists, which can supposedly lead to a reduction of hepatic lipids and inflammation, are under investigation, including lanifibranor (a pan-PPAR agonist, phase 3, NCT04849728), saroglitazar (a dual PPAR-α/γ agonist, phase 2, NCT03863574), and elafibranor (a dual PPAR-α/δ agonist, phase 3, NCT02704403) [3]. Despite promising data obtained in early clinical trials, it was recently reported in an interim analysis of the phase 3 'RESOLVE-IT' trial that elafibranor did not meet its primary endpoint, namely 'NASH resolution without worsening fibrosis' [8,9]. However, no mechanistic explanation for the failed trial was provided. Since no clinical data (including transcriptomics data) or study samples are available from NASH patients treated with elafibranor, we were prompted to perform whole genome transcriptomics analyses of human-based in vitro NASH models exposed to elafibranor in order to investigate its molecular effects. By comparing the obtained results to transcriptomics datasets of liver samples from NAFLD patients after bariatric surgery, we aimed at gaining insights in the mechanisms that could possibly lay at the basis of the negative outcome of the interim analysis of the phase 3 'RESOLVE-IT' trial. Bariatric surgery impacts several hepatic metabolic pathways even before weight loss occurs, leading to a reduction of liver triglyceride levels as a major cause for the reduction of NASH [10,11]. Therefore, it was considered as a valid reference.

2. Materials and Methods

2.1. hSKP-HPC Cell Cultures

hSKP were isolated from foreskin samples after circumcision of 2–10 year old boys after obtaining informed consent from the parents and authorization from the medical ethical committee of the UZ Brussel. hSKP were isolated, cryopreserved, cultured, and differentiated to hSKP-HPC as earlier described [12]. Briefly, hSKP were recovered from liquid nitrogen at passage 6 and plated onto 75 cm^2 culture flasks (BD Falcon, Erembodegem, Belgium) in a humidified incubator with 5% (*v/v*) CO_2 at 37 °C. Hereafter, hSKP were split twice using TrypLE Express reagent (Thermo Fisher Scientific, Waltham, MA, USA). Hepatic differentiation was initiated when the culture reached 90 to 95% confluence in rat tail collagen type I (Corning Incorporated, Corning, NY, USA)-coated 75 cm^2 culture flasks (BD Falcon, Erembodegem, Belgium) in a humidified incubator with 5% (*v/v*) CO_2 at 33 °C. At day 20, the cells were detached using TrypLE Express reagent (Thermo Fisher Scientific, Waltham, MA, USA), transferred onto rat tail collagen type I (Corning Incorporated, Corning, NY, USA)-coated 24-multiwell plates (BD Falcon, Erembodegem, Belgium), and kept in culture for 7 more days. Then, hSKP-HPC were exposed for 24 h in a humidified incubator with 5% (*v/v*) CO_2 at 37 °C to 'NASH'-inducing factors, consisting of 100 nM insulin, 65 μM sodium oleate, 45 μM palmitic acid, 4.5 mg/mL glucose (all Sigma-Aldrich, St. Louis, MO, USA), 50 ng/mL tumor necrosis factor (TNF)-α (Prospec, Rehovot, Israel), 25 ng/mL interleukin (IL)-1β, and 8 ng/mL transforming growth factor (TGF)-β (both Peprotech, Cranbury, NJ, USA), as earlier documented [6].

2.2. HepaRG Cell Cultures

Differentiated HepaRG cells were purchased from Biopredic International (St.-Grégoire, France) and cultured according to the supplier's instructions. Briefly, the cryopreserved differentiated HepaRG cells were plated onto rat tail collagen type I (Corning Incorporated, Corning, NY, USA)-coated 24-multiwell plates (BD Falcon, Erembodegem, Belgium) using basal hepatic medium (Biopredic International, St.-Grégoire, France) supplemented with 'thawing/plating/general purpose medium supplement with antibiotics' (Biopredic International, St.-Grégoire, France) in a humidified incubator with 5% (*v/v*) CO_2 at 37 °C. After 24 h, the medium was replaced by basal hepatic medium (Biopredic International, St.-Grégoire, France) supplemented with 'maintenance/metabolism medium supplement with antibiotics' (Biopredic International, St.-Grégoire, France) and changed every 2–3 days for one week. 'NASH' inducers were added to William's E medium (Thermo Fisher Scientific, Waltham, MA, USA) supplemented with 10% (*v/v*) FBS (Hyclone Laboratories, Logan, UT, USA), 1% (*v/v*) PenStrep (Thermo Fisher Scientific, Waltham, MA, USA), 2 mM L-glutamine (Sigma-Aldrich, St. Louis, MO, USA), 872.69 nM insulin (Sigma-Aldrich, St. Louis, MO, USA), and 0.5 nM hydrocortisone. The 'NASH' inducers were identical to those described under 'hSKP-HPC culture', with the exception of the insulin concentration, which was raised to 8726.92 nM.

2.3. Primary Human Hepatocyte Cell Cultures

Fresh primary human hepatocytes (PHH) cultures were purchased from Biopredic International (St.-Grégoire, France). The cells were isolated and plated freshly onto rat tail collagen type I-coated 24-multiwell plates. After isolation, the cells were maintained in long term culture medium (Biopredic International, St.-Grégoire, France) and shipped at 37 °C. Four days after isolation, the cells were exposed for 24 h to conditions identical to those described for HepaRG cell cultures.

2.4. Compound Preparation

Elafibranor (Adooq Bioscience, Irvine, CA, USA) was dissolved in dimethylsulfoxide (DMSO) (Sigma–Aldrich, St. Louis, MO, USA) at a concentration of 60 mM. Elafibranor was added to the cultures at 60 μM concomitantly with the 'NASH' triggers.

2.5. RNA Extraction

After exposure for 24 h, the cell cultures were lysed using RNA-lysis buffer containing 1% *v/v* β-mercaptoethanol (Sigma-Aldrich, St. Louis, MO, USA) and stored in RNAse-free tubes at −80 °C. Total RNA was extracted and purified using the GenElute Mammalian Total RNA Purification Miniprep Kit (Sigma-Aldrich, St. Louis, MO, USA). RNA quantification was performed using a Nanodrop spectrophotometer (Thermo Fisher Scientific, Waltham, MA, USA).

2.6. Microarray

Whole genome microarrays were run using GeneChip™ Human Genome U133 Plus 2.0 Arrays (Thermo Fisher Scientific, Waltham, MA, USA) and reagents. 50 to 100 ng RNA was amplified using the GeneChip 3′ IVT PLUS Reagent Kit. aRNA was purified using magnetic particles before fragmentation. Fragmented aRNA was hybridized to Affymetrix Human Genome U133 plus 2.0 arrays and put in a Genechip Hybridization Oven-645 (Thermo Fisher Scientific, Waltham, MA, USA) at 45 °C and rotated at 60 rpm for 16 h. Then, the arrays were washed using a Genechip Fluidics Station-450 (Thermo Fisher Scientific, Waltham, MA, USA) and stained with an Affymetrix HWS kit. The chips were scanned using a GeneChip™ Scanner 3000 7G (Thermo Fisher Scientific, Waltham, MA, USA). Background adjustment, normalization (quantile), and summarization (median polish) of the raw data was done using Robust Multichip Average (RMA) Express. Volcano plots of all probesets (a probeset represents one or multiple hybridization probe pairs used together to interrogate a sequence corresponding to a specific gene) were made using Transcriptome Analysis Console (TAC) software (Thermo Fisher Scientific, Waltham, MA, USA) (eBayes, FDR fold change 2 and $p < 0.025$). Pathway analyses were performed using Ingenuity Pathway Analysis (IPA) (Qiagen, Hilden, Germany, version summer release 2020). A two-sided t-test with unequal variance was used for calculating the p-values before loading the data in IPA. Genes with a fold change <-2 or $>+2$ and $p < 0.025$ were included for further analysis. Activation z-scores of $\geq+2$ and ≤-2 were considered as significant for a likely activation or inhibition status, respectively. All specific terms related to IPA and used in the manuscript have been explained in Table S1. The raw .CEL files and normalized data of PHH (control-'NASH'-'NASH' + elafibranor), HepaRG ('NASH'-'NASH' + elafibranor), and hSKP-HPC ('NASH' + elafibranor) have been deposited in the Gene Expression Omnibus (GEO) database with accession number GSE166186.

2.7. Publicly Available Datasets

Microarray datasets of HepaRG control samples were obtained from Rodrigues et al. (GSE74000) [13]. Microarray datasets of hSKP-HPC control samples and hSKP-HPC 'NASH' samples were obtained from Boeckmans et al. (GSE126484) [6]. For comparative pathway analysis, IPA-deposited transcriptomics analyses (accessed October 2020–April 2021) of human liver biopsies were assessed in IPA-Analysis Match (Analysis names, disease states, dataset IDs, contributors, and references: '3-obesity [liver] NA 9506' (obesity (n = 16) vs. normal control (n = 12), GSE48452, Ahrens et al. [14]), '1- fatty liver [liver] NA 9504' (steatosis (n = 9) vs. normal control (n = 12), GSE48452, Ahrens et al. [14]), '2-nonalcoholic steatohepatitis (NASH) [liver] NA 9505' (NASH (n = 17) vs. normal control (n = 12), GSE48452, Ahrens et al. [14]), '3-nonalcoholic steatohepatitis (NASH) [liver] NA 3543' (NASH (n = 7) vs. normal control (n = 19), E-MEXP-3291, Lake et al. [15]), '2-nonalcoholic steatohepatitis (NASH) [liver] NA 10520' (NASH (n = 24) vs. normal control (n = 37), GSE61260, Horvath et al. [16]), '2-nonalcoholic steatohepatitis (NASH) [liver] NA 2178' (NASH (n = 15) vs. normal control (n = 14), GSE126848, Suppli et al. [17]), '5-obesity [liver] NA 9508' (obesity + bariatric surgery (n = 11) vs. none (n = 4), GSE48452, Ahrens et al. [14]), '4-fatty liver [liver] NA 9507' (steatosis + bariatric surgery (n = 5) vs. none (n = 7), GSE48452, Ahrens et al. [14]), '1-nonalcoholic fatty liver disease (NAFLD); overweight [liver] NA 12329' (NAFLD + bariatric surgery (n = 23) vs. none (n = 28), GSE83452, Lefebvre et al. and '3-nonalcoholic steatohepatitis (NASH);overweight [liver] NA 12331' (NASH + bariatric surgery (n = 15) vs. baseline (n = 55), GSE83452, Lefebvre et al. [18])).

2.8. Statistical Analysis

Z-scores were calculated using IPA. Other statistical analyses were performed using GraphPad Prism 7.0 (San Diego, CA, USA). The data are presented as the mean ± standard deviation. A one-way ANOVA with post hoc Bonferroni correction was used when comparing selected groups (control with 'NASH' and 'NASH' with 'NASH' + elafibranor). Three independent replicates were used.

3. Results

3.1. Transcriptomes of Human Hepatic In Vitro NASH Models Exhibit Similarities with Those of Human NASH Liver

Exposure of PHH, HepaRG, and hSKP-HPC cultures to 'NASH' triggers results in total probeset modulations of 6.30 %, 6.36 %, and 10.06 % in TAC, respectively, indicating that hSKP-HPC most sensitively responds to steatogenic and inflammatory triggers (Figure 1A and S1). These transcriptional alterations result in 2001, 2277, and 3213 analysis-ready molecules/genes for PHH, HepaRG, and hSKP-HPC in IPA, respectively. Comparative analysis of 'upstream regulators' (URs) and 'diseases and functions' between the in vitro datasets and NASH patient datasets reveals up to 50% similarity, depending on the model used and the reference clinical dataset, and shows NASH-specific transcriptional alterations in the in vitro models exposed to 'NASH' triggers (Figure 1B, top panels). NASH-triggered PHH exhibits the highest overall similarity z-score (with the dataset of Lake et al.), suggesting that PHH most closely mimics this patient-specific transcriptome (Figure 1B). Hierarchical clustering of URs of the 4 NASH patient datasets with the in vitro 'NASH' datasets shows that HepaRG cultures are closer to PHH cultures than hSKP-HPC cultures (Figure 1C top and bottom panels). NASH-triggered hSKP-HPC are, however, closer to the patient dataset of Suppli et al. than to the PHH and HepaRG in vitro NASH models (Figure 1C bottom panel). This indicates that all three in vitro NASH models exhibit a number of NASH-specific transcriptional features. Exposure of the in vitro NASH models to elafibranor results in modulation of 0.67%, 0.15%, and 8.19% of probesets in PHH, HepaRG, and hSKP-HPC, respectively, indicating that the latter model most sensitively responds to this compound (Figure 1D and S2). Similarity analysis of elafibranor-exposed PHH, HepaRG, and hSKP-HPC (based on 106, 45, and 2857 analysis-ready molecules/genes, respectively) indicates reversion of NASH-specific transcriptional responses, which is the most pronounced for hSKP-HPC (Figure 1E).

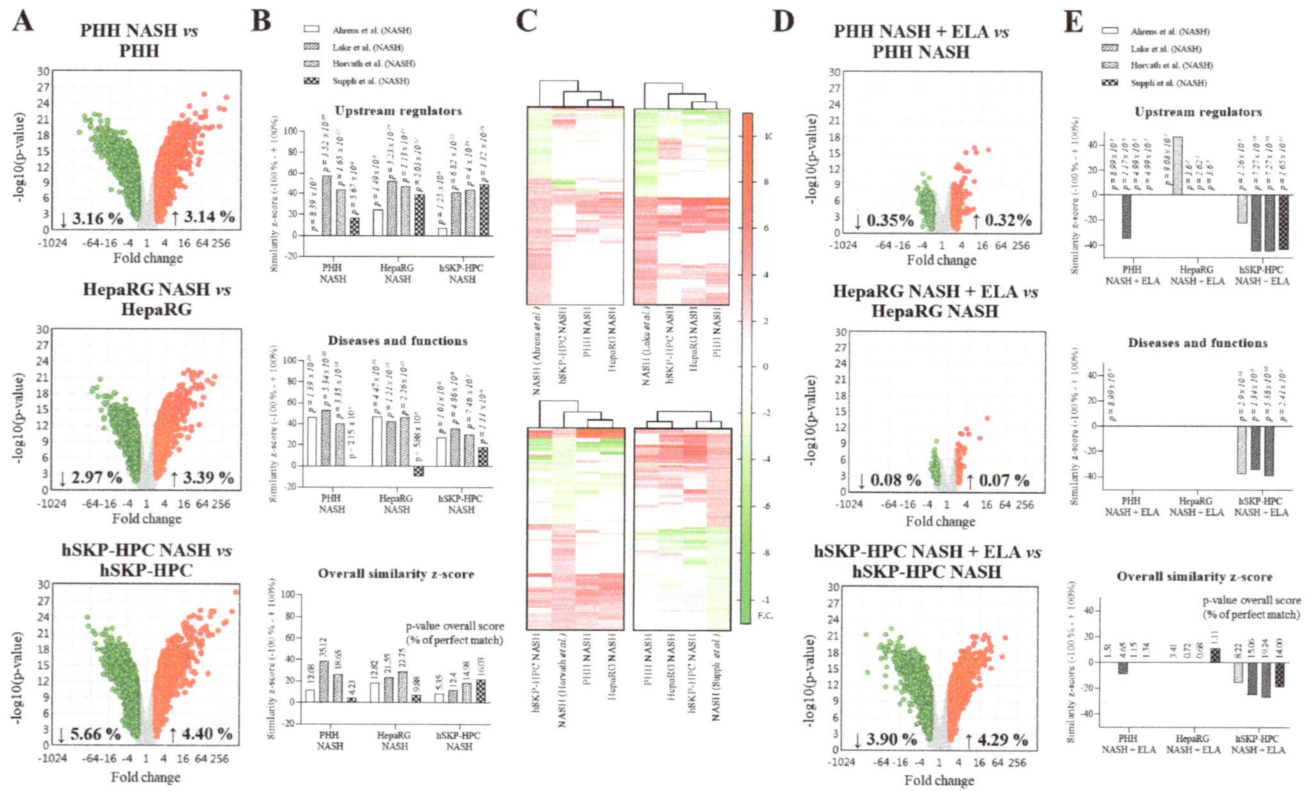

Figure 1. Transcriptional responses of human hepatic in vitro models exposed to 'NASH' triggers correspond with those of human NASH liver biopsies. (**A**) Microarray volcano plots of differentially modulated probesets in PHH, HepaRG, and hSKP-HPC cultures exposed to 'NASH' triggers. (**B**) Similarity analysis (IPA z-score) of in vitro NASH models and human liver NASH datasets. (**C**) Hierarchical clustering of upstream regulators in the in vitro NASH models with human liver NASH datasets. (**D**) Volcano plots of probesets of PHH-, HepaRG- and hSKP-HPC-'NASH' cultures exposed to elafibranor. (**E**) Similarity analysis (IPA z-score) of in vitro NASH models exposed to elafibranor with human liver NASH datasets [Volcano plots exclusion criteria: eBayes, FDR fold change <−2 or >+2 and $p < 0.025$; IPA analysis: Fisher's Exact Test, fold change cutoff: <−2 or >+2 and $p < 0.025$].

3.2. *Elafibranor Reduces Toll-like Receptor-Dependent Inflammation in PHH and hSKP-HPC In Vitro NASH Models*

One specific mechanism by which hepatic inflammation is initiated in NASH occurs via Toll-like receptors (TLRs) and MyD88 signaling [19]. Free fatty acids, such as palmitic acid, and damage-associated molecular patterns (DAMPs), such as high mobility group box 1 (HMGB1), are able to activate TLRs to initiate the inflammatory response. Therefore, we investigated the involvement of TLR-2, -4, and -9 signaling on the inflammatory response in the in vitro models exposed to NASH triggers and elafibranor. Pathway analysis shows that TLR-2, -4, and -9 signaling are activated in NASH-triggered PHH, but only TLR-2 and -9 or TLR-4 and -9 contribute to the inflammatory response in HepaRG and hSKP-HPC, respectively (Figure 2). These data suggest that the NASH-triggered hepatic in vitro models exhibit subtle differences in inflammation-promoting mechanisms. However, exposure to elafibranor seems to attenuate the inflammatory response by intervening in TLR-2, -4 as -9 signaling in PHH and hSKP-HPC (Figure 2). This effect is, however, absent in elafibranor-exposed HepaRG 'NASH' cultures, which do not show inhibition of TLR signaling, nor inhibition of the inflammatory response.

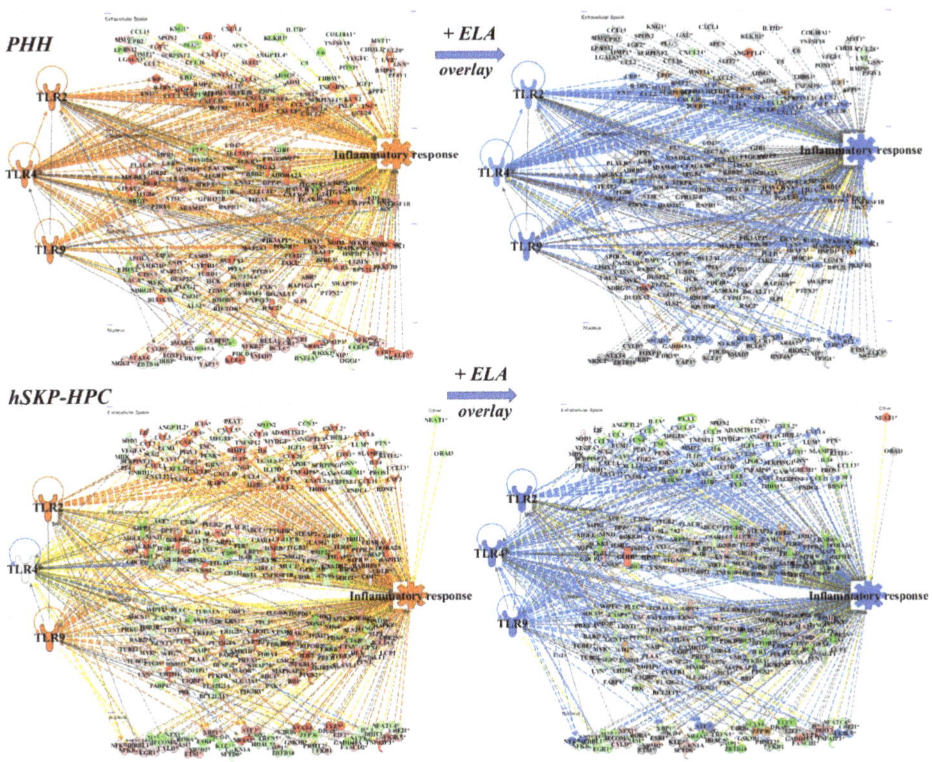

Figure 2. *Cont.*

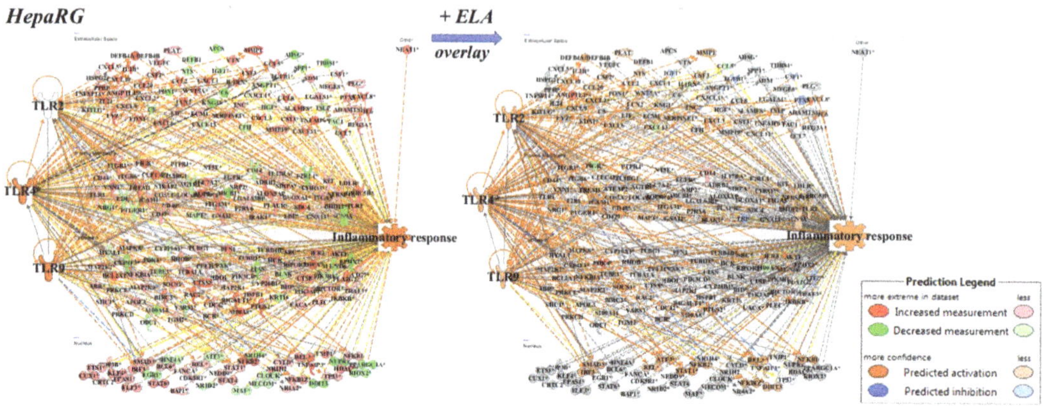

Figure 2. Pathway analysis predicts a reduction of TLR-2/4/9-dependent inflammatory responses by elafibranor in PHH and hSKP-HPC 'NASH' cultures, but not in HepaRG 'NASH' cultures. Prediction chart based on TLR signaling shows an activation of inflammatory responses in PHH, hSKP-HPC and HepaRG cell cultures exposed to NASH triggers (left panels). When PHH and hSKP-HPC 'NASH' cultures are further exposed to elafibranor, this inflammatory response is inhibited in PHH and hSKP-HPC cell cultures, but not in HepaRG cultures (right panels) [orange line = 'leads to activation'; blue line = 'leads to inhibition'; yellow line = 'findings inconsistent with state of downstream molecule'; grey line = 'effect not predicted'].

3.3. *Similarities in NASH-Related Upstream Regulators and Canonical Pathways in Hepatic In Vitro NASH Models Exposed to Elafibranor and Liver Samples of Patients after Bariatric Surgery*

Z-score determination of URs in NASH-triggered PHH, HepaRG, and hSKP-HPC cultures shows the predicted activation of prototypical mediators of inflammation during NASH, which is the most pronounced for NFkB (complex). However, p38 (mitogen-activated protein kinase) MAPK and Jun are also predicted as being activated in all three in vitro NASH models, indicating that multiple mechanisms lay at the basis of inflammatory signaling. Of note, these mediators are also predicted as being activated in the NASH patient datasets, indicating the clinical relevance of the inflammatory response in the in vitro NASH models at the transcriptional level. Inflammatory cytokines that lay at the origin of inflammation in human NASH are also highly involved in the in vitro NASH datasets, suggesting that the activation of the inflammatory transcription factors and kinases also occurs in a similar way. The TLR and MyD88 signaling are also activated in both in vitro and patient datasets. Of note, these NASH-inducing transcriptional alterations that occur in vitro can only be observed in specific NASH patient datasets, and not in datasets derived from livers obtained from patients suffering from obesity or steatosis alone. Exposure of the NASH-triggered hepatic in vitro models to elafibranor reverses the prediction statuses of most URs, which is the least pronounced for the HepaRG cultures. Furthermore, the elafibranor-induced reversion of prediction statuses of NASH-specific URs also occurs in the livers of patients that underwent bariatric surgery, which suggests that some of the weight loss-induced transcriptional effects can be observed in vitro through pharmacological intervention (Figure 3A).

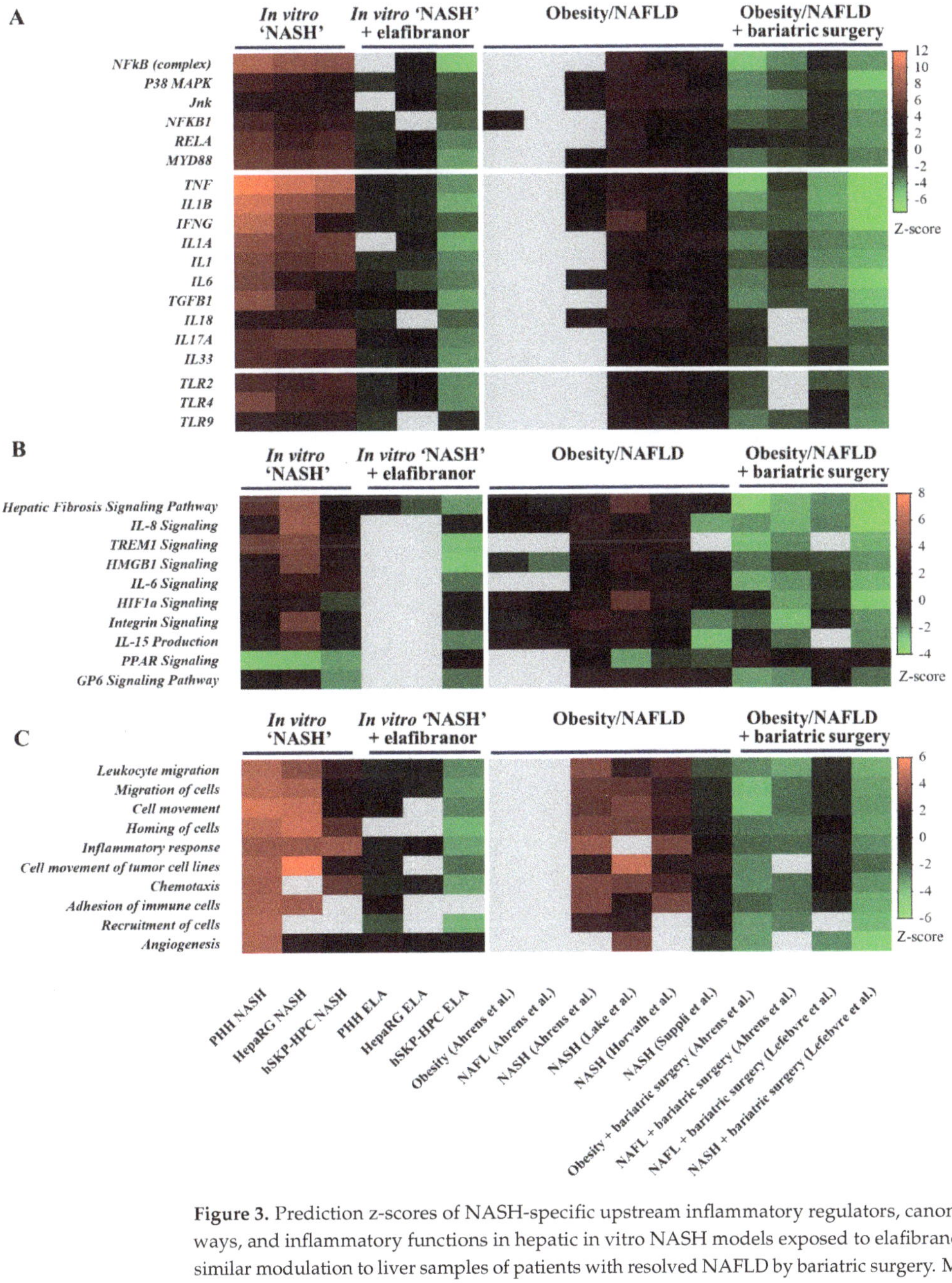

Figure 3. Prediction z-scores of NASH-specific upstream inflammatory regulators, canonical pathways, and inflammatory functions in hepatic in vitro NASH models exposed to elafibranor show a similar modulation to liver samples of patients with resolved NAFLD by bariatric surgery. Microarray analysis of upstream inflammatory mediators and Toll-like receptors (**A**), canonical pathways (**B**), and inflammatory functions (**C**) in PHH, HepaRG, and hSKP-HPC cultures exposed to 'NASH' triggers in the presence or absence of elafibranor is compared to data sets from NASH patients and NASH patients that underwent bariatric surgery [z-score of ≥ 2 indicates a significant activation; z-score of ≤ -2 indicates a significant inhibition; grey fields indicate no available z-score].

Exposure of PHH, HepaRG, and hSKP-HPC to 'NASH' triggers results in predicted activation of NASH-specific canonical pathways, including HMGB1 and triggering receptor expressed on myeloid cells-1 (TREM1)-signaling that are predicted as being activated in all three in vitro NASH datasets and at least one NASH patient dataset (Figure 3B). These canonical pathways are only predicted as being inhibited in the elafibranor-treated hSKP-HPC NASH samples, yet also in liver samples obtained from patients that underwent bariatric surgery, pointing to the clinical relevance of the latter model. Further, PPAR-signaling is impaired in all in vitro NASH datasets, which is also seen in human NASH liver samples. However, a recovery of the PPAR signaling is only observed in hSKP-HPC NASH model exposed to elafibranor and the clinical datasets from patients with resolved NASH upon bariatric surgery (Figure 3B).

Furthermore, NASH-triggered PHH, HepaRG, and hSKP-HPC were found to exhibit increased prediction statuses of functions related to chemotaxis and inflammatory cell migration (Figure 3C). These prominent inflammatory responses can be partially ascribed to the increased mRNA levels of a series of inflammatory chemokines (Figure S3). The cell migration-promoting functions that are activated in the in vitro models are also found as being activated in the NASH patient liver samples, suggesting that the similarities between the in vitro and patient datasets mainly rely on cell migratory processes. The addition of elafibranor to NASH-stimulated cell cultures robustly induces negative prediction statuses of these functions only in hSKP-HPC, which indicates that this model most sensitively responds to this compound to alter NASH-specific disease mechanisms. Moreover, these alterations are also present in liver biopsies obtained from patients with resolved NAFLD upon undergoing bariatric surgery, which further illustrates the relevance of hSKP-HPC.

3.4. PPARGC1A, PPARA, and SIRT1 Regulate Shared Elafibranor-Induced Genes in NASH-Triggered PHH, HepaRG, and hSKP-HPC

In order to detect biologically relevant changes in gene expression in response to elafibranor, we opted to combine the data of the three different human-based cell systems (PHH, HepaRG, and hSKP-HPC). Comparison analysis of NASH-triggered PHH, HepaRG, and hSKP-HPC cultures exposed to elafibranor unveils that Peroxisome Proliferator Activated Receptor Alpha (PPARA), PPARG Coactivator 1 Alpha (PPARGC1A), and Sirtuin 1 (SIRT1) are the highest expressed common upstream regulators, suggesting that this triad of regulators control the expression of an important subset of genes involved in PPAR-α/δ signaling (Figure 4A). The mechanistic networks of PPARGC1A, PPARA, and SIRT1 and their downstream modulators based on the datasets of PHH-, HepaRG-, and hSKP-HPC-NASH exposed to elafibranor, unveils common regulation of CCL5, ANGPTL4, PDK4, and PLIN2 (Figure 4B). In addition, an unfiltered comparison analysis of the datasets obtained from the three in vitro NASH models exposed to elafibranor shows that elafibranor commonly modulates 6 genes, namely TDO2, ARG2, and also the previously identified CCL5, ANGPTL4, PDK4, and PLIN2 (Figure 4C). Considering that the level of expression of TDO2 is very low in hSKP-HPC and that the modulations of ARG2 are not consistent among the different in vitro models (Figure S4), only CCL5, ANGPTL4, PDK4, and PLIN2 are regulated by shared top-upstream regulators, and we selected these 4 genes for further investigation.

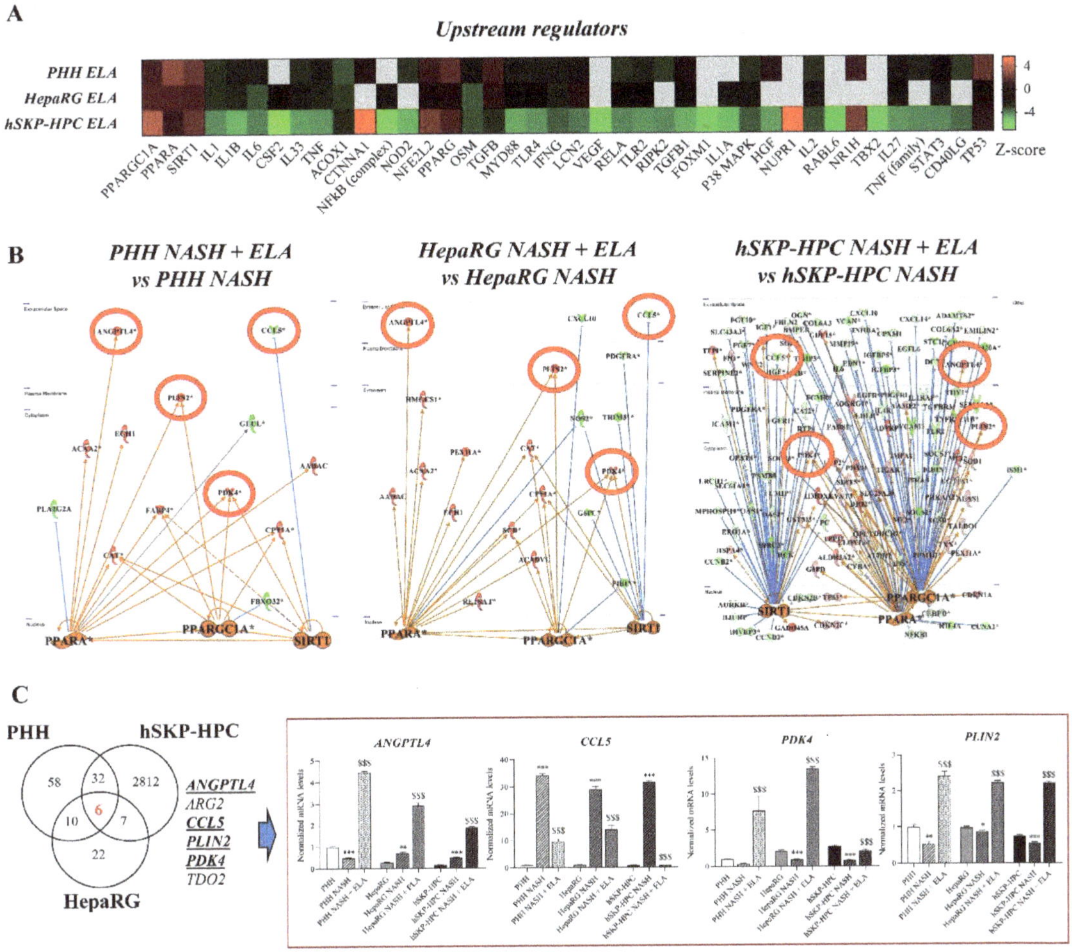

Figure 4. PPAR-α/δ-induced commonly modulated genes among hepatic in vitro NASH models are regulated by shared upstream regulators. (**A**) Top-upstream regulators in PHH, HepaRG, and hSKP-HPC NASH models exposed to elafibranor. (**B**) Mechanistic networks of hepatic in vitro NASH models exposed to elafibranor with connected SIRT1, PPARGC1A1, and PPARA top upstream regulators. (**C**) Number of commonly modulated genes induced by elafibranor in PHH-, HepaRG-, and hSKP-HPC NASH models. (**A**,**B**): fold change: <−2 or >+2 and Fisher's Exact Test with $p < 0.025$; (**C**) One-way ANOVA with post hoc Bonferroni correction vs. control condition (*, ** and ***; $p \leq 0.05$, $p \leq 0.01$, and $p \leq 0.001$) and vs. 'NASH' condition ($$$; $p \leq 0.001$)].

3.5. Elafibranor-Mediated Induction of ANGPTL4, PDK4, and PLIN2 In Vitro Predicts Induction of Steatosis, Contrarily to Anti-NASH Bariatric Surgery

The robust induction of ANGPTL4, PDK4, and PLIN2 by elafibranor brings up the question of whether or not these lipid metabolism-related genes induce NAFLD-promoting effects. Pathway analysis revealed that these 3 genes have a downstream effect in functions related to hepatic steatosis and inflammation in the three investigated in vitro NASH models (Figure 5A). As such, elafibranor-induced modulation of ANGPTL4, PDK4, and PLIN2 leads to a predicted increase in multiple pro-steatogenic functions, including 'accumulation and formation of lipid droplets', 'fatty acid metabolism', and 'concentration of triacylglycerol'. However, the induction of this set of genes is absent in the clinical samples of patients suffering from obesity/fatty liver/NASH that underwent bariatric surgery, which is considered as an intervention that improves NAFLD/NASH (Figure 5B). The discordant effect of elafibranor in in vitro NASH models in comparison with the clinical datasets should be further evaluated as a potential molecular explanation for the failure of elafibranor during the phase 3 'RESOLVE-IT' trial.

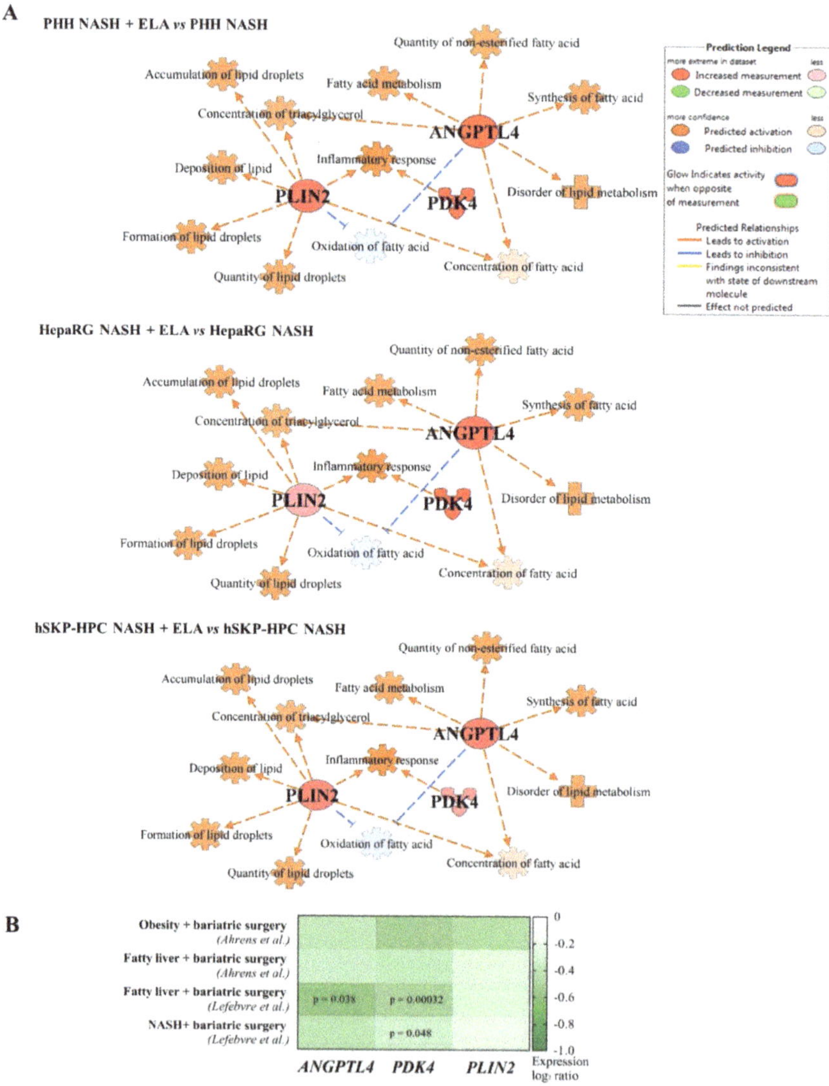

Figure 5. Elafibranor-mediated upregulation of ANGPTL4, PDK4, and PLIN2 is predicted to induce steatosis functions in in vitro NASH models, contrarily to observed downregulation in patients with resolved NASH due to bariatric surgery. (**A**) Regulation of steatosis- and inflammation-related functions by PLIN2, ANGPTL4, and PDK4 in PHH, HepaRG, and hSKP-HPC cultures exposed to 'NASH' triggers and elafibranor. (**B**) Differential expression of ANGPTL4, PDK4, and PLIN2 in liver tissue from obesity/NAFLD patients that underwent bariatric surgery.

4. Discussion

NASH is a severe form of NAFLD and is regarded as an important stage during disease progression towards fibrosis, cirrhosis, and cancer. In recent years, several phase 3 NASH clinical trials have failed due to lack of efficacy of the compound under study [20,21]. Therefore, preclinical screening for novel anti-NASH compounds is still highly relevant. However, it is of utmost importance to learn from those failed large clinical studies to improve future pharmacological strategies.

NASH is traditionally studied in animal models. However, due to the complexity of NASH and its human-specificity, NASH-related research has been largely directed towards human-based technologies in recent years [22]. To that extent, we recently developed a methodology that can be employed to study NASH in vitro and screen for potential anti-NASH compounds [6,7]. In the present study, we confirmed the clinical relevance of the thus-developed in vitro disease models by whole genome-based pathway analysis using four publicly available datasets of patients suffering from NASH. Mechanistically, triggering of human hepatic in vitro models resulted in the activation of NASH-specific upstream regulators, including soluble inflammatory mediators and TLRs, as well as transcription factors and kinases known to be activated in the livers of NASH patients, such as NFκB, Jun, and p38 MAPK [23,24]. Furthermore, there were canonical pathways, among which HMGB1- [25] and TREM1-signaling pathways [26] that could be modulated in these in vitro systems, along with inflammation- and cell migration-related processes. Hence, apart for drug screening purposes, these models could also be used for the mechanistic investigation of NASH-promoting processes.

PPAR agonism is considered a promising treatment option for NASH [3]. Elafibranor, a dual modulator of PPAR-α/δ, is one of the developed anti-NASH compounds under evaluation that proved to be effective in a phase 2 study [27]. Nonetheless, an interim analysis of the phase 3 elafibranor trial showed that elafibranor is not more efficacious than placebo [8]. This prompted us to investigate the potential opposite effects to the reduction of NASH characteristics that elafibranor can induce. To this end, we investigated the effects of elafibranor using in vitro NASH models and relied on datasets of patients with resolved NASH upon bariatric surgery as reference. Since no clinical information of elafibranor is available, including a lack of transcriptomic liver data, and since bariatric surgery is the only clinical therapy proven to resolve NASH, among others inducing hepatic metabolic pathways leading to the reduction of liver triglycerides, we believe that these clinical datasets are appropriate to use in our analysis. In a first step, we found that PPARG1A1, PPARA, and SIRT1 are shared elafibranor-induced URs that are predicted as being activated in PHH-, HepaRG-, and hSKP-HPC-derived NASH models. These regulators are, however, known for their NASH-resolving properties, since PPARGC1A1 (i.e., PGC-1α) is a master regulator of mitochondrial biogenesis [28], PPARA expression inversely correlates with histological severity of NASH and recovers upon NASH improvement [29] and deletion of SIRT1, a NAD+-dependent protein deacetylase important for energy homeostasis, results in impaired PPAR-α signaling and decreased fatty acid β-oxidation [30]. In contrast, these three regulators also dictate the increased expression of ANGPTL4, PDK4, and PLIN2, which are known PPAR-α target genes [31]. Although PPAR agonism has been considered as a promising treatment strategy for NASH, these genes are also known for their NASH-promoting effects. Indeed, PLIN2, a lipid droplet protein, promotes obesity and progressive fatty liver disease in mice [32,33], and loss of PLIN2 lessens diet-induced hepatic steatosis, inflammation, and fibrosis [34]. PDK4, which is an inhibitor of the pyruvate dehydrogenase complex [35], is higher expressed on the protein and gene level in NASH liver samples, and PDK4 deficiency in mice leads to reduced steatosis [36]. The effect of ANGPTL4, an exercise-induced hepatokine, is less straightforward. ANGPTL4 is an inhibitor of lipoprotein lipase that stimulates lipolysis but inhibits triglyceride plasma clearance [37]. ANGPTL4 overexpression leads to reduced adiposity at the expense of increased liver steatosis and can therefore be considered as having both positive and negative effects on

NASH [38,39]. Notwithstanding, ANGPTL4 has also been earlier proposed as a biomarker for drug-induced steatosis [40].

5. Conclusions

By using in vitro NASH models, this study elucidates elafibranor-induced transcriptomic modulations that predict a number of opposing mechanistic effects to those observed in patients with improved NAFLD after bariatric surgery. These ambiguities seem to be related with the modulation of the functions of ANGPTL4, PDK4, and PLIN2 that play prominent roles in lipid metabolism. However, whether these effects truly contributed to the failure of the clinical phase 3 elafibranor trial should be further explored when clinical datasets or samples become available.

Supplementary Materials: The following are available online at https://www.mdpi.com/article/10.3390/cells11050893/s1, Figure S1: Venn Diagram of in vitro NASH disease models, Figure S2: PCA plots of hepatic in vitro models exposed to 'NASH' triggers and elafibranor, Figure S3: Modulation inflammatory chemokines in hepatic in vitro models exposed to 'NASH' triggers and elafibranor relative to the PHH control samples, Figure S4: Gene expression of TDO2 and ARG2 in human hepatic in vitro models exposed to 'NASH' triggers and elafibranor, Table S1: Terms related to Ingenuity Pathway Analysis used in the manuscript.

Author Contributions: Conceptualization, J.B. and R.M.R.; methodology, J.B. and R.M.R.; software, J.B.; formal analysis, J.B., A.G. and R.M.R.; investigation, J.B., A.G., R.M.R., T.V., A.H., M.R., V.R. and J.D.K.; resources, T.V., V.R., R.M.R. and J.B.; data curation, J.B.; writing—original draft preparation, J.B., A.G. and R.M.R.; writing—review and editing, J.B., A.G., R.M.R., T.V., A.H., M.R., V.R. and J.D.K.; visualization, J.B. and R.M.R.; supervision, R.M.R. and T.V.; project administration, T.V. and R.M.R.; funding acquisition, T.V., R.M.R. and J.B. All authors have read and agreed to the published version of the manuscript.

Funding: This work was funded by grants of Research Foundation Flanders (12H2219N, 1S73019N and G042019N), and the Research Chair Mireille Aerens for the development of Alternatives to Animal Testing.

Institutional Review Board Statement: Ethics approval for the use of human skin-derived precursors (hSKP) was obtained from the medical ethical committee of the UZ Brussel (Belgium).

Informed Consent Statement: Not applicable.

Data Availability Statement: Raw microarray data have been deposited in NCBI's Gene Expression Omnibus and are accessible through GEO Series accession number GEO: GSE166186. Public available datasets have been obtained as described above under 'Materials and Methods'.

Conflicts of Interest: The authors declare no conflict of interest.

References

1. Cotter, T.G.; Rinella, M. Nonalcoholic Fatty Liver Disease 2020: The State of the Disease. *Gastroenterology* **2020**, *158*, 1851–1864. [CrossRef] [PubMed]
2. Younossi, Z.; Anstee, Q.M.; Marietti, M.; Hardy, T.; Henry, L.; Eslam, M.; George, J.; Bugianesi, E. Global Burden of NAFLD and NASH: Trends, Predictions, Risk Factors and Prevention. *Nat. Rev. Gastroenterol. Hepatol.* **2018**, *15*, 11–20. [CrossRef] [PubMed]
3. Boeckmans, J.; Natale, A.; Rombaut, M.; Buyl, K.; Rogiers, V.; De Kock, J.; Vanhaecke, T.; Rodrigues, R.M. Anti-NASH Drug Development Hitches a Lift on PPAR Agonism. *Cells* **2020**, *9*, 37. [CrossRef] [PubMed]
4. Vilar-Gomez, E.; Martinez-Perez, Y.; Calzadilla-Bertot, L.; Torres-Gonzalez, A. Weight Loss via Lifestyle Modification Significantly Reduces Features of Nonalcoholic Steatohepatitis. *Gastroenterology* **2015**, *149*, 367–378. [CrossRef]
5. Klebanoff, M.J.; Corey, K.E.; Chhatwal, J.; Kaplan, L.M.; Chung, R.T.; Hur, C. Bariatric Surgery for Nonalcoholic Steatohepatitis: A Clinical and Cost-Effectiveness Analysis. *Hepatology* **2017**, *65*, 1156–1164. [CrossRef]
6. Boeckmans, J.; Buyl, K.; Natale, A.; Vandenbempt, V.; Branson, S.; De Boe, V.; Rogiers, V.; De Kock, J.; Rodrigues, R.M.; Vanhaecke, T. Elafibranor Restricts Lipogenic and Inflammatory Responses in a Human Skin Stem Cell-Derived Model of NASH. *Pharmacol. Res.* **2019**, *144*, 377–389. [CrossRef]
7. Boeckmans, J.; Natale, A.; Rombaut, M.; Buyl, K.; Cami, B.; De Boe, V.; Heymans, A.; Rogiers, V.; De Kock, J.; Vanhaecke, T.; et al. Human Hepatic in Vitro Models Reveal Distinct Anti-NASH Potencies of PPAR Agonists. *Cell Biol. Toxicol.* **2020**, *37*, 293–311. [CrossRef]

8. *GENFIT Announces Results from Interim Analysis of RESOLVE-IT Phase 3 Trial of Elafibranor in Adults with NASH and Fibrosis*; Genfit: Loos, France, 2020; pp. 1–5.
9. Harrison, S.A.; Ratziu, V.; Bedossa, P.; Dufour, J.-F.; Kruger, F.; Schattenberg, J.M.; Francque, S.M.; Arrese, M.; George, J.; Bugianesi, E.; et al. RESOLVE-IT Phase 3 Trial of Elafibranor in NASH: Final Results of the Week 72 Interim Surrogate Efficacy Analysis. Proceedings of AASLD 2020, Boston, MA, USA, 13–16 November 2020.
10. Lefere, S.; Onghena, L.; Vanlander, A.; van Nieuwenhove, Y.; Devisscher, L.; Geerts, A. Bariatric Surgery and the Liver-Mechanisms, Benefits, and Risks. *Obes. Rev.* **2021**, *22*. [CrossRef]
11. Chambers, A.P.; Jessen, L.; Ryan, K.K.; Sisley, S.; Wilsonpérez, H.E.; Stefater, M.A.; Gaitonde, S.G.; Sorrell, J.E.; Toure, M.; Berger, J.; et al. Weight-Independent Changes in Blood Glucose Homeostasis after Gastric Bypass or Vertical Sleeve Gastrectomy in Rats. *Gastroenterology* **2011**, *141*, 950–958. [CrossRef]
12. Rodrigues, R.M.; De Kock, J.; Branson, S.; Vinken, M.; Meganathan, K.; Chaudhari, U.; Sachinidis, A.; Govaere, O.; Roskams, T.; De Boe, V.; et al. Human Skin-Derived Stem Cells as a Novel Cell Source for In Vitro Hepatotoxicity Screening of Pharmaceuticals. *Stem Cells Dev.* **2014**, *23*, 44–55. [CrossRef]
13. Rodrigues, R.M.; Heymans, A.; De Boe, V.; Sachinidis, A.; Chaudhari, U.; Govaere, O.; Roskams, T.; Vanhaecke, T.; Rogiers, V.; De Kock, J. Toxicogenomics-Based Prediction of Acetaminophen-Induced Liver Injury Using Human Hepatic Cell Systems. *Toxicol. Lett.* **2016**, *240*, 50–59. [CrossRef] [PubMed]
14. Ahrens, M.; Ammerpohl, O.; von Schönfels, W.; Kolarova, J.; Bens, S.; Itzel, T.; Ahrens, M.; Ammerpohl, O.; von Scho, W.; Teufel, A.; et al. DNA Methylation Analysis in Nonalcoholic Fatty Liver Disease Suggests Distinct Disease-Specific and Remodeling Signatures after Bariatric Surgery. *Cell Metab.* **2013**, *18*, 296–302. [CrossRef] [PubMed]
15. Lake, A.D.; Novak, P.; Fisher, C.D.; Jackson, J.P.; Hardwick, R.N.; Billheimer, D.D.; Klimecki, W.T.; Cherrington, N.J. Analysis of Global and Absorption, Distribution, Metabolism, and Elimination Gene Expression in the Progressive Stages of Human Nonalcoholic Fatty Liver Disease. *Drug Metab. Dispos.* **2011**, *39*, 1954–1960. [CrossRef]
16. Horvath, S.; Erhart, W.; Brosch, M.; Ammerpohl, O.; Von Schönfels, W.; Ahrens, M.; Heits, N.; Bell, J.T.; Tsai, P.C.; Spector, T.D.; et al. Obesity Accelerates Epigenetic Aging of Human Liver. *Proc. Natl. Acad. Sci. USA* **2014**, *111*, 15538–15543. [CrossRef] [PubMed]
17. Suppli, M.P.; Rigbolt, K.T.G.; Veidal, S.S.; Heebøll, S.; Eriksen, P.L.; Demant, M.; Bagger, J.I.; Nielsen, J.C.; Oró, D.; Thrane, S.W.; et al. Hepatic Transcriptome Signatures in Patients with Varying Degrees of Nonalcoholic Fatty Liver Disease Compared with Healthy Normal-Weight Individuals. *Am. J. Physiol. Gastrointest. Liver Physiol.* **2019**, *316*, G462–G472. [CrossRef]
18. Lefebvre, P.; Lalloyer, F.; Baugé, E.; Pawlak, M.; Gheeraert, C.; Dehondt, H.; Vanhoutte, J.; Woitrain, E.; Hennuyer, N.; Mazuy, C.; et al. Interspecies NASH Disease Activity Whole-Genome Profiling Identifies a Fibrogenic Role of PPARα-Regulated Dermatopontin. *JCI Insight* **2017**, *2*. [CrossRef] [PubMed]
19. Miura, K.; Seki, E.; Ohnishi, H.; Brenner, D.A. Role of Toll-like Receptors and Their Downstream Molecules in the Development of Nonalcoholic Fatty Liver Disease. *Gastroenterol. Res. Pract.* **2010**, *2010*, 362847. [CrossRef] [PubMed]
20. Lambrecht, J.; van Grunsven, L.A.; Tacke, F. Current and Emerging Pharmacotherapeutic Interventions for the Treatment of Liver Fibrosis. *Expert Opin. Pharmacother.* **2020**, *21*, 1637–1650. [CrossRef]
21. Neuschwander-Tetri, B.A. Therapeutic Landscape for NAFLD in 2020. *Gastroenterol* **2020**, *158*, 1984–1998.e3. [CrossRef]
22. Boeckmans, J.; Natale, A.; Buyl, K.; Rogiers, V.; De Kock, J.; Vanhaecke, T.; Robim, M. Human-Based Systems: Mechanistic NASH Modelling Just around the Corner? *Pharmacol. Res.* **2018**, *134*, 257–267. [CrossRef]
23. Frades, I.; Andreasson, E.; Mato, J.M.; Alexandersson, E.; Matthiesen, R.; Martínez-Chantar, M.L. Integrative Genomic Signatures Of Hepatocellular Carcinoma Derived from Nonalcoholic Fatty Liver Disease. *PLoS ONE* **2015**, *10*, e0124544. [CrossRef] [PubMed]
24. Schuster, S.; Cabrera, D.; Arrese, M.; Feldstein, A.E. Triggering and Resolution of Inflammation in NASH. *Nat. Rev. Gastroenterol. Hepatol.* **2018**, *15*, 349–364. [CrossRef]
25. Gaskell, H.; Ge, X.; Nieto, N. High-Mobility Group Box-1 and Liver Disease. *Hepatol. Commun.* **2018**, *2*, 1005–1020. [CrossRef] [PubMed]
26. Rao, S.; Huang, J.; Shen, Z.; Xiang, C.; Zhang, M.; Lu, X. Inhibition of TREM-1 Attenuates Inflammation and Lipid Accumulation in Diet-Induced Nonalcoholic Fatty Liver Disease. *J. Cell. Biochem.* **2019**, *120*, 11867–11877. [CrossRef] [PubMed]
27. Ratziu, V.; Harrison, S.A.; Francque, S.; Bedossa, P.; Lehert, P.; Serfaty, L.; Romero-Gomez, M.; Boursier, J.; Abdelmalek, M.; Caldwell, S.; et al. Elafibranor, an Agonist of the Peroxisome Proliferator-Activated Receptor-α and -δ, Induces Resolution of Nonalcoholic Steatohepatitis Without Fibrosis Worsening. *Gastroenterology* **2016**, *150*, 1147–1159. [CrossRef]
28. Scarpulla, R.C. Metabolic Control of Mitochondrial Biogenesis through the PGC-1 Family Regulatory Network. *Biochim. Biophys. Acta* **2011**, *1813*, 1269–1278. [CrossRef]
29. Francque, S.; Verrijken, A.; Caron, S.; Prawitt, J.; Paumelle, R.; Derudas, B.; Lefebvre, P.; Taskinen, M.R.; Van Hul, W.; Mertens, I.; et al. PPAR-α Gene Expression Correlates with Severity and Histological Treatment Response in Patients with Non-Alcoholic Steatohepatitis. *J. Hepatol.* **2015**, *63*, 164–173. [CrossRef]
30. Purushotham, A.; Schug, T.T.; Xu, Q.; Surapureddi, S.; Guo, X.; Li, X. Hepatocyte-Specific Deletion of SIRT1 Alters Fatty Acid Metabolism and Results in Hepatic Steatosis and Inflammation. *Cell Metab.* **2009**, *9*, 327–338. [CrossRef]
31. Rakhshandehroo, M.; Knoch, B.; Müller, M.; Kersten, S. Peroxisome Proliferator-Activated Receptor Alpha Target Genes. *PPAR Res.* **2010**, *2010*, 393– 416. [CrossRef]

32. Najt, C.P.; Lwande, J.S.; McIntosh, A.L.; Senthivinayagam, S.; Gupta, S.; Kuhn, L.A.; Atshaves, B.P. Structural and Functional Assessment of Perilipin 2 Lipid Binding Domain(s). *Biochemistry* **2014**, *53*, 7051–7066. [CrossRef]
33. Orlicky, D.J.; Libby, A.E.; Bales, E.S.; McMahan, R.H.; Monks, J.; La Rosa, F.G.; McManaman, J.L. Perilipin-2 Promotes Obesity and Progressive Fatty Liver Disease in Mice through Mechanistically Distinct Hepatocyte and Extra-Hepatocyte Actions. *J. Physiol.* **2019**, *597*, 1565–1584. [CrossRef] [PubMed]
34. Najt, C.P.; Senthivinayagam, S.; Aljazi, M.B.; Fader, K.A.; Olenic, S.D.; Brock, J.R.L.; Lydic, T.A.; Jones, A.D.; Atshaves, B.P. Liver-Specific Loss of Perilipin 2 Alleviates Diet-Induced Hepatic Steatosis, Inflammation, and Fibrosis. *Am. J. Physiol.-Gastrointest. Liver Physiol.* **2016**, *310*, G726–G738. [CrossRef]
35. Zhang, S.; Hulver, M.W.; McMillan, R.P.; Cline, M.A.; Gilbert, E.R. The Pivotal Role of Pyruvate Dehydrogenase Kinases in Metabolic Flexibility. *Nutr. Metab.* **2014**, *11*, 1–9. [CrossRef] [PubMed]
36. Zhang, M.; Zhao, Y.; Li, Z.; Wang, C. Pyruvate Dehydrogenase Kinase 4 Mediates Lipogenesis and Contributes to the Pathogenesis of Nonalcoholic Steatohepatitis. *Biochem. Biophys. Res. Commun.* **2018**, *495*, 582–586. [CrossRef] [PubMed]
37. Takahashi, H.; Kotani, K.; Tanaka, K.; Egucih, Y.; Anzai, K. Therapeutic Approaches to Nonalcoholic Fatty Liver Disease: Exercise Intervention and Related Mechanisms. *Front. Endocrinol.* **2018**, *9*, 588. [CrossRef] [PubMed]
38. Mandard, S.; Zandbergen, F.; Van Straten, E.; Wahli, W.; Kuipers, F.; Müller, M.; Kersten, S. The Fasting-Induced Adipose Factor/Angiopoietin-like Protein 4 Is Physically Associated with Lipoproteins and Governs Plasma Lipid Levels and Adiposity. *J. Biol. Chem.* **2006**, *281*, 934–944. [CrossRef]
39. Xu, A.; Lam, M.C.; Chan, K.W.; Wang, Y.; Zhang, J.; Boo, R.L.C.; Xu, J.Y.; Chen, B.; Chow, W.S.; Tso, A.W.K.; et al. Angiopoietin-like Protein 4 Decreases Blood Glucose and Improves Glucose Tolerance but Induces Hyperlipidemia and Hepatic Steatosis in Mice. *Proc. Natl. Acad. Sci. United States Am.* **2005**, *102*, 6086–6091. [CrossRef]
40. Sahini, N.; Selvaraj, S.; Borlak, J. Whole Genome Transcript Profiling of Drug Induced Steatosis in Rats Reveals a Gene Signature Predictive of Outcome. *PLoS ONE* **2014**, *9*, e114085. [CrossRef]

cells

MDPI

Article

Selective PPARα Modulator Pemafibrate and Sodium-Glucose Cotransporter 2 Inhibitor Tofogliflozin Combination Treatment Improved Histopathology in Experimental Mice Model of Non-Alcoholic Steatohepatitis

Kentaro Murakami [1,2,†], Yusuke Sasaki [1,2,†], Masato Asahiyama [2,†], Wataru Yano [2], Toshiaki Takizawa [2], Wakana Kamiya [1], Yoshihiro Matsumura [3], Motonobu Anai [1], Tsuyoshi Osawa [4], Jean-Charles Fruchart [5], Jamila Fruchart-Najib [5], Hiroyuki Aburatani [6], Juro Sakai [3,7], Tatsuhiko Kodama [1] and Toshiya Tanaka [1,*,†]

1 Department of Nuclear Receptor Medicine, Laboratories for Systems Biology and Medicine (LSBM) at the Research Center for Advanced Science and Technology (RCAST), The University of Tokyo, Tokyo 153-8904, Japan; k-murakm@lsbm.org (K.M.); y-sasaki@kowa.co.jp (Y.S.); kamiya@lsbm.org (W.K.); anai@lsbm.org (M.A.); kodama@lsbm.org (T.K.)
2 Pharmaceutical Division, Kowa Company, Ltd., Tokyo 189-0022, Japan; m-asahiy@kowa.co.jp (M.A.); wyano@kowa.com (W.Y.); ttakizaw@kowa.co.jp (T.T.)
3 Division of Metabolic Medicine, Laboratories for Systems Biology and Medicine (LSBM) at the Research Center for Advanced Science and Technology (RCAST), The University of Tokyo, Tokyo 153-8904, Japan; matsumura-y@lsbm.org (Y.M.); jmsakai-tky@umin.ac.jp (J.S.)
4 Division of Integrative Nutriomics and Oncology, Laboratories for Systems Biology and Medicine (LSBM) at the Research Center for Advanced Science and Technology (RCAST), The University of Tokyo, Tokyo 153-8904, Japan; osawa@lsbm.org
5 R3i Foundation, Picassoplatz 8, 4010 Basel, Switzerland; jean-charles.fruchart@r3i.org (J.-C.F.); jamila.fruchart@yahoo.fr (J.F.-N.)
6 Genome Science Division, Laboratories for Systems Biology and Medicine (LSBM) at the Research Center for Advanced Science and Technology (RCAST), The University of Tokyo, Tokyo 153-8904, Japan; haburata-tky@umin.ac.jp
7 Tohoku University Graduate School of Medicine, Division of Molecular Physiology and Metabolism, Sendai 980-8575, Japan
* Correspondence: tanaka@lsbm.org; Tel./Fax: +81-3-5452-5025
† These authors contributed equally to this work.

Citation: Murakami, K.; Sasaki, Y.; Asahiyama, M.; Yano, W.; Takizawa, T.; Kamiya, W.; Matsumura, Y.; Anai, M.; Osawa, T.; Fruchart, J.-C.; et al. Selective PPARα Modulator Pemafibrate and Sodium-Glucose Cotransporter 2 Inhibitor Tofogliflozin Combination Treatment Improved Histopathology in Experimental Mice Model of Non-Alcoholic Steatohepatitis. *Cells* **2022**, *11*, 720. https://doi.org/10.3390/cells11040720

Academic Editors: Kay-Dietrich Wagner and Nicole Wagner

Received: 25 January 2022
Accepted: 15 February 2022
Published: 18 February 2022

Abstract: Ballooning degeneration of hepatocytes is a major distinguishing histological feature of non-alcoholic steatosis (NASH) progression that can lead to cirrhosis and hepatocellular carcinoma (HCC). In this study, we evaluated the effect of the selective PPARα modulator (SPPARMα) pemafibrate (Pema) and sodium-glucose cotransporter 2 (SGLT2) inhibitor tofogliflozin (Tofo) combination treatment on pathological progression in the liver of a mouse model of NASH (STAM) at two time points (onset of NASH progression and HCC survival). At both time points, the Pema and Tofo combination treatment significantly alleviated hyperglycemia and hypertriglyceridemia. The combination treatment significantly reduced ballooning degeneration of hepatocytes. RNA-seq analysis suggested that Pema and Tofo combination treatment resulted in an increase in glyceroneogenesis, triglyceride (TG) uptake, lipolysis and liberated fatty acids re-esterification into TG, lipid droplet (LD) formation, and *Cidea/Cidec* ratio along with an increased number and reduced size and area of LDs. In addition, combination treatment reduced expression levels of endoplasmic reticulum stress-related genes (*Ire1a*, *Grp78*, *Xbp1*, and *Phlda3*). Pema and Tofo treatment significantly improved survival rates and reduced the number of tumors in the liver compared to the NASH control group. These results suggest that SPPARMα and SGLT2 inhibitor combination therapy has therapeutic potential to prevent NASH-HCC progression.

Keywords: SPPARMα; SGLT2; ballooning; ER stress

1. Introduction

Non-alcoholic steatosis (NASH) is a severe form of non-alcoholic fatty liver disease (NAFLD), which is closely linked to type 2 diabetes and metabolic syndrome [1–3]. NASH is defined as the presence of steatosis, inflammation, ballooning degeneration of hepatocytes with or without fibrosis, and the eventual development of cirrhosis and hepatocellular carcinoma (HCC). In particular, higher grades of steatosis, inflammation, and ballooning degeneration are important steps in the pathogenesis of cirrhosis and HCC and are strongly associated with morbidity and mortality of liver disease [4,5]. However, the mechanism by which lipid accumulation in hepatocytes affects NASH progression is unclear. In addition, no effective therapeutic agents have been approved for treating NASH. Therefore, the development of a therapeutic approach for NASH is urgently needed.

Lipid droplets (LDs) are storage organelles that store neutral lipids such as triglycerides (TGs) and sterol esters during excess energy states and serve as a reservoir of energy supplies during the fasting state [6–8]. Importantly, not only role in maintaining lipid homeostasis but also buffering function of toxic lipid species have emerged with respect to LD biology. Dysregulated LDs homeostasis is considered to induce toxic lipid release and trigger cell death through prolonged activation of signaling pathways, such as the unfolded protein response (UPR) [9,10]. However, extensive LD accumulation in hepatocytes is not always in accordance with cellular dysfunction [11,12]. Although accumulation of LDs in hepatocytes is a prerequisite step for NASH development, changes in the composition of the lipids and proteins of LDs may play an important role in the progression from NAFLD to NASH [13,14]. Thus, the investigation of LD biogenesis and degradation, as well as the regulation of hepatic fatty acid and TG metabolism by a balance of de novo lipogenesis (DNL), glyceroneogenesis, VLDL assembly and secretion, lipolysis, and fatty acid oxidation (FAO) at the transcriptional and post-transcriptional levels, is important for understanding NASH development.

Pemafibrate (Pema) is the first clinically available selective PPARα modulator (SPPARMα); it is used to improve dyslipidemia and reduce macrovascular and microvascular complications [15–19]. We have reported that activation of PPARα by Pema induces the expression of a series of genes involved in TG hydrolysis, fatty acid uptake, fatty acid β-oxidation, and ketogenesis in the liver, supporting its ability to reduce plasma TG [20,21]. In our previous study using STAM NASH model mice, we reported that Pema treatment prevents NASH development by reducing myeloid cell recruitment without reducing hepatic TG content [22]. Therefore, we suggest that the combination of Pema and drugs that enhance the excretion or inhibit the absorption of carbohydrates and/or lipids has the potential to alleviate LD accumulation in hepatocytes and impede NASH development.

Sodium-glucose cotransporter-2 (SGLT2) inhibitors are a class of lower blood glucose drugs that increase urinary glucose excretion by inhibiting glucose reabsorption at the proximal tubule in the kidney [23–25]. Recent studies suggested that SGLT2 inhibitor treatment can reduce hepatic lipid levels and alleviate NAFLD, and has blood glucose-lowering effects [26,27]. The hepatic lipid-lowering effect of SGLT2 inhibitors has been suggested, in part, based on their ability to lower circulating glucose and insulin levels, which reduces DNL. In this study, we evaluated the therapeutic potential of the combination of Pema and the SGLT2 inhibitor tofogliflozin (Tofo) in STAM NASH model mice at two time points (onset of NASH progression and HCC survival).

2. Materials and Methods

2.1. Reagents

Pema and Tofo were kindly provided by Kowa Co., Ltd. (Nagoya, Japan). Streptozotocin (STZ) was purchased from Sigma-Aldrich (St. Louis, MO, USA) and Arabic gum from Wako Pure Chemical Industries (Osaka, Japan).

2.2. Animal Treatment

2.2.1. Progression Prevention Study

STAM mice were generated as previously described [22]. Pathogen-free pregnant C57BL/6J mice were obtained from CLEA Japan (Tokyo, Japan). All mice were housed in a temperature-controlled (24 °C) facility with a 12-h light/12-h dark cycle (08:00–20:00 h) and ad libitum access to food and water, except for the drug treatment period. Two days after birth, male mice received a subcutaneous injection of 200 μg STZ (Sigma, St. Louis, MO, USA) and were fed HFD32 (32% fat, CLEA Japan) ad libitum after 4 weeks of age of weaning. Two weeks after HFD32 feeding, mice were randomly divided into four groups: STAM control group fed HFD32 with vehicle treatment, Pema-treated group fed HFD32 with Pema (0.1 mg/kg), Tofo-treated group fed a HFD32 with Tofo (10 mg/kg), and Pema and Tofo combination (Pema 0.1 mg/kg and Tofo 10 mg/kg) for 3 weeks (6–9 weeks). Drugs were administered at 5 mL/kg body weight by oral intubation in 3% Arabic gum daily between 09:30 and 10:00 h. HFD32 was fed in a pair-feeding manner (2.3–2.8 g/mouse/day). In the drug treatment groups, animals were fed the same amount of HFD32 diet as that consumed by the control group over the preceding 24 h. In addition, normal diet (CE-2; 5% fat, CLEA Japan) fed normal group was orally administered vehicle for 3 weeks. Four hours after final administration, mice were sacrificed, and serum parameters measurement, histology, TG content determination, and gene expression analysis of liver were carried out. The study protocol was approved in accordance with the relevant guidelines and regulations of the Animal Care and Use Committee of the University of Tokyo (RAC12011, RAC170001).

2.2.2. Survival Study

Male STAM mice were purchased from SMC Laboratories (Tokyo, Japan) at 5 weeks of age and were fed HFD32. C57BL/6J normal mice were purchased from Japan SLC (Shizuoka, Japan) and fed a normal diet, CE-2. All mice were housed at 23 ± 3 °C and 55 ± 15% relative humidity (RH) under a 12-h light/dark cycle (07:00–19:00 h) and provided with food and water ad libitum. At 6 weeks of age, STAM mice were divided into four groups based on body weight: control, Pema (0.00008% equivalent to 0.1 mg/kg), Tofo (0.015% equivalent to 10 mg/kg) [28,29], and Pema and Tofo combination (n = 20 each). C57BL/6J mice with normal chow were assigned to the normal group (n = 8). Pema and/or Tofo were mixed in the diet and administered to each group. The study protocol was approved in accordance with the relevant guidelines and regulations of the Animal Care and Use Committee of Tokyo New Drug Research Laboratories, Kowa Company, Ltd. (Tokyo, Japan).

2.3. Blood Parameter

Serum total cholesterol (TC), TG, glucose, non-esterified fatty acids (NEFA), AST, ALT, phospholipids (PL), and creatinine (CRN) levels were determined using a Labospect 003 autoanalyzer (Hitachi High-Technologies Corporation, Tokyo, Japan).

2.4. Histology

A histological study was performed as previously described [22]. For immunohistochemistry, blocking of endogenous peroxidase activity was performed using 0.03% H_2O_2 in methanol. Obtained liver sections were treated with the anti-ER-TR7 (Abcam, Cambridge, MA, USA) antibodies overnight at 4 °C. After treatment with secondary antibodies, the substrate reaction was performed using 3,3′-diaminobenzidine (Dojindo, Kumamoto, Japan) solution.

According to Kleiner et al. [30], the NAFLD activity score (NAS) was calculated. Quantitative five grades assessment of Oil Red O staining was carried out by scoring of positive areas. Quantitative estimations of ER-TR7 and Sirius-red positive areas were carried out of the positive areas in five fields. Briefly, for each animal, bright field images of stained sections were captured around the central veins at 400-fold magnification using a digital camera (DP72, Olympus, Tokyo, Japan), and quantitatively estimated using WinROOF image processing software (Mitani, Tokyo, Japan). The results were shown as the mean of five different fields in each section.

2.5. RNA-Sequencing

For genome-wide transcriptome analysis, RNA-sequencing (RNA-Seq) was performed as previously described [22]. Briefly, sequencing of the RNA libraries was carried out using 150-bp paired-end mode of the TruSeq Rapid PE Cluster Kit and TruSeq Rapid SBS kit (Illumina) on the Illumina HiSeq 2500 platform. RNA-seq reads were mapped onto the reference mouse genome (NCBI37/mm9) and transcriptome (UCSC gene), respectively, using Burrows-Wheeler Aligner. Transcript coordinates were converted to genomic positions, and then an optimal mapping result was chosen either via transcript or genome mapping by comparing the minimal edit distance to the reference. Local realignment was implemented within an in-house short read aligner with a smaller k-mer size (k = 11). Eventually, fragments per kilo base of exon per million fragments mapped (fpkm) values were calculated for each UCSC gene while considering strand-specific information.

2.6. Quantitative Real-Time PCR (qPCR)

qPCR was performed as previously described [22,31,32]. *Ppia* mRNA was used as an independent control. All primers used for qPCR are listed in Supplementary Table S1.

2.7. LD Analysis

LD evaluation was performed as previously described [22]. For hepatic LD analysis, "Image J" imaging software (https://imagej.nih.gov/ij/download.html (accessed on 7 August 2016) was applied. H&E staining images were opened with Image J software and converted into grayscale (8 bit). Then, the lipid drop areas were extracted by using the threshold (Min: 220, Max: 255). After eliminating blood vessels, LD areas were analyzed and quantified using the "Analyze particles" function. Quantified LD area data were firstly obtained by pixel, and then they were converted into μm^2 (1 μm = 3 pixels, determined by scale bar size). LD diameter was also analyzed. Data were shown as the mean values from three different images of each animal. A histogram was created with Microsoft Excel spreadsheet software.

2.8. Statistical Analyses

All data are presented as the mean ± SEM. Homogeneity invariance was evaluated by Bartlett's test followed by parametric or non-parametric Dunnett's multiple comparison test (two-sided). * $p < 0.05$, ** $p < 0.01$. In the survival study, data were analyzed using the multiple log-rank test and Cox proportional hazard model.

3. Results

3.1. Pema and Tofo Combination Prevents Ballooning Degeneration of Hepatocytes

To investigate the effects of Pema and Tofo combination on NASH development in the STAM mouse model, each drug and combination was administered for three weeks. STAM mice showed significant hyperglycemia and hypertriglyceridemia; higher phospholipid, FGF21, and AST levels; lower body weight; and higher liver weight, compared to normal C57BL/6J mice (Table 1). Pema significantly reduced serum TG levels, but it did not alter serum glucose (Figure 1A,B), AST, and ALT levels. In addition, Pema significantly increased liver weight, which is a well-known effect of PPARα stimulation in rodents [33,34]. Tofo significantly reduced serum TG and glucose levels (Figure 1A,B). Pema and Tofo combination treatment effectively reduced serum TG and glucose levels and increased FGF21 levels.

Table 1. Effects of Pema, Tofo, and Pema and Tofo combination on body and liver weight, biochemical parameters in the serum, immunohistochemical analysis, and NAS.

	Normal	STAM			
		Vehicle	Pemafibrate	Tofogliflozin	Pemafibrate/Tofogliflozin
n	6	6	5	7	6
Body weight (g)	23.49 ± 0.35 **	18.65 ± 0.45	18.15 ± 0.82	18.46 ± 0.49	18.25 ± 0.1
Liver weight (g)	1.31 ± 0.08 **	0.97 ± 0.03	1.43 ± 0.15 **	1.34 ± 0.03 **	1.65 ± 0.08 **
TC (mg/dL)	97.3 ± 0.7 **	171.3 ± 10.6	234.4 ± 12.1 **	194.6 ± 10.7	225.2 ± 10.7 **
PL (mg/dL)	207.7 ± 3.5 **	330.2 ± 12.9	380.4 ± 10.3 *	323.6 ± 15.7	331.3 ± 13.4
NEFA (mEq/L)	0.77 ± 0.03	0.85 ± 0.10	0.71 ± 0.06	0.91 ± 0.07	0.75 ± 0.07
β-hydroxybutylate (nmol)	17.5 ± 2.5 **	140.2 ± 8.1	149.1 ± 0.7	152.8 ± 1.6	152.7 ± 2.9
FGF21 (pg/mL)	156.1 ± 55.4 *	1579.6 ± 458.8	2449.0 ± 500.1	1430.4 ± 212.9	2759.6 ± 143.5 *
CRN (mg/dL)	0.108 ± 0.004	0.122 ± 0.005	0.100 ± 0.011	0.114 ± 0.006	0.093 ± 0.013
AST (U/L)	128.0 ± 16.4 *	181.8 ± 9.6	192.2 ± 17.3	189.6 ± 15.4	192.7 ± 10.6
ALT (U/L)	38.3 ± 7.9	71.2 ± 7.6	76.6 ± 13.2	62.9 ± 5.5	84.8 ± 12.3
Oil Red O score	0.3 ± 0.2 **	2.8 ± 0.3	3.2 ± 0.2	2.6 ± 0.2	3.2 ± 0.2
ER-TR-7 (% area)	1.667 ± 0.037 **	3.718 ± 0.451	3.074 ± 0.205	3.347 ± 0.319	3.096 ± 0.143
Sirius Red (% area)	0.257 ± 0.019 **	1.160 ± 0.205	1.096 ± 0.146	1.301 ± 0.127	1.062 ± 0.136
Steatosis	0.00 **	2.2 ± 0.3	2.2 ± 0.2	1.7 ± 0.2	2.0 ± 0.0
Inflammation	0.00 **	1.17 ± 0.17	1.20 ± 0.20	1.29 ± 0.18	1.50 ± 0.22
NAS	0.00 **	4.67 ± 0.56	4.00 ± 0.32	3.71 ± 0.36	3.67 ± 0.33

TC: total cholesterol, PL: phosphorlipids, NEFA: non-esterified fatty acid, CRN: creatinine, AST: aspartate aminotransferase, ALT: alanine aminotransferase, NAS: NAFLD activity score. Error bars show s.e.m. * $p < 0.05$; ** $p < 0.01$: Significantly difference from STAM control group by Dunnett's multiple comparison test.

H&E staining clarified that STAM control mouse liver owned liver nodules, macro- and micro-vesicular lipid accumulation, inflammatory cell infiltration, and ballooning degeneration of hepatocytes, unlike normal mouse liver (Figure 1C). Pema-treated mouse liver included less macrovesicular lipid accumulation, less ballooning degeneration, and a tendency to reduce the NAS compared to STAM control mice (Table 1). Tofo treatment reduced macrovesicular lipid accumulation and ballooning degeneration. The Pema and Tofo combination treatment significantly reduced ballooning degeneration (Figure 1D).

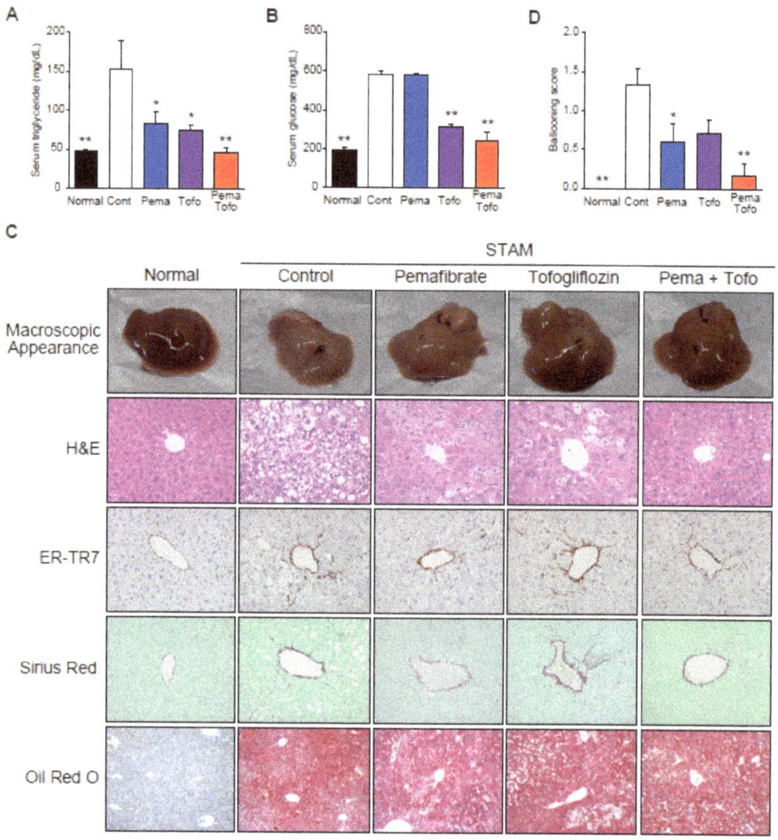

Figure 1. Pemafibrate and Tofogliflozin combination improves hypertriglyceridemia, hyperglycemia, macrovesicular steatosis, and ballooning score in STAM mice liver. (**A**) Serum triglyceride, (**B**) Serum glucose, (**C**) Representative gross morphology of liver, H&E stained, ER-TR7 stained, Sirius-red stained, and Oil red O stained liver section, and (**D**) Ballooning score of normal, control, pemafibrate, tofogliflozin, and pemafibrate and tofogliflozin combination-treated STAM mice. Error bars show s.e.m. * $p < 0.05$; ** $p < 0.01$: Significantly difference from STAM control group by Dunnett's multiple comparison test.

3.2. Pema and Tofo Combination Treatment Induces Lipolysis and Re-Esterification Cycles of TG in STAM Mouse Livers

Although Pema, Tofo, and combination treatment resulted in decreased macrovesicular steatosis, this reduction was not reflected in the steatosis and Oil Red O staining scores (Table 1). To verify the effect of Pema and Tofo combination treatment on STAM mouse livers, we performed a global gene expression analysis by RNA-seq using liver tissues collected from normal, STAM control, Pema-treated, Tofo-treated, and Pema and Tofo combination-treated STAM mice. We identified 125 upregulated and 68 downregulated genes in the Pema and Tofo combination treatment compared with the STAM control group according to our stringent criteria (Supplementary Figure S1A). These genes included almost all Pema- and/or Tofo-regulated genes. In particular, the most upregulated genes by combination treatment were PPARα target genes involved in lipid metabolism (Supplementary Figure S1B).

To understand the effect of Pema and Tofo combination treatment on STAM mouse livers, lipid and carbohydrate metabolism-related gene expression levels were analyzed. We found that the expression levels of genes related to TG hydrolysis, fatty acid uptake, fatty acid activation, fatty acid binding, peroxisomal and mitochondrial oxidation, and ketogenesis were increased in the STAM control group than in the normal group (Supplementary Figure S2). Tofo and Pema monotherapy upregulated the expression of these genes, and the combination treatment upregulated their expression further. Importantly, the combination of Pema and Tofo dramatically increased the *Pdk4* gene expression level, indicating that it mediates the inhibition of glucose oxidation and preferential activation of FAO (Supplementary Figure S2).

Increased glucose and fructose uptake in hepatocytes accelerate glycolysis and DNL to generate TG. Especially, the glycerolipid synthesis pathway (glyceroneogenesis) and the monoacylglycerol pathway are key players in TG synthesis (Figure 2A). The STAM control mouse livers exhibited higher levels of glycolysis-related gene expression than the normal mouse livers (Supplementary Figure S3 and Figure 2B). In addition, we found that glyceroneogenesis and re-esterification of 2-monoacylglycerol were induced in STAM control livers, in addition to simultaneous TG uptake and hydrolysis. The Pema and Tofo combination treatment did not affect glycolysis-related gene expression, but it significantly induced a series of genes involved in TG synthesis from glycerol 3-phosphate and re-esterification from monoacylglycerols and diacylglycerols generated by TG hydrolysis in STAM mouse livers (Figure 2B). These results suggest that the Pema and Tofo combination enhances the uptake and oxidation of fatty acids, TG synthesis from glycerol 3-phosphate, and the re-esterification of glycerol generated by TG hydrolysis in STAM mouse livers.

3.3. *Pema and Tofo Combination Increases Small LDs in STAM Mouse Livers*

To better understand the effect of Pema and Tofo combination treatment on steatosis in STAM mice, we measured the TG concentration in the liver (Figure 3A). The STAM control group showed a significantly increased TG content in the liver. Pema significantly increased, and Tofo and Pema and Tofo combination tended to decrease, the TG content in STAM mouse livers (Figure 3A). Because the combination of Pema and Tofo markedly improved macrovesicular steatosis (Figure 1C), we evaluated LD counts and size distributions. Pema and Tofo treatments increased the droplet number and decreased the LD area (Figure 3B,C). Furthermore, this drug combination treatment increased the percentage of cells representing small LDs (<1 μm) from 29.40% in the control to 49.65% and decreased the percentage of cells representing large LDs (>3 μm) from 39.36% in the STAM control to 7.87% (Figure 3D).

LDs consist of an inner core of neutral lipids, including TG and sterol esters, a phospholipid monolayer, and LD-associated proteins (LDAPs) [6–8]. Because LDAPs affect LD function and dynamics [6–8], we evaluated the effect of Pema and Tofo combination on LDAP expression (Figure 3E). The Pema and Tofo combination group showed increased expression of genes related to LD inner core lipid synthesis (*Agpat6*, *Dgat1*, and *Acat1*), formation (*Agpat6*, *Acsl3*, *Mettl7b*, and *Plin2*), budding (*Fitm2* and *Bscl2*), stabilization (*Plin4* and *Plin5*), lipolysis (*Pnpla2*, *Hsd17b11*, *Pcyt1a*, and *Abhd5*), expansion (*Agpat3*, *Pex3*, and *Tcp1*), and fusion (*Cidea* and *Cidec*). Among these genes, the Pema and Tofo combination induced *Cidea* expression. Recently, Sans et al. suggested that hepatic CIDEA and CIDEC correlated negatively and positively, respectively, with steatohepatitis and liver injury in mice, as well as steatosis and NASH in obese humans [35]. In addition, suppression of CIDEC has been reported to reduce LD size and stimulate lipolysis [36]. Pema, Tofo, and combination treatments induced expression of *Cidea* and *Cidec*, and combination treatment strongly induced *Cidea* gene expression, thereby increasing the *Cidea*/*Cidec* ratio. These results may contribute to the reduction in LD size and stimulation of lipolysis by combination treatment.

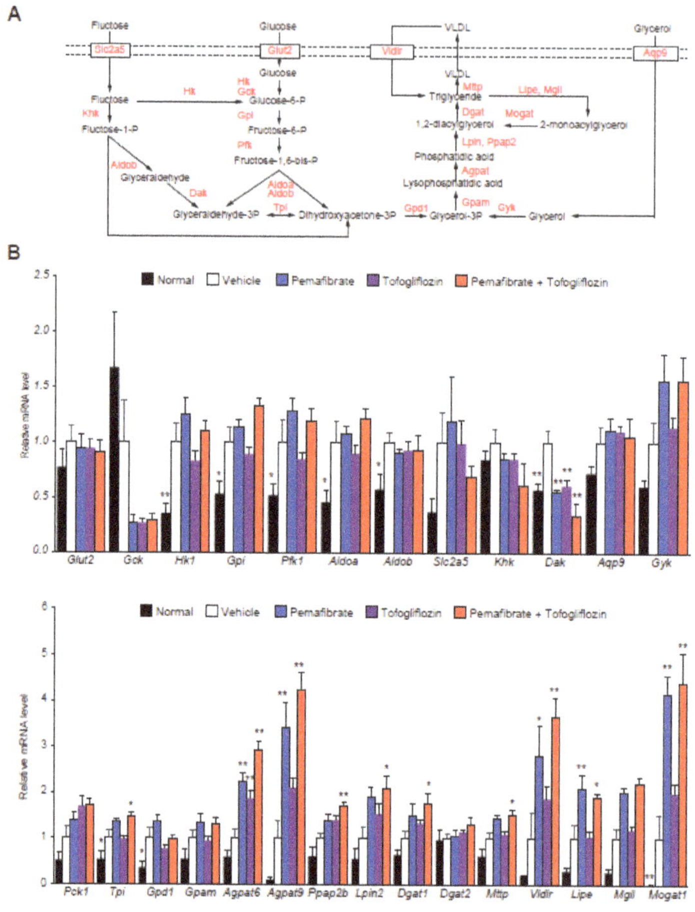

Figure 2. Pemafibrate and Tofogliflozin combination induces lipolysis and fatty acid re-esterification genes expression in STAM mice liver. (**A**) Schematic representation of the glycolytic and TG synthesis pathways in the liver. (**B**) qPCR validation of glycolytic and TG metabolism-related genes expression of normal, control, pemafibrate, tofogliflozin, and pemafibrate and tofogliflozin combination-treated STAM mice liver. Error bars show s.e.m. * $p < 0.05$; ** $p < 0.01$: Significantly difference from STAM control group by Dunnett's multiple comparison test.

3.4. *Pema and Tofo Combination Inhibits the IRE1α-XBP1-PHLDA3 Pathway*

LD biogenesis and enhanced esterification of fatty acids play a key role in buffering toxic lipid species [6–8]. Several reports have indicated that fatty acid accumulation in hepatocytes can lead to cell dysfunction and cell death through endoplasmic reticulum (ER) stress [37,38]. UPR signaling is mainly driven by three sensors mediated by inositol requiring enzyme 1 α (IRE1α), protein kinase RNA-like endoplasmic reticulum (ER) kinase (PERK), and activating transcription factor 6 (ATF6) [39,40]. In addition, the luminal side of each UPR sensor interacts with chaperones of immunoglobulin-binding protein/glucose regulatory protein 78 (BiP/GRP78) [39–41]. Livers from the STAM control group showed enhanced UPR sensors (*Ire1a*, *Perk*, and *Atf6*), chaperones (*Grp78* and *Pdi1a*), antioxidant defense-regulated transcription factor (*Nrf2*), IRE1α interaction protein form apoptosis mediator complex (*Traf2*), and proapoptotic BCL-2 protein family (*Bak1* and *Bax*). Pema and Tofo combination significantly reduced *Ire1a*, *Grp78*, *Xbp1*, and *Phlda3* expression levels (Figure 4). Recently, ER stress in hepatocytes has been reported to induce PHLDA3 via the IRE1–Xbp1s pathway, which facilitates liver injury by inhibiting Akt [42]. These data suggest that the combination of Pema and Tofo prevents liver injury by inhibiting the IRE1α-XBP1-PHLD3A pathway.

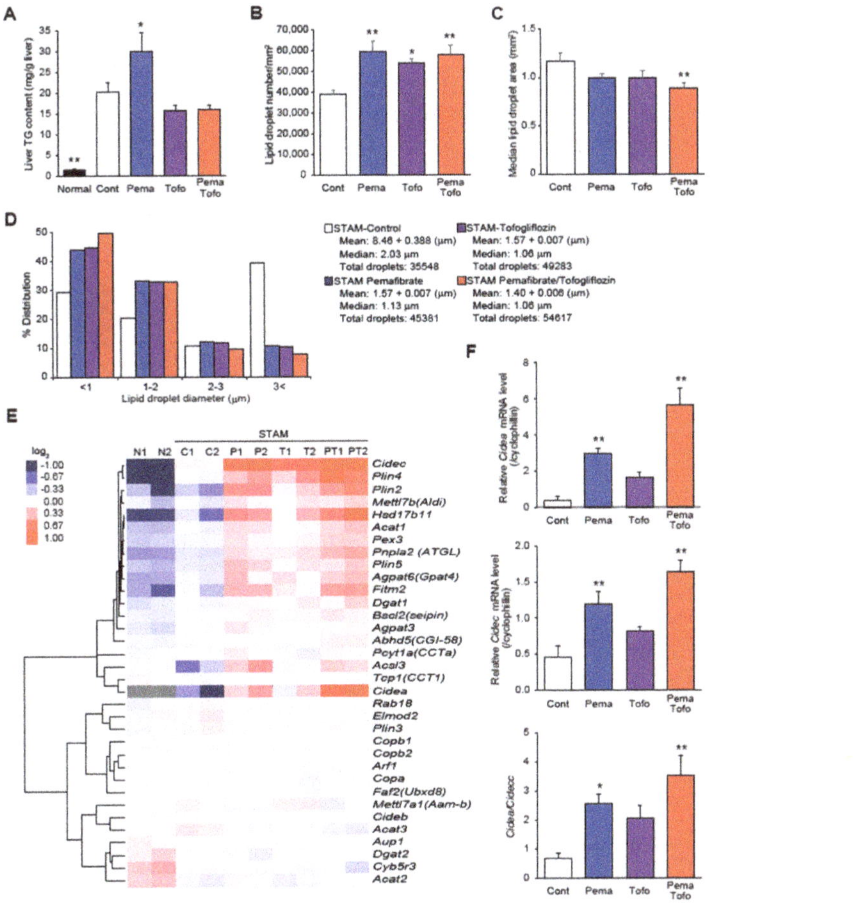

Figure 3. Pemafibrate and Tofogliflozin combination induces lipid droplets formation. (**A**) Liver TG content, (**B**) Lipid droplet number, (**C**) Median lipid droplet, (**D**) Lipid droplet sizes distribution, (**E**) Heatmap of hierarchical clustering of LDAPs and formation-related genes, and (**F**) *Cidea*, *Cidec*, and *Cidea*/*Cidec* ratio of control, pemafibrate, tofogliflozin, and pemafibrate and tofogliflozin combination-treated STAM mice. * $p < 0.05$; ** $p < 0.01$: Significantly difference from STAM control group by Dunnett's multiple comparison test.

3.5. Pema and Tofo Combination Improves HCC-Related Survival

Finally, to determine whether Pema and Tofo combination can prevent the progression of NASH to HCC, we treated STAM mice for 16 weeks. As observed in the 3-week drug treatment on NASH progression, the combination of Pema and Tofo treatment resulted in a significant decrease in serum TG and blood glucose levels (Figure 5A,B). Levels of serum AFP, an oncofetal protein that is used as a tumor marker, significantly increased in the STAM vehicle control group and decreased in the combination treatment group (Figure 5C). Kaplan-Meier survival curves show that the survival rate of the control group decreased to 10%. Pema in the diet showed a tendency to increase the survival rate (30%), and the combination of Pema and Tofo significantly improved the survival rates (50%) compared to the control group. In addition, each drug-treated group had a markedly reduced number of tumors in the liver (Figure 5E,F).

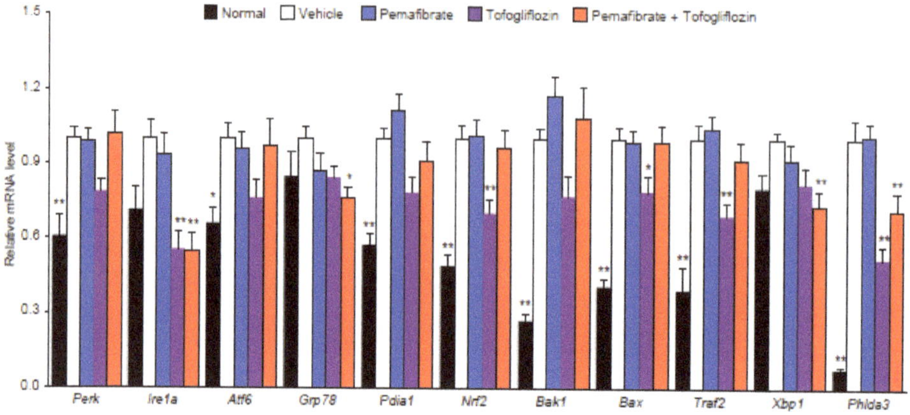

Figure 4. Pemafibrate and Tofogliflozin combination improves ER stress genes expression in STAM mice liver. qPCR validation of ER stress-related genes expression of normal, control, pemafibrate, tofogliflozin, and pemafibrate and tofogliflozin combination-treated STAM mice liver. * $p < 0.05$; ** $p < 0.01$: Significantly difference from STAM control group by Dunnett's multiple comparison test.

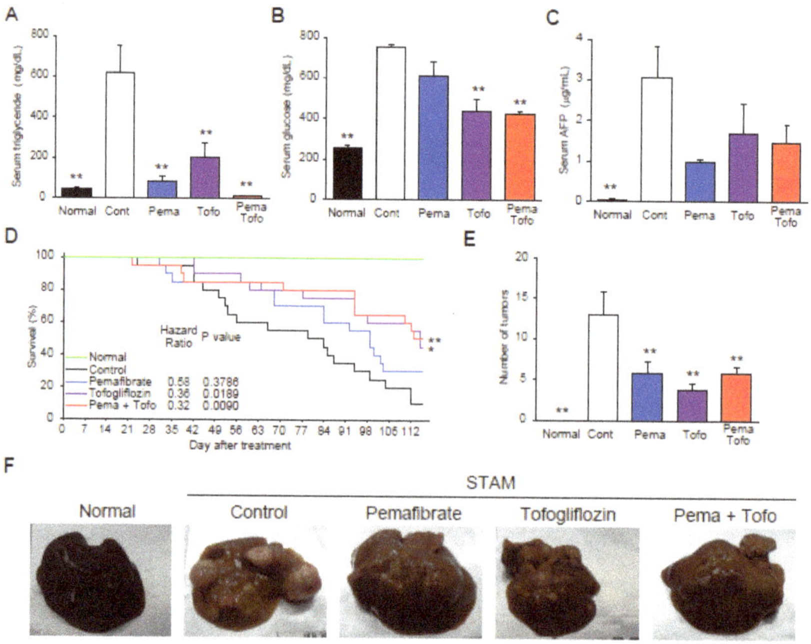

Figure 5. Pemafibrate and Tofogliflozin combination improves survival rate in STAM mice liver. **(A)** Serum triglyceride, **(B)** Serum glucose, **(C)** Serum AFP, **(D)** Kaplan-Meier survival curves, **(E)** Number of tumors, and **(F)** Representative gross morphology of liver from normal, control, pemafibrate, tofogliflozin, and pemafibrate and tofogliflozin combination-treated STAM mice. Log-rank p-value and hazard ratio were shown in the survival curve figure. Error bars show s.e.m. * $p < 0.05$; ** $p < 0.01$: Significantly difference from STAM control group by Dunnett's multiple comparison test.

4. Discussion

Hepatic TG accumulation has been suggested to play a central role in NAFLD and NASH, which can progress to cirrhosis and liver failure [1–6]. The mechanisms underlying the pathogenesis of NASH in a subset of patients with steatosis have not been clarified, but several proposed hypotheses suggest that steatosis with additional factors, such as insulin resistance, oxidative stress, ER stress, and mitochondrial dysfunction, may be involved [1–3]. Our previous study revealed that Pema prevents NASH development without reducing the TG content in the liver [22]. We also revealed that although Pema improves macrovesicular steatosis by enhancing TG hydrolysis while simultaneously enhancing esterification of fatty acids for TG and LD biogenesis, it may not result in a sufficient TG reduction in STAM mouse livers. Based on these observations, we hypothesized that the combination of Pema with a drug that enhances the excretion of carbohydrates from the kidney via SGLT2 inhibition has the potential to improve TG accumulation and NASH development. The combination of Pema and Tofo significantly improved ballooning degeneration of hepatocytes and reduced hepatic TG accumulation. In addition, the combination of Pema and Tofo specifically reduced *Ire1a-Xbp1-Phld3a* gene expression in the NASH liver. These results suggest that the combination of SPPARMα and SGLT2 inhibitors has therapeutic potential for NASH and NASH-related HCC via reduction of ER stress-induced liver injury.

LDs are organelles that store neutral lipids, such as TG and sterol esters, as sources of energy and cell membrane synthesis [6–8]. When cells face excess neutral lipids, such as fatty acids and sterols, they synthesize LDs and disperse them into the cytoplasm. This process is also important for protecting cells from the toxicity associated with an excess of lipids such as fatty acids, glycerolipids, and sterols [6–8]. Therefore, control of LD biogenesis and consumption plays a key role in the pathogenesis of NASH. In this study, we found that the Pema and Tofo combination induced expression of genes involved in TG uptake, lipolysis, fatty acid uptake, fatty acid β-oxidation and esterification, and ketogenesis, as well as PDK4, which inhibits glucose oxidation. These results are consistent with our previous results in Pema-treated STAM mice liver [22] and were enhanced by the combination treatment used in this study. We found that Tofo improved hyperglycemia and serum TG levels and reduced TG content in the liver of STAM mice, whereas Pema reduced serum TG levels but did not reduce liver TG content. The STAM model is characterized by hyperglycemia and reduced body weight with reduced *Gck* expression, which is exclusively regulated by insulin signaling [43]. Therefore, insulin-stimulated DNL gene regulation mediated through sterol regulatory element-binding protein 1c (SREBP1c) is unlikely to contribute to these effects, suggesting that carbohydrate response element-binding protein (ChREBP) signaling regulates glycolytic and DNL genes in this model. Although PPARα activation by Pema did not affect hyperglycemia, glycolytic genes, or *G6pc* expression, which is associated with ChREBP activation by fructose, Tofo tended to reduce the expression levels of these genes. These results suggest that SGLT2 inhibitors reduce the influx of the substrate for DNL and reduce TG content in the liver of STAM mice. Our transcriptome analysis also shows that PPARα activation by Pema induced FAO and ketogenesis, but in STAM mice with a high concentration of β-hydroxybutyrate in the blood stream, it led to re-esterification of fatty acids released from the TG and sterol esters by lipolysis and uptake into the TG for LD synthesis in the liver.

In addition, we found that Pema and Tofo combination significantly increased LD number, reduced LD size, and improved macrovesicular steatosis. Consistent with the increased number of LDs, Pema and Tofo combination also induced expression of genes involved in LD inner core lipid synthesis and formation (*Agpat6*, *Dgat1*, and *Acat1*), budding (*Fitm2* and *Bscl2*), and fusion (*Cidea* and *Cidec*) proteins. Although the biological role of LD diversification has not been clarified yet, increased numbers of small LDs may explain the protection against lipotoxicity [6]. LDs have been suggested to protect against lipotoxicity under a variety of stressful conditions such as lipid overload, hypoxia, oxidative stress, autophagic flux, and dysfunctional lipolysis [6–8]. In fact, DGAT1-dependent LD biogenesis has been suggested to prevent lipotoxic mitochondrial dysfunction [44]. In addition, Becuwe et al. have reported that Fit2, encoded by *Fitm2*, is an evolutionarily conserved fatty acyl-CoA diphosphatase that maintains the ER structure, protects against ER stress, and enables normal lipid storage in LDs [45]. Furthermore, the differential expression of cell death-inducing DFF45-like effector (CIDE) family members CIDEC and CIDEA, recognized as regulators of LD growth, has been reported to be linked to NAFLD progression and liver injury, and CIDEA expression level decreases with NAFLD severity [35]. CIDEA and CIDEC are strongly expressed in brown adipocytes and white adipocytes, respectively, and are associated with the formation of multilocular small LDs (that are prone to lipolysis) and the storage form of unilocular LDs [36]. In this study, we found that *Cidea* was the most highly induced gene among LDAPs, and the *Cidea*/*Cidec* ratio was significantly increased by the Pema and Tofo combination. These results suggest that the combination of Pema and Tofo promotes fatty acid catabolism via lipolysis and β-oxidation while promoting re-esterification of excess fatty acids and LD biogenesis, thereby preventing lipotoxicity.

Hepatic steatosis has a risk of steatohepatitis, fibrosis, cirrhosis, liver failure, and HCC, and it was reported that dysregulation of microbial metabolites such as aromatic and branched-chain amino acid (AAA and BCAA) or of iron metabolism are related to liver fat accumulation and facilitate steatosis [46,47]; however, additional factors, such as insulin resistance, oxidative stress, ER stress, and mitochondrial dysfunction, are also involved in a disease progression [37,38]. Because the sequential esterification of fatty acids into a glycerol backbone to generate TG and budding as nascent LDs occur at the ER membrane, LDs are closely associated with ER homeostasis. Induction of UPR sensors (*Ire1a, Perk, and Atf6*), chaperones (*Grp78* and *Pdi1a*), antioxidant defense-regulated transcription factor (*Nrf2*), IRE1α interaction protein form apoptosis mediator complex (*Traf2*), and proapoptotic BCL-2 protein family (*Bak1* and *Bax*) genes were observed in the liver of STAM mice. Although numerous reports have indicated that the ER stress response plays a key role in NASH development, it is unknown which UPR sensor signaling contributes to the development of this disorder [39,40]. Pema and Tofo combination selectively reduced *Ire1a, Grp78, Xbp1*, and *Phlda3* expression levels in STAM mouse livers. Among the UPR sensors, IRE1 is the most evolutionarily conserved, implying that it plays a crucial role in cell fate determination under ER stress conditions. It has been indicated that IRE1 is capable of inducing cell fate by two distinct pathways through XBP1-mediated gene regulation and interaction with TNF receptor-associated factor 2 (TRAF2) to initiate the apoptosis signal-regulating kinase 1 (ASK1) and c-Jun N-terminal kinase (JNK) signaling cascades [48]. However, because ASK1 is activated by stress responses to ER stress, as well as ROS and TNFα, the impact of ASK1 activation by the ER stress pathway on NASH development remains unclear. A recent study reported a positive and negative effect of hepatic ASK1 ablation on NASH development in HFD-fed mice [49]. Similarly, treatment with the ASK1 inhibitor selonsertib in patients with NASH yielded controversial results [50]. However, because the effects of ASK1 inhibitors are not limited to the liver and include effects on other tissues, further studies investigating the role of ASK1 in NASH development are warranted. Recently, several reports indicated that pleckstrin homology-like domain family A member 3 (PHLDA3) functions as an AKT inhibitor and plays a crucial role in the cell fate of cancer cells [42]. In addition, PHLDA3 overexpression causes tissue injury, and the IRE1-Xbp1 pathway induces PHLDA3 overexpression, which facilitates liver injury [51]. Therefore, these results and reports suggest that the Pema and Tofo combination prevents liver injury by inhibiting the lipotoxicity-induced IRE1α-XBP1-PHLD3A pathway, thereby controlling toxic lipid esterification, LD biogenesis, and the lipolysis cycle.

Epidemiological studies have shown that NASH is closely linked to type 2 diabetes and metabolic syndrome [1–3]. However, the STAM mouse is recognized as a type 1 diabetes-related NASH model with hyperglycemia, reduced body weight gain, and lack of insulin secretion and fatty acid mobilization from adipose tissue. In general, storage TG in hepatocytes requires both fatty acids and glycerol and has been suggested to be mainly regulated by the pool size of fatty acid [52]. Although fat accumulation in the liver with type 1 diabetes has been reported, much less attention could be attributed to NASH prevalence of type 1 diabetes as compared to type 2 diabetes and metabolic syndrome. However, a recent report has suggested that NAFLD prevalence in patients with type 1 diabetes is considerable in meta-analysis [53], and several hypotheses have been proposed to explain the pathogenesis of liver steatosis in type 1 diabetes. These include insufficient TG secretion from the liver as VLDL, SREBP1c, and ChREBP induced DNL and conversion of sugar into fat [54]. On the other hand, the importance of circulating fatty acid influx has been suggested to contribute to increased hepatic lipid accumulation in type 2 diabetes [55], and circulating NEFA, dietary fat, and DNL have been reported to account for 59, 15, and 26% of the TG content in hepatocytes, respectively [56]. From these observations, adipose tissue-derived fatty acid influx and DNL have been suggested to play a crucial role in hepatic TG accumulation in type 2 diabetes-related NASH. In fact, it is well known that DNL is stimulated by insulin via SREBP1c activation and by influx glucose via ChREBP [57]. Insulin also activates LXRα, which in turn induces SREBP1c expression. In addition, impaired lipoprotein metabolism (VLDL export) and mitochondrial function (fatty acid entry and oxidation) have been suggested in the hepatic TG accumulation under insulin resistance [58]. Therefore, increased fatty acid influx, enhanced DNL, impaired TG secretion as VLDL, and mitochondrial dysfunction have been linked to human type 2 diabetic-related NASH. In the present study, we showed that there were no significant changes in serum NEFA in the STAM mouse compared to the normal mouse. In addition, our RNA-seq analysis indicated that impaired VLDL secretion and SREBP1c mediated DNL is unlikely to be the cause of hepatic steatosis in the STAM mouse model because VLDL assembly regulated *Mttp* was induced, insulin and SREBP1c target gene of *Gck* was reduced, and *Pck1*, which is negatively regulated by insulin, was induced. From several reports and our observations, this model may not completely reflect the human NASH liver metabolic state and may be a model in which the effect of SGLT2 inhibitors is more likely to be effective. Thus, additional studies using other NASH models with obesity and insulin resistance are warranted to evaluate the effect of the Pema and Tofo combination treatment on human NASH development.

In addition, although Pema and Tofo combination treatment significantly induced fatty acid catabolism, fatty acid re-esterification, and LD biogenesis; impeded the IRE1α-XBP1-PHLD3A pathway; and alleviated ballooning degeneration of hepatocytes, the precise underlying mechanism is still largely unknown. It is well known that hepatocytes are not equally responsible for liver metabolism, and the existence of so-called metabolic zonation based on oxygen tension has been proposed [59,60]. For example, gluconeogenesis, fatty acid β-oxidation, cholesterol synthesis, and ureagenesis are mainly considered to be performed by hepatocytes in the periportal region, where the oxygenated blood is transported via hepatic arteries, whereas glycolysis, DNL, bile acid synthesis, and xenobiotic detoxification occur in the pericentral region, which is relatively hypoxic [59,60]. Dysregulation of metabolic zonation is considered to lead to the development of lifestyle-related diseases such as obesity, diabetes, and NAFLD [61,62]. In fact, NAFLD is considered to begin with pericentral steatosis and inflammation with periportal inflammation and fibrosis considered late-occurring histological lesions [63]. However, the periportal disease has been associated with worse metabolic outcomes and more adverse hepatic fibrosis than pericentral disease [64]. In addition, interactions between hepatocytes and sinusoidal endothelial cells, Kupffer cells, and stellate cells are known to be involved in the pathogenesis of NASH [65]. Although Kupffer cells, T lymphocytes, and dendritic cells are more abundant in the periportal regions, infiltrating macrophages have been observed both in the periportal and pericentral regions [66]. Furthermore, preferential effects on the periportal and pericentral regions have been suggested for vitamin E and cysteamine, and PPARγ and FXR agonists, respectively, as per several randomized clinical trials [67–69]. Therefore, to understand the pathogenesis of NASH and the mechanism of therapeutic efficacy of the Pema and Tofo combination, it will be necessary to explore the spatial gene expression profile of hepatocytes and non-parenchymal cells using scRNA-seq and slide seq technologies [70–72].

5. Conclusions

In conclusion, the combination of SPPARMα and SGLT2 inhibitor treatment prevented ballooning degeneration of hepatocytes and HCC progression. Our global gene expression analysis gives evidence of the liver protective effect of the combination therapy by inhibiting the lipotoxicity-induced IRE1α-XBP1-PHLD3A pathway. Taken together with our previous report that SPPARMα treatment prevents NASH development by reducing myeloid cell recruitment without reducing hepatic TG content [22], the combination of SPPARMα and SGLT2 inhibitor presents a promising new therapy for NASH. Our results presented in this report using the NASH mouse model gives reason to hope that the combination of SPPARMα and SGLT2 inhibitor will be synergistic. Therefore, this combination is much more effective in human NASH than monotherapy and could become an ideal strategy for long-term treatment for NASH-HCC progression.

Supplementary Materials: The following supporting information can be downloaded at: https://www.mdpi.com/article/10.3390/cells11040720/s1, Supplementary Figure S1: Pemafibrate and tofogliflozin combination regulated genes in STAM mouse liver. Supplementary Figure S2: Effect of pemafibrate and tofogliflozin combination on fatty acid metabolism-related genes expression in STAM mouse liver. Supplementary Figure S3: Effect of pemafibrate and tofogliflozin combination on glucose metabolism and triglyceride synthesis-related genes expression in STAM mouse liver. Supplementary Table S1: qPCR Primer lists.

Author Contributions: Conceptualization, T.T. (Toshiya Tanaka), J.S. and T.K. designed the studies. T.T. (Toshiya Tanaka), K.M. and Y.S. wrote the manuscript. K.M., T.T. (Toshiaki Takizawa), Y.S., M.A. (Masato Asahiyama), W.Y. and T.T. (Toshiya Tanaka) performed experiments. W.K. assisted with the animal experiments and qPCR. M.A. (Motonobu Anai) performed the histological experiments. H.A. contributed to the RNA-sequencing data analysis. Y.M., M.A. (Motonobu Anai), T.O., J.-C.F. and J.F.-N. read and commented on the manuscript. All authors have read and agreed to the published version of the manuscript.

Funding: This work was supported in part by a collaborative research fund from Kowa Co. Kowa Company, Ltd.

Institutional Review Board Statement: Not applicable.

Informed Consent Statement: Not applicable.

Data Availability Statement: The data are available upon request from the corresponding author.

Conflicts of Interest: T.K. is an advisory board member and a recipient of support from a collaborative research fund from Kowa Company, Ltd. J.-C.F., and J.F.-N. are consultants of Kowa Company, Ltd. Kowa Co. had no role in the design of the study; in the collection, analyses, or interpretation of data; in the writing of the manuscript, or in the decision to publish the results. K.M., Y.S., M.A. (Masato Asahiyama), W.Y. and T.T. (Toshiaki Takizawa) are employees of Kowa Company, Ltd. Rest of the authors declare no conflict of interest.

References

1. Sheka, A.C.; Adeyi, O.; Thompson, J.; Hameed, B.; Crawford, P.A.; Ikramuddin, S. Nonalcoholic Steatohepatitis: A Review. *JAMA* **2020**, *323*, 1175–1183. [CrossRef] [PubMed]
2. Tilg, H.; Moschen, A.R.; Roden, M. NAFLD and diabetes mellitus. *Nat. Rev. Gastroenterol. Hepatol.* **2017**, *14*, 32–42. [CrossRef] [PubMed]
3. Suzuki, A.; Diehl, A.M. Nonalcoholic Steatohepatitis. *Annu. Rev. Med.* **2017**, *68*, 85–98. [CrossRef]
4. Haas, J.T.; Francque, S.; Staels, B. Pathophysiology and Mechanisms of Nonalcoholic Fatty Liver Disease. *Annu. Rev. Physiol.* **2016**, *78*, 181–205. [CrossRef]
5. Taylor, R.S.; Taylor, R.J.; Bayliss, S.; Hagström, H.; Nasr, P.; Schattenberg, J.M.; Ishigami, M.; Toyoda, H.; Wai-Sun Wong, V.; Peleg, N.; et al. Association Between Fibrosis Stage and Outcomes of Patients With Nonalcoholic Fatty Liver Disease: A Systematic Review and Meta-Analysis. *Gastroenterology* **2020**, *158*, 1611–1625. [CrossRef] [PubMed]

6. Walther, T.C.; Chung, J.; Farese, R.V., Jr. Lipid Droplet Biogenesis. *Annu. Rev. Cell Dev. Biol.* **2017**, *33*, 491–510. [CrossRef] [PubMed]
7. Olzmann, J.A.; Carvalho, P. Dynamics and functions of lipid droplets. *Nat. Rev. Mol. Cell Biol.* **2019**, *20*, 137–155. [CrossRef]
8. Scorletti, E.; Carr, R.M. A new perspective on NAFLD: Focusing on lipid droplets. *J. Hepatol.* 2021; *in press*. [CrossRef]
9. Fu, S.; Watkins, S.M.; Hotamisligil, G.S. The role of endoplasmic reticulum in hepatic lipid homeostasis and stress signaling. *Cell Metab.* **2012**, *15*, 623–634. [CrossRef]
10. Colgan, S.M.; Hashimi, A.A.; Austin, R.C. Endoplasmic reticulum stress and lipid dysregulation. *Expert Rev. Mol. Med.* **2011**, *13*, e4. [CrossRef] [PubMed]
11. Mashek, D.G. Hepatic lipid droplets: A balancing act between energy storage and metabolic dysfunction in NAFLD. *Mol. Metab.* **2021**, *50*, 101115. [CrossRef]
12. Thiam, A.R.; Beller, M. The why, when and how of lipid droplet diversity. *J. Cell Sci.* **2017**, *130*, 315–324. [CrossRef] [PubMed]
13. Alonso, C.; Fernández-Ramos, D.; Varela-Rey, M.; Martínez-Arranz, I.; Navasa, N.; Van Liempd, S.M. Metabolomic identification of subtypes of nonalcoholic steatohepatitis. *Gastroenterology* **2017**, *152*, 1449–1461. [CrossRef]
14. Chaurasia, B.; Tippetts, T.S.; Monibas, R.M.; Liu, J.; Li, Y.; Wang, L. Targeting a ceramide double bond improves insulin resistance and hepatic steatosis. *Science* **2019**, *365*, 386–392. [CrossRef]
15. Yamazaki, Y.; Abe, K.; Toma, T.; Nishikawa, M.; Ozawa, H.; Okuda, A.; Araki, T.; Oda, S.; Inoue, K.; Shibuya, K.; et al. Design and synthesis of highly potent and selective human peroxisome proliferator-activated receptor agonists. *Bioorg. Med. Chem. Lett.* **2007**, *17*, 4689–4693. [CrossRef]
16. Fruchart, J.C. Peroxisome proliferator-activated receptor-α (PPARα): At the crossroads of obesity, diabetes and cardiovascular disease. *Atherosclerosis* **2009**, *205*, 1–8. [CrossRef]
17. Fruchart, J.C. Selective peroxisome proliferator-activated α receptor modulators (SPPARM): The next generation of peroxisome proliferator-activated receptor α-agonists. *Cardiovasc. Diabetol.* **2013**, *12*, 82. [CrossRef]
18. Fruchart, J.C. Pemafibrate (K-877), a novel selective peroxisome proliferator-activated receptor α modulator for management of atherogenic dyslipidaemia. *Cardiovasc. Diabetol.* **2017**, *16*, 124. [CrossRef] [PubMed]
19. Takei, K.; Han, S.I.; Murayama, Y.; Satoh, A.; Oikawa, F.; Ohno, H.; Osaki, Y.; Matsuzaka, T.; Sekiya, M.; Iwasaki, H.; et al. Selective peroxisome proliferator-activated receptor-α modulator K-877 efficiently activates the peroxisome proliferator-activated receptor-α pathway and improves lipid metabolism in mice. *J. Diabetes Investig.* **2017**, *8*, 446–452. [CrossRef]
20. Raza-Iqbal, S.; Tanaka, T.; Anai, M.; Inagaki, T.; Matsumura, Y.; Ikeda, K.; Taguchi, A.; Gonzalez, F.J.; Sakai, J.; Kodama, T. Transcriptome Analysis of K-877 (a Novel Selective PPARα Modulator (SPPARMα))-Regulated Genes in Primary Human Hepatocytes and the Mouse Liver. *J. Atheroscler. Thromb.* **2015**, *22*, 754–772. [CrossRef]
21. Sasaki, Y.; Raza-Iqbal, S.; Tanaka, T.; Murakami, K.; Anai, M.; Osawa, T.; Matsumura, Y.; Sakai, J.; Kodama, T. Gene Expression Profiles Induced by a Novel Selective Peroxisome Proliferator-Activated Receptor α Modulator (SPPARMα) Pemafibrate. *Int. J. Mol. Sci.* **2019**, *20*, 5682. [CrossRef] [PubMed]
22. Sasaki, Y.; Asahiyama, M.; Tanaka, T.; Yamamoto, S.; Murakami, K.; Kamiya, W.; Matsumura, Y.; Osawa, T.; Anai, M.; Fruchart, J.C.; et al. Pemafibrate, a selective PPARα modulator, prevents non-alcoholic steatohepatitis development without reducing the hepatic triglyceride content. *Sci. Rep.* **2020**, *10*, 7818. [CrossRef]
23. Ferrannini, E.; Solini, A. SGLT2 inhibition in diabetes mellitus: Rationale and clinical prospects. *Nat. Rev. Endocrinol.* **2012**, *8*, 495–502. [CrossRef]
24. Scheen, A.J. Pharmacodynamics, efficacy and safety of sodium-glucose co-transporter type 2 (SGLT2) inhibitors for the treatment of type 2 diabetes mellitus. *Drugs* **2015**, *75*, 33–59. [CrossRef]
25. Fonseca-Correa, J.I.; Correa-Rotter, R. Sodium-Glucose Cotransporter 2 Inhibitors Mechanisms of Action: A Review. *Front. Med.* **2021**, *8*, 777861. [CrossRef] [PubMed]
26. Yoshioka, N.; Tanaka, M.; Ochi, K.; Watanabe, A.; Ono, K.; Sawada, M.; Ogi, T.; Itoh, M.; Ito, A.; Shiraki, Y.; et al. The sodium-glucose cotransporter-2 inhibitor Tofogliflozin prevents the progression of nonalcoholic steatohepatitis-associated liver tumors in a novel murine model. *Biomed. Pharmacother.* **2021**, *140*, 111738. [CrossRef]
27. Honda, Y.; Imajo, K.; Kato, T.; Kessoku, T.; Ogawa, Y.; Tomeno, W.; Kato, S.; Mawatari, H.; Fujita, K.; Yoneda, M.; et al. The Selective SGLT2 Inhibitor Ipragliflozin Has a Therapeutic Effect on Nonalcoholic Steatohepatitis in Mice. *PLoS ONE* **2016**, *11*, e0146337. [CrossRef] [PubMed]
28. Ito, S.; Hosaka, T.; Yano, W.; Itou, T.; Yasumura, M.; Shimizu, Y.; Kobayashi, H.; Nakagawa, T.; Inoue, K.; Tanabe, S.; et al. Metabolic effects of Tofogliflozin are efficiently enhanced with appropriate dietary carbohydrate ratio and are distinct from carbohydrate restriction. *Physiol. Rep.* **2018**, *6*, e13642. [CrossRef]
29. Obata, A.; Kubota, N.; Kubota, T.; Iwamoto, M.; Sato, H.; Sakurai, Y.; Takamoto, I.; Katsuyama, H.; Suzuki, Y.; Fukazawa, M.; et al. Tofogliflozin Improves Insulin Resistance in Skeletal Muscle and Accelerates Lipolysis in Adipose Tissue in Male Mice. *Endocrinology* **2016**, *157*, 1029–1042. [CrossRef]
30. Kleiner, D.E.; Brunt, E.M.; Van Natta, M.; Behling, C.; Contos, M.J.; Cummings, O.W.; Ferrell, L.D.; Liu, Y.C.; Torbenson, M.S.; Unalp-Arida, A.; et al. Nonalcoholic Steatohepatitis Clinical Research Network. Design and validation of a histological scoring system for nonalcoholic fatty liver disease. *Hepatology* **2005**, *41*, 1313–1321. [CrossRef]

31. Tanaka, T.; Yamamoto, J.; Iwasaki, S.; Asaba, H.; Hamura, H.; Ikeda, Y.; Watanabe, M.; Magoori, K.; Ioka, R.X.; Tachibana, K.; et al. Activation of peroxisome proliferator-activated receptor delta induces fatty acid beta-oxidation in skeletal muscle and attenuates metabolic syndrome. *Proc. Natl. Acad. Sci. USA* **2003**, *100*, 15924–15929. [CrossRef]
32. Tanaka, T.; Tahara-Hanaoka, S.; Nabekura, T.; Ikeda, K.; Jiang, S.; Tsutsumi, S.; Inagaki, T.; Magoori, K.; Higurashi, T.; Takahashi, H.; et al. PPARβ/δ activation of CD300a controls intestinal immunity. *Sci. Rep.* **2014**, *4*, 5412. [CrossRef] [PubMed]
33. Gonzalez, F.J.; Shah, Y.M. PPARα: Mechanism of species differences and hepatocarcinogenesis of peroxisome proliferators. *Toxicology* **2008**, *246*, 2–8. [CrossRef]
34. Misra, P.; Reddy, J.K. Peroxisome proliferator-activated receptor-α activation and excess energy burning in hepatocarcinogenesis. *Biochimie* **2014**, *98*, 63–74. [CrossRef] [PubMed]
35. Sans, A.; Bonnafous, S.; Rousseau, D.; Patouraux, S.; Canivet, C.M.; Leclere, P.S.; Tran-Van-Nhieu, J.; Luci, C.; Bailly-Maitre, B.; Xu, X.; et al. The Differential Expression of Cide Family Members is Associated with Nafld Progression from Steatosis to Steatohepatitis. *Sci. Rep.* **2019**, *9*, 7501. [CrossRef]
36. Gao, G.; Chen, F.J.; Zhou, L.; Su, L.; Xu, D.; Xu, L.; Li, P. Control of lipid droplet fusion and growth by CIDE family proteins. *Biochim. Biophys. Acta Mol. Cell Biol. Lipids* **2017**, *1862*, 1197–1204. [CrossRef] [PubMed]
37. Parthasarathy, G.; Revelo, X.; Malhi, H. Pathogenesis of Nonalcoholic Steatohepatitis: An Overview. *Hepatol. Commun.* **2020**, *4*, 478–492. [CrossRef] [PubMed]
38. Peng, C.; Stewart, A.G.; Woodman, O.L.; Ritchie, R.H.; Qin, C.X. Non-Alcoholic Steatohepatitis: A Review of Its Mechanism, Models and Medical Treatments. *Front. Pharmacol.* **2020**, *11*, 603926. [CrossRef] [PubMed]
39. Koo, J.H.; Han, C.Y. Signaling Nodes Associated with Endoplasmic Reticulum Stress during NAFLD Progression. *Biomolecules* **2021**, *11*, 242. [CrossRef] [PubMed]
40. Flessa, C.M.; Kyrou, I.; Nasiri-Ansari, N.; Kaltsas, G.; Papavassiliou, A.G.; Kassi, E.; Randeva, H.S. Endoplasmic Reticulum Stress and Autophagy in the Pathogenesis of Non-alcoholic Fatty Liver Disease (NAFLD): Current Evidence and Perspectives. *Curr. Obes. Rep.* **2021**, *10*, 134–161. [CrossRef] [PubMed]
41. Malhi, H.; Kaufman, R.J. Endoplasmic reticulum stress in liver disease. *J. Hepatol.* **2011**, *54*, 795–809. [CrossRef]
42. Han, C.Y.; Lim, S.W.; Koo, J.H.; Kim, W.; Kim, S.G. PHLDA3 overexpression in hepatocytes by endoplasmic reticulum stress via IRE1-Xbp1s pathway expedites liver injury. *Gut* **2016**, *65*, 1377–1388. [CrossRef]
43. Iynedjian, P.B. Molecular physiology of mammalian glucokinase. *Cell Mol. Life Sci.* **2009**, *66*, 27–42. [CrossRef] [PubMed]
44. Chitraju, C.; Mejhert, N.; Haas, J.T.; Diaz-Ramirez, L.G.; Grueter, C.A.; Imbriglio, J.E.; Pinto, S.; Koliwad, S.K.; Walther, T.C.; Farese, R.V., Jr. Triglyceride Synthesis by DGAT1 Protects Adipocytes from Lipid-Induced ER Stress during Lipolysis. *Cell Metab.* **2017**, *26*, 407–418. [CrossRef] [PubMed]
45. Becuwe, M.; Bond, L.M.; Pinto, A.F.M.; Boland, S.; Mejhert, N.; Elliott, S.D.; Cicconet, M.; Graham, M.M.; Liu, X.N.; Ilkayeva, O.; et al. FIT2 is an acyl-coenzyme A diphosphatase crucial for endoplasmic reticulum homeostasis. *J. Cell Biol.* **2020**, *219*, e202006111. [CrossRef] [PubMed]
46. Hoyles, L.; Fernandez-Real, J.M.; Federici, M.; Serino, M.; Abbott, J.; Charpentier, J.; Heymes, C.; Luque, J.L.; Anthony, E.; Barton, R.H.; et al. Molecular phenomics and metagenomics of hepatic steatosis in non-diabetic obese women. *Nat. Med.* **2018**, *24*, 1070–1080. [CrossRef] [PubMed]
47. Mayneris-Perxachs, J.; Cardellini, M.; Hoyles, L.; Latorre, J.; Davato, F.; Moreno-Navarrete, J.M.; Arnoriaga-Rodríguez, M.; Serino, M.; Abbott, J.; Barton, R.H.; et al. Iron status influences non-alcoholic fatty liver disease in obesity through the gut microbiome. *Microbiome* **2021**, *9*, 104. [CrossRef] [PubMed]
48. Zheng, Q.Y.; Li, P.P.; Jin, F.S.; Yao, C.; Zhang, G.H.; Zang, T.; Ai, X. Ursolic acid induces ER stress response to activate ASK1-JNK signaling and induce apoptosis in human bladder cancer T24 cells. *Cell Signal.* **2013**, *25*, 206–213. [CrossRef]
49. Challa, T.D.; Wueest, S.; Lucchini, F.C.; Dedual, M.; Modica, S.; Borsigova, M.; Wolfrum, C.; Blüher, M.; Konrad, D. Liver ASK1 protects from non-alcoholic fatty liver disease and fibrosis. *EMBO Mol. Med.* **2019**, *11*, e10124. [CrossRef]
50. Harrison, S.A.; Wong, V.W.; Okanoue, T.; Bzowej, N.; Vuppalanchi, R.; Younes, Z.; Kohli, A.; Sarin, S.; Caldwell, S.H.; Alkhouri, N.; et al. Selonsertib for patients with bridging fibrosis or compensated cirrhosis due to NASH: Results from randomized phase III STELLAR trials. *J. Hepatol.* **2020**, *73*, 26–39. [CrossRef]
51. Chen, Y.; Ohki, R. p53-PHLDA3-Akt Network: The Key Regulators of Neuroendocrine Tumorigenesis. *Int. J. Mol. Sci.* **2020**, *21*, 4098. [CrossRef] [PubMed]
52. Roden, M. Mechanisms of Disease: Hepatic steatosis in type 2 diabetes—pathogenesis and clinical relevance. *Nat. Clin. Pract. Endocrinol. Metab.* **2006**, *2*, 335–348. [CrossRef] [PubMed]
53. de Vries, M.; Westerink, J.; Kaasjager, K.H.; de Valk, H.W. Prevalence of Nonalcoholic Fatty Liver Disease (NAFLD) in Patients with Type 1 Diabetes Mellitus: A Systematic Review and Meta-Analysis. *J. Clin. Endocrinol. Metab.* **2020**, *105*, 3842–3853. [CrossRef] [PubMed]
54. Regnell, S.E.; Lernmark, A. Hepatic steatosis in type 1 diabetes. *Rev. Diabet. Stud.* **2011**, *8*, 454–467. [CrossRef] [PubMed]
55. Ravikumar, B.; Carey, P.E.; Snaar, J.E.; Deelchand, D.K.; Cook, D.B.; Neely, R.D.; English, P.T.; Firbank, M.J.; Morris, P.G.; Taylor, R. Real-time assessment of postprandial fat storage in liver and skeletal muscle in health and type 2 diabetes. *Am. J. Physiol. Endocrinol. Metab.* **2005**, *288*, E789–E797. [CrossRef] [PubMed]

56. Donnelly, K.L.; Smith, C.I.; Schwarzenberg, S.J.; Jessurun, J.; Boldt, M.D.; Parks, E.J. Sources of fatty acids stored in liver and secreted via lipoproteins in patients with nonalcoholic fatty liver disease. *J. Clin. Investig.* **2005**, *115*, 1343–1351. [CrossRef] [PubMed]
57. Tamura, S.; Shimomura, I. Contribution of adipose tissue and de novo lipogenesis to nonalcoholic fatty liver disease. *J. Clin. Investig.* **2005**, *115*, 1139–1142. [CrossRef]
58. Grefhorst, A.; van de Peppel, I.P.; Larsen, L.E.; Jonker, J.W.; Holleboom, A.G. The Role of Lipophagy in the Development and Treatment of Non-Alcoholic Fatty Liver Disease. *Front. Endocrinol.* **2021**, *11*, 601627. [CrossRef] [PubMed]
59. Soto-Gutierrez, A.; Gough, A.; Vernetti, L.A.; Taylor, D.L.; Monga, S.P. Pre-clinical and clinical investigations of metabolic zonation in liver diseases: The potential of microphysiology systems. *Exp. Biol. Med.* **2017**, *242*, 1605–1616. [CrossRef] [PubMed]
60. Kietzmann, T. Metabolic zonation of the liver: The oxygen gradient revisited. *Redox Biol.* **2017**, *11*, 622–630. [CrossRef] [PubMed]
61. Cunningham, R.P.; Porat-Shliom, N. Liver Zonation—Revisiting Old Questions With New Technologies. *Front. Physiol.* **2021**, *12*, 732929. [CrossRef] [PubMed]
62. Steinman, J.B.; Salomao, M.A.; Pajvani, U.B. Zonation in NASH—A key paradigm for understanding pathophysiology and clinical outcomes. *Liver Int.* **2021**, *41*, 2534–2546. [CrossRef]
63. Bedossa, P. Histological Assessment of NAFLD. *Dig. Dis. Sci.* **2016**, *61*, 1348–1355. [CrossRef] [PubMed]
64. Brunt, E.M.; Kleiner, D.E.; Wilson, L.A.; Unalp, A.; Behling, C.E.; Lavine, J.E.; Neuschwander-Tetri, B.A. NASH Clinical Research NetworkA list of members of the Nonalcoholic Steatohepatitis Clinical Research Network can be found in the Appendix. Portal chronic inflammation in nonalcoholic fatty liver disease (NAFLD): A histologic marker of advanced NAFLD-Clinicopathologic correlations from the nonalcoholic steatohepatitis clinical research network. *Hepatology* **2009**, *49*, 809–820. [PubMed]
65. Kumar, S.; Duan, Q.; Wu, R.; Harris, E.N.; Su, Q. Pathophysiological communication between hepatocytes and non-parenchymal cells in liver injury from NAFLD to liver fibrosis. *Adv. Drug Deliv. Rev.* **2021**, *176*, 113869. [CrossRef]
66. Gadd, V.L.; Skoien, R.; Powell, E.E.; Fagan, K.J.; Winterford, C.; Horsfall, L.; Irvine, K.; Clouston, A.D. The portal inflammatory infiltrate and ductular reaction in human nonalcoholic fatty liver disease. *Hepatology* **2014**, *59*, 1393–1405. [CrossRef] [PubMed]
67. Schwimmer, J.B.; Lavine, J.E.; Wilson, L.A.; Neuschwander-Tetri, B.A.; Xanthakos, S.A.; Kohli, R.; Barlow, S.E.; Vos, M.B.; Karpen, S.J.; Molleston, J.P.; et al. In children with non-alcoholic fatty liver disease, cysteamine bitartrate delayed release improves liver enzymes but does not reduce disease activity scores. *Gastroenterology* **2016**, *151*, 1141–1154. [CrossRef]
68. Park, S.R.; Cho, C.-S.; Xi, J.; Kang, H.M.; Lee, J.H. Holistic characterization of single-hepatocyte transcriptome responses to high-fat diet. *Am. J. Physiol. Endocrinol. Metab.* **2021**, *320*, E244–E258. [CrossRef] [PubMed]
69. Brosch, M.; Kattler, K.; Herrmann, A.; von Schönfels, W.; Nordström, K.; Seehofer, D.; Damm, G.; Becker, T.; Zeissig, S.; Nehring, S.; et al. Epigenomic map of human liver reveals principles of zonated morphogenic and metabolic control. *Nat. Commun.* **2018**, *9*, 4150. [CrossRef]
70. Halpern, K.B.; Shenhav, R.; Matcovitch-Natan, O.; Tóth, B.; Lemze, D.; Golan, M.; Massasa, E.E.; Baydatch, S.; Landen, S.; Moor, A.E.; et al. Single-cell spatial reconstruction reveals global division of labour in the mammalian liver. *Nature* **2017**, *542*, 352–356. [CrossRef] [PubMed]
71. Xiong, X.; Kuang, H.; Ansari, S.; Liu, T.; Gong, J.; Wang, S.; Zhao, X.Y.; Ji, Y.; Li, C.; Guo, L.; et al. Landscape of Intercellular Crosstalk in Healthy and NASH Liver Revealed by Single-Cell Secretome Gene Analysis. *Mol. Cell.* **2019**, *75*, 644–660. [CrossRef] [PubMed]
72. Hildebrandt, F.; Andersson, A.; Saarenpää, S.; Larsson, L.; Van Hul, N.; Kanatani, S.; Masek, J.; Ellis, E.; Barragan, A.; Mollbrink, A.; et al. Spatial Transcriptomics to define transcriptional patterns of zonation and structural components in the mouse liver. *Nat. Commun.* **2021**, *12*, 7046. [CrossRef] [PubMed]

cells

Review

Potential Therapeutic Effects of PPAR Ligands in Glioblastoma

Rossella Basilotta [1], Marika Lanza [1], Giovanna Casili [1], Giulia Chisari [2], Stefania Munao [2], Lorenzo Colarossi [2], Laura Cucinotta [1], Michela Campolo [1], Emanuela Esposito [1,*] and Irene Paterniti [1]

1 Department of Chemical, Biological, Pharmaceutical and Environmental Sciences, University of Messina, Viale Ferdinando Stagno D'Alcontres, 98166 Messina, Italy; robasilotta@unime.it (R.B.); marika.lanza@unime.it (M.L.); gcasili@unime.it (G.C.); laura.cucinotta@unime.it (L.C.); campolom@unime.it (M.C.); irene.paterniti@unime.it (I.P.)

2 Istituto Oncologico del Mediterraneo, Via Penninazzo 7, 95029 Viagrande, Italy; giulia.chisari@grupposamed.com (G.C.); stefania.munao@grupposamed.com (S.M.); lorenzo.colarossi@grupposamed.com (L.C.)

* Correspondence: eesposito@unime.it; Tel.: +39-090-676-5208

Abstract: Glioblastoma (GB), also known as grade IV astrocytoma, represents the most aggressive form of brain tumor, characterized by extraordinary heterogeneity and high invasiveness and mortality. Thus, a great deal of interest is currently being directed to investigate a new therapeutic strategy and in recent years, the research has focused its attention on the evaluation of the anticancer effects of some drugs already in use for other diseases. This is the case of peroxisome proliferator-activated receptors (PPARs) ligands, which over the years have been revealed to possess anticancer properties. PPARs belong to the nuclear receptor superfamily and are divided into three main subtypes: PPAR-α, PPAR-β/δ, and PPAR-γ. These receptors, once activated by specific natural or synthetic ligands, translocate to the nucleus and dimerize with the retinoid X receptors (RXR), starting the signal transduction of numerous genes involved in many physiological processes. PPARs receptors are activated by specific ligands and participate principally in the preservation of homeostasis and in lipid and glucose metabolism. In fact, synthetic PPAR-α agonists, such as fibrates, are drugs currently in use for the clinical treatment of hypertriglyceridemia, while PPAR-γ agonists, including thiazolidinediones (TZDs), are known as insulin-sensitizing drugs. In this review, we will analyze the role of PPARs receptors in the progression of tumorigenesis and the action of PPARs agonists in promoting, or not, the induction of cell death in GB cells, highlighting the conflicting opinions present in the literature.

Keywords: glioblastoma; cancer; PPARs; brain; neuro-oncology

Citation: Basilotta, R.; Lanza, M.; Casili, G.; Chisari, G.; Munao, S.; Colarossi, L.; Cucinotta, L.; Campolo, M.; Esposito, E.; Paterniti, I. Potential Therapeutic Effects of PPAR Ligands in Glioblastoma. *Cells* **2022**, *11*, 621. https://doi.org/10.3390/cells11040621

Academic Editors: Kay-Dietrich Wagner and Nicole Wagner

Received: 17 December 2021
Accepted: 8 February 2022
Published: 10 February 2022

1. Introduction

Glioblastoma (GB) is the most common and aggressive subtype of malignant brain tumors. This tumor belongs to the large family of gliomas and is also known as grade IV astrocytoma [1]. GB originates from glial cells and astrocytes, which play supporting roles within the central nervous system (CNS) and is characterized by abnormal angiogenesis, apoptosis alteration, and invasiveness. This tumor manifests itself with nonspecific signs and symptoms, which vary according to the size and location of the tumor. Patients often present symptoms of increased intracranial pressure, including focal or progressive headache and neurologic deficits, personality changes, and seizures [2]. Among the genetic risk factors, a set of single nucleotide genetic polymorphisms (SNPs) have been identified, located on different genes (NF1, NF2, IDH1/IDH2, TERT, EGFR, CCDC26, CDKN2B, PHLDB1, TP53, RTEL1), which seem to contribute to gliomagenesis and to the development of all grades and histologies of gliomas [3,4]. The glioblastomas are divided into two broad categories: primary and secondary [5]. Primary GB accounts for 90% of total cases, is more frequent in the elderly population, and has a worse prognosis than its secondary counterpart. Primary GB onco-markers include overexpression of the epidermal growth factor receptor (EGFR) and mutations in the tumor suppressor gene of phosphatase and tensin homolog (PTEN) and the telomerase reverse transcriptase promoter (TERT) [6]. Secondary GB constitutes 5% of cases and develops from astrocytomas with a lower degree of malignancy, affects younger patients and is related to mutations in isocitrate dehydrogenase 1 and 2 (IDH 1 and 2) and tumor protein 53 (P53) [7]. It is a tumor characterized by an extraordinary intra-tumoral heterogeneity which often results in the inability of traditional therapies to obtain long-term remissions. Glioblastomas, in fact, differ in phenotypic properties, including transient quiescence, self-renewal, adaptation to hypoxia, and resistance to therapy-induced DNA damage. For this reason, the development of new personalized treatment strategies for GB represents both a preclinical and clinical challenge [8]. To date, the causes and physiopathology of GB are unknown. Current therapy of choice consists of surgical resection or biopsy, followed by radiotherapy and concomitant chemotherapy with temozolamide (TMZ). TMZ, the gold standard anticancer drug for the treatment of GB, belongs to the class of orally administered alkylating agents and improves prognosis by increasing median patient survival. The approved standard therapy consists of a daily dose of 150 to 200 mg per square meter of body surface area for 5 days of each 28-day cycle. Daily therapy at a dose of 75 mg per square meter for up to seven weeks is safe; at these doses, TMZ depletes the DNA repair enzyme MGMT, resulting in tumor tissue shrinkage, an effect associated with longer survival among glioblastoma patients [9,10]. Considering the aggressiveness of GB and the low efficacy of therapeutic strategies, it is currently necessary to identify new therapeutic targets able to reduce or arrest the progression of GB. GB aggressiveness appears to be related to the presence of tumor stem cell populations called GSCs that contribute to GB malignancy by promoting tumor growth, angiogenesis, and therapeutic resistance. Unlike well-differentiated tumor cells, which show limited replicative potential, GSCs can proliferate indefinitely and spread to tissues and organs distant from the primary tumor site, becoming responsible for initiation, growth of metastases and resistance to therapy. According to some studies, the preferential overexpression of nicotinamide N-methyltransferase (NNMT) in GSC, a cytosolic enzyme involved in the biotransformation of many xenobiotics, causes the exhaustion of the methyl donor S-adenosyl methionine (SAM), with consequent hypomethylation of the GB DNA, causing the translation of the tumor towards a mesenchymal phenotype and accelerating its growth [11,12]. The main mechanism of resistance to TMZ therapy is related to the overexpression of O-6-methylguanine-DNA methyltransferase (MGMT), a gene located on chromosome 10q26 that codes for a DNA repair protein, which removes alkyl groups from the O-6 position of guanine, an important alkylation site. High levels of MGMT in tumor cells, therefore, induce the formation of a resistant phenotype by reducing the efficacy of alkylating agents such as TMZ, thus leading to therapeutic failure. MGMT in its methylated and therefore inactive form represents a molecular marker of clinical relevance, associated with the response to alkylating chemotherapy and survival of patients with GB [13–15].

Peroxisome proliferator-activated receptors (PPARs) are ligand-inducible transcription factors that belong to the superfamily of proteins called nuclear hormone receptors (Nrs), to which steroid, thyroid, and retinoid receptors also belong. Three isoforms of PPAR have been identified: PPAR-α, PPAR-β/δ, and PPAR-γ, differently expressed based on the physiological role, tissue distribution, and specificity of the ligands. Each of the isoforms activates or suppresses different genes involved primarily in the metabolism and homeostasis of fats and carbohydrates, as well as in proliferation and cell differentiation, inflammation, and cancer [16].

All three isotypes of PPAR are co-expressed in the central nervous system (CNS), but their function in this tissue is still poorly understood. Some studies show that, at the level of CNS, PPARs are involved in lipid metabolism, neuronal differentiation, and death, as well as inflammation and neurodegeneration. Observations both in vitro and in vivo show that PPAR-β/δ is the prevalent isoform in neurons of different brain areas, while PPAR-α is expressed at very low levels, predominantly in astrocytes, and appears to be involved in the neurotransmission of excitatory amino acids [17]. The expression of PPAR-γ in the brain has been extensively studied in relation to inflammation and neurodegeneration [18].

Although PPAR receptors are known for their role in lipid metabolism and glucose homeostasis, a lot of research in the literature has demonstrated the contribution of these receptors in tumors and GB biology [19,20].

Preclinical and clinical studies have shown beneficial effects of PPAR agonists against GB growth, inhibiting the invasion and motility of glioma cells and thus increasing the chance of survival [21]. The purpose of this review is to summarize the role of PPARs in GB, focusing on the antitumor action of their synthetic and natural ligands, in order to consider them as potential additional treatments to conventional therapies.

2. Peroxisome Proliferator-Activated Receptors

Peroxisome proliferator-activated receptors (PPARs) are ligand-activated transcription factors involved in various processes at the cellular level and in the regulation of lipid, carbohydrate, and amino acid metabolism. PPARs belong to the superfamily of nuclear hormone receptors (Nrs), which after interacting with specific ligands (synthetic or non-synthetic), translocate to the nucleus where they modify their conformation and regulate gene transcription through the differential recruitment of cofactors and enzymes modifying the histone [22]. Once translocated to the nucleus, PPARs interacts with retinoid X receptors (RXR), peroxisome proliferator-activated receptor gamma-coactivators (PGC), steroid receptor coactivators (SRC), and CREB binding protein (CBP/p300), then bind to the sequences of the peroxisome proliferating receptor element (PPRE) and consequently initiate the transcription of target genes involved in various physiological processes [23]. PPARs, in fact, control gene expression involved in energy homeostasis, lipid metabolism, and adipogenesis: they represent the main receptors for dietary fats such as oleic and linolenic acids and for many lipid metabolites, for example, prostaglandin J2, 8S-hydroxyheicosatetraenoic acid (8-HETE), and oxidized phospholipids. The altered expression of PPARs is also related to the onset of many diseases such as type 2 diabetes, dyslipidemia, obesity, atherosclerosis, and metabolic syndrome [21]. PPARs are also expressed in the cardiovascular system (endothelial cells, vascular smooth muscle cells, monocytes, and macrophages), and many clinical and preclinical studies have shown the significant role of PPARs also in cardiovascular diseases [24]. Today, increasing attention has been paid to the critical role of PPARs in inflammation and cancer; numerous studies have, in fact, highlighted the overexpression of PPARs in many human solid tumors [25].

Studies based on X-ray crystallography and molecular modeling have shown that the structure of PPARs consists of six functional domains, from A to F [26]. The N-terminal portion of PPARs exhibits the ligand-independent transactivation domain (or A/B domain), also called activation function 1 (AF-1) responsible for transcriptional activation, followed by the C domain, also called DNA binding domain (DBD), involved in recognition of the DNA sequence in the promoter region of genes known as peroxisome proliferator response element (PPRE). The C-terminal of the PPAR receptor, on the other hand, contains the D domain, which thanks to its flexibility acts as a docking site for the cofactors and the E/F (or LBD) domain, which is responsible for the specificity of the ligand and for dimerization of the receptor with the retinoid X receptors (RXR). The dimerization domain is critical for the formation of heterodimers with the retinoic acid receptor α (RXRα), an important prerequisite for PPARs to bind DNA in regions containing the DNA sequence AGGTCANAGGTCA. The C-terminal also possesses the AF-2 activation domain, that after binding with the ligand, synergizes with AF-1 and undergoes conformational modifications allowing the recruitment of co-activating proteins p300, CREB binding protein (CBP), or coactivator steroid receptor 1 (SRC1), important for the transcriptional activation of their target genes (Figure 1) [27].

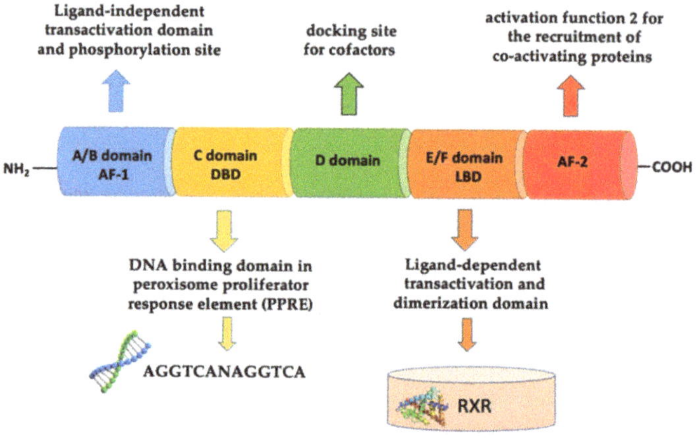

Figure 1. Schematic representation of the structure of PPAR receptor.

PPARs Isoforms: Tissue Distribution and Biological Activity

The PPAR family includes three different subtypes which differ in terms of tissue distribution, affinity for ligands, and biological activity. They are called PPAR-α, PPAR-β/δ, and PPAR-γ. All isoforms participate differently in lipid homeostasis and glucose regulation (energy balance), but each of them is capable of suppressing or activating different genes [28].

Based on their tissue distribution, PPAR-α receptors are mainly expressed in brown adipose tissues, skeletal muscle, kidney, heart, liver, and intestinal mucosa and are involved in glucose metabolism and homeostasis and in the oxidation of fatty acids [29]. The PPAR-α receptor is activated by natural ligands, including saturated, monounsaturated, and polyunsaturated fatty acids and their metabolites such as 8S-HETE and 8-HEPE, leukotrienes B4 (LTB4), oxidized phospholipids, and lipolytic lipoprotein products. Among these omega-3 fatty acids, being highly polyunsaturated, oxidize easily and stimulate PPAR-α, causing a decrease in lipid levels and the elimination of triglycerides from the plasma with a consequent increase in the levels of high-density lipoprotein cholesterol (HDL) and reduction of inflammation and arteriosclerosis in the cardiovascular system. Furthermore, an important anti-inflammatory effect also derives from the inhibition of the oxidation of omega-3 fatty acids, mediated by NF-κB in a PPAR-α-dependent pathway [30]. Synthetic PPAR-α ligands are represented by fibrates (clofibrate, gemfibrozil, fenofibrate, and bezafibrate), a class of lipid-lowering drugs that are used in the treatment of hypertriglyceridemia. Through the activation of PPAR-α, they cause an increase in gene expression involved in the β-oxidation of fatty acids leading to the reduction of triglyceride-rich lipoproteins in the serum and to the increase of HDL cholesterol, slowing the progression of arteriosclerosis and reducing cardiovascular events (Figure 2) [31].

The PPAR-β/δ receptor consists of 441 amino acids, with a molecular weight of 49.9 kDa and is ubiquitously expressed in almost all tissues, including the liver, intestines, kidneys, abdominal fat, skeletal muscle, brain, and pancreas. Like the other members of the PPAR family, it mainly intervenes in the metabolism of lipids, participating in the oxidation of fatty acids, both at the level of adipose tissue, reducing adiposity, and consequently preventing the development of obesity, both at the level of skeletal muscles and heart and regulating the concentrations of cholesterol and blood glucose [32]. Moreover, many studies have revealed a large expression of PPAR-β/δ in the central nervous system (CNS), in particular at the level of neurons, astrocytes, oligodendrocytes and in microglia cells, suggesting the role of these receptors as targets for neuroinflammation and neurodegeneration [33]. Although PPAR-β/δ has a smaller binding domain (LBD) than other members of the PPAR family, it has the ability to bind many endogenous ligands, but with relatively low selectivity. Natural ligands include polyunsaturated fatty acids (arachidonic and linoleic acids) and their metabolites like prostacyclin PGI2, 13S-hydroxyoctadecadienoic acid (13S-HODE), and 15S-hydroxyheicosatetraenoic acid (15S-HETE), which could have promising applications in cardiomyopathy diabetic. Synthetic agonists (GW501516, GW0742, L-165041, and MBX-802) that have been developed and proposed as treatments for obesity and metabolic syndrome have not been used for clinical trials due to their carcinogenic effects, so none of them have been approved for clinical use to date (Figure 2) [34]. Despite controversial data on its pro-tumorigenic versus anti-tumorigenic action, few results suggest that PPAR-β/δ activation can reduce the growth of neuroectodermal tumors, including glioblastomas [35].

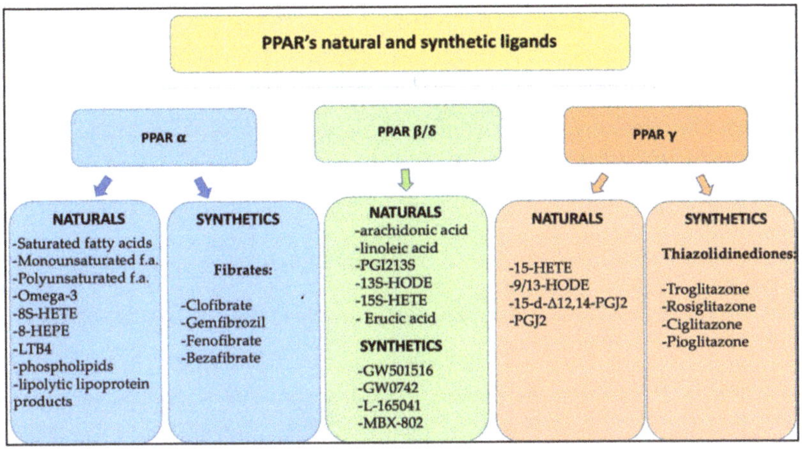

Figure 2. Classification of natural and synthetic ligands of PPARs receptors.

The PPAR-γ receptor is expressed in many human tissues and has three different isoforms: PPAR-γ1, γ2, and γ3. PPAR-γ1 is ubiquitously expressed in all human cells, PPAR-γ2 has been predominantly detected in white and brown adipose tissue, as well as in the large intestine and spleen, while PPAR-γ3 expression is limited. PPAR-γ is poorly expressed in the central nervous system (CNS), but it was found in different cell types such as neurons, astrocytes, oligodendrocytes, and microglia. Its physiological role includes the regulation of adipogenesis and the levels of adipokines such as adiponectin, TNFα, MCP-1, and resistin, and it is also involved in the energy balance and lipid biosynthesis. Thanks to the size of its binding cavity, the PPAR-γ receptor is able to bind a large variety of natural or synthetic lipophilic acids. The natural modulators of PPAR-γ are mainly unsaturated fatty acids and their metabolites, including 15- hydroxyeicosatetraenoic acid (15-HETE), 9- and 13-hydroxyoctadecadienoic acid (9/13-HODE), 15-deoxy- Δ12,14-prostaglandin J2 (15-d-Δ12,14-PGJ2), and prostaglandin PGJ2, whose physiological role is still to be clarified [36]. Synthetic PPAR-γ ligands belong to the thiazolidinediones class (TZD), including troglitazone, rosiglitazone, ciglitazone, and pioglitazone, and they are known as insulin-sensitizing drugs. By activating the PPAR-γ receptor, in fact, they reduce the hepatic production of glucose and prolong the function of pancreatic cells, preventing apoptosis of β cells (Figure 2). TZD are used in the treatment of type 2 diabetes, as they increase insulin sensitivity and improve glucose control, but their use is limited by important adverse events, including the risk of bone fracture and congestive heart failure [37] (Figure 3).

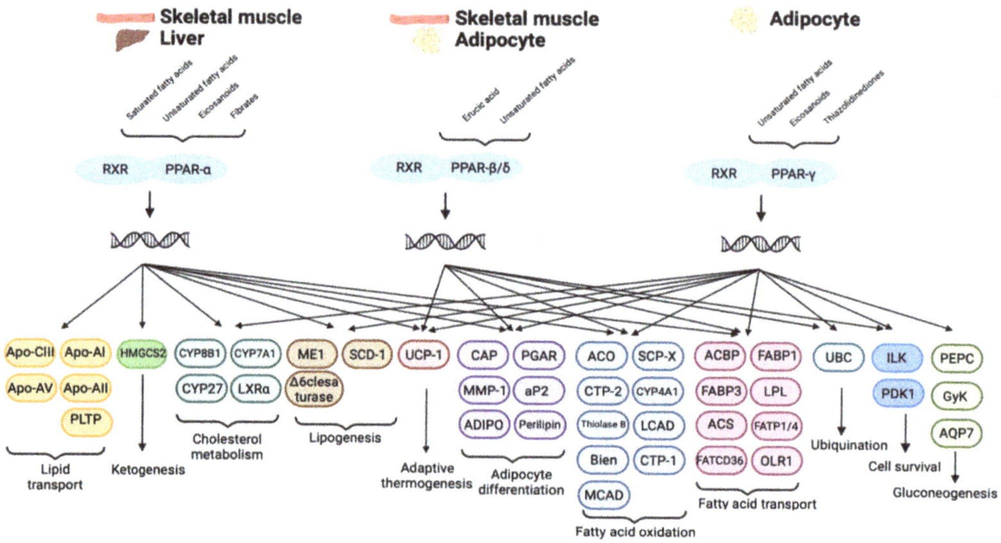

Figure 3. A schematic representation at cellular level of PPARs signaling.

3. Role of PPARs in Tumors

Certain PPAR-related metabolic alterations, such as obesity and type 2 diabetes, have been identified as risk factors for cancer cell proliferation and thus tumor progression. Hence, research is currently focused on using PPARs as targets for cancer therapy, and a few studies have focused on understanding their role in human cancer and in the antitumor activity of their natural and synthetic agonists. Furthermore, current studies have revealed conflicting results on the role of the different isoforms of PPARs in various tumor types; most investigations have shown that PPAR-β/δ activation is linked to tumor progression, while PPAR-α and PPAR-γ are associated with an antitumor action [38]. Some in vitro studies on breast cancer cells SUM149PT and SUM1315MO2 have shown interesting results in the context of nuclear PPAR-α receptor signaling, demonstrating that its activation by clofibrate agonist suppresses the inflammatory activity of cyclooxygenase-2 (COX-2) and 5-lipoxygenase (5-LO) and determines the decrease in the secretion of prostaglandin-E2 (PGE2) and leukotrienes-B4 (LB4), effectively inhibiting cell survival and cell cycle-related kinases [39]. Another study conducted in vivo showed that the combination of clofibric acid (PPAR-α agonist) and pioglitazone (PPAR-γ agonist) reduces angiogenesis, induces apoptosis, significantly decreasing the expression of COX-2 and VEGF, through inhibition of activator protein-1 (AP-1) and suppressing the growth of solid ovarian tumors [40]. Evidence suggests that certain PPAR-α ligands, including bezafibrate and fenofibrate, may act as potential chemopreventive agents in colon carcinogenesis by reducing intestinal polyp formation in Apc-deficient mice, by inhibiting AOM/DSS-induced colon carcinogenesis. However, the exact mechanism by which PPAR-α activation suppresses colon carcinogenesis is still unclear [40]. Moreover, only a few studies support the pro-carcinogenic role of PPAR-α. For example, some investigations carried out on human breast cancer cell lines (MDA-MB-231) have shown that inhibition of PPAR-α by the antagonist GW6471 leads to an impairment of the mevalonate pathway and a substantial reduction in cholesterol and lipid droplets. This causes a consequent perturbation of lipid metabolism and cell death, influencing the pathways involved in the control of proliferation, such as the pathways involving the Rho Family and YAP/TAZ and Wnt/β-catenin signaling [41]. The conflicting results presented in the literature regarding the action of PPAR-α on tumor progression could be due to the different evaluation conditions, cell lines used, the stage of differentiation, the cellular context, and the microenvironment.

The role of PPAR-β/δ on cell proliferation, induction of angiogenesis, and cell death has been extensively investigated. Although there are conflicting opinions on the effects of PPAR-β/δ activation for cancer progression, most reports suggest that its stimulation could have pro-tumorigenic effects [42]. PPAR-β/δ exerts proangiogenic effects, directly or indirectly modulating downstream proinflammatory or proangiogenic molecules that act on multiple cell types in the tumor microenvironment, promoting cancer progression and metastasis [43]. The increased expression of PPAR-β/δ mRNA in colon cancers has been attributed to APC-β-catenin-TCF4-mediated transcription, similar to the well-known target gene β-catenin-TCF4 CCND1, which encodes cyclin D1. This supports the hypothesis that PPAR-β/δ regulates genes that increase cell proliferation and promote colon carcinogenesis [44]. Consistent with these results, PPAR-β/δ has been shown to strongly potentiate aberrant activation of β-catenin in mouse genetic models of human CRC, with representative APC mutations and overexpression or deletion of PPAR-β/δ in intestinal epithelial cells (IEC), activating pro-invasive pathways to promote CRC tumorigenesis. These results have demonstrated that PPAR-β/δ strongly accelerates APC mutation-driven CRC progression and invasion through multiple important pro-tumorigenic pathways, including BMP7/TAK1/-catenin, PDGFRβ, AKT1, EIF4G1, and CDK1 [45]. The role of PPAR-β/δ in enhancing cell proliferation was supported by a further study performed on human liposarcoma cells (SW872, T778), in which an increase in cell proliferation was observed in response to PPAR-β/δ activation by the agonist GW0742, which appears to be caused by leptin repression, suggesting the potential therapeutic use of PPAR-β/δ antagonists for the treatment of unresectable liposarcomas [46]. PPAR-β/δ seems to be overexpressed in breast cancer cells and the elevated levels appear to correlate with greater migratory and metastatic properties. PPAR-β/δ mediates these effects by mechanisms including increased expression of antioxidant proteins such as catalase and increased AKT-mediated survival signaling after prolonged nutrient deprivation [47].

Among the three isoforms, PPAR-γ is certainly the most studied in tumors. It is, in fact, expressed in a wide variety of cancers and its role in cancer initiation/progression has long been debated. The literature suggests that PPAR-γ plays a key role in tumorigenesis as a tumor suppressor. Indeed, PPAR-γ activation by many agonists has shown antiproliferative and proapoptotic actions in colon, esophageal, thyroid, breast, lung, and prostate cancers [48–50]. The mechanism by which it induces tumor cell growth arrest appears to be related to the PPAR-γ-dependent upregulation of the tumor suppressor gene and of the homologous tensin phosphatase (PTEN), which inhibits the phosphorylation of PI3-kinase and AKT by reducing cell migration and proliferation. To the antiproliferative effects is also added the downregulation of the anti-apoptotic protein B-cells/lymphoma 2 (Bcl-2), the anti-angiogenic activity through the inhibition of VEGF and its receptors in various cells and anti-inflammatory properties through NFκB-mediated inhibition of gene transcription. Furthermore, PPAR-γ appears to hinder the formation of metastases through the inhibition of the epithelial–mesenchymal transition (EMT), a process by which epithelial cells lose their cell polarity and cell–cell adhesion, and gain migratory and invasive properties. Consistent with these findings, growing evidence suggests that PPAR-γ overexpression has important suppressive activities in colorectal cancer growth. In fact, preclinical studies currently analyze new PPAR-γ agonists, capable of inhibiting the Wnt/β-catenin pathway, acting as modulators of PPAR-γ signaling, and interfering with related pathways in order to provide new therapies for CRC [51]. PPAR-γ thiazolidinediones (TZD) ligands have been shown to counteract the stimulatory effects of leptin on breast cancer growth in vivo and in vitro models. The results show that PPAR-γ activation inhibited cell proliferation and prevented the development of leptin-induced MCF-7 tumor xenografts [52].

PPAR-γ expression was studied in patients with esophageal squamous cell carcinoma (ESCC), on which the antiproliferative effect and mechanism of action of the PPAR-γ agonist, ephatutazone, were investigated. Ephatutazone has been shown to cause a 49.6%

reduction in the proliferation of xenotransplanted ESCC cells through a mechanism that involves the regulation of p21Cip1 protein levels in the nucleus by inactivating the Akt signal and dephosphorylating p21 to Thr145, without altering the transcriptional activity of p21Cip1 [53]. Another study found a significant increase in PPAR-γ expression in NSCLC cell lines by both immunohistochemistry and Western blotting. In addition, significant antitumor activity was observed both in vivo and in vitro in response to treatment with troglitazone or pioglitazone, with a correlated reduction in metastases [54].

Although many results strongly support the role of PPAR-γ as a tumor suppressor, other studies, on the contrary, argue that it plays a role as a tumor promoter. Indeed, a recent animal experiment of prostate orthotransplantation identified the involvement of PPAR-γ in the upregulation of the AKT3-PGC1α axis. PPAR-γ seems to increase the regulation of AKT3, destabilizing CRM1 and favoring the localization of PGC1α in the nucleus with a consequent increase in mitochondrial function and ATP levels. The high levels of ATP appear to be related to promoting tumor growth and metastasis [55]. Further evidence reveals that PPAR-γ activation reduces TXNIP expression in human melanoma cells (A375 and C8161), affecting the expression of proteins of particular relevance to melanoma cell invasiveness, such as integrin alpha-v/beta-3 and TIMP-2, resulting in melanoma progression to a metastatic phenotype [56].

3.1. Role of PPAR-α Agonists in GB

Conflicting results emerged regarding the role of PPAR-α in GB tumorigenesis. Studies have examined the expression of PPAR-α protein and PPAR-α mRNA in primary wild-type human IDH1 GB, arguing that their overexpression in GB is related to the degree of glioma malignancy [57]. This overexpression appears to be accompanied by the significant increase in 30-hydroxy-30-methylglutaryl-CoA reductase (HMGCR), the cholesterol biosynthesis-limiting enzyme (CHO), that catalyzes the formation of mevalonate (MVA). These results indicate that PPAR-α, by regulating CHO metabolism, is involved in the strong alteration of lipid homeostasis observed in gliomas and could therefore drive the tumorigenesis process. In fact, the use of a compound derived from N-phenylsulfonyl (AA452), capable of blocking the activation of PPAR-α, has determined a strong effect on cell viability, reducing cell proliferation and migration and therefore decreasing tumor invasiveness [58]. Other evidence discusses that the expression of PPAR-α receptors is negatively correlated to the degree of malignancy of the glioma, in fact, its activation suppresses the proliferation of tumor cells, delays the cell cycle to the G1 phase and induces apoptosis and the accumulation of species reactive oxygen (ROS) in U87 cells [59,60]. The anticancer effects of the PPAR-α agonist fenofibrate have been demonstrated in several cell lines of colon, breast, endometrial, skin, medulloblastoma, and melanoma cancers [61,62]. Among the PPAR-α ligands tested in GB, fenofibrate received the most attention, due to its capacity to cross the blood brain barrier (BBB) and has an established anti-inflammatory activity and limited toxicity and a better side effect profile [63]. In order to activate the PPAR-α receptor, fenofibrate must first be converted into fenofibric acid (FA) by blood and tissue esterases. Once converted FA then binds to the PPAR-α receptor and triggers the expression of numerous metabolic enzymes involved in the oxidation of fatty acids and reduces glucose uptake by repressing the insulin-dependent glucose transporter GLUT4. This metabolic switch could explain the mechanism by which fenofibrate initiates a gradual decline in energy metabolism in cancer cells. However, the anticancer effects of fenofibrate are more pronounced than other PPAR agonists and may be due to its accumulation in the mitochondrial fraction of human GB cells, which respond with a sudden and severe inhibition of mitochondrial respiration and an immediate increase but transient glycolysis, effectively triggering an energy catastrophe in GB cells with significantly reduced toxicity in normal astrocytes [64]. The anticancer effects of fenofibrate are therefore very complex and cannot be explained simply by activation of PPAR-dependent transcription, but it is, therefore, necessary to consider PPAR-independent mechanisms. Some investigations have also confirmed that fenofibrate is a neuroprotective agent; the results obtained in vitro on high-grade glioma

(HGG) cell lines U87 and U343 (p53 wild-type), U251 and T98 (p53 mutant), confirmed the antiproliferative and pro-apoptotic effects of fenofibrate demonstrating inhibition of NF-κB expression, cyclin D1, and Akt (Figure 4). Akt is well established as a therapeutic target for HGG, and there are a number of Akt inhibitors being evaluated as an adjunct treatment for HGG [65]. The transcription factor Fork-head box O1 (FoxO1) is an Akt substrate that plays a key role in tumor suppression by promoting transcriptional activation of p27kip. A recent study showed that fenofibrate by activating the PPARα/FoxO1/p27kip pathway could actually induce the death of human U87MG glioma cells, causing cell cycle arrest in the G0/G1 phase [66]. The activation of the PPAR-α receptor by the agonist fenofibrate attenuates the signaling responses of the IGF-IR, a factor that helps to support malignant growth and invasion of glioma cells and causes the accumulation of reactive species of oxygen (ROS), loss of mitochondrial membrane potential and a deficit in ATP production, which together may explain the severe impairment of glioma cell motility [67]. A recent study demonstrated that fenofibrate modulates the expression of hypoxia-inducible factor-1 alpha (HIF-1α), an overexpressed transcription factor under hypoxic conditions in human GB samples. HIF-1α is involved in the transactivation of genes involved in the altered metabolism, causing the accumulation of metabolites in the tumor environment, thus contributing to the growth and increase of the aggressiveness of GB. Fenofibrate inhibits the expression of HIF-1α by activation of HO-1 via the AMPK pathway. Furthermore, the activation of HO-1 involves the upregulation of SIRT1, causing its translocation into the nucleus, with consequent deacetylation of HIF-1α and inhibition of transcriptional activity [68].

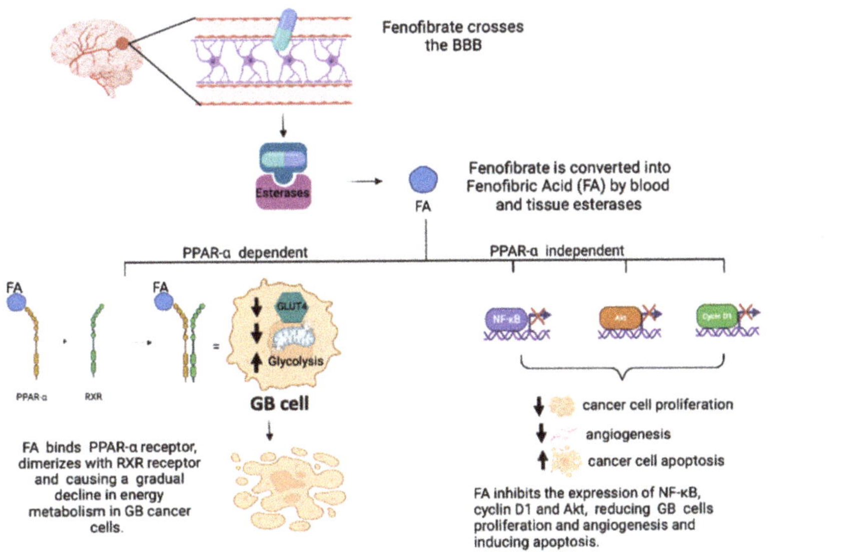

Figure 4. An exemplified overview of fenofibrate's PPAR-dependent and PPAR-independent mechanisms of action in GB tumoral cells.

A dual agonist PPAR-α/PPAR-γ, also called TZD18, inhibited cell growth and induced apoptosis in human GB T98G cells, through the activation of caspase-3 and down-regulation of the expression of Bcl-2, suggesting that TZD18 may have a therapeutic role in the treatment of human GB [69].

Moreover, PPAR-α and PPAR-γ agonists have the ability to selectively upregulate catalase expression on human astrocytes. In fact, both C6 GB cells and normal astrocytes were treated with PPAR-α and PPAR-γ agonists, showing a significant increase in catalase expression only in normal astrocytes, while, on the contrary, they failed to increase catalase expression in glioma cells. These results are promising because current data support the concept that selective manipulation of catalase gene expression and/or activity can provide greater protection of astrocytes from H_2O_2-induced damage and consequently can improve normal tissue survival during radiotherapy [70].

3.2. *Role of PPAR-β/δ in GB*

Although PPAR-β/δ agonists have been shown to cross the blood–brain barrier and modulate oxidative stress and proinflammatory responses associated with acute and chronic CNS disorders, overall, there are only a few studies evaluating the action of the PPAR-β/δ receptor in brain tumors [71]. The effects of a PPAR-β/δ ligand, namely erucic acid (EA), an omega-9 fatty acid, were investigated. EA has been shown to block the growth of C6 glioma cells and also reduce the cardiotoxicity of doxorubicin, thus suggesting that the combination of systemic EA with DOX-chemotherapy can reduce DOX concentrations in the systemic circulation, hinder toxic interactions and induce selective killing of glioma cells [72]. However, these results are too small to evaluate the role of the receptor PPAR-β/δ in brain tumors, especially considering the prevalence of the results in the literature that instead supports a procarcinogenic action in other types of tumors.

3.3. *Role of PPAR-γ Agonists in GB*

In addition to well-defined metabolic actions, PPAR-γ agonists exhibit various anti-neoplastic effects and induce cell death by apoptosis in various brain tumor cell lines. There are several possible mechanisms by which PPAR-γ agonists inhibit cell proliferation, such as induction of cell cycle arrest in the G0/G1 phase, a reduction in MYC levels upstream of the S phase transition, as well as possible down-regulation of CCND1 (cyclin D1) and associated cyclin-dependent kinases, but also the upregulation of cyclin-dependent kinase inhibitors CDKN1A, CDKN1B, and CDKN2B [21]. PPAR-γ agonists also blocking the Janus kinase/signal transducer and transcription activator (JAK/STAT) pathway, inhibit the expansion of CD133 + brain tumor stem cells (BTSC is also called tumor-initiating brain cells). In vivo and in vitro models have shown that inhibition of JAK2 (upstream regulator of STAT3) by the agonist PPAR-γ troglitazone promotes the slowing of the progression of GB disease, causing the phosphorylation of STAT3 tyrosine 705 and leading to the down-regulation of CCND1 and BCL2L1 (B-cell lymphoma extra-large protein 2) [73].

Agonists PPAR-γ, PGJ2, and rosiglitazone, have been shown to inhibit the proliferation of GB cell lines (U87-MG) through G2/M arrest and promoting the induction of programmed cell death [74]. These results are consistent with another study, in which evaluated growth inhibition and induction of apoptosis by another TZD, ciglitazone. Ciglitazone in addition to inducing cell cycle arrest, causes a significant reduction in the activity of telomerase, an enzyme that is constitutively active in most tumor cells, in human GB cell lines U-87 MG and U-118 MG [75]. Other data suggest that ciglitazone is able to induce PPAR-γ-independent apoptotic cell death in human T98G glioma cells by down-regulation of Akt and reduction of mitochondrial membrane potential (MMP), an effect that was accompanied by a down-regulation of Bcl-2 expression and an increase in Bid cleavage [76]. A study shows that the PPAR-γ receptor is an important positive regulator of the expression of CIDEA, a member of cell death-inducing DFFA-like effector (CIDE) protein family. In fact, it has been shown that PPAR-γ inhibition improves CIDEA expression, triggering glioma cell apoptosis and a decrease in HIF-1α activation, justifying further investigations aimed at evaluating the efficacy of PPAR-γ inhibitors as an effective anti-glioma therapeutic strategy [77]. Another TZD, Pioglitazone, showed anticancer efficacy on human glioma cells (U87MG, T98G, and U251MG) in vitro. Pioglitazone-induced inhibition of glioma cell proliferation and invasion occurred in a PPAR-γ-dependent manner and is in agreement with its ability to dramatically reduce β-catenin expression and transcriptional activity, resulting in decreased cell proliferation, migration, and apoptosis. These results indicate that PPAR-γ activation induces suppression of glioma cell turnover [78]. The agonist PPAR-γ Pioglitazone has also been shown to increase the functional expression of the glutamate transporter EAAT2 in glioma cells, preventing excitotoxic damage and glutamate-mediated seizures related to glioma [79]. In addition to the antineoplastic and anticonvulsant effects of pioglitazone, there is also the demonstration of its ability to cross the blood–brain barrier after oral and intracerebral administration in a human glioma xenograft model, suggesting its possible use as an additional candidate in the current regimen for double mechanistic efficacy in subtherapeutic doses to avoid associated adverse effects [80].

Clinical studies suggest a potential protective effect of the PPAR-γ agonists pioglitazone or rosiglitazone in diabetic patients with GB. The results show that diabetic patients with GB who had been treated with PPAR-γ agonists showed an increase in median survival of 19 months compared to patients who received standard treatment only [81]. In addition, a Phase 1 clinical study conducted on patients with primary and metastatic brain tumors treated with radiotherapy highlighted the protective role of pioglitazone in the prevention of radiation-induced cognitive decline (RICD) and its good tolerability at the 45 mg dose, not showing dose-limiting toxicity (DLT), which may be suggested for efficacy studies [82] (Table 1).

Table 1. This table summarizes the studies that analyze the effects of PPAR ligands.

Drug	Target	Effects	References
Fenofibrate	PPAR-α	-Repression of GLUT4 -Inhibition of NF-κB, cyclin D1 and Akt expression	[63–68]
TZD18	PPAR-α/PPAR-γ	-Activation of caspase-3 -Reduction of Bcl-2 expression	[69]
PGJ2 Rosiglitazone	PPAR-γ	-Induction of G2/M arrest	[74]
Ciglitazone	PPAR-γ	-Reduction of telomerase activity -Reduction Akt and Bcl-2 expression	[75,76]
Pioglitazone	PPAR-γ	-Reduction β-catenin expression -Increase EAAT2 expression	[78–80]

4. Conclusions and Future Prospects

In this review, a detailed analysis was carried out in order to summarize the role of the PPAR receptor family in GB, a brain tumor characterized by high aggression. The PPARs family includes three different subtypes PPAR-α, PPAR-β/δ, and PPAR-γ, which differ in terms of tissue distribution, affinity for ligands, and biological activity, and which participate differently in the maintenance of lipid and glucose homeostasis [83]. Their role in the progression and differentiation of cancer cells in different types of solid tumors is also widely studied, even if often the data present in the literature report conflicting opinions. Nevertheless, most investigations have shown that PPAR-β/δ activation is related to tumor progression, whereas PPAR-α and PPAR-γ are associated with antitumor action [84]. Based on these findings, considerable interest has been shown in PPAR ligands as potential therapeutic agents in the treatment of gliomas; however, the molecular mechanisms underlying the suppression of carcinogenesis in gliomas, determined by PPAR activation have not yet been fully elucidated. Particularly among the fibrates, a PPAR-α ligands, fenofibrate has received the most attention due to its capacity to reduce the proliferation of GB cells through both PPAR-dependent and PPAR-independent mechanisms [85]. The thiazolidinediones (TZD) class, PPAR-γ ligands known as insulin-sensitizing drugs, including troglitazone, ciglitazone, rosiglitazone, and pioglitazone have also been shown to interact with several pathways involved in the induction of cell death in GB cells [86]. On the other hand, little attention has been paid to the PPAR-β/δ ligands, probably due to the conflicting evidence in the literature regarding its pro-carcinogen action, so further studies would be needed to clarify its function in this context. Finally, there are clinical studies aimed at evaluating the efficacy and safety of these ligands in patients with GB, but further results are certainly needed in order to be able to suggest PPAR ligands as potential treatments in the therapy of GB.

Author Contributions: R.B. drew up the manuscript; I.P., M.C. and E.E. were involved in the design and intellectual concept of the study; G.C. (Giulia Chisari), S.M., L.C. (Lorenzo Colarossi) and L.C. (Laura Cucinotta) performed the literature search; M.L. and G.C. (Giovanna Casili) supervised the study; I.P. and E.E. critically revised the manuscript. All authors have read and agreed to the published version of the manuscript.

Funding: This research received no external funding.

Institutional Review Board Statement: Not applicable.

Informed Consent Statement: Not applicable.

Data Availability Statement: Not applicable.

Conflicts of Interest: The authors declare no conflict of interest.

References

1. Kapoor, M.; Gupta, V. *Astrocytoma*; StatPearls: Treasure Island, FL, USA, 2021.
2. Le Rhun, E.; Preusser, M.; Roth, P.; Reardon, D.A.; Van den Bent, M.; Wen, P.; Reifenberger, G.; Weller, M. Molecular targeted therapy of glioblastoma. *Cancer Treat. Rev.* **2019**, *80*, 101896. [CrossRef] [PubMed]
3. Rajaraman, P.; Melin, B.S.; Wang, Z.; McKean-Cowdin, R.; Michaud, D.S.; Wang, S.S.; Bondy, M.; Houlston, R.; Jenkins, R.B.; Wrensch, M.; et al. Genome-wide association study of glioma and meta-analysis. *Hum. Genet.* **2012**, *131*, 1877–1888. [CrossRef] [PubMed]
4. Wrensch, M.; Jenkins, R.B.; Chang, J.S.; Yeh, R.F.; Xiao, Y.; Decker, P.A.; Ballman, K.V.; Berger, M.; Buckner, J.C.; Chang, S.; et al. Variants in the CDKN2B and RTEL1 regions are associated with high-grade glioma susceptibility. *Nat. Genet.* **2009**, *41*, 905–908. [CrossRef] [PubMed]
5. Ardizzone, A.; Scuderi, S.A.; Giuffrida, D.; Colarossi, C.; Puglisi, C.; Campolo, M.; Cuzzocrea, S.; Esposito, E.; Paterniti, I. Role of Fibroblast Growth Factors Receptors (FGFRs) in Brain Tumors, Focus on Astrocytoma and Glioblastoma. *Cancers* **2020**, *12*, 3825. [CrossRef]
6. Stoyanov, G.S.; Dzhenkov, D.L. On the Concepts and History of Glioblastoma Multiforme-Morphology, Genetics and Epigenetics. *Folia Med.* **2018**, *60*, 48–66. [CrossRef]
7. Sasmita, A.O.; Wong, Y.P.; Ling, A.P.K. Biomarkers and therapeutic advances in glioblastoma multiforme. *Asia Pac. J. Clin. Oncol.* **2018**, *14*, 40–51. [CrossRef]
8. Patel, A.P.; Tirosh, I.; Trombetta, J.J.; Shalek, A.K.; Gillespie, S.M.; Wakimoto, H.; Cahill, D.P.; Nahed, B.V.; Curry, W.T.; Martuza, R.L.; et al. Single-cell RNA-seq highlights intratumoral heterogeneity in primary glioblastoma. *Science* **2014**, *344*, 1396–1401. [CrossRef]
9. Stupp, R.; Mason, W.P.; Van den Bent, M.J.; Weller, M.; Fisher, B.; Taphoorn, M.J.; Belanger, K.; Brandes, A.A.; Marosi, C.; Bogdahn, U.; et al. Radiotherapy plus concomitant and adjuvant temozolomide for glioblastoma. *N. Eng. J. Med.* **2005**, *352*, 987–996. [CrossRef]
10. Stupp, R.; Hegi, M.E.; Mason, W.P.; Van den Bent, M.J.; Taphoorn, M.J.; Janzer, R.C.; Ludwin, S.K.; Allgeier, A.; Fisher, B.; Belanger, K.; et al. Effects of radiotherapy with concomitant and adjuvant temozolomide versus radiotherapy alone on survival in glioblastoma in a randomised phase III study: 5-year analysis of the EORTC-NCIC trial. *Lancet Oncol.* **2009**, *10*, 459–466. [CrossRef]
11. Jung, J.; Kim, L.J.; Wang, X.; Wu, Q.; Sanvoranart, T.; Hubert, C.G.; Prager, B.C.; Wallace, L.C.; Jin, X.; Mack, S.C.; et al. Nicotinamide metabolism regulates glioblastoma stem cell maintenance. *JCI Insight* **2017**, *2*, e90019. [CrossRef]
12. Pozzi, V.; Salvolini, E.; Lucarini, G.; Salvucci, A.; Campagna, R.; Rubini, C.; Sartini, D.; Emanuelli, M. Cancer stem cell enrichment is associated with enhancement of nicotinamide N-methyltransferase expression. *IUBMB Life* **2020**, *72*, 1415–1425. [CrossRef] [PubMed]
13. Hegi, M.E.; Diserens, A.C.; Gorlia, T.; Hamou, M.F.; De Tribolet, N.; Weller, M.; Kros, J.M.; Hainfellner, J.A.; Mason, W.; Mariani, L.; et al. MGMT gene silencing and benefit from temozolomide in glioblastoma. *N. Eng. J. Med.* **2005**, *352*, 997–1003. [CrossRef] [PubMed]
14. Riemenschneider, M.J.; Hegi, M.E.; Reifenberger, G. MGMT promoter methylation in malignant gliomas. *Target. Oncol.* **2010**, *5*, 161–165. [CrossRef]
15. Poon, M.T.C.; Keni, S.; Vimalan, V.; Ip, C.; Smith, C.; Erridge, S.; Weir, C.J.; Brennan, P.M. Extent of MGMT promoter methylation modifies the effect of temozolomide on overall survival in patients with glioblastoma: A regional cohort study. *Neurooncol. Adv.* **2021**, *3*, vdab171. [CrossRef]
16. Derosa, G.; Sahebkar, A.; Maffioli, P. The role of various peroxisome proliferator-activated receptors and their ligands in clinical practice. *J. Cell. Physiol.* **2018**, *233*, 153–161. [CrossRef]

17. Wojtowicz, S.; Strosznajder, A.K.; Jezyna, M.; Strosznajder, J.B. The Novel Role of PPAR Alpha in the Brain: Promising Target in Therapy of Alzheimer's Disease and Other Neurodegenerative Disorders. *Neurochem. Res.* **2020**, *45*, 972–988. [CrossRef]
18. Villapol, S. Roles of Peroxisome Proliferator-Activated Receptor Gamma on Brain and Peripheral Inflammation. *Cell. Mol. Neurobiol.* **2018**, *38*, 121–132. [CrossRef] [PubMed]
19. Cimini, A.; Cristiano, L.; Colafarina, S.; Benedetti, E.; Di Loreto, S.; Festuccia, C.; Amicarelli, F.; Canuto, R.A.; Ceru, M.P. PPARgamma-dependent effects of conjugated linoleic acid on the human glioblastoma cell line (ADF). *Int. J. Cancer* **2005**, *117*, 923–933. [CrossRef]
20. Fidoamore, A.; Cristiano, L.; Laezza, C.; Galzio, R.; Benedetti, E.; Cinque, B.; Antonosante, A.; d'Angelo, M.; Castelli, V.; Cifone, M.G.; et al. Energy metabolism in glioblastoma stem cells: PPARalpha a metabolic adaptor to intratumoral microenvironment. *Oncotarget* **2017**, *8*, 108430–108450. [CrossRef]
21. Ellis, H.P.; Kurian, K.M. Biological Rationale for the Use of PPARgamma Agonists in Glioblastoma. *Front. Oncol.* **2014**, *4*, 52. [CrossRef] [PubMed]
22. Chan, L.S.; Wells, R.A. Cross-Talk between PPARs and the Partners of RXR: A Molecular Perspective. *PPAR Res.* **2009**, *2009*, 925309. [CrossRef]
23. Yousefnia, S.; Momenzadeh, S.; Seyed Forootan, F.; Ghaedi, K.; Nasr Esfahani, M.H. The influence of peroxisome proliferator-activated receptor gamma (PPARgamma) ligands on cancer cell tumorigenicity. *Gene* **2018**, *649*, 14–22. [CrossRef]
24. Mirza, A.Z.; Althagafi, I.I.; Shamshad, H. Role of PPAR receptor in different diseases and their ligands: Physiological importance and clinical implications. *Eur. J. Med. Chem.* **2019**, *166*, 502–513. [CrossRef] [PubMed]
25. Yasui, Y.; Kim, M.; Tanaka, T. PPAR Ligands for Cancer Chemoprevention. *PPAR Res.* **2008**, *2008*, 548919. [CrossRef] [PubMed]
26. Lazennec, G.; Canaple, L.; Saugy, D.; Wahli, W. Activation of peroxisome proliferator-activated receptors (PPARs) by their ligands and protein kinase A activators. *Mol. Endocrinol.* **2000**, *14*, 1962–1975. [CrossRef]
27. Brunmeir, R.; Xu, F. Functional Regulation of PPARs through Post-Translational Modifications. *Int. J. Mol. Sci.* **2018**, *19*, 1738. [CrossRef]
28. Grygiel-Gorniak, B. Peroxisome proliferator-activated receptors and their ligands: Nutritional and clinical implications—A review. *Nutr. J.* **2014**, *13*, 17. [CrossRef]
29. Takeyama, K.; Kodera, Y.; Suzawa, M.; Kato, S. Peroxisome proliferator-activated receptor (PPAR)—Structure, function, tissue distribution, gene expression. *Nihon Rinsho* **2000**, *58*, 357–363. [PubMed]
30. Mishra, A.; Chaudhary, A.; Sethi, S. Oxidized omega-3 fatty acids inhibit NF-kappaB activation via a PPARalpha-dependent pathway. *Arterioscler. Thromb. Vasc. Biol.* **2004**, *24*, 1621–1627. [CrossRef]
31. Han, L.; Shen, W.J.; Bittner, S.; Kraemer, F.B.; Azhar, S. PPARs: Regulators of metabolism and as therapeutic targets in cardiovascular disease. Part I: PPAR-alpha. *Future Cardiol.* **2017**, *13*, 259–278. [CrossRef]
32. Wagner, K.D.; Wagner, N. Peroxisome proliferator-activated receptor beta/delta (PPARbeta/delta) acts as regulator of metabolism linked to multiple cellular functions. *Pharmacol. Ther.* **2010**, *125*, 423–435. [CrossRef]
33. Strosznajder, A.K.; Wojtowicz, S.; Jezyna, M.J.; Sun, G.Y.; Strosznajder, J.B. Recent Insights on the Role of PPAR-beta/delta in Neuroinflammation and Neurodegeneration, and Its Potential Target for Therapy. *Neuromolecular. Med.* **2021**, *23*, 86–98. [CrossRef]
34. Liu, Y.; Colby, J.K.; Zuo, X.; Jaoude, J.; Wei, D.; Shureiqi, I. The Role of PPAR-delta in Metabolism, Inflammation, and Cancer: Many Characters of a Critical Transcription Factor. *Int. J. Mol. Sci.* **2018**, *19*, 3339. [CrossRef]
35. Altinoz, M.A.; Elmaci, I.; Hacimuftuoglu, A.; Ozpinar, A.; Hacker, E.; Ozpinar, A. PPARdelta and its ligand erucic acid may act anti-tumoral, neuroprotective, and myelin protective in neuroblastoma, glioblastoma, and Parkinson's disease. *Mol. Aspects Med.* **2021**, *78*, 100871. [CrossRef] [PubMed]
36. Marion-Letellier, R.; Savoye, G.; Ghosh, S. Fatty acids, eicosanoids and PPAR gamma. *Eur. J. Pharmacol.* **2016**, *785*, 44–49. [CrossRef] [PubMed]
37. Takada, I.; Makishima, M. Peroxisome proliferator-activated receptor agonists and antagonists: A patent review (2014–present). *Expert Opin. Ther. Pat.* **2020**, *30*, 1–13. [CrossRef] [PubMed]
38. Cheng, H.S.; Yip, Y.S.; Lim, E.K.Y.; Wahli, W.; Tan, N.S. PPARs and Tumor Microenvironment: The Emerging Roles of the Metabolic Master Regulators in Tumor Stromal-Epithelial Crosstalk and Carcinogenesis. *Cancers* **2021**, *13*, 2153. [CrossRef]
39. Chandran, K.; Goswami, S.; Sharma-Walia, N. Implications of a peroxisome proliferator-activated receptor alpha (PPARalpha) ligand clofibrate in breast cancer. *Oncotarget* **2016**, *7*, 15577–15599. [CrossRef]
40. Shigeto, T.; Yokoyama, Y.; Xin, B.; Mizunuma, H. Peroxisome proliferator-activated receptor alpha and gamma ligands inhibit the growth of human ovarian cancer. *Oncol. Rep.* **2007**, *18*, 833–840.
41. Castelli, V.; Catanesi, M.; Alfonsetti, M.; Laezza, C.; Lombardi, F.; Cinque, B.; Cifone, M.G.; Ippoliti, R.; Benedetti, E.; Cimini, A.; et al. PPARalpha-Selective Antagonist GW6471 Inhibits Cell Growth in Breast Cancer Stem Cells Inducing Energy Imbalance and Metabolic Stress. *Biomedicines* **2021**, *9*, 127. [CrossRef]
42. Wagner, N.; Wagner, K.D. PPAR Beta/Delta and the Hallmarks of Cancer. *Cells* **2020**, *9*, 1133. [CrossRef] [PubMed]
43. Du, S.; Wagner, N.; Wagner, K.D. The Emerging Role of PPAR Beta/Delta in Tumor Angiogenesis. *PPAR Res.* **2020**, *2020*, 3608315. [CrossRef]
44. He, T.C.; Chan, T.A.; Vogelstein, B.; Kinzler, K.W. PPARdelta is an APC-regulated target of nonsteroidal anti-inflammatory drugs. *Cell* **1999**, *99*, 335–345. [CrossRef]

45. Liu, Y.; Deguchi, Y.; Tian, R.; Wei, D.; Wu, L.; Chen, W.; Xu, W.; Xu, M.; Liu, F.; Gao, S.; et al. Pleiotropic Effects of PPARD Accelerate Colorectal Tumorigenesis, Progression, and Invasion. *Cancer Res.* **2019**, *79*, 954–969. [CrossRef] [PubMed]
46. Wagner, K.D.; Benchetrit, M.; Bianchini, L.; Michiels, J.F.; Wagner, N. Peroxisome proliferator-activated receptor beta/delta (PPARbeta/delta) is highly expressed in liposarcoma and promotes migration and proliferation. *J. Pathol.* **2011**, *224*, 575–588. [CrossRef]
47. Wang, X.; Wang, G.; Shi, Y.; Sun, L.; Gorczynski, R.; Li, Y.J.; Xu, Z.; Spaner, D.E. PPAR-delta promotes survival of breast cancer cells in harsh metabolic conditions. *Oncogenesis* **2016**, *5*, e232. [CrossRef]
48. Elix, C.; Pal, S.K.; Jones, J.O. The role of peroxisome proliferator-activated receptor gamma in prostate cancer. *Asian J. Androl.* **2018**, *20*, 238–243. [CrossRef] [PubMed]
49. Ferrari, S.M.; Materazzi, G.; Baldini, E.; Ulisse, S.; Miccoli, P.; Antonelli, A.; Fallahi, P. Antineoplastic Effects of PPARgamma Agonists, with a Special Focus on Thyroid Cancer. *Curr. Med. Chem.* **2016**, *23*, 636–649. [CrossRef] [PubMed]
50. Ammu, R.K.; Garikapati, K.K.; Krishnamurthy, P.T.; Chintamaneni, P.K.; Pindiprolu, S.K.S. Possible role of PPAR-gamma and COX-2 receptor modulators in the treatment of Non-Small Cell lung carcinoma. *Med. Hypotheses* **2019**, *124*, 98–100. [CrossRef]
51. Sabatino, L.; Pancione, M.; Votino, C.; Colangelo, T.; Lupo, A.; Novellino, E.; Lavecchia, A.; Colantuoni, V. Emerging role of the beta-catenin-PPARgamma axis in the pathogenesis of colorectal cancer. *World J. Gastroenterol.* **2014**, *20*, 7137–7151. [CrossRef]
52. Catalano, S.; Mauro, L.; Bonofiglio, D.; Pellegrino, M.; Qi, H.; Rizza, P.; Vizza, D.; Bossi, G.; Ando, S. In vivo and in vitro evidence that PPARgamma ligands are antagonists of leptin signaling in breast cancer. *Am. J. Pathol.* **2011**, *179*, 1030–1040. [CrossRef] [PubMed]
53. Sawayama, H.; Ishimoto, T.; Watanabe, M.; Yoshida, N.; Sugihara, H.; Kurashige, J.; Hirashima, K.; Iwatsuki, M.; Baba, Y.; Oki, E.; et al. Small molecule agonists of PPAR-gamma exert therapeutic effects in esophageal cancer. *Cancer Res.* **2014**, *74*, 575–585. [CrossRef]
54. Keshamouni, V.G.; Reddy, R.C.; Arenberg, D.A.; Joel, B.; Thannickal, V.J.; Kalemkerian, G.P.; Standiford, T.J. Peroxisome proliferator-activated receptor-gamma activation inhibits tumor progression in non-small-cell lung cancer. *Oncogene* **2004**, *23*, 100–108. [CrossRef] [PubMed]
55. Galbraith, L.C.A.; Mui, E.; Nixon, C.; Hedley, A.; Strachan, D.; MacKay, G.; Sumpton, D.; Sansom, O.J.; Leung, H.Y.; Ahmad, I. PPAR-gamma induced AKT3 expression increases levels of mitochondrial biogenesis driving prostate cancer. *Oncogene* **2021**, *40*, 2355–2366. [CrossRef] [PubMed]
56. Meylan, P.; Pich, C.; Winkler, C.; Ginster, S.; Mury, L.; Sgandurra, M.; Dreos, R.; Frederick, D.T.; Hammond, M.; Boland, G.M.; et al. Low expression of the PPARgamma-regulated gene thioredoxin-interacting protein accompanies human melanoma progression and promotes experimental lung metastases. *Sci. Rep.* **2021**, *11*, 7847. [CrossRef]
57. Haynes, H.R.; White, P.; Hares, K.M.; Redondo, J.; Kemp, K.C.; Singleton, W.G.B.; Killick-Cole, C.L.; Stevens, J.R.; Garadi, K.; Guglani, S.; et al. The transcription factor PPARalpha is overexpressed and is associated with a favourable prognosis in IDH-wildtype primary glioblastoma. *Histopathology* **2017**, *70*, 1030–1043. [CrossRef]
58. Benedetti, E.; d'Angelo, M.; Ammazzalorso, A.; Gravina, G.L.; Laezza, C.; Antonosante, A.; Panella, G.; Cinque, B.; Cristiano, L.; Dhez, A.C.; et al. PPARalpha Antagonist AA452 Triggers Metabolic Reprogramming and Increases Sensitivity to Radiation Therapy in Human Glioblastoma Primary Cells. *J. Cell. Physiol.* **2017**, *232*, 1458–1466. [CrossRef]
59. Gao, Y.; Han, D.; Sun, L.; Huang, Q.; Gai, G.; Wu, Z.; Meng, W.; Chen, X. PPARalpha Regulates the Proliferation of Human Glioma Cells through miR-214 and E2F2. *Biomed. Res. Int* **2018**, *2018*, 3842753. [CrossRef]
60. Shi, Y.; Luan, W.; Tao, T.; Wang, J.; Qian, J.; Dong, Q.; Liu, N.; You, Y. Expression of peroxisome proliferators-activated receptor in glioma and its effect on the growth of human glioma cells. *Zhonghua Yi Xue Yi Chuan Xue Za Zhi* **2014**, *31*, 317–321. [CrossRef] [PubMed]
61. Grabacka, M.; Plonka, P.M.; Urbanska, K.; Reiss, K. Peroxisome proliferator-activated receptor alpha activation decreases metastatic potential of melanoma cells in vitro via down-regulation of Akt. *Clin. Cancer Res.* **2006**, *12*, 3028–3036. [CrossRef] [PubMed]
62. Urbanska, K.; Pannizzo, P.; Grabacka, M.; Croul, S.; Del Valle, L.; Khalili, K.; Reiss, K. Activation of PPARalpha inhibits IGF-I-mediated growth and survival responses in medulloblastoma cell lines. *Int. J. Cancer* **2008**, *123*, 1015–1024. [CrossRef]
63. Stalinska, J.; Zimolag, E.; Pianovich, N.A.; Zapata, A.; Lassak, A.; Rak, M.; Dean, M.; Ucar-Bilyeu, D.; Wyczechowska, D.; Culicchia, F.; et al. Chemically Modified Variants of Fenofibrate with Antiglioblastoma Potential. *Transl. Oncol.* **2019**, *12*, 895–907. [CrossRef]
64. Wilk, A.; Wyczechowska, D.; Zapata, A.; Dean, M.; Mullinax, J.; Marrero, L.; Parsons, C.; Peruzzi, F.; Culicchia, F.; Ochoa, A.; et al. Molecular mechanisms of fenofibrate-induced metabolic catastrophe and glioblastoma cell death. *Mol. Cell. Biol.* **2015**, *35*, 182–198. [CrossRef]
65. Binello, E.; Mormone, E.; Emdad, L.; Kothari, H.; Germano, I.M. Characterization of fenofibrate-mediated anti-proliferative pro-apoptotic effects on high-grade gliomas and anti-invasive effects on glioma stem cells. *J. Neurooncol.* **2014**, *117*, 225–234. [CrossRef]
66. Han, D.F.; Zhang, J.X.; Wei, W.J.; Tao, T.; Hu, Q.; Wang, Y.Y.; Wang, X.F.; Liu, N.; You, Y.P. Fenofibrate induces G0/G1 phase arrest by modulating the PPARalpha/FoxO1/p27 kip pathway in human glioblastoma cells. *Tumor Biol.* **2015**, *36*, 3823–3829. [CrossRef] [PubMed]

67. Drukala, J.; Urbanska, K.; Wilk, A.; Grabacka, M.; Wybieralska, E.; Del Valle, L.; Madeja, Z.; Reiss, K. ROS accumulation and IGF-IR inhibition contribute to fenofibrate/PPARalpha-mediated inhibition of glioma cell motility in vitro. *Mol. Cancer* **2010**, *9*, 159. [CrossRef] [PubMed]
68. Lin, C.; Lai, S.W.; Shen, C.K.; Chen, C.W.; Tsai, C.F.; Liu, Y.S.; Lu, D.Y.; Huang, B.R. Fenofibrate inhibits hypoxia-inducible factor-1 alpha and carbonic anhydrase expression through activation of AMP-activated protein kinase/HO-1/Sirt1 pathway in glioblastoma cells. *Environ. Toxicol.* **2021**, *36*, 2551–2561. [CrossRef] [PubMed]
69. Liu, D.C.; Zang, C.B.; Liu, H.Y.; Possinger, K.; Fan, S.G.; Elstner, E. A novel PPAR alpha/gamma dual agonist inhibits cell growth and induces apoptosis in human glioblastoma T98G cells. *Acta Pharmacol. Sin.* **2004**, *25*, 1312–1319.
70. Khoo, N.K.; Hebbar, S.; Zhao, W.; Moore, S.A.; Domann, F.E.; Robbins, M.E. Differential activation of catalase expression and activity by PPAR agonists: Implications for astrocyte protection in anti-glioma therapy. *Redox Biol.* **2013**, *1*, 70–79. [CrossRef]
71. Schnegg, C.I.; Kooshki, M.; Hsu, F.C.; Sui, G.; Robbins, M.E. PPARdelta prevents radiation-induced proinflammatory responses in microglia via transrepression of NF-kappaB and inhibition of the PKCalpha/MEK1/2/ERK1/2/AP-1 pathway. *Free Radic. Biol. Med.* **2012**, *52*, 1734–1743. [CrossRef] [PubMed]
72. Altinoz, M.A.; Bilir, A.; Elmaci, I. Erucic acid, a component of Lorenzo's oil and PPAR-delta ligand modifies C6 glioma growth and toxicity of doxorubicin. Experimental data and a comprehensive literature analysis. *Chem. Biol. Interact.* **2018**, *294*, 107–117. [CrossRef]
73. Spagnolo, A.; Grant, E.N.; Glick, R.; Lichtor, T.; Feinstein, D.L. Differential effects of PPARgamma agonists on the metabolic properties of gliomas and astrocytes. *Neurosci. Lett.* **2007**, *417*, 72–77. [CrossRef]
74. Morosetti, R.; Servidei, T.; Mirabella, M.; Rutella, S.; Mangiola, A.; Maira, G.; Mastrangelo, R.; Koeffler, H.P. The PPARgamma ligands PGJ2 and rosiglitazone show a differential ability to inhibit proliferation and to induce apoptosis and differentiation of human glioblastoma cell lines. *Int. J. Oncol.* **2004**, *25*, 493–502.
75. Strakova, N.; Ehrmann, J.; Dzubak, P.; Bouchal, J.; Kolar, Z. The synthetic ligand of peroxisome proliferator-activated receptor-gamma ciglitazone affects human glioblastoma cell lines. *J. Pharmacol. Exp. Ther.* **2004**, *309*, 1239–1247. [CrossRef]
76. Lee, M.W.; Kim, D.S.; Kim, H.R.; Kim, H.J.; Yang, J.M.; Ryu, S.; Noh, Y.H.; Lee, S.H.; Son, M.H.; Jung, H.L.; et al. Cell death is induced by ciglitazone, a peroxisome proliferator-activated receptor gamma (PPARgamma) agonist, independently of PPARgamma in human glioma cells. *Biochem. Biophys. Res. Commun.* **2012**, *417*, 552–557. [CrossRef]
77. Chatterjee, A.; Mondal, P.; Ghosh, S.; Mehta, V.S.; Sen, E. PPARgamma regulated CIDEA affects pro-apoptotic responses in glioblastoma. *Cell Death Discov.* **2015**, *1*, 15038. [CrossRef] [PubMed]
78. Wan, Z.; Shi, W.; Shao, B.; Shi, J.; Shen, A.; Ma, Y.; Chen, J.; Lan, Q. Peroxisome proliferator-activated receptor gamma agonist pioglitazone inhibits beta-catenin-mediated glioma cell growth and invasion. *Mol. Cell. Biochem.* **2011**, *349*, 1–10. [CrossRef] [PubMed]
79. Ching, J.; Amiridis, S.; Stylli, S.S.; Bjorksten, A.R.; Kountouri, N.; Zheng, T.; Paradiso, L.; Luwor, R.B.; Morokoff, A.P.; O'Brien, T.J.; et al. The peroxisome proliferator activated receptor gamma agonist pioglitazone increases functional expression of the glutamate transporter excitatory amino acid transporter 2 (EAAT2) in human glioblastoma cells. *Oncotarget* **2015**, *6*, 21301–21314. [CrossRef] [PubMed]
80. Grommes, C.; Karlo, J.C.; Caprariello, A.; Blankenship, D.; Dechant, A.; Landreth, G.E. The PPARgamma agonist pioglitazone crosses the blood-brain barrier and reduces tumor growth in a human xenograft model. *Cancer Chemother. Pharmacol.* **2013**, *71*, 929–936. [CrossRef] [PubMed]
81. Grommes, C.; Conway, D.S.; Alshekhlee, A.; Barnholtz-Sloan, J.S. Inverse association of PPARgamma agonists use and high grade glioma development. *J. Neurooncol.* **2010**, *100*, 233–239. [CrossRef]
82. Cramer, C.K.; Alphonse-Sullivan, N.; Isom, S.; Metheny-Barlow, L.J.; Cummings, T.L.; Page, B.R.; Brown, D.R.; Blackstock, A.W., Jr.; Peiffer, A.M.; Strowd, R.E.; et al. Safety of pioglitazone during and after radiation therapy in patients with brain tumors: A phase I clinical trial. *J. Cancer Res. Clin. Oncol.* **2019**, *145*, 337–344. [CrossRef]
83. Korbecki, J.; Bobinski, R.; Dutka, M. Self-regulation of the inflammatory response by peroxisome proliferator-activated receptors. *Inflamm. Res.* **2019**, *68*, 443–458. [CrossRef] [PubMed]
84. Bundscherer, A.; Reichle, A.; Hafner, C.; Meyer, S.; Vogt, T. Targeting the tumor stroma with peroxisome proliferator activated receptor (PPAR) agonists. *Anticancer Agents Med. Chem.* **2009**, *9*, 816–821. [CrossRef]
85. Giordano, A.; Macaluso, M. Fenofibrate triggers apoptosis of glioblastoma cells in vitro: New insights for therapy. *Cell Cycle* **2012**, *11*, 3154. [CrossRef] [PubMed]
86. Tapia-Perez, J.H.; Kirches, E.; Mawrin, C.; Firsching, R.; Schneider, T. Cytotoxic effect of different statins and thiazolidinediones on malignant glioma cells. *Cancer Chemother. Pharmacol.* **2011**, *67*, 1193–1201. [CrossRef] [PubMed]

cells MDPI

Article

PPARγ Regulates Triclosan Induced Placental Dysfunction

Jing Li [1,2,*,†], Xiaojie Quan [1,2,†], Yue Zhang [1,2,†], Ting Yu [1], Saifei Lei [3], Zhenyao Huang [1], Qi Wang [1], Weiyi Song [1], Xinxin Yang [1] and Pengfei Xu [3,4,*]

1 School of Public Health, Xuzhou Medical University, 209 Tong-Shan Road, Xuzhou 221002, China; quanxiaojie7@gmail.com (X.Q.); Feb_20th@163.com (Y.Z.); yuting20211213@163.com (T.Y.); huangzhenyao@gmail.com (Z.H.); wangqixzmu2018@163.com (Q.W.); songweiyi798@xzhmu.edu.cn (W.S.); xxyang@xzhmu.edu.cn (X.Y.)
2 Key Laboratory of Human Genetics and Environmental Medicine, Xuzhou Medical University, Xuzhou 221004, China
3 Center for Pharmacogenetics and Department of Pharmaceutical Sciences, University of Pittsburgh, Pittsburgh, PA 15261, USA; sal208@pitt.edu
4 Beijing Key Laboratory of Gene Resource and Molecular Development, College of Life Sciences, Beijing Normal University, Beijing 100875, China
* Correspondence: 100002008046@xzhmu.edu.cn (J.L.); pex9@pitt.edu (P.X.)
† These authors contributed equally to this work.

Abstract: Exposure to the antibacterial agent triclosan (TCS) is associated with abnormal placenta growth and fetal development during pregnancy. Peroxisome proliferator-activated receptor γ (PPARγ) is crucial in placenta development. However, the mechanism of PPARγ in placenta injury induced by TCS remains unknown. Herein, we demonstrated that PPARγ worked as a protector against TCS-induced toxicity. TCS inhibited cell viability, migration, and angiogenesis dose-dependently in HTR-8/SVneo and JEG-3 cells. Furthermore, TCS downregulated expression of PPARγ and its downstream viability, migration, angiogenesis-related genes *HMOX1*, *ANGPTL4*, *VEGFA*, *MMP-2*, *MMP-9*, and upregulated inflammatory genes *p65*, *IL-6*, *IL-1β*, and *TNF-α* in vitro and in vivo. Further investigation showed that overexpression or activation (rosiglitazone) alleviated cell viability, migration, angiogenesis inhibition, and inflammatory response caused by TCS, while knockdown or inhibition (GW9662) of PPARγ had the opposite effect. Moreover, TCS caused placenta dysfunction characterized by the significant decrease in weight and size of the placenta and fetus, while PPARγ agonist rosiglitazone alleviated this damage in mice. Taken together, our results illustrated that TCS-induced placenta dysfunction, which was mediated by the PPARγ pathway. Our findings reveal that activation of PPARγ might be a promising strategy against the adverse effects of TCS exposure on the placenta and fetus.

Keywords: triclosan; PPARγ; placenta toxicity; cell migration; angiogenesis; inflammation

Citation: Li, J.; Quan, X.; Zhang, Y.; Yu, T.; Lei, S.; Huang, Z.; Wang, Q.; Song, W.; Yang, X.; Xu, P. PPARγ Regulates Triclosan Induced Placental Dysfunction. *Cells* **2022**, *11*, 86. https://doi.org/10.3390/cells11010086

Academic Editors: Kay-Dietrich Wagner and Nicole Wagner

Received: 9 November 2021
Accepted: 24 December 2021
Published: 28 December 2021

1. Introduction

Triclosan (TCS) is a synthetic broad-spectrum antimicrobial and exposure mainly occurs in dermal application (soaps, hand sanitizers, toothpaste, cosmetics, antiperspirants, and bedclothes) and oral use of consumer products (water, food products) [1–3]. TCS, as a kind of exogenous biological signal termed Endocrine disruptors (EDCs), can mimic endogenous estrogenic hormones, and interferes with the maintenance of homeostasis and the regulation of developmental processes [4]. Previous epidemiological research demonstrated that TCS has been distinguished in mothers' milk (1–13.6 ng/mL), urine (2.5–107 ng/mL), and cord blood samples [5–7]. In addition, high urinary TCS levels in patients with spontaneous abortion have been reported [8]. A Denmark Odense Child Cohort study stated that median unadjusted urinary TCS was 0.88 ng/mL in pregnant women, and high maternal urinary TCS levels were associated with reduced head and abdominal circumference at birth [9]. Animal experiments suggested that the exposure of pregnant mice to TCS reduced fetal body weight and viability [8]. TCS has been detected with a concentration of up to 478 ng/L and 1329 ng/g in surface waters and sediment, respectively, in China rivers [10]. Therefore, a further understanding of the toxicity of TCS is important for the inhibition of TCS pollution.

The nuclear receptors, peroxisome proliferator-activated receptors (PPARs) are activated by binding natural ligand-inducible transcription factors such as glitazones [11,12]. PPARs are responsible for metabolism and cell function and have three subtypes including PPARα, PPARβ/δ, and PPARγ [13,14]. PPARγ is highly expressed in reproductive tissues (ovary, testis, uterus, prostate, mammary gland) and the trophoblast labyrinthine zone in rodent placentas and human placentas [15]. PPAR has been suggested to be involved in regulating cell trophoblast proliferation, inflammatory reactions oxidative response, and nutrient transport to mediate placenta development [16–18]. Additionally, research has illustrated that PPARγ agonists enhanced, while PPARγ antagonists reduced, proliferative and migratory capabilities of endothelial cells [19]. The vasculature defects were observed in placentas at embryonic GD 9.5 in PPARγ-null mice [16]. Similar animal experiments implicated embryonic lethality induced by defects in placental vascularization in PPARγ-null mice [20,21]. The downregulated protein level of PPARγ was associated with placental disorders [22]. PPARγ expression levels were also found to be decreased in pregnant mice or zebrafish when exposed to TCS [23,24].

The effect of prenatal TCS exposure on placental toxicity and its role in fetal growth and development are still unclear. Whether the effect of prenatal TCS exposure on placental toxicity is associated with PPARγ remains to be revealed. Our research aims to clarify the potential character of prenatal TCS exposure on placenta function and its potential mechanism of PPARγ in the process of placental exposure to TCS.

2. Materials and Methods

2.1. Reagents

TCS (CAS No. 3380-34-5, purity > 98% pure) was obtained from Sigma–Aldrich (Oakville, ON, Canada). DMEM/F12 medium and MEM medium were purchased from KeyGEN BioTECH (Jiangsu, China). RPMI 1640 medium was purchased from GIBCO (Grand Island, N.Y.). Dimethyl sulfoxide (DMSO), fetal bovine serum (FBS), the Cell Counting Kit-8 (CCK-8), and corn oil used was obtained from Vicmed (Busan, Korea) and Macklin (Shanghai, China), respectively, and were analytical grades. The pcDNA-*PPAR*γ vector and *si-PPAR*γ were purchased from GenePharma (Shanghai, China). Moreover, the rosiglitazone (a specific PPARγ agonist) and GW9662 (a specific PPARγ antagonist) were obtained from MedChemExpress (Shanghai, China).

2.2. Cell Culture and Transfection

Thanks to Dr. Xinru Wang from Nanjing Medical University (Nanjing, China) for sending HTR-8/SVneo and JEG-3 cells as a gift to us. HTR-8/SVneo and JEG-3 cells were seeded in DMEM/F12 and MEM medium, respectively, and supplemented with 10% FBS in a 5% CO_2 humidified atmosphere at 37 °C. Based on the manufacturer's instructions, these two cell lines were transfected with small interfering RNA (siRNA) (*si-PPAR*γ; 20 nM), the control siRNA (*si-Con*; 20 nM), pcDNA-*PPAR*γ (2 μg), and pcDNA 3.1 (2 μg) targeting PPARγ by Lipofectamine 2000 (Invitrogen, Carlsbad, CA, USA). The medium was replaced after 4 h of cell transfection. After cell transfection for 24 h, they were incubated with triclosan for 24 h. For PPARγ activation or deactivation, cells were treated with rosiglitazone or GW9662 with 10 μM for 24 h and pre-treated for 0.5 h before incubating with TCS.

2.3. Animal Treatment

Institute of Cancer Research (ICR) mice, at 10 weeks old and weighing about 30 g, were obtained under a protocol approved by Xuzhou Medical University Animal Center. On day GD 7.5, after the vaginal plug (GD 0.5) was observed in pregnant mice, they were randomly separated into five groups (n = 8 per group). All pregnant mice were orally managed with

0, 10, 50, and 100 mg/kg/day TCS and from GD 7.5 to GD 17.5, while control pregnant mice accepted the same volume of corn oil. Firstly, rosiglitazone was dissolved in DMSO, and the ratio of DMSO to corn oil was 1:1000, and pregnant mice were orally administered with rosiglitazone (20 mg/kg/day) together with TCS (100 mg/kg/day). Mice placentas were isolated, put into liquid nitrogen to be quickly frozen and stored at −80 °C. According to manufacturers of Xuzhou Medical University Animal Ethics Committee, the entire experiment was performed in a Specified Pathogen-Free (SPF) environment (protocols 201605w025, 25 May 2016 and 202106A237, 25 June 2021).

2.4. Cell Viability Assay

HTR-8/SVneo or JEG-3 cells (5×10^3 cells/well) were seeded into sterile plates of 96 wells in complete medium (DMEM/F12, MEM, 10% FBS). Cells were incubated and cultured with TCS of different concentrations for 24 h. An amount of 10 μL CCK-8 (Vicmed, Busan, Korea) solution was added to each well and incubated for 0.5–4 h at 37 °C in a 5% CO_2 atmosphere. The plate was detected at 450 nm on a microplate reader Spark (TECAN, Austria, 30086376).

2.5. Cell Migration Assay

HTR-8/SVneo or JEG-3 cells were grown in 6-well cell plates overnight to obtain approximately 70% confluency. After corresponding treatment, HTR-8/SVneo and JEG-3 cells were starved in 1% serum medium. Scratch wounds were made via a sterile 10 μL pipette tip to obtain three parallel lines. Multiple photographs were taken of the wound using an inverted microscope (KS400; Carl Zeiss Imaging GmbH), and the migration distance was measured by Image J analysis software (National Institutes of Health, v1.8.0). The whole wound closure distance was calculated and the distance of newly covered cells was measured at 0 and 24 h to evaluate migration.

2.6. Tube Formation Assay

Approximately 50 μL Matrigel (10 mg/mL) (BD Biosciences, San Diego, CA, USA) was decked into 96-well plates and allowed to completely solidify at 37 °C for 1 h to form a gel. Approximately 5×10^3/mL HUVECs were suspended in a conditioned medium (the conditioned medium was derived from HTR-8/SVneo and JEG-3 cells and added into the wells of the Matrigel solidified plate. Tube formation in each well was monitored and imaged using an inverted microscope after the plate was incubated at 37 °C for 5 h. We repeated the experiment three times independently. Tube branch length was measured with six duplicates per well, while average branch length was taken from three random microscopic fields per well. Each assay was done in triplicate and quantification was conducted with Image J software (National Institutes of Health, v1.8.0).

2.7. Real Time PCR (RT-PCR)

Placental tissues, trophoblast cells, and Trizol (Vicmed, Busan, Korea) were used to isolate total RNA. RNA concentration was detected by Nanodrop 2000 (Thermo, Scientific). The reverse transcription was conducted using SYBR Green qPCR SuperMix and 500 ng of the total RNA was reverse transcribed to cDNA. The quantitative RT-PCR was carried out using SYBR Green Master Mix on a 7500 fast real-time PCR System (Applied Biosystem, Foster, CA, USA) according to the manufacturer's instructions. The mRNA levels were analyzed using the comparative cycle threshold method ($2^{-\Delta\Delta CT}$), and the relative levels were normalized to the level of GAPDH. The primers were designed by Sangon Biotech (Shanghai, China), and primer sequences were listed in Supplementary Table S1.

2.8. Western Blot Analysis

Total cellular protein was extracted and lysed with RIPA lysis buffer with protease inhibitors (KeyGEN, Nanjing, China) and phosphatase inhibitors (KeyGEN, Nanjing, China). Proteins were separated with 10% SDS-PAGE gels and transferred to PVDF membranes

(Merck Millipore, MA, USA). Membranes were blocked with 5% non-fat milk for 1h and then incubated with primary antibodies of PPARγ, ANGPTL4, MMP-2, IL-1β, GAPDH, and tubulin β (Bioword, Beijing, China) overnight at 4 °C. After being washed five times for five minutes, the membranes were incubated with the corresponding secondary antibodies for 1h at room temperature. The protein bands were visualized with an Enhanced Chemiluminescence (ECL) detection kit (Amersham, NJ, USA). The protein bands were analyzed using Image Lab Software (Bio-Rad, CA, USA).

2.9. Statistical Analysis

The above assays were carried out three independent times and all data were carried out by GraphPad Prism 8.3 software (San Diego, CA, USA), and data were exhibited as the mean ± SEM. The comparison of the data was subjected to a one-way analysis of variance among each independent group. The *t*-test or one-way ANOVA was implemented using the SPSS 19.0 (IBM, Armonk, NY, USA). p-value < 0.05 and p-value < 0.01 were considered a statistically significant difference and higher significance, respectively.

3. Results

3.1. PPARγ Is Crucial in Cell Viability Inhibition Induced by TCS

The cell viability of HTR-8/SVneo and JEG-3 cells exposed to TCS at different concentrations was determined by the CCK-8 assay. TCS dose-dependently inhibited cell viability of HTR-8/SVneo and JEG-3 cells, significantly decreasing cell viability at 20 μM, 30 μM, or 40 μM (Figure 1A). Of interest, TCS significantly inhibited *PPARγ* mRNA expression levels in those two cell lines especially HTR-8/SVneo cells (Figure 1B).

To investigate whether PPARγ was involved in the TCS-induced inhibition of cell viability, cell viability was analyzed when PPARγ was overexpressed and activation (rosiglitazone) or knockdown and inhibition (GW9662), and then exposed to TCS. Rosiglitazone and GW9662 with the concentration of 10 μM were not toxic in HTR-8/SVneo and JEG-3 cells (Figure S1A). The efficiency of PPARγ overexpression and knockdown is shown in Figure S1B–D. PPARγ mRNA and protein expression level was increased when PPARγ was overexpressed (Figure S1B,C) but decreased after knockdown compared to the control in HTR-8/SVneo and JEG-3 cells (Figure S1B,D). The results exhibited that pcDNA-*PPARγ* and rosiglitazone alleviated TCS-elicited cell viability inhibition in HTR-8/SVneo and JEG-3 cells (Figure 1C,D). In contrast, GW9662 and *si-PPARγ* aggravated the inhibition in these two cell lines (Figure 1C,E). In addition, treatment with rosiglitazone increased the expression of PPARγ-regulated genes, such as *HMOX1*, *ANGPTL4*, *VEGFA*, *MMP-2*, and *MMP-9*, and decreased the expression of *p65*, *IL-6*, *IL-1β*, and *TNF-α* in vitro. On the other hand, treatment with GW9662 has an opposite trend (Figure S1E).

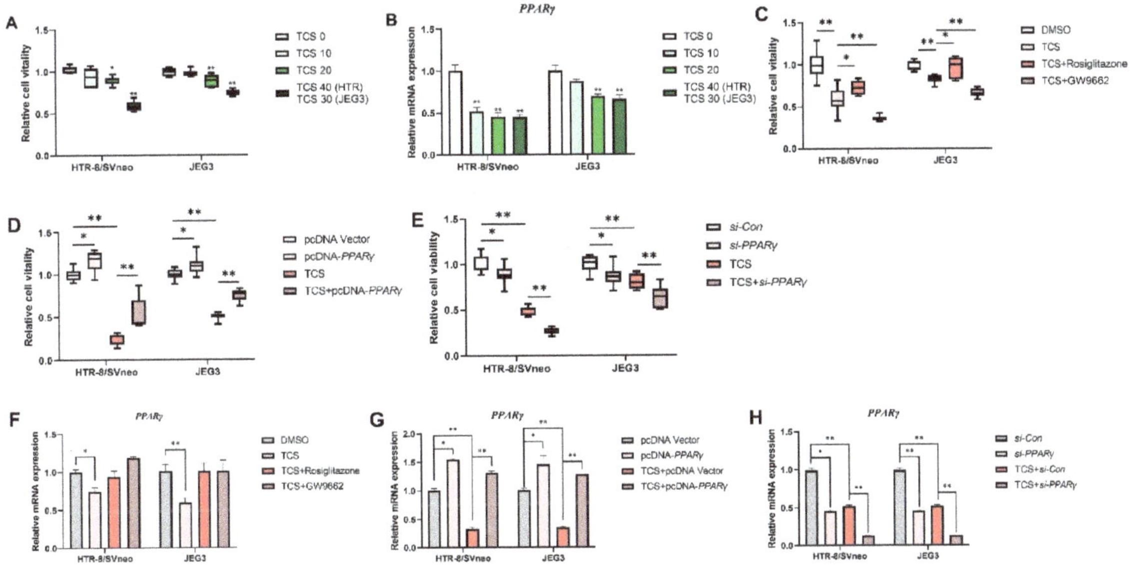

Figure 1. The viability of TCS on HTR-8/SVneo and JEG-3 cells was detected by Cell Counting Kit-8 assay for 24 h. (**A**) Cell vitality in HTR-8/SVneo and JEG-3 cells exposed to TCS. (**B**) Expression of PPARγ was determined by RT-PCR in HTR-8/SVneo and JEG-3 cells exposed to indicated TCS for 24 h. (**C**) Cell vitality was detected when exposed to TCS (40 μM for HTR-8/SVneo, 30 μM for JEG-3) in the absence or presence of rosiglitazone or GW9662 in the two cell lines. PPARγ was overexpressed (**D**) and knockdown (**E**) in the two cell lines and co-treated with TCS in HTR-8/SVneo and JEG-3 cells while cell vitality was analyzed. (**F**) HTR-8/SVneo and JEG-3 cells exposed to TCS while in the absence or presence of rosiglitazone (10 μM) and GW9662 (10 μM). HTR-8/SVneo and JEG-3 cells exposed to TCS during PPARγ overexpression (**G**) or knockdown (**H**). The data are shown as the means ± S.E.M. * $p < 0.05$; ** $p < 0.01$; compared with the indicated group, $n = 3$.

3.2. *PPARγ Is Involved in TCS-Elicited Impaired Migration*

The wound healing assay confirmed that the migration ability of HTR-8/SVneo and JEG-3 cells were inhibited after concentration dependent on exposure to TCS (Figure 2A). To assess the impact of TCS on cell migration influenced by PPARγ in vitro, the migration assays were conducted in the absence or presence of rosiglitazone or GW9662. Our results in Figure 2B demonstrated that treatment with rosiglitazone significantly increased cell migration, while GW9662 decreased the cell migration compared to the TCS-exposed group in HTR-8/SVneo and JEG-3 cells. PPARγ displayed overexpression and knockdown in HTR-8/SVneo and JEG-3 cells. Our results indicated that PPARγ overexpression alleviated the migration inhibition induced by TCS in HTR-8/SVneo and JEG-3 cells (Figure 2C). Moreover, PPARγ knockdown aggravated the migration inhibition induced by TCS in these two cell lines (Figure 2D).

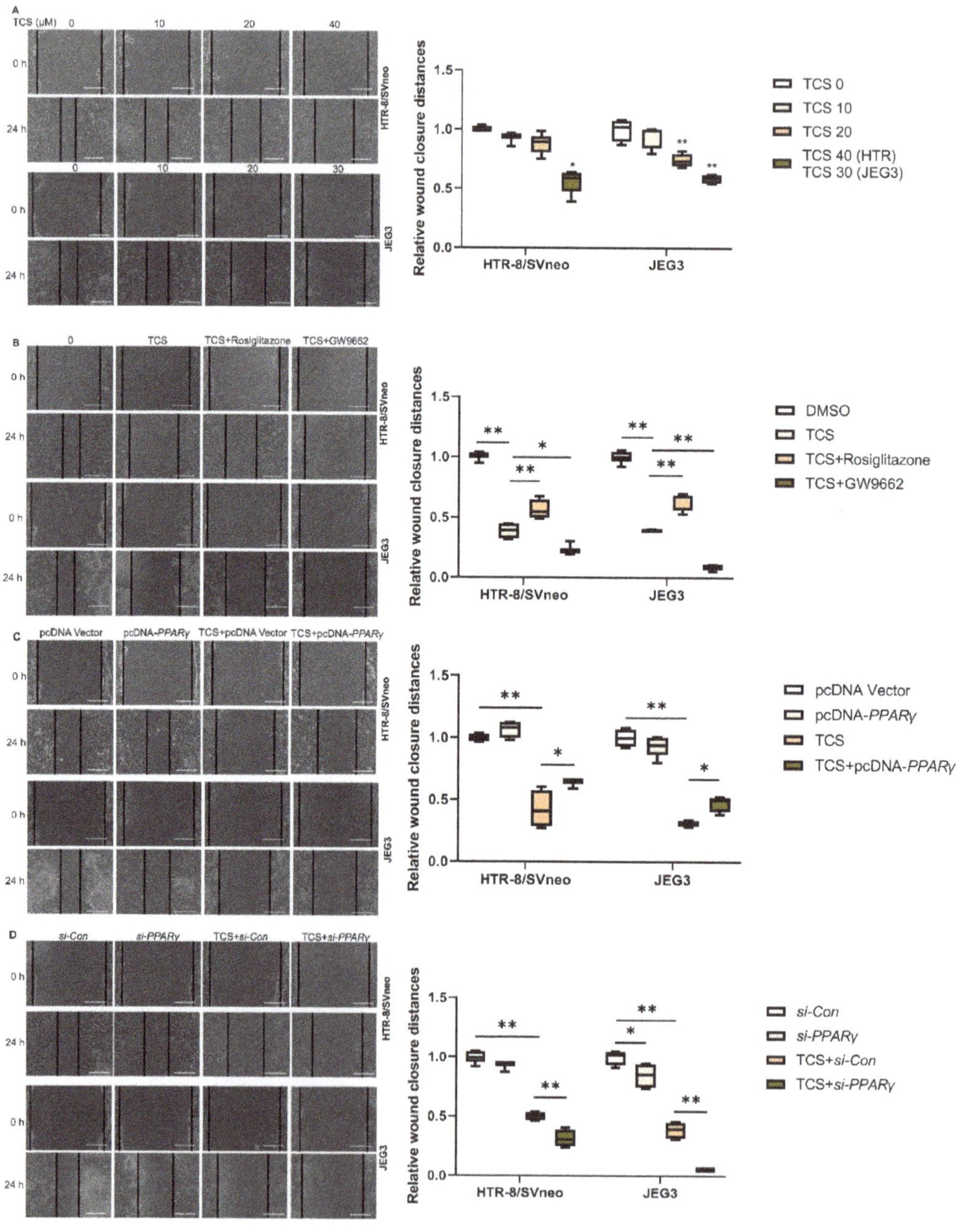

Figure 2. Role of PPARγ in migration exposed to TCS in vitro. The migration distances were measured after exposure to TCS at 0 h and 24 h. (**A**) HTR-8/SVneo and JEG-3 cell migration distance was decreased with increasing concentration of TCS. Co-treated with TCS in the absence or presence of rosiglitazone or GW9662 in HTR-8/SVneo and JEG-3 cells (**B**). The migration distance in response to PPAR overexpression (**C**) and knockdown (**D**) when exposed to TCS. Scale bar: 200 μm. The relative wound closure distances are shown on the right. The data are shown as the means ± S.E.M. * $p < 0.05$; ** $p < 0.01$; compared with the indicated group, $n = 3$.

3.3. PPARγ Alleviates TCS-Elicited Angiogenesis Inhibition

An angiogenesis assay was implemented to describe the influence of TCS on tube branch length. The HTR8/SVneo and JEG-3 cell lines were treated with different concentrations of TCS. TCS showed significantly decreased tube branch length compared to the control (DMSO) at 30 μM or 40 μM in the two cell lines, respectively (Figure 3A). The impact of PPARγ on TCS-elicited tube branch length is displayed in Figure 3. The existing results illustrated that PPARγ overexpression or rosiglitazone co-treatment alleviated the inhibition of tube branch length induced by TCS in vitro (Figure 3B,C). In contrast, PPARγ knockdown or GW9662 co-treatment aggravated the inhibition of tube branch length induced by TCS in HTR8/SVneo and JEG-3 cells (Figure 3B,D).

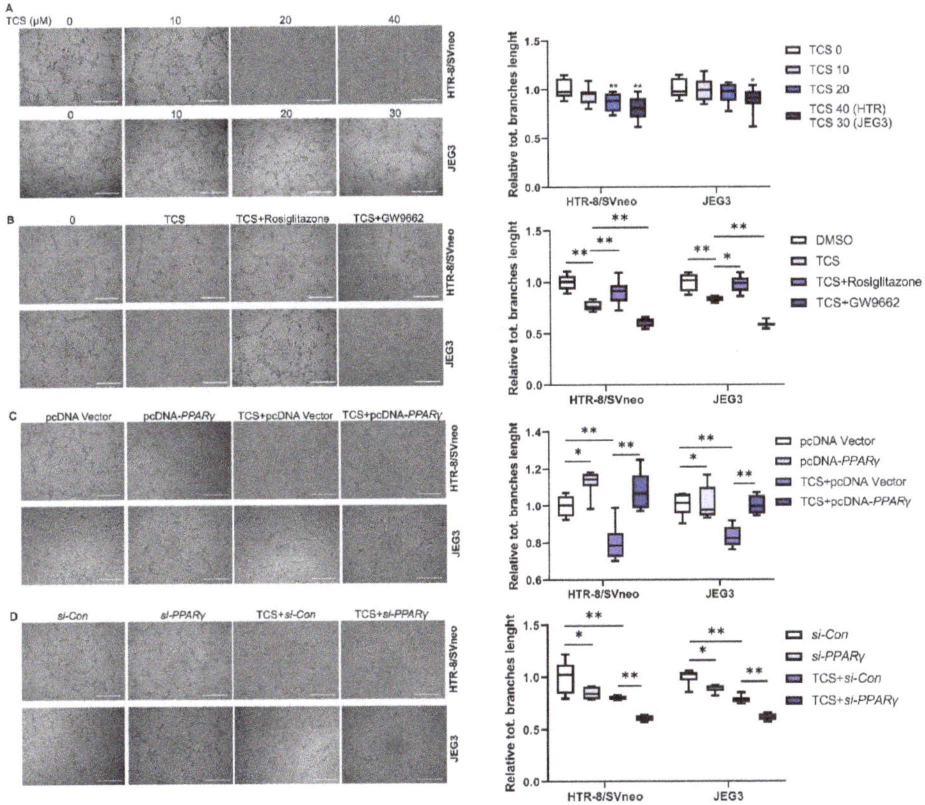

Figure 3. TCS reduced HTR-8/SVneo and JEG-3 cells angiogenesis through PPARγ pathway. (**A**) TCS decreased the branch length of HTR-8/SVneo and JEG-3 cells. (**B**) Co-treated with TCS (40 μM for HTR-8/SVneo, 30 μM for JEG-3) in the absence or presence of rosiglitazone and GW9662 in these two cells. (**C**,**D**) PPARγ overexpression alleviated, while (**C**) PPARγ knockdown exacerbated. (**D**) TCS-induced cell angiogenesis inhibition of HTR-8/SVneo and JEG-3 cells. Scale bar: 400 μm. The data are shown as the means ± S.E.M. * $p < 0.05$; ** $p < 0.01$; compared with the indicated group, $n = 3$.

3.4. TCS Alters the Expression of PPARγ Target Genes Associated with Viability, Angiogenesis, and Migration

The *HMOX1*, *ANGPTL4*, *VEGFA*, *MMP-2*, and *MMP-9* genes were determined as the target genes of PPARγ and play considerable roles in cell viability, angiogenesis, and migration [25–27]. In comparison with the control group, the levels of *HMOX1*, *ANGPTL4*,

VEGFA, *MMP-2*, and *MMP-9* mRNA were significantly decreased in the TCS-exposed group (Figure 4A,B). In addition, the protein levels of *ANGPTL4* and *MMP-2* were significantly decreased in the TCS-exposed groups (Figure 4I,J). We also assessed the influence of *PPARγ* on the above genes in the absence or presence of TCS. The results revealed that rosiglitazone enhanced the expression of *HMOX1*, *ANGPTL4*, *VEGFA*, *MMP-2*, and *MMP-9* levels, although TCS decreased them in vitro (Figure 4C,D). The *HMOX1*, *ANGPTL4*, *VEGFA*, *MMP-2*, and *MMP-9* mRNA levels were upregulated after overexpression of PPARγ in vitro (Figure 4E,F). In contrast, PPARγ knockdown downregulated the cell viability, angiogenesis, and migration-related genes in HTR-8/SVneo and JEG-3 cells (Figure 4G,H). Therefore, TCS inhibited the PPARγ pathway, leading to the downregulation of *HMOX1*, *ANGPTL4*, *VEGFA*, *MMP-2*, and *MMP-9*.

3.5. TCS Changes Expression Level of PPARγ-Regulated Inflammation Genes

The effect of TCS on PPARγ-regulated inflammatory genes was investigated, and results demonstrated that TCS was able to upregulate genes involved in inflammation such as *p65*, *IL-6*, *IL-1β*, and *TNF-α* in HTR-8/SVneo and JEG-3 cells (Figure 5A,B). The protein levels of *IL-1β* were significantly increased in TCS-exposed groups (Figure 4I). However, PPARγ overexpression or rosiglitazone co-treatment significantly alleviated the level of these inflammatory cytokines elicited by TCS in vitro (Figure 5C–F). In addition, PPARγ knockdown or GW9662 co-treatment enhanced the expression levels of those genes elicited by TCS in HTR-8/SVneo and JEG-3 cells (Figure 5C–H). These results emphasized that TCS upregulated inflammatory gene expression through the PPARγ pathway.

3.6. TCS Induces Placenta Dysfunction through PPARγ Pathway in Mice

To reveal the potential toxicity of TCS on the placenta and fetal development, pregnant ICR mice were treated with TCS daily by gavage at doses of 0, 10, 50, and 100 mg/kg/day from gestation day GD7.5 to GD17.5. Uterus, placenta, and fetus were collected for analysis in our study. As the results showed, the size of uterus (Figure 6A), fetus weight (Figure 6B), and placenta (Figure 6C) in TCS-exposed mice were decreased obviously in the 50 and 100 mg/kg/day gavage group compared to the vehicle group, indicating the serious placenta toxicity of TCS. Moreover, rosiglitazone (20 mg/kg/day) administration prevented the decrease of fetus weight (Figure 6B) and placenta diameter (Figure 6C) induced by TCS.

3.7. TCS Alters Expression of PPARγ-Regulated Genes in Mice Placenta

The effects of gestational TCS exposure on viability, migration, angiogenesis, and inflammatory genes in mice placenta were analyzed. As described in Figure 7, placental *Pparγ* (Figure 7A) and *Pparγ*-regulated genes *Homx1*, *Angptl4*, *Vegfa*, *Mmp-2*, and *Mmp-9* mRNA levels (Figure 7B) were decreased, while inflammatory genes *p65*, *Il-6*, *Il-1β*, and *Tnf-α* mRNA levels (Figure 7C) were significantly elevated in TCS-exposed pregnant mice. The protein expression of ANGPTL4, MMP-2, and IL-1β were consistent with the gene expression trends in the placenta of GD17.5 mice treated with or without TCS (Figure 7D). In addition, treatment with rosiglitazone reversed the PPARγ-regulated viability, migration, angiogenesis, and inflammatory gene expression induced by TCS (Figure 7B,C).

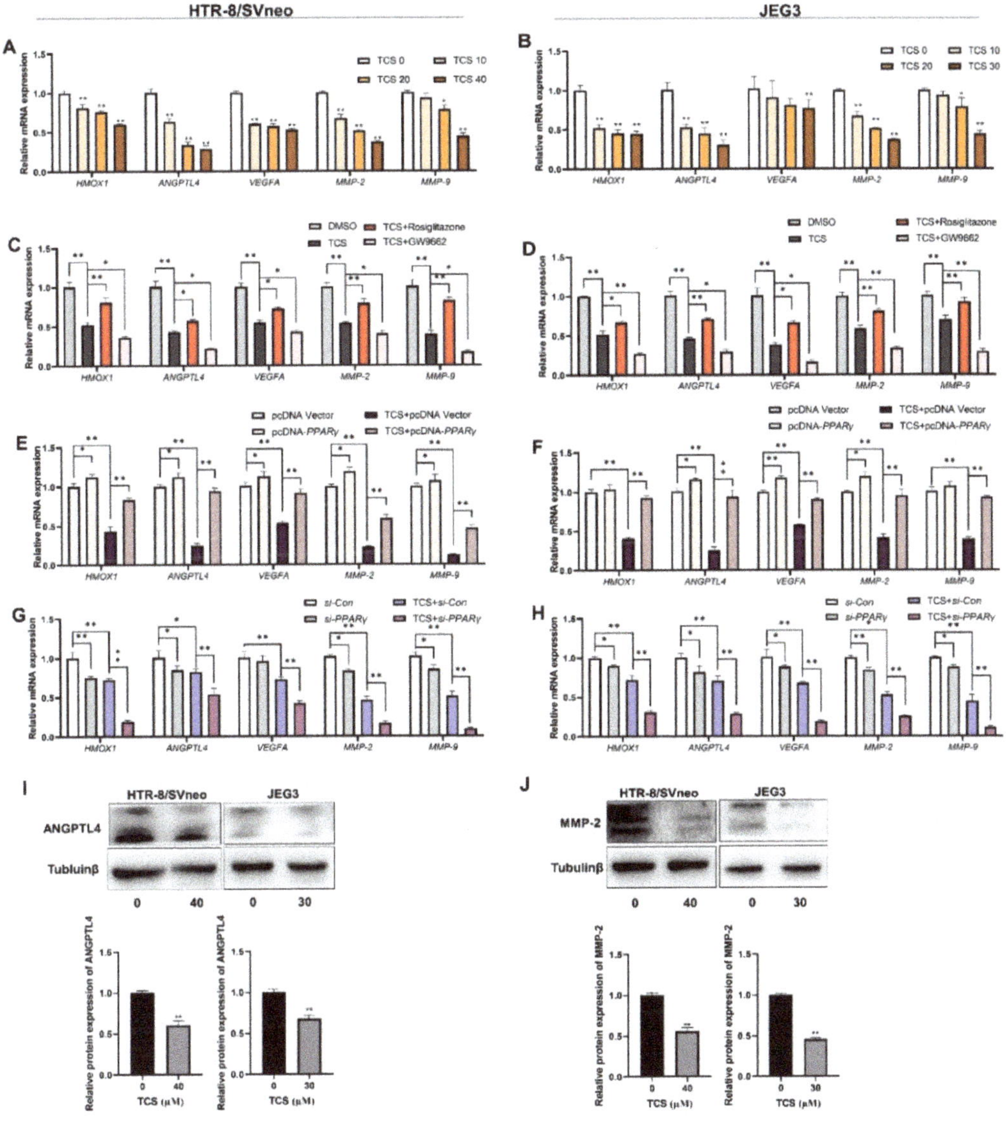

Figure 4. TCS inhibited expression of PPARγ target genes related to cell vitality, migration, and angiogenesis via PPARγ pathway in vitro. (**A**,**B**) The mRNA expression was analyzed by RT-PCR exposed to the indicated dose of TCS in HTR-8/SVneo and JEG-3 cells. (**C**,**D**) HTR-8/SVneo and JEG-3 cells treated with or without rosiglitazone and GW9662 for 24 h in HTR-8/SVneo and JEG-3 cells exposed to TCS (40 μM for HTR-8/SVneo, 30 μM for JEG-3). (**E**–**H**) PPARγ was overexpressed (**E**,**F**) or displayed knockdown (**G**,**H**) and co-treated with TCS in HTR-8/SVneo and JEG-3 cells. (**I**,**J**) The protein expression of ANGPTL4 (**I**) and MMP-2 (**J**) was analyzed by Western blot in HTR-8/SVneo and JEG-3 cells exposed to TCS. The relative values of protein expression are shown below. The data are shown as the means ± S.E.M. * $p < 0.05$; ** $p < 0.01$; compared with the indicated group, $n = 3$.

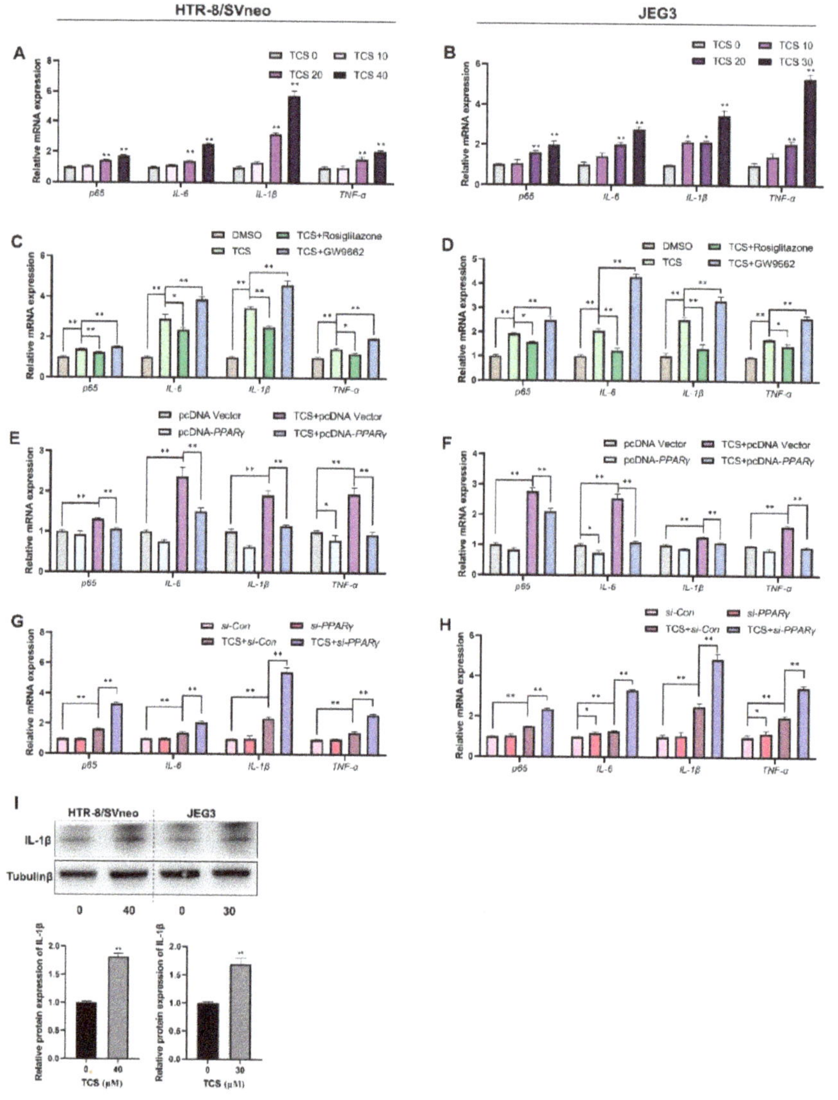

Figure 5. TCS increased expression of PPARγ-regulated genes related to inflammation through PPARγ pathway in vitro. (**A**,**B**) The mRNA expression level of the inflammatory genes exposed to indicated dose of TCS in HTR-8/SVneo and JEG-3 cells. (**C**,**D**) HTR-8/SVneo and JEG-3 cells treated with or without rosiglitazone and GW9662 for 24 h in HTR-8/SVneo and JEG-3 cells co-treated with TCS (40 μM for HTR-8/SVneo, 30 μM for JEG-3). (**E**–**H**) PPARγ was overexpressed (**E**,**F**) or displayed knockdown (**G**,**H**) and co-treated with TCS in HTR-8/SVneo and JEG-3 cells. (**I**) The protein expression of IL-1β was analyzed by Western blot in HTR-8/SVneo and JEG-3 cells exposed to TCS. The relative value of protein expression is shown below. The data are shown as the means ± S.E.M. * $p < 0.05$; ** $p < 0.01$; compared with the indicated group, $n = 3$.

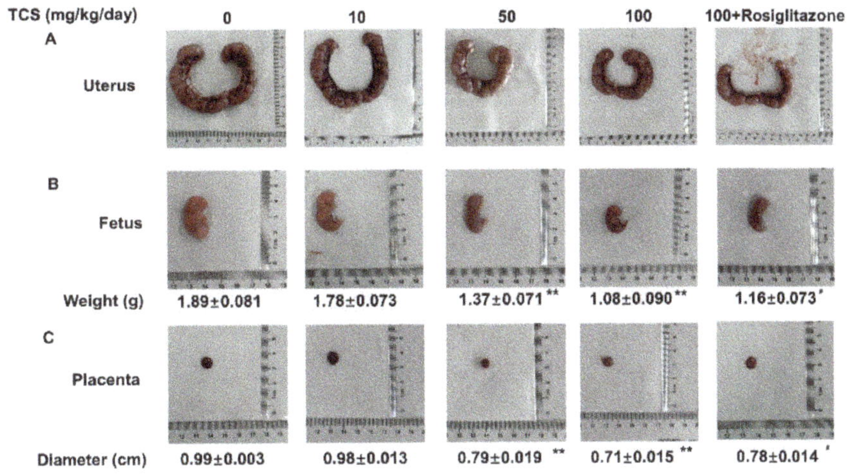

Figure 6. PPARγ participated in placental and fetal development toxicity of TCS. Representative picture of uterus (**A**), fetus (**B**), and placenta (**C**) of GD17.5 mice exposed to indicated doses of TCS. The data are shown as the means ± S.E.M; * $p < 0.05$ and ** $p < 0.01$, compared with control group; # $p < 0.05$, compared with TCS (100 mg/kg/day) group; $n = 8$.

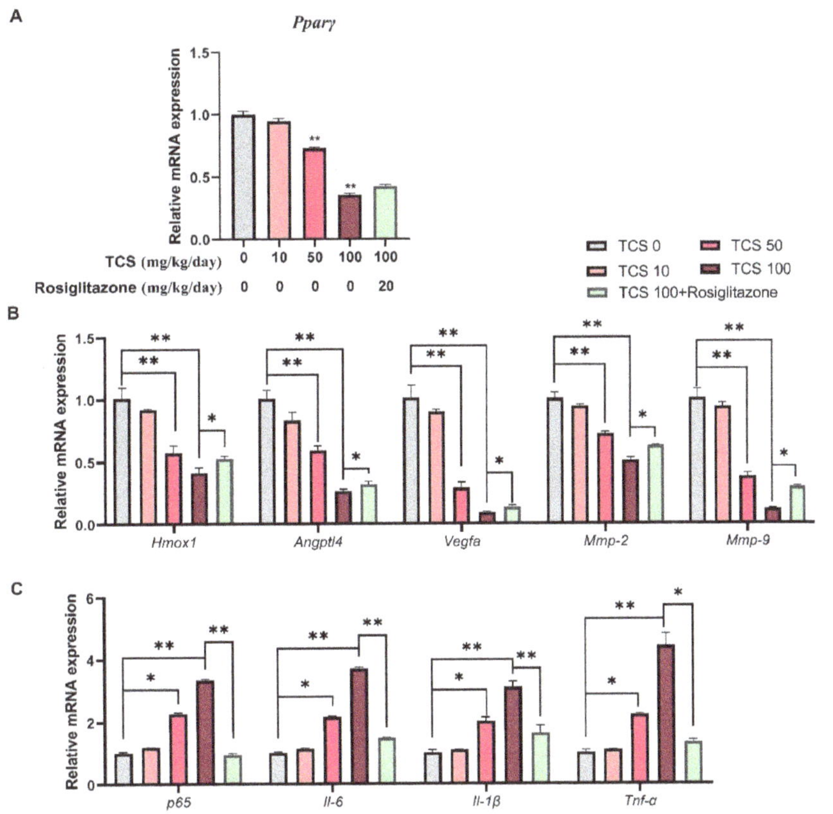

Figure 7. *Cont.*

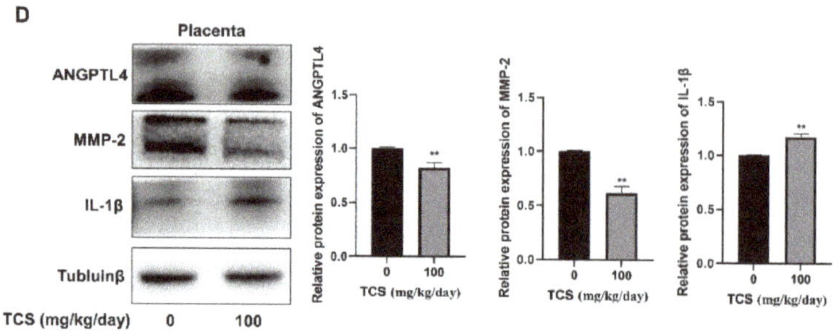

Figure 7. The gene and protein expression of PPARγ-regulated genes in placentas of gestational mice exposed to TCS. (**A**) The PPARγ expression level was detected by RT-PCR in GD17.5 mice placentas. (**B**,**C**) The mRNA expression of cell growth, angiogenesis, migration (**B**), and inflammation (**C**)-related genes in the placenta of GD17.5 mice treated with indicated doses of TCS and with or without rosiglitazone. (**D**) The protein expression of ANGPTL4, MMP-2, and IL-1β were analyzed by Western blot in placenta of GD17.5 mice treated with or without TCS (100 mg/kg/day). The relative values of protein expression are shown on the right. The data are shown as the means ± S.E.M. * $p < 0.05$; ** $p < 0.01$; compared with the indicated group, $n = 8$.

4. Discussion

In spite of numerous developmental toxicities shown to be triggered by TCS, the mechanism of TCS-elicited severe placental dysfunction has not been well elaborated. Here, our research revealed that TCS dose-dependently inhibited cell growth, migration, angiogenesis in HTR-8/SVneo and JEG-3 cells, and impaired placental development in mice. The mechanism of TCS-induced effects was studied. PPARγ was partly involved in the toxicity of TCS by regulating placental cell growth, migration, angiogenesis, and inflammatory responses in vitro and in vivo.

Abnormal placental cell proliferation may induce pregnancy complications such as fetal growth retardation, miscarriage, preeclampsia, and macrosomia [28]. Previous studies have verified that TCS had cytotoxic impression, elicited apoptosis, and inhibited cell vitality of human placental trophoblasts [29,30]. Placenta angiogenesis works for placental transport, endocrine, metabolic, and immune function regulation. It is essential for embryogenesis and is involved in the reproductive cycle and wound healing [16]. The intra-placental vascular lesions may result in preeclampsia, fetal growth restriction, or birthweight decrease [31]. Previous research has reported that TCS exposure decreased fetal viability and fetal body weight through placental thrombosis [32]. Some evidence has also demonstrated that TCS stimulated vascular branch disappearance and vascular injury in zebrafish [33]. In our animal research, placental diameter and fetal weight significantly decreased in pregnant mice in the TCS high-dose group. Previous studies have illustrated that the placenta had the greatest bio-accumulation of TCS, and reduced uterine weight and abortion were observed in the TCS (600 mg/kg/day) group in rats [34]. Following TCS exposure, reduction of gravid uterine weight and the occurrence of abortion was observed in pregnant rats [34]. PPARγ-deficient mice have placental abnormalities and defects in trophoblast differentiation and vascular development [20]. In addition, rosiglitazone increased placental vascularization and trophoblast migration and invasion in villous cytotrophoblast cells (VCT) and placental explants [35]. Our results also proved that PPARγ was inhibited in the mice placentas after treatment with TCS, and rosiglitazone prevented TCS-induced placenta toxicity by activation of PPARγ. It will be interesting to know the mechanism by which TCS affects PPARγ expression. Our results demonstrated that TCS inhibited cell growth, migration, and angiogenesis and influenced placenta development through the PPARγ pathway in vitro and in vivo.

Vascular endothelial growth factors A (*VEGFA*) and Heme oxygenase 1 (*HO-1*) are key modules of the angiogenesis process [36]. Angiopoietin-like protein 4 (*ANGPTL4*) is a secretory glycoprotein member of the angiopoietin family, which participates in the regulation of migration and angiogenesis [37,38]. Furthermore, functional studies found that PPARγ targeted *ANGPTL4* in placental development and angiogenesis, mediating the survival, proliferation, migration, and invasion of HTR8/SVneo cells [39,40]. Our results indicated that PPARγ activation alleviated the decrease of *HMOX1*, *ANGPTL4*, and *VEGFA* expression caused by TCS. MMPs have been demonstrated to mediate the migration and invasion of trophoblast cells, and *MMP-2* and *MMP-9* expression levels were decreased in preeclamptic placental tissues [41,42]. Similarly, our findings suggested that PPARγ mediated TCS-induced migration inhibition via *MMP-2* and *MMP-9* expression. In this study, cell growth, migration, and angiogenesis inhibition by TCS was partly ameliorated by PPARγ elevation or activation. These findings prove that TCS affects cell viability, migration, and angiogenesis through the PPARγ pathway.

Pregnancy is a process of dynamic inflammatory phases, and the inflammatory state is present almost throughout pregnancy and towards the end [43]. Activation of PPARγ is reported to suppress inflammation through NF-kappaB and TNF-α [44]. Indeed, pro-inflammatory conditions have been associated with pregnancy complications such as intrauterine growth restriction and preeclampsia [45]. TCS increased the inflammatory response by promoting *TNF-a* and *IL-6* expression in HUVEC [32]. Interestingly, PPARγ has a critical role in regulating inflammatory cytokines including *TNF-α* and *IL-6*, which were linked to preterm labor, miscarriage, and preeclampsia [22]. Moreover, PPAR-γ has been proved for its anti-inflammatory effects and downregulation of the expression of pro-inflammatory cytokines, such as *IL-6*, *IL-8*, and *TNF-α*, in human gestational tissue [46,47]. Furthermore, activation of PPARγ by rosiglitazone reversed the LPS-mediated effects on inflammatory cytokine release and proliferation inhibition in HTR-8/SVneo cells [48]. Similarly, we demonstrated that TCS can induce the expression of inflammatory genes in two cell lines. Rosiglitazone or PPARγ overexpression alleviated PFOS-induced cell growth, migration, angiogenesis inhibition, and the release of inflammatory cytokines in HTR-8/SVneo and JEG-3 cells [49]. Previous studies indicated that decreased PPARγ expression or the inhibition of PPARγ activity led to mitochondrial fission, hyperpolarization, and increased oxidative stress [50]. The PPARγ pathway was one mechanism of triclosan-induced mitochondria-targeted effects, regulating the function of these organelles and the permeability of their membranes [51,52]. Our research showed that PPARγ activation or overexpression mitigated, while PPARγ inhibition or silence aggravated the increase of inflammatory gene expression caused by TCS. It is interesting that PPARγ prevented TCS-induced toxic phenotypes, while treatments with a PPARγ agonist or antagonist alone had no effects, suggesting some new mechanisms related to PPARγ can be initiated by TCS, which is worth more future study.

All the results in this study illustrated that TCS caused significant abnormal functions of HTR-8/SVneo and JEG-3 cells, including viability, migration, angiogenesis, and inflammation response. In addition, treatment with rosiglitazone or overexpression of PPARγ almost completely prevented the abnormal cell function changes in vitro. On the other hand, treatment with GW9662 or *si-PPARγ* aggravated the toxicity. Similarly, animal studies also indicated that rosiglitazone mitigated the decrease of placental diameter and fetal weight caused by TCS. In addition, curcumin, a natural compound and a modulator of PPARγ, has been known to have beneficial effects on pregnancy outcomes [53,54]. It will be interesting to know the effect of curcumin or other natural PPARγ modulators on TCS placental exposure in the future. Our results suggested that TCS placental exposure had adverse effects in vitro and in vivo through the PPARγ pathway (Figure 8), and activation or increased expression of PPARγ is a potential strategy to protect against the placenta toxicity induced by TCS.

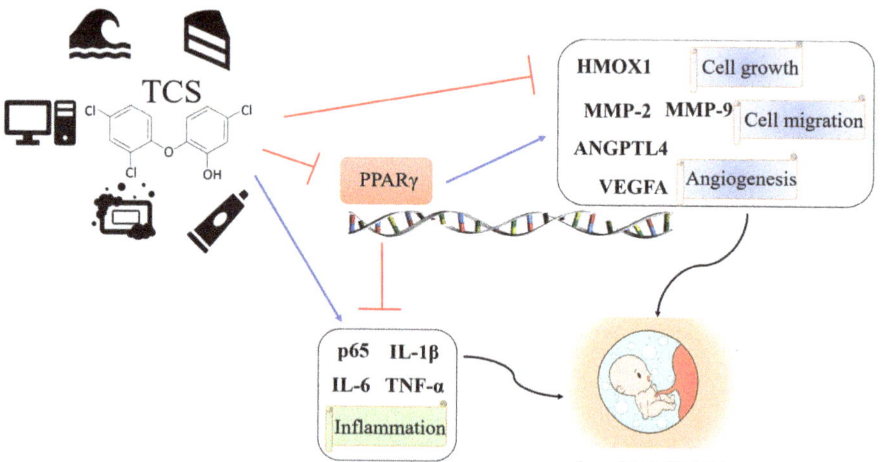

Figure 8. Schematic model depicting TCS-induced placental dysfunction and low birth-weight infants through PPARγ signaling pathways. TCS inhibited cell growth, angiogenesis, and migration and promoted inflammation of placenta via PPARγ-regulated genes *HMOX1, ANGPTL4, VEGFA, MMP-2, MMP-9, p65, IL-6, IL-1β*, and *TNF-α*.

Supplementary Materials: The following supporting information can be downloaded at: https://www.mdpi.com/article/10.3390/cells11010086/s1, Figure S1: The efficiency of PPARγ overexpression and knockdown; Table S1: RT-PCR primers.

Author Contributions: Conceptualization, J.L. and P.X.; methodology, X.Q.; formal analysis, J.L. and S.L.; investigation, J.L. and Q.W.; data curation, X.Q., Y.Z. and T.Y.; writing—original draft preparation, X.Q. and Z.H.; writing—review and editing, X.Q., S.L. and P.X.; supervision, J.L., X.Y. and P.X.; project administration, W.S., P.X. and J.L. All authors have read and agreed to the published version of the manuscript.

Funding: This research was funded by the National Natural Science Foundation of China [No. 81703260]; the Science and Technology Department of Jiangsu Province [No. BK20160227]; the China Postdoctoral Science Foundation funded project [No. 2016M601892]; the Priority Academic Program Development of Jiangsu Higher Education Institutions (PAPD), and Jiangsu Overseas Visiting Scholar Program for University Prominent Young and Middle-aged Teachers and Presidents, the Postgraduate Research and Practice Innovation Program of Jiangsu Province (No. KYCX21_2725).

Institutional Review Board Statement: The animal study protocol was approved by the Xuzhou Medical University Animal Ethics Committee, the entire experiment was per-formed in a Specified Pathogen-Free (SPF) environment (protocols 201605w025, 25 May 2016 and 202106A237, 25 June 2021).

Informed Consent Statement: Not applicable.

Data Availability Statement: The data sets generated and/or analyzed during the current study are available from the corresponding author on reasonable request.

Conflicts of Interest: The authors declare no conflict of interest.

References

1. Weatherly, L.M.; Gosse, J.A. Triclosan exposure, transformation, and human health effects. *J. Toxicol. Environ. Health B. Crit. Rev.* **2017**, *20*, 447–469. [CrossRef]
2. Alfhili, M.A.; Lee, M.-H. Triclosan: An Update on Biochemical and Molecular Mechanisms. *Oxid. Med. Cell Longev.* **2019**, *2019*, 1607304. [CrossRef]
3. Rodricks, J.V.; Swenberg, J.A.; Borzelleca, J.F.; Maronpot, R.R.; Shipp, A.M. Triclosan: A critical review of the experimental data and development of margins of safety for consumer products. *Crit. Rev. Toxicol.* **2010**, *40*, 422–484. [CrossRef]

4. Ribeiro, E.; Ladeira, C.; Viegas, S. EDCs Mixtures: A Stealthy Hazard for Human Health? *Toxics* **2017**, *5*, 5. [CrossRef]
5. Bever, C.S.; Rand, A.A.; Nording, M.; Taft, D.; Kalanetra, K.M.; Mills, D.A.; Breck, M.A.; Smilowitz, J.T.; German, J.B.; Hammock, B.D. Effects of triclosan in breast milk on the infant fecal microbiome. *Chemosphere* **2018**, *203*, 467–473. [CrossRef]
6. Etzel, T.; Muckle, G.; Arbuckle, T.E.; Fraser, W.D.; Ouellet, E.; Séguin, J.R.; Lanphear, B.; Braun, J.M. Prenatal urinary triclosan concentrations and child neurobehavior. *Environ. Int.* **2018**, *114*, 152–159. [CrossRef]
7. Ashley-Martin, J.; Dodds, L.; Arbuckle, T.E.; Marshall, J. Prenatal triclosan exposure and cord blood immune system biomarkers. *Int. J. Hyg. Environ. Health* **2016**, *219*, 454–457. [CrossRef]
8. Wang, X.; Chen, X.; Feng, X.; Chang, F.; Chen, M.; Xia, Y.; Chen, L. Triclosan causes spontaneous abortion accompanied by decline of estrogen sulfotransferase activity in humans and mice. *Sci. Rep.* **2015**, *5*, 18252. [CrossRef]
9. Lassen, T.H.; Frederiksen, H.; Kyhl, H.B.; Swan, S.H.; Main, K.M.; Andersson, A.-M.; Lind, D.V.; Husby, S.; Wohlfahrt-Veje, C.; Skakkebæk, N.E.; et al. Prenatal Triclosan Exposure and Anthropometric Measures Including Anogenital Distance in Danish Infants. *Environ. Health Perspect.* **2016**, *124*, 1261–1268. [CrossRef]
10. Zhao, J.-L.; Zhang, Q.-Q.; Chen, F.; Wang, L.; Ying, G.-G.; Liu, Y.-S.; Yang, B.; Zhou, L.-J.; Liu, S.; Su, H.-C.; et al. Evaluation of triclosan and triclocarban at river basin scale using monitoring and modeling tools: Implications for controlling of urban domestic sewage discharge. *Water Res.* **2013**, *47*, 395–405. [CrossRef]
11. Hong, F.; Xu, P.; Zhai, Y. The Opportunities and Challenges of Peroxisome Proliferator-Activated Receptors Ligands in Clinical Drug Discovery and Development. *Int. J. Mol. Sci.* **2018**, *19*, 2189. [CrossRef]
12. Xu, P.; Zhai, Y.; Wang, J. The Role of PPAR and Its Cross-Talk with CAR and LXR in Obesity and Atherosclerosis. *Int. J. Mol. Sci.* **2018**, *19*, 1260. [CrossRef]
13. Hong, F.; Pan, S.; Guo, Y.; Xu, P.; Zhai, Y. PPARs as Nuclear Receptors for Nutrient and Energy Metabolism. *Molecules* **2019**, *24*, 2545. [CrossRef]
14. Xu, P.; Hong, F.; Wang, J.; Wang, J.; Zhao, X.; Wang, S.; Xue, T.; Xu, J.; Zheng, X.; Zhai, Y. DBZ is a putative PPARγ agonist that prevents high fat diet-induced obesity, insulin resistance and gut dysbiosis. *Biochim. Biophys. Acta Gen. Subj.* **2017**, *11 Pt A*, 2690–2701. [CrossRef]
15. Xi, Y.; Zhang, Y.; Zhu, S.; Luo, Y.; Xu, P.; Huang, Z. PPAR-Mediated Toxicology and Applied Pharmacology. *Cells* **2020**, *9*, 200. [CrossRef]
16. Nadra, K.; Quignodon, L.; Sardella, C.; Joye, E.; Mucciolo, A.; Chrast, R.; Desvergne, B. PPARgamma in placental angiogenesis. *Endocrinology* **2010**, *151*, 4969–4981. [CrossRef]
17. Fournier, T.; Tsatsaris, V.; Handschuh, K.; Evain-Brion, D. PPARs and the placenta. *Placenta* **2007**, *28*, 65–76. [CrossRef]
18. Debril, M.B.; Renaud, J.P.; Fajas, L.; Auwerx, J. The pleiotropic functions of peroxisome proliferator-activated receptor gamma. *J. Mol. Med.* **2001**, *79*, 30–47. [CrossRef]
19. Wagner, N.; Wagner, K.-D. PPARs and Angiogenesis-Implications in Pathology. *Int. J. Mol. Sci.* **2020**, *21*, 5723. [CrossRef]
20. Barak, Y.; Nelson, M.C.; Ong, E.S.; Jones, Y.Z.; Ruiz-Lozano, P.; Chien, K.R.; Koder, A.; Evans, R.M. PPAR gamma is required for placental, cardiac, and adipose tissue development. *Mol. Cell* **1999**, *4*, 585–595. [CrossRef]
21. Kubota, N.; Terauchi, Y.; Miki, H.; Tamemoto, H.; Yamauchi, T.; Komeda, K.; Satoh, S.; Nakano, R.; Ishii, C.; Sugiyama, T.; et al. PPAR gamma mediates high-fat diet-induced adipocyte hypertrophy and insulin resistance. *Mol. Cell* **1999**, *4*, 597–609. [CrossRef]
22. Kadam, L.; Kohan-Ghadr, H.R.; Drewlo, S. The balancing act—PPAR-γ's roles at the maternal-fetal interface. *Syst. Biol. Reprod. Med.* **2015**, *61*, 65–71. [CrossRef]
23. Hua, X.; Cao, X.-Y.; Wang, X.-L.; Sun, P.; Chen, L. Exposure of Pregnant Mice to Triclosan Causes Insulin Resistance via Thyroxine Reduction. *Toxicol. Sci.* **2017**, *160*, 150–160. [CrossRef] [PubMed]
24. Szychowski, K.A.; Skóra, B.; Wójtowicz, A.K. Triclosan affects the expression of nitric oxide synthases (NOSs), peroxisome proliferator-activated receptor gamma (PPARγ), and nuclear factor kappa-light-chain-enhancer of activated B cells (NF-κB) in mouse neocortical neurons in vitro. *Toxicol. Vitro* **2021**, *73*, 105143. [CrossRef]
25. La Paglia, L.; Listì, A.; Caruso, S.; Amodeo, V.; Passiglia, F.; Bazan, V.; Fanale, D. Potential Role of ANGPTL4 in the Cross Talk between Metabolism and Cancer through PPAR Signaling Pathway. *PPAR Res.* **2017**, *2017*, 8187235. [CrossRef]
26. Shyu, L.-Y.; Chen, K.-M.; Lu, C.-Y.; Lai, S.-C. Regulation of Proinflammatory Enzymes by Peroxisome Proliferator-Activated Receptor Gamma in Astroglia Infected with Toxoplasma gondii. *J. Parasitol.* **2020**, *106*, 564–571. [CrossRef]
27. Shen, B.; Zhao, C.; Wang, Y.; Peng, Y.; Cheng, J.; Li, Z.; Wu, L.; Jin, M.; Feng, H. Aucubin inhibited lipid accumulation and oxidative stress via Nrf2/HO-1 and AMPK signalling pathways. *J. Cell Mol. Med.* **2019**, *23*, 4063–4075. [CrossRef]
28. Kelleher, A.M.; DeMayo, F.J.; Spencer, T.E. Uterine Glands: Developmental Biology and Functional Roles in Pregnancy. *Endocr. Rev.* **2019**, *40*, 1424–1445. [CrossRef]
29. Zhang, N.; Wang, W.; Li, W.; Liu, C.; Chen, Y.; Yang, Q.; Wang, Y.; Sun, K. Inhibition of 11β-HSD2 expression by triclosan via induction of apoptosis in human placental syncytiotrophoblasts. *J. Clin. Endocrinol. Metab.* **2015**, *100*, E542–E549. [CrossRef]
30. Honkisz, E.; Zieba-Przybylska, D.; Wojtowicz, A.K. The effect of triclosan on hormone secretion and viability of human choriocarcinoma JEG-3 cells. *Reprod. Toxicol.* **2012**, *34*, 385–392. [CrossRef]
31. Burton, G.J.; Jauniaux, E. Pathophysiology of placental-derived fetal growth restriction. *Am. J. Obstet. Gynecol.* **2018**, *218*, S745–S761. [CrossRef]
32. Zhang, M.; Zhu, R.; Zhang, L. Triclosan stimulates human vascular endothelial cell injury via repression of the PI3K/Akt/mTOR axis. *Chemosphere* **2020**, *241*, 125077. [CrossRef]

33. Zhang, Y.; Liu, M.; Liu, J.; Wang, X.; Wang, C.; Ai, W.; Chen, S.; Wang, H. Combined toxicity of triclosan, 2,4-dichlorophenol and 2,4,6-trichlorophenol to zebrafish (*Danio rerio*). *Environ. Toxicol. Pharmacol.* **2018**, *57*, 9–18. [CrossRef]
34. Feng, Y.; Zhang, P.; Zhang, Z.; Shi, J.; Jiao, Z.; Shao, B. Endocrine Disrupting Effects of Triclosan on the Placenta in Pregnant Rats. *PLoS ONE* **2016**, *11*, e0154758. [CrossRef] [PubMed]
35. Garnier, V.; Traboulsi, W.; Salomon, A.; Brouillet, S.; Fournier, T.; Winkler, C.; Desvergne, B.; Hoffmann, P.; Zhou, Q.-Y.; Congiu, C.; et al. PPARγ controls pregnancy outcome through activation of EG-VEGF: New insights into the mechanism of placental development. *Am. J. Physiol. Endocrinol. Metab.* **2015**, *309*, E357–E369. [CrossRef]
36. Shibuya, M. Vascular endothelial growth factor and its receptor system: Physiological functions in angiogenesis and pathological roles in various diseases. *J. Biochem.* **2013**, *153*, 13–19. [CrossRef]
37. Hato, T.; Tabata, M.; Oike, Y. The role of angiopoietin-like proteins in angiogenesis and metabolism. *Trends Cardiovasc. Med.* **2008**, *18*, 6–14. [CrossRef]
38. Goh, Y.Y.; Pal, M.; Chong, H.C.; Zhu, P.; Tan, M.J.; Punugu, L.; Tan, C.K.; Huang, R.-L.; Sze, S.K.; Tang, M.B.Y.; et al. Angiopoietin-like 4 interacts with matrix proteins to modulate wound healing. *J. Biol. Chem.* **2010**, *285*, 32999–33009. [CrossRef]
39. Liu, L.; Zhuang, X.; Jiang, M.; Guan, F.; Fu, Q.; Lin, J. ANGPTL4 mediates the protective role of PPARγ activators in the pathogenesis of preeclampsia. *Cell Death Dis.* **2017**, *8*, e3054. [CrossRef]
40. Woeckel, V.J.; Bruedigam, C.; Koedam, M.; Chiba, H.; van der Eerden, B.C.J.; van Leeuwen, J.P.T.M. 1α,25-dihydroxyvitamin D3 and rosiglitazone synergistically enhance osteoblast-mediated mineralization. *Gene* **2013**, *512*, 438–443. [CrossRef]
41. Tian, F.-J.; Qin, C.-M.; Li, X.-C.; Wu, F.; Liu, X.-R.; Xu, W.-M.; Lin, Y. Decreased stathmin-1 expression inhibits trophoblast proliferation and invasion and is associated with recurrent miscarriage. *Am. J. Pathol.* **2015**, *185*, 2709–2721. [CrossRef]
42. Yang, Y.; Zhang, J.; Gong, Y.; Liu, X.; Bai, Y.; Xu, W.; Zhou, R. Increased expression of prostasin contributes to early-onset severe preeclampsia through inhibiting trophoblast invasion. *J. Perinatol.* **2015**, *35*, 16–22. [CrossRef]
43. Vannuccini, S.; Clifton, V.L.; Fraser, I.S.; Taylor, H.S.; Critchley, H.; Giudice, L.C.; Petraglia, F. Infertility and reproductive disorders: Impact of hormonal and inflammatory mechanisms on pregnancy outcome. *Hum. Reprod. Update* **2016**, *22*, 104–115. [CrossRef]
44. Remels, A.H.V.; Langen, R.C.J.; Gosker, H.R.; Russell, A.P.; Spaapen, F.; Voncken, J.W.; Schrauwen, P.; Schols, A.M.W.J. PPARgamma inhibits NF-kappaB-dependent transcriptional activation in skeletal muscle. *Am. J. Physiol. Endocrinol. Metab.* **2009**, *297*, E174–E183. [CrossRef] [PubMed]
45. Challis, J.R.; Lockwood, C.J.; Myatt, L.; Norman, J.E.; Strauss, J.F.; Petraglia, F. Inflammation and pregnancy. *Reprod. Sci.* **2009**, *16*, 206–215. [CrossRef]
46. Jiang, C.; Ting, A.T.; Seed, B. PPAR-gamma agonists inhibit production of monocyte inflammatory cytokines. *Nature* **1998**, *391*, 82–86. [CrossRef]
47. Lappas, M.; Permezel, M.; Georgiou, H.M.; Rice, G.E. Regulation of proinflammatory cytokines in human gestational tissues by peroxisome proliferator-activated receptor-gamma: Effect of 15-deoxy-Delta(12,14)-PGJ(2) and troglitazone. *J. Clin. Endocrinol. Metab.* **2002**, *87*, 4667–4672. [CrossRef] [PubMed]
48. Kadam, L.; Kilburn, B.; Baczyk, D.; Kohan-Ghadr, H.R.; Kingdom, J.; Drewlo, S. Rosiglitazone blocks first trimester in-vitro placental injury caused by NF-κB-mediated inflammation. *Sci. Rep.* **2019**, *9*, 2018. [CrossRef] [PubMed]
49. Li, J.; Quan, X.; Lei, S.; Huang, Z.; Wang, Q.; Xu, P. PFOS Inhibited Normal Functional Development of Placenta Cells via Signaling. *Biomedicines* **2021**, *9*, 276. [CrossRef] [PubMed]
50. Yeligar, S.M.; Kang, B.-Y.; Bijli, K.M.; Kleinhenz, J.M.; Murphy, T.C.; Torres, G.; San Martin, A.; Sutliff, R.L.; Hart, C.M. PPARγ Regulates Mitochondrial Structure and Function and Human Pulmonary Artery Smooth Muscle Cell Proliferation. *Am. J. Respir. Cell Mol. Biol.* **2018**, *58*, 648–657. [CrossRef] [PubMed]
51. Teplova, V.V.; Belosludtsev, K.N.; Kruglov, A.G. Mechanism of triclosan toxicity: Mitochondrial dysfunction including complex II inhibition, superoxide release and uncoupling of oxidative phosphorylation. *Toxicol. Lett.* **2017**, *275*, 108–117. [CrossRef]
52. Belosludtsev, K.N.; Belosludtseva, N.V.; Tenkov, K.S.; Penkov, N.V.; Agafonov, A.V.; Pavlik, L.L.; Yashin, V.A.; Samartsev, V.N.; Dubinin, M.V. Study of the mechanism of permeabilization of lecithin liposomes and rat liver mitochondria by the antimicrobial drug triclosan. *Biochim. Biophys. Acta Biomembr.* **2018**, *1860*, 264–271. [CrossRef]
53. Tossetta, G.; Fantone, S.; Giannubilo, S.R.; Marzioni, D. The Multifaced Actions of Curcumin in Pregnancy Outcome. *Antioxidants* **2021**, *10*, 126. [CrossRef]
54. Blanquicett, C.; Kang, B.-Y.; Ritzenthaler, J.D.; Jones, D.P.; Hart, C.M. Oxidative stress modulates PPAR gamma in vascular endothelial cells. *Free Radic. Biol. Med.* **2010**, *48*, 1618–1625. [CrossRef] [PubMed]

MDPI
St. Alban-Anlage 66
4052 Basel
Switzerland
Tel. +41 61 683 77 34
Fax +41 61 302 89 18
www.mdpi.com

Cells Editorial Office
E-mail: cells@mdpi.com
www.mdpi.com/journal/cells

www.ingramcontent.com/pod-product-compliance
Lightning Source LLC
Chambersburg PA
CBHW041937260926
51850CB00018B/213

9783036581101